WITHDRAWN
NDSU

Carbon Forms and Functions
in Forest Soils

Carbon Forms and Functions in Forest Soils

Editors
William W. McFee and J. Michael Kelly

Based on the papers presented at the Eighth North American Forest Soils Conference; and sponsored by the Soil Science Society of America, Society of American Foresters, Canadian Institute of Forestry, and Canadian Soil Science Society. Financial support provided by USDA Forest Service–Global Climate Change Program, ITT– Rayonier, Inc., Packaging of America, Georgia Pacific, Container Corporation of America–Jefferson Smurfit Corporation, Champion International Corporation, International Paper Company, and University of Florida–Institute of Food and Agricultural Sciences; Gainsville, Florida, May 1993.

Organizing Committee

D.H. Alban	M.F. Jurgensen
J.R. Boyle	J.M. Kelly
R.D. Briggs	W.W. McFee
J.A. Burger	I.K. Morrison
D.W. Cole	D.G. Neary
N.B. Comerford	P.E. Pope
R.F. Fisher	R.F. Powers
N.W. Foster	D.H. Van Lear
J.D. Joslin	

Editor–in–Chief SSSA
Jerry M. Bigham

Managing Editor
Jon M. Bartels

Soil Science Society of America, Inc.
Madison, Wisconsin USA
1995

Cover Design: Mary Lou Jones
Cover Artwork: Mary Lou Jones

Copyright © 1995 by the Soil Science Society of America, Inc.

ALL RIGHTS RESERVED UNDER THE U.S. COPYRIGHT ACT OF 1976 (P.L. 94-553)

Any and all uses beyond the limitations of the "fair use" provision of the law require written permission from the publisher(s) and/or the author(s); not applicable to contributions prepared by officers or employees of the U.S. Government as part of their official duties.

Soil Science Society of America, Inc.
677 South Segoe Road, Madison, WI 53711 USA

Library of Congress Cataloging-in-Publication Data

North American Forest Soils Conference (8th : 1993 : Gainesville, Fla.)
  Carbon forms and functions in forest soils / editors, William W. McFee and J. Michael Kelly : organizing committee, D.H. Alban ... [et al.] : editor-in-chief SSSA, Jerry B. Bigham : managing editor, Jon M. Bartels.
     p.   cm.
  "Based on the papers presented at the Eighth North American Forest Soils Conference; and sponsored by the Soil Science Society of America ... [et al.]; Gainesville, Florida, May 1993."
  Includes bibliographical references.
  1. Forest soils—Congresses. 2. Soils—Carbon content—Congresses. 3. Carbon cycles (Biogeochemistry)—Congresses.
I. McFee, W. W. (William Warren), 1935-   . II. Kelly, James Michael, 1944-   . III. Soil Science Society of America.
IV. Title.
  SD390.N67   1993
  631.4'17'09152—dc20   94-35401
                                                            CIP

Printed in the United States of America

DEDICATION

Charles (Chuck) B. Davey, Professor Emeritus, North Carolina State University, has had a distinguished career in forest soils. Since receiving his Ph.D. from the University of Wisconsin in 1955, he has directed the work of 50 some graduate students, been named outstanding teacher, served as President of the Soil Science Society of America, been consultant to forestry organizations in ten countries, produced over 120 refereed publications and been a prominent leader among forest soil scientists. We are proud to dedicate this volume to an effective educator, distinguished scientist, and esteemed colleague.

Charles (Chuck) B. Davey

CONTENTS

Dedication... v

Foreword.. xi

Preface... xiii

Contributors.. xv

1 Soil Organic Matter: Clue or Conundrum?
   Richard F. Fisher.. 1

2 Forest Soil Organic Matter: Characterization and Modern Methods of Analysis
   John G. McColl and Noam Gressel... 13

3 Fractionation of Soil Organic Matter with Supercritical Freon
   Felipe G. Sanchez and Gregory A. Ruark... 33

4 The Influence of Low-Molecular-Weight Organic Acids on Properties and Processes in Forest Soils
   Thomas R. Fox... 43

5 Characterization of Dissolved and Colloidal Organic Matter in Soil Solution: A Review
   Bruce E. Herbert and Paul M. Bertsch.. 63

6 Chemistry of Carbon Decomposition Processes in Forests as Revealed by Solid-State Carbon-13 Nuclear Magnetic Resonance
   J.A. Baldock and C.M. Preston... 89

7 Management-Induced Changes in the Actively Cycling Fractions of Soil Organic Matter
   B.H. Ellert and E.G. Gregorich... 119

8 Long-Term Changes in Organic Matter in Soils Receiving Applications of Municipal Biosolids
   Robert B. Harrison, Charles L. Henry, Dale W. Cole, and Dongsen Xue... 139

9 Soil Carbon, Soil Formation, and Ecosystem Development
   Keith Van Cleve and Robert F. Powers... 155

10 Soil Organic Carbon in the Missouri Forest-Prairie Ecotone
   R. David Hammer, Gray S. Henderson, Ranjith P. Udawatta and Donna K. Brandt.. 201

11 Carbon Cycling in a Loblolly Pine Forest: Implications for the Missing Carbon Sink and for the Concept of Soil
   D.D. Richter, D. Markewitz, C.G. Wells, H.L. Allen, J.K. Dunscombe, K. Harrison, P.R. Heine, A. Stuanes, B. Urrego, and G. Bonani.. 233

12 Toward a New Theory of Podzolization
   Bryant A. Browne.. 253

13 A Model of Soil Organic Matter and its Function in Temperate Forest Soil Development
   Robert C. Santore, Charles T. Driscoll, and Michael Aloi.............. 275

14 Role of Carbon in the Cycling of Other Nutrients in Forested Ecosystems
   Dale W. Johnson.. 299

15 Carbon Controls on Spodosol Nitrogen, Sulfur, and Phosphorus Cycling
   Mark B. David, George F. Vance, and Anna J. Krzyszowska......... 329

16 Carbon and Nitrogen Cycling within Mid- and Late-Rotation Jack Pine
   Neil W. Foster, Ian K. Morrison, Paul W. Hazlett, Gary D. Hogan, and Maria I. Salerno... 355

17 Carbon Chemistry and Nutrient Supply in Cedar–Hemlock and Hemlock–Amabilis Fir Forest Floors
   Cindy E. Prescott, L.E. Demontigny, C.M. Preston, Rodney J. Keenan, and Gordon F. Weetman... 377

18 Belowground Responses to Atmospheric Carbon Dioxide in Forests
   Richard J. Norby, E.G. O'Neill, and Stan D. Wullschleger............ 397

19 Soil Organic Matter: A Link Between Forest Management and Productivity
   Gray S. Henderson... 419

20 Soil Carbon in Northern Forested Wetlands: Impacts of Silvicultural Practices
   Carl C. Trettin, Martin F. Jurgensen, Margaret R. Gale, and James W. McLaughlin... 437

21 Carbon Dynamics Following Clear-Cutting of a Northern Hardwood Forest
    Chris E. Johnson, Charles T. Driscoll, Timothy J. Fahey, Thomas G.
    Siccama, and Jeffery W. Hughes.. 463

22 Distribution of Carbon in a Piedmont Soil as Affected by Loblolly Pine
    Management
    D.H. Van Lear, P.R. Kapeluck, and Melissa M. Parker.................... 489

23 The Role of Forest Soils in the Global Carbon Cycle
    Alex F. Bouwman and Rik Leemans... 503

24 Comparison of Carbon Accumulation in Douglas Fir and Red Alder Forests
    Dale W. Cole, Jana E. Compton, Peter S. Homann, R.L. Edmonds, and
    Helga Van Miegroet.. 527

25 Modeling Carbon and Nitrogen Dynamics in Western Red Cedar and
    Western Hemlock Forests
    Rodney J. Keenen, J.P (Hamish) Kimmins, and John Pastor.......... 547

26 Carbon and Nitrogen Dynamics in Oak Stands along an Urban-Rural
    Gradient
    Richard V. Pouyat, Mark J. McDonnell, S.T.A. Pickett, Peter M.
    Groffman, M.M. Carreiro, Robert W. Parmelee, Kimberly E. Medley,
    and W.C. Zipperer... 569

27 A Perspective on the Evolution of Forest Soil Science
    Robert F. Chandler, Jr.. 589

FOREWORD

There is a renewed public awareness of the importance of forests and forest lands for agriforestry, biodiversity, ecosystem management, global climatic change, resource conservation, timber production, wildlife habitat, recreation, and sustainability of the biosphere. The balance between societal needs and utilization of forest resources is a continuing public debate. This publication serves a critical niche in capturing the state of knowledge on forest soils relative to several of the above national and international priorities. It targets organic matter and the role of carbon and nitrogen cycling as driving functions of forest ecosystems. Further, the publication forges a bridge and continued synergistic interaction among forest science professionals of North America.

The sponsors, editors, and authors are to be commended for their individual and collective efforts in bringing this publication to fruition. It provides a valuable treatise at a time when the public mandates increased sensitivity and attention to sustainability of the biosphere.

LARRY P. WILDING, *President*
Soil Science Society of America

PREFACE

Forest productivity and the factors which influence it have been the underlying theme uniting the seven previous North American Forest Soil Conferences which have occurred at 5-year intervals since the series was initiated in 1958. While that theme continues into this eighth conference, the focus has been directed to the single factor that in many ways describes, defines, and delineates the study of forest soils as a unique niche in the broader continuum of soil science.

In their preface to the proceedings of the third North American Forest Soils Conference C. T. Youngberg and C.B. Davey noted that "the growth of the tree and the productivity of the forest are directly affected by the quantity, quality and location of the soil." Soil carbon in its myriad forms, perhaps as no other factor, influences soil–plant relationships through its direct and indirect impacts on mineral solubility, exchange capacity, nutrient availability, moisture supply, aggregate formation, and soil erosion. Hans Jenny, in his classic equation describing the factors of soil formation, placed organic matter in a pivotal role in the development of the soil and soil properties as well as the succession of associated plant communities.

Since the previous conference many changes in our approach to the management and preservation of the forest resource have occurred. Most significant of these developments has been the growing emphasis placed on understanding and manipulating forests as ecological units. Paralleling and to some extent driving this move toward an ecological focus has been the renewed realization of the important and dynamic nature of the interaction between the physical and biological resources. The role of forest soils, as well as the aboveground portion of the forest, as significant buffers against the impacts of anthropogenic activities on atmospheric carbon dioxide and the earth's energy balance has driven home the necessity of continuing to develop and refine our knowledge of the interactions of soils, plants, and the environment in which they exist.

Three generations of forest soil scientists have now contributed to these volumes. During that time our knowledge has continued to expand and gain clearer focus. These gains have been the result of the collective work and interaction of the individual researcher and the peer community. Each of us has played a role in this procession and realizing that there is much yet to be done, it seems only appropriate that we collect our thoughts at these five year intervals and celebrate our accomplishments.

CONTRIBUTORS

H.L. Allen	Department of Forestry, North Carolina State University, Raleigh, NC
Michael Aloi	Research Assistant, Department of Civil & Environmental Engineering Syracuse University, Syracuse, NY 13244–1190
J.A. Baldock	Research Scientist, Canadian Forest Service, Petawawa National Forestry Institute, Chalk River, Ontario, Canada K0J 1J0
Paul M. Bertsch	Associate Professor, Soil Physical Chemistry, Division of Biogeochemistry, Savannah River Ecology Laboratory, University of Georgia, Aiken, SC 29802
G. Bonani	ETH, Zurich, Switzerland
Alex F. Bouwman	National Institute of Public Health and Environmental Protection, Bilthoven, the Netherlands
Donna K. Brandt	Research Agronomist, Rose Lake Plant Materials Center, USDA–SCS, East Lansing, MI 48823
Bryant A. Browne	Assistant Professor of Soil and Water Resources, College of Natural Resources, University of Wisconsin, Stevens Point, WI 54481
M.M. Carreiro	Assistant Professor, Louis Calder Center, Fordham University, Armonk, NY 10504
Robert F. Chandler, Jr.	Director, Emeritus, International Rice Research Institute, Clermont, FL 34711
Dale W. Cole	Professor of Forest Soils, College of Forest Resources, University of Washington, Seattle, WA 98195
Jana E. Compton	Research Assistant, College of Forest Resources, University of Washington, Seattle, WA 98185; current position and address is Research Associate, Harvard Forest, Harvard University, Petersham, MA 01366
Mark B. David	Associate Professor of Forest Soils, Department of Forestry, University of Illinois, Urbana, IL 61801
L.E. deMontigny	Research Silviculturalist, Research Branch, Ministry of Forests, Victoria, BC, Canada V8W 3E7
Charles T. Driscoll	Distinguished Professor of Civil & Environmental Engineering, Department of Civil & Environmental Engineering, Syracuse University, Syracuse, NY 13244–1190

CONTRIBUTORS

J.K. Dunscomb — School of the Environment, Duke University, Durham, NC 27708–0328

R.L. Edmonds — Professor, College of Forest Resources, University of Washington, Seattle, WA 98195

B.H. Ellert — Research Scientist, Agriculture and Agri-Food Canada Research Centre, Lethbridge, Alberta, Canada T1J 4B1

Timothy J. Fahey — Associate Professor, Department of Natural Resources, Cornell University, Ithaca, NY 14853

Richard F. Fisher — Professor of Forest Soils, Forest Science Department, Texas A&M University, College Station, TX 77843–2135

Neil W. Foster — Research Scientist, Canadian Forest Service, Sault Ste. Marie, Ontario, Canada P6A 5M7

Thomas R. Fox — Forest Research Group Leader, Forest Research Center, Rayonier, Inc., Yulee, FL 32097

Margaret R. Gale — Associate Professor, School of Forestry and Wood Products, Michigan Technological University, Houghton, MI 49931

E.G. Gregorich — Research Scientist, Centre for Land and Biological Resources Research, Agriculture and Agri–Food Canada, Central Experimental Farm, Ottawa, Ontario, Canada K1A 0C6

Noam Gressel — Graduate Student Researcher, Environmental Science, Policy and Management Department, Division of Ecosystem Sciences, University of California, Berkeley, CA 94720

Peter M. Groffman — Associate Scientist, Institute of Ecosystem Studies, Millbrook, NY 12545

R. David Hammer — Associate Professor of Soil Science, School of Natural Resources, University of Missouri, Columbia, MO 65211

Robert B. Harrison — Associate Professor of Soil Chemistry, University of Washington, Seattle, WA 98195

K. Harrison — Columbia University, New York, NY 10027

Paul W. Hazlett — Research Assistant, Canadian Forest Service, Sault Ste. Marie, Ontario, Canada P6A 5M7

P.R. Heine — School of the Environment, Duke University, Durham, NC 27708–0328

Gray S. Henderson — Professor of Forest Soils, School of Natural Resources, University of Missouri, Columbia, MO 65211

Charles L. Henry — Research Assistant Professor of Forest Soils, University of Washington, Seattle, WA 98195

Bruce E. Herbert — Assistant Professor of Environmental Geochemistry, Department of Geology & Geophysics, Texas A & M University, College Station, TX 77843

CONTRIBUTORS

Gary D. Hogan	Research Scientist, Canadian Forest Service, Sault Ste. Marie, Ontario, Canada P6A 5M7
Peter S. Homann	Research Associate, Department of Forest Science, Oregon State University, Corvallis, OR 97331–7501
Jeffrey W. Hughes	Assistant Professor, Department of Botany, University of Vermont, Burlington, VT 05405
Chris E. Johnson	Assistant Professor, Department of Civil and Environmental Engineering, Syracuse University, Syracuse, NY 13244
Dale W. Johnson	Assistant Professor, Biological Sciences Center, Desert Research Institute, Reno NV 89506; and Environmental Resource Science, University of Nevada, Reno, NV 98512
Martin F. Jurgensen	Professor of Forest Soils, School of Forestry and Wood Products, Michigan Technological University, Houghton, MI 49931
P.R. Kapeluck	Research Forester, Department of Forest Resources, Clemson University, Clemson, SC 29634–1003
Rodney J. Keenan	Senior Research Scientist, Queensland Forest Research Institute, Department of Primary Industries, Atherton, Australia
J. Michael Kelly	Senior Technical Specialist, Tennessee Valley Authority, Norris, TN 37828
J.P. (Hamish) Kimmins	Faculty of Forestry, University of British Columbia, Vancouver, BC, Canada V6T 1Z4
Anna J. Krzyszowska	Research Scientist, Department of Plant, Soil and Insect Sciences, University of Wyoming, Laramie, WY 82071
Rik Leemans	National Institute of Public Health and Environmental Protection, Bilthoven, the Netherlands
D. Markewitz	School of the Environment, Duke University, Durham, NC, 27708–0328
John G. McColl	Professor, Environmental Science, Policy, and Management Department, Division of Ecosystem Sciences, University of California, Berkeley, CA 94720
Mark J. McDonnell	Director, Bartlett Arboretum, University of Connecticut, Stamford, CT 06903
William W. McFee	Professor, Department of Agronomy, Purdue University, West Lafayette, IN 47907
James W. McLaughlin	Assistant Professor, Department of Applied Ecology and Environmental Sciences, University of Maine, Orono, ME 14469
Kimberly E. Medley	Assistant Professor of Geography, Department of Geography, Miami University, Oxford, OH 45056

CONTRIBUTORS

Ian K. Morrison	Research Scientist, Canadian Forest Service, Sault Ste. Marie, Ontario, Canada, P6A 5M7
Richard J. Norby	Research Scientist, Environmental Sciences Division, Oak Ridge National Laboratory, Oak Ridge, TN 37831–6034
E.G. O'Neill	Research Scientist, Environmental Sciences Division, Oak Ridge National Laboratory, Oak Ridge, TN 37831–6034
Melissa M. Parker	Conservation Scientist, Texas Parks and Wildlife Department, Nacogdoches, TX 75962
Robert W. Parmelee	Research Scientist, Department of Entomology, Ohio State University, Columbus, OH 43210
John Pastor	University of Minnesota, Duluth, MN
S.T.A. Pickett	Scientist, Institute of Ecosystem Studies, Millbrook, NY 12545
Richard V. Pouyat	Research Forester, USDA–FS, NEFES, c/o Institute of Ecosystem Studies, Millbrook, NY 12545
Robert F. Powers	Principal Research Silviculturalist, Pacific Southwest Research Station, USDA–FS, Redding, CA 96001
Cindy E. Prescott	Research Associate, Faculty of Forestry, University of British Columbia, Vancouver, BC, Canada V6T 1Z4
C.M. Preston	Research Scientist, Canadian Forest Service, Pacific Forestry Centre, Victoria, BC, Canada V82 1M5
D.D. Richter	Professor, School of the Environment, Duke University, Durham, NC 27708–0328
Gregory A. Ruark	Research Administrator, USDA–FS, Washington, DC 20250
Maria I. Salerno	Research Assistant, Convenio Ministeries de Asuntas Agrarios, Universidad Nacional de la Plata, La Plata, Argentina
Felipe G. Sanchez	Research Chemist, USDA–FS, SEFES, Research Triangle Park, NC 27709
Robert C. Santore	Research Assistant, Department of Civil and Environmental Engineering, Syracuse, NY 13244–1190
Thomas G. Siccama	Lecturer and Director of Field Studies, School of Forestry and Environmental Studies, Yale University, New Haven, CT 06511
A. Stuanes	Norwegian Forest Research Institute, AAS–NLH, Norway
Carl C. Trettin	Research Soil Scientist, Environmental Sciences Division, Oak Ridge National Laboratory, Oak Ridge, TN 37831–6038, current address is Center for Forested Wetlands Research, 2730 Savannah Hwy., Charleston, SC 29414

CONTRIBUTORS

Ranjith P. Udawatta	Graduate Research Assistant, University of Missouri, Columbia, MO 65211
B. Urrego	Department of Forestry, North Carolina State University, Raleigh, NC
George F. Vance	Associate Professor of Soil Chemistry, Department of Plant, Soil and Insect Sciences, University of Wyoming, Laramie, WY 82071
Keith Van Cleve	Professor of Forest Soils, Forest Soils Laboratory, University of Alaska, Fairbanks, AK 99775
D.H. Van Lear	Bowen Professor of Forestry, Department of Forest Resources, Clemson University, Clemson, SC 29634–1003
Helga Van Miegroet	Assistant Professor of Wetland Soils and Biogeochemical Cycling, Department of Forest Resources, Utah State University, Logan, UT 84322–5215
Gordon F. Weetman	Professor of Silviculture, Faculty of Forestry, University of British Columbia, Vancouver, BC, Canada V6T 1Z4
C.G. Wells	Southeastern Forest Experimental Station, USDA–FS, Research Triangle Park, NC 27709
Stan D. Wullschleger	Research Scientist, Environmental Sciences Division, Oak Ridge National Laboratory, Oak Ridge, TN 37831–6034
Dongsen Xue	Research Associate, College of Forest Resources, University of Washington, Seattle, WA 98195
W.C. Zipperer	Research Forester, USDA–FS, Syracuse, NY 13210

1 Soil Organic Matter: Clue or Conundrum?

Richard F. Fisher
Texas A&M University
College Station, Texas

This chapter is not intended to be an exhaustive review of what we know about soil organic matter, nor of how we use that knowledge in determining soil quality or managing the soil resource. The chapters that follow will address those topics in detail. This chapter addresses the questions "is soil organic matter an important determinant of soil quality?" and if it is "why are formal relationships between soil organic matter and soil quality so poor?" It also considers how we might explain the role of soil organic matter in determining soil quality, and how we might form strong relationships that predict that role.

Soil organic matter content is an easily observable, easily quantifiable soil property. "Black dirt" suggests a productive site, and mankind has long used soil color as an index of the worth of land for plant production (Kononova, 1961). Soil organic matter or organic C is routinely measured in the laboratory by wet, Walkley-Black, and dry, loss on ignition, combustion methods (Page et al., 1982). Although dark, organic matter-rich soils are generally more productive than pale, organic matter-poor soils, there is seldom a strong statistical correlation between organic matter content and plant growth. We have a powerful intuitive sense that organic matter is important for tree growth and the maintenance of site productivity, but we have few strong statistical relationships to support that sense. Thus we are often ambivalent when it comes to sacrificing yield or profit to maintain soil organic matter levels.

There are many reasons for the imperfect relationship between soil organic matter content and tree growth. A cardinal reason is that in wetter soils growth often declines as soil organic matter content increases with increased wetness while in drier soils the reverse is generally true. The decline in growth on wetter soils is usually related to decreases in the availability of O_2 in the root zone, and the increase in growth on drier soils is related to increased water and nutrient holding capacity. Nonetheless, the relationship of soil organic matter content to growth is confounded (Pritchett & Fisher, 1987).

A more difficult to understand, but probably more cogent, reason for the lack of a strong linear relationship between total soil organic C and tree growth lies in the large difference between "total" and "active" organic C. Mineral soils vary in

---

Copyright © 1995 Soil Science Society of America, 677 S. Segoe Rd., Madison, WI 53711, USA.
Carbon Forms and Functions in Forest Soils.

their total organic C content from as little as 10 000 mg/kg to as much as 100 000 mg/kg or more. This equates to from 20 000 to 200 000 kg of C in the upper 20 cm of soil with a bulk density of 1000 kg/m^3. However the amount of active organic C, that which is in solution or exchangeable at any one time, is on the order of 10 to 100 mg/kg. This means that less than 1% of the total organic C is active, and large changes in the quantity and nature of this active fraction can take place without being reflected in the size of the total organic C pool. There may be a statistically significant relation between tree growth and some particular fraction of soil C, but we are yet to explain that relationship.

At an ecological level, the relationship of tree growth to soil organic matter may be extraordinary. There is a complex detrital trophic structure in the forest floor and soil. The intricate food web of detrital feeders may be important in producing the organic compounds that allow maximum tree growth rates, or they may modulate the rate of production of critical organic compounds. This complex web has a vital function in controlling the rate of mineralization so important to plant growth. We remain surprisingly ignorant of the extent to which the structure and function of the detrital system influences tree growth and long-term soil quality.

THE IMPORTANCE OF SOIL ORGANIC MATTER

Just how important is soil organic matter to the life of the soil and the plants it supports? Soil organic matter is the fuel that runs the soil's engine. The storage and decomposition of plant and animal remains in soil are basic biological processes and are essential parts of the C cycle. During this process C is recirculated to the atmosphere as carbon dioxide (CO_2), nitrogen (N) is made available as ammonium (NH^{4+}) and nitrate (NO^{3-}); and other associated elements, principally phosphorus (P), sulphur (S), and micronutrients such as iron (Fe), manganese (Mn), copper (Cu), boron (B), molybdenum (Mo), and zinc (Zn), appear in forms required by higher plants. Soil organic matter plays other important roles in the soil. Humus, one portion of soil organic matter, provides buffering over a broad range of soil acidity, and much of the exchange capacity of the soil may be due to colloidal humic substances. Humus greatly affects soil physical properties such as aeration, aggregate stability, water-holding capacity, and permeability. Soil organic compounds both large and small are important in binding metallic ions thus increasing the availability of some nutrients and reducing toxic effects of others (Stevenson, 1982). Externally supplied humic acids can directly affect plant growth; however, it is unlikely that these humic substances have a major, direct affect on plants under natural conditions (Wiersum, 1974).

THE NATURE OF SOIL ORGANIC MATTER

Organic materials in the soil are diverse in nature and often chemically complex. Soil organic matter is derived from litter fall and root turnover and the bodies of the organisms that participate in the detrital food chain. In forests a large portion of the material is resident for varying periods of time on the soil surface. Although a voluminous literature exists on the classification of these surface layers, our knowledge of the complex decomposition processes that occur within them remains sketchy (Green et al., 1993). Materials mixed or leached from these surface layers contribute to the soil organic matter, but a large share of soil organic matter comes from roots, soil fauna, and microorganisms that live and die in the soil. The annual input of organic C to forest soils varies more or less linearly with latitude (Fig. 1–1). This is, of course, an oversimplification since large site-to-site differences exist within any latitudinal belt due to precipitation, altitude, and other local factors.

In forest ecosystems, soil organic matter is added both continuously and episodically. The more or less continuous additions of organic material maintain an endemic level of decomposer populations, but these populations fluctuate rather widely with the episodic additions of large amounts of organic material that occur in most forest soils. The dynamics of both the organic additions and the decomposer populations are vital attributes of forest ecosystems, and quite likely should attract greater research attention.

The terms humus and soil organic matter are most often used interchangeably and generally refer to the organic constituents in the soil, excluding undecayed plant and animal tissue, partial decomposition products, and the soil biomass. Soil organic matter thus includes: (i) simple substances such as amino acids, sugars, and other small molecules, (ii) identifiable high-molecular-weight compounds such as polysaccharides, proteins, and lipids, and (iii) humic substances, a series of high-molecular-weight, yellow to black substances formed by secondary synthesis reactions.

The first of these groups of compounds provides a ready source of energy for soil organisms. The second group is decomposed to yield members of the first group and to provide building blocks for the third group. It is this latter group of

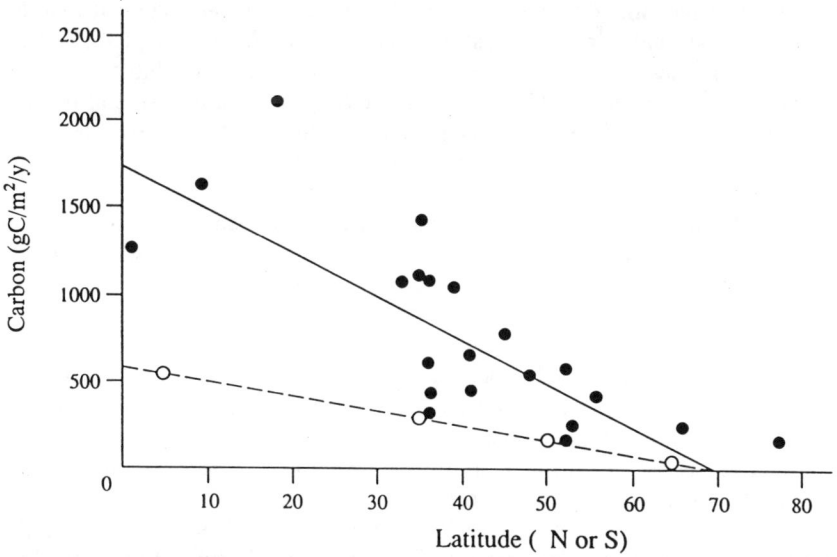

Fig. 1–1. Latitudinal trend in mean annual litter fall (dashed line) and respirational loss of CO_2 including root respiration (solid line) from Schlesinger, 1977.

compounds that gives soil organic matter most of its unique properties. Humic compounds are true products of the soil. They are condensations of smaller organic radicals produced by catabolism of plant and animal products within the soil. Our knowledge of the structure and function of these humic compounds remains rudimentary.

Humic substances are defined as naturally occurring, biogenic, heterogeneous organic compounds that can generally be characterized as being yellow to black in color, of high molecular weight, and refractory (Aiken et al., 1985). They are commonly placed into three fractions: humic acid, soluble in water but not below pH 2; fulvic acid, soluble in water at any pH; and humin, insoluble in water at any pH. These fractions are an extraordinarily complex mixture of organic compounds. While virtually every chemical separation technique has been applied to humic substances, no one has yet succeeded in separating them into discrete components (Stevenson, 1982; MacCarthy et al., 1990).

THE SOIL TROPHIC SYSTEM

A complex trophic system composed of numerous species of organisms representing many phyla are involved in the decay process. The rate and kind of processes that occur in forest soils are closely related to the populations of soil dwelling organisms that compose this system. Soil quality is closely linked to these soil processes, and, consequently, to the soil biota. The community of soil organisms that influences soil processes and soil quality includes: macrofauna, mesofauna, microfauna, microorganisms, and plant roots (Lal, 1988; Schlesinger, 1991).

Macrofauna are species large enough to disrupt soil structure through their burrowing or feeding activities (2–20 mm) and include small mammals, earthworms, termites, ants, and other insects (Dindal, 1990; Stork & Eggleton, 1992; Berry, 1994). These organisms create macropores that influence infiltration rate and promote gas exchange. They are responsible for a great deal of soil mixing and can profoundly influence horizonation and rooting patterns. They often consume a very high C/N ratio organic material and produce a much lower C/N ratio detritus.

Mesofauna are those soil dwelling biota that are large enough (0.1–2 mm) to escape the surface tension of water and move freely in the soil, but small enough not to disrupt soil structure through their movement. They include some mollusks, springtails, mites, wood lice, and other small arthropods. These organisms are extremely important in decomposition and mineralization (Crossley et al., 1992). Although a great deal has been written about them, it is largely taxonomic rather than functional in nature. We are still quite unclear as to the total function and the essentiality of this group of organisms in soil organic matter dynamics.

Microfauna are organisms that are less than 100 μ in size and are usually confined to water films on soil particles. This group includes protozoa, turbellaria worms, rotifers, nematodes, and gastrotricha worms. Nematodes are the main component of this class. They are often thought of solely as root parasites, but they also are important decomposers. They are important predators on fungi and

Table 1–1. Biomass and respiration of soil fauna in various forest types (after Kitazawa, 1971).

Forest type	Biomass kJ/m²	Respiration
	kj/m²	kj/m²/yr
Alpine coniferous	117	715
Temperate deciduous	109	694
Subtropical rain	109	2048
Tropical rain	38	865
Equatorial wet	380	1580
Equatorial dry	50	1141

bacteria. In this way they regulate microbial populations, and they are important in mineralization.

The biomass and metabolism of the soil fauna have been estimated for a variety of sites (Table 1–1). The biomass of soil fauna may exceed the biomass of other terrestrial fauna on a per area basis, and their respiration can produce a significant loss of soil C. Large differences in the ratio of biomass to respiration in Table 1–1 are due to temperature, moisture, and the specific make up of the soil faunal population.

An incredible array of microorganisms live in the soil (Table 1–2). These estimates of bacteria, actinomycetes, fungi, and yeasts are based on plate counts and are probably conservative (Stevenson, 1986). Many forest soils are rich in fungi and contain more than a million per gram. Estimates of microbial biomass vary widely. Total live weight of bacteria in soil is about 0.15 to 1.5 g/kg of soil and the weight of fungal tissue ranges from 0.24 to 2.4 g/kg of soil.

As organic material passes through the soil food web, some C is converted to CO_2, some is incorporated into the soil biomass, and some is converted into stable humus. The humus fraction of the soil is constantly being mineralized (or decomposed); thus, although large quantities of organic residues may be returned to the soil annually, decay does not necessarily increase soil humus content. Maintenance of steady-state levels of organic matter requires a return of CO_2 to the atmosphere through soil respiration approximately equal to gains from photosynthesis. Thus, the organic matter content of most undisturbed soils remains essentially constant from year to year.

Several stages can be delineated in the decay of organic remains in the soil. Earthworms and other soil animals play an important role in reducing the size of

Table 1–2. Approximate numbers of microorganisms commonly found in soils (Stevenson, 1986).

Organism	Estimated number
	(per gram of soil)
Bacteria	3 000 000 – 500 000 000
Actinomycetes	1 000 000 – 20 000 000
Fungi	5 000 – 900 000
Yeasts	1 000 – 100 000
Algae	1 000 – 500 000
Protozoa	1 000 – 500 000
Nematodes	50 – 200

fresh plant material. Enzymes produced by soil microorganisms also are important in this phase of decomposition. The initial phase of microbial attack is characterized by the rapid loss of readily decomposable organic substances such as simple sugars, amino acids, most proteins, and some polysaccharides. In subsequent stages, soil biomass and more resistant compounds such as lignin and cellulose are decomposed, and humic substances are synthesized.

Stages in the microbial decomposition of organic residues are depicted in Fig. 1–2. During each stage some C is converted to CO_2, some is incorporated into soil biomass, and some is converted to stable humus. In most temperate zone forest soils, about one-third of the annual C addition will remain in the soil at the end of the growing season (Stevenson, 1986). The amount of C in microbial biomass has generally been considered insignificant in comparison to the amount of C in stable humus and therefore has been ignored by soil scientists in considering the C balance of the soil.

Although the C content of the microbial biomass is low, it is another "active fraction" of the soil C pool. Interest in soil microbial biomass has risen with the realization that microorganisms play a major role in the retention and release of nutrients (Paul & Voroney, 1980; Jenkinson & Ladd, 1981; Smith, 1994). Microbial biomass varies from 0.4 to 4 g/kg soil, admittedly a small fraction of the total soil organic C. However, from 0.5 to 15% of the soil N and from 2 to

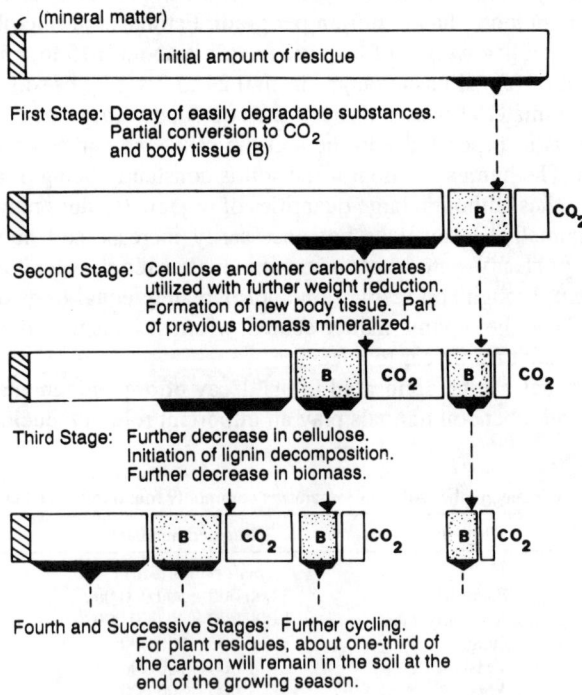

Fig. 1–2. Stages in the microbial decomposition of organic residue in soil. The letter B denotes microbial biomass from Stevenson, 1986.

20% of soil P is contained in this biomass (Paul & Juma, 1981; Brookes et al., 1982). The growth of trees may be more closely related to the amount and/or turnover rate of soil microbial biomass than to total soil organic C.

CONFRONTING THE CONUNDRUM

There are many areas of potential relationship between tree growth and soil organic matter in which our knowledge is imperfect. Some of them have already been mentioned. I would like to concentrate on three areas that might hold the key to the soil organic matter conundrum: the role of organic compounds as chelating agents, the role of extractable or solution C, and the tropnic system.

Despite all our efforts in humus layer classification and humic substance characterization, our understanding of the ecological role of humus is very poor. We know that humic compounds are powerful chelators and that they form very stable metal-organic complexes (MacCarthy et al., 1990). But exactly how and to what extent does this protect against metal toxicity or increase nutrient availability? Are these chelating or complexing compounds catabolized in the soil and must they constantly be synthesized if they are to play an effective role in the soil environment? Are these metal-organic complexes simply chemical curiosities or are they significant in promoting aggregate stability and enhancing soil structure?

Are simple organic coatings more valuable than complexes in promoting soil structure? Are such coatings short- or long-lived in the soil environment? Must they be constantly augmented with compounds newly derived from fresh organic material? Is the humic fraction of soil organic matter a source or a sink for N, P, S? Does it shift roles during succession or following disturbance? These are but a few of the questions we must answer before we can adequately address the ecological role of humus.

A second area in which we lack sufficient understanding is the role of C in soil solution, or water-extractable organic C, in soil fertility. We know that many compounds in this tiny fraction of soil organic matter can chelate metallic ions. This can promote nutrient availability and moderate toxic effects of Al (Ae et al., 1990; Fox & Comerford, 1992). These compounds should be easily decomposed in the soil, so their effect may be ephemeral if they are not continuously produced. What regulates their production?

Water-extractable organic C may contain most of the compounds that are the source of mineralizable N and P. What modulates the movement of N- and P-rich compounds into the soluble pool? If most of these soluble organic compounds form a ready energy source for soil microorganisms, to what extent does this fraction of soil organic matter promote immobilization rather than mineralization? The composition of the soluble fraction of soil organic matter likely controls which of these two processes is favored. What regulates that composition?

What role does water soluble organic C play in developing soil structure and influencing other soil physical properties? We know that disturbance can drastically alter the amount of soluble organic C in the soil. To what degree does this influence soil-plant interactions, short-term soil quality, and long-term productivity? We are only beginning to understand the important ecological roles of soluble organic C in the soil system.

The final area is the most complex and least well understood. Our concern for it is so recent that it does not have an accepted name. I have called it the soil trophic system. There is an intricate food web and a complex trophic structure involved in the decomposition of organic material and the creation of soil organic matter in the forest ecosystem. This is the most complex, diverse, and complicated portion of nearly all forest ecosystems, and it is the portion of the system about which we know least. Our understanding of the complexity and importance of the detrital food chain in aquatic systems has steadily advanced over the past three decades. At the same time our knowledge of the terrestrial detrital food chain has remained minimal. If this portion of the system is as important in forests as it is in lakes, then it is no wonder that soil organic matter is more of a conundrum than a clue.

We know that soil organisms are important in promoting aeration, aggregate stability, mineral mixing, creating macropores, narrowing the C/N ratio and releasing nutrients, and in the production of humic substances. We do not know how important soil organisms are to each other. For example, how important is it that all nodes in a food web are occupied by stable populations of organisms? How important is it that all of the niches are filled? How dependent is one trophic level on another for the appropriate substrate to function properly?

It seems that human activity is most likely to alter the structure and balance of the detrital food chain. Is this the way in which we impact soil quality and long-term productivity? The chances are very good that it is. Such disruption of the ecosystem might have little immediate effect in agronomic systems where intensive management often overrides natural function, but in less intensively managed forests such disruption of the ecosystem might have profound effects.

CONCLUSION

Robert Peters, in *A Critique for Ecology* (1991) decries the status of ecological science, while lauding the ability of forestry and other applied ecological initiatives to "provide a highly sophisticated set of predictive relations." He faults ecologists for using tenuous models, posing unanswerable questions, employing spurious or trivial correlations, and in general failing to properly use the power of hypothetico-deductive science. He employs several examples to illustrate the "poor health" of ecology concluding by saying that evidently ecology is a weak science because those environmental problems that ecology should be able to solve are instead worsening, growing more imminent and more monstrous each day. Criticism is an essential part of science, and Peters' critique should prove to be healthy for ecology. Forest science also should pay attention to the points Peters raises.

In forestry, we also see the problems we should be able to solve becoming more monstrous and intractable each day. We also find our science weak in the face of the demands placed upon it by an ever-questioning public. There may be a few outstanding examples of "sophisticated predictive relations" in forest pest management, genetics, etc., but we pose many unanswerable questions, employ spurious or trivial correlations, invoke tenuous models, and ignore the power of the hypothetico-deductive approach to science. Ecological research may too often be only "scholastic puzzle-solving," but forestry research is too often "empirical hegemony."

We should take a more theoretical approach in our attempt to understand the role of soil organic matter in specific and the maintenance of "soil quality" in general. We hope to develop "sophisticated predictive relations," but the road to such relations lies in the development of cogent theories rather than the development of giant regression models. We would be well served by more judicious application of the hypothetico-deductive method in our research (Fig. 1–3). We should pay close attention to Peters' criteria for judging science.

Science should be judged at both the analytical (scientific) and the social (public) level. Peters proposes the following analytical criteria: **Goal**—is the goal of the study well and properly defined? **Relevance**—is the goal relevant to a "need to know"? **Immediacy**—the theory involving fewest intermediates is best. **Operationalism**—will the research develop something that can be made operational? **Accuracy**—can we reliably measure the accuracy of the theory or model? **Generality**—to what degree can results of the work be generalized? **Precision**—who much error is there in predictions from the theory? **Quantification**—is the theory quantitative as proposed to qualitative? **Economy**—is there an easier, valid way to do it? Peters also suggests that science should be judged with respect to its social dimension. **Practicability**—theories should be easy to put into service. **Simplicity**—theories should be as conceptually simple as possible.

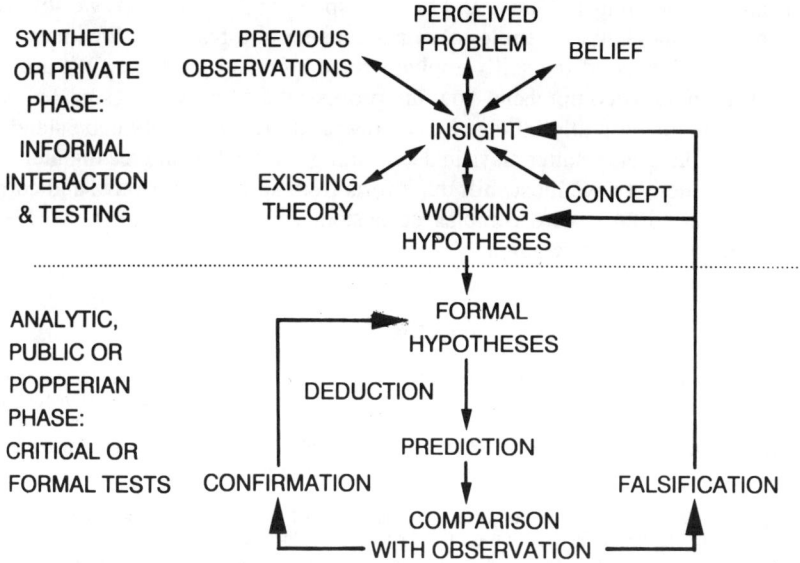

Fig. 1–3. A schematic diagram of the hypothetico-deductive method, indicating the separation of private and public phases of theory building, from Peters, 1991.

Consistency—theories should be described within the common norms of science and in plain English. **Heuristic power**—a scientific construct with weak predictive power may be valuable if it stimulates others to build on it.

If we apply Peters' criteria to our science and emphasize the hypothetico-deductive approach, we are positioned to make rapid advances in our understanding of the relation of soil organic matter to soil quality, tree growth, and long-term site productivity. As we have seen in this chapter, and as will be reiterated in following chapters we have inductively established many valid questions about these relationships. We have at our disposal the theories, classifications, techniques, technologies, etc., to generate testable hypotheses, create and perform critical experiments, derive predictive models and, in short, to push our science forward a quantum leap, and make it strong.

SUMMARY

Soil organic matter content is an easily observable, easily quantifiable soil property. "Black dirt" suggests a productive site, and people have long used soil color as an index of the worth of land for growing crops. Soil organic matter or organic C is routinely measured in the laboratory by a variety of methods. But, although dark, organic matter-rich soils are generally more productive than pale, organic matter-poor soils, there is seldom a strong statistical correlation between total organic matter content and crop yield. Soil scientists have devoted countless hours to discussing the importance and classification of surface organic layers and soil organic substances. They have worked diligently at describing, classifying, and enumerating soil fauna and flora. In spite of all this activity, we still have a poor understanding of the roles of organic matter in forest soils. We know little about the influence of the soil's trophic structure on soil processes. We are also only beginning to comprehend how the processes driven by or associated with soil organic matter really influence tree growth. If we are to truly understand the part that soil organic matter plays in determining soil quality and the maintenance of soil productivity, we must shift our efforts from simple descriptive and correlative studies toward more complex ecosystem level studies and process level hypothetico-deductive research.

REFERENCES

Ae, N., J. Arihara, K. Okada, T. Yoshihara, and C. Johansen. 1990. Phosphorus uptake by pigeon pea and its role in cropping systems of the Indian subcontinent. Science (Washington, DC) 248:477–480.

Aiken, G.R., D.M. McKnight, R.L. Wershaw, and P. MacCarthy (ed.). 1985. Humic substances in soil, sediment, and water: Geochemistry, isolation, and characterization. Wiley-Interscience, New York.

Berry, E.C. 1994. Earthworms and other fauna in the soil. p. 61–90. *In* J.L. Hatfield, and B.A. Stewart (ed.) Soil biology: Effects on soil quality. Lewis Publ., Boca Raton, FL.

Brookes, P.C., D.S. Powlson, and D.S. Jenkinson. 1982. Measurement of microbial biomass phosphorus in soil. Soil Biol. Biochem. 14:319–329.

Crossley, D.A., B.R. Mueller, and J.C. Perdue. 1992. Biodiversity of microarthropods in agricultural soils: relations to processes. Agric. Ecosyst. Environ. 40:37–46.

Dindal, D.L. 1990. Soil biology guide. John Wiley & Sons, New York.

Fox, T.R., and N.B. Comerford. 1992. Influence of oxalate loading on phosphorus and aluminum solubility in Spodosols. Soil Sci. Soc. Am. J. 56:290–294.
Green, R.N., R.L. Trowbridge, and K. Klinka, 1993. Towards a taxonomic classification of humus forms. For. Sci. Monogr. 29. Soc. Am. For., Bethesda, MD.
Jenkinson, D.S., and J.N. Ladd. 1981. Microbial biomass in soil: Measurement and turnover. p. 415–472. In E.A. Paul and J.N. Ladd (ed.) Soil biochemistry. Vol. 5. Dekker, New York.
Kitazawa, Y. 1971. Metabolism in forest litter. p. 485–498. In P. Duvigneaud (ed.) Productivity of forest ecosystems. UNESCO, Paris.
Kononova, M.M. 1961. Soil organic matter, its nature, its role in soil formation and in fertility. Pergamon Press, Oxford, England.
Lal, R. 1988. Effects of microfauna on soil properties in tropical ecosystems. Agric. Ecosyst. Environ. 24:101–116.
MacCarthy, P., C.E. Clapp, R.L. Malcolm, and P.R. Bloom (ed.). 1990. Humic substances in soil and crop sciences: Selected readings. ASA and SSSA, Madison, WI.
Page, A.L., R.H. Miller, and D.R. Keeney (ed.). 1982. Methods of soil analysis. Part 2. Vol. 2. ASA and SSSA, Madison, WI.
Paul, E.A., and R.P. Vorney. 1980. Nutrient and energy flows through soil microbial biomass. p. 215–237. In D.C. Ellwood et al. (ed.) Contemporary microbial ecology. Acad. Press, New York.
Paul, E.A., and N.G. Juma. 1981. Mineralization and immobilization of soil nitrogen by microorganisms. p. 179–195. In F.E. Clark and T. Rosswall (ed.) Terrestrial nitrogen cycles: Processes, ecosystems, strategies, and management impacts. Ecol. Bull. 33. Swedish Natural Sci. Res. Council, Stockholm.
Peters, R.H. 1991. A critique for ecology. Cambridge Univ. Press, New York.
Pritchett, W.L., and R.F. Fisher. 1987. Properties and management of forest soils. John Wiley & Sons, New York.
Schlesinger, W.H. 1977. Carbon balance in terrestrial detritus. Annu. Rev. Ecol. System. 8:51–81.
Schlesinger, W.H. 1991. Biogeochemistry: An analysis of global change. Acad. Press, San Diego, CA.
Smith, J.L. 1994. Cycling of nitrogen through microbial activity. In J.L. Hatfield and B.A. Stewart (ed.) Soil biology: Effects on soil quality. Lewis Publ., Boca Raton, FL.
Stevenson, F.J. 1982. Humus chemistry: Genesis, composition, reactions. Wiley-Interscience, New York.
Stevenson, F.J. 1986. Cycles of soil, carbon, nitrogen, phosphorus sulphur, micronutrients. Wiley-Interscience, New York.
Stork, N.E., and P. Eggleton. 1992. Invertebrates as determinants of soil quality. Am. J. Altern. Agric. 7:38–47.
Wiersum, L.K. 1974. p. 123–129. In The activity of specific growth stimulating substances in the soil in relation to the application of organic matter. Trans. Int. Congr. Soil Sci. Conf. 3, 10. Int. Soc. Soil Sci. Moscow, USSR.

2 Forest Soil Organic Matter: Characterization and Modern Methods of Analysis

John G. McColl and Noam Gressel

University of California
Berkeley, California

IMPORTANCE OF ORGANIC MATTER

Forests are long-lived and, in contrast to many seasonal agricultural crops, forests themselves contribute in a major way to the development and maintenance of the supporting soil. The longevity of forests leads to nutrient cycling processes that are in an approximate state of equilibrium, with little natural disturbance from year to year, and an intact litter layer and soil organic matter (OM) content that determine many of the physical, chemical and biological properties of the forest soil.

Litter accumulates on the surface mineral soil, forming the O horizon, which is a protective layer modifying the water and temperature regimes of the profile. It is a major source of humus in the mineral soil, supplies both inorganic and organic leachates to lower horizons, and provides a rich substrate for microorganisms and macroinvertebrates. Soil OM is the sum of all C-containing substances in soils (other than carbonates), consisting of a mixture of plant and animal residues in various stages of decomposition, of substances synthesized microbially or chemically from the breakdown products, and of the remains and living bodies of soil microorganisms (Schnitzer, 1991). Reviews of the classification, nature and reactions of soil OM are given by Aiken et al. (1985), Huang and Schnitzer (1986), MacCarthy et al. (1990), Tate (1987), and Wilson (1991). It is gratifying to note that recent texts in general ecology include chapters on soil OM and its importance in cycling processes (e.g., Aber & Melillo, 1991).

Changes in long-term forest productivity are closely related to changes in OM including the forest floor and in mineral soil in particular (Powers et al., 1990). Forest manipulations that alter the addition of OM to the soil or change decomposition rates will alter the C balance. There is concern that forest harvest

---

Copyright © 1995 Soil Science Society of America, 677 S. Segoe Rd., Madison, WI 53711, USA.
Carbon Forms and Functions in Forest Soils.

practices may remove OM or understory vegetation which otherwise would contribute to soil fertility and maintenance of site productivity (Powers et al., 1990). This concern may be justified, depending on a number of factors, including logging residues, soil disturbance, and presence of N-fixing plants.

The fertility of a soil is intimately related to the organic C content. Commonly the soil OM of a forest A-horizon accounts for 75 to 90% of total cation exchange capacity (McColl, 1994). Thus the amount of readily available nutrient cations derived from mineral weathering, is largely dependent on the amount of soil OM present. Soil OM, also contains essentially all of the N, P and S which are mineralized through microbial oxidation processes. Manipulation of soils also results in changes in soluble soil organics, which affect microbial dynamics, mineral weathering rates, and availability of nutrient cations. It is obvious therefore, that changes in OM affect both physical and chemical soil characteristics and subsequently have profound effects on the availability of water and nutrients essential to sustain forest health (Stevenson, 1986; Powers, 1991).

ROLE OF SOIL ORGANISMS

A healthy forest supports a diverse community of microorganisms and mesofauna which mix and decompose OM. Forest soils generally have greater species diversity of decomposer organisms than grassland, although the organism biomass and density are usually lower, as shown by data from the humid Tropics (Lavelle et al., 1992). Temperate forests of the Pacific Northwestern USA may harbor greater diversity of species (mostly decomposer soil organisms) than even tropical forests that are generally considered to be among the most diverse of ecosystems (Moldenke, 1990), but quantitative proof of this is lacking. In tropical rainforests, litter is rapidly decomposed by bacteria and fungi, but in old-growth forests of Oregon arthropods appear to be the key decomposers of forest litter. There are hundreds of invertebrate species and about 75 species of oribatid mites per square meter of soil in these forests, and up to 150 different kinds of mycorrhizae on the roots of a single mature Douglas fir [*Pseudotsuga menzesii* (Mirb.) Franco] (Moldenke, 1990).

Forest soils are usually more acid than grassland soils (Jenny, 1980), and conifer forest litter is usually more acid than that of hardwoods (Pritchett, 1979). Fungi have a wide tolerance to soil pH, but bacteria prefer neutral to alkaline soils (Brady, 1990); this favors fungi in conifer soils and bacteria in hardwood soil in temperate zones. In addition to decomposer fungi and bacteria, there is a wide array of other decomposer organisms in forest soils that break down plant litter and dead roots (Gosz, 1984). These include earthworms, centipedes, fly larvae, mites, termites and other mesofauna. Decomposer organisms help maintain soil structure and fertility, and are important in nutrient cycling processes. Quantitative data on their role is scarce (e.g., Moldenke, 1992). Papers presented at the recent International Workshop on "Soil Structure/Soil Biota Interrelationships," provide a good review of the role of soil organisms, both in mediating transformations of important nutrients in soil and also as key players in determining physical characteristics of soil (e.g., Ladd et al., 1993; Oades, 1993).

PHYSICAL AND CHEMICAL CHARACTERISTICS OF SOIL ORGANIC MATTER

This review of the nature of soil OM in forest soils focuses on its chemical characteristics, but this is not to imply that some major effects of OM in soil are not physical, affecting aggregation, aeration, porosity, moisture characteristics and related properties. In fact, many physical effects of soil OM are intimately related to chemical effects, and are difficult to separate. For example, we have found that laboratory storage decreased the dispersibility of forest soil samples, thus implying an increase in aggregate stability (Jersak et al., 1992). This physical effect was accompanied by a decrease in the chemical extractability of inorganic Fe, Al and Si and their organic complexes, thought to be due to inorganic phases becoming more stable, and/or due to increased organic-metallic hydrophobicity as the samples dried during storage.

Soil OM occurs in the solid, colloidal and soluble states. The surface area per unit volume of colloidal material is very large and many important chemical and biological reactions (particularly sorption and partitioning processes) occur at the solid–liquid interface. Much of the soil OM under field conditions is insoluble and bound: as insoluble macromolecular complexes; as macromolecular complexes bound together by di- and trivalent cations, such as $Ca(II)$, $Fe(III)$, $Al(III)$; and in various combinations with clay minerals, including polyvalent cation bridging (clay–metal–humus), H bonding, and van der Waal's forces (Stevenson, 1985).

Broadly speaking, humus of forest soil (e.g., Alfisols, Spodosols, and Ultisols) is characterized by a greater content of the more-soluble fulvic acid, whereas that of peat (Histosols) and grassland soil (e.g., Mollisols) contains more of the less-soluble humic acid (Stevenson, 1985). Although comprising only a small fraction of total OM, the soluble organic fraction in forest soil is very important as it is responsible for weathering of primary minerals, via chelation and acidification. Polyphenols, organic acids, and other complexing substances leaching from litter and from soil OM are influential in the formation of Spodosols, through their dissolution of Fe and Al sesquioxides (Stobbe & Wright, 1959). Low-molecular-weight organics (LMWO) from a wide array of overstory and understory leaf-litter from California mixed-conifer forests have been identified and found to be largely controlling the dissolution of soil Al, Fe, Mn and Mg (McColl et al., 1990; Tam & McColl, 1990, 1991).

Fresh plant material decomposes, releasing C, H, O, N, S, P and cationic elements such as Ca, Mg, and K. Nutrient elements are recycled by microorganisms in the soil and litter, and often made available for plant uptake. The elemental composition and functional group content of a humic acid from a Mollisol Ah horizon, and of a fulvic acid from a Spodosol Bh horizon are given in Table 2–1 (Schnitzer, 1991). Interactions between the inorganic elements and the organic fraction are complex but can have important effects on the availability of nutrients for plant uptake. For example, soluble organics can affect P availability through chelation; oxalate, malate and other organic acids are very effective in preventing the precipitation of phosphate by Fe and Al oxides (Struthers & Sieling, 1950; Fox et al., 1990).

Table 2–1. Analytical characteristics of HA and FA (from Schnitzer, 1991).

	HA	FA
Element	g kg^{-1}	
C	564	509
H	55	33
N	41	7
S	11	3
O	339	448
Functional group	cmol kg^{-1}	
COOH	450	910
Phenolic OH	210	330
Alcohol OH	280	360
Quinonoid C=O	250	60
Ketonic C=O	190	250
OCH$_3$	30	10
E$_4$/E$_6$ ratio	4.3	7.1

ORGANIC FRACTION CHARACTERIZATION

The organic constituents are comprised of carbohydrates, amino acids and proteins, nucleic acids, lipids, lignins and humus, listed in approximate order of increasing resistance to decay. The following is a description of the organic components of OM and is largely abstracted from the recent text by Tan (1993).

The carbohydrates are polyhydroxy aldehydes, ketones or substances yielding these compounds on hydrolysis, and include simple and complex sugars, starches, hemicellulose and cellulose. Amino acids are the fundamental structural units of protein; N occurs in the amino group attached to the C chain, and the carboxyl group exhibits acidic properties. In acid soils, most of the amino acids occur as positively charged compounds. Nucleic acids include plant-derived deoxyribonucleic acid (DNA) and ribonucleic acid (RNA) as long chains of alternating sugar and phosphate residues. Lipids are heterogeneous compounds of fatty acids, waxes and oils, without particular chemical structures, but are soluble in fat solvents including ether, chloroform, or benzene; phospholipids contain P. Lipids have low solubility in water and exhibit hydrophobic characteristics. Lignins are complex thermoplastic, highly aromatic polymers, derived from coniferyl alcohol or guaiacol propane monomers, and are an important source of aromatic components in humus. Lignin and lipids are highly resistant to microbial decomposition and are insoluble in water, many organic solvents, and strong acids. Lignins form large, polymeric molecules of complex structure, difficult to describe, although Tan (1993) suggests that a systematic arrangement of the basic monomers into lignin is a possibility.

The composition of organic compounds in fresh oak and pine litter (Table 2–2), is taken from the classic study of Waksman and Tenney (1928). Tree litter typically has higher lignin and lower protein contents than succulent plants and most agricultural crops, and pines have very high contents of fats, waxes and

Table 2–2. Composition of fresh leaf litter of oak and pine (from Waksman & Tenney, 1928).

Component	Oak	Pine
	g kg^{-1}	
Water soluble	130	73
Hemicelluolse	180	171
Cellulose	128	148
Lignin	248	219
Protein	43	21
Fats, waxes, resins	98	240
Ash	51	25

resins. Consequently, tree litter, especially that of pines, usually takes longer to decompose than that of seasonal agricultural crops.

Soil humus is defined as "that more or less stable fraction of the soil OM remaining after the major portions of added plant and animal residues have decomposed. Usually it is dark in color" (Brady, 1990). Humus consists of a heterogeneous mixture of compounds for which no single structural formula can be given. Humus has three main fractions, operationally defined by their solubilities: humin is insoluble in water at any pH, humic acid (HA) is insoluble in water under acid conditions (<pH 2), fulvic acid (FA) is soluble in water under all pH conditions. Humic acids have higher molecular weights than FA, and FA contain higher O but lower C contents and have more acidic functional groups, particularly COOH (Table 2–1).

Decomposition of OM is characterized by gradual changes in C functional groups. Changes can be detected as a function of decomposition time (Blair, 1988; Inbar et al., 1989), down a gradient of litter and soil horizons (Kögel, 1986; Tam et al., 1991) or as a function of size and density fractions (Baldock et al., 1992). The major changes that occur are a decrease in carbohydrates, an increase in the relative proportion of alkyl (aliphatic chains) and carboxyl C, and partial breakdown of lignin (Stevenson, 1982; Zech et al., 1990). Mineralization of important inorganic nutrient elements also occurs as OM decomposes, although such relationships are not well researched. For example, although there is clear evidence of differential changes in total P, organic P, and microbial P as litter decomposes (Blair, 1988; Hart & Firestone, 1991; Schoenau et al., 1989; Walbridge et al., 1991), little is known of the relationship between changes in P species with changes in associated functional C groups.

Considerable work has been done in describing and classifying the soluble fractions of OM in various aqueous solutions (Junk et al., 1974; Keefer et al., 1984; Leenheer & Huffman, 1976). Pohlman and McColl (1988) developed a fractionation scheme for soluble litter organics because of their importance in solubilizing metals and thus facilitating metal movement from forest litter and in the mineral soil profile. Hydrophobic acids and bases, hydrophilic acids and bases, and hydrophobic and hydrophilic neutrals were identified, and significant correlations were found between certain organic fractions and dissolved Al and Fe (Table 2–3).

Table 2–3. Correlation coefficients for litter organics vs. soil dissolved Al and Fe content of leachates (from Pohlman & McColl, 1988).

Organics	Al	Fe
Hydrophobics		
Acid C	0.91***	0.71**
Base C	0.82***	0.54*
Neutral C	0.57*	0.32
Hydrophilics		
Acid C	0.93***	0.85***
Base C	0.90***	0.82***
Neutral C	0.95***	0.79***
Polyphenolic C	0.94***	0.89***

*,**,*** r values significant at $P<0.05$, 0.01, and 0.001, respectively; $n = 13$.

NEW METHODS AND RELEVANT APPLICATIONS

Historical perspectives of research on soil humus, and interactions of HA and FA with soil mineral components, are discussed by Stevenson (1985). MacCarthy and Rice (1985) emphasize that the fundamental chemical nature of humic substances is still largely a mystery and that, "the study of humic substances is the study of complicated, ill-defined mixtures, and many researchers appear to overlook this basic fact." Rice and Lin (1993) recently published a unique article demonstrating that it is possible to describe the nature of humic materials using the concepts of fractal geometry. They noted that, "on the basis of the characteristics that humic materials must possess in order to perform their functions in natural systems, it has been proposed that the fundamental chemical characteristic of humic material is not a discrete chemical structure but a profound lack of order on a molecular level."

Classification of soil OM and descriptions of its components have historically been predicated on fractionation and extraction procedures which destroy the nature of the original soil OM as found in field conditions. New methods which attempt to preserve the natural integrity of soil OM are gradually replacing the older chemical methods, and are antiquating the solubility-based, operational classification of soil OM into FA, HA and humin fractions. Concepts of soil OM, its components, functions and formation, change with the introduction of new technologies—providing new insights and facilitating new theories. Descriptions of some of the newer methods of analysis and their applications to forest soil OM research follow.

Spectroscopy in Infrared and Ultraviolet-Visible Regions

Fourier transform infrared spectroscopy (FTIR) is extremely useful for qualitative and semiquantitative analysis of soil OM. Stretching and bending vibrations of chemical bonds can be increased by infrared light, and bonds from different functional groups will resonate at different frequencies, thus absorbing at characteristic wavelengths.

The spectra of water-soluble OM extracted from litter and soil horizons under oak (*Quercus*) in a Mediterranean forest are shown in Fig. 2–1. With an increase in soil depth, increases in absorption were observed for peaks at 1410 cm^{-1} (COO$^-$, phenolic OH and aliphatic bonds), 1125 cm^{-1} (sulfone S-O) and 870 cm^{-1} (carbonates), and decreases were observed for peaks at 1600 cm^{-1} (aromatic C=C, COOH and ketone C=O) and 1080 cm^{-1} (polysaccharide C-O); peak assignments follow Sposito et al. (1976), Stevenson (1982) and Baes and Bloom (1989). These changes indicate increasing humification with soil depth. Further confirmation of this trend is apparent from the C/N ratios of the bulk samples (Fig. 2–1).

Compounds that contain double bonds or nonbonded electrons on O, halogens and S atoms will absorb in the ultraviolet (UV; 200–400 nm) and visible (400–800 nm) regions. A chromophore is any part of a molecule that contains one or more of the chemical species listed above. Absorption occurs when electrons in σ, π and η-orbitals are excited from their ground state energy level to higher energy levels.

The diversity of chromophores in soil OM usually results in featureless absorption spectra in the UV and visible spectral regions. As a result, little information can be gained about specific functional groups in soil OM using this

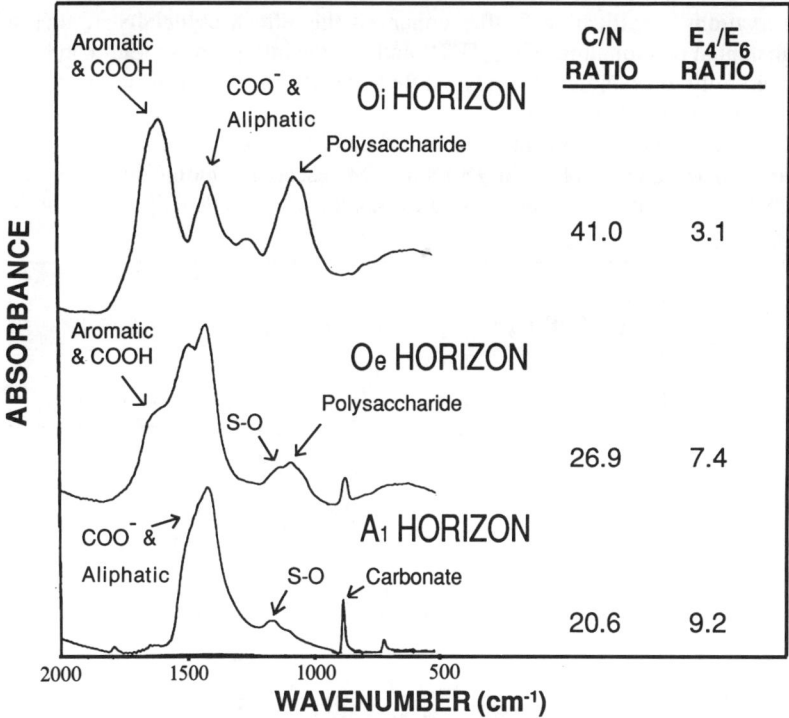

Fig. 2–1. Changes in organic compounds with decomposition; aqueous extracts from oak litter and the A horizon of a calcareous forest Haploxeroll in Israel, as shown in FTIR spectra (modified from Gressel et al., 1994).

spectroscopic technique (Stevenson, 1982). However, the ratio of absorption at 465 nm and 665 nm (the E_4/E_6 ratio) can provide characteristic information about the chromophoric nature of OM in solution. Chen et al. (1977) found the E_4/E_6 ratio of humic substances to be governed by molecular weight, with correlation to contents of O, C, COOH and total acidity. Thus, smaller molecules, higher total acidity and higher COOH content correspond to larger E_4/E_6 ratios. Ratios of 3.0 to 5.7 have been reported for HA, and ratios of 6.0 to 10.0 for FA, as shown by Schnitzer (1991) in Table 2–1, and reported elsewhere (Stevenson, 1982; Chen et al., 1977). We have reported E_4/E_6 ratios of 3.3 to 10.5 for aqueous extracts of forest litter and 7.3 to 11.7 for aqueous extracts of forest soil (Gressel & McColl, 1992; Gressel et al., 1994); the ratio increased with soil depth, which is in agreement with FTIR spectral data of the extracts, and with C/N ratios (Fig. 2–1).

We also have used data from FTIR and E_4/E_6 ratios to investigate effects of understory vegetation and fertilization on the nature of aqueous extracts of pine litter (Gressel & McColl, 1992), as part of a comprehensive study of the primary factors controlling the growth of *Pinus ponderosa* (Dougl. ex Laws. var. ponderosa) plantations in California (Powers et al., 1992). Reduction of understory vegetation with herbicides resulted in relatively greater oxidation and aliphatic content in the litter extracts compared to nontreated plots (Fig. 2–2). Higher E_4/E_6 ratios were measured in litter from herbicided plots, in agreement with the FTIR observations. Fertilization further enhanced this effect, especially at sites with high annual precipitation. Thus, FTIR and UV-visible spectroscopies can be used successfully to examine relatively subtle and early changes resulting from recent forest management practices.

Developments in infrared spectroscopic techniques are currently being examined for their applicability to soil OM research. Among them are diffuse reflectance infrared Fourier transform spectroscopy (DRIFT) and cylindrical

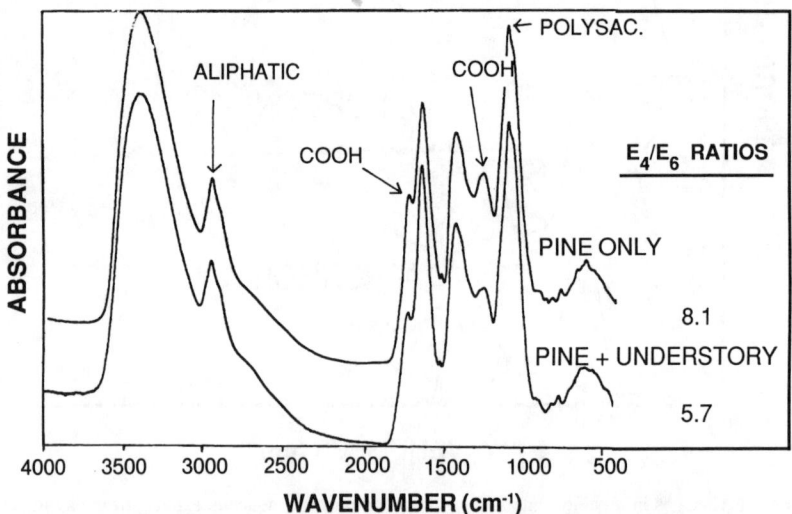

Fig. 2–2. Organics from FTIR spectra and E_4/E_6 ratio of litter from a 6-yr-old Ponderosa pine plantation on a Ultisol in northern California, with and without understory vegetation (from Gressel & McColl, 1992).

internal reflectance (CIR). The DRIFT requires minimal sample preparation for powdered samples of soil, litter and OM, replacing the older method of compression into pellets with potassium bromide (Stevenson, 1982). Cylindrical IR enables direct analysis of OM in solution without need for drying, which may alter the chemistry of soluble OM.

Fluorescence spectra of dissolved organic carbon (DOC) in UV-visible wavelengths provide greater spectral detail than the absorption spectra. Effects of pH and metals are apparent, although more research is needed in interpretation of results. For example, Cabaniss (1992) studied the effects of metal binding with OM. Figure 2–3 shows the effect of CU(II) on a FA. The difference between the spectra with and without the metal is called the "quenching spectrum," exhibiting a strong peak centered at 465 nm. Three distinct types of quenching spectra were obtained for divalent transition metals (Cu(II), Pb(II), Ni(II), Co(II) and Mn(II)), Mg(II), and Al(III) (Cabaniss, 1992). This suggests that synchronous quenching spectra may be used for qualitative study of metal binding sites in OM.

Chromatography and Extraction Techniques

Different chromatographic techniques have been used to separate chemical compounds in solutions and gaseous mixtures. A sample is generally run through a separating gel or column, which will slow down certain compounds while enabling other compounds to pass through rapidly. Separation is achieved through ionic or other specific chemical reactions between the column resin and the compounds, or through pore-size specificity. Detectors sensitive to the compounds of interest (usually electroconductivity, UV absorbance or fluorescence detectors) determine the amounts leaving the column at any given time. Comparison of the retention times of samples to those of standards provides

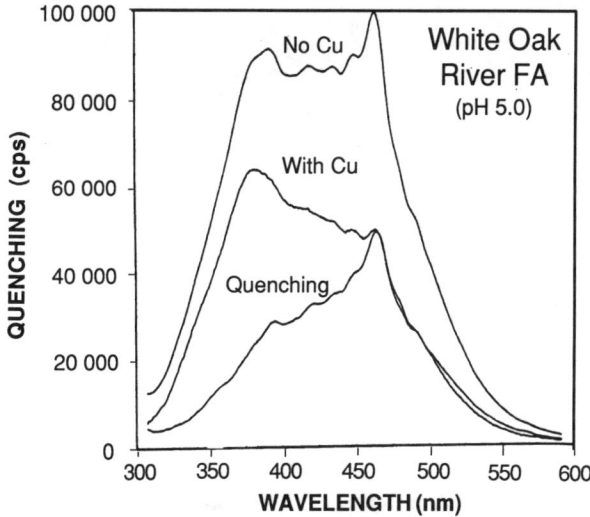

Fig. 2–3. Synchronous fluorescence spectra of DOC of White Oak River FA. Quenching spectra (smoothed difference) illustrates Cu binding (modified from Cabaniss, 1992).

quantitative identification. Further identification of the separated compounds is typically achieved by chemical analysis or mass spectroscopy.

Gas chromatography (GC), ion chromatography (IC) and high performance liquid chromatography (HPLC) are often used to study forest OM. Gas chromatography can be applied to study emissions from forest floors of volatile organic compounds and greenhouse gases, and to study N transformations. The HPLC and IC are used to identify low molecular-weight organics (LMWO) and their reactions in soils, and to monitor the reactions of xenobiotics with litter, leachates and soil components.

Pohlman and McColl (1988) used HPLC to identify specific LMWO compounds classified in the categories of acid, base and neutral organics in DOC, as mentioned earlier (Table 2–3). Relationships between DOC and dissolved Al and Fe contents have deterministic effects on pedogenic processes, and on availability of nutrients such as P. Phosphorus may be released as a result of competitive absorption with LMWO. Fox et al. (1990) found a close relationship between the inorganic P released from a spodic (Bh) horizon and the Al stability constants (K_{Al}) of the LMWO used to leach the soil (Fig. 2–4). The organic P released as a result of competition with LMWO was not related to the Al stability constants of the organics. However, organic P is 75% of the total P in the soil, though much of it may be in highly recalcitrant forms.

Degradation of pesticides and their absorption on minerals and soil OM can be determined using HPLC. Incorporation of xenobiotics into soil OM is considered the major reaction for many xenobiotics in soils. Certain forest practices, including removal or burning of litter after harvesting, may reduce OM at a site. During the first few years of regrowth, pesticides are often used to eradicate weed species, and soil OM and litter amounts are low. Under these conditions, reactions with metal oxides may be of major importance. For example, we have

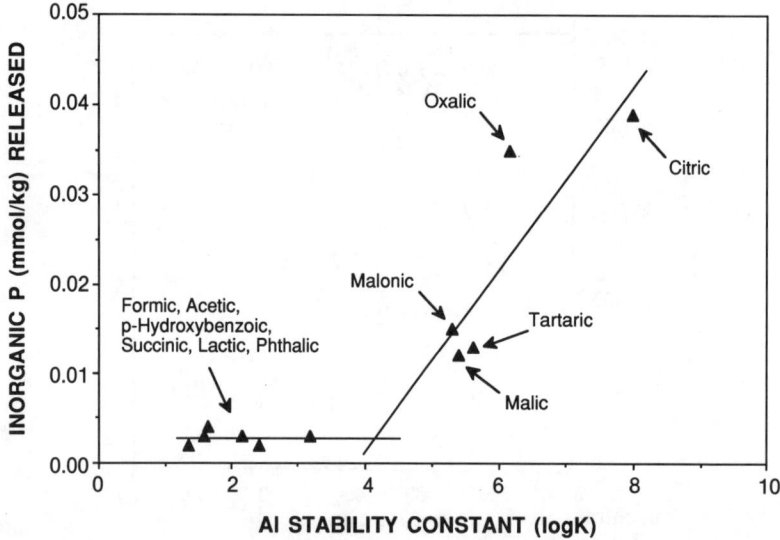

Fig. 2–4. Inorganic P release vs. Al stability constant ($logK_{Al}$) of organic acids in Bh horizon of a Florida pine forest Spodosol, showing a threshold $logK_{Al}$ of 4.1 (modified from Fox et al., 1990).

demonstrated differences in the loss of common herbicides from forest soils in solution, with and without OM or metal oxides (McGrath & McColl, 1992). Herbicide concentrations remaining in solution were determined using HPLC.

Other chromatographic and extraction techniques are currently being developed, providing detection of specific components within the continuum of soil OM. Supercritical Gas Extraction (SGE) has been used effectively by Schulten and Schnitzer (1991) to extract aliphatic acids from two soils. Carbon dioxide was used since it requires relatively mild conditions (supercritical conditions for CO_2 are 7.35 MPa and 31°C). Long-chain alkanes, alkenes, saturated and unsaturated fatty acids, alcohols, ketones and alkyl-esters were identified. Previous studies used n-pentane, ethanol, distilled water and various mixtures as solvents, all yielding different ratios of aliphatic and aromatic compounds (Schnitzer et al., 1986; Schnitzer & Preston, 1987; Schnitzer et al., 1991).

Nuclear Magnetic Resonance Spectroscopy of Soil Organic Matter

Nuclear magnetic resonance (NMR) is a form of spectroscopy based on the properties of nuclides that possess a nonzero spin and associated magnetic moment so that a resonance occurs at a characteristic frequency. This resonance can be detected by an appropriate experimental arrangement: included are a static magnetic field (H_0) necessary for polarizing the nuclear spins, an oscillating magnetic field at the characteristic frequency to induce resonance (H_1), and a receiving coil for detecting the NMR signal. Molecules may have magnetic nuclei precessing at slightly different frequencies representative of their different chemical environments. The resulting time-dependent NMR signal is complex, but a Fourier transformation and computer processing of the data provide an NMR spectrum. A detailed review of the principals and techniques is provided by Wilson (1987). Nondestructive solid-state ^{13}C-NMR techniques involving cross-polarization and magic-angle-spinning (CPMAS ^{13}C-NMR) have been used extensively over the past decade to study soil C. Unfortunately, research has focused primarily on temperate agricultural soils (Preston, 1991). A typical CPMAS ^{13}C-NMR spectrum of a Histosol from Québec, demonstrating the general peak designations, is presented in Fig. 2–5 (Preston et al., 1987). Since there is overlap between the peak areas of C-species, these divisions are somewhat arbitrary.

A recent study of a large variety of soils using CPMAS ^{13}C-NMR has enabled Baldock et al. (1992) to present a simple model describing OM decomposition. For example, data from an Australian Mollisol shows increasing decomposition with decreasing size fractions which corresponds to a decrease of the C/N ratio (Fig. 2–6). Carbohydrates and proteins are degraded readily in the litter followed by slow oxidation of lignin, leaving relatively large concentrations of the recalcitrant alkyl (aliphatic) structures in the smaller particle-size fractions.

Preston et al. (1990) studied fallen logs at various decay stages in old-growth Douglas fir (*P. menzesii*) forests using ^{13}C-NMR. Results showed relative increases in concentrations of lignin, waxes and resins, and overall loss of cellulose and hemicellulose during the first 50 yr of decomposition (Fig. 2–7). There was little chemical alteration after the initial 50 yr and little change of the lignin structure with decomposition.

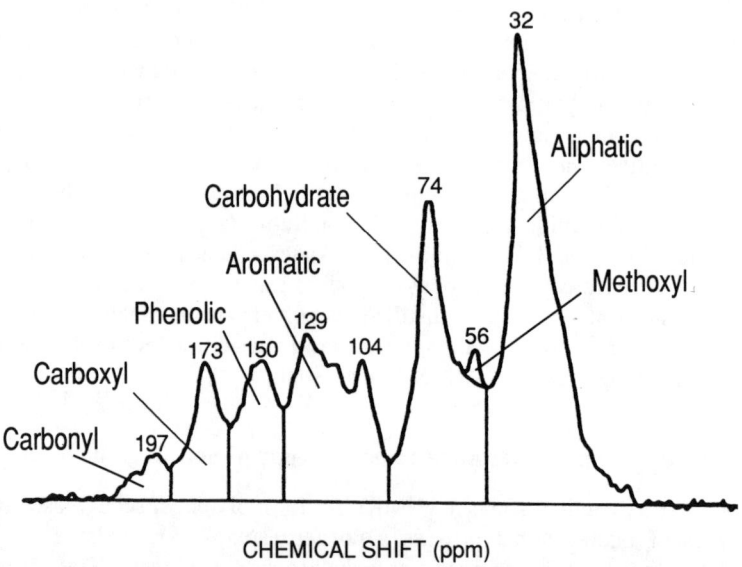

Fig. 2–5. Different C types at 95-cm depth in virgin soil at a site in transition to a bog in Quebec, shown in a CPMAS ^{13}C NMR spectrum (modified from Preston et al., 1987).

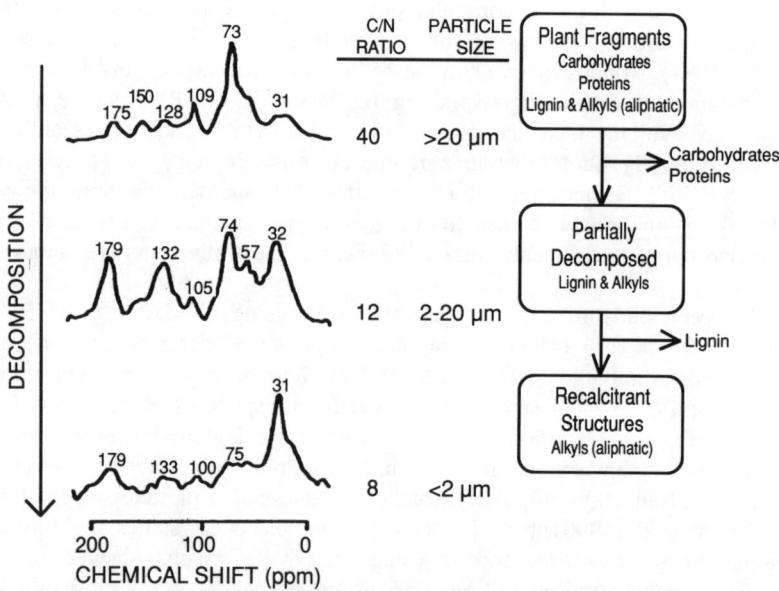

Fig. 2–6. Decomposition of soil OM in the A horizon of a South Australia wheat (Triticum aestivum) pasture Mollisol as a function of particle size. Parallel changes are shown in CPMAS ^{13}C NMR spectra and the C/N ratio (modified from Baldock et al., 1992).

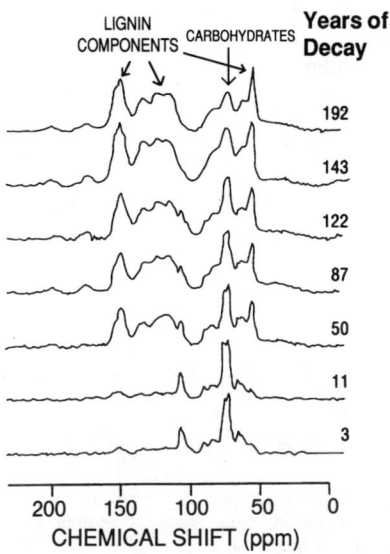

Fig. 2–7. Decomposition of heartwood in fallen boles of Douglas fir from sites in Washington and Oregon, as shown in CPMAS ^{13}C NMR spectra (modified from Preston et al., 1990).

A number of other nuclei can produce NMR signals apart from ^{13}C; ^{1}H, ^{15}N and ^{31}P are of major interest for soil OM research (Wilson, 1987). ^{1}H-NMR is often used in conjunction with ^{13}C-NMR for structural analysis of OM. The ^{31}P nuclei provides a good probe, since all P in nature is comprised of this isotope. Low concentrations of P in the soil counteract this advantage for direct NMR probing of soils. Thus, ^{31}P-NMR spectroscopy of soils usually involves extraction with NaOH solution or exchange resins. Adams and Byrne (1989) present ^{31}P-NMR spectra from *Eucalyptus* forest soils in Western Australia. The forms of P apparent in the spectra are inorganic P, monoester-P (mainly inositol phosphates), diester P (nucleic acids and phospholipids), phosphonates (with C-P bonds), polyphosphates and pyrophosphates (Fig. 2–8). Gil-Sotres et al. (1990) found correlations between bicarbonate-extractable organic P and the amounts of diester-P ($r = 0.87$) and nonmonoester-P ($r = 0.97$), as determined by ^{31}P-NMR. These results confirm that labile organic P consists mainly of diester P and explains the poor correlation between labile organic P and measurements of phosphomonoesterase activity.

Nitrogen-15 NMR can be used effectively only after appropriate enrichment with ^{15}N, a relatively rare isotope in natural abundance. This drawback can be used as an advantage to study interactions of N fertilizers or xenobiotics with soil components. Enrichment of the compound of interest with ^{15}N enables good spectral resolution with minimal interference (noise) from background soil N, which is not enriched. This technique has been used by Ginwalla and Mikita (1992) to study the reaction of Suwannee River FA (a standard FA) with chloramine, a compound being considered for disinfection of municipal water; there was similarity between the reactions of ammonia and chloramine with the FA (Fig. 2–9). The signals can be assigned to primary amides, formed by reactions of esters with

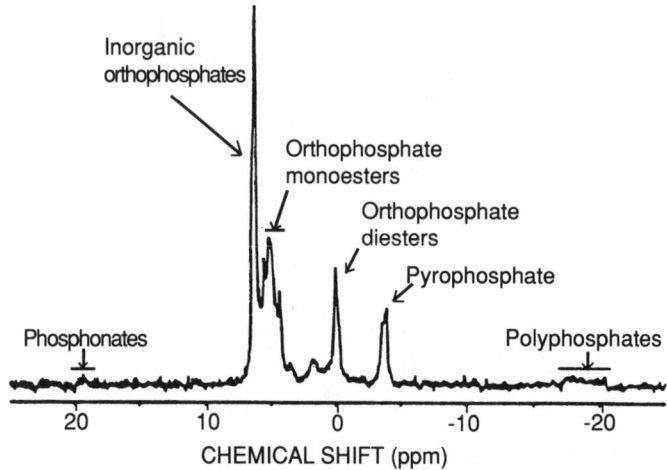

Fig. 2–8. Phosphorus compounds in concentrated extracts of highly weathered, red earth soils supporting old-growth Karri forest at Pemberton, Western Australia, as shown in 31P NMR spectra (modified from Adams & Byrne, 1989).

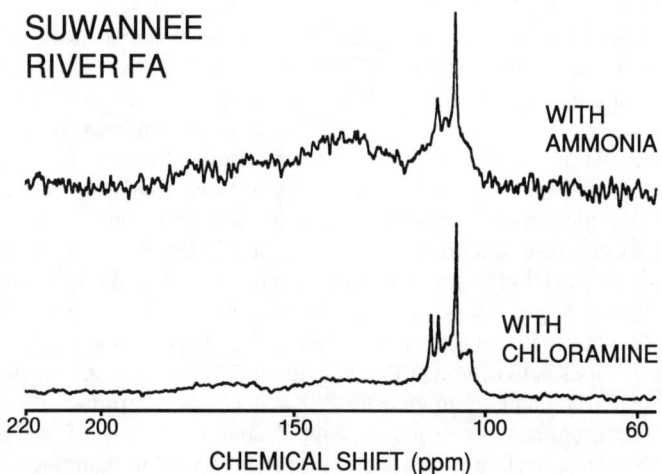

Fig. 2–9. Identification of N-containing compounds in water with Suwannee River FA following addition of ammonia and chloroamine, as shown in 15N NMR spectra (modified from Ginwalla & Mikita, 1992).

chloramine or, more likely, with its coproduct ammonia. A recent study by Almendros et al. (1991) combined ^{13}C and ^{15}N-NMR to examine changes that occurred during composting. *Lolium rigidum* plants were enriched with ^{15}N and then composted for 70 d. While the ^{13}C-NMR spectra exhibited a typical pattern of humification, no new ^{15}N peaks were formed during the 70-d period. These results suggest that assigning a significant fraction of HA N to heteroaromatic structures may be wrong. Further investigation is required before final conclusions can be drawn, but this study provides a good example of how new techniques lead us to question conventional theories.

Pyrolysis Techniques

Chemical degradation of macrostructures such as lignin and humic substances provide limited information due to incomplete degradation and artifacts. Pyrolysis of organic compounds can yield better information of their chemical building blocks. In this technique, a pulse of thermal energy is applied to a sample, splitting the molecules at the weaker bonds, creating an array of typical products for each molecular structure. Pyrolysis products are then passed through a gas chromatograph or through field ionization, before identification on a mass spectrometer (Py-GC-MS and Py-FIMS, respectively). The resulting mass signals are compared to data sets of peaks of assigned compounds. Examples of such studies are reported by Kögel et al. (1988), and Schulten and Schnitzer (1992).

Pyrolysis products of litter, whole soils and humic substances have been studied in forest soil research, but more research is needed before reliable quantitative information can be gained using this technique.

Electron Spin Resonance

Organic free radicals and metals such as Cu, Fe, and Mn can be detected using electron spin resonance (ESR). Senesi and Sposito (1989) determined the modes of complexation of OM with these metals. Similarly, Tam et al. (1991) examined litter horizons from a ponderosa pine plantation in California. The three principal features of the ESR spectra are a resonance at $g = 4.3$ [Fe(III) in inner-sphere complexes], a six-lined resonance at $g = 2.0$ [Mn(II) in outer-sphere coordination] and a sharp resonance at $g = 2.0$ arising from semiquinone-type free radicals, suggesting an increase in humification with increasing litter depth (Fig. 2–10). Extractability of Mn and Fe from these forest floor layers was consistent with the coordination chemistry observed by ESR.

SUMMARY AND DIRECTIONS FOR FUTURE RESEARCH

Soil OM has important interactions with clays and metals, contributes to the soil cation exchange capacity and nutrient availability, is influential in the formation of aggregates, aids in retardation of erosion and enhancement of the moisture regime, and is a major factor in soil formation processes.

Past research has largely been restricted by the inability to study soil OM in an undisturbed state. Methods of analyses employed traditional wet-chemistry techniques, involving destructive extraction and fractionation of OM, using many solvents or other techniques that destroy the solid state of the OM and arbitrarily divide it into portions that may have little relevance to its original state and functionality. Many of the current research techniques are moving away from these approaches, by employing modern analytical equipment that allow analysis of

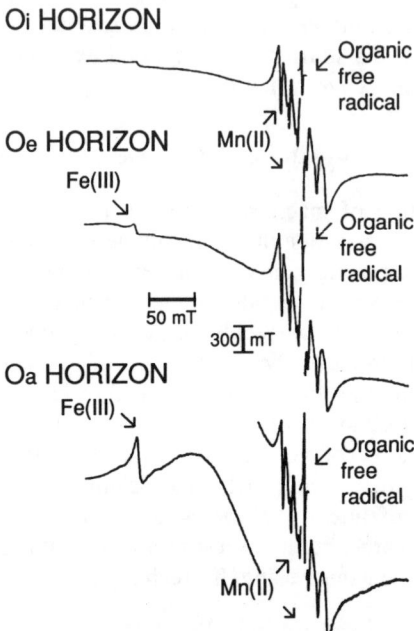

Fig. 2–10. Increases of organic-free radicals with increasing decomposition of Ponderosa pine litter in California, as shown in ESR spectra; spectra also indicate that Fe(III) was bound as inner-sphere complexes, whereas Mn(II) was bound as outer-sphere complexes (modified from Tam et al., 1991).

OM in its natural forms. As Schnitzer (1991) stated so well in his recent paper entitled "Soil organic matter—the next 75 years," there will be "no toxic odors of organic solvents, corrosive liquids, explosive methylating reagents" etc., in the modern analytical laboratory researching the nature of soil OM. Modern methods include the use of high performance liquid chromatography (HPLC), supercritical gas extraction (SGE), and pyrolysis-mass spectroscopy (Py/MS). These techniques supplement the nondestructive spectroscopic methods of NMR, FTIR, ESR, and fluorescence spectroscopy (FS). Studies using combinations of these techniques are elucidating the structure and functionality of OM in forest soils, without the necessity of destructive extraction procedures that are typical of older, wet-chemistry techniques.

The obvious, overwhelming characteristic of soil OM which is the main obstacle in arriving at a valid concept of its structure, is its heterogeneity. Future research, fueled by recent availability of more-sophisticated equipment, will start to unfold the complex structures of OM, and will lead to a better understanding of interactions between OM and other soil components.

Many of the organic compounds containing N, P and S have yet to be identified; until this is done, the processes controlling the availability of these elements will remain largely as a mystery! Metal- and clay-humic interactions also will be researched in the future using the modern analytical equipment, and thus will greatly enhance knowledge of the processes of soil genesis and nutrient availability, and an understanding of soil structure and related physical properties.

Interactions between soil OM and xenobiotics also will be elucidated by modern techniques. Such findings will be of great value to understand and predict the persistence, degradation, bioavailability, leaching and volatility of contaminants in forest soils.

Apart from the use of modern analytical techniques, there is a renewed interest in the role of mesofauna in the breakdown and cycling of soil OM in forests. This type of work is usually laborious and time-consuming but should be encouraged, as many of the important soil mesofauna are still to be classified, let alone understood in terms of their particular roles in molding both the physical and chemical characteristics of forest soils.

Recent concern over carbon dioxide build-up in the atmosphere and the need for better inputs in models of global cycling of C, dictate the need to know more about the structures, functions and amounts of soil OM in forest soils. On more-local scales, better information also is needed about the role of soil OM in maintaining the long-term productivity of forests, and the effect of different forest management practices on soil OM.

Use of new analytical techniques is essential to answer current and critical applied problems. Forest researchers also must reduce the lag-time between discovery of new techniques and their applicability. Many disciplines, including soil chemistry, organic chemistry, ecology, forest management and microbiology, must be brought together in future research on soil OM, as it is such a complex, but important component of forest ecosystems.

ACKNOWLEDGMENTS

This research review was largely supported by Hatch Project no. CA-B-SSC-5607.

REFERENCES

Aber, J.D., and J.M. Melillo. 1991. Terrestrial ecosystems. Saunders College Publ., Philadelphia, PA.

Adams, M.A., and L.T. Byrne. 1989. ^{31}P-NMR analysis of phosphorus compounds in extracts of surface soils from selected Karri (*Eucalyptus diversicolor* F. Muell.) forests. Soil Biol. Biochem. 21:523–528.

Aiken, G.R., D.M. McKnight, R.L. Wershaw, and P. MacCarthy (ed.). 1985. Humic substances in soil, sediment, and water: geochemistry, isolation, and characterization. Wiley-Interscience, New York.

Almendros, G., R. Fründ, F.J. Gonzalez-Vila, K.M. Haider, H. Knicker, and H.-D. Lündermann. 1991. Analysis of ^{13}C and ^{15}N CPMAS NMR-spectra of soil organic matter and composts. FEBS Lett. 282:119–121.

Baes, A.U., and P.R. Bloom. 1989. Diffuse reflectance and transmission Fourier transform infrared (DRIFT) spectroscopy of humic and fulvic acids. Soil Sci. Soc. Am. J. 53:695–700.

Baldock, J.A., J.M. Oades, A.G. Waters, X. Peng, A.M. Vassallo and M.A. Wilson. 1992. Aspects of the chemical structure of soil organic materials as revealed by solid-state ^{13}C NMR spectroscopy. Biogeochemistry 16:1–42.

Blair, J.M. 1988. Nitrogen, sulfur and phosphorus dynamics in decomposing deciduous leaf litter in the southern Appalachians. Soil Biol. Biochem. 20:693–701.

Brady, N.C. 1990. The nature and properties of soils. 10th ed. MacMillan, New York.

Cabaniss, S.E. 1992. Synchronous fluorescence spectra of metal-fulvic acid complexes. Environ. Sci. Technol. 26:1133–1139.

Chen, Y., N. Senesi, and M. Schnitzer. 1977. Information provided on humic substances by E_4/E_6 ratios. Soil Sci. Soc. Am. J. 41:352–358.

Fox, T.R. N.B. Comerford, and W.W. McFee. 1990. Phosphorus and aluminum release from a spodic horizon mediated by organic acids. Soil Sci. Soc. Am. J. 54:1763–1767.

Gil-Sotres, F., W. Zech, and H.G. Alt. 1990. Characterization of phosphorus fractions in surface horizons of soils from Galicia (N.W. Spain) by ^{31}P NMR spectroscopy. Soil Biol. Biochem. 22:75–79.

Ginwalla, A.S., and M.A. Mikita. 1992. Reaction of Suwannee River fulvic acid with chloramine: characterization of products via N-15 NMR. Environ. Sci. Technol. 26:1148–1150.

Gosz, J.R. 1984. Biological factors influencing nutrient supply in forest soils. p. 119–146. *In* G.D. Bowen and E.K.S. Nambiar (ed.) Nutrition of plantation forests. Acad. Press, New York.

Gressel, N., and J.G. McColl. 1992. Organic matter in forest soil: Carbon and phosphorus fractions. p. 347. *In* Agronomy abstracts. ASA, Madison, WI.

Gressel, N., Y. Inbar, A. Singer, and Y. Chen. 1994. Chemical and spectroscopic properties of leaf litter and decomposed organic matter in the Carmel Range, Israel. Soil Biol. Biochem. (In press.)

Hart, S.C., and M.K. Firestone. 1991. Forest floor—mineral interactions in the internal nitrogen cycle of an old-growth forest. Biogeochemistry 12:103–127.

Huang, P.M., and M. Schnitzer (ed.). 1986. Interactions of soil minerals with natural organics and microbes. SSSA Spec. Publ. 17. SSSA, Madison, WI.

Inbar, Y., Y. Chen, and Y. Hadar. 1989. Solid-state carbon-13 nuclear magnetic resonance and infrared spectroscopy of composted organic matter. Soil Sci. Soc. Am. J. 53:1695–1701.

Jenny, H. 1980. The soil resource: Origin and behavior. Springer-Verlag, New York.

Jersak, J.M., J.G. McColl, and J.F. Hetzel. 1992. Changes in extractability of iron, aluminum and silicon and dispersibility by storage of California forest soils. Commun. Soil Sci. Plant Anal. 23:993–1018.

Junk, G.A., J.J. Richard, M.D. Grieser, D. Witiak, J.L. Witiak, M.D. Arguello, E. Vick, H.J. Svec, J.S. Fritz, and G.V. Calder. 1974. Use of macroreticular resins on the analysis of water for trace organic contaminants. J. Chromatogr. 99:745–762.

Keefer, R.F., E.E. Codling, and R.W. Singh. 1984. Fractionation of metal-organic components extracted from a sludge-amended soil. Soil Sci. Soc. Am. J. 48:1054–1059.

Kögel, I. 1986. Estimation and decomposition pattern of the lignin component in forest humus layers. Soil Biol. Biochem. 18:589–594.

Kögel, I., R. Hempfling, W. Zech, P.G. Hatcher, and H.-R. Schulten. 1988. Chemical composition of the organic matter in forest soils: 1. Forest litter. Soil Sci. 146:124–136.

Ladd, J.M., R.C. Foster, and J.O. Skjemstad. 1993. Soil structure: Carbon and nitrogen metabolism. Geoderma 56:401–434.

Lavelle, P., E. Blanchant, A. Martin, A.V. Spain, and S. Martin. 1992. Impact of soil fauna on the properties of soils in the humid tropics. p. 157–185. *In* R. Lal and P.A. Sanchez (ed.) Myths and science of the tropics. SSSA Publ. 29. SSSA and ASA, Madison, WI.

Leenheer, J.A., and E.W.D. Huffman, Jr. 1976. Classification of organic solutes in water by using macroreticular resins. J. Res. U.S. Geol. Surv. 4:737–751.

MacCarthy, P., and J.A. Rice. 1985. Spectroscopic methods (other than NMR) for determining functionality in humic substances. p. 527–560. *In* G.R. Aiken et al. (ed.) Humic substances in soil, sediment, and water: Geochemistry, isolation, and characterization. Wiley-Interscience, New York.

MacCarthy, P., C.E. Clapp, R.L. Malcom, and P.R. Bloom (ed.). 1990. Humic substances in soil and crop sciences: Selected reading. SSSA and ASA, Madison, WI.

McColl, J.G. 1994. Forest soils. *In* C.W. Finkl Jr. (ed.). Encyclopedia of soil science and technology. Chapman and Hall, New York. (In press.)

McColl, J.G., A.A. Pohlman, J.M. Jersak, S.-C Tam, and R.R. Northup. 1990. Organic and metal solubility in California forest soils. p. 178–195. *In* S.P. Gessel et al. (ed.) Sustained productivity of forest soils. Proc. 7th N. Am. For. Soils Conf., Vancouver, Canada. 24–28 July 1988. Univ. British Columbia. Faculty For Publ., Vancouver, Canada.

McGrath, A.E., and J.G. McColl. 1992. Sorption and surface reactions of herbicide on ferrihydrite, birnessite, and forest soils. p. 47. *In* P.M. Huang (ed.) Impacts of interactions of inorganic, organic and microbiological soil components on environmental quality. ISSS Working Group MO, 1st Workshop. Univ. of Alberta, Edmonton, Canada.

Moldenke, A.R. 1990. One hundred twenty thousand little legs. Wings. Summer, p. 11–14.

Moldenke, A.R. 1992. Non-target impacts of management practices on the soil arthropod community of Ponderosa pine plantations. p. 78–103. *In* K. Harcksen (ed.) Proc. 13th Annu. For. Vegetation Manage. Conf., Eureka, CA. 14–16 January. For. Veg. Manage. Conf., Redding, CA.

Oades, J.M. 1993. The role of biology in the formation, stabilization and degradation of soil structure. Geoderma 56:377–400.

Pohlman, A.A., and J.G. McColl. 1988. Soluble organics from forest litter and their role in metal dissolution. Soil Sci. Soc. Am. J. 52:265–271.

Powers, R.F. 1991. Are we maintaining the productivity of forest lands? Establishing guidelines through a network of long-term studies. p. 70–81. *In* A.E. Harvey and L.P. Neuenschwander (ed.) Proc. Manage. Productivity of Western-Montane Forest Soils, Boise, ID. 10–12 April 1990. Gen. Tech. Rep. INT-280. USDA-FS, Intermountain Res. Stn., Ogden, UT.

Powers, R.F., D.H. Alban, D.H. Miller, A.E. Tiarks, C.G. Wells, P.E. Avers, R.G. Cline, N.S. Jr. Loftus and R.O. Fitzgerald. 1990. Sustaining productivity in North American forests: Problems and prospects. p. 49–79. *In* S.P. Gessel et al. (ed.) Sustained productivity of forest soils. Proc. 7th N. Am. For. Soils Conf., Vancouver, Canada. 21–28 July 1988. Univ. British Columbia. Faculty For. Publ., Vancouver, Canada.

Powers, R.F., G.T. Ferrell, and T.W. Koerber. 1992. The Garden of Eden experiment: Four year growth of Ponderosa pine plantations. p. 46–63. *In* K. Harcksen (ed.) Proc. 13th Annu. For. Veg. Manage. Conf., Eureka, CA. 14–16 January. For. Veg. Manage. Conf., Redding, CA.

Preston, C.M. 1991. Using NMR to characterize the development of soil organic matter with varying climate and vegetation. p. 27–36. *In* Stable isotopes in plant nutrition, soil fertility and environmental studies. Proc. Int. Symp., Vienna, Austria. 1–5 October 1990. IAEA, Vienna, Austria.

Preston, C.M., S.-E. Shipitalo, R.L. Dudley, C.A. Fyfe, S.P. Mathur, and M. Levesque. 1987. Comparison of ^{13}C CPMAS NMR and chemical techniques for measuring the degree of decomposition in virgin and cultivated peat profiles. Can. J. Soil Sci. 67:187–198.

Preston, C.M., P. Sollins, and B.G. Sayer. 1990. Changes in organic components for fallen logs in old-growth Douglas-fir forests monitored by ^{13}C nuclear magnetic resonance spectroscopy. Can. J. For. Res. 20:1382–1391.

Pritchett, W.L. 1979. Properties and management of forest soils. John Wiley & Sons, New York.

Rice, J.A., and J.-S. Lin. 1993. Fractal nature of humic materials. Environ. Sci. Technol. 27:413–414.

Schnitzer, M. 1991. Soil organic matter—the next 75 years. Soil Sci. 151:41–48.

Schnitzer, M., D.A. Hindle, and M. Meglic. 1986. Supercritical gas extraction of alkanes and alkanoic acids from soils and humic materials. Soil Sci. Soc. Am. J. 50:913–919.

Schnitzer, M., and C.M. Preston. 1987. Supercritical gas extraction of soil with solvents of increasing polarities. Soil Sci. Soc. Am. J. 51:639–646.

Schnitzer, M., H.-R. Schulten, P. Schuppli, and D.A. Angers. 1991. Extraction of organic matter from soils with water at high pressures and temperatures. Soil. Sci. Soc. Am. J. 55:102–108.

Schoenau, J.J., J.W.B. Stewart, and J.R. Bettany. 1989. Forms and cycling of phosphorus in prairie and boreal forest soils. Biogeochemistry 8:223–237.

Schulten, H.-R., and M. Schnitzer. 1991. Supercritical carbon dioxide extraction of long-chain aliphatics from two soils. Soil Sci. Soc. Am. J. 55:1603–1611.

Schulten, H.-R., and M. Schnitzer. 1992. Structural studies on soil humic acids by Curie-point pyrolysis-gas chromatography/mass spectroscopy. Soil Sci. 153:205–224.

Senesi, N., and G. Sposito. 1989. Characterization and stability of transition metal complexes of chestnut (*Castanea sativa* L.) leaf litter. J. Soil Sci. 40:461–472.

Sposito, G., K.M. Holtzclaw, and J. Baham. 1976. Analytical properties of the soluble, metal-complexing fractions in sludge-soil mixtures: II. Comparative structural chemistry of fulvic acid. Soil Sci. Soc. Am. J. 40:691–697.

Stevenson, F.J. 1982. Humus chemistry: Genesis, composition, reactions. Wiley-Interscience, New York.

Stevenson, F.J. 1985. Geochemistry of soil humic substances, p. 13–52. *In* G.R. Aiken et al. (ed.) Humic substances in soil, sediment, and water: Geochemistry, isolation, and characterization. Wiley-Interscience, New York.

Stevenson, F.J. 1986. Cycles of soil. C, N, P, S, micronutrients. Wiley-Interscience, New York.

Stobbe, P.C., and J.R. Wright. 1959. Modern concepts of the genesis of podsols. Soil Sci. Soc. Am. Proc. 23:161–164.

Struthers, P.H., and D.H. Sieling. 1950. Effects of organic anions on phosphate precipitation by iron and aluminum as influenced by pH. Soil Sci. 62:205–213.

Tam, S.-C., and J.G. McColl. 1990. Aluminum and calcium-binding affinities of organic ligands in acidic conditions. J. Environ. Qual. 19:514–520.

Tam, S.-C., and J.G. McColl. 1991. Aluminum-binding ability of soluble organics in Douglas fir litter and soil as examined by chromatography. Soil Sci. Soc. Am. J. 55:1421–1427.

Tam, S.-C., G. Sposito, and N. Senesi. 1991. Spectroscopic and chemical evidence of variability across a pine litter layer. Soil Sci. Soc. Am. J. 55:1320–1325.

Tan, K.H. 1993. Principles of soil chemistry. 2nd ed. Marcel Dekker, New York.

Tate, R.L., III. 1987. Soil organic matter: Biological and ecological effects. John Wiley & Sons, New York.

Waksman, S.A., and F.G. Tenney. 1928. Composition of natural organic materials and their decomposition in the soil: III. The influence of nature of plant upon the rapidity of its decomposition. Soil Sci. 26:155–171.

Walbridge, M.R., C.J. Richardson, and W.T. Swank. 1991. Vertical distribution of biological and geochemical phosphorus subcycles in two southern Appalachian forest soils. Biogeochemistry 13:61–85.

Wilson, M.A. 1987. NMR techniques and applications in geochemistry and soil chemistry. Pergamon, New York.

Wilson, W.S. (ed.). 1991. Advances in soil organic matter research: the impact of agriculture and the environment. Spec. Publ. 90. R. Soc. Chem., Cambridge, England.

Zech, W., R. Hempfling, L. Haumaier, H.-R. Schulten, and K. Haider. 1990. Humification in subalpine rendzinas: Chemical analysis, IR and ^{13}C NMR spectroscopy and pyrolysis-field ionization mass spectroscopy. Geoderma 47:123–138.

3 Fractionation of Soil Organic Matter with Supercritical Freon

Felipe G. Sanchez
USDA-FS
Research Triangle Park, North Carolina

Gregory A. Ruark
USDA-FS
Washington, District of Columbia

Soil organic matter (SOM) in the surface mineral soil is considered a major determinant of forest ecosystem productivity because it affects water retention, soil structure, and nutrient cycling (Powers et al., 1989). Forest productivity may be sensitive to climate variations that alter SOM decomposition rates, particularly those associated with the highly labile fractions that directly control nutrient cycling (Ruark & Blake, 1991). In general, nutrient turnover rates are more closely related to forest productivity than the total amount of SOM (Cole & Rapp 1981). Thus, in equations for predicting N-fertilization responses and N uptake, variations in the factors regulating decomposition (C/N ratio, moisture, temperature) are often of greater importance than total SOM (Edmonds & Hsiang, 1987, Binkley & Hart, 1987).

In the national efforts to monitor the health of forest ecosystems, many of the indicators of site quality are soil parameters, including total SOM. The recommended procedure is to report SOM values in units of total elemental C (Nelson & Sommers, 1986). During a 1-yr study in five forest stands in Georgia, monthly estimates of total SOM in the surface mineral horizons differed by more than 100% (Haines & Cleveland, 1981). The large short-term changes probably were caused by episodes of fine root mortality in response to soil moisture depletion. To track the status of forest productivity over time it is imperative to identify a baseline measure that represents the relatively recalcitrant SOM fraction. If this baseline measure is independent of season of sampling, it should reflect the long-term status of SOM. The size of the labile pool—the difference between total SOM and the baseline—may be equally important as an indicator of the status of nutrient cycling.

Copyright © 1995 Soil Science Society of America, 677 S. Segoe Rd., Madison, WI 53711, USA.
Carbon Forms and Functions in Forest Soils.

SUPERCRITICAL FLUIDS

Supercritical fluids (SCFs) are compounds that have been raised past their critical temperature and pressure. The critical temperature is the temperature above which a gas will not condense, regardless of the pressure applied. The critical pressure is the pressure above which a liquid will not vaporize, regardless of the temperature applied. Its state enables an SCF to diffuse and transport like a gas and to dissolve materials like a liquid, hence supercritical fluid extraction (SFE) is an attractive alternative to previous SOM extraction procedures. In other attempts to analyze SOM through SFE (Schnitzer et al., 1986; Schulten & Schnitzer, 1990; Capriel et al., 1990; Spiteller, 1985), no fractionation was attempted; however, Spiteller (1985) reported achieving four to five times better recovery than with conventional NaOH extraction. Recent advances in instrumentation and the introduction of within-matrix derivatization techniques for the extraction of polar and ionic compounds (Miller et al., 1991) allow for considerable improvements in the extraction and fractionation of SOM.

In this report we describe fractionation of SOM into labile and recalcitrant pools via SFE with freon-22 (chlorodifluoromethane). We also describe the trends these pools exhibit in relation to measures of productivity (leaf biomass and height increment) for different stands. Finally, we will describe the limitations of this procedure for the extraction of certain soils.

MATERIALS AND METHODS

A major need in the development of a standardized laboratory procedure for the determination of the labile and recalcitrant SOM pools is determining when the labile pool has been sufficiently removed. To accomplish this we drew on the SOM's inherent seasonal variability. We collected samples from each of three soil series [Lakeland (thermic, coated Typic Quartzipsamment); Fuquay (loamy, siliceous, thermic Arenic Plinthic Kandiudult); and Orangeburg (fine-loamy, siliceous, thermic Typic Paleudult)] that span a moisture/texture gradient at the Savannah River Station (SRS) in Aiken, South Carolina. Fifteen samples were collected at random from the surface 15 cm of the Ap horizon at each site every month from April 1990 to May 1991. The collected soil volume was 1750 cm^3 for each sample. The sites were occupied by second rotation, 15-yr-old (10-yr-old for the Orangeburg soil) loblolly pine (*Pinus taeda* L.) plantations planted on abandoned farm land where the previous agricultural practices left a homogenous plow horizon. The samples were air dried and passed through a 2-mm mesh sieve. Carbon content of each sample was measured on a Carlo Erba NA 1500 N/C/S analyzer [1](Fisons Instruments, Inc., Beverly, MA).

The SOM extraction scheme is shown in Fig. 3–1 and a complete description is presented herein. A 5-g subset of each sample was shaken overnight on a wrist action shaker with 50 cm^3 of 1 M HCl. The SOM and fine mineral components were separated from the coarse mineral fraction by decanting the acid, SOM, and fine mineral fraction into a Buchner funnel and filtering off the acid. Residual

---

[1]Product name is indicated for the benefit of the reader and does not imply preferential endorsement.

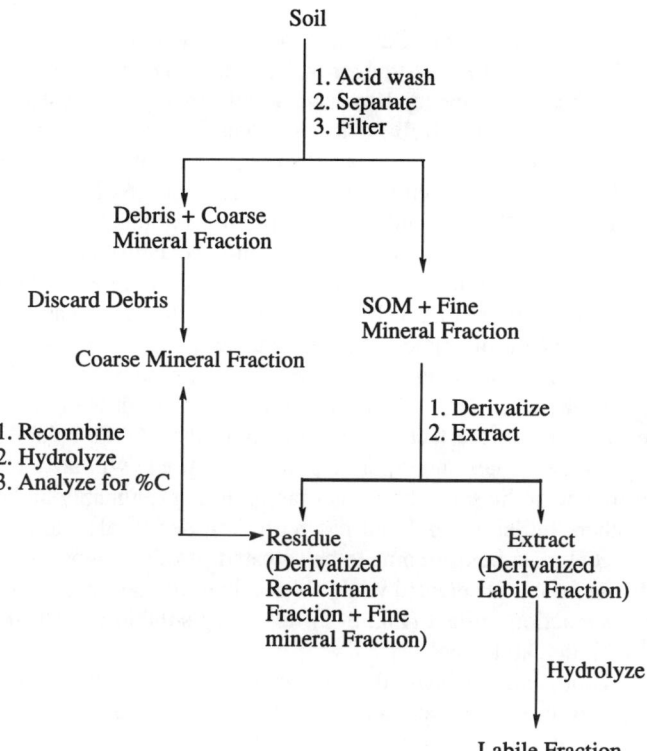

Fig. 3–1. Soil extraction procedure.

debris (i.e., small roots) was removed from the coarse mineral fraction by adding water and decanting off the lighter debris.

The Orangeburg soil required an additional step of clay removal prior to the separation procedure described above. This step is necessary because the clay tends to clog the extraction lines of the extractor described below. The procedure for clay removal was adapted from the particle size fractionation procedure described by Baldock et al. (1990). Briefly, a 5-g subset of each sample was added to 50 cm^3 of deionized water in a 150-cm^3 beaker and sonified for 5 min using a Branson 8200 Sonifier[1] (Branson Ultrasonics Corp., Danbury, CT) operating at 50% power. Ice was packed around the beaker to prevent sample heating. The dispersed sample was passed through a 53-µm sieve and the clay fraction (≤2-µm diam. particles) was removed by gravitational sedimentation in deionized water. Materials retained on the 53-µm sieve were added to the >2-µm diam. fraction. The >2-µm diam. fraction was further separated as described in the preceding paragraph, the clay was combined with the coarse mineral fraction.

The SOM and fine mineral fraction were air dried, weighed and subjected to off-line SFE on a Suprex MPS 225 supercritical fluid chromatograph/extractor[1] (Suprex Corp., Pittsburgh, PA). Freon-22 was selected as the solvent because of

---

[1]Product name is indicated for the benefit of the reader and does not imply preferential endorsement.

its demonstrated superiority to CO_2 and NO_2 for the extraction of materials from soils (Hawthorne et al., 1992). Stronger SCF solvents (i.e., methanol and acetone) were avoided because they tend to have relatively high critical temperatures (240.0 and 235.5°C, respectively) that could lead to the thermal degradation of some compounds. Freon (96°C) is the strongest SCF solvent we could use and still maintain moderate extraction conditions. The inert SCF freon atmosphere minimizes unwanted SOM chemical transformations at this temperature.

Our extraction strategy included increasing the labile pool's solubility in SCF freon by converting the strong polar and ionic sites present in both the mineral and SOM fractions to relatively weak nonpolar sites. This was accomplished by using 10% (v/v) of trimethylsilylimidazole (TMSI), a strong silylating reagent, as an additive to freon. The general reaction scheme for TMSI and polar compounds is shown in Fig. 3–2. We hypothesized that, in the absence of polar and ionic interactions, the factors that determine the SOM's solubility in SCF freon (i.e., Van der Waals and dipole-dipole interactions) will also influence the SOM's residency in the soil. The recalcitrant pool is presumably higher molecular weight than the labile pool and has more Van der Waals and dipole-dipole interactions that must be overcome to be solvated. If this assumption is accurate, then the labile pool was solvated while the recalcitrant pool remained in the soil. Imidazole, a reaction artifact (Fig. 3–2), is highly soluble in SCF freon and is extracted with the labile pool.

The SFE procedure consisted of a 10-min static run followed by a 10-min dynamic run at an oven temperature of 110°C and pressure of 40.5 MPa. The oven temperature was lowered to 100°C and the dynamic run was continued for another 10 min. The average flow rate was 1.5 cm^3/min during the dynamic mode and less than 0.2 cm^3/min for the static mode. The extraction procedure was done in triplicate in order to derivatize and remove all the material. This was confirmed when the fourth extraction did not yield any material detectable by gas chromatography.

The derivatized extract was converted to its original form by mild acid hydrolysis (silyl derivatives are very sensitive to hydrolysis). The hydrolysis procedure involved dissolving the extract in 5 cm^3 of 1 M HCl, neutralizing the acid with 1 M NaOH, and centrifuging the mixture. The liquid was decanted off and the residue washed three times with deionized water. The material was filtered and the solid residue was air dried. Imidazole is highly water soluble and is eliminated from the labile pool in this step. The residue was dissolved in

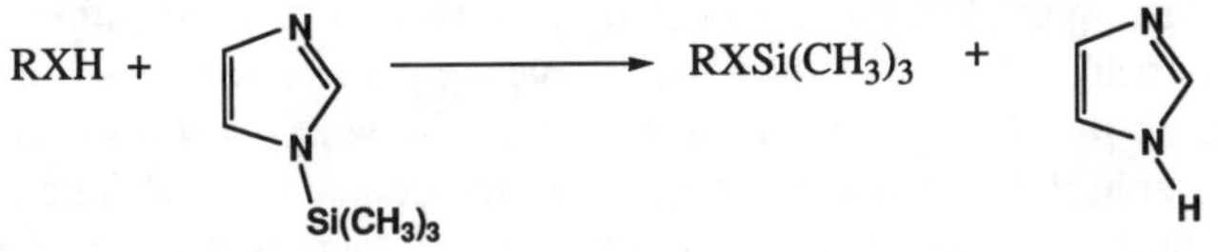

RXH + [imidazole-Si(CH$_3$)$_3$] ⟶ RXSi(CH$_3$)$_3$ + [imidazole-H]

X = O, N, S   TMSI   Imidazole

Fig. 3–2. General derivatization scheme.

N,N-dimethylformamide-d_7 and a liquid state ^1H nuclear magnetic resonance (NMR) spectrum was obtained on a Bruker AM × 300 spectrometer (Bruker Instruments, Inc., Billerica, MA).

The remaining SOM and fine mineral fraction was recombined with the coarse mineral fraction. Residual derivative was removed by adding 5 cm^3 of 1 M HCl and thoroughly mixing the sample on a vortex mixer. The acid was neutralized with 1 M NaOH, the mixture was centrifuged and the liquid decanted off. The mixture was washed three times with deionized water and filtered. The solid material was air dried, finely ground, and analyzed for total C.

Leaf biomass and height increment were measured as indicators of forest productivity. Leaf biomass was assessed by tracking five randomly arranged 1-m^2 litter traps monthly for 2 yr. Tree diameters at DBH (diameter at breast height) and height increment of five dominant and codominant trees were measured annually. Foliar samples were taken from the upper crown each month to assess nutrition.

RESULTS AND DISCUSSION

Figures 3–3 and 3–4 show the labile and recalcitrant SOM fractions for the Lakeland (sand) and Fuquay (loamy sand) soils, respectively. The seasonal variations in SOM values obtained for the soils before SFE are consistent with the results of Haines and Cleveland (1981). Each of the data points obtained before SFE represents the mean of 15 replicates. The data obtained after SFE represents a smaller data set (6 replicates per data point), but we were able to identify a baseline value for each stand. If this baseline measure accurately depicts the separation between the labile and recalcitrant pools, the area between the total SOM and

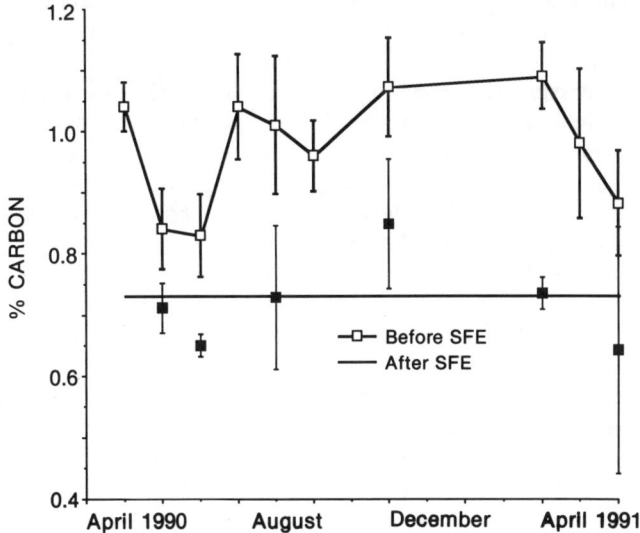

Fig. 3–3. Percentage C, before and after SFE, for the Lakeland soil.

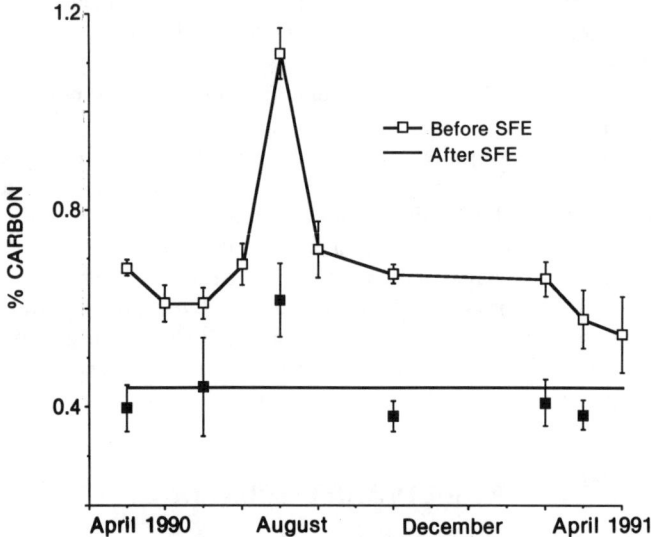

Fig. 3–4. Percentage C, before and after SFE, for the Fuquay soil.

the baseline may be an indication of the nutrient cycling dynamics of the site. We calculated this area and found that it was essentially the same for the two sites (53 and 52 Mg of C/Ha/yr for the Lakeland and Fuquay sites, respectively).

We were unsuccessful in determining a baseline for the Orangeburg site. The percentage C values obtained after extraction could not be statistically differentiated from those obtained before extraction. It is possible that some of the labile pool was removed with the clay. This would result in only a fraction of the labile pool being subjected to extraction. Attempted extraction of the Orangeburg soil (approximately 10% clay in the Ap horizon) without prior clay removal was equally unsuccessful. The extractions were characterized by slow flow rates (<0.5 cm^3/min) in the dynamic mode. The SFE extraction vessels are equipped with 2-μm frits designed to prevent particulate matter from clogging the extraction lines. Clay particles will effectively clog this frit thus hampering solvent flow. No attempt was made to determine the minimum clay concentration that would allow for successful extraction of the labile pool.

The labile pool has been estimated by several researchers to have a mean residence time (MRT) of a few decades. Our analysis indicates changes occurring in the SOM during 1 yr. It is probable that some labile SOM components are tied into the humic macromolecule and have not yet been "broken off" by microbial degradation to be soluble by our extraction method. This would result in the seasonal variation of the baseline that we observe. However, if we assume that the ecosystem is at a steady state, then the baseline measured is a good estimation of the actual baseline of the system.

Preliminary spectral characterization of a Lakeland soil extract through liquid state ^{1}H NMR is shown in Fig. 3–5. The spectrum indicates the presence of phenolic (7–8 ppm), aromatic (8–9 ppm) and carboxyl (11–13 ppm) functional

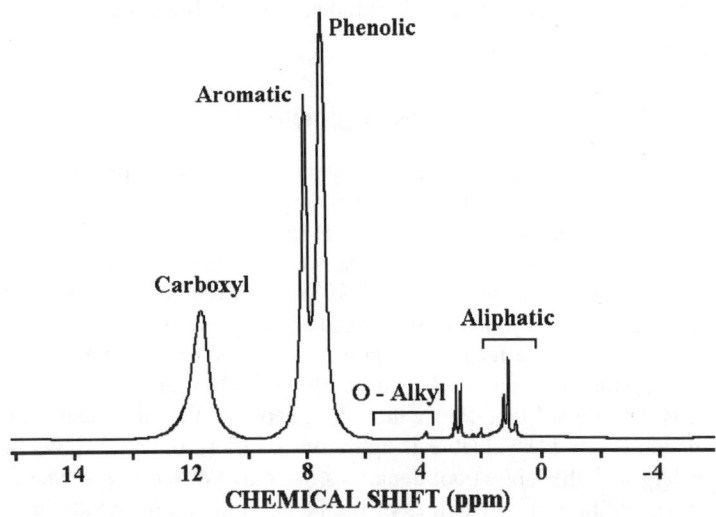

Fig. 3–5. Proton NMR of a composite Lakeland soil extract.

groups. Aliphatic material (0–2 ppm), which is resistant to microbial degradation, was virtually absent. An area of concern is the absence of an appreciable O-alkyl signal (4–6 ppm). A signal in this region would indicate the presence of carbohydrate-type materials which are generally considered labile. This material may have been lost during extraction, tied into a polymeric form (i.e., cellulose) which may resist extraction, or was not present in the sample. To determine what occurred, 0.5-g samples of cellulose were subjected to the extraction procedure. The experiment was repeated in triplicate and in all cases the derivatized material was highly soluble (>70% by weight) in SCF freon. This leads us to believe that this material was not present in the sample. Since carbohydrate-type materials are the most labile SOM components (Baldock et al., 1990), it is possible that they could have already been microbially consumed before the soil sample was collected.

The Fuquay and Lakeland stands are similar in age and species composition. The primary difference among the sites is the soil type. Total C and the SOM baseline values are higher for the Lakeland soil (Fig. 3–3) than for the Fuquay soil (Fig. 3–4). Based on these C measurements alone, we would intuitively expect the Lakeland site to be more productive. This expectation conflicts with the actual net productivity of the stands. The average height of five dominant and codominant trees was determined in 1990 for these stands and this measurement (11.7 and 15.3 m for the Lakeland and Fuquay stands, respectively) identified the Fuquay stand as more productive. However, measures of annual productivity for the sites indicate that they are essentially equal. Mean leaf biomass production for 1990 and 1991 was 2.2 and 2.1 kg for the Lakeland and Fuquay stands, respectively, and total leaf biomass production for 1990 was the same (2.0 kg) for both sites. The areas of the labile pool for the Lakeland and Fuquay sites (see previous section), which may be an indication of the nutrient cycling dynamics, reflect this observation.

CONCLUSIONS

Our attempts to fractionate the Orangeburg soil into labile and recalcitrant pools are continuing. In addition, other questions remain to be answered. Foremost among these is: Have we actually extracted the labile pool? Although this study provides indications that this may be the case, final confirmation by C dating methods is forthcoming. Hsieh (1993) has developed a means of estimating the age of relatively young (a few decades) samples. Another question is: Are we extracting part of the recalcitrant pool? This question also may be answered by calculating the mean age of the extract through C dating.

If it is determined that our method does separate the labile and recalcitrant pools, it could provide valuable insights on a variety of issues. By modeling weather data with the labile pool dynamics, we may be able to estimate fine root turnover. In addition, the dynamics of the labile pool could provide insight into nutrient cycling dynamics. Finally, the baseline value could provide information on the issue of belowground C sequestering.

ACKNOWLEDGMENT

The authors of this paper would like to thank the Department of Energy and the Southern Global Climate Change Program for their financial support. Additional gratitude is expressed to David Josephus, Robert Eaton, Sandra Kelly and Robbie Barham for their excellent technical support.

REFERENCES

Baldock, J., J.M. Oades, A.M. Vassallo, and M.A. Wilson. 1990. Solid state CP/MAS ^{13}C N.M.R. analysis of particle size and density fractions of a soil incubated with uniformly labelled ^{13}C-glucose. Aust. J. Soil Res. 28:193–212.

Binkley, D., and S.C. Hart. 1989. The components of nitrogen availability assessments in forest soils. Adv. Soil Sci. 10:57–112.

Capriel, P., T. Beck, H. Borchert, and P. Harter. 1990. Relationship between soil aliphatic fraction extracted with supercritical hexane, soil microbial biomass and soil aggregate stability. Soil Sci. Soc. Am. J. 54:415–420.

Cole, D.W., and M. Rapp. 1981. Elemental cycling. p. 341–409. *In* D.E. Riechle (ed.) Dynamic properties of forest ecosystems. Cambridge Univ. Press, Cambridge, England.

Edmonds, R.L., and T. Hsiang. 1987. Forest floor and soil influences on response of Douglas-fir to urea. Soil Sci. Soc. Am. J. 51:1332–1337.

Haines, S.G., and G. Cleveland. 1981. Seasonal variation in properties of five forest soils in southwest Georgia. Soil Sci. Soc. Am. J. 45:139–143.

Hawthorne, S.B., J.J. Langenfeld, D.J. Miller, and M.D. Burford. 1992. Comparison of supercritical $CHClF_2$, N_2O, and CO_2 for the extraction of polychlorinated biphenyls and polycyclic aromatic hydrocarbons. Anal. Chem. 64:1614–1622.

Hsieh, Y.P. 1993. Radiocarbon signatures of turnover rates in active soil organic matter pools. Soil Sci. Soc. Am. J. 57:1020–1023.

Miller, D.J., S.B. Hawthorne and J.J. Langenfeld. 1991. SFE with chemical derivatization for the recovery of polar and ionic analytes. p. 55–156. *In* M. Lee (ed.) Proc. Int. Symp. Supercritical Fluid Chromatography and Extraction, Park City UT. 15–17 January. Brigham Young Univ., Provo, UT.

Nelson, D.W., and L.E. Sommers. 1986. Total carbon, organic carbon, and organic matter. p. 539–580. *In* A.L. Page et al. (ed.) Methods of soil analysis. Part 2. 2nd ed. Agron. Monogr. 9. ASA and SSSA, Madison, WI.

Powers, R.F., D.H. Alban, R.E. Miller, A.E. Tiarks, C.G. Wells, P.E. Avers, R.G. Cline, R.O. Fitzgerald, and N.S. Loftus, Jr. 1990. Sustaining site productivity in North American Forest: Problems and prospects. p. 49–79. *In* S.P. Gessel et al. (ed.) Sustaining productivity of forest soils. N. Am. For. Soils Conf., 7th, Vancouver, Canada. 24–28 July 1988. Faculty For., Univ. British Columbia, Vancouver.

Ruark, G.A., and J.I. Blake. 1991. Conceptual stand model of plant carbon allocation with a feedback linkage to soil organic matter maintenance. p. 187–198. *In* W.J. Dyck and C.A. Mees (ed.) Long-term field trails to assess environmental impacts of harvesting. For. Res. Inst. 161. For. Res. Inst. Rotorua, New Zealand.

Schnitzer, M., C.A. Hindle, and M. Meglic. 1986. Supercritical gas extraction of alkanes and alkanoic acids from soils and humic materials. Soil Sci. Soc. Am. J. 50:913–919.

Schulten, H.R., and M. Schnitzer. 1990. Aliphatics in soil organic matter in fine-clay fractions. Soil Sci. Soc. Am. J. 54:98–105.

Spiteller, M. 1985. Extraction of soil organic matter by supercritical fluids. Organ. Geochem. 8:111–113.

4   The Influence of Low-Molecular-Weight Organic Acids on Properties and Processes in Forest Soils

Thomas R. Fox

Rayonier, Inc.
Yulee, Florida

The presence of low-molecular-weight (LMW) organic acids alters both chemical and biological processes in soils. Low-molecular-weight organic acids function as ligands and through complexation reactions in solution and ligand exchange reactions at mineral surfaces, they affect metal solubility and speciation (Stumm & Morgan, 1981; Martell et al., 1988). Therefore, they play an important role in mineral weathering and soil genesis (Huang & Schnitzer, 1986). Interest in the role of LMW organic acids on solubility and speciation of metals such as Pb, Cd, and Al has increased in recent years in response to environmental issues such as atmospheric deposition of acids and the land application of municipal and industrial wastes (Inskeep & Baham, 1983; McColl & Pohlman, 1986; Tam, 1987). The bioavailability of plant nutrients such as P and K increases in the presence of LMW organic acids which may improve plant nutrition (Marschner, 1986). Organic acids also may indirectly affect plant nutrition in acid soils by complexing Al, thus alleviating Al toxicity and improving the growth and physiologic function of roots (Hue et al., 1986). Conversely, the incomplete decomposition of organic matter may lead to the accumulation of toxic concentrations of organic acids that detrimentally affects plant growth (Wang et al., 1967; Jalal & Read, 1983a,b).

The importance of soil C, especially the dissolved organic carbon (DOC) fraction, in regulating and influencing ecosystem processes is well recognized. Much of the recent work has emphasized homologous groups of compounds, such as hydrophobic and hydrophilic acids and bases, rather than individual compounds because most of the soil organic matter can not be specifically identified with current analytical techniques (Buffle, 1990). In contrast, the chemical and physical properties of LMW organic acids are well characterized, including the number, type and arrangement of functional groups in the molecule. Consequently, unlike the more complex forms of soil organic matter, specific

---

Copyright © 1995 Soil Science Society of America, 677 S. Segoe Rd., Madison, WI 53711, USA. Carbon Forms and Functions in Forest Soils.

LMW organic acids can be separated and identified using simple chromatographic techniques (Lee & Lord, 1986; Howe et al., 1990; Lilieholm et al., 1992).

Although LMW organic acids usually comprise less than 10% of the DOC in soils (Pohlman & McColl, 1988; Fox et al., 1990a), they have a large impact on soil processes (Stevenson, 1967). They disproportionately influence soil processes because of their small size, relatively high solubility and ability to form strong complexes with metals. In addition, the concentration of LMW organic acids in the rhizosphere and other localized sites in the soil is frequently much greater than in the bulk soil (Curl & Truelove, 1986). Low-molecular-weight compounds also form the structural components of larger humic and fulvic acids (Stevenson, 1982). Therefore, LMW organic acids can serve as simple, albeit limited, models of the more complex but less well-characterized forms of organic matter in soil (Buffle, 1990; Robert & Berthelin, 1986). Studies using LMW organic acids have greatly increased our understanding of the impact of organic matter on soil properties and processes (Stumm, 1986).

The purpose of this paper is to review the role of LMW organic acids in forest soils and to update the information in the thorough review of organic acids in soils by Stevenson (1967). The origin and distribution of LMW organic acids and their impact on soil processes, particularly Al and P solubility, will be covered. Emphasis will be placed on the mechanisms by which LMW organic acids influence soil processes. Additional information on various aspects of this topic can be found in Stevenson (1967, 1982), Flaig (1971), Hodgkinson (1977), Stumm & Morgan (1981), Thurman (1985), Huang & Schnitzer (1986) and Buffle (1990).

CLASSIFICATION OF LOW-MOLECULAR-WEIGHT ORGANIC ACIDS

Size and Molecular Weight

As a group, LMW organic acids are well-characterized molecules generally less than 1.0 nm in diameter. They are smaller than most other DOC such as proteins, polysaccharides and humic compounds, although some of the smaller fulvic acids may be of similar size (Buffle, 1990). A considerable range exists in the molecular weight of organic acids (Table 4–1), and no well-defined upper limit has been established separating low- and high-molecular-weight compounds. Formic acid (HCOOH), the simplest organic acid found in soil, has a molecular weight of 46. At the other extreme, complex long chain fatty acids like dihydroxystearic $[CH_3(CH_2)_7CHOH\text{-}CHOH(CH_2)_7COOH]$ and cerotic $[CH_3(CH_2)_{24}COOH]$, with molecular weights approaching 400, have been identified as free acids in soils (Stevenson, 1982). Most of the more common LMW organic acids in soils have molecular weights of less than 300 (Table 4–1).

Chemical Structure

A more informative classification uses the chemical structure of LMW organic acids, including the basic structure of the molecule and the type and arrangement of functional groups present (Fig. 4–1). These features determine

Table 4-1. Properties of selected LMW organic acids.

Organic acid & chemical formula	Molecular weight	pK_a's	$LogK_{Al}$
Acetic CH_3COOH	60.05	4.75	1.57
Aconitic $(CH_2)(OH)_3COOH$	174.11	2.80; 4.45	?
Benzoic $(C_6H_5)COOH$	122.13	4.19	?
Cinnamic $(C_6H_5)CHCHCOOH$	148.17	3.89	?
Citric $HOC(CH_2COOH)_2COOH$	192.12	3.14; 4.77; 6.39	7.98
Formic $HCOOH$	46.03	3.75	1.36
Fumaric $HOOCCHCHCOOH$	116.07	3.03; 4.44	?
Gallic $(C_6H_2)(OH)_3COOH$	169.11	4.41; 9.1; 11	?
Lactic $CH_3CHOHCOOH$	90.08	3.08	2.38
Maleic $HOOCCHHCHCOOH$	116.07	1.83; 6.07	5.48
Malic $HOOCCHOHCH_2COOH$	134.09	3.40; 5.11	5.34
Malonic $HOOCCH_2COOH$	104.06	2.83; 5.69	5.24
p–hydroxybenzoic $(C_6H_4)OHCOOH$	138.12	4.48; 9.32	1.66
Phthalic $(C_6H_4)COOHCOOH$	166.13	2.89; 5.51	3.18
Protocatechuic $(C_6H_3)OHOHCOOH$	166.13	4.35; 8.84	?
Oxalic $HOOCCOOH$	90.04	1.23; 4.19	6.10
Salicylic $(C_6H_4)OHCOOH$	138.12	2.97; 13.4	4.44
Succinic $HOOCCH_2CH_2COOH$	118.09	4.16; 5.61	2..09
Tartaric $HOOCCHOHCHOHCOOH$	150.09	2.98; 4.34	5.62
Vanilliic $(C_6)OHOCH_3COOH$	168.15	?	?

solubility and acidity of LMW organic acids as well as their ability to complex metals. The presence or absence of a benzene ring structure in the molecule has a large impact on the solubility and acidity of the molecule (Thurman, 1985). Aromatic compounds tend to be less soluble in aqueous solutions than aliphatic compounds. However, they are frequently stronger acids than aliphatic compounds.

The functional groups that occur in organic acids include acidic groups such as carboxylic (R-COOH), enolic (R-CH=CH-OH), phenolic OH (Ar-OH), and quinones (Ar=O); neutral groups such as alcoholic OH (R-CH_2OH), ethers (R-CH_2-O-CH_2-R), ketones [R-C=O(-R)], aldehydes [R-C=O(-H)], and esters

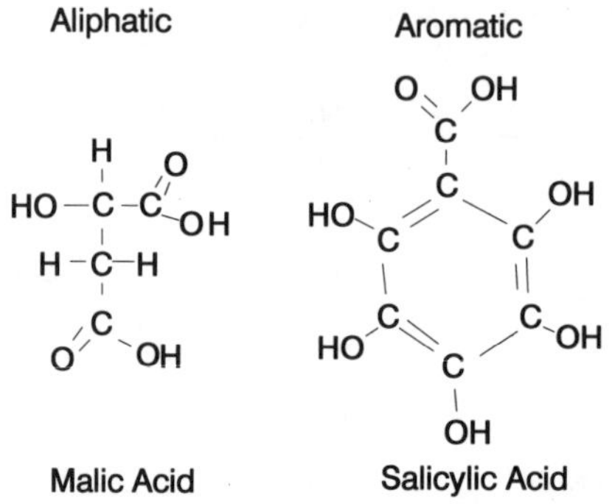

Fig. 4–1. Chemical structure and arrangement of functional groups on selected LMW organic acids.

[R-C=O(-OR)]; and neutral groups such as amines (R-CH$_2$-NH$_2$) and amides [R-C=O(NH-R)].

The carboxylic acid functional group is by far the most important functional group because of its large contribution to the aqueous solubility and acidity of LMW organic acids (Thurman, 1985). Because the carboxyl group is a weakly acid functional group that dissociates in aqueous solutions, LMW organic acids generally occur in soils in the anionic form. (Nevertheless, for simplicity the term "acid" will be used throughout this paper whether referring to the dissociated or undissociated molecule.) The various forms of carboxylic acids include simple aliphatic and aromatic acids, hydroxyacids and aliphatic and aromatic di- and tricarboxylic acids. The addition of a second carboxylic acid group lowers the pK_a of the first (Table 4–1). For example the pK_a's of oxalic acid are 1.2 and 4.2 compared with a pK_a of 3.7 in formic acid. Hydroxyl groups, both alcoholic and phenolic also are important functional groups in LMW organic acids. The phenol functional group of aromatic compounds is acidic while alcoholic OH groups in aliphatic compounds are neutral (Bailey & Bailey, 1981). Hydroxyl groups contribute to the aqueous solubility of LMW organic acids through hydrogen bonding with water (Thurman, 1985).

ORIGIN OF LOW-MOLECULAR-WEIGHT ORGANIC ACIDS IN FOREST SOILS

Low-molecular-weight organic acids are synthesized in many metabolic processes of animals, plants and microorganisms. They also are produced as secondary metabolites during decomposition of more complex forms of organic matter. The major sources of LMW organic acids in forest soils are root exudation, release from soil fungi, leaching from decomposing litter in the forest floor, and the decomposition of organic matter in the soil. Substantial amounts of volatile fatty acids also may enter forest soils in rainfall.

Root Exudates

Plants release a large number of LMW organic acids into the rhizosphere. Considerable work has been done with crop plants including maize (*Zea mays* L.), barley (*Hordeum vulgare* L.), wheat (*Triticum aestivum* L.), rape (*Brassica napus* L.) as well as vegetables and legumes (Vancura, 1964; Vancura & Hovadik, 1965; Vancura & Hanzlikova, 1972; Kraffczyk et al., 1984; Gardner et al., 1983; Foy & Lee, 1987; Hoffland et al., 1989b; Hoffland, 1992). Although a wide variety of LMW organic acids have been identified in the root exudates of these species, oxalic, citric and malic seem to be the most abundant. The information available on LMW organic acids in root exudates of forest trees is limited. Smith (1969) examined the root exudates from seedlings of four species of pine (*Pinus radiata* D. Don, *P. Lambertiana* Dougl., *P. banksiana* Lamb., and *P. rigida* Mill.) as well as black locust (*Robina pseudoacacia* L.) under axenic conditions. Oxalic and acetic acids were the dominant LMW organic acids identified, although succinic, fumaric, malonic and glycolic also were detected. In a subsequent study Smith (1976) examined the root exudates of mature *Betula alleghaniensis* Britton, *Fagus grandifolia* Ehrh. and *Acer saccharum* Marsh. The LMW organic acids identified were acetic, aconitic, citric, fumaric, malic, malonic, oxalic and succinic.

The amount of organic C released into the soil from roots can be quite large. From 20 to 40% or more of the C translocated to the roots may be released into the soil (Barber & Martin, 1976; Johansson, 1992). Although much of this C is released as CO_2 during root respiration, considerable amounts of more complex organic materials, including LMW organic acids are released. For example, the total quantity of LMW organic acids released by maize roots over a 32-d period was nearly 16 g kg^{-1} root dry matter (Kraffczyk et al., 1984). Similar amounts of LMW organic acids were observed in the root exudates of several tree species (Smith, 1976).

The LMW organic acids released as root exudates depends on both the species and the physiologic status of the plant. The dominant LMW organic acids in the root exudates of *B. alleghaniensis* were acetic, citric, fumaric, malic, and oxalic while in *A. saccharum* only acetic, citric and malonic acids were detected (Smith, 1976). The total quantity of LMW organic acids released also was greater in yellow birch compared to sugar maple, 8 g kg^{-1} dry root vs. 3 mg g^{-1} dry root over 14-d period. The amount of root exudates generally increases in response to stress in the plant. In a series of papers, Gardner and coworkers (1982a,b, 1983) reported substantial release of citric acid into the rhizosphere by white lupine (*Lupinus alpus* L.) in response to P deficiencies. Increased release of organic acid in response to P stress has been observed in several other plants including rape, wheat, maize, and a variety of legumes (Ratnayake et al., 1978; Hirata et al., 1982; Lipton et al., 1987; Ohwaki & Hirata, 1992; Hoffland et al., 1989b). In a similar manner, the amount of LMW organic acids released from the roots of maize increased when the supply of K to the plant was reduced (Kraffczyk et al., 1984).

Soil Fungi and Bacteria

Both free-living and mycorrhizal fungi may directly release LMW organic acids to the soil. In fungi, oxalic acid is by far the most abundant LMW organic acid (Sollins et al., 1981). Calcium oxalate may comprise up to 45% of the dry weight of the mycelium in some fungi (Hodgkinson, 1977). Production of oxalic acid plays a critical role in the pathogenicity of many fungi, often leading to the production of relatively large amounts of oxalate (Pierson & Rhodes, 1992). For example, the pathogenic fungi *Sclerotium rolfsii* produced over 1 g of oxalic for each gram of dry weight of hyphae (Bateman & Beer, 1965; Maxwell & Bateman, 1968). Oxalic acid also is produced by mycorrhizal fungi in forest ecosystems (Sollins et al., 1981; Malajczuk & Cromack, 1982). Large perennial mats of the hyphae of the ectomycorrhizal fungus *Hyserangium setchellii* (Fisher) often form in the forest floor under Douglas fir [*Pseudotsuga menziesii* (Mirb.) Franco] in the western USA and contain massive amounts of calcium oxalate (Cromack et al., 1979; Graustein et al., 1977). The calcium oxalate content of the A horizon of forest soils containing these fungal mats can exceed 850 kg ha^{-1} (Sollins et al., 19891). Citric, malic, lactic, fumaric and succinic also are common but less abundant metabolic products of fungi (Stevenson, 1967).

In contrast to fungi, bacteria tend to produce more of the simple volatile fatty acids such as formic and acetic (Stevenson, 1967). Metabolic products originating from the microbial degradation of higher-weight organic compounds include many LMW organic acids (Robert & Berthelin, 1986).

Leaching From Litter in the Forest Floor

Large quantities of LMW organic acids are synthesized by plants in metabolic pathways such as the tricarboxylic acid cycle and the shikimic acid cycle. These compounds can be leached from litter as it decomposes. The mixture of LMW organic acids identified in extracts of the forest floor is much more varied than that reported from root exudates or other sources. Analysis of leachates from forest floor material often reveals the presence of a substantial amount of aromatic compounds.

Oxalic, citric and formic acids were identified in the extracts of litter from spruce (*Picea* genus) and birch (*Betula* genus) forests (Kaurichev et al., 1963). Muir and coworkers (1964) identified malic, citric, shikimic, and quinic acid in the aqueous extracts of Scots pine (*P. sylvestris* L.) needles. In a classic study of the extracts of litter from several forest trees including *Quercus petraea* (Mattuschka) Liebl., *Fagus silvatica* L., and *P. silvestris* L., Bruckert (1970a) identified a wide variety of LMW organic acids including the aromatic compounds vanillic, protocatechuic, *p*-hycroxybenzoic, *p*-coumaric, caffeic, ferulic, and gallic. He also found aliphatic LMW organic acids such as citric, malic, oxalic, acetic, malonic, lactic, and succinic. Kuiters and Sarink (1986) studied litter leachates from 14 species of forest trees, both coniferous and deciduous, and identified benzoic, salicylic, *p*-hydroxybenzoic, vanillic, protocatechuic, gallic, resorcylic, gentisic, syringic, cinnamic, ferulic, caffeic, *p*-coumaric, and *o*-coumaric acids.

McColl and coworkers (1990) have studied the LMW organic acids leached from the forest floor under several tree species in the western USA including Douglas fir, ponderosa pine (*P. ponderosa* Laws.), and incense cedar [*Calocedrus decurrens* (Torr.) Florin.] (McColl & Pohlman, 1986; Pohlman & McColl, 1988; Tam & McColl, 1991). They detected aliphatic acids such as oxalic, maleic, aconitic, malic, fumaric, succinic, and citric; and aromatic acids including trans-cinnamic, benzoic, protocatechuic, *p*-hydroxybenzoic, vanillic, *p*-coumaric, shikimic, gallic, phthalic, and salicylic.

In the southern USA, Fox and Comerford (1990) identified oxalic, formic, citric, acetic, malic, lactic and aconitic acids in litter extracts form slash and long leaf pine stands. Muck and coworkers (1991) recently reported on the LMW organic acids in pasture grasses. Citric, malic, malonic, oxalic, succinic and palmitic acids were the most common. Given the relatively high concentrations reported, the potential for large amounts of LMW organic acids to enter the soil from grass and other understory species in forest ecosystems should not be overlooked.

Decomposition of Organic Matter

Soils contain large amounts of high-molecular-weight compounds such as long-chain fatty acids, polysaccharides, and proteins which can degrade to LMW organic compounds (Schnitzer & Khan, 1978; Stevenson, 1982). For example, long-chain fatty acids such as polygalacturonic acid, may be microbially degraded into simple uronic acids (Stevenson, 1982). Complex forms of soil organic matter include lignin, humic and fulvic acids and other operationally defined homologous groups of compounds (Buffle, 1990) which also decompose to some degree and release LMW organic acids into the soil (Schnitzer & Khan, 1978; Stevenson, 1982; Tan, 1986). Oxidative degradation of humic and fulvic acids releases aliphatic and aromatic LMW organic acids (Schnitzer, 1986). The most abundant aliphatic compounds resulting from this decomposition are fatty acids and the di- and tricarboxylic acids. The dominant aromatic degradation products contain from one to three OH groups and one to five COOH groups per benzene ring (Schnitzer, 1986). Large amounts of the aromatic acids tend to be released during decomposition of organic matter in waterlogged soils (Stevenson, 1967). Fermentation processes in anaerobic soils also leads to the production of large amounts of volatile aliphatic acids such as acetic and butyric (Stevenson, 1967).

Rainfall

Low-molecular-weight organic acids, particularly the volatile fatty acids may contribute from 16 to 35% of the total acidity in rainfall in the USA (Avery et al., 1991). Formic and acetic acids have been shown to make a significant contribution to the acidity of rainwater in both urban and rural environments (Kawamura & Kaplan, 1983, 1984; Avery et al., 1991; Wiley & Wilson, 1993). Kawamur and Kaplan (1984) found up to 5.4 mg L^{-1} of volatile fatty acids in rainfall. Automotive exhaust and the photochemical decomposition of plant produced terpenes are the probable sources (Kawamura et al., 1985; Martin et al., 1991).

DEGRADATION OF LOW-MOLECULAR-WEIGHT ORGANIC ACIDS IN SOILS

The concentration of LMW organic acids in soils is determined by the balance between production and degradation. Most LMW organic acids are rapidly degraded by microorganisms and thus have a transitory existence in the soil. A large number of microorganisms are capable of utilizing even very simple C1 and C2 compounds such as oxalate and formate as their sole energy source (Harder, 1973). McColl and coworkers (1990) demonstrated rapid degradation of the organic acids in litter extracts from Douglas fir. Concentrations of gallic, fumaric, protocatechuic, vanillic, and syringic decreased by more than 50% within 48 h while concentrations of gensitic and ferulic acids decreased from 5.8 and 5.9 mM to 0 within 24 h. Nearly complete degradation of oxalate and formate was reported within 73 h in Spodosols by Fox and coworkers (1990a). In anaerobic environments, fatty acids can be directly oxidized by sulfate reducing bacteria (Monetti & Scranton, 1992). Hodgkinson (1977) details a multitude of degradation pathways for oxalate, many of which are undoubtedly active in the soil. The formation of metal-organic acid complexes often retards the microbial degradation of organic acid in soil (Boudot, 1992).

Low-molecular-weight organic acids such as malic, formic, oxalic also participate in the reduction of metals (i.e., the reduction of MnO_2 to Mn^{+2}) by acting as electron acceptors during which they are oxidized and degraded (Jauregui & Reisenauer, 1982). It also has been demonstrated that some plants may reabsorb LMW organic acids that are present in the rhizosphere (Jones & Darrah, 1992).

DISTRIBUTION AND ABUNDANCE OF LOW-MOLECULAR-WEIGHT ORGANIC ACIDS IN FOREST SOILS

Low-molecular-weight organic acids appear to be ubiquitous in forest soils (Stevenson, 1967). The LMW organic acids that have been identified in soil include aliphatic compounds such as formic, acetic, malic, malonic, maleic, lactic, aconitic, fumaric, succinic, citric and oxalic acids; and aromatic compounds such as benzoic, cinnamic, protocatechuic, vanillic; p-hydroxybenzoic, phthalic, salicylic, p-coumaric, gallic, ferulic, shikimic, and syringic (Table 4–1) (Schwartz et al., 1954; Kaurichev et al., 1963, Muir et al., 1964; Whitehead, 1964; Bruckert, 1970a; Graustein et al., 1977; Cromack et al., 1979; Gardner et al., 1982a; Jalal & Read, 1983a; Hue et al., 1986; Kuiters & Sarink, 1986; McColl & Pohlman, 1986; Rozycki & Strzelczyk, 1986; Pohlman & McColl, 1988; Fox & Comerford, 1990; Tam & McColl, 1991).

Concentrations of LMW organic acids in soils range from less than 10^{-6} to greater than 10^{-3} mole per liter (Stevenson, 1982). Quantitative analysis of the LMW organic acids present in soil almost universally follows the sequence aliphatic > aromatic > amino acids (Robert & Berthelin, 1986). Factors affecting the types and amounts of organic acids present in the soil include the vegetation present, soil type, depth in the profile, proximity to roots, and soil aeration.

In general, concentrations of LMW organic acids are greater in forest soils than agricultural soils (Stevenson, 1967). For example, Hue et al. (1986) observed that malic, succinic and lactic acid concentrations were 117, 282 and 52 µmol L^{-1} in the BE horizon of a Dothan series soil (Plinthic Paleudults) supporting an oak and pine forest in Alabama but were not detected in the same horizon of a cultivated soil. Organic acid concentrations also are generally higher in the litter layer compared to the mineral horizons of forest soils (Bruckert, 1970a; McColl & Pohlman, 1986).

The pattern of LMW organic acid distribution among mineral soil horizons is less clear. Fox and Comerford (1990) found higher concentrations of oxalic and formic acid in the Bh horizon compared to the A horizons of Spodosols in the Lower Coastal Plain. In contrast, Bruckert (1970a), found higher concentrations in the A horizon compared to the Bh horizon in a series of podzols in France.

Consistently higher concentrations of LMW organic acids occur in the rhizosphere of plant roots (Gardner et al., 1983; Hoffland et al., 1989b; Marschner, 1991). The concentration of oxalic acid in the rhizosphere of slash pine (*P. elliottii* Engelm.) exceeded 1 g kg^{-1} soil (Fox & Comerford, 1990). High concentrations of LMW organic acids also are associated with fungal hyphae in the soil (Graustein et al., 1977). Calcium oxalate content in fungal mats of *Hysterangium setchellii*, a mycorrhizal associate of Douglas fir, may exceed 80 g m^{-2} (Cromack et al., 1979). Similarly, high concentrations of calcium oxalate were associated with ectomycorrhizal roots of *P. radiata* and *Eucalyptus marginata* in Australia (Malajczuk & Cromack, 1982).

REACTION MECHANISMS

Before discussing the impacts of LMW in soils, a brief review of coordination chemistry at soil surfaces and in aqueous solutions is necessary. More detailed discussions of this topic are provided in Stumm et al. (1983), Sposito (1984), Stumm (1986).

Low-molecular-weight organic acids function as organic ligands and affect soil properties and processes by (i) chelating metals in solution and (ii) ligand exchange at metal-hydroxide surfaces (Stumm, 1986). The formation of surface and solution complexes with metals also extends the domain of congruent dissolution of minerals thus allowing higher concentrations of soluble metals to exist in solution before a new mineral phase precipitates (Stumm, 1986). For example, Shotyk and Nesbitt (1992) demonstrated increased congruent dissolution of plagioclase feldspar in the presence of oxalate. The impact of LMW organic acids on soil processes depends on the acid present and their concentration. Those organic acids that form stable complexes with metals, both in solution and at mineral surfaces, will have a greater impact on soil processes.

Metals in aqueous solutions and at the soil surfaces in contact with aqueous solutions, including those in phylosillicate clays and metal oxides, are coordinated with surface hydroxyl groups. The central metals ions acts as a Lewis acid site with the hydroxyl groups containing the complex forming O donor groups (Sposito, 1984, Stumm, 1986). Low-molecular-weight organic acids form inner sphere complexes with these metals (Fig. 4–2). The formation of stable metal complexes depends on the type and arrangement of functional groups on the organic acid (Martell et al., 1988). Functional groups which contain unshared pairs of electrons can form covalent bonds between the functional group and the central metal ion (Sposito, 1984). The order of decreasing affinity of functional groups for metal ions is given by Stevenson (1982) as

$$O^- > NH_2 > N=N > COO^- > -O- \gg C=O$$

Carboxylate and enolate are the most important functional groups present in LMW organic acids. The stability of metal-ligand complexes depends on properties of both the metal and the ligand (Stevenson, 1982). The number of atoms that form a bond with the metal is important because stable complexes tend to involve five- and six-membered ring structures. The stability of the complex increases as the number of these ring structures increases. The pH of the system also strongly affects the stability of the complexes formed.

The affinity of a metal ion for a ligand is indicated by the equilibrium stability constant ($LogK$) of the metal-ligand complex in solution (Martell et al., 1988). The Al stability constant for various LMW organic acids is listed in Table 4–1. Although ligands with higher stability constants form stronger metal complexes, many metals tend to undergo complex hydrolytic reactions in natural aqueous solutions which may complicate the use of simple, thermodynamic ligand stability constants (Motekaitis & Martell, 1984; Thomas et al., 1991). Conditional stability constants must often be calculated for the system of interest. Stumm and coworkers (1983) demonstrated that the tendency of a ligand to form surface complexes is closely related to the tendence to form complexes in solution. However, stearic effects may have a large impact on the formation of complexes at mineral surfaces because of geometric restrictions imposed by the solid surface (Stumm, 1986).

Low-molecular-weight organic acids affect soil processes through (i) ligand exchange at oxide surfaces and the subsequent formation of surface complexes, and (ii) complexation of metals in solution.

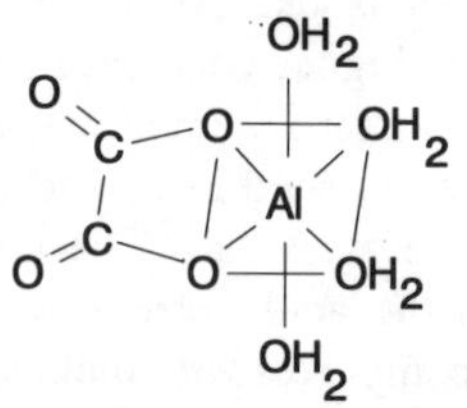

Fig. 4–2. Proposed structure of an Al-oxalate complex illustrating a stable five-membered ring structure.

In a ligand exchange reaction, the nucleophilic ligand binds to a metal center, displacing another ligand from the surface forming an inner-sphere complex (Fig. 4–3). Frequently the ligand displaced from the metal is an OH group, although other ligands such as SO_4 and PO_4 also may be involved in the ligand exchange reaction at mineral surfaces (Parfitt et al., 1977; Goldberg & Sposito, 1985; Hingston et al., 1974; Fox et al., 1990a).

The formation of surface complexes promotes the dissolution of metal-oxide surfaces due to a change in the coordination sphere of the metal following ligand exchange (Stumm et al., 1983; Stumm, 1986; Furrer & Stumm, 1986; Zinder et al., 1986). In a ligand exchange reaction, the nucleophilic ligand binds to a metal center by replacing an OH group. This polarizes the remaining Me-O bonds of that metal center. The formation of surface complexes also is usually accompanied by a decrease in surface charge of the metal oxide. These two factors weaken the Me-O bonds and enhance the detachment of the metal, either as a Me-H_2O or Me-L group. The detachment of metals usually occurs at the so called "kink" or "step" sites of the oxide surface. Similar mechanisms appear to be at work in the ligand promoted dissolution of minerals such as oligoclase, labrodorite, and anorthite (Amrhein & Suarez, 1988; Casey et al., 1988, 1989; Mast & Drever, 1987).

Complexation reactions in solution are similar to those at mineral surfaces. Ligands replace the coordinated hydroxyls from metals in solution and form inner sphere complexes where the metal and ligand share pairs of electrons in a covalent bond. The formation of solution complexes lowers the activity of the free metal in solution which causes dissolution of metal oxides through equilibrium solubility reactions. The formation of stable solution complexes between

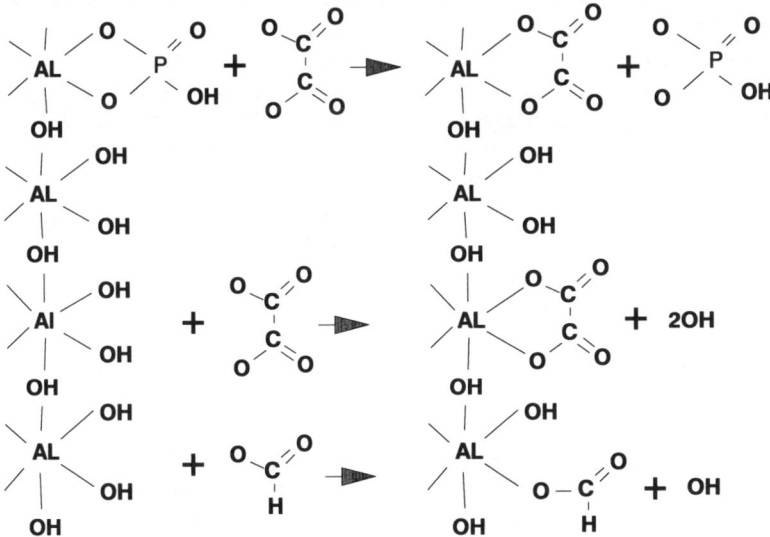

Fig. 4–3. Ligand exchange reactions between oxalate and formate at an Al-oxide surface.

LMW organic acids and metals also retards the precipitation of metal oxides (Huang & Violante, 1986).

EFFECT OF LOW-MOLECULAR-WEIGHT ORGANIC ACIDS ON METAL SOLUBILITY IN FOREST SOILS

The mobilization of metals such as Fe and Al by organic acids leached from the forest floor was recognized as a dominant feature of many forest soils by early soil scientists (Buurman, 1984). Although more complex organic compounds such as humic and fulvic acids are thought to play the dominant role, LMW compounds also are important in this process (Bloomfield, 1953; Muir et al., 1964; Bruckert, 1970b). Because this topic was recently reviewed elsewhere (Huang & Schnitzer, 1986), only some of the more recent work directly related to forest soils will be discussed here.

The impact of organic acids on Al solubility in forest soils of the northern California has been extensively investigated and reported on in a series of papers by McColl and coworkers (McColl & Pohlman, 1986; Pohlman & McColl, 1986, 1988; Tam, 1987; Tam & McColl, 1990, 1991; McColl et al., 1990). This work very clearly indicates the large impact of LMW organic acids entering the soil in litter leachates on metal solubility, particularly Al. A large number of LMW organic acids were identified in litter extracts and their Al-binding strengths were investigated. Those organic acids with carboxylic and phenolic functional groups arranged so that five- and six-membered ring structures could be formed with the metal, had a much greater impact on metal solubility. These compounds also strongly affected the kinetics of Al release. The importance of aromatic compounds such as p-hydroxybenzoic acid and protocatechuic acid in the systems under study by this group is a significant finding. More work is needed on the impact of aromatic compounds in other forest ecosystems.

In the southern USA, several groups have recently investigated the role of LMW organic acids in Al and Fe solubility and speciation. Miller et al. (1986) studied the impact of 13 organic acids on the dissolution of iron oxides. They found that oxalic, malonic, malic, citric and tartaric completely dissolved a noncrystalline iron oxide while acetic, lactic, salicylic, and pthalic acids were much less effective. Hue et al. (1986) examined the impact of 11 organic acids on reducing Al toxicity in subsoils of Ultisols and placed the organic acids into three groups: strong detoxifiers (citric, oxalic, tartaric), moderate detoxifiers (malic, malonic, salicylic), and weak detoxifiers (succinic, lactic, formic, acetic, and pthalic). Evans and Zelazny (1990) found that the kinetics of Al release from forest soils also was strongly affected by the addition of certain LMW organic acids. Oxalic and malonic had a relatively large effect while tartaric, malic and succinic had a lesser impact. The release of Al from the Bh horizon of a Spodosol increased in the presence of oxalate but not formate (Fox et al., 1990a). Fox and Comerford (1992) also showed that the amount of Al released from this soil increased as the amount of oxalate added to the soil increased.

These results clearly indicate that those LMW organic acids able to form stable metal-ligand complexes have a larger impact on Al or Fe solubility. Fox and coworkers (1990b) observed an exponential relationship between the Al-stability constant of the LMW organic acid and Al solubility. This suggests that the LogK_{Al} value can be used to predict the impact of LMW organic acids on metal solubility in soils. However, other factors affect the impact of LMW organic acids on metal solubility and it may not always be possible to accurately predict the impact of LMW organic acids from published thermodynamic data. This problem is clearly demonstrated by salicylic acid. The arrangement of functional groups in this molecule (Fig. 4–1) suggests it will form six-membered rings with a metal cation. This is confirmed by the relatively high thermodynamic metal-ligand stability constant of salicylic acid (LogK_{Al} = 12.9). However, in numerous instances it has been shown that salicylic acid does not impact metal solubility to the extent expected (Low & Black, 1950; Miller et al., 1986; Hue et al., 1986; Pohlman & McColl, 1986; Fox et al., 1990b). The general conclusion is that stearic effects at mineral surfaces inhibit the formation of stable metal-salicylate complexes. Recent work suggests that a more appropriate conditional stability constant for salicylic acid with Al in soil is between 4.5 and 5.0 (Hue et al., 1986; Fox et al., 1990b).

EFFECT OF LOW-MOLECULAR-WEIGHT ORGANIC ACIDS ON PHOSPHORUS AVAILABILITY IN FOREST SOILS

The presence of LMW organic acids increases plant nutrient availability in soils (Marschner et al., 1986). Availability of nutrients such as S, K, Fe, Zn and other micronutrients has been shown to increase in the presence of LMW organic acids (Hodgson, 1969; Elgawhary et al., 1970; Graustein et al., 1977; Cline et al., 1982; Reid et al., 1985; Inskeep & Comfort, 1986; Song & Huang, 1988; Marschner, 1991; Evans, 1991; Gerke, 1992b). However, the majority of the work in this area has focused on the effects of LMW organic acids on P availability.

Increased release of P following the addition of certain LMW organic acids to soils was clearly demonstrated by the middle of this century (Dean & Rubins, 1947; Swenson et al., 1949; Low & Black, 1950; Struthers & Sieling, 1950). Recent work has demonstrated that only those LMW organic acids that form stable complexes with metals increase P availability. In a study of P sorption on synthetic Fe and Al gels as well as natural soils, citric acid had a much larger effect than tartaric while acetic acids had no effect (Earl et al., 1979). Similar results were obtained in a montmorillonitic soil where P solubility was greatest in the presence of citrate and decreased in the presence of tartrate and formate (Traina et al., 1986a). Citric acid forms more stable complexes with metals than tartrate while formic acid and acetic acid do not form stable complexes.

The impact of 16 LMW organic acids and water on the release of P from a spodic horizon was examined by Fox et al. (1990b), who found an extremely wide range in the amount of P released following the addition of the LMW organic acids. A significant portion of the variability in P release from this soil could be explained by the Al stability constant of the organic acids. Those LMW organic acids with Al stability constants less than approximately 4.1 were no more effective than water in increasing P solubility. This suggests that in this soil, the stability of Al-P surface complexes was equivalent to a $\text{Log}K_{Al}$ of approximately four.

The impact of LMW organic acids on P release increases with their concentration in solution. In a system dominated by Al-oxide surfaces, P release increased monotonically when increasing amounts of oxalic acid were added to the soil (Fox & Comerford 1992). In a similar manner, the solubility of P in an Oxisol, Luvisol and a Podzol increased as the concentration of citric acid in solution increased (Gerke, 1992a).

The presence of LMW organic acids also affects the kinetics of P release from soils with faster release occurring in the presence of LMW organic acids that form stable metal complexes. Kuo and Lotse (1974) reported that the rate of P desorption from gibbsite was faster in the presence of EDTA than in the presence of formate. Likewise, Traina et al. (1987) observed a more rapid initial release of P from an acid, montmorillonitic soil in the presence of citrate compared with formate. Fox and coworkers (1990a), examined the kinetics of P release from a Spodosol in Florida in the presence of oxalate and formate. The addition of oxalate greatly increased the rate of P release from the soil whereas the rate of P release in the presence of formic acid was no different than in water.

Low-molecular-weight organic acids affect P availability in a number of ways, all of which depend on the ability of the LMW organic acid to complex metals. The first mechanism is ligand exchange of P sorbed to Lewis acid sites on both clays and oxide surfaces that directly releases P (Sposito, 1984; Goldberg & Sposito, 1985; Fox et al., 1990a). The second mechanism whereby LMW organic acids affect P release from soils is through the dissolution of oxides surfaces. Significant amounts of P in soils may be contained within these oxide layers rather than at the surface (Van Riemsdijk et al., 1984) which is released as the oxide coating dissolve in the presence of LMW organic acids. The third mechanism whereby LMW organic acids affect P availability is by decreasing the P adsorption capacity of the soil. Surface complexation of LMW organic acids at oxide surfaces blocks P sorption sites on mineral surfaces and dissolution of the oxide surface decreases sorbing surfaces in the soil. The presence of "excess" organic acids in solution to chelate the solublized metals is important to maintaining the elevated levels of P in solution. In the absence of excess organic acids, metal-P complexes will form in solution and eventually reduce P availability as they polymerize and precipitate (Grossl & Inskeep, 1991). A fourth mechanism whereby LMW organic acids may affect P availability is through the formation of organic-metal-phosphate complexes (Bloom, 1981; Arp & Myer, 1985; Gerke, 1992a) which decrease the activity of P in solution.

The impact of LMW organic acids in plant-soil systems has been confirmed in several instances (Marschner et al., 1986). Gardner and coworkers (1983) attributed improved P uptake in white lupine to the large amounts of citrate observed in its root exudates. They postulated that citrate forms ferric-hydroxy-phosphate polymers which diffuse to the root. At the root surface they are degraded by reducing agents which releases the P which is subsequently taken up by the plant. Release of citric and malic acid by P stressed rape plants has been shown to increase P availability in this species (Hoffland, 1992; Hoffland et al., 1989a,b; 1990). Comerford and Skinner (1989) demonstrated that incremental loading of oxalic and citric acid increased P solubility in a New Zealand forest soil supporting radiata pine 14 yr after P fertilization. This may explain some of the longevity in the response to P fertilization in radiate pine since Malajczuk and Cromack (1982) have found large amounts of oxalate in the ectomycorrhizal roots of this species. Increased nutrient availability and soil weathering has been shown under the hyphal mats of soil fungi, both in the Pacific Northwest and in eastern North America which is clearly related to the high concentrations of LMW organic acids in these locations (Fisher, 1972; Cromack et al., 1979; Entry et al., 1992).

SUMMARY

Low-molecular-weight organic acids are ubiquitous in forest soils. Although they generally comprise only a small fraction of the total soil C, LMW have a disproportionate influence on soil processes because they are some of the most reactive forms of C in soil. In addition, high concentrations of LMW organic acids occur in numerous localized microsites in the soil. Low-molecular-weight organic acids also are structural components and decomposition products of larger, more complex forms of soil organic matter. The major sources of LMW organic acids in forest soils include plant root exudates, forest floor leachates, soil fungi and bacteria, organic matter decomposition, and rainfall. Concentrations of LMW organic acids range from less than 10^{-6} to greater than 10^{-3} mole per liter. The final concentration of LMW organic acids is a balance between the rates of production and degradation in the soil environment. Since most LMW organic acids are rapidly degraded, the nearly ubiquitous nature of these compounds in forest soils suggests a constant rate of production. Although the concentration in bulk soil is often quite low ($<10^{-6}$ M), extremely high concentrations ($>10^{-3}$M) can exist at specific microsites in the soil, such as in the root rhizosphere or near hyphal mats of soil fungi.

Low-molecular-weight organic acids alter chemical process in soils through complexation reactions with metals in solution and via ligand exchange reactions at soil surfaces. The impact of a specific LMW organic acid on soil processes depends to a large extent on its ability to form stable complexes with metal. This depends on the number and arrangement of functional groups on the molecule. Carboxylic, phenolic and alcoholic functional groups are the most important in LMW organic acids. Those organic acids with functional groups arranged in such a manner that they can form stable five- and six-membered ring structures via covalently bonded inner-sphere complexes can significantly affect metal solubility and speciation. Metal-ligand stability complexes can be used to estimate impact of LMW organic acids on soil processes. Low-molecular-weight organic acids can significantly affect metal solubility and speciation and thus impact mineral weathering and soil genesis. The presence of LMW organic acids also alters nutrient availability in forest soils. This has been definitively demonstrated for P. Because they function as organic ligands, LMW organic acids increase P availability by releasing P through ligand exchange reactions at mineral surfaces and by accelerating the dissolution of these mineral surfaces. Therefore, LMW organic acids affect the mineral nutrition of plants, particularly those growing in nutrient poor soil.

REFERENCES

Amrhein, C., and D.L. Suarez. 1988. The use of a surface complexation model to describe the kinetics of ligand-promoted dissolution of anorthite. Geochim. Cosmo. Acta 52:2785–2793.

Arp, P.A., and W.L. Meyer. 1985. Formation constants for selected organo-metal (Al^{+3}, Fe^{+3})- phosphate complexes. Can. J. Chem. 63:3357–3366.

Avery, G.B., Jr., J.D. Willey, and C.A. Wilson. 1991. Formic and acetic acids in coastal North Carolina rainwater. Environ. Sci. Technol. 25:1875–1880.

Bailey, P.S., Jr., and C.A. Bailey. 1981. Organic chemistry. 2nd ed. Allyn and Bacon, Inc., Boston.

Barber, D.A., and J.K. Martin. 1976. The release of organic substances by cereal roots into soil. New Phytol. 76:69–80.

Bateman, D.F., and S.V. Beer. 1965. Simultaneous production and synergistic action of oxalic acid and polygalacturonase during pathogenesis by Sclerotium rolfsii. Phytopathology 55:204–211.

Bloom, P.R. 1981. Phosphorus adsorption by an aluminum-peat complex. Soil Sci. Soc. Am. J. 45:267–272.

Bloomfield, C. 1953. A study of podzolization. Part 1. The mobilization of iron and aluminum by Scots pine needles. J. Soil Sci. 4:5–17.

Boudot, J.P. 1992. Relative efficiency of complexed aluminum, noncrystalline Al hydroxide, allophane and imogolite in retarding the biodegradation of citric acid. Geoderma 52:29–39.

Bruckert, S. 1970a. Influences des composes organiques solubles sur la pedogenese en milieu acide. I. Etudes de terrain. Ann. Agron. 21:421–452.

Bruckert, S. 1970b. Influences des composes organiques solubles sur la pedogenese en milieu acide. II. Experiences de laboratoire modalites d'action des agents complexants. Ann. Agron. 21:725–757.

Buffle, J. 1990. Complexation reactions in aquatic systems. An analytical approach. Ellis Horwood, New York.

Buurman, P. (ed.). 1984. Podzols. Van Nostrand Reinhold Co., New York.

Casey, W.H., H.R. Westrich, and G.W. Arnold. 1988. Surface chemistry of labradorite feldspar reacted with aqueous solutions at pH = 2,3, and 12. Geochim. Cosmochim. Acta 52:2795–2807.

Casey, W.H., H.R. Westrich, G.W. Arnold, and J.F. Banfield. 1989. The surface chemistry of dissolving labradorite feldspar. Geochim. Cosmochim. Acta 53:821–832.

Cline, G.R., P.E. Powell, P.J. Szaniszlo, and C.P.P. Reid. 1982. Comparison of the abilities of hydroxamic, synthetic, and other natural organic acids to chelate iron and other ions in nutrient solution. Soil Sci. Soc. Am. J. 46:1158–1164.

Comerford, N.B., and M.F. Skinner. 1989. Residual phosphorus solubility for an acid, clayey, forested soil in the presence of oxalate and citrate. Can. J. Soil. Sci. 69:111–117.

Cromack, K., Jr., P. Sollins, W.C. Graustein, K. Speidel, A.W. Todd, G. Spycher, C.Y. Li, and R.L. Todd. 1979. Calcium oxalate accumulation and soil weathering in mats of the hypogeous fungus *Hysterangium crassum*. Soil Biol. Biochem. 11:463–468.

Curl, E.A., and B. Truelove. 1986. The rhizosphere. Springer Verlag, Berlin.

Dean, L.A., and E.J. Rubins. 1947. Anion exchange in soils: I. Exchangeable phosphorus and the anion retention capacity. Soil Sci. 63:377–387.

Earl, K.D., J.K. Syers, and J.R. McLaughlin. 1979. Origin of the effects of citrate, tartrate, and acetate on phosphate sorption by soils and synthetic gels. Soil Sci. Soc. Am. J. 43:674–678.

Elgawhary, S.M., W.L. Lindsay, and W.D. Inskeep. 1970. Effect of complexing agents and acids on the diffusion of zinc to a simulated root. Soil Sci. Soc. Am. Proc. 34:211–214.

Entry, J.A., C.L. Rose, and K. Cromack. 1992. Microbial biomass and nutrient concentrations in hyphal mats of the ectomycorrhizal fungus of *Hysterangium setchellii* in a coniferous forest soil. Soil Biol. Biochem. 24:447–453.

Evans, A., Jr. 1991. Influence of low molecular weight organic acids on zinc distribution within micronutrient pools and zinc uptake by wheat. J. Plant Nutr. 14:1307–1318.

Evans, A., Jr. and L.W. Zelazny. 1990. Kinetics of aluminum and sulfate release from forest soil by mono- and diprotic aliphatic acids. Soil Sci. 149:324–330.

Fisher, R.F. 1972. Spodosol development and nutrient distribution under *Hydnaceae* fungal mats. Soil Sci. Soc. Am. Proc. 36:492–495.

Flaig, W. 1971. Organic compounds in soils. Soil Sci. 111:19–33.
Fox, T.R., and N.B. Comerford. 1990. Low-molecular-weight organic acids in selected forest soils of the southeastern USA. Soil Sci. Soc. Am. J. 54:1139–1144.
Fox, T.R., and N.B. Comerford. 1992. Influence of oxalate loading on phosphorus and aluminum solubility in Spodosols. Soil Sci. Soc. Am. J. 56:290–294.
Fox, T.R., N.B. Comerford, and W.W. McFee. 1990a. Kinetics of phosphorus release from Spodosols: Effects of oxalate and formate. Soil Sci. Soc. Am. J. 54:1441–1447.
Fox, T.R., N.B. Comerford, and W.W. McFee. 1990b. Phosphorus and aluminum release from a spodic horizon mediated by organic acids. Soil Sci. Soc. Am. J. 54:1763–1767.
Foy, C.D., and E.H. Lee. 1987. Differential aluminum tolerances of two barley cultivars related to organic acids in their roots. J. Plant Nutr. 10:1089–1101.
Furrer, G., and W. Stumm. 1986. The coordination chemistry of weathering: 1. Dissolution kinetics of Al_2O_3 and BeO. Geochim. Cosmochim. Acta 50:1847–1860.
Gardner, W.K., D.G. Parbery, and D.A. Barber. 1982a. The acquisition of phosphorus by *Lupinus albus* L. I. Some characteristics of the soil/root interface. Plant Soil 68:19–32.
Gardner, W.K., D.G. Parbery, and D.A. Barber. 1982b. II. The effect of varying phosphorus supply and soil type on some characteristics of the soil/root interface. Plant Soil 68:33–41.
Gardner, W.K., D.A. Barber, and D.G. Parbery. 1983. The acquisition of phosphorus by *Lupinus albus* L. III. The probable mechanism by which phosphorus movement in the soil/root interface is enhanced. Plant Soil 70:107–124.
Gerke, J. 1992a. Orthophosphate and organic phosphate in the soil solution of four sandy soils in relation to pH—Evidence for humic-Fe-(Al)-phosphate complexes. Commun. Soil Sci. Plant Anal. 23:601–612.
Gerke, J. 1992b. Phosphate, aluminum and iron in the soil solution of three different soils in relation to varying concentrations of citric acid. Z. Pflanzenernahr. Bodenk. 155:339–343.
Goldberg, S., and G. Sposito. 1985. On the mechanism of specific phosphate adsorption by hydroxylated mineral surfaces: A review. Commun. Soil Sci. Plant Anal. 16:801–821.
Graustein, W.C., K. Cromack, Jr., and P. Sollins. 1977. Calcium oxalate: Occurrence in soils and effect on nutrient and geochemical cycles. Science (Washington, DC) 198:1252–1254.
Grossl, P.R., and W.P. Inskeep. 1991. Precipitation of dicalcium phosphate dihydrate in the presence of organic acids. Soil Sci. Soc. Am. J. 55:670–675.
Harder, W. 1973. Microbial metabolism of organic C1 and C2 compounds. Antonie van Leewenhoek. 39:650–652.
Hingston, F.J., A.M. Posner, and J.P. Quirk. 1974. Anion adsorption by goethite and gibbsite. II. Desorption of anions from hydrous oxide surfaces. J. Soil Sci. 25:16–26.
Hirata, H., H. Hisaka, and A. Hirata. 1982. Effects of phosphorus and potassium deficiency treatment on root secretion of wheat and rice seedlings. Soil Sci. Plant. Nutr. 28:543–552.
Hodgkinson, A. 1977. Oxalic acid in biology and medicine. Acad. Press, London.
Hodgson, J.R. 1969. Contribution of metal-organic complexing agents to the transport of metals to roots. Soil Sci. Soc. Am. Proc. 33:68–75.
Hoffland, E. 1992. Quantitative evaluation of the role of organic acid exudation in the mobilization of rock phosphate by rape. Plant Soil 140:279–289.
Hoffland, E., G.R. Findenegg, and J.A. Nelemans. 1989a. Solubilization of rock phosphate by rape. I. Evaluation of the role of nutrient uptake pattern. Plant Soil 113:155–160.
Hoffland, E., G.R. Findenegg, and J.A. Nelemans. 1989b. Solubilization of rock phosphate by rape. II. Local root exudation of organic acids as a response to P starvation. Plant Soil 113:161–165.
Hoffland, E., J.A. Nelemans, and G.R. Findenegg. 1990. Origin of organic acids exuded by roots of phosphorus stressed rape (*Brassica napus*) plants. p. 179–183. *In* M.L. van Beusichem (ed.) Plant nutrition—physiology and applications. Kluwer Acad. Publ. New York.
Nowe, T.G., L.M. Dudley, and J.J. Jurinak. 1990. Determination of oxalic acid (ethanedioic acid) in soil extracts using high performance liquid chromatography. Commun. Soil Sci. Plant Anal. 21:2371–2378.
Huang, P.M., and M. Schnitzer (ed.). 1986. Interactions of soil minerals with natural organics and microbes. SSSA Spec. Publ. 17. SSSA, Madison, WI.
Huang, P.M., and A. Violante. 1986. Influence of organic acids on crystallization and surface properties of precipitation products of aluminum. p. 159–222. *In* P.M. Huang and M. Schnitzer (ed.) 1986. Interactions of soil minerals with natural organics and microbes. SSSA Spec. Publ. 17. SSSA, Madison, WI.
Hue, N.V., G.R. Craddock, and F. Adams. 1986. Effect of organic acids on aluminum toxicity in subsoils. Soil Sci. Soc. Am. J. 50:28–34.

Inskeep, W.P., and J. Baham. 1983. Competitive complexation of Cd(III) and Cu(II) by water soluble organic ligands and Na-montmorillonite. Soil Sci. Soc. Am. J. 47:1109–1115.

Inskeep, W.P., and S.D. Comfort. 1986. Thermodynamic predictions for the effects of root exudates on metal speciation in the rhizosphere. J. Plant Nutr. 9:567–586.

Jalal, M.A.F., and D.J. Read. 1983a. The organic acid composition of Calluna heathland soil with special reference to phyto- and fungitoxicity. I. Isolation and identification of organic acids. Plant Soil 70:257–272.

Jalal, M.A.F., and D.J. Read. 1983b. The organic acid composition of Calluna heathland soil with special reference to phyto- and fungitoxicity. II. Monthly quantitative determination of the organic acid content of Calluna and spruce dominated soils. Plant Soil 70:273–286.

Jauregui, M.A., and H.M. Reisenauer. 1982. Dissolution of oxides of manganes and iron by root exudate compounds. Soil Sci. Soc. Am. J. 46:314–317.

Johansson, G. 1992. Release of organic C from growing roots of meadow fescue (*Festuca pratensis* L.). Soil Biol. Biochem. 24:427–433.

Jones, D.L., and P.R. Darrah. 1992. Re-sorption of organic components by roots of *Zea mays* L. and its consequences in the rhizosphere. I. Re-sorption of ^{14}C labelled glucose, mannose and citric acid. Plant Soil 143:259–266.

Kaurichev, I.S., T.N. Ivanova, and Y.M. Nozdrunova. 1963. Low molecular organic acid content of water-soluble organic matter in soil. Sov. Soil Sci. 1963:223–229.

Kawamura, W.A., and I.R. Kaplan. 1983. Organic compounds in the rainwater of Los Angeles. Environ. Sci. Technol. 17:497–501.

Kawamura, W.A., and I.R. Kaplan. 1984. Capillary gas chromatography determination of volatile organic acids in rain and fog samples. Anal. Chem. 56:1616–1620.

Kawamura, W.A., L.L. Ng, and I.R. Kaplan. 1985. Determination of organic acids (C_1–C_{10}) in the atmosphere, motor exhausts, and engine oils. Environ. Sci. Technol. 19:5122–5130.

Kraffczyk, I., G. Trolldenier, and H. Beringer. 1984. Soluble root exudates of maize: Influence of potassium supply and rhizosphere microorganisms. Soil Biol. Biochem. 4:315–322.

Kuiters, A.T., and H.M. Sarink. 1986. Leaching of phenolic compounds from leaf and needle litter of several deciduous and coniferous trees. Soil Biol. Biochem. 18:475–480.

Kuo, S., and E.G. Lotse. 1974. Kinetics of phosphate adsorption and desorption by hematite and gibbsite. Soil Sci. 116:400–406.

Lee, D.P., and A.D. Lord. 1986. A high performance phase for the organic acids. LC-GC 5:261–265.

Lilieholm, B.C., L.M. Dudley, and J.J. Jurinak. 1992. Oxalate determination in soils using ion chromatography. Soil Sci. Soc. Am. J. 56:324–326.

Lipton, D.S., R.W. Blanchar, and D.G. Blevins. 1987. Citrate, malate and succinate concentration in exudates from P-sufficient and P-stressed *Medicago sativa* L. seedlings. Plant Physiol. 85:315–317.

Low, P.F., and C.A. Black. 1950. Reactions of phosphate with kaolinite. Soil Sci. 70:273–290.

Malajczuk, N., and K. Cromack, Jr. 1982. Accumulation of calcium oxalate in the mantle of ectomycorrhizal roots of *Pinus radiata* and *Eucalyptus marginate*. New Phytol. 92:527–531.

Marschner, H. 1986. Mineral nutrition of higher plants. Acad. Press, London.

Marschner, H. 1991. Root-induced changes in the availability of micronutrients in the rhizosphere. p. 503–528. *In* Y. Waisel et al. (ed.) Plant roots. The hidden half. Marcel Dekker, Inc., New York.

Marschner, H., V. Romhels, W.J. Horst, and P. Martin. 1986. Root induced changes in the rhizosphere: Importance for the mineral nutrition of plants. Z. Pflanzenernaehr. Bodenk. 149:441–456.

Martell, A.E., R.J. Motekaitis, and R.M. Smith. 1988. Structure-stability relationships of metal complexes and metal speciation in environmental aqueous solutions. Environ. Tox. Chem. 7:417–434.

Martin, R.S., H. Westber, E. Allwine, L. Ashman, J.C. Farmer, and B. Lamb. 1991. Measurement of isoprene and its atmospheric oxidation products in a central Pennsylvania deciduous forest. J. Atmos. Chem. 13:1–32.

Mast, M.A., and J.I. Drever. 1987. The effect of oxalate on the dissolution rates of oligoclase and tremolite. Geochim. Cosmochim. Acta 2559–2568.

Maxwell, D.P., and D.F. Bateman. 1968. Oxalic acid biosynthesis by Sclerotium rolfsii. Phytopathology 58:1635–1642.

McColl, J.G., and A.A. Pohlman. 1986. Soluble organic acids and their chelating influence on Al and other metal dissolution from forest soils. Water Air Soil Pollut. 31:917–927.

McColl, J.G., A.A. Pohlman, J.M. Jersak, S.C. Tam, and R.R. Northup. 1990. Organics and metal solubility in California forest soils. p. 178 195. *In* S.P. Gessel et al. (ed.) Sustained productivity of forest soils. Proc. North Am. For. Soils Conf., 7th, Vancouver, Canada. 24–28 July 1990. Univ.British Columbia, Faculty For. Publ. Vancouver, Canada.

Miller, W.P., L.W. Zelazny, and D.C. Martens. 1986. Dissolution of synthetic crystalline and noncrystalline iron oxides by organic acids. Geoderma 37:1–13.

Monetti, M.A., and M.I. Scranton. 1992. Fatty acid oxidation in anoxic marine sediments: The importance of hydrogen sensitive reactions. Biogeochemistry. 17:23–47.

Motekaitis, R.J., and A.E. Martell. 1984. Complexes of aluminum(III) with hydroxy carboxylic acids. Inorg. Chem. 23:18–23.

Muck, R.E., R.K. Wilson, and P. O'Kiely. 1991. Organic acid content of permanent pasture grasses. Ir. J. Agric. Res. 30:143–152.

Muir, J.W., R.I. Morrison, C.J. Brown, and J. Logan. 1964. The mobilization of iron by aqueous extracts of plants. I. Composition of the amino acid and organic acid fractions of an aqueous extract of pine needles. J. Soil Sci. 15:220–225.

Ohwaki, Y., and H. Hirata. 1992. Differences in carboxylic acid exudation among P-starved leguminous crops in relation to carboxylic acid contents in plant tissues and phospholipid level in roots. Soil Sci. Plant Nutr. 38:235–243.

Parfitt, R.L., A.R. Fraser, J.D. Russell, and V.C. Farmer. 1977. Adsorption on hydrous oxides. II. Oxalate, benzoate and phosphate on gibbsite. J. Soil Sci. 28:40–47.

Pierson, P.E., and L.H. Rhodes. 1992. Effect of culture medium on the production of oxalic acid by *Sclerotinia trifoliorum*. Mycologia 84:467–469.

Pohlman, A.A., and J.G. McColl. 1986. Kinetics of metal dissolution from forest soils by soluble organic acids. J. Environ. Qual. 15:86–92.

Pohlman, A.A., and J.G. McColl. 1988. Soluble organics from forest litter and their role in metal dissolution. Soil Sci. Soc. Am. J. 52:265–271.

Ratnayake, M., R.T. Leaonald, and J.A. Menge. 1978. Root exudation in relation to supply of phosphorus and its possible relevance to mycorrhizal formation. New Phytol. 81:543–552.

Reid, R.K., C.P.P. Reid, and P.J. Szaniszlo. 1985. Effects of synthetic and microbially produced chelates on the diffusion of iron and phosphorus to a simulated root in soil. Biol. Fert. Soils 1:45–52.

Robert, M., and J. Berthelin. 1986. Role of biological and biochemical factors in soil weathering. p. 453–495. *In* P.M. Huang and M. Schnitzer (ed.) Interactions of soil minerals with natural organics and microbes. SSSA Spec. Publ. 17. SSSA, Madison, WI.

Rozycki, H., and F. Strzelczyk. 1986. Organic acid production by *Streptomyces* spp. isolated from soil, rhizosphere and mycorrhizosphere of pine (*Pinus sylvestris* L.). Plant Soil 96:337–345.

Schnitzer, M. 1986. Binding of humic substances by soil mineral colloids. p. 77–102. *In* P.M. Huang and M. Schnitzer (ed.) Interactions of soil minerals with natural organics and microbes. SSSA Spec. Publ. 17. SSSA, Madison, WI.

Schnitzer, M., and S.U. Khan (ed.). 1978. Soil organic matter. Elsevier Sci. Publ. Co., New York.

Schwartz, S.M., J.E. Varner, and W.P. Martin. 1954. Separation of organic acids form several dormant and incubated Ohio soils. Soil Sci. Soc. Am. Proc. 18:174–177.

Shotyk, W., and H.W. Nesbitt. 1992. Incongruent and congruent dissolution of plagioclase feldspar: Effect of feldspar composition and ligand complexation. Geoderma 55:55–78.

Smith, W.H. 1969. Release of organic materials from the roots of tree seedlings. For. Sci. 15:138–143.

Smith, W.H. 1976. Character and significance of forest tree root exudates. Ecology 57:324–331.

Sollins, P., K. Cromack, Jr., and C.Y. Li. 1981. Role of low-molecular-weight organic acids in the inorganic nutrition of fungi and higher plants. p. 607–619. *In* D.T. Wicklow and G.C. Carroll (ed.) The fungal community. Marcel Dekker, New York.

Song, S.K., and P.M. Huang. 1988. Dynamics of potassium release from potassium bearing minerals as influenced by oxalic and citric acids. Soil Sci. Soc. Am. J. 52:383–390.

Sposito, G. 1984. The surface chemistry of soils. Oxford Univ. Press, New York.

Stevenson, F.J. 1967. Organic acids in soil. p. 119–146. *In* A.D. McLaren and G.H. Peterson (ed.) Soil biochemistry. Vol. 1. Marcel Dekker, New York.

Stevenson, F.J. 1982. Humus chemistry. John Wiley & Sons, New York.

Struthers, P.H., and D.H. Sieling. 1950. Effects of organic anions on phosphate precipitation by iron and aluminum as influenced by pH. Soil Sci. 69:205–213.

Stumm, W. 1986. Coordination interactions between soil solids and water: An aquatic chemists' point of view. Geoderma 38:19–30.

Stumm, W., and J.J. Morgan. 1981. Aquatic chemistry. John Wiley & Sons, New York.

Stumm, W., G. Furrer, and B. Kunz. 1983. The role of surface coordination in precipitation and dissolution of mineral phases. Croat. Chem. Acta 56:593–611.

Swenson, R.M., C.V. Cole, and D.H. Sieling. 1949. Fixation of phosphate by iron and aluminum and replacement by organic and inorganic ions. Soil Sci. 67:3–22.

Tam, S.C. 1987. Simulated acid rain and the importance of organic ligands on the availability of aluminum in soil. Water Air Soil Pollut. 36:193–206.

Tam, S.C., and J.G. McColl. 1990. Aluminum and calcium binding affinities of some organic ligands in acidic conditions. J. Environ. Qual. 19:514–520.

Tam, S.C., and J.G. McColl. 1991. Aluminum binding ability of soluble organics in Douglas-fir litter and soil. Soil Sci. Soc. Am. J. 55:1421–1427.

Tan, K.H. 1986. Degradation of soil minerals by organic acids. p. 1–28 *In* P.M. Huang and M. Schnitzer (ed.) Interactions of soil minerals with natural organics and microbes. SSSA Spec. Publ. 17. SSSA, Madison, WI.

Thomas F., A. Masion, J.Y. Bottero, J. Rouiller, F. Genevrier, and D. Boudot. 1991. Aluminum(III) speciation with acetate and oxalate. A potentiometric and ^{27}Al NMR study. Environ. Sci. Technol. 25:1553–1559.

Thurman, E.M. 1985. Organic geochemistry of natural waters. Marinus Nijhoff/Dr. W. Junk Publ., Dordrecht, the Netherlands.

Traina, S.J., G. Sposito, D. Hesterberg, and U. Kafkafi. 1986a. Effects of pH and organic acids on orthophosphate solubility in an acidic, montmorillonitic soil. Soil Sci. Soc. Am. J. 50:45–52.

Traina, S.J., G. Sposito, G.R. Bradford, and U. Kafkafi. 1987. Kinetic study of citrate effects on orthophosphate solubility in an acid, montmorillonitic soil. Soil Sci. Soc. Am. J. 51:1483–1487.

Vancura, V. 1964. Root exudates of plants. I. Analysis of root exudates of barley and wheat in their initial phases of growth. Plant Soil 21:231–248.

Vancura, V., and A. Hanzlikova. 1972. Root exudates of plants. IV. Differences in chemical composition of seed and seedling exudates. Plant Soil 36:271–282.

Vancura, V., and A. Hovadik. 1965. Root exudates of plants. II. Composition of root exudates of some vegetables. Plant Soil 22:21–32.

Van Riemsdijk, W.H., L.J.M. Bourans, and F.A.M. De Haan. 1984. Phosphate sorption by soils: I. A model for phosphate reaction with metal-oxides in soil. Soil Sci. Soc. Am. J. 41:870–876.

Wang, T.S.C., S. Cheng, and H. Tung. 1967. Dynamics of soil organic acids. Soil Sci. 104:138–144.

Whitehead, D.C. 1964. Identification of *p*-hydroxybenzoic, vanillic, *p*-coumaric and ferulic acids in soils. Nature (London) 202:417–418.

Wiley, J.D., and C.A. Wilson. 1993. Formic and acetic acids in atmospheric condensate in Wilmington, North Carolina. J. Atmos. Chem. 16:123–133.

Zinder, B., G. Furrer, and W. Stumm. 1986. The coordination chemistry of weathering: II. Dissolution of Fe(III) oxides. Geochim. Cosmochim. Acta 50:1861–1869.

5 Characterization of Dissolved and Colloidal Organic Matter in Soil Solution: A Review

Bruce E. Herbert

*Texas A&M University
College Station, Texas*

Paul M. Bertsch

*University of Georgia
Savannah River Ecology Laboratory
Aiken, South Carolina*

Dissolved organic matter (DOM) in soil solution is important in many biogeochemical processes and can have a strong influence on the fate of contaminants in soils, surface waters, and groundwaters. Dissolved organic matter plays an important role in the cycling of nutrients, such as C, N, P, and S, within soil, and the export of these nutrients from soil systems to streams and groundwaters (McDowell & Wood, 1984; Malcolm, 1985; Murphy et al., 1989a; Homann et al., 1990; Qualls et al., 1991; Vaithiyanathan & Corell, 1992). It also is a substrate for microbial growth (Zsolnay & Steindl, 1991; Qualls & Haines, 1992). Dissolved organic matter mediates the transport of inorganics, such as Al and other metals, within soils (Saar & Weber, 1982; Inskeep & Baham, 1983; Krug & Isaacson, 1984) and is important in the translocation of oxides, humus, and silicate clays in soils (Wright & Schnitzer, 1963; Dawson et al., 1978, 1981; Jenny, 1980). Chelation by organic acid functional groups in DOM also is responsible for enhancing mineral weathering (Antweiler & Drever, 1983; Pohlman & McColl, 1988).

The fate and transport of organic contaminants, including polycyclic aromatic hydrocarbons, chlorinated hydrocarbons, pesticides and herbicides in the environment can be influenced by the presence of DOM. Many natural systems can be considered three-phase systems, where organic contaminants sorb to the solid matrix, to DOM, or are freely dissolved. The relative proportions of contaminant in each of the phases affect its fate and transport. Dissolved organic matter extracted from lakes and streams, soils, and lake sediments increases the solubility of organic contaminants, such as DDT [1,1,1-trichloro-2,2 bis(*p*-chlorophenyl)ethane], n-alkanes, 1,2,3 trichlorobenzene, and lindane (Wershaw et al., 1969; Boehm & Quinn, 1973; Carter & Suffet, 1982; Means &

Copyright © 1995 Soil Science Society of America, 677 S. Segoe Rd., Madison, WI 53711, USA.
Carbon Forms and Functions in Forest Soils.

Wijayaratne, 1982; Chiou et al., 1986). Dissolved organic matter also decreases the volatilization of certain organic contaminants (Callaway et al., 1984; Hassett & Milicic, 1985) and may facilitate the transport of these compounds through soils and groundwaters (McDowell-Boyer et al., 1986; McCarthy & Zachara, 1989). Dissolved organic matter can decrease the abiotic hydrolysis of pesticides and other organic contaminants (Perdue & Wolfe, 1982; Macalady et al., 1989) photosensitize transformations of toxic organic compounds (Zepp et al., 1981; Zafiriou et al., 1984; Ross & Crosby, 1985; Zepp et al., 1985; Simmons & Zepp, 1986; Mills & Schwind, 1990). Finally, DOM in lake water and sediment pore water decreases the bioaccumulation of organic contaminants to organisms such as *Daphnia*, fish and amphipods (McCarthy et al., 1985; McCarthy & Jimenez, 1985; Landrum et al., 1987; Kukkonen et al., 1990).

Many of these studies quantified the influence of DOM on organic contaminant fate using organic material isolated from aquatic systems or alkali-extracted humic materials isolated from soils. There is reasonable cause to question whether these materials are adequate surrogates of soil solution DOM. In fact, relatively few studies have characterized soil solution DOM, its biogeochemistry, and its influence on the fate of contaminants in soils. We review the present state of knowledge on soil solution DOM characteristics and present several areas of needed research in the isolation and characterization of soil solution DOM. We also suggest that organic matter collected in water extracts of soil may be better analogs of soil solution DOM than aquatic DOM or alkali-extracted humic solutions. In particular, the objectives of this review are to: (i) evaluate different techniques for collecting, extracting, and fractionating soil solution DOM; (ii) compile data on the characteristics of soil solution DOM; and (iii) review the biogeochemistry of DOM.

DEFINITIONS OF DISSOLVED ORGANIC MATTER AND WATER SOLUBLE ORGANIC MATTER

Dissolved organic matter is a complex mixture of many organic compounds and therefore must be operationally defined (MacCarthy & Suffet, 1989). It is composed of both humic materials and other organic compounds, such as amino acids, carbohydrates, aliphatic and aromatic acids, and hydrocarbons. The sizes and molecular weights (MW) of common organic species found in soil solution are illustrated in Fig. 5–1, along with other common biotic and abiotic components. A large portion of the organic matter in soil solution is present as humic substances with average molecular weights on the order of 1000 MW (Thurman, 1985a). Some organic colloidal material also may be present as humic and lipid aggregates of 2000 to 100000 MW (2–50 nm), or associated with inorganic colloids or organic particulates (Thurman, 1985b; McCarthy & Zachara, 1989). Soil solution DOM is usually defined as the organic matter that passes a 0.45-μm filter (Thurman, 1985a). Other size filters, especially 0.2 μm, also have been used to define DOM (Buffle et al., 1982), because colloidal material may not be retained on the 0.45-μm filters (Riley, 1973; Meadows et al., 1978). Unless solution free of colloidal material is required, it is best to define DOM as the organic matter dissolved in soil solution which passes a 0.45-μm filter. In this way, the

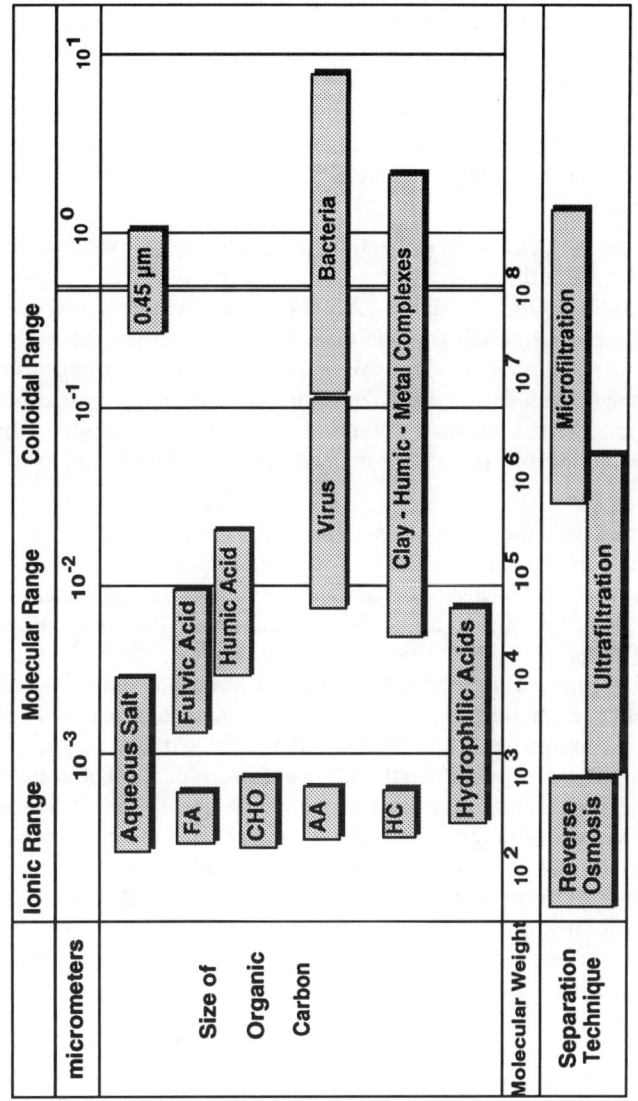

Fig. 5–1. Relative size of dissolved and colloidal organic species in soil solution (Thurman, 1985a). The FA, CHO, AA, HC refers to fatty acids, carbohydrates, amino acids, and hydrocarbons, respectively.

isolated material would be comparable to the greatest body of published literature.

Soil organic matter extracts have been used as surrogates for soil solution DOM. Organic matter extracted from soil is operationally defined based on the solvent used in the extraction (Hayes, 1985). Water soluble organic matter (WSOM) is the fraction of soil organic matter extracted with water or a dilute salt solution that passes a 0.45 µm or other appropriate filter. Several studies have used this material to represent soil solution DOM (Bremner & Lees, 1949; Hayes et al., 1975; Candler & van Cleve, 1982; Inskeep & Baham, 1983; Krug & Isaacson, 1984; Candler et al., 1988; Traina et al., 1989; Herbert et al., 1993). This review uses the following definitions for soil solution DOM, WSOM and humic materials:

1. Soil Solution DOM = Dissolved organic matter in soil solution that can pass a 0.45-µm filter.
2. Water Soluble Organic Matter (WSOM) = Soil organic matter extracted with water or dilute salt solution that can pass a 0.45-µm filter.
3. Humic Substances = Large MW macromolecules resulting from the microbial degradation of plant and animal remains. Soil humic substances are defined based on their solubility in acidic and alkaline solutions. Aquatic humic substances are defined based on their sorption to XAD resins.

Many studies have used aqueous solutions of alkaline-extracted soil organic matter as analogs of soil solution DOM. For example, several studies quantified DOM-organic contaminant interactions using alkaline-extracted organic matter (Chiou et al., 1986; Gauthier et al., 1987; Lee & Farmer, 1989), or commercial humic acids (Carter & Suffet, 1982; Hassett & Milicic, 1985; McCarthy & Jimenez, 1985; Gauthier et al., 1987; MacCarthy & Malcolm, 1989). Use of these results to predict interactions between organic contaminants and soil solution DOM may be inappropriate if the properties of alkaline-extracted organic matter and soil solution DOM are significantly different (Malcolm & MacCarthy, 1986). Likewise, false conclusions may be reached in studies of the importance of allochthonous sources of organic C to streams and rivers if the characteristics of alkaline-extracted soil humics are compared to dissolved humic substances from aquatic systems (Malcolm, 1985). It is a central premise of this paper that the extraction of soil organic matter with water or dilute salt solutions results in an organic matter fraction that is more representative of soil solution DOM than alkaline-extracted fulvic or humic acids.

COLLECTION AND EXTRACTION OF DISSOLVED ORGANIC MATTER AND WATER SOLUBLE ORGANIC MATTER

Collection of Soil Solution Dissolved Organic Matter

Among the better methods for sampling soil solution DOM are tension and zero-tension porous cup lysimeters, though there are problems associated with the use of these instruments (Barbee & Brown, 1986; Litaor, 1988). The volume of soil solution, and hence the mass of DOM, collected by lysimeters can be limited depending on soil conditions. Soil heterogeneity also may limit the ability of the lysimeter to sample a representative volume of soil (Haines et al., 1982; Barbee & Brown, 1986). This has led to the recommendation that larger samplers be used (Radulovich & Sollins, 1987). Finally, tension and zero-tension lysimeters sample different solutions (Haines et al., 1982; Cronan et al., 1990). Tension lysimeters preferentially sample soil solution in smaller pores, while zero-tension lysimeters sample soil water at near saturated conditions (Haines et al., 1982; Litaor, 1988). The difference in the soil solution sampled by each lysimeter may affect the characteristics of DOM collected.

Extraction of Water Soluble Organic Matter

By definition, WSOM is extracted with water or aqueous solution. Batch extractions of soil organic matter have been conducted using water or dilute salt solutions (Forsyth, 1947; Khan, 1970; Linehan, 1977; Candler & van Cleve, 1982; Inskeep & Baham, 1983; Madhun et al., 1986; Candler et al., 1988; Traina et al., 1989; Novak & Bertsch, 1991; Herbert et al., 1993), and salt solutions of sodium pyrophosphate, $NaNO_3$, and $NaSO_4$ (Hayes et al., 1975; David et al., 1989). Leaching soils with water and aqueous salt solutions, particularly NaCl, $NaNO_3$ and $(Na)_2SO_4$ salts, also has been used to isolate WSOM (Cronan et al., 1978; Chang & Alexander, 1984; Krug & Isaacson, 1984; Cronan, 1985; Evans et al., 1988; Vance & David, 1989; Dahlgren et al., 1990). Novak and Bertsch (1991) found that 90% of WSOM was extracted after three sequential batch extractions during a series of eight sequential extractions. Likewise, continuous leaching studies indicate labile and nonlabile pools of leachable organic matter exist in soils (Chang & Alexander, 1984; Krug & Isaacson, 1984). The existence of a nonlabile pool of DOM may not affect sequential batch extractions of WSOM, but could influence dynamic extractions using leaching techniques.

The amount of WSOM extracted is dependent on solution chemistry of the extractant, including ionic strength, pH, and the dominant anion present. Variation of pH, in the range of 3.5 to 5.5, has a small effect on the amount of WSOM extracted (Chang & Alexander, 1984; Cronan, 1985; Hay et al., 1985; David et al., 1989; Vance & David, 1989). Only a small amount of WSOM was extracted with solutions at pH 2, while at a pH of less than 2, a large amount of WSOM was released due to mineral dissolution (David et al., 1989). Solutions at pH greater than 8 solubilize large amounts of WSOM (Schnitzer, 1978). For instance, anhydrous and aqueous ammonia, phosphate based fertilizers, and combinations of these fertilizers have been found to increase DOM concentrations in solution due to the high pH of the soil solution after their application (Nemec & Vopenka, 1971; Tomasiewicz & Henry, 1985; Bell & Black, 1970; Giordano et al., 1971; Myers & Thien, 1988). Ionic strength and the dominant anion present in solution also influence the amount of WSOM extracted from soils. A low ionic strength solution (0.2 mmol L^{-1}) leached more WSOM than solutions of higher ionic strengths (Evans et al., 1988). The presence of SO_4^{2-} increased the amount of WSOM extracted at constant ionic strength and pH, compared to extractants containing Cl^- or NO_3^- (Vance & David, 1989).

Solution characteristics also determine qualitative differences in WSOM (Cronan, 1985; David et al., 1989; Vance & David, 1989). Decreasing the pH of the extraction solution resulted in a decrease in the hydrophobic acid content of WSOM extracted from forest soils. The pH effect was largest in WSOM leached from O horizons and least in WSOM leached from B horizons. These studies also showed that the influence of solution chemistry on the amount and characteristics of WSOM is secondary to the influence of the soil horizon.

Several studies have compared the amount and characteristics of WSOM and soil organic matter extracted with other solvents such as NaOH (Forsyth, 1947; Bremner & Lees, 1949; Hayes et al., 1975; Novak & Bertsch, 1991). In general, aqueous solutions extract less organic matter from soils than solutions of organic chelating agents, dipolar aprotic solvents, or NaOH (Novak & Bertsch, 1991). Forsyth (1947) and Linehan (1977) used infrared spectoscopy, elemental analysis, and acidity measurements to show that the chemical characteristics of WSOM were similar to NaOH extracted fulvic acid. In contrast, ^{13}C NMR spectroscopy indicated significant differences in chemical characteristics of WSOM, fulvic acid, and humic acid extracted from soils along a toposequence in South Carolina (Novak & Bertsch, 1991).

On the other hand, comparisons between WSOM and soil solution DOM collected from the same soil have not been made. Since solution chemistry influences the amount and characteristics of WSOM, it is logical that an extractant solution adjusted to the pH, ionic strength, Na/Ca ratio, and type of dominant anion of the soil solution of interest would result in WSOM that best represents soil solution DOM at that site. This is a critical research need if the usefulness of WSOM extractions is to be validated.

Isolation of Dissolved Organic Matter from Soil Solution or Water Extracts

Several techniques used to isolate and concentrate DOM from aquatic systems may be useful to concentrate soil solution DOM or WSOM. These techniques include ultrafiltration, reverse osmosis, and sorption chromatography. Ultrafiltration, which has been used to collect soil solution (Jackson et al., 1976; Levin & Jackson, 1977), should work well for DOM or WSOM isolation. Ultrafiltration separates dissolved and colloidal species by passage through molecular filters under hydrostatic pressure (Aiken, 1984). Filters are available with nominal molecular weight cutoffs (NMWCO) of 500 to 300000. Different techniques include flat membrane, field-flow, and hollow fiber ultrafiltration (Aiken, 1985). These techniques have isolated DOM from aquatic sources (Gjessing, 1970; Wheeler, 1976; Buffle et al., 1978, 1982). The field flow and hollow fiber membranes are particularly useful for the rapid filtration of samples without fouling the membranes. Dissolved organic matter or WSOM of MW greater than 200 to 500 is isolated by ultrafiltration using filters of 500 or 1000 NMWCO. Ultrafiltration allows significant amounts of lower MW, nonhumic organics to pass through the membranes.

Reverse osmosis also has been used to isolate aquatic DOM (Aiken, 1985; Serkiz & Perdue, 1990). This technique has the ability to process large volumes and, as opposed to ultrafiltration, also will collect most of the smaller organic compounds. Of course, inorganics also will be collected by reverse osmosis. This may induce precipitation of the organics or introduce contamination. Reverse osmosis equipment also is more expensive than ultrafiltration.

Sorption chromatography utilizing alumina, activated C, ion exchange resin, nonionic macroreticular resin (XAD resins), and C-18 bonded resin (solid phase extraction, SPE), has been used to isolate aquatic organics (Mills & Quinn, 1981; Aiken, 1985; Amador et al., 1990). Fu and Pocklington (1983) found that a combination of XAD-2 and activated charcoal was able to extract greater than 90% of the DOM in sea water. Aiken et al. (1992) isolated aquatic DOM on a preparative scale using a two column array of XAD-8 and XAD-4 resins. The percentage of total dissolved organic carbon (DOC) retained on the XAD-8 resin varied between 23 to 58%, while the percentage retained on the XAD-4 was 7 to 25%. Solid phase extraction also has been used to extract DOM from seawater (Mills et al., 1982; Amador et al., 1990). Similar amounts of seawater DOM are isolated with SPE and macroreticular resins (Amador et al., 1990). An advantage of using SPE instead of XAD resin for DOM isolation is that the resin is not subject to the extensive pretreatment steps required by the macroreticular resins.

All three classes of techniques should be useful for the isolation of soil solution DOM or WSOM, though they need to be tested. The major limitation of these techniques is that they generally isolate only a fraction of total DOM, and therefore may best be suited for DOM fractionation.

FRACTIONATION OF DISSOLVED ORGANIC MATTER AND WATER SOLUBLE ORGANIC MATTER

Due to the complexity of DOM and WSOM, separation into fractions based on physiochemical characteristics is useful (Buffle et al., 1982; Hayes, 1985; Swift, 1985; Thurman, 1985a). These characteristics include solubility, size or MW, and the ability to sorb to different materials (Swift, 1985). When studying DOM biogeochemistry, it is useful to choose a fractionation technique based on important physiochemical properties related to the biogeochemical process of interest. For instance, DOM fractionation by sorption chromatography may provide insight into DOM sorption mechanisms to clay minerals or soil particles.

Fractionation by Solubility

Solubility or organic matter in different extractants has often been used as the basis of fractionation (Swift, 1985). Alkali-extracted soil organic matter has classically been fractionated into humic and fulvic acids based on solubility in dilute solutions of NaOH and HCl. This approach represents the standard soil organic matter fractionation technique of the International Humic Substances Society (Stevenson, 1985; Schnitzer & Preston, 1986). This technique also has been used to fractionate aquatic humics (Thurman & Malcolm, 1981; Leenheer, 1985; Abbt-Braun et al., 1989), but has not been applied to WSOM. Fractionation of WSOM based on solubility in acid and base has the advantage that the fractions would be directly comparable to the vast amount of literature concerned with soil humic and fulvic acids.

Fractionation by Molecular Size

Three techniques have been used to fractionate soil derived DOM and WSOM according to molecular size: gel permeation chromatography, ultrafiltration, and centrifugation. The fractionation of DOM based on size is important because of the strong correlation between the size of organic molecules and their chemical and physical properties (Swift & Posner, 1972; Thurman et al., 1982).

Gel permeation chromatography is based on the preferential size exclusion of large MW organics from the interior of cross-linked macroreticular resins. This creates an inverse relationship between the elution pattern and molecular size. The Sephadex resins are most often used (Hine & Bursill, 1984), though other resins such as the Bio-Gel P-series also have been used (Hayes et al. 1975; Buffle et al., 1978). Eluants used include distilled water (Khan, 1970; Candler & van Cleve, 1982), TRIS buffer (2-amino 2-hydroxymethyl-1,3-propanediol) (Hayes et al. 1975; Dawson et al., 1981), and 0.1 M NaOH (Dawson et al., 1981). The method is calibrated with macromolecules of known MW in order to estimate the MW of natural organic molecules.

Problems with gel permeation chromatography involve interpreting the estimated average MW of the collected fractions. Organic solutes can interact with the resin gel through hydrophobic and electrostatic bonds (Swift & Posner, 1971; Hine & Bursill, 1984; Haumaier et al., 1990). This invalidates the interpretation of the elution pattern based solely on molecular size. The fractionation process also is influenced by the column size and eluant composition (Hine & Bursill, 1984). For example, Dawson et al. (1981) eluted samples with TRIS buffer and NaOH. The two eluants gave incongruous results because of different gel-solute interactions. The NaOH eluant resulted in separation based on ion exclusion, while the TRIS buffer minimized ion exclusion. These problems limit the use of gel permeation chromatography as a technique of absolute DOM MW determination and fractionation.

As described above, ultrafiltration can fractionate dissolved and colloidal organic species based on size (Aiken, 1984). It has been found that the shape of the molecules affect retention; a linear molecule may pass through a filter, while a globular molecule of the same MW may be retained (Batley, 1989). Large MW organic molecules change conformation depending on solution chemistry, which will affect retention (Ghosh & Schnitzer, 1980). The pores of the filters themselves are not exactly uniform which results in some diversity of retained molecular size (Wershaw & Aiken, 1985). These limitations indicate that ultrafiltration, like gel permeation chromatography, is a semiquantitative technique for DOM molecular size fractionation providing only relative size ranges under a given set of solution characteristics. On the other hand, the much lower surface area of the filters compared to gel permeation resins limits sorption problems during fractionation. The newer ultrafiltration techniques of tangential flow fractionation using hollow fiber or spiral wound filters limit the problem of concentration polarization exhibited by flat ultrafiltration membranes and result in faster filtration times

(Kwak et al., 1977; Buffle et al., 1978; Aiken, 1984). These ultrafiltration techniques should be useful for the isolation and fractionation of DOM based on relative molecular size.

Buffle et al. (1978) tested the ability of ultracentrifugation to fractionate radio-labeled Cd(II) and Pb(II) complexes of aquatic and soil organic matter, and aqueous leaf letter extracts. The solutions were centrifuged at 40000 rpm, and the application of the Svedberg equation gave an order-of-magnitude estimate for the MW of the metal complexes. Reid at al. (1990) recently reported the fractionation and MW determination of three humic samples using analytical [ultraviolet (UV) scanning] ultracentrifugation. Though this technique gave reasonable results for average MW of the humics, the determination of absolute MW is dependent on the assumed relationship between the extinction coefficient of DOM and its MW. Overall, ultracentrifugation is rarely used for DOM fractionation and is more suited to particulate and colloidal studies (Lammers, 1967; Batley, 1989).

The disadvantages of fractionation techniques based on molecular size include the possibility of confounded results due to conformational changes of DOM during the fractionation (Batley, 1989). The reproducibility of fractionation results are maximized if DOM aggregation is kept to a minimum. The advantages of these techniques lie in the observation that the molecular size of DOM is often correlated with other molecular properties. These techniques also allow the fractionation and concentration of DOM with minimal addition of chemical reagents.

Fractionation by Sorption Chromatography

The fractionation of DOM by sorption chromatography is useful because fractionation is based on important DOM properties, including hydrophobicity and acidity, which regulate DOM interaction with organic contaminants and soil surfaces, and it can operate at both the analytical and preparative scale (Leenheer, 1985). The most common technique for the fractionation of aquatic DOM (Aiken & Malcom, 1979; Leenheer, 1981) is based on the sorption of DOM to nonionic (Amberlite XAD-8, Rohm and Haas, Co., Philadelphia, PA) and ion-exchange resins (Biorad AG-MP-50 Bio-Rad Laboratories, Life Science Group, Richmond, CA; Duolite A-7 Rohn and Haas, Co., Philadelphia, PA). Several studies have used this approach to fractionate soil solution DOM and WSOM. Fractions based on the method of Leenheer (1981) were used to follow soil solution DOM cycling over space and time, and relate these observations to mineral dissolution rates (Antweiller & Drever, 1983). Changes in the hydrophobic/hydrophilic ratios of DOM as a function of acidic inputs also have been studied (Cronan, 1985; Cronan & Aiken, 1985; David et al., 1989; Vance & David, 1989). The sorption of different soil solution DOM fractions to soil has been analyzed by this method (Jardine et al., 1989a). A related technique has been used to isolate the hydrophobic and hydrophilic acid fractions of DOM (Aiken et al., 1992). This technique uses a sequence of XAD-8 and XAD-4 resins to fractionate DOM adjusted to pH 2. The XAD-8 resin retains the hydrophobic acids, while the XAD-4 retains the hydrophilic acids.

Disadvantages of this technique include its labor intensiveness, the potential for resin bleed from XAD resins (Thurman & Malcolm, 1981), and the extensive cleanup of the XAD resin that is required (van Rossum & Webb, 1978). Aggregation of DOM and undesirable DOM-adsorbent interactions also are possible, which could confound the results (Leenheer, 1985). Qualls and Haines (1991) have compiled a list of organic compounds expected to be isolated in each fraction (Table 5–1).

Reverse phase chromatography utilizing C-18 Sep-Pak cartridges (Waters Associates, Milford, MA) has been used to fractionate natural marine DOM, lake DOM and WSOM, and may represent a new fractionation technique for soil solution DOM (Mills & Quinn, 1981; Mills et al., 1982, 1987, 1989; Amador et al., 1990; David & Vance, 1991; Herbert et al., 1993). At natural solution pH, Sep-Paks extracted 10 to 30% of total marine DOM (Mills & Quinn, 1981). The hydrophobic fraction retained on the cartridges was further fractionated by eluting the organic material from the cartridges with 50:50 v/v water-methanol solutions, and then 100% methanol (Mills et al., 1982, 1987). The relative polarities of size fractionated WSOM, humic acid, and fulvic acid were measured using the

Table 5–1. Substance found in specific fractions of DOM fractionated by sorption chromatography using nonionic and ion-exchange resins (reprinted with permission from Qualls & Haines, 1991).

Fraction	Compounds
Hydrophobic neutrals	Hydrocarbons†
	Chlorophyll
	Cartenoids
	Phospholipids
	Humics with <1 ionic or phenolic group per 13 C atoms‡
Weak (phenolic) hydrophobic acids	Tannins‡
	Flavonoids
	Other polyphenols (<1 carboxyl group per 13 C atoms)‡
	Vanillin
Strong (carboxylic) hydrophobic acids	Fulvic acid and humic acid‡
	Humic-bound amino acids and peptides
	Humic-bound carbohydrates
	Aromatic acids (including phenolic and carboxlic acids)§
	Oxidized polyphenols (>1 carboxyl group per 12 C atoms)‡
	Long-chain fatty acids (>C5)§
Hydrophilic acids	Humic-like substances with lower molecular size and higher COOH/C ratios
	Oxidized carbohydrates with COOH groups
Hydrophilic neutrals	Small carboxylic acids¶
	Inositol and other sugar phosphates
	Simple neutral sugars
	Non humic-bound polysaccharides†
	Alcohols (<C4)§
Bases	Proteins
	Free amino acids and peptides
	Aromatic amines
	Amino-sugar polymers (such as from microbial cell walls)

†Substances indicated by infrared spectra of fractions of river-water DOM.
‡Thurman, 1985a.
§Determined by retention of model compounds by XAD-8 resin (Thurman et al., 1978).
¶Fractionation of model compound, oxalic acid (David et al., 1989).

C-18 Sep-Pak cartridges (Herbert et al., 1993). The solutions of organic matter were adjusted to pH 2 and passed through pretreated Sep-Paks. David and Vance (1991) observed similar retention of lake DOM on XAD-8 resin and reverse phase C-18 resin. Retention of DOM on reverse phase C-18 resin and XAD-8 resin should be similar because of similarities in the basic sorption mechanisms (Amador et al., 1990). Relative characteristics of DOM isolated by reverse phase C-18 and the macroreticular XAD-8 resins need to be compared in greater detail. Natural organics also have been fractionated using reverse phase, high performance liquid chromatography (Mills et al., 1982, 1987; Saleh et al., 1989; Saleh & Chang, 1987). This method is limited because of problems in detection of all organic solutes in DOM with UV absorption (Mills et al., 1982, 1987). Other detectors may offer some promise, such as electrochemical detection, though these detectors are subject to fouling (Novotny, 1989).

Careful thought should be given to the choice of fractionation technique because methodology affects observed fraction characteristics. The greatest amount of information is gained from DOM fractionation when the chosen technique is based on a DOM characteristic related to important biogeochemical processes. For instance, comparisons of DOM from soil and aquatic systems can best be made if the same fractionation scheme for all isolated organic material is used (Malcolm, 1985). It also is possible that DOM should not be fractionated in certain cases. Recent results suggest the properties of DOM fractions may not correspond to the properties of the original sample because of the importance of interactions between fractions (Kukkonen et al., 1990).

DISSOLVED ORGANIC MATTER AND WATER SOLUBLE ORGANIC MATTER CHARACTERISTICS

Dissolved Organic Matter Concentrations in Soil Solution

Several processes in the vadose zone mediate DOM concentration in soil solution, including microbial degradation, leaching, and sorption (Fig. 5–2). Most observations of soil solution DOM concentrations have been made in forest soils, especially Spodosols. Dissolved organic matter concentrations in these soils typically decrease from 20 mg L^{-1} in the O and A horizons to about 2 mg L^{-1} or less in the C horizon. Average DOC concentrations as a function of soil order and horizon are reported in Table 5–2. Temporal variations in DOM concentrations occur on two scales: seasonal and by storm event. Large fluxes of DOM have been observed after snowmelt, in the spring after plant growth is initiated, and in the fall after leaf fall (Herbauts, 1980; Antweiler & Drever, 1983; Meyer & Tate, 1983). In general, DOM in soil solution is higher in summer than in winter (Cronan & Aiken, 1985). Dissolved organic matter concentrations in soils also vary during and after storm events, depending on the water flux through the vadose zone (McDowell & Wood, 1984). There is much less variation in B and C horizon DOM concentrations than in the A horizon. This is probably a function of the strong DOM sorption to the solid phase in the lower horizons (Meyer & Tate, 1983; McDowell & Wood, 1984; Cronan & Aiken, 1985).

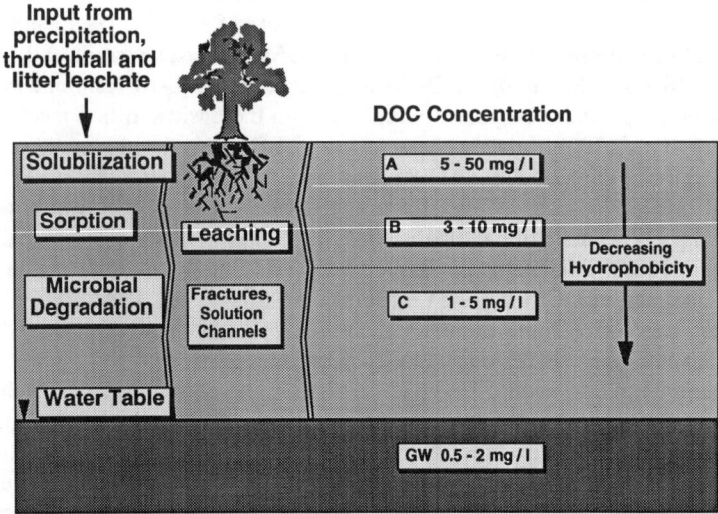

Fig. 5–2. The biogeochemistry of DOM in soil. Average DOC concentrations in soil horizons and groundwater are given, along with the relative change in DOM hydrophobicity as DOM leaches through the soil profile.

Dissolved Organic Matter and Water Soluble Organic Matter Chemical Characteristics

Soil solution DOM is composed of about 50% humic substances, most of which are fluvic acids, 30% macromolecular hydrophilic acids, and about 20% identifiable organic compounds such as carbohydrates, carboxylic acids, amino acids, and hydrocarbons (Thurman, 1985a; McDowell & Likens, 1988; Qualls & Haines, 1991). Average MW of DOM is approximately 1000 Da, though this is dependent on sample type, methodology, and treatment (Buffle et al., 1978; Dawson et al., 1978, 1981; Candler & van Cleve, 1982; Homann & Grigal, 1992). Titratable acidity ($\leq$ pH 7) measured for a number of soil DOM samples collected with lysimeters varied between 6.5 and 11.5 eq kg^{-1} C (Cronan & Aiken, 1985; Litaor & Thurman, 1988). The acidity of WSOM was 14.9 eq kg^{-1} C, as measured by direct titrations to pH 10 (Herbert et al., 1993) (Table 5–3). The acidity of WSOM size fractions retained on ultrafiltration filters with NMWCO of 100000, 30000, 10000, and 1000, varied between 4.5 and 7.95 eq kg^{-1} C. The higher acidity of the unfractionated WSOM reflected the presence of simple organic acids in this sample. These organic acids were not retained on the filters with the fractionated organic material.

Dissolved organic matter also has been characterized using nuclear magnetic resonance (NMR) spectroscopy. Novak and Bertsch (1991) used ^{13}C CPMAS (Cross polarization, magic angle spinning) NMR to compare the characteristics of WSOM, fulvic acid, and humic acid extracted from several soils collected along a hillslope (Table 5–4). Their work indicated that WSOM had a higher portion of O-alkyl-C, reflecting the presence of polysaccharides and aliphatic acids

Table 5–2. Average observed DOC concentrations in soil solution as a function of soil type, horizon and collection method. Dissolved organic matter concentrations were analyzed using automated persulfate oxidation.

Soil	Horizon	DOC	Method†	Source
		(mg/L)		
Spodosol	A, B	10–53	t. lysimeter	Cronan & Aiken, 1985
	B, C	2–12		
Spodosol	<50 cm	15–69	t. lysimeter	Wallis et al., 1981
	GW	2–16	piezometer	
	stream	2.2		
Spodosol	A	28.1	t. lysimeter	McDowell & Wood, 1984
	Bs	5.91		
	B	2.96		
	stream	2.16		
Spodosol	O	14.0	0 t. lysimeter	Cronan et al., 1990
	B	7.4		
	BC	2.8		
	stream	3.8		
Ultisol	O	36	t. lysimeter	Dawson et al., 1981
	BA	22		
	B	10		
	GW	7		
Ultisol	B	2–13	t. lysimeter	Meyer & Tate, 1983
	GW	0.2–0.7		
Ultisol	O	13.7	0 t. lysimeter	Cronan et al., 1990
	Bt	2.1	t. lysimeter	
	C	0.78		
	stream	1.38		
Andisol	O	22	t. lysimeter	Dawson et al., 1981
	A	23		
	B	11		
	stream	7		
Inceptisol	O1	32.5	0 t. lysimeter	Qualls et al., 1991
	Oa	32.5		

†Methods include tension lysimeters (t. lysimeter) and zero-tension lysimeters (0 t. lysimeter).

and a lower portion of aromatic C than fulvic and humic acid. The differences in O-alkyl-C content between WSOM and the humics were accentuated in the bottomland samples. Due to their saturated condition, microbial degradation in bottomland soils is reduced which leads to the accumulation of polysaccharides and other less recalcitrant organic compounds. These results are consistent with the characterization of DOM from an Inceptisol formed under mixed-forest vegetation in Germany (Candler et al., 1988) and in DOM collected in pore waters of marine and fresh water sediments (Orem & Hatcher, 1987).

Several studies have measured the concentrations of specific identifiable organic compounds in DOM. Polyphenolic compounds composed between 10 to

Table 5–3. Acidities and hydrophobic content of WSOM, fulvic acid, and humic acid extracted from a Cumulic Humaquept, Aiken County, South Carolina. The WSOM was fractionated using ultrafiltration†.

Sample	Fraction	NMWCO‡	Acidity	Hydrophobic fraction
			eq kg^{-1} OC	%
WSOM	Unfractionated	NA§	14.9 (0.31)	8.6 (1.0)
	R1	100 000	ND¶	23.8 (3.4)
	R2	30 000	4.47 (0.04)	7.7 (0.8)
	R3	10 000	7.95 (1.00)	18.4 (1.4)
	R4	1 000	7.93 (0.64)	22.2 (0.4)
Fulvic acid		NA	NA	19.5 (0.34)	34.4 (0.5)
Humic acid		NA	NA	12.1 (0.79)	75.3 (6.5)

†Data taken from Herbert et al. (1993). Standard errors are in parentheses.
‡Nominal molecular weight cut off.
§Not applicable.
¶Not determined.

23% of total WSOM extracted from an O horizon (Pohlman & McColl, 1988). Five aliphatic and six aromatic acids also were detected, including oxalic, malic, and gallic acids. Phenolic compounds and carbohydrates comprised about 5% of total DOM collected with tension lysimeters from the E and B horizons of a forest soil (McDowell & Likens, 1988). Lipids, including alkanes, fatty acids, and alcohols, were quantified in WSOM isolated from the A horizons of soils along a hillslope (Herbert & Mills, 1993). The dominant lipid compounds in the water extracts were the C15 to C18 fatty acids which were present at concentrations of 50 to 200 µg kg^{-1} extracted dry soil. Characterization of DOM through the quantification of specific organic classes is limited because only a small percentage of total DOM is actually studied (Qualls & Haines, 1991). On the other hand, characterizing DOM through quantification of identifiable organic compounds may be best suited when the organic compounds are directly related to particular biogeochemical processes of interest. These compounds can act as molecular markers which aid in identifying specific soil microbial and other biochemical processes.

Most studies of soil DOM properties focused on relative amounts and characteristics of hydrophobic and hydrophilic fractions separated by sorption chromatography as described above. These techniques fractionate DOM based on properties that also regulate its interaction with soil surfaces (Qualls & Haines, 1991). The hydrophobic/hydrophilic ratios of several DOM samples collected from terrestrial and aquatic sources are given in Table 5–5. Dissolved organic matter is typically dominated by the hydrophobic and hydrophilic acid fractions. Specific organic compounds found in these fractions include humic materials, fatty acids, and polyphenols (Table 5–1). Dissolved organic matter generally contains between 25 to 60% hydrophobic material, primarily hydrophobic acids (Antweiller & Drever, 1983; Cronan & Aiken, 1985; Qualls & Haines, 1991). The relative proportion of hydrophobic acid decreases with depth in the soil profile (Cronan & Aiken, 1985). This indicates that the hydrophilic fractions, primarily the acid fraction, can be transported preferentially through the soil. Similar

Table 5–4. Average distribution of C as determined by ^{13}C CPMAS NMR in WSOC, fulvic acids, and humic acids extracted from an upland-bottomland sequence.†

Sample	Landscape position	Aliphatic-C 0–50 ppm‡	O-Akyl-C 50–110 ppm‡	C Species Aromatic C 110–160 ppm‡	COOH-C 160–190 ppm‡	Carbonyl C 190–240 ppm‡
			%			
WSOC	Upland	26	53	6	11	6
	Bottomland	15	65	6	12	4
Fulvic Acid	Upland	22	38	15	20	6
	Bottomland	20	48	9	19	5
Humic Acid	Upland	27	26	31	13	4
	Bottomland	23	23	34	15	5

† Data taken from Novak & Bertsch (1991).
‡ Chemical shift region used for integration.

Table 5–5. Percentage of hydrophobic/hydrophilic fractions in DOC from froest soils and streams.†

Sample	Hydrophobic acids	Hydrophilic acids	Bases and neutrals
		%	
Panther outlet	34	39	27
Woods outlet	25	36	39
Sagamore outlet	40	45	15
Hardwood O/A	47	37	16
Hardwood B	25	56	19
Conifer O/A	43	46	11
Conifer B	30	50	20

†DOC fractionated by Leenheer (1981) method (data taken from Cronan & Aiken, 1985).

hydrophobic contents, as measured by reverse-phase chromatography, was observed for size-fractionated WSOM (Table 5–3). The amount of material retained on the cartridges varied between 7.7 to 23.8% for the WSOM fractions, 34.4% for fulvic acid, and 75.3% for humic acid (Herbert et al., 1993) showing that the hydrophobic fraction content of WSOM is much less than the base extracted humic materials.

While a large fraction of soil solution DOM is composed of humic materials, especially fulvic acid, soil solution DOM is chemically distinct from base extracted humic and fulvic acid. Observed increases in DOM in the spring and fall indicate that nonhumic and partially humified organic compounds also are an important fraction of soil solution DOM, but their existence is more transitory. The use of sorption chromatography to characterize DOM has been very useful. The relative amounts of hydrophobic and hydrophilic material in A horizon soil solution DOM are approximately equal. The hydrophobic fraction decreases with depth in the soil profile suggesting preferential sorption to soil surfaces. Future work on the temporal variation of the hydrophobic fraction of DOM, along with the spatial and temporal characterization of other DOM properties is needed.

BIOGEOCHEMISTRY OF DISSOLVED ORGANIC MATTER

Transport of Dissolved Organic Matter in the Vadose Zone

Both field and laboratory observations have shown that DOM is mobile in soil solutions. Nutrient export to streams and groundwaters from soils, transport of metals and sesquioxides to lower horizons, and the facilitated transport of organic contaminants can result from soil DOM mobilization and transport. Column transport studies of humic acid (Enfield et al., 1989; Abdul et al., 1990), and DOM (Dunnivant et al., 1992), along with studies of DOM transport through forest pedons (Jardine et al., 1989b, 1990) have focused on the coupled physical and chemical processes which control DOM mobility. Generally, retardation of DOM is observed in column studies with the preferential sorption of the hydrophobic fractions of DOM to the solid matrix (Abdul et al., 1990; Dunnivant et al., 1992). Size exclusion of DOM, resulting in early breakthrough compared to the transport of tritium, has been reported. These observations may have been the result of the high MW and large molecular size of the humic substances used (Enfield et al., 1989). While most DOM leached from surface horizons is attenuated in the soil profile, a significant fraction can be transported to groundwaters or surface waters. Using the isotopic composition of groundwater DOM, Murphy et al. (1989b) and Wassenaar et al. (1991) have shown that soil is a source of groundwater DOM. In particular, Murphy et al. (1989b) showed that the large MW organics in groundwater, as isolated by sorption to XAD-8 resin, originated from the soil environment.

Several studies at both the column and field scale have shown that DOM transport can be strongly time dependent. Local equilibrium of DOM with the solid matrix was not achieved at high pore water velocities (Jardine et al., 1992). Subsurface flow through preferential flow paths during storm events can increase the flux of DOM through forest pedons by limiting DOM sorption (Jardine et al., 1989b, 1990). Significant transport of the hydrophobic fraction of DOM also has been observed under these conditions, even though this fraction selectively sorbs to soil surfaces (Jardine et al., 1989a, 1990). The importance of transient flow in increasing DOM mobility during storm events has been shown at the watershed scale (Meyer & Tate, 1983; Cantrell, 1989).

Dissolved organic matter mobility is a major factor affecting cycling of nutrients in the soil system and the export of nutrients from soils to surface waters. This is particularly true for N and P, as the organic forms of these nutrients can be a significant fraction of total dissolved concentrations (Qualls et al., 1991). In a mountain watershed in the southern Appalachians, Qualls et al. (1991) observed fluxes of C, N, and P from the O horizons of an Inceptisol of 40.5, 1.0, and 0.029 g m^{-2} yr^{-1}, respectively. These fluxes represented 12.5, 17.3, and 4.9% net leaching from the forest floor for C, N, and P, respectively. A similar organic C flux (26.3 g m^{-2} yr^{-1}) was observed at the Hubbard Brook forest in New Hampshire (McDowell & Likens, 1988). These fluxes represent important inputs of nutrients to mineral soil horizons and have been used to explain the amount of mineralizable N in soil below 36 cm (Cassman & Munns, 1980).

Hydrology can have a strong influence on nutrient fluxes. In a Western lodgepole forest, most leaching of organic nutrients occurred during snowmelt, with a much higher net N leaching than found at the Coweeta Hydrologic Laboratory, North Carolina (Yavitt & Fahey, 1986; Qualls et al., 1991). Soil-derived dissolved and particulate organic nutrients also can act as important sources of nutrients for aquatic systems. The export of total organic P by an agricultural and forest watershed to a river system was reported to be 15.33 and 3.63 g P ha^{-1} wk^{-1}, respectively (Vaithiyanathan & Corell, 1992).

Sorption of Dissolved Organic Matter in the Vadose Zone

Mobility of DOM through a soil profile is controlled in part by sorption to soil surfaces (Cronan & Aiken, 1985). Dissolved organic matter sorbs to many different surfaces including gibbsite (Davis, 1980; Davis & Gloor, 1981), lab-synthesized iron oxides (Tipping, 1981a,b), phyllosilicates (Davis, 1982; Murphy et al., 1990), and soils (Sibanda & Young, 1986; Jardine et al., 1989b). The degree of DOM sorption to these surfaces is strongly dependent on pH, the nature of the surface, and the average molecular weight of DOM, and is less dependent on ionic strength and the source of DOM (Davis, 1980; Davis & Gloor, 1981; Jardine et al., 1989a).

Jardine et al. (1989b) studied the influence of several factors on the sorption of DOM to minerals and soils. Dissolved organic matter sorption was strongly pH dependent, with higher sorption observed at lower pH. Greater DOM sorption to B horizon material, along with the preferential removal of hydrophobic fractions of DOM, also was observed. Removal of the initial soil organic matter from soil surfaces resulted in increased DOM sorption, while removal of iron oxides dramatically decreased sorption. The preferential sorption of the hydrophobic fraction is consistent with greater sorption of higher MW DOM fraction observed by Davis and Gloor (1981). This can explain the lower DOM concentrations in B and C horizons and the lower concentrations of hydrophobic material in these samples (Cronan & Aiken, 1985; Qualls & Haines, 1991).

Dissolved organic matter sorbs to natural surfaces through several different mechanisms due its complex nature (Greenland, 1971). Competitive sorption between DOM and PO_4^{3-} and increases in pH during DOM sorption is evidence for ligand exchange reactions between carboxylate groups and mineral surface hydroxyls (Parfitt et al., 1977; Tipping, 1981a,b; Murphy et al., 1990). Anion-exchange reactions also are important in DOM sorption as indicated by the decrease in DOM sorption in the presence of SO_4^{2-}, which binds to mineral surfaces primarily through electrostatic interactions (Inoue & Wada, 1971a,b; Jardine et al., 1989a). Finally, the importance of physical adsorption is shown by preferential sorption of the higher molecular weight fraction of DOM, increased sorption at pH's less than 5.5, and the weak temperature dependence of sorption (Jardine et al., 1989a). Recent studies of polymer adsorption mechanisms show that system conditions can subtly control which adsorption mechanism dominates (van de Steeg et al., 1992). This may explain the contrasting conclusions drawn by recent studies concerning the dominate DOM adsorption mechanism operating in soils (Murphy et al., 1990; Jardine et al., 1989a).

Microbial Degradation of Dissolved Organic Matter in the Vadose Zone

Microbial mineralization and immobilization of soil organic matter are an integral part of nutrient cycling in soils and soil organic matter dynamics (Mazzarino et al., 1993). Biodegradation can mediate DOM transport in soils, exportation to aquatic systems, and influence nutrient cycling. Microbes can degrade DOM in soil solution (Zsolnay & Steindl, 1991; Qualls & Haines, 1992), DOM leached from litterfall (Lock & Hynes, 1976), and DOM in streams and rivers (Meyer et al., 1987). Comparisons have been made of the biodegradability of DOM from several sources, including throughfall, forest floor, soil solution, and stream sources (Qualls & Haines, 1992; Meyer et al., 1987). The average loss of DOC due to biodegradation over time scales of 3 to 143 d was 14 to 33%, except for DOC from throughfall and leaf leachate, which varied between 35 and 70%. The biodegradability of DOM also is a function of its depth in soil (Qualls & Haines, 1992). Water soluble organic matter extracted from the A horizon was the least degradable compared to material from the O and B horizons.

Several studies have attempted to correlate chemical composition of DOM to its biodegradability. One conceptual model considers DOM to be composed of two fractions, a labile fraction and a recalcitrant fraction (Zsolnay & Steindl, 1991). The recalcitrant fraction roughly corresponds to the humic fraction in DOM. This simple model can explain the relative biodegradability of leaf litter leachate vs. soil solution DOM, however, it is unable to explain the more complex findings reported by Meyer et al. (1987) and Qualls and Haines (1992). Meyer et al. (1987) fractionated blackwater stream DOM with ultrafiltration filters of <1000, 1000 to 10000, and >10000 NMWCO. The midsize fraction had the lowest biodegradability. The greater biodegradabilities of the smallest and largest DOM fractions were attributed to the lability of low MW compounds and the presence of these compounds complexed with the large MW DOM. Qualls and Haines (1992) fractionated soil solution DOM by sorption chromatography and observed that DOM degradation and the percentage of humic materials, as defined by the hydrophobic acid fraction, was uncorrelated. This indicates that humic material may not be totally recalcitrant. Dissolved organic matter degradation was correlated with the percentage of hydrophilic neutrals, which contain significant amounts of carbohydrates. This points out the limitation in using gross fractionation techniques to explore processes such as microbial degradation which are intimately tied to concentrations of specific organic compounds.

Evidence from several studies conducted at the Coweeta Hydrologic Laboratory indicates that microbial decomposition has a relatively minor role in removing DOM from soil solution (Qualls & Haines, 1991, 1992). The dominant mechanism in the B and C horizons appears to be sorption to mineral surfaces, resulting in a relatively constant DOM concentration in the soil solution in these horizons. This hypothesis also can explain the increase in DOM biodegradability in soil solution collected below the A horizon. In the B and C horizons, most of the recalcitrant DOM is sorbed by the mineral surfaces, leaving the labile organic compounds, such as carbohydrates, in solution and available for degradation.

SUMMARY

Soil solution DOM is intimately tied to many biogeochemical processes in soils such as mineral weathering, translocation of oxides, humus, and silicate clays, and nutrient cycling. In addition, many processes controlling the fate and transport of organic contaminants in soils, such as sorption, chemical degradation, bioavailability, and transport, are mediated by the presence of DOM in soil solution. Because of the difficulty in collecting soil solution DOM, WSOM has been used as a surrogate for soil solution DOM. While there are probably some differences in their respective properties, WSOM is a better surrogate for soil solution DOM than base-extracted fulvic or humic acids.

Soil solution DOM is typically collected by lysimeters, though reverse osmosis may be a good technique in saturated systems. Water soluble organic matter is extracted with dilute salt solutions adjusted to the pH, ionic strength, and Na/Ca ratio of the soil. A number of techniques can be used to fractionate DOM including ultrafiltration and sorption chromatography. The use of macroreticular and ion-exchange resins to fractionate DOM into hydrophobic and hydrophilic fractions is increasing. Based on these techniques, DOM has been shown to be about 50% hydrophobic acids and neutrals. Concentrations of these fractions decrease with depth in the soil. Humic material, particularly fulvic acid, makes up a large percentage of DOM. Transient amounts of identifiable organics, such as lipids, carbohydrates, and organic acids, also make up a significant fraction of total DOM in soil solution. Characterization of WSOM by NMR spectroscopy has shown that this material is significantly different from base-extracted fulvic acid. Water soluble organic matter, and presumably soil solution DOM, have higher proportions of O-alkyl-C species typical of polysaccharides compared to base-extracted humics.

The ability of DOM to facilitate the transport of organic contaminants through soils is dependent on the hydrophobicity of organic contaminants, the specific characteristics of DOM, and DOM mobility. One of the primary processes attenuating DOM transport is sorption to soil surfaces, which selectively sorb the large MW and hydrophobic fractions. However, DOM transport also is influenced by preferential flow and transient seasonal and hydrologic processes as evidenced by the transport of large MW and hydrophobic organic matter through forest pedons to groundwater. The rapid transport of these fractions of DOM through soils, coupled with their strong sorption of nonionic organic contaminants, may represent an important facilitated transport mechanism of organic contaminants transport in soils.

ACKNOWLEDGMENTS

Special acknowledgment goes to Dr. Gary Mills and Dr. Carl Strogan of the Savannah River Ecology Laboratory, University of Georgia, and Shannon Garcia of the Department of Geology, Texas A&M University for their contributions and reviews. The authors would also like to thank two anonymous reviewers for their contributions. This research was partially funded by contract DE-AC09-76SR00819 between the University of Georgia and the U.S. Department of Energy.

REFERENCES

Abbt-Braun, G., F.H. Frimmel, and H. Schulten. 1989. Structural investigations of aquatic humic substances by pyrolysis-field ionization mass spectrometry and pyrolysis-gas chromatography/mass spectrometry. Water Resour. 23:1579–1591.

Abdul, A.S., T.L. Gilson, and D.N. Rai. 1990. Use of humic acid solution to remove organic contaminants from hydrogeologic system. Environ. Sci. Technol. 24:328–333.

Aiken, G.R. 1984. Evaluation of ultrafiltration for determining molecular weight of fulvic acid. Environ. Sci. Technol. 18:978–981.

Aiken, G.R. 1985. Isolation and concentration techniques for aquatic humic substances. p. 363–385. *In* G.R. Aiken et al. (ed.) Humic substances in soil, sediment, and water. John Wiley, New York.

Aiken, G.R., and R.L. Malcolm. 1979. Comparison of XAD macroporous resins for the concentration of fulvic acid from aqueous solution. Anal. Chem. 51:1799–1803.

Aiken, G.R., D.M. McKnight, K.A. Thorn, and E.M. Thurman. 1992. Isolation of hydrophilic organic acids from water using nonionic macroporous resins. Org. Geochem. 18:567–573.

Amador, J.A., P.J. Milne, C.A. Moore, and R.G. Zika. 1990. Extraction of chromophoric humic substances from seawater. Mar. Chem. 29:1–17.

Antweiler, R.C., and J.I. Drever. 1983. The weathering of a late Tertiary volcanic ash: Importance of organic solutes. Geochim. Cosmochim. Acta 47:623–629.

Barbee, G.C., and K.W. Brown. 1986. Comparison between suction and free-draining soil solution samplers. Soil Sci. 141:149–154.

Batley, G.E. 1989. Physiochemical separation methods for trace element speciation in aquatic samples. p. 43–76. *In* G. E. Batley (ed.) Trace element speciation analytical methods and problems. CRC Crit. Rev. Environ. Control, Boca Raton, FL.

Bell, L.C., and C.A. Black. 1970. Comparison of methods for identifying crystalline phosphates produced by interactions of phosphate fertilizers with soils. Soil Sci. Soc. Am. Proc. 3:579–582.

Boehm, P.D., and J.G. Quinn. 1973. Solubilization of hydrocarbons by the dissolved organic matter in sea water. Geochim. Cosmochim. Acta 37:2459–2477.

Bremner, J.M., and H. Lees. 1949. Studies on soil organic matter. Part II. The extraction of organic matter from soil by neutral reagents. J. Agric. Sci. 39:274–279.

Buffle, J., P. Deladoey, and W. Haerdi. 1978. The use of ultrafiltration for separation and fractionation of organic ligands in fresh waters. Anal. Chim. Acta 101:339–357.

Buffle, J., P. Deladoey, J. Zumstein, and W. Haerdi. 1982. Analysis and characterization of natural organic matters in freshwaters. I. Study of analytical techniques. Schweiz. Z. Hydrol. 44:327–362.

Callaway, J.Y., K.V. Gabbita, and V.L. Vilker. 1984. Reduction of low molecular weight halocarbons in the vapor phase above concentrated humic acid solutions. Environ. Sci. Technol. 18:890–893.

Candler, R., and K. van Cleve. 1982. A comparison of aqueous extracts from the B horizon of a birch and aspen forest in interior Alaska. Soil Sci. 134:176–180.

Candler, R., W. Zech, and H.G. Alt. 1988. Characterization of water-soluble organic substances from a Typic Dystrochrept under spruce using GPC, IR, ^{1}H NMR and ^{13}C NMR spectroscopy. Soil Sci. 146:445–452.

Cantrell, K.J. 1989. Role of soil organic and carbonic acids in the acidification of forest streams and soils. Ph.D. diss. Georgia Inst. Technol., Atlanta (DAI-B51/01).

Carter, C.W., and I.H. Suffet. 1982. Binding of DDT to dissolved humic materials. Environ. Sci. Technol. 16:735–740.

Cassman, K.G., and D.N. Munns. 1980. Nitrogen mineralization as affected by soil moisture, temperature, and depth. Soil Sci. Soc. Am. J. 44:1233–1237.

Chang, F.H., and M. Alexander. 1984. Effects of simulated acid precipitation on decomposition and leaching of organic carbon in forest soils. Soil Sci. 138:226–234.

Chiou, C.T., R.L. Malcolm, T.I. Brinton, and D.E. Kile. 1986. Water solubility enhancement of some organic pollutants and pesticides by dissolved humic and fulvic acids. Environ. Sci. Technol. 20:502–508.

Cronan, C.S. 1985. Comparative effects of precipitation acidity on three forest soils: Carbon cycling responses. Plant Soil 88:101–112.

Cronan, C.S., and G.R. Aiken. 1985. Chemistry and transport of soluble humic substances in forested watersheds of the Adirondack Park, New York. Geochim. Cosmochim. Acta 49:1697–1705.

Cronan, C.S., C.T. Driscoll, R.M. Newton, J.M. Kelly, C.L. Schofield, R.J. Bartlett, and R. April. 1990. A comparative analysis of aluminum biogeochemistry in a northern and a southeastern forested watershed. Water Resour. Res. 26:1413–1430.

Cronan, C.S., W.A. Reiners, R.C. Reynolds, Jr., and G.E. Lang. 1978. Forest floor leaching: Contributions from mineral, organic, and carbonic acids in New Hampshire subalpine forests. Science (Wasington, DC) 200:309–311.

Dahlgren, R.A., D.C. McAvoy, and C.T. Driscoll. 1990. Acidification and recovery of a Spodosol Bs horizon from acid deposition. Environ. Sci. Technol. 24:531–537.

David, M.B., and G.F. Vance. 1991. Chemical character and origin of organic acids in streams and seepage lakes of central Maine. Biogeochemistry 12:17–41.

David, M.B., G.F. Vance, J.M. Rissing, and F.J. Stevenson. 1989. Organic carbon fractions in extracts of O and B horizons from a New England Spodosol: Effects of acid treatment. J. Environ. Qual. 18:212–217.

Davis, J.A. 1980. Adsorption of natural organic matter from freshwater environments by aluminum oxide. p. 179–305. *In* R.A. Baker (ed.) Contaminants and Sediments. Vol. 2. Ann Arbor Sci., Ann Arbor, MI.

Davis, J.A. 1982. Adsorption of natural dissolved organic matter at the oxide/water interface. Geochim. Cosmochim. Acta 46:2381–2393.

Davis, J.A., and R. Gloor. 1981. Adsorption of dissolved organics in lake water by aluminum oxide. Effect of molecular weight. Environ. Sci. Technol. 15:1223–1229.

Dawson, H.J., B.F. Hrutfiord, R.J. Zasoski, and F. C. Ugolini. 1981. The molecular weight and origin of yellow organic acids. Soil Sci. 132:191–199.

Dawson, H.J., F.C. Ugolini, B.F. Hrutfiord, and J. Zachara. 1978. Role of soluble organics in the soil processes of a Podzol, central Cascades, Washington. Soil Sci. 126:290–296.

Dunnivant, F.M., P.M. Jardine, D.L. Taylor, and J.F. McCarthy. 1992. Transport of naturally occurring dissolved organic carbon in laboratory columns containing aquifer material. Soil Sci. Soc. Am. J.56:437–444.

Enfield, C.G., G. Bengtsson, and R. Lindqvist. 1989. Influence of macromolecules on chemical transport. Environ. Sci. Technol. 23:1278–1286.

Evans, Jr., A., L.W. Zelazny, and C.E. Zipper. 1988. Solution parameters influencing dissolved organic carbon levels in three forst soils. Soil Sci. Soc. Am. J. 52:1789–1792.

Forsyth, W.G.C. 1947. Studies on the more soluble complexes of soil organic matter. Biochem. J. 41:176–181.

Fu, T., and R. Pocklington. 1983. Quantitative adsorption of organic matter from seawater on solid matrices. Mar. Chem. 13:255–264.

Gauthier, T.D., W.R. Seitz, and C.L. Grant. 1987. Effects of structural and compositional variations of dissolved humic materials on pyrene Koc values. Environ. Sci. Technol. 21:243–248.

Ghosh, K., and M. Schnitzer. 1980. Macromolecular structures of humic substances. Soil Sci. 129:266–276.

Giordano, P.M., E.C. Sample, and J.J. Mortvedt. 1971. Effect of ammonia ortho- and pyrophosphate on Zn and P in soil solution. Soil Sci. 111:101–106.

Gjessing, E.T. 1970. Ultrafiltration of aquatic humus. Environ. Sci. Technol. 4:437–438.

Greenland, D.J. 1971. Interactions between humic and fulvic acids and clays. Soil Sci. 111:34–39.

Haines, B.L., J.B. Waide, and R.L. Todd. 1982. Soil solution nutrient concentrations sampled with tension and zero-tension lysimeters: report of discrepancies. Soil Sci. Soc. Am. J. 46:658–661.

Hassett, J.P., and E. Milicic. 1985. Determination of equilibrium and rate constants for binding of a polychlorinated biphenyl congener by dissolved humic substances. Environ. Sci. Technol. 19:638–643.

Haumaier, L.,W. Zech, and G. Franke. 1990. Gel permeation chromatography of water-soluble organic matter with deionized water as eluent-I. Examination of the method. Org. Geochem. 15:413–417.

Hay, G.W., J.H. James, and G.W. van Loon. 1985. Solubilization effects of simulated acid rain on the organic matter of forest soil; preliminary results. Soil Sci. 139:422–430.

Hayes, M.H.B. 1985. Extraction of humic substances from soil. p. 329–362. *In* G.R. Aiken et al. (ed.) Humic substances in soil, sediment, and water. John Wiley, New York.

Hayes, M.H.B., R.S. Swift, R.E. Wardle, and J.K. Brown. 1975. Humic materials from an organic soil: A comparison of extractants and of properties of extracts. Geoderma 13:231–245.

Herbauts, J. 1980. Direct evidence of water soluble organic matter leaching in brown earths and slightly podzolized soils. Plant Soil 54:317–321.

Herbert, B.E., P.M. Bertsch, and J.M. Novak. 1993. Pyrene sorption to water-soluble organic carbon. Environ. Sci. Technol. 27:398–403.

Herbert, B.E., and G.L. Mills. 1993. Water extractable lipids in soils. p. 251. *In* Agronomy abstracts. ASA, Madison, WI.

Hine, P.T., and D.B. Bursill. 1984. Gel permeation chromatography of humic acid. Problems associated with Sephadex gel. Water Res. 18:1461–1465.

Homann, P.S., M.J. Mitchell, H.V. Miegroet, and D.W. Cole. 1990. Organic sulfur in throughfall, stem flow, and soil solutions from temperate forests. Can. J. For. Res. 20:1535–1539.

Homann, P.S., and D.F. Grigal. 1992. Molecular weight distribution of soluble organics from laboratory-manipulated surface soils. Soil Sci. Soc. Am. J. 56:1305–1310.

Inoue, T., and K. Wada. 1971a. Reactions between humified clover extracts and imogolite as a model of humus clay interaction. Part 1. Clay Sci. 4:61–70.

Inoue, T., and K. Wada. 1971b. Reactions between humified clover extracts and imogolite as a model of humus clay interaction. Part 2. Clay Sci. 4:71–80.

Inskeep, W.P., and J. Baham. 1983. Competitive complexation of Cd(II) and Cu(II) by water soluble organic ligands and Na-Montmorillonite. Soil Sci. Soc. Am. J. 47:1109–1115.

Jackson, D.R., F.S. Brinkley, and E.A. Bondietti. 1976. Extraction of soil water using cellulose-acetate hollow fibers. Soil Sci. Soc. Am. Proc. 4:327–329.

Jardine, P.M., F.M. Dunnivant, H.M. Selim, and J.F. McCarthy. 1992. Comparisons of models for describing the transport of dissolved organic carbon in aquifer columns. Soil Sci. Soc. Am. J. 56:437–444.

Jardine, P.M., G.V. Wilson, R.J. Luxmore, and J.F. McCarthy. 1989a. Transport of inorganic and natural organic tracers through an isolated pedon in a forest watershed. Soil Sci. Soc. Am. J. 53:317–323.

Jardine, P.M., N.L. Weber, and J.F. McCarthy. 1989b. Mechanisms of dissolved organic carbon adsorption on soil. Soil Sci. Soc. Am. J. 53:1378–1385.

Jardine, P.M., G.V. Wilson, J.F. McCarthy, R.J. Luxmore, D.L. Taylor, and L.W. Zelazny. 1990. Hydrogeochemical processes controlling the transport of dissolved organic carbon through a forested hillslope. J. Contam. Hydrol. 6:3–19.

Jenny, H. 1980. The soil resource. Origin and behavior. Springer-Verlag, New York.

Khan, S.U. 1970. Organic matter association with soluble salts in the water extract of a black Solonetz soil. Soil Sci. 109:227–228.

Krug, E.C., and P.J. Isaacson. 1984. Comparison of water and dilute acid treatment on organic and inorganic chemistry of leachate from organic-rich horizons of an acid forest soil. Soil Sci. 137:370–378.

Kukkonen, J., J.F. McCarthy, and A. Oikari. 1990. Effects of XAD-8 fractions of dissolved organic carbon on the sorption and bioavailability of organic micropollutants. Arch. Environ. Contam. Toxicol. 19:551–557.

Kwak, J.C.T., R.W.P. Nelson, and D.S. Gamble. 1977. Ultrafiltration of fulvic and humic acids, a comparison of stirred cell and hollow fiber techniques. Geochim. Cosmochim. Acta 41:993–996.

Lammers, W.T. 1967. Separation of suspended and colloidal particles from natural water. Environ. Sci. Technol. 1:52–57.

Landrum, P.F., S.R. Nihart, B.J. Eadie, and L.R. Herche. 1987. Reduction in bioavailability of organic contaminants to "Pantoporelia hoyi" by dissolved organic matter of sediment interstitial waters. Environ. Toxicol. Chem. 6:11–20.

Lee, D.Y., and W.J. Farmer. 1989. Dissolved organic matter interaction with naproamide and four other nonionic pesticides. J. Environ. Qual. 18:468–474.

Leenheer, J.A. 1981. Comprehensive approach to preparative isolation and fractionation of dissolved organic carbon from natural waters and wastewaters. Environ. Sci. Technol. 15:578–587.

Leenheer, J.A. 1985. Fractionation techniques for aquatic humic substances, p. 409–429. In G.R. Aiken, D.M. McKnight, R.L. Wershaw, and P. MacCarthy (ed.) Humic substances in soil, sediment, and water. John Wiley, New York.

Levin, M.J., and D.R. Jackson. 1977. A comparison of in situ extractors for sampling soil water. Soil Sci. Soc. Am. J. 41:535–536.

Linehan, D.J. 1977. A comparison of the polycarboxylic acids extracted by water from an agricultural top soil with those extracted by alkali. J. Soil Sci. 28:369–378.

Litaor, M.I. 1988. Review of soil solution samplers. Water Resour. Res. 24:727–733.

Litaor, M.I., and E.M. Thurman. 1988. Acid neutralizing processes in an alpine watershed from range, Colorado, U.S.A.-1: buffering capacity of dissolved organic carbon in soil solutions. Appl. Geochem. 3:645–652.

Lock, M.A., and H.B.N. Hynes. 1976. The fate of "dissolved" organic carbon derived from autumn-shed maple leaves (*Acer saccharum*) in a temperate hard-water stream. Limnol. Oceanogr. 21:436–443.

Macalady, D.L., P.G. Tratnyek, and N.L. Wolfe. 1989. Influences of natural organic matter on the abiotic hydrolysis of organic contaminants in aqueous systems. p. 323–331. In I.H. Suffet and P. MacCarthy (ed.) Aquatic humic substances: Influence on fate and treatment of pollutants. ACS, Washington, DC.

MacCarthy, P., and R.L. Malcolm. 1989. The nature of commercial humic acids. p. 55–63. In I.H. Suffet and P. MacCarthy (ed.) Aquatic humic substances: Influence on fate and treatment of pollutants. ACS, Washington, DC.

MacCarthy, P., and I.H. Suffet. 1989. Introduction: Aquatic humic substances and their influence on the fate and transport of pollutants. p. xvii–xxx. In I.H. Suffet and P. MacCarthy (ed.) Aquatic humic substances: Influence on fate and treatment of pollutants. ACS, Washington, DC.

Madhun, Y.A., J.L. Young, and V.H. Freed. 1986. Binding of herbicides by water-soluble organic materials from soil. J. Environ. Qual. 15:64–68.

Malcolm, R.L. 1985. Geochemistry of stream fulvic and humic substances. p. 181–209. In G.R. Aiken et al. (ed.) Humic substances in soil, sediment, and water. John Wiley, New York.

Malcolm, R.L., and P. MacCarthy. 1986. Limitations in the use of commercial humic acids in water and soil research. Environ. Sci. Technol. 20:904–911.

Mantoura, R.F.C., and J.P. Riley. 1975. The analytical concentration of humic substances from natural waters. Anal. Chim. Acta 76:97–106.

Mazzarino, M.J., L. Szott, and M. Jimenez. 1993. Dynamics of soil total C and N, microbial biomass, and water-soluble C in tropical agroecosystems. Soil Biol. Biochem. 25:205–214.

McCarthy, J.F., and B.D. Jiminez. 1985. Reduction in bioavailability to bluegills of polycyclic aromatic hydrocarbons bound to dissolved humic material. Environ. Toxicol. Chem. 4:511–521.

McCarthy, J.F., B.D. Jimenez, and T. Barbee. 1985. Effect of dissolved humic materials on accumulation of polycyclic aromatic hydrocarbons: Structure-activity relationships. Aquat. Toxicol. 7:15–24.

McCarthy, J.F., and J.M. Zachara. 1989. Subsurface transport of contaminants. Environ. Sci. Technol. 23:496–502.

McDowell, W.H., and G.E. Likens. 1988. Origin, composition, and flux of dissolved organic carbon in the Hubbard Brook Valley. Ecol. Monogr. 58:177–195.

McDowell, W.H., and T. Wood. 1984. Podozolization: Soil processes control dissolved organic carbon concentrations in stream water. Soil Sci. 137:23–32.

McDowell-Boyer, L.M., J.R. Hunt, and N. Sitar. 1986. Particle transport through porous media. Water. Resour. Res. 22:1901–1921.

Meadows, J.W.T., C.F. Smith, D.G. Coles, L. Maynard, and J. Dellis. 1978. Sampling natural waters: Are filtered samples true solutions. p. 253–264. *In* D.C. Adriano and I.L. Brisbin, Jr. (ed.) Environmental chemistry and cycling processes. Proc. Symp. Augusta, Georgia. 28 April May 1976. Technol. Info. Center, U.S. Dep. Energy, Washington, DC.

Means, J.C., and R. Wijayaratne. 1982. Role of natural colloids in the transport of hydrophobic pollutants. Science (Washington, DC) 215:968–970.

Meyer, J.L., R.T. Edwards, and R. Risley. 1987. Bacterial growth on dissolved organic carbon from a blackwater river. Microb. Ecol. 13:13–29.

Meyer, J.L., and C.M. Tate. 1983. The effects of watershed disturbance on dissolved organic carbon dynamics of a stream. Ecology 64:33–44.

Mills, G.L., G.S. Douglas, and J.G. Quinn. 1989. Dissolved organic copper isolated by C18 reverse-phase extraction in an anoxic basin located in the Pettaquamscutt River estuary. Mar. Chem. 26:277–288.

Mills, G.L., A.K. Hanson, Jr., J.G. Quinn, W.R. Lammela, and N.D. Chasteen. 1982. Chemical studies of copper-organic complexes isolated from estuarine waters using C18 reverse-phase liquid chromatography. Mar. Chem. 11:355–377.

Mills, G.L., E. McFadden, and J.G. Quinn. 1987. Chromatographic studies of dissolved organic matter and copper-organic complexes isolated from estuarine waters. Mar. Chem. 20:313–325.

Mills, G.L., and J.G. Quinn. 1981. Isolation of dissolved organic matter and copper-organic complexes from estuarine waters using reverse-phase liquid chromatography. Mar. Chem. 10:93–102.

Mills, G.L., and D. Schwind. 1990. Photochemical degradation rates of tetraphenylborate and diphenylboric acid sensitized by dissolved organic matter in stream water. Environ. Toxicol. Chem. 9:569–574.

Murphy, E.M., J.M. Zachara, and S.C. Smith. 1990. Influence of mineral-bound humic substances on the sorption of hydrophobic organic compounds. Environ. Sci. Technol. 24:1507–1516.

Murphy, E.M., S.N. Davis, A. Long, D. Donahue, and A.J.T. Jull. 1989a. Characterization and isotopic composition of organic and inorganic carbon in the Milk River aquifer. Water Resour. Res. 25:1893–1905.

Murphy, E.M., S.N. Davis, A. Long, D. Donahue, and A.J.T. Jull. 1989b. 14C in fractions of dissolved organic carbon in ground water. Nature (London) 337:153–155.

Myers, R.G., and S.J. Thien. 1988. Organic matter solubility and soil reaction in an amonium and phosphorus application zone. Soil Sci. Soc. Am. J. 52:516–522.

Nemec, A., and L. Vopenka. 1971. The effect of anhydrous ammonia on humus substances and cations in soils. Agrochimica 15:321–327.

Novak, J.M., and P.M. Bertsch. 1991. The influence of topography on the nature of humic substances in soil organic matter at a site in the Atlantic Coastal Plain of South Carolina. Biogeochemistry 15:111–126.

Novotny, M.V. 1989. Recent developments in analytical chromatography. Science (Washington, DC) 246:51–57.

Orem, W.H., and P.G. Hatcher. 1987. Solid-state 13C NMR studies of dissolved organic matter in pore waters from different depositional environments. Org. Geochem. 11:73–82.

Parfitt, R.L., A.R. Fraser, and V.C. Farmer. 1977. Adsorption on hydrous oxides III. Fulvic acid and humic acid on goethite, gibbsite and imogolite. J. Soil Sci. 28:289–296.

Perdue, E.M., and N.L. Wolfe. 1982. Modification of pollutant hydrolysis kinetics in the presence of humic substances. Environ. Sci. Technol. 16:847–852.

Pohlman, A.A., and J.G. McColl. 1988. Soluble organics from forest litter and their role in metal dissolution. Soil Sci. Soc. Am. J. 52:265–271.
Qualls, R.G., and B.L. Haines. 1991. Geochemistry of dissolved organic nutrients in water percolating through a forest ecosystem. Soil Sci. Soc. Am. J. 55:1112–1123.
Qualls, R.G., and B.L. Haines. 1992. Biodegradability of dissolved organic matter in forest throughfall, soil solution, and stream water. Soil Sci. Soc. Am. J.56:578–586.
Qualls, R.G., B.L. Haines, and W.T. Swank. 1991. Fluxes of dissolved organic nutrients and humic substances in a deciduous forest. Ecology 72:254–266.
Radulovich, R., and P. Sollins. 1987. Improved performance of zero-tension lysimeters. Soil Sci. Soc. Am. J. 51:1386–1388.
Reid, P.M., A.E. Wilkinson, E. Tipping, and M.N. Jones. 1990. Determination of molecular weights of humic substances by analytical (UV scanning) ultracentrifugation. Geochim. Cosmochim. Acta 54:131–138.
Riley, G.A. 1973. Particulate and dissolved organic carbon in the oceans. p. 204–220. *In* G.M. Woodwell and E.V. Pecan (ed.) Carbon and the biosphere. NTIS, U.S. Dep. of Commerce. Springfield, VA.
Ross, R.D., and D.G. Crosby. 1985. Photooxidant activity in natural waters. Environ. Toxicol. Chem. 4:773–778.
Saar, R.A., and J.H. Weber. 1982. Fulvic acid: Modifier of metal-ion chemistry. Environ. Sci. Technol. 16:510A–517A.
Saleh, F.Y., and D.Y. Chang. 1987. Retention behavior of Suwannee river fulvic acid components in RP-HPLC. Sci. Total Environ. 62:67–74.
Saleh, F.Y., W.A. Ong, and D.Y. Chang. 1989. Structural features of aquatic fulvic acids. Analytical and preparative reversed-phased high performance liquid chromatography with photodiode array detection. Anal. Chem. 61:2792–2800.
Schnitzer, M. 1978. Humic substances: Chemistry and reactions. p. 1–64. *In* M. Schnitzer and S.U. Khan (ed.) Soil organic matter. Elsevier, New York.
Schnitzer, M., and C.M. Preston. 1986. Analysis of humic acids by solution and solid-state carbon-13 nuclear magnetic resonance. Soil Sci. Soc. Am. J. 50:326–331.
Serkiz, S.M., and E.M. Perdue. 1990. Isolation of dissolved organic matter from the Suwannee River using reverse osmosis. Water Res. 24:911–916.
Sibanda, H.M., and S.D. Young. 1986. Competitive adsorption of humus acids and phosphate on goethite, gibbsite and two tropical soils. J. Soil Sci. 37:197–204.
Simmons, M.S., and R.G. Zepp. 1986. Influence of humic substances on photolysis of nitroaromatic compounds in aqueous systems. Water Res. 20:899–904.
Stevenson, F.J. 1985. Geochemistry of soil humic substances. p. 13–52. *In* G.R. Aiken et al. (ed.) Humic substances in soil, sediment, and water. John Wiley & Sons, New York.
Swift, R.S. 1985. Fractionation of soil humic substances. p. 387–408. *In* G.R. Aiken et al. (ed.) Humic substances in soil, sediment, and water. John Wiley & Sons, New York.
Swift, R.S., and A.M. Posner. 1971. Gel chromatography of humic acid. J. Soil Sci. 22:237–249.
Swift, R.S., and A.M. Posner. 1972. Nitrogen, phosphorus and sulfur contents of humic acids fractionated with respect to molecular weight. J. Soil Sci. 23:50–57.
Thurman, E.M. 1985a. Organic geochemistry of natural waters. Martinus Nijhoff/Dr. W. Junk, Dordrecht, the Netherlands.
Thurman, E.M. 1985b. Humic substances in groundwater. p. 87–103. *In* G.R. Aiken et al. (ed.) Humic substances in soil, sediment, and water. John Wiley, New York.
Thurman, E.M., and R.L. Malcolm. 1981. Preparative isolation of aquatic humic substances. Environ. Sci. Technol. 15:463–466.
Thurman, E.M., R.L. Malcolm, and G.R. Aiken. 1978. Prediction of capacity factors for aqueous solutes on a porous acrylic resin. Anal. Chem. 50:775–779.
Thurman, E.M., R.L. Wershaw, R.L. Malcolm, and D.J. Pinckney. 1982. Molecular size of aquatic humic substances. Org. Geochem. 4:27–37.
Tipping, E. 1981a. Adsorption by goethite (a-FeOOH) of humic substances from three different lakes. Chem. Geol. 33:81–89.
Tipping, E. 1981b. The adsorption of aquatic humic substances by iron oxides. Geochim. Cosmochim. Acta 45:191–199.
Tomasiewicz, D.J., and J.L. Henry. 1985. The effect of an anhydrous ammonia application on the solubility of soil organic carbon. Can. J. Soil Sci. 65:737–747.
Traina, S.J., D.A. Spontak, and T.J. Logan. 1989. Effects of cations on complexation of naphthalene by water-soluble organic carbon. J. Environ. Qual. 18:221–227.

Vaithiyanathan, P., and D.L. Correll. 1992. The Rhode River watershed: Phosphorus distribution and export in forest and agricultural soils. J. Environ. Qual. 21:280–288.

van de Steeg, H.G.M., M.A. Cohen Stuart, A. de Keizer, and B.H. Bijsterbosch. 1992. Polyelectrolyte adsorption: A subtle balance of forces. Langmuir 8:2538–2546.

van Rossum, P., and R.G. Webb. 1978. Isolation of organic water pollutants by XAD resins and carbon. J. Chromatogr. 150:381–392.

Vance, G.F., and M.B. David. 1989. Effect of acid treatment on dissolved organic carbon retention by a spodic horizon. Soil Sci. Soc. Am. J. 53:1242–1247.

Wallis, P.M., H.B.N. Hynes, and S.A. Telang. 1981. The importance of groundwater in the transportation of allochthonous dissolved organic matter to the streams draining a small mountain basin. Hydrobiologia 79:77–90.

Wassenaar, L.I., R. Aravena, P. Fritz, and J.F. Barker. 1991. Controls on the transport and carbon isotopic composition of dissolved organic carbon in a shallow groundwater system, Central Ontario, Canada. Chem. Geol. 87:39–57.

Wershaw, R.L., and G.R. Aiken. 1985. Molecular size and weight measurements of humic substances. p. 477–492. *In* G.R. Aiken et al. (ed.) Humic substances in soil, sediment, and water. John Wiley, New York.

Wershaw R.L., P.J. Burcar, and M.C. Goldberg. 1969. Interaction of pesticides with natural organic material. Environ. Sci. Technol. 3:271–273.

Wheeler, J.R. 1976. Fractionation by molecular weight of organic substances in Georgia coastal waters. Limnol. Oceanogr. 21:846–852.

Wright, J.R., and Schnitzer. 1963. Metallo-organic interactions associated with podzolization. Soil Sci. Soc. Am. Proc. 27:171–176.

Yavitt, J.B., and T.J. Fahey. 1986. Long-term litter decay and leaching from the forest in *Pinus contorta* (lodgepole pine) ecosystems. J. Ecol. 74:525–545.

Zafiriou, O.C., J. Joussot-Dubien, R.G. Zepp, and R.G. Zika. 1984. Photochemistry of natural waters. Environ. Sci. Technol. 18:358A–371A.

Zepp, R.G., G.L. Baugham, and P.F. Schlotzhauer. 1981. Comparison of photochemical behavior of various humic substances in water: I. Sunlight induced reactions by humic substances. Chemosphere 10:109–117.

Zepp, R.G., P.F. Schlotzhauer, and R.M. Sink. 1985. Photosensitized transformations involving electronic energy transfer in natural waters: Role of humic substances. Environ. Sci. Technol. 19:74–81.

Zsolnay, A., and H. Steindl. 1991. Geovariability and biodegradability of the water-extractable organic material in an agricultural soil. Soil Biol. Biochem. 23:1077–1082.

6 Chemistry of Carbon Decomposition Processes in Forests as Revealed by Solid-State Carbon-13 Nuclear Magnetic Resonance

J. A. Baldock

Canadian Forest Service
Natural Resources Canada
Petawawa National Forestry Institute
Chalk River, Ontario, Canada

C. M. Preston

Canadian Forest Service
Natural Resources Canada
Pacific Forestry Centre
Victoria, British Columbia, Canada

The forest floor is a complex mixture of plant, microbial, insect, and animal residues. The nature and magnitude of decomposition processes operating within this layer control rates of nutrient release and carbon dioxide evolution and influence forest productivity and sustainability. Traditional approaches to studying decomposition in forests have used litter bags to quantify decomposition rates. Determination of changes in chemical composition during decomposition has relied on the use of wet chemical techniques capable of extracting specific compounds from the decomposing residues. The introduction of high-resolution solid-state ^{13}C nuclear magnetic resonance (NMR) spectroscopy, however, has made it possible to determine the chemical composition of a diverse range of decomposing forest organic materials including wood (Hedges et al., 1985; Spiker & Hatcher, 1987; Preston et al., 1990; Bates & Hatcher, 1989; Hortling et al., 1992; Bates et al., 1991; Martinez et al., 1991), forest litter and soil organic matter (Zech et al., 1992; Baldock et al., 1992), and peat (Hammond et al., 1985; Preston et al., 1987, 1989). The objectives of this paper are as follows:

Copyright © 1995 Soil Science Society of America, 677 S. Segoe Rd., Madison, WI 53711, USA.
Carbon Forms and Functions in Forest Soils.

1. To describe the types of information obtained using solid-state ^{13}C NMR spectroscopy.
2. To review studies which have utilized this technique to characterize chemical changes associated with the decomposition of forest organic materials including wood, forest litter, and peat.
3. To present a general model which describes the chemical changes associated with decomposition in forests.
4. To determine if the chemical changes induced by decomposition can be used to assess the extent to which forest organic materials have been decomposed.

INTERPRETATION OF SOLID-STATE CARBON-13 NUCLEAR MAGNETIC RESONANCE

In this section we present an overview of the information obtainable from conventional solid-state ^{13}C NMR analyses. A detailed presentation of the theory of solid-state NMR can be obtained from texts by Wilson (1987) and Pfeffer and Gerasimowicz (1989).

An example of a solid-state ^{13}C NMR spectrum acquired for decomposing needles under a red pine (*Pinus resinosa*) plantation is presented in Fig. 6–1. Carbon contained in different chemical structures is differentiated on the basis of chemical shift values (horizontal axis on Fig. 6–1). Each chemically distinct type of C has a characteristic chemical shift value. VanderHart (1976) showed that the chemical shift values of the terminal CH_3 C, the CH_2 C adjacent to terminal CH_3, and the $(CH_2)_n$ C in a polymethylene structure (*n*-eicosane) were 17, 27, and 35 ppm, respectively. Signals having chemical shift values in the vicinity of 30 to 35 ppm are thus ascribed to alkyl C in long chain polymethylene, $(CH_2)_n$, type structures (e.g., fatty acids, waxes, and resins). A decrease in the chemical shift of this signal towards 25 ppm indicates a reduction in the amount of long chain polymethylene material and an increase in short chain material. Signals in the vicinity of 17 ppm arise from methyl C.

Signal intensity associated with the 45 to 95 ppm region has typically been assigned to carbohydrate O-alkyl structures such as cellulose (Fig. 6–2a) or hemicelluloses; however, the oxygenated C_α, C_β, C_γ, and methoxyl C of the phenylpropane lignin structural units (Fig. 6–2b) and signals derived from amine C also appear in this region. Signals derived from cellulose include the C_2, C_3, and C_5 ring C atoms (72–75 ppm), the crystalline (65 ppm) and noncrystalline (62 ppm) components of C_6, the crystalline (89 ppm) and noncrystalline (84 ppm) components of C_4, and the dioxygenated anomeric C_1 (105 ppm) (Earl & VanderHart, 1981; Maciel et al., 1982; VanderHart & Atalla, 1984). Signals derived from hemicelluloses are often contained within the cellulose peaks, but weak signals at 22 ppm and 173 ppm due to the methyl and carboxyl C in the acetate groups of hemicelluloses have been observed in samples of undecomposed wood (Hedges et al., 1985; Bates & Hatcher, 1989; Bates et al., 1991). The oxygenated C_α, C_β, and C_γ C atoms of lignin appear over the range of 60 to 90 ppm (Nimz et al.,

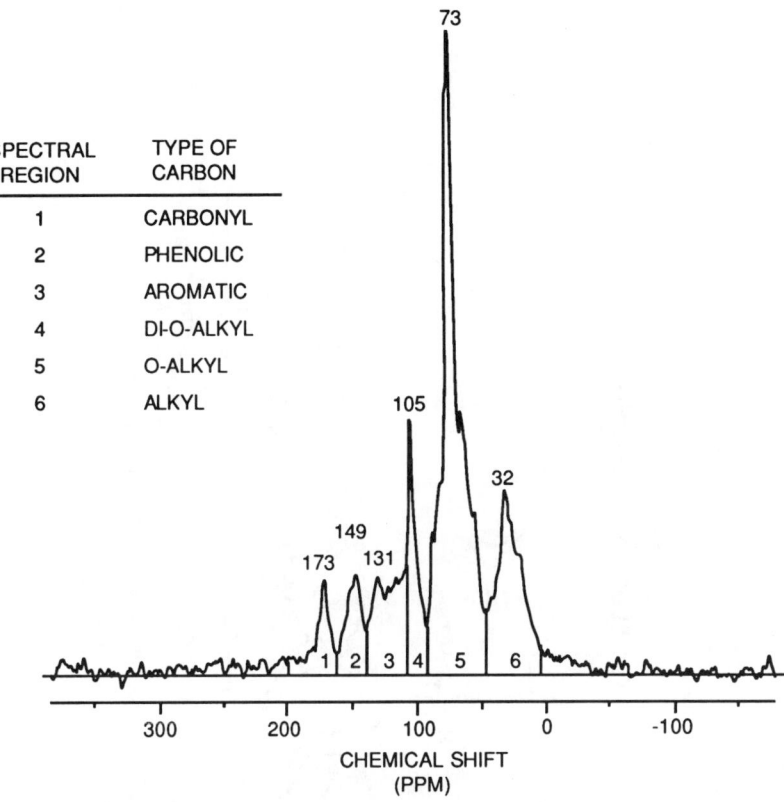

Fig. 6–1. Solid-state ^{13}C NMR spectrum of decomposing red pine needles showing chemical shift values of the major types of C present and the six regions into which the acquired spectral intensity has been divided.

1981; Hatfield et al., 1987; Hatcher, 1987). Methoxyl C signals appear at 56 ppm and can be separated from carbohydrate signals in some samples. Amine C signals appear over the range of 45 to 65 ppm (Duncan, 1987) but may not be resolved completely from the more intense carbohydrate signals.

Signals for C- or H-substituted aromatic C occur over the range of 110 to 145 ppm while those arising from O-substituted aromatic C appear near between 145 and 160 ppm. In fresh undecomposed plant residues these signals are derived dominantly from lignin structures (Fig. 6–2b). Data presented by Nimz et al. (1981) have been used to determine the chemical shift values associated with the various lignin C atoms. Syringyl C_2 and C_6 C atoms are observed between 100 and 108 ppm and thus occur at similar chemical shift values to the anomeric C_1 of carbohydrate structures. Guaiacyl C_2, C_5, and C_6 C atoms are observed between 115 and 120 ppm. Signals in the vicinity of 133 ppm are derived from guaiacyl and syringyl C_1 C atoms and the phenolic syringyl C_4 C. The phenolic guaiacyl C_4 C signal appears near 153 ppm along with the syringyl C_3 and C_5 C atoms. The C_3 of guaiacyl is observed at 148 ppm.

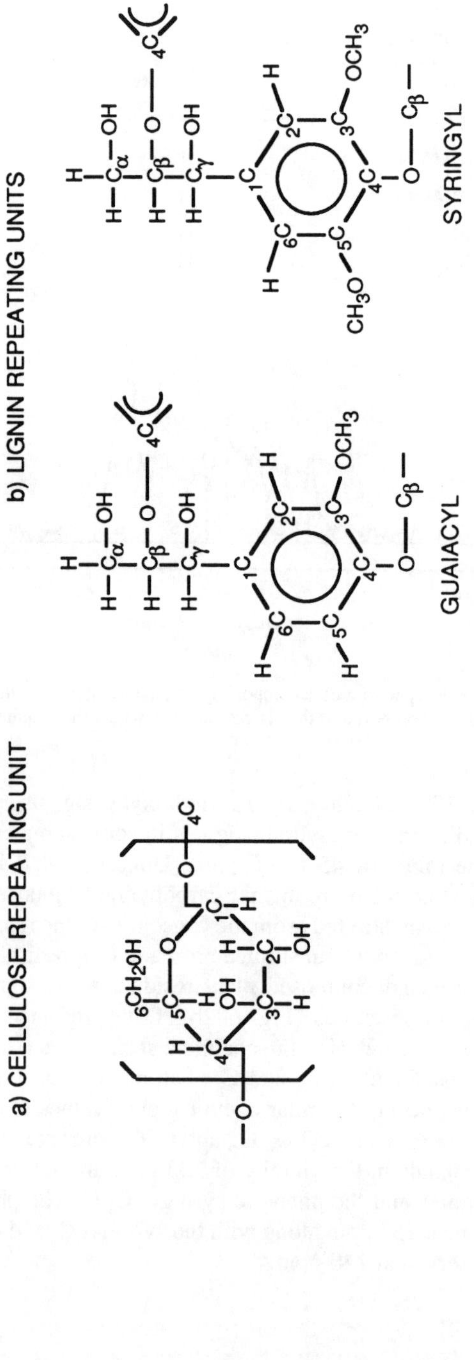

Fig. 6–2. Chemical structure of the repeating monomeric units in (*a*) cellulose and (*b*) guaiacyl and syringyl lignin.

Signals around 175 ppm can originate from carboxylic, amide, or ester C and ketone and aldehyde C signals are observed near 200 ppm. Duncan (1987) has compiled a list of solid-state ^{13}C NMR chemical shift values of C atoms contained in many known chemical structures.

The signal intensity (vertical axis in Fig. 6–1) observed for a given chemical shift value, expressed as a fraction of the total signal intensity acquired, represents the proportion of that type of C present in a sample. Spectra can be broken up into chemical shift regions in which the chemistry of the C atoms within each region is similar. By integrating the signal intensity contained within each chemical shift region, the proportion of a given type of C atom can be calculated. The spectrum presented in Fig. 6–1 has been divided into six spectral regions; however, for spectra with poor signal to noise ratios it may be possible to delineate only four regions corresponding to alkyl (10–45 ppm), O-alkyl (45–110 ppm), aromatic(110–160 ppm), and carbonyl (160–200 ppm) C atoms. The name given to each region is general and thought to be indicative of the dominant form of C present. Based on the previous discussion of chemical shift values, it should be apparent that the proportions of each type of C obtained by this approach are only approximate as there would be undoubtedly some overlap between adjacent regions.

Another important factor which must be considered when interpreting solid-state ^{13}C NMR spectra is whether or not the spectra reflect quantitatively the actual distribution of C types present in the sample. Conventional solid-state ^{13}C NMR analyses, which incorporate cross-polarization and magic-angle spinning, are only quantitative when the following conditions are met: (i) the rate of sample spinning is fast enough that no spinning side bands are produced; (ii) the recycle delay period inserted between repetitive pulses is long enough to allow all protons to relax fully; (iii) the rates of signal development (T_{CH}) and decay ($T_{1p}H$) during the contact time are similar for all types of C in the sample; and, (iv) $T_{CH} \ll T_{1p}H$ (Wilson, 1987; Pfeffer & Gerasimowicz, 1989). The first two conditions can be evaluated by performing some simple preliminary NMR experiments prior to acquiring a spectrum for a sample or series of similar samples. Conditions 3 and 4 can only be evaluated by determining the relationship between signal intensity and the length of the contact time used in the pulse sequence.

WOOD DECOMPOSITION

Preston et al. (1990) utilized solid-state ^{13}C NMR to quantify the chemical changes associated with the decomposition of fallen large diameter (<15 cm) Douglas-fir (*Pseudotsuga menziesii*), western hemlock (*Tsuga heterophylla*), and western red cedar (*Thuja plicata*) logs from old growth forests located on the west slope of the Cascade Range in Oregon and Washington. Logs were classified according to the system presented in Table 6–1 and dated as described by Sollins et al. (1987). Changes observed in the NMR spectra with increasing decay class and residence time for Douglas-fir wood indicated a loss of O-alkyl C and an increase in the content of methoxyl, aromatic, and phenolic C atoms and, to a lesser extent, alkyl and carbonyl C atoms (Fig. 6–3 and 6–4). Such spectral

Table 6–1. Log decay classification system developed by R. Fogel, M. Ogawa, and J. Trappe (personal communication) and used by Preston et al. (1990) to rank fallen decomposing logs.

Class	Description
L	Dried wood which was living when sampled
I	Logs freshly fallen, bark and all wood sound, current-year twigs attached
II	Sapwood decayed but present, bark and heartwood mainly sound, twigs absent
III	Logs still support own weight, sapwood decayed but some still present, bark sloughs, heartwood decayed but structurally sound
IV	Logs do not support own weight, sapwood and bark mainly absent, heartwood not structurally sound, branch stubs can be removed
V	Heartwood mainly fragmented, forming ill-defined, elongated mounds on forest floor sometimes invisible from surface

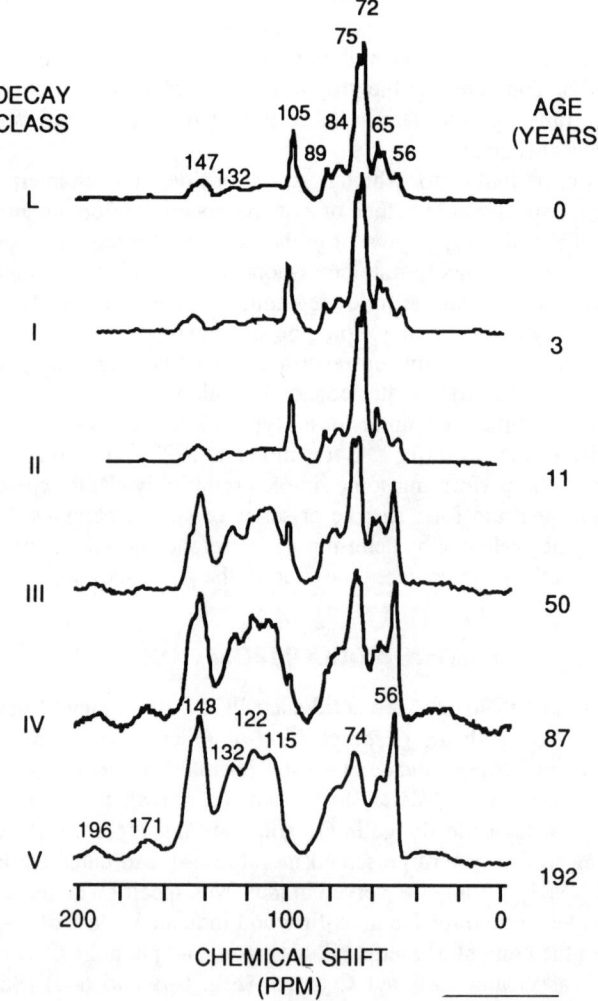

Fig. 6–3. Solid-state ^{13}C NMR spectra acquired for Douglas-fir wood of increasing decay class and residence time (Preston et al., 1990).

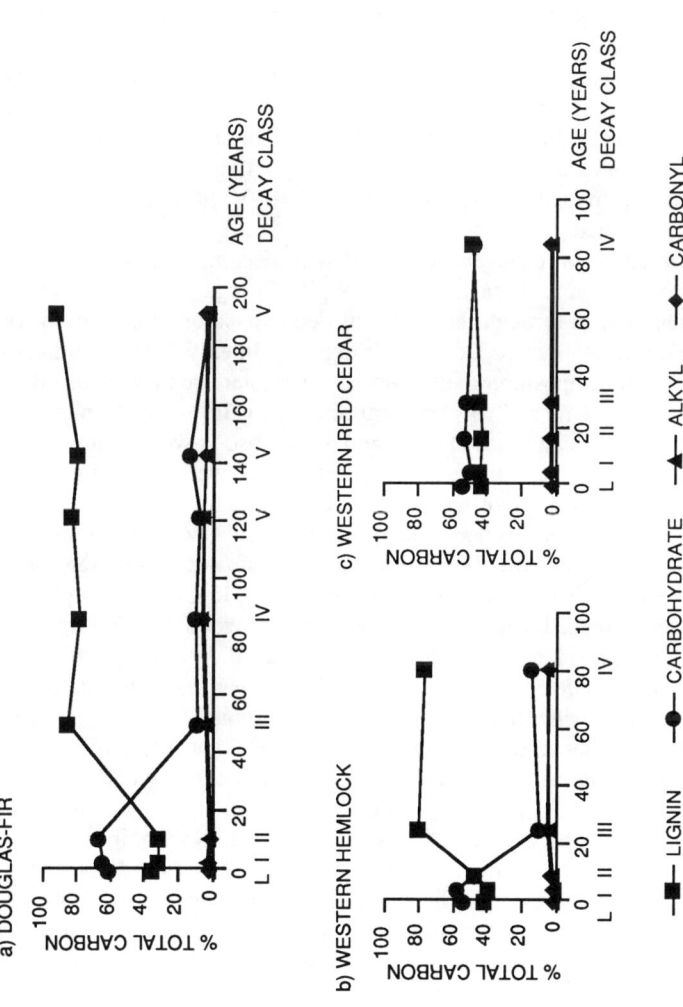

Fig. 6–4. Changes in the chemical composition of large diameter (>15 cm) (*a*) Douglas-fir, (*b*) western hemlock, and (*c*) western red cedar heartwood with increasing residence time and extent of decomposition (Preston et al., 1990).

changes are consistent with a loss of cellulose and hemicellulose, a concentration of lignin, waxes, and resins and an increase in the extent of wood oxidation.

Preston et al. (1990) suggested that Douglas-fir wood passes through three distinct phases of decomposition. Phase I, corresponding to a progression from fresh wood to decay Class II, is characterized by a small increase in carbohydrate C and likely reflects an enhancement in cellulose content as hemicellulose is preferentially degraded (Kirk & Highley, 1973). Increases in cellulose content during the early stages of decay also have been noted for logs of amabilis fir (*Abies amabalis*) (Hope, 1987) and balsam fir (*A balsamea*) (Lambert et al., 1980). In Phase 2 (decay Class II–III) an extensive loss of carbohydrate C and a concentration of lignin C occurs. Small increases in alkyl and carbonyl C contents also occur presumably due to the recalcitrant nature of waxes and resins (Paul & vanVeen, 1978) and the increased oxidation of organic molecules, respectively. In the third stage (decay Class III–V) little change in chemical composition is noted.

The chemical changes associated with decomposition of western hemlock logs were similar to those noted for Douglas-fir logs (Fig. 6–4), but western hemlock logs passed through the stages of decomposition more rapidly. Stage three was reached in 50 yr for Douglas-fir logs, but in only 25 yr for western hemlock logs. The decomposition pattern noted for western red cedar logs differed from that of Douglas-fir and western hemlock logs (Fig. 6–4). Despite a progression from fresh to decay Class IV logs and a decrease in wood density, alterations to the chemical structure of western red cedar wood were minor, indicating no preferential decomposition of specific classes of organic compounds.

In recent studies we have used solid-state ^{13}C NMR to examine the chemical changes associated with the decomposition of small diameter logs (<15 cm) in a red pine and a Douglas-fir plantation (Fig. 6–5). In the red pine plantation, samples were collected from the sapwood and heartwood of living trees, fallen logs, and thinning residues exhibiting characteristics of decomposition Classes L to III described in Table 6–1. All wood samples had been decomposing for ≤15 yr. The contents of carbohydrate, lignin, alkyl, and carbonyl C atoms were calculated in a manner similar to that of Preston et al. (1990). In progressing from living to Class III, a progressive decrease in the content of carbohydrate C and an increase in the content of lignin C was observed for sapwood. Small increases in the contents of alkyl and carbonyl C also were noted. No significant changes were observed in the chemical composition of heartwood as the extent of decomposition increased, indicating that microbial colonization of the heartwood was limited by the short length of time which the logs had decomposed and/or by the presence of compounds which limited microbial activity. In the Douglas-fir plantation, wood was collected from living trees and the residues of thinning operations to produce samples that had decomposed for 0 to 19 yr. Little if any change in composition was noted for the sapwood after 4 yr of decomposition; however, from 10 yr onwards a loss of carbohydrate and an accumulation of lignin structures were evident. For the Douglas-fir heartwood no compositional changes were noted up to 11 yr of decomposition but, after 19 yr, a loss of carbohydrate and an accumulation of lignin structures were evident. Changes in the chemical composition of small diameter logs as a result of microbial decomposition occurred at a faster rate than noted for large diameter logs by Preston et al. (1990). This observation is suspected to result from the smaller surface area to

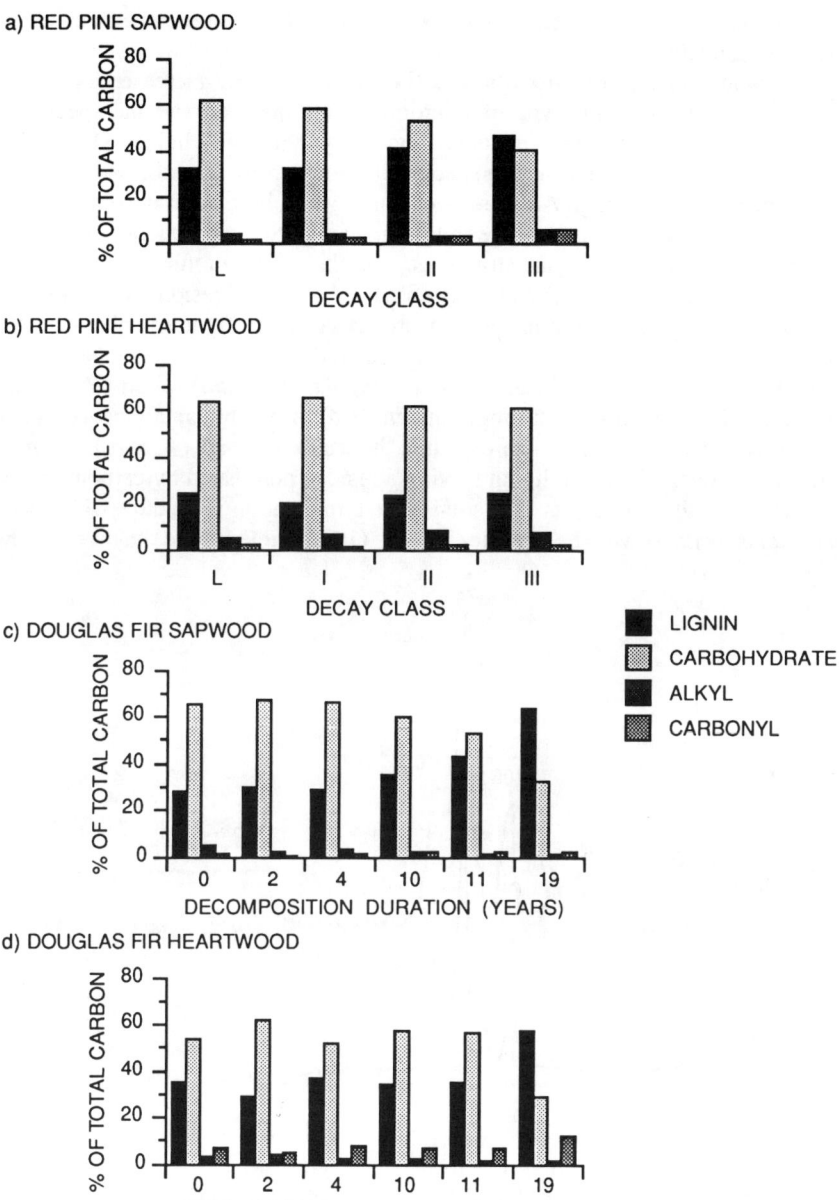

Fig. 6–5. Changes in the chemical composition of small diameter (<15 cm) red pine (*a*) sapwood and (*b*) heartwood and Douglas-fir (*c*) sapwood and (*d*) heartwood as the extent of decomposition increased.

volume ratio and the shorter distance which microorganisms must penetrate in smaller diameter logs.

Changes in wood composition as the extent of decay increases have been shown to depend on the type of microorganisms involved and the species of wood. In an in vitro wood decomposition study, Kirk and Highley (1973) followed the decomposition of five species of softwood by three brown-rot fungi and three white-rot fungi. All brown-rot fungi were observed to decompose carbohydrate structures in preference to lignin, leading to the production of a wood residue dominated by lignin structures. For the white-rot fungi, no selective decomposition of carbohydrate C was noted and the wood residue was similar in composition to the starting material. Martínez et al. (1991) used solid-state ^{13}C NMR to determine the influence of brown-rot and white-rot fungi on the composition of decomposed Chilean ulmo (*Eucryphia cordifolia*) wood (Fig. 6–6). Decomposition by brown-rot fungi (unidentified) resulted in an almost complete removal of carbohydrate structures, while the white rot fungus (*Ganoderma australe*) selectively degraded lignin leaving a residue dominated by carbohydrates. Using the white rot fungus in a solid-state fermentation procedure with beech (*Fagus sylvatica*) wood, Martínez et al. (1991) noted little change in the

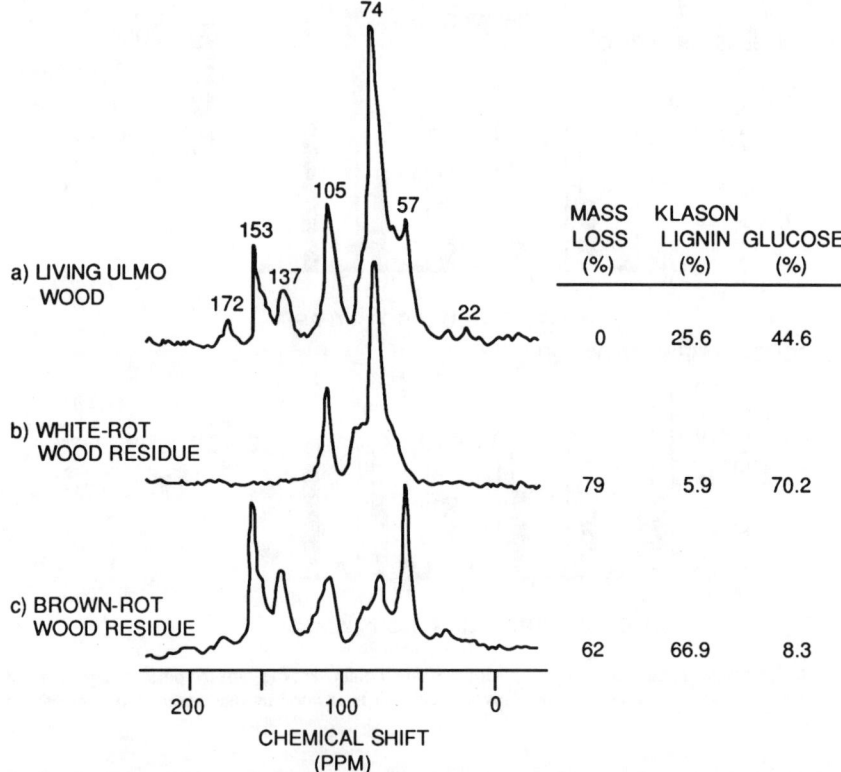

Fig. 6–6. Chemical composition of (*a*) living Chilean ulmo wood, (*b*) ulmo wood decomposed by white-rot fungi, and (*c*) ulmo wood decomposed by brown-rot fungi (Martinez et al., 1991).

chemical composition of the decomposed wood despite a 36% loss of mass. Hortling et al. (1992) used solid-state ^{13}C NMR to investigate the decomposition of birch wood by white-rot fungi and observed little change in chemical composition despite an estimated 75% reduction in wood density. These results indicate that the pattern of decomposition induced by white rot fungi depends on an interaction between the species of fungus and wood. Similar results were obtained in an ultrastructural study where selective delignification of *Laurelia philippiana* wood by the white-rot fungus *Phlevia chrysocrea* was noted, but decomposition of the same wood by *G. australe* resulted in increased lignin contents (Barrassa et al., 1992). The selective degradation of carbohydrates by brown-rot fungi appears to occur independent of the fungal or wood species involved. However, the presence of a selective or nonselective degradation process for white-rot fungi depends on the species of fungus and wood.

Decomposition processes can be influenced by environmental parameters such as temperature, nutrient status, and water content. Temperature and nutrient status exert their major influence by controlling decomposition rates, but also have been shown to affect lignin decomposition by fungi. Under conditions of low N availability, lignin degradation by white-rot fungi was stimulated, whereas under high N conditions, polysaccharide degradation was stimulated (Blanchette, 1991). Low temperatures in conjunction with high humidity also have been shown to stimulate lignin degradation (Blanchette, 1991). The exclusion of O_2 under saturated conditions limits the activity of obligate aerobes such as wood decomposing fungi (Nicholas, 1973; Liese, 1975). Bacteria are thought to be the major decomposer organisms present under anaerobic conditions (Hedges et al., 1985). Recent studies of modern and buried wood using solid-state ^{13}C NMR have shown that decomposition under anaerobic conditions involves a loss of carbohydrate structures and a preservation of lignin and alkyl C (Hedges et al., 1985; Spiker & Hatcher, 1987; Bates & Hatcher, 1989; Bates et al., 1991).

Hedges et al. (1985) acquired solid-state ^{13}C NMR spectra for modern and buried white oak (*Quercus alba*) and red alder (*Alnus rubra*) wood. Spectra acquired for the modern and buried woods (Fig. 6–7) indicated that, under the anaerobic conditions induced by burial, decomposition resulted in a preferential and almost complete loss of carbohydrate C and a selective preservation of lignin and alkyl C. These findings were supported by wet chemical methods used to determine total carbohydrate or lignin content quantitatively. Based on the contents of monomeric species contained in the major wood biopolymers, Hedges et al. (1985) developed the following stability series for the biopolymers found in the alder and oak wood investigated: guaiacyl lignin > syringyl lignin > pectin > cellulose > hemicellulose. On the basis of this series, one would predict that gymnosperm wood, which contains guaiacyl lignin exclusively, would be more stable to decomposition than angiosperm wood which contains a mixture of guaiacyl and syringyl lignin. Hedges et al. (1985) observed that buried gymnosperm (*Picea sitchensis*) wood found along side the angiosperm (*Alnus* spp.) wood exhibited little evidence of degradation. The observation that gymnosperm wood is more resistant to decomposition than angiosperm wood, at least under the anaerobic conditions of burial, agrees with results of Kohara (1956) and Nicholas (1973).

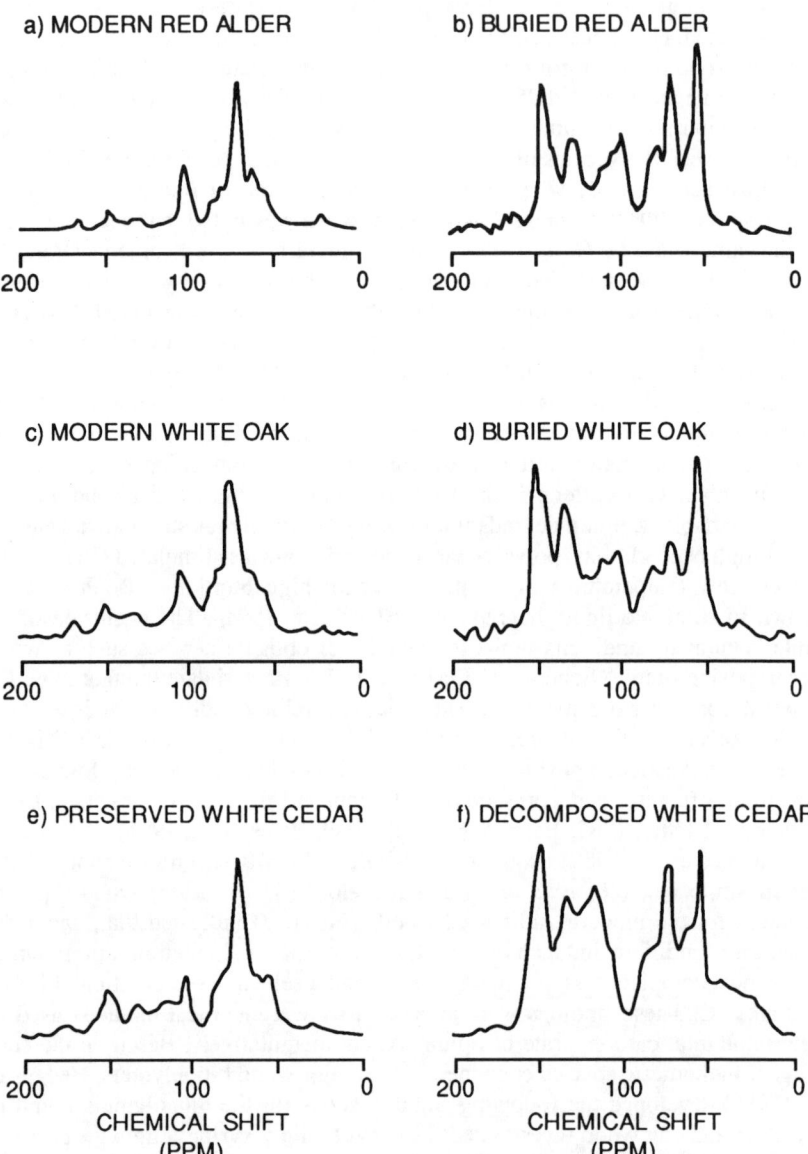

Fig. 6–7. Solid-state ^{13}C NMR spectra acquired for modern and buried oak and alder (Hedges et al., 1985) and preserved and decomposed buried white cedar (Spiker & Hatcher, 1987).

Decomposition of a white cedar (*Chamaecyparis thyoides*) log buried in peat was investigated by Spiker and Hatcher (1987). Most of the wood in the log was well preserved; however, the inner portion was decomposed by microorganisms. The preserved wood was dominated by O-alkyl and acetal C signals indicative of cellulose and hemicellulose structures (Fig. 6–7). In contrast, the decomposed wood was dominated by signals originating from the aromatic, phenolic, and methoxyl C atoms typical of that found in lignin structures (Fig. 6–7). The large decrease in carbohydrate signal intensity suggested that carbohydrate structures were of minor importance in the decomposed wood and that the remaining O-alkyl intensity originated from the propane side chains found in lignin structures (Fig. 6–2). The strong resemblance between the decomposed wood spectrum and that of lignin isolated from the preserved wood indicated a preferential decomposition of carbohydrate structures and a preservation of lignin structures during decomposition. Changes in the $\partial^{13}C$ values of C contained in these samples also suggested a preferential decomposition of carbohydrate C.

Bates and Hatcher (1989) used solid-state ^{13}C NMR to characterize the chemical changes associated with samples taken from a cross-section of a buried gymnosperm (Araucariaceae) log (Fig. 6–8). The extent of decomposition of wood in the log increased progressing from its center to its periphery. The ^{13}C NMR spectra acquired for samples collected from the center of the log resembled that acquired for modern *A. excelsa* wood and were dominated by signals arising from carbohydrate structures; however, in progressing towards the periphery of the log, the composition of the wood changed from being dominated by carbohydrate structures to being dominated by lignin structures, indicating a selective decomposition of carbohydrates and preservation of lignin structures. Similar results were obtained by Bates et al. (1991) for cross-sectional samples taken from a buried angiosperm (*Sapotaceae*) log found 3 m below the surface of a peat bog (Fig. 6–8).

FOREST LITTER DECOMPOSITION

Two approaches have been used to characterize compositional changes associated with forest litter decomposition using solid-state ^{13}C NMR. The first approach involves collecting fresh litter and placing it in litter bags which are returned to the environment for various lengths of time. In such studies, solid-state ^{13}C NMR analyses are performed on original fresh litter and samples are collected periodically to monitor changes in chemical composition as decomposition proceeds. In the second approach, the organic materials contained in forest litter layers are separated into fresh litter (L horizon), partially decomposed litter (F horizon), and well-decomposed litter (H and Ah horizons) by sampling along a depth profile through the litter layer and into the mineral soil. An underlying assumption in this second approach is that the well-humified organic materials were derived from litter having a similar composition to that of the fresh litter.

One of the first studies which utilized solid-state ^{13}C NMR to monitor chemical changes during the decomposition of litter placed in litter bags was that of Wilson et al. (1983), who followed decomposition of beech (*Fagus* genus) leaves and pine needles for 2 yr (Table 6–2). The O-alkyl content of both beech leaves and pine needles decreased with age, consistent with a decrease in carbohydrate content. Associated with this decrease was an increase in the content of aromatic C for the pine needles and alkyl C for the beech leaves. The lack of an accumulation of aromatic structures in the decomposing beech litter indicated that beech lignin was more susceptible to decomposition than pine lignin, which supports

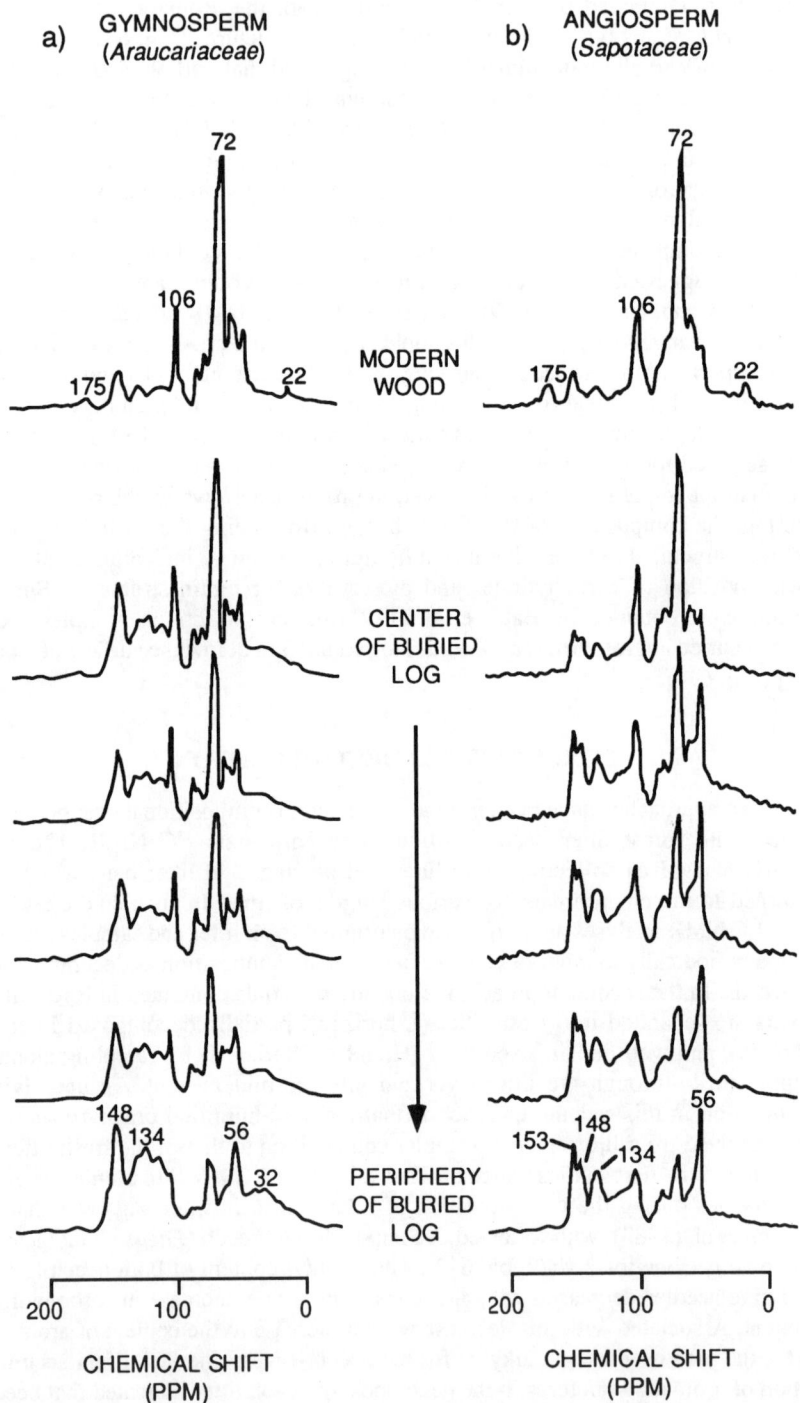

Fig. 6–8. Solid-state ^{13}C NMR spectra acquired for buried (a) gymnosperm (Bates & Hatcher, 1989) and (b) angiosperm wood (Bates et al., 1991)

Table 6–2. Changes in chemical composition of beech leaves and pine needles left to decompose for 2 yrs (Wilson et al., 1983).

		Content of each type of organic C				
		Carbonyl	Aromatic	Acetal	O-alkyl	Alkyl
				%		
Beech leaves	1976 (undecomposed)	6	14	19	42	18
	1977	6	11	14	35	29
	1978	6	15	14	37	28
Pine needles	1976 (undecomposed)	4	12	16	48	19
	1977	7	20	15	44	19
	1978	9	21	15	42	18

the suggestion of Hedges et al. (1985) that angiosperm lignin is more susceptible to microbial decomposition than gymnosperm lignin. Results of Zech et al. (1987) indicated that, after 3 yr of decomposition, Scots pine and Norway spruce needles were depleted in O-alkyl C and enriched in alkyl, aromatic, and carboxyl C atoms. The accumulation of alkyl C was greater for the Scots pine than Norway spruce [P. abies (L.) Karsten] needles. A similar result was obtained by Nordén and Berg (1990) for Scots pine (*P. sylvestris* L.) needles left to decompose for approximately 4 yr. Using chemical composition data acquired for experiments of this type with mass and C balance data, it should be possible to calculate decomposition or accumulation rates for each of the various types of C provided that the spectra obtained reflect the distribution of C types quantitatively. However, this approach does not appear to have been utilized.

Solid-state ^{13}C NMR spectra which we have acquired recently for living needles, fresh litter, litter horizons, and the mineral Ah horizon from a red pine plantation are presented in Fig. 6–9. In progressing from living needles to well-decomposed organic materials contained in the mineral Ah horizon, a loss of O-alkyl C and an accumulation of alkyl, aromatic, and carbonyl C atoms was noted (Table 6–3). Such compositional changes are similar to those noted for litterbag studies and consistent with a preferential utilization of carbohydrate structures, an accumulation of the more recalcitrant waxes, resins, and lignin, and a synthesis of carboxyl structures as a product of oxidative decomposition.

Zech et al. (1992) performed solid-state ^{13}C NMR analyses on the litter and A horizons of seven German soils representative of each of the major humus forms: mull, moder, mor, and tangelmor. Irrespective of humus form, the compositional differences noted by these researchers as decomposition proceeded were similar to those noted by us for the red pine plantation with the exception that an accumulation of aromatic C was not observed. Other studies utilizing solid-state ^{13}C NMR to follow decomposition of forest litter have noted no accumulation of aromatic C (Hempfling et al., 1987; Zech et al., 1985, 1987, 1990; Kögell-Knabner et al., 1988); however, Oades et al. (1987, 1988) and Baldock et al. (1992) observed an accumulation of aromatic C during the initial stages of decomposition of organic materials in mineral soil horizons. The observation of Zech et al. (1992) that CuO-lignin content decreased from the L to A horizon, even though total aromatic C content remained constant, indicated that lignin

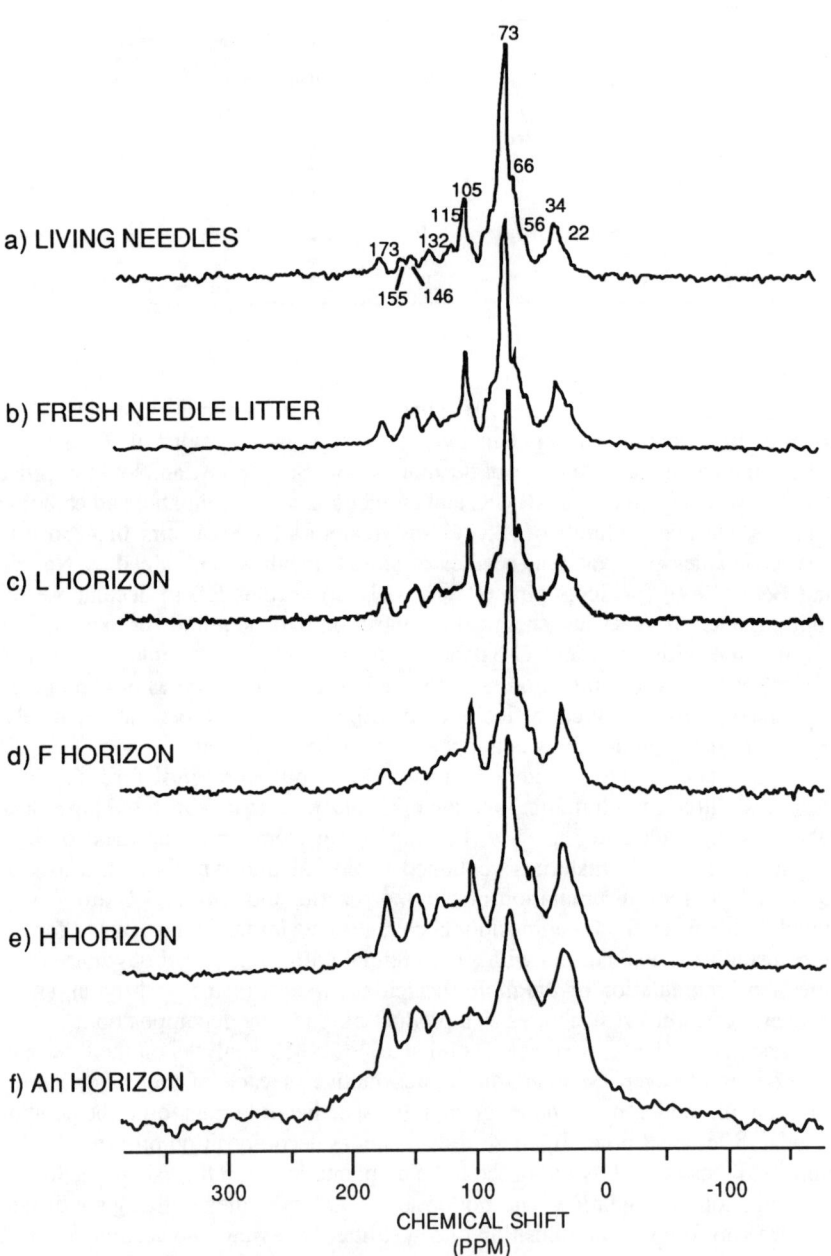

Fig. 6–9. Solid-state ^{13}C NMR spectra acquired for living needles, fresh needle litter, litter horizons, and mineral Ah horizon under a red pine plantation.

Table 6–3. Chemical composition of organic C contained in living needles, fresh litter, and organic and mineral soil horizons of a red plantation.

Horizon	C_{org}	N_{tot}	Content of each type of organic C			
			Carbonyl	Aromatic	O-alkyl	Alkyl
			———————— % ————————			
Living needles	50.6	1.24	5	17	64	14
Fresh litter	51.2	0.58	5	20	59	16
L			5	19	60	16
F	50.1	1.42	4	17	60	18
H	38.1	1.22	8	25	48	19
Ah	9.3	0.29	11	24	43	21

molecules were decomposed and nonlignin aromatic structures, or at least significantly altered lignin, accumulated with progressive decomposition. Whether or not aromatic C accumulates during decomposition is thought to be a function of the relationships between the chemical nature of the litter, the composition of the decomposer community, and the environmental conditions as outlined in the discussion of wood decomposition. In a system containing white-rot fungi an accumulation of aromatic materials may not be observed because these fungi will decompose lignin; however, in a system dominated by bacteria and/or brown-rot fungi, lignin may persist, leading to an accumulation of aromatic C. An accumulation of aromatic C also may be induced in wet environments where fungal activity is suppressed by anaerobic conditions.

The loss of O-alkyl C as the extent of decomposition increases was investigated further by Zech et al. (1992). Using a sulphuric acid digestion procedure to determine total polysaccharide content, they noted that the decrease in O-alkyl C was not as large as the decrease in total polysaccharide content, indicating that nonpolysaccharide O-alkyl C also accumulated during decomposition. A portion of this nonpolysaccharide O-alkyl C may be the oxygenated C atoms of the propane side chain and methoxyl groups of lignin monomers (Fig. 6–2).

An accumulation of alkyl C has been noted in all solid-state ^{13}C NMR studies of litter decomposition and in studies where the decomposition of organic materials in mineral soils has been characterized (Oades et al., 1987, 1988; Baldock et al., 1989, 1992). A list of rate constants for the decomposition of various classes of organic compounds in mineral soils has been presented by Paul and vanVeen (1978). The lowest rate constants were obtained for alkyl compounds, indicating that an accumulation of alkyl C during decomposition may result from the selective degradation of other forms of C and a preservation of the more recalcitrant alkyl structures. The increased content of alkyl C with increasing extent of decomposition and humification may result from one or more of the following possible mechanisms: (i) a selective preservation of recalcitrant alkyl structures present in the original plant material; (ii) a synthesis of alkyl materials as a metabolic product of the organisms decomposing the organic materials; and/or (iii) a transformation of the original or synthesized alkyl structures into a more recalcitrant form.

The accumulation of alkyl C as decomposition proceeds has been investigated by Kögel-Knabner et al. (1989, 1992a,b) using a combination of solid-state

^{13}C NMR and pyrolysis-gas chromatography-mass spectrometry. The importance of plant-derived alkyl structures was assessed by determining the content of hydroxy fatty acids derived from cutin and/or suberin (cutin acids) in litter and mineral horizons (Kögel-Knabner et al., 1989). The cutin acid content of the organic materials contained in litter horizons was found to be greater than that of the more humified organic materials contained in the corresponding mineral horizons, indicating a net loss of cutin structures during decomposition. The susceptibility of cutin to decomposition also was demonstrated by Kögel-Knabner et al. (1992a) for decomposing litter in a long-term litterbag field experiment and by Ziegler and Zech (1990) and Goñi and Hedges (1991). Therefore, the increase in alkyl C content as decomposition proceeds is not suspected to result from an accumulation of unaltered alkyl structures derived from plant materials.

Solid-state ^{13}C NMR has been used to demonstrate the ability of algae, bacteria, and fungi to synthesize alkyl C structures as metabolic products of decomposition processes (Zelibor et al., 1988; Baldock et al., 1990; Kögel-Knabner et al., 1992b). Kögel-Knabner et al. (1992b) investigated the contribution of alkyl structures synthesized by microorganisms to the accumulation of alkyl C during decomposition using dipolar dephasing ^{13}C NMR experiments. Such experiments allow the alkyl signal intensity to be separated into two components: mobile and rigid alkyl structures. Fresh litter was shown to contain approximately equal quantities of mobile and rigid alkyl structures; however, as decomposition progressed mobile structures were preferentially lost, presumably by either selective degradation or a conversion to more rigid structures. Microbial alkyl C has been shown to contain both mobile and rigid structures (Baldock et al., 1990; Kögel-Knabner et al. (1992b). The rigid microbial alkyl C may contribute to the accumulation of alkyl C directly, but it is unlikely that unaltered mobile C makes a significant contribution.

Kögel-Knabner et al. (1992b) have suggested that an increase in the degree of cross linking may be responsible for a conversion of mobile to rigid alkyl structures. Any other mechanism which reduces the molecular motion exhibited by alkyl structures also may be of importance. For example, adsorption onto pieces of organic debris or soil mineral components could reduce molecular motion and shift dipolar dephased solid-state ^{13}C NMR signal intensity towards a more rigid structure.

PEAT DECOMPOSITION

Attempts to characterize the extent of decomposition of peat focused initially on various physical properties (e.g., rubbed fiber content and bulk density) and chemical characteristics determined using wet chemical methods (e.g., carbohydrate analysis, functional group analysis, and pyrophosphate index). Lévesque and Mathur (1979) found that the degree of decomposition of peat was best estimated from its rubbed fiber content. More recently, solid-state ^{13}C NMR has been used to quantify changes in the chemical composition of peat as decomposition proceeds by examining the depth profiles of peat deposits, peat particle size fractions, and the influence of cultivation.

Table 6–4. Changes in the chemical composition of samples collected from depth profiles of three peats (Hammond et al., 1985).

Peat	Depth	Rubbed fiber	Content of each type of organic C			
			Carbonyl	Aromatic	O-alkyl	Alkyl
	cm			%		
Ormstown	0–10	90	6	15	60	19
	10–17	48	6	18	51	25
	17–33	41	6	15	54	25
	85–92	23	6	27	35	32
Clair	30–40	59	5	18	56	21
	40–70	26	5	18	47	30
	70–95	26	5	22	44	29
	95–125	30	6	22	46	26
	125–140	–	6	21	48	25
	140–165	55	7	19	50	24
St-Chrysostome	0–25	27	8	21	41	30
	25–65	21	9	28	36	27
	65–85	34	8	32	37	23
	85–125	19	9	33	34	24

Hammond et al. (1985) performed solid-state ^{13}C NMR analyses on depth profiles of three Canadian peats (Table 6–4). As rubbed fiber content decreased and the extent of decomposition increased, changes in chemical composition observed by Hammond et al. (1985) for each type of peat were similar. The content of O-alkyl C decreased while that of alkyl and aromatic C increased. No significant changes in the carbonyl region, which includes carboxyl C, were noted. Such changes are consistent with a selective degradation of carbohydrate structures and an accumulation of alkyl and aromatic C. The lack of an accumulation of carboxyl C suggests that oxidative decomposition processes were limited presumably by anaerobic conditions. Under anaerobic conditions, the activity of lignin-degrading fungi would be reduced (Alexander, 1977), which may account for the observed accumulation of aromatic C structures.

Preston et al. (1987) used solid-state ^{13}C NMR to determine changes in the chemical composition of samples collected from the depth profiles of a peat subjected to 0, 5, and 10 yr of cultivation. The changes noted with depth were equivalent to those observed by Hammond et al. (1985); O-alkyl C content decreased, alkyl and aromatic C content increased, and carbonyl C content remained constant. However, the compositional changes induced by cultivation in the surface 0- to 5-cm layer included a loss of both O-alkyl and aromatic C, an accumulation of alkyl C, and little change in carbonyl C content. The loss of aromatic C induced by cultivation indicated that mixing and aerating this peat enhanced the decomposition of aromatic structures. Strong correlations between aromatic, phenolic, and methoxyl C suggest that changes in aromatic C content may be the result of an increased decomposition of lignin structures. Increased activity of lignin decomposing fungi in response to increased O_2 levels was demonstrated by Reid and Seifert (1982).

Preston et al. (1989) and Nordén et al. (1992) used solid-state ^{13}C NMR to investigate compositional changes during decomposition of peats by acquiring spectra from particle size fractions. The extent of decomposition of finer size fractions isolated from peat was found to be greater than that of coarser size fractions (Bracewell et al., 1980; Lévesque & Dinel, 1977, 1982; Morita & Lévesque, 1980; Williams, 1983). Preston et al. (1989) isolated six particle size fractions (>2000, 1000–2000, 500–1000, 150–500, 75–150, and <75 μm) from six different peat profiles selected to include a variety of landform, plant type, and degree of decomposition. Solid state ^{13}C NMR data from highly decomposed and poorly decomposed peats studied by Preston et al. (1989) are presented in Fig. 6–10. The finer particle size fractions of all six peats were observed to contain less O-alkyl and more alkyl C than the coarser fractions. Aromatic C contents tended to be lower in finer particles for all but the Gatineau peat, but changes were not as large as that observed for O-alkyl or alkyl C. Carboxyl C content did not increase as particle size decreased and the extent of decomposition increased, suggesting that the operative decomposition processes were taking place in a dominantly anaerobic environment. The magnitude of the compositional changes was related to the extent of decomposition exhibited by the unfractionated peat samples. The highly decomposed peats showed a greater change in chemical composition in progressing from the coarse to fine particle size fractions than that observed for the poorly decomposed peats (Fig. 6–10). The solid-state ^{13}C NMR results obtained by Nordén et al. (1992) for particle size fractions isolated from two well-decomposed *Carex* peats and three less decomposed *Sphagnum* peats also showed a progressive loss and accumulation of O-alkyl and alkyl C atoms, respectively, with decreasing particle size. Based on the spectra presented by Nordén et al. (1992), little if any changes in aromatic and carbonyl C content were noted.

COMPOSITIONAL CHANGES RESULTING FROM DECOMPOSITION PROCESSES

Changes in chemical composition of wood, forest litter, and peat induced by decomposition have been summarized in Fig. 6–11. Under aerobic conditions, the decomposition of wood is dominated by fungi (Hedges et al., 1985) which can be classified into two groups, white-rot and brown-rot fungi. Brown-rot fungi have been shown to degrade carbohydrate structures selectively resulting in residues with high aromatic C contents (Kirk & Highley, 1973; Martinez et al., 1991). The influence of white-rot fungi is more variable than that observed for brown-rot fungi and depends on an interaction between the fungal and wood species involved. The line with double arrowheads under white-rot fungi in Fig. 6–11a indicates that white-rot fungi have been observed to induce structural changes ranging from a selective delignification to no compositional change as decomposition progressed. Different chemical changes have been induced by exposing several species of wood to a single white-rot fungus (Martinez et al., 1991) and by exposing a single type of wood to different species of white rot (Blanchette, 1991; Barassa et al., 1992). Under anaerobic conditions wood decomposition processes are thought to be dominated by bacteria (Hedges et al., 1985) and

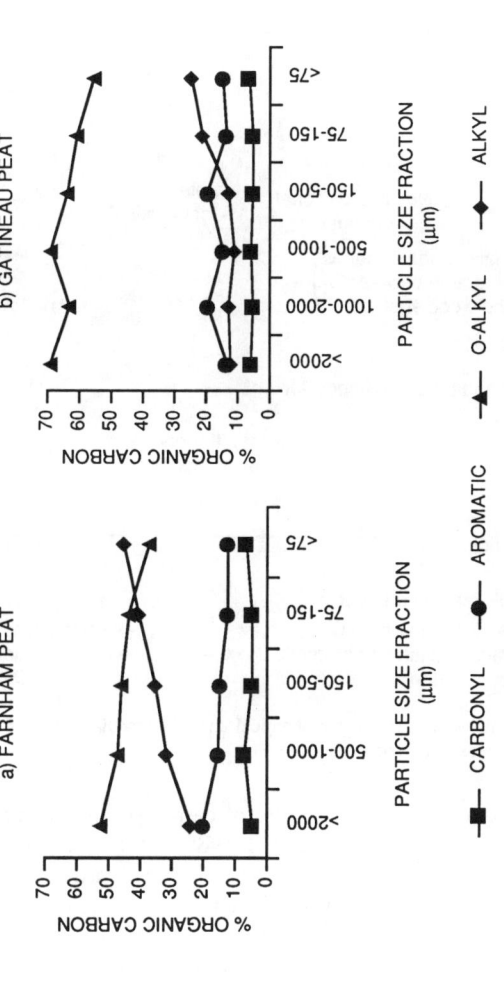

Fig. 6–10. Chemical composition of particle size fractions isolated from (a) the highly decomposed Farnham peat and (b) the poorly decomposed Gatineau peat (Preston et al., 1989).

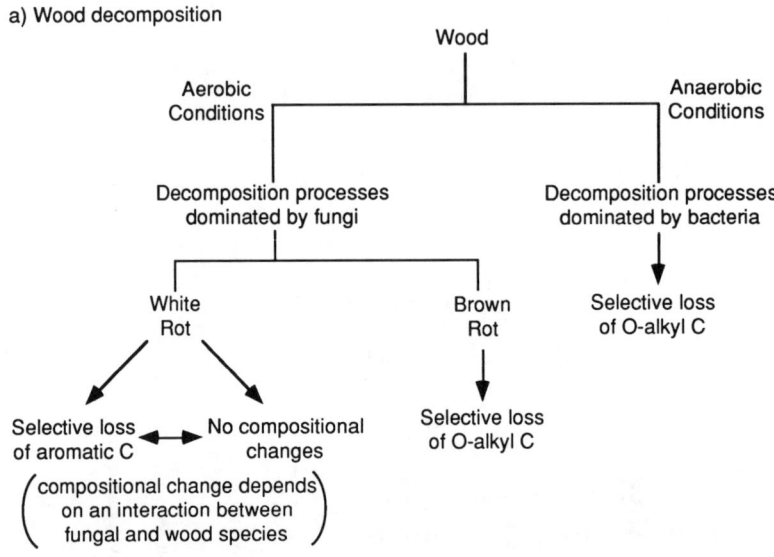

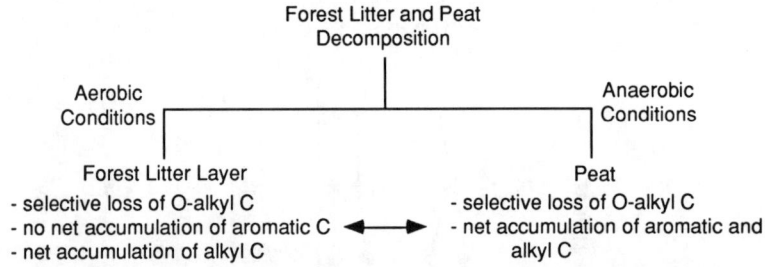

Fig. 6–11. A model summarizing the chemical changes associated with the decomposition of (*a*) wood and (*b*) forest litter and peat.

invariably resulted in a selective loss of carbohydrate structures and an accumulation of aromatic C (Hedges et al., 1985; Spiker & Hatcher, 1987; Bates & Hatcher, 1989; Bates et al., 1991). The lack of a significant accumulation of alkyl C during wood decomposition under aerobic or anaerobic conditions is thought to be a function of the low alkyl content of undecomposed wood and an inability of wood decomposing organisms to synthesize biologically stable alkyl structures.

Compared to wood decomposition, little is known about forest litter and peat decomposition processes. Experiments designed to assess the influence of various types of microorganisms (e.g., white-rot fungi vs. brown-rot fungi or bacteria) on compositional changes induced by decomposition of litter and peat have not been performed. However, the compositional changes observed for forest litter and peat are consistent with results of wood decomposition and allow some inferences to be made. Under aerobic conditions, deposition of forest litter results in the formation of litter layers on the mineral soil surface and decomposition processes appear to result in a selective loss of carbohydrate structures, little change in aromatic C content, and an accumulation of alkyl C (Zech et al., 1992). The absence of an accumulation of aromatic C is indicative of the presence of white-rot fungi in addition to other forms of microorganisms. Under anaerobic conditions, the decomposition rate of forest litter is reduced, organic materials accumulate, and peat deposits are formed. In peats, carbohydrates are degraded selectively and lignin accumulates in a manner similar to that observed for the anaerobic decomposition of wood (Hammond et al., 1985; Preston et al., 1987, 1989; Nordén et al., 1992). In forest litter layers which experience intermittent periods of waterlogging, an accumulation of aromatic structures, as we observed in the red pine plantation, may result due to a periodic suppression of fungal activity. Significant accumulations of alkyl C are observed during aerobic and anaerobic decomposition of forest litter and peat (Hammond et al., 1985; Preston et al., 1987, 1989; Nordén et al., 1992; Zech et al., 1992).

A DECOMPOSITION INDEX BASED ON CHEMICAL COMPOSITION

Lignin has often been considered to be the most recalcitrant component of decomposing forest organic materials; however, in the presence of white-rot fungi, lignin may be preferentially degraded with little or no loss of carbohydrate or other organic materials. The results presented in this paper demonstrate clearly that the traditional idea of an accumulation of aromatic C in response to decomposition and humification of organic materials is not accurate under all conditions. As a result, using aromaticity (content of aromatic C expressed as a fraction of total organic C content) as a general indicator of the extent to which organic materials have been decomposed appears inappropriate.

Alkyl C containing polymethylene type structures, $(CH_2)_n$, offer the most stability against microbial attack and have been noted to accumulate during aerobic and anaerobic decomposition of all forest organic materials, especially forest litter and peat. The small accumulation of alkyl C during wood decomposition, relative to that observed for forest litter and peat, is suspected to be a function of the low alkyl content of undecomposed wood and an inability of wood decomposing organisms to synthesize biologically stable alkyl structures. In an attempt to correlate the extent of decomposition of samples collected from three peats, as assessed by rubbed fiber content, with chemical composition, Hammond et al. (1985) noted a much better correlation with alkyl C content ($R^2 = 0.70$, $P = 0.0003$) than with aromaticity ($R^2 = 0.39$, $P = 0.0229$). For forest litter and peat, increases in alkyl C content were always associated with decreases in O-alkyl (carbohydrate) C content. It would, therefore, appear that a more sensitive index of decomposition based on chemical characteristics could be obtained using the ratio of alkyl to O-alkyl C contents. As the magnitude of the alkyl/O-alkyl ratio increases the extent of decomposition also increases. The correlation between rubbed fiber and the alkyl/O-alkyl ratio for the data of Hammond et al. (1985) ($R^2 = 0.74$, $P = 0.0002$) was approximately the same as that obtained using alkyl C content alone when all three peats were considered together. When the analysis was performed on the samples collected from individual peats, the alkyl/O-alkyl ratio was more strongly correlated with rubbed fiber content than aromaticity or alkyl C content.

Aromaticity and alkyl/O-alkyl ratio values have been plotted as a function of particle size fraction for the six peats studied by Preston et al. (1989) (Fig. 6–12). As mentioned previously, the extent of decomposition of the organic materials contained in finer peat fractions is greater than that contained in coarser fractions. For all of the peats, except the Gatineau peat where little change in aromaticity was noted, a decrease in aromaticity was observed in progressing from coarser to finer particle size fractions. Such a trend contradicts the notion that aromaticity increases with an increase in the extent of decomposition. The value of the alkyl/O-alkyl ratio increased progressively as particle size decreased for each peat, which is consistent with an increase in extent of decomposition.

Aromaticity and alkyl/O-alkyl ratio values also were calculated for the forest litter and mineral horizons of the seven soils studied by Zech et al. (1992) (Fig. 6–13). In progressing down the soil profile from fresh litter to humified organic materials contained in mineral horizons, changes in aromaticity values were variable. There was no consistent increase in aromaticity as the extent of decomposition increased. However, with the exception of the Of horizon of the Lithic Udorthent, a gradual increase in the alkyl/O-alkyl ratio was observed with increasing soil depth and extent of organic matter decomposition.

The alkyl/O-alkyl ratio appears to give a better estimation of the extent to which organic materials contained in forest soil horizons and peats are decomposed than does the traditional value of aromaticity. As a result of the low contents of alkyl C in wood and the selective nature which decomposition of wood can take, it seems improbable that the alkyl/O-alkyl ratio, or any other parameter related to chemical composition, would adequately assess the extent of decomposition of wood or woody residues. Differences in the magnitude of the changes in alkyl/O-alkyl ratio across the particle size fractions of the peats studied by Preston et al. (1989) and the soil horizons studied by Zech et al. (1992) may have resulted from two factors: (i) variations in the parameters controlling decomposition rates and compositional changes; and/or (ii) differences in the chemical composition of the original organic materials. As a result, the alkyl/O-alkyl ratio should only be used to assess the extent of decomposition of organic materials, where the organic materials are known to be derived from starting materials having a similar composition.

CONCLUSIONS

Solid-state ^{13}C NMR has been used successfully in numerous studies to characterize changes in the chemistry of decomposing forest organic materials. Results acquired from these studies have indicated that changes in chemical composition of wood, forest litter, and peat induced by decomposition result from an interaction between the composition of the original unaltered substrate and the species composition of the decomposer microbial community. Under different environmental conditions, decomposition of the same substrate can result in the

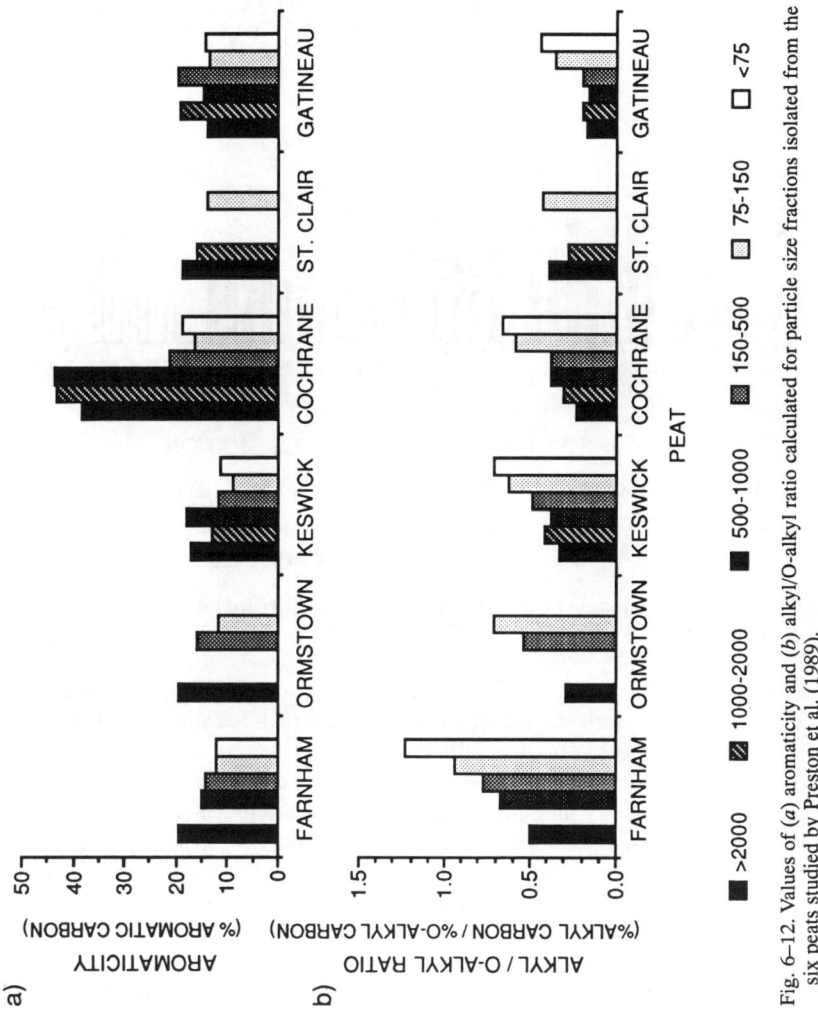

Fig. 6–12. Values of (*a*) aromaticity and (*b*) alkyl/O-alkyl ratio calculated for particle size fractions isolated from the six peats studied by Preston et al. (1989).

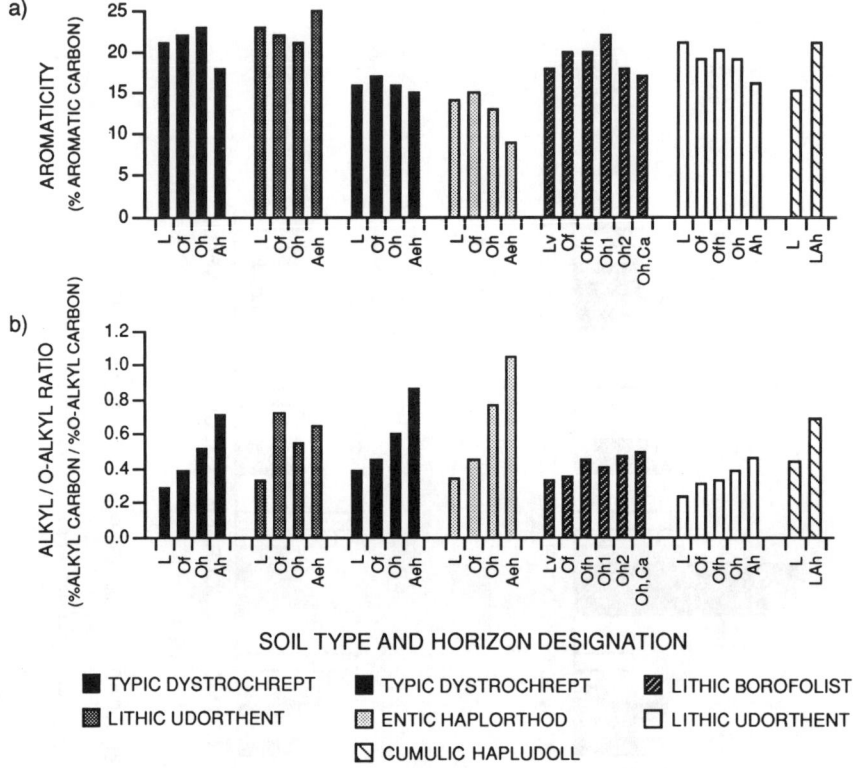

Fig. 6–13. Values of (*a*) aromaticity and (*b*) alkyl/O-alkyl ratio calculated for litter and mineral horizons of the seven forest soils studied by Zech et al. (1992).

formation of residues with a different chemical structure. White-rot fungi have been observed to degrade carbohydrate and lignin structures simultaneously or to delignify wood selectively depending on the species of fungus and wood involved. Brown-rot fungi have been shown to utilize wood carbohydrate structures selectively, thereby inducing accumulations of lignin. Bacteria, which dominate decomposition processes under anaerobic conditions, also utilize wood carbohydrate structures selectively. Although results pertaining to the activity of specific types of fungi and bacteria were related to decomposition of wood, it would seem likely that similar processes will occur during the decomposition of forest litter and peat.

The traditional view that aromaticity increases as the extent of decomposition proceeds has not been confirmed by the solid-state ^{13}C NMR results presented. It is thought that the lack of a relationship between aromaticity and extent of decomposition is a function of the selective nature which decomposition processes may take in the presence of different microbial decomposer communities. For organic materials contained in forest soil horizons and in peat deposits, the ratio of alkyl (polymethylene) to O-alkyl (carbohydrates) C content appears to provide a good index of the extent of decomposition; however, use of this ratio to characterize decomposition processes should be restricted to situations where the origin of the organic materials is similar.

In this paper, the majority of the solid-state ^{13}C NMR results presented used conventional solid-state ^{13}C NMR analyses. Many other analyses (pulse sequences) that can provide more detailed structural information are also available. For example, the degree of protonation of aromatic and aklyl structures and/or their molecular mobility can be assessed using the dipolar dephasing pulse sequence described by Wilson (1987). In addition, recent technical advances, such as improved electronics and computer systems and the ability to use larger sample sizes, will facilitate the acquisition of improved spectra using shorter acquisition times.

REFERENCES

Alexander, M. 1977. Introduction to soil microbiology. 2nd ed. John Wiley & Sons, New York.

Baldock, J.A., J.M. Oades, A.M. Vassallo, and M.A. Wilson. 1989. Incorporation of uniformly labelled ^{13}C-glucose carbon into the organic fraction of a soil. Carbon balance and CP/MAS ^{13}C NMR measurements. Aust. J. Soil Res. 27:725–746.

Baldock, J.A., J.M. Oades, A.M. Vassallo, and M.A. Wilson. 1990. Solid-state CP/MAS ^{13}C NMR analysis of bacterial and fungal cultures isolated from a soil incubated with glucose. Aust. J. Soil Res. 28:213–225.

Baldock, J.A., J.M. Oades, A.G. Waters, X. Peng, A.M. Vassallo, and M.A. Wilson. 1992. Aspects of the chemical structure of soil organic materials as revealed by solid-state ^{13}C NMR spectroscopy. Biogeochemistry 16:1–42.

Barassa, J.M., A.E. González, and A.T. Martinez. 1992. Ultrastructural aspects of fungal delignification of Chilean wood by *Ganoderma australe* and *Phelbia chrysocrea*. A study of natural and invitro degradation. Holzforschung 46:1–8.

Bates, A.L., P.G. Hatcher, H.E. Lerch, C.B. Cecil, S.G. Neuzil, and Supardi. 1991. Studies of a peatified angiosperm log cross section from Indonesia by nuclear magnetic resonance spectroscopy and analytical pyrolysis. Org. Geochem. 17:37–45.

Bates, A.L., and P.G. Hatcher. 1989. Solid-state ^{13}C NMR studies of a large fossil gymnosperm from the Yallourn Open Cut, Latrobe Valley, Australia. Org. Geochem. 14:609–617.

Bracewell, J.M., G.W. Robertson, and B.L. Williams. 1980. Pyrolysis-mass spectrometry studies of humification in a peat and peaty podzol. J. Anal. Appl. Pyrol. 2:53–62.

Blanchette, R.A. 1991. Delignification by wood-decay fungi. Ann. Rev. Phytopathol. 29:381–398.

Duncan, T.M. 1987. ^{13}C chemical shielding in solids. J. Phys. Chem. Ref. Data 16:125–151.

Earl, W.L., and D.L. VanderHart. 1981. Observations by high-resolution carbon-13 nuclear magnetic resonance of cellulose I related to morphology and crystal structure. Macromolecules 14:570–574.

Goñi, M.A., and J.I. Hedges. 1991. The diagenetic behavior of cutin acids in buried conifer needles and sediments from a coastal marine environment. Geochim. Cosmochim. Acta 54:3083–3093.

Hammond, T.E., D.G. Cory, W.M. Ritchey, and H. Morita. 1985. High resolution solid state ^{13}C NMR of Canadian peats. Fuel 64:1687–1695.

Hatcher, P.G. 1987. Chemical structural studies of natural lignin by dipolar dephasing solid-state ^{13}C nuclear magnetic resonance. Org. Geochem. 11:31–39.

Hatfield, G.R., G.E. Maciel, O. Erbatur, and G. Erbatur. 1987. Qualitative and quantitative analysis of solid lignin samples by carbon-13 nuclear magnetic resonance spectrometry. Anal. Chem. 59:172–179.

Hedges, J.I., G.L. Cowie, J.R. Ertel, R.J. Barbour, and P.G. Hatcher. 1985. Degradation of carbohydrates and lignins in buried woods. Geochim. Cosmochim. Acta 49:701–711.

Hempfling, R., F. Ziegler, W. Zech, and H.-R. Schulten. 1987. Litter decomposition and humification in acidic forest soils studied by chemical degradation, IR and NMR spectroscopy and pyrolysis field ionization mass spectrometry. Z. Pflanzenernähr. Bodenk. 150:179–186.

Hope, S.M. 1987. Classification of decayed *Abies amabilis* logs. Can. J. For. Res. 17:559–564.

Hortling, B., I. Forsskåhl, J. Janson, J. Sundquist, and L. Viikari. 1992. Investigations of fresh and biologically decayed birch. Holzforschung 46:9–19.

Kirk, T.K., and T.L. Highley. 1973. Quantitative changes in structural components of conifer woods during decay by white- and brown-rot fungi. Phytopathology 63:1338–1342.

Kögel-Knabner, I., W. Zech, and P.G. Hatcher. 1988. Chemical composition of the organic matter in forest soils: The humus layer. Z. Pflanzenernähr. Bodenk. 151:331–340.

Kögel-Knabner, I., P.G. Hatcher, E.W. Tegelaar, and J.W. de Leeuw. 1992a. Aliphatic components of forest soil organic matter as determined by solid-state ^{13}C NMR and analytical pyrolysis. Sci. Total Environ. 113:89–106.

Kögel-Knabner, I., J.W. de Leeuw, and P.G. Hatcher. 1992b. Nature and distribution of alkyl carbon in forest soil profiles: implication for the origin and humification of aliphatic biomacromolecules. Sci. Total Environ. 117/118:175–185.

Kögel-Knabner, I., F. Ziegler, M. Riederer, and W. Zech. 1989. Distribution and decomposition pattern of cutin and suberin in forest soils. Z. Pflanzenernähr. Bodenk. 152:409–413.

Kohara, J. 1956. Studies of Japanese old timbers—XX. Chemical analyses of unearthed woods. Mokuzai Gakkaishi 19:195–200.

Lambert, R.L., G.E. Lang, and W.A. Reiners. 1980. Loss of mass and chemical change in decaying boles of a subalpine balsam fir forest. Ecology 61:1460–1473.

Lévesque, M., and H. Dinel. 1977. Fiber content, particle-size distribution and some related properties of four peat materials in eastern Canada. Can. J. Soil Sci. 57:187–195.

Lévesque, M., and H. Dinel. 1982. Some morphological and chemical aspects of peats applied to the characterization of histosols. Soil Sci. 133:324–332.

Lévesque, M., and S.P. Mathur. 1979. A comparison of various means of measuring the degree of decomposition of virgin peat materials in the context of their relative biodegradability. Can. J. Soil Sci. 59:397–400.

Liese, W. 1975. Transformation of woods by microorganisms. Intl. Congr. Plant Path., 2nd. Springer-Verlag, New York.

Maciel, G.E., W.L. Kolodziejski, M.S. Bertran, and B.E. Dale. 1982. ^{13}C NMR and order in cellulose. Macromolecules 15:686–687.

Martínez, A.T., A.E. González, M. Valmaseda, B.E. Dale, M.J. Lambregts, and J.F. Haw. 1991. Solid-state NMR studies of lignin and plant polysaccharide degradation by fungi. Holzforschung 45:49–54.

Morita, J., and M. Lévesque. 1980. Monosaccharide composition of peat fractions based on particle size. Can. J. Soil Sci. 60:285–289.

Nicholas, D.D. 1973. Wood deterioration and its prevention by preservative treatments. Vol. 1. Syracuse Univ. Press, Syracuse, NY.

Nim H.H., D. Robert, O. Faix, and M. Nemr. 1981. Carbon-13 NMR spectra of lignins. 8. Structural differences between lignins of hardwoods, softwoods, grasses and compression wood. Holzforschung 35:16–26.

Nordén, B., E. Bohlin, M. Nilsson, Å. Albano, and C. Röckner. 1992. Characterization of particle size fractions of peat. An integrated biological, chemical and spectroscopic approach. Soil Sci. 153:382–396.

Nordén, B., and B. Berg. 1990. A non-destructive method (solid-state ^{13}C NMR) for determining organic chemical components of decomposing litter. Soil Biol. Biochem. 22:271–275.

Oades, M.J., A.M. Vassallo, A.G. Waters, and M.A. Wilson. 1987. Characterization of organic matter in particle size and density fractions from a red-brown earth by solid-state ^{13}C N.M.R Aust. J. Soil Res. 25:71–82.

Oades, J.M., A.G. Waters, A.M. Vassallo, M.A. Wilson, and G.P. Jones. 1988. Influence of management on the composition of organic matter in a red-brown earth as shown by ^{13}C nuclear magnetic resonance. Aust. J. Soil Res. 26:289–299.

Paul, E.A., and H. VanVeen. 1978. The use of tracers to determine the dynamic nature of organic matter. p. 61–102. *In* Cong. Int. Soc. Soil Sci., 11th, Edmonton, Canada. 19 27 June. Vol. 3. Int. Soc. Soil Sci.

Pfeffer, P.E., and W.V. Gerasimowicz. (ed.). 1989. Nuclear magnetic resonance in agriculture. CRC Press, Boca Raton, FL.

Preston, C.M., D.E. Axelson, M. Lévesque, S.P. Mathur, H. Dinel, and R.L. Dudley. 1989. Carbon-13 NMR and chemical characterization of particle-size separates of peats differing in degree of decomposition. Org. Geochem. 13:393–403.

Preston, C.M., S.-E. Shipitalo, R.L. Dudley, C.A. Fyfe, S.P. Mathur, and M. Lévesque. 1987. Comparison of ^{13}C CPMAS NMR and chemical techniques for measuring the degree of decomposition in virgin and cultivated peat profiles. Can. J. Soil Sci. 67:187–198.

Preston, C.M., P. Sollins, and B.G. Sayer. 1990. Changes in organic components for fallen logs in old-growth Douglas-fir forests monitored by ^{13}C nuclear magnetic resonance spectroscopy. Can. J. For. Res. 20:1382–1391.

Reid, I.D., and K.A. Seifert. 1982. Effect of an atmosphere of oxygen on growth, respiration and lignin degradation by white rot fungi. Can. J. Bot. 60:252–260.

Sollins, P., S.P. Cline, T. Verhoeven, D. Sachs, and G. Spycher. 1987. Patterns of log decay in old-growth Douglas-fir forests. Can. J. For. Res. 17:1585–1595.

Spiker, E.C., and P.G. Hatcher. 1987. The effects of early diagenesis on the chemical and stable carbon isotopic composition of wood. Geochim. Cosmochim. Acta 51:1385–1391.

VanderHart, D.L. 1976. Characterization of the methylene carbon-13 chemical shift tensor in the normal alkane n-eicosane. J. Chem. Phys. 64:830–834.

VanderHart, D.L., and R.H. Attala. 1984. Studies of microstructure in native celluloses using solid-state ^{13}C NMR. Macromolecules 17:1465–1472.

Williams, B.L. 1983. The nitrogen content of particle size fractions separated from peat and its rate of mineralization during incubation. J. Soil Sci. 34:113–125.

Wilson, M.A. 1987. NMR Techniques and applications in geochemistry and soil chemistry. Pergamon Press, Oxford, England.

Wilson, M.A., S. Heng, K.M. Goh, R.J. Pugmire, D.M. Grant. 1983. Studies of litter and acid insoluble soil organic matter fractions using ^{13}C-cross polarization nuclear magnetic resonance spectroscopy with magic angle spinning. J. Soil Sci. 34:83–97.

Zech, W., R. Hempfling, L. Haumaier, H.-R. Schulten, and K. Haider. 1990. Humification in subalpine rendzinas: chemical analyses, IR and ^{13}C NMR spectroscopy and pyrolysis-field ionization mass spectrometry. Geoderma 47:123–138.

Zech, W., M.-B. Johansson, L. Haumaier, R.L. Malcolm. 1987. CPMAS ^{13}C NMR and IR spectra of spruce and pine litter and of the Klason lignin fraction at different stages of decomposition. Z. Pflanzenernähr. Bodenk. 150:262–265.

Zech, W., I. Kögel, A. Zucker, and H. Alt. 1985. CP-MAS-13C-NMR-Spektren organischer Lagen einer Tangelrendzina. Z. Pflanzenernähr. Bodenk. 148:481–488.

Zech, W., F. Ziegler, I. Kögel-Knabner, and L. Haumaier. 1992. Humic substances distribution and transformation in forest soils. The science of the total environment. 117/118:155–174.

Zelibor, J.L., L. Romankow, P.G. Hatcher, and R.R. Colwell. 1988. Comparative analysis of the chemical composition of mixed and pure cultures of green algae and their decomposed residues by ^{13}C nuclear magnetic resonance spectroscopy. Appl. Environ. Microbiol. 54:1051–1060.

Zielger, F., and W. Zech. 1990. Decomposition of beech litter cutin under laboratory conditions. Z. Pflanzenernähr. Bodenk. 153:373–374.

7 Management-Induced Changes in the Actively Cycling Fractions of Soil Organic Matter

B. H. Ellert

Agriculture and Agri-Food Canada Research Centre
Lethbridge, Canada

E. G. Gregorich

Centre for Land and Biological Resources Research
Ottawa, Canada

Soil organic matter consists of compounds or fractions that vary widely in susceptibility to decomposition (Fig. 7–1). The decomposability of organic matter fractions likely varies along a continuum, but current methods identify at least two fractions of organic matter that decompose at different rates. The fractions may be categorized as either "actively cycling," or "resistant to decomposition" during one decade.

Actively cycling fractions of soil organic matter receive inputs of fresh plant and microbial residues and are susceptible to rapid decomposition. These fractions release mobile compounds that may function as plant nutrients, water pollutants, precursors for more stable soil organic matter, or precursors for gases that are released to the atmosphere. Consequently, changes in the actively cycling fractions may foreshadow shifts in the quantity and overall composition of whole soil organic matter.

Humans have a profound influence on terrestrial ecosystems through management, which alters the structure and function of ecosystems to achieve specific goals. In a traditional sense, management refers to agronomic and silvicultural practices designed to optimize production in agriculture and forestry. In a broader sense, management also includes: interconversion of forest and agricultural land, range management, mitigation of pollution, and management of parks and nature preserves. Holling (1986) views management activities as external pressures that elicit changes in ecosystem structure and function, thereby exposing the inner workings of the ecosystem. Comparative studies of soils under contrasting management regimes help to develop sound management practices and to elucidate the inner workings of soils and terrestrial ecosystems. The distinction between

---

Copyright © 1995 Soil Science Society of America, 677 S. Segoe Rd., Madison, WI 53711, USA.
Carbon Forms and Functions in Forest Soils.

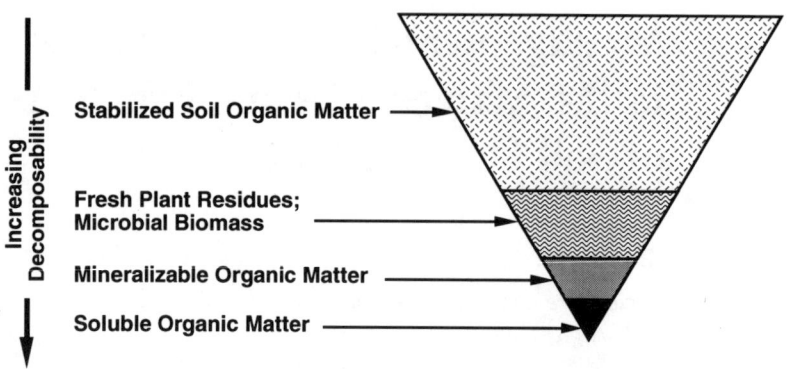

Fig. 7–1. Fractions of soil organic matter with different turnover times, and size of the actively cycling fractions in relation to total soil organic matter.

actively cycling and more stable fractions of soil organic matter may be useful to assess management impacts, because the active fractions are expected to be more sensitive to management.

The objective of this chapter is to explore the relationships between management practices and the actively cycling fractions of soil organic matter. An understanding of the actively cycling fractions is crucial to predict whether a given management practice will alter nutrient cycling or cause soil degradation. The most promising methods (excluding isotopic techniques) of measuring actively cycling organic matter are reviewed, because there is a lack of consensus on how to assess the active fraction.

The chemical and physical diversity of constituents in active soil organic matter ensures that no single technique can separate this fraction from soil organic matter with longer turnover times (Paul, 1989). Several studies have considered the influence of agronomic or silvicultural practices on soil organic matter, but few have described the changes that occur when forests are cleared and tilled to produce agricultural crops. This chapter reviews the results from separate studies on management-induced changes in the active organic matter of forest or agricultural soils, but we also present data from our comparative studies of organic matter in forest and agricultural soils. Results from separate studies illustrate the decomposability and amounts of actively cycling organic matter in forest or agricultural soils, but comparisons among studies are hampered by differences in methodology and the lack of statistical tests. A broad overview of actively cycling soil organic matter is presented to illustrate how management-induced changes in soil organic matter may be monitored.

BIOLOGICAL INDICES OF ACTIVELY CYCLING ORGANIC MATTER

Biological evaluation of the actively cycling fraction usually involves measuring decomposition rates or estimating the soil microbial and faunal biomass that performs decomposition. Incubations to measure decomposition rates may be conducted under controlled conditions in the laboratory or under ambient conditions in the field.

Laboratory Incubations

The actively cycling fraction of soil organic matter may be equated with CO_2 or inorganic N produced under controlled conditions in the laboratory. Laboratory incubations essentially are bioassays to determine the fraction of soil organic matter that is susceptible to decomposition. Rather than directly defining the decomposable organic constituents, laboratory incubations let the soil microorganisms and fauna define what is decomposable. Mineralizable C and N are measured as CO_2 or inorganic N released from soils incubated in the laboratory (Anderson, 1982; Campbell et al., 1993). Many authors use the first-order kinetic model to describe the time sequence of mineralization and obtain extrapolative estimates of mineralizable C and N present initially (Campbell et al., 1993). The conceptual basis of using kinetic models to estimate mineralizable C and N is appealing, but the cumulative amounts of C and N released during incubation often are more meaningful than coefficients estimated for kinetic models.

Several studies have indicated that mineralizable C and N in agricultural soils are sensitive to management practices such as crop rotation and tillage methods (Gregorich et al., 1994), but the cumulative amounts of both C and N rarely have been reported for forest soils. In four studies that included at least one forest soil, the rates of mineralization in mineral horizons ranged from 13 to 30 mg C kg^{-1} d^{-1} and from 0.03 to 5.0 mg N kg^{-1} d^{-1} (Table 7–1). Comparisons among studies, however, are hampered by differences in incubation conditions, especially the total duration, and by changes in mineralization rates during the incubation. Rates of C and N mineralization in the organic layers were more than 100 times those in the mineral soils, but the differences decreased when horizon thickness and bulk density were used to express mineralization on an area basis (Table 7–1). Perhaps the most meaningful and reliable method of assessing the actively cycling fraction of soil organic matter from mineralization data is to calculate the proportion of total soil C or N that is mineralizable. This proportion indicates the relative decomposability of the organic matter without being excessively influenced by the concentrations of total soil C or N.

Forest soils and native grasslands often have greater reserves of total soil C and N than agricultural soils under annual crops, but the influence of land use and agronomic or silvicultural practices on mineralizable C and N is less clear. Greater proportions of soil C and N were in the mineralizable fractions of forest litter layers than were in mineral soils (Table 7–1). The O and E horizons of the Cryoboralfs at the Saskatchewan site were mixed to form the Ap horizon in the cleared field. The proportions of mineralizable C and N in the cleared Ap were intermediate between the forest O and E horizons and similar to the cultivated Ap horizon (Table 7–1). The proportions of mineralizable C and N appeared to be insensitive to both the shift from forest to agricultural crops on the Hapludalfs in Ontario, and the shift from pine to heath vegetation on the Ranker soils (Haplumbrepts) in Spain (Table 7–1).

Laboratory incubations are useful to assess actively cycling soil organic matter, but are unreliable to estimate rates of mineralization occurring in the field.

Table 7-1. Potentially mineralizable C and N estimated by incubating soils in the laboratory.

Soil classification, location, incubation conditions, & reference	Management, horizon, & thickness	Rate†		Mineralization Areal rate‡		Active/totals§	
		C	N	C	N	C	N
		mg kg⁻¹ d⁻¹		g m⁻² d⁻¹		%	
Cryoboralfs (Star City, SK) 9 wk, 30°C (Ellert, 1990)	Forest O, 10 cm	330	14.83	6.6	0.30	12	11
	Forest E, 13 cm	na¶	0.03	na	0.01	na	1
	Cleared Ap, 15 cm	18	1.24	3.6	0.24	6	7
	Cultivated Ap, 15 cm	13	0.83	2.5	0.16	6	5
Hapludalfs (Plainfield, ON) 22 wk, 30°C# (Ellert, 1992, unpublished)	Forest Ah, 15 cm	28	1.40	5.4	0.27	6	5
	Pasture Ah, 15 cm	17	1.15	3.3	0.22	4	3
	Cultivated Ap, 15 cm	15	1.19	2.9	0.23	7	7
Cryorthods (Petawawa, ON) 3 wk, 30°C (Hendrickson & Robinson, 1984)	Forest O, 5 cm	868	14.36	8.7	0.14	38	na
	Forest Ah, 0–5 cm	30	0.71	2.0	0.05	17	na
	Forest E, 5–10 cm	19	0.41	1.2	0.03	12	na
Haplumbrepts (Galicia, Spain) 1.4 wk, 25°C (Díaz-Raviña et al., 1993)	Heath Ah, 15 cm	17	5.05	3.3	0.98	1	6
	Pine Ah, 15 cm	18	2.58	3.5	0.50	3	10

† Linear rate calculated as the cumulative amount of C or N mineralized divided by the duration of the incubation.
‡ Calculated under the assumptions that bulk density was 1.3 Mg m⁻³ for mineral soils and 0.20 Mg m⁻³ for organic layers.
§ Proportion of C or N in the mineralizable fraction = daily rate of C or N mineralization times 100 d divided by total soil organic C or N.
¶ Not available.
# Mineralizable C and N determined by the method in Campbell et al. (1993), rates were nearly linear throughout 22 wk of incubation.

Field rates are difficult to infer from laboratory assays because soil moisture, temperature, structure, bulk density and aeration are modified when soils are removed from the field. Laboratory incubations usually are conducted under nearly optimal conditions to decrease the time required to measure production of CO_2 or inorganic N. Laboratory incubations represent potential rates that may never be reached in the field, because environmental conditions in situ invariably are suboptimal. Suboptimal conditions in the field may even favor the build-up of actively cycling organic matter that persists until the soil is brought into the laboratory and incubated under nearly optimal conditions.

Simulation models are useful to explore the influence of suboptimal environmental conditions on laboratory estimates of C and N mineralization. The responses of C and N mineralization to temperature and moisture may be assessed in laboratory incubations, and rates of mineralization in situ can be predicted from field measurements of soil temperature and moisture. The transformation of mineralizable soil organic matter to CO_2 often is viewed as a first-order reaction in which the rate depends on the instantaneous amount of mineralizable C.

An elementary simulation model for a hypothetical soil, containing 3000 µg mineralizable C per gram of soil, was constructed to illustrate how temperature and moisture might influence organic matter dynamics (Fig. 7–2). Typical values for soil temperature and the percentage of water-filled soil pores in the temperate zone were used as driving variables (Fig. 7–3). Multiplicative reduction factors were used to adjust k, the first-order rate coefficient, downwards to account for suboptimal temperature and moisture. Soil moisture was represented by the percentage of water-filled pores, because this variable reflects both water content and soil aeration (Linn & Doran, 1984). Simulated rates of C mineralization varied with soil temperature and moisture, and often were considerably less than potential rates typically observed in laboratory incubations (Fig. 7–4, Table 7–1). Although the model oversimplifies soil biological processes and more data are required to define the parameters, it clarifies the importance of soil environmental conditions. The model suggests that in situ mineralization may be equally or more sensitive to management-induced changes in soil environmental conditions than to management-induced changes in the amounts of potentially mineralizable organic matter.

Field Incubations

Field incubations provide information on the dynamics of actively cycling soil organic matter under realistic environmental conditions. Litter decay studies yield data on the quantity and composition of inputs to the actively cycling fractions, whereas studies of in situ C and N mineralization estimate outputs from the actively cycling fraction. The amount of organic matter in the actively cycling fraction may be inferred from field estimates of CO_2 evolution or inorganic N production. Soil respiration may be calculated from micrometeorological estimates of CO_2 fluxes above soil surfaces, from the amounts of CO_2 evolved into chambers placed on the soil surface, or from diffusion coefficients and CO_2 gradients in soil profiles (Nakayama, 1990). In situ N mineralization usually is inferred from changes in the concentrations of inorganic N in field soils.

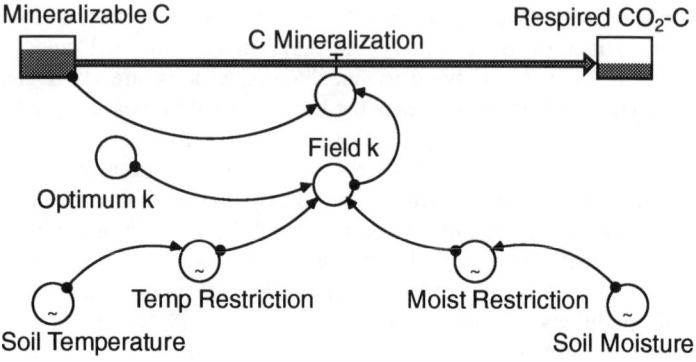

Initial Mineralizable Carbon = 3000 µg C g^{-1}

C Mineralization, µg g^{-1} d^{-1} = $\frac{-d[\text{Mineralizable C}]}{d\text{Time}}$ = [Mineralizable C] • Field k

Field k = Optimum k • Temperature Restriction • Moisture Restriction

Optimum k = first order rate coefficient = 0.01 d^{-1}

Reduction Factors for Temperature and Moisture Restrictions Defined by:

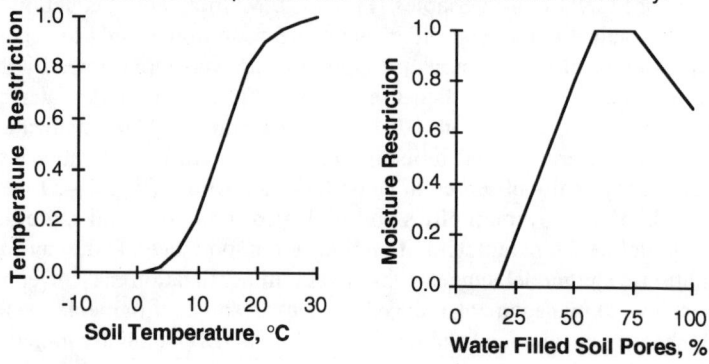

Fig. 7–2. System dynamics diagram of an elementary first-order model to simulate release of CO_2-C from the mineralizable fraction of soil organic matter.

Litter Decay Studies

Litter decay studies have been used extensively to assess the loss of mass and specific chemical constituents from plant residues that are enclosed in mesh bags or otherwise labeled for retrieval from the field after defined time intervals. Litter decay studies expose the changes that occur when plant material is converted to soil organic matter, and are useful to assess residue persistence and turnover in contrasting environments. The loss of litter mass often is described by a first-order decay curve and the first-order rate coefficient is used as a biological index of litter decay (Wieder & Lang, 1982). Litter decay data typically are plotted to show the proportion of original dry matter or element mass remaining after vari-

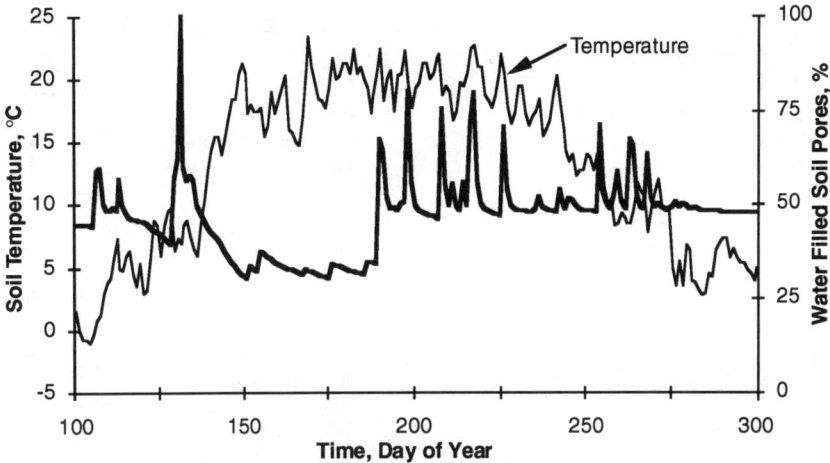

Fig. 7–3. Soil temperatures and percentages of water-filled pores used as driving variables to simulate C mineralization.

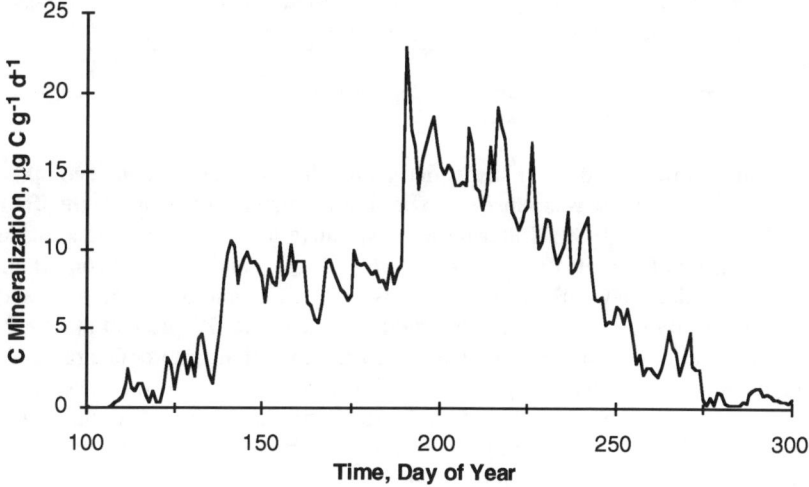

Fig. 7–4. Simulated rates of C mineralization for a hypothetical soil, containing 3000 μg mineralizable C per gram, in the temperate zone.

ous periods of incubation. More realistic descriptions of litter decay sometimes can be obtained by describing mass loss as a function of degree days or heat units. Soil degree days, calculated by summing mean daily soil temperatures that were above 0°C, successfully accounted for the influence of temperature on the decay of forest litter in Alaskan soils (Sparrow et al., 1992).

Litter decay studies have been used to assess the impact of management on the persistence of forest and crop residues (Table 7–2). In a semiarid environment, wheat (*Triticum aestivum* L.) straw that was buried at 15 cm decomposed

Table 7–2. Litter decomposition as influenced by management or residue placement in various ecosystems.

Residue‡, location, and reference	Treatment	k, yr⁻¹
Wheat straw (Pendleton, OR) (Douglas et al., 1980)	Buried residues	–0.73
	Surface-placed residues	–0.18
	Suspended above surface	–0.11
Loblolly pine needles (FL) (Polglase et al., 1992)	Reference	–0.40
	Weeds controlled	–0.34
	Fertilized	–0.45
	Fertilized & weeds controlled	–0.36
Maple leaves (Coweeta Basin, NC) (Blair & Crossley, 1988)	WS† 7, precut	–0.53
	WS 2, precut reference	–0.48
	WS 7, 1 yr postcut	–0.37
	WS 2, uncut reference	–0.62
	WS 7, 8 yr postcut	–0.55
	WS 2, uncut reference	–0.64
Ponderosa pine needles (CA) (Hart et al., 1992)	Old-growth conifers	–0.08
	Conifer plantation	–0.15

†WS = watershed.
‡Loblolly pine = *Pinus taeda* L., maple = *Acer* spp., Ponderosa pine = P. *Ponderrosa* Douglas ex P. Lawson & Lawson wheat = *Triticum aestivum.*.

faster than straw placed on or suspended above the soil surface, and decomposition of the buried straw was sensitive to both soil moisture and temperature (Table 7–2; Douglas et al., 1980). Litter decay is sensitive to changes in residue placement caused by implements used to prepare soils for planting. Weed control and fertilization also may influence litter decay rates, but microclimatic and landscape variability may hamper interpretation of field data (Polglase et al., 1992). Tree removal often makes microenvironment conditions more favorable for decomposition, but litter decay rates were lower in the clear-cut watershed compared to the uncut forests at Coweeta, North Carolina (Blair & Crossley, 1988; Table 7–2). Decay coefficients vary from <0.10 yr⁻¹ for conifers in harsh environments to >2.00 yr⁻¹ for tropical vegetation with residue pools that turn over several times each year (Singh & Gupta, 1977). Litter decay studies represent the dynamics of actively cycling organic matter, but inaccuracies may arise from the loss of fragmented material from mesh bags and the exclusion of fauna larger than mesh openings.

Soil Respiration

Soil respiration includes CO_2 from heterotrophic decomposition of the actively cycling fractions and CO_2 from autotrophic respiration (mainly plant roots and rhizomes). Partitioning soil respiration between roots and heterotrophic decay remains a difficult task, and estimates for the percentage derived from roots range from 20 to 60% depending on the plant community and growth stage (Singh & Gupta, 1977; de Jong & Paul, 1979). Perhaps the most important information gained from attempts to partition soil respiration is the dependence of heterotrophic decay on photosynthetic inputs to the soil. Plant growth and soil biological processes are inextricably linked in the rhizosphere (Williams, 1985; Perry et al., 1990; Dormaar, 1991).

Soil respiration has been studied for several decades, but the measurements still are method-dependent. Increasingly sophisticated instruments have been used to measure soil respiration, but most measurements have been based on the absorption of CO_2 by alkali traps in respiration chambers. Most estimates of soil respiration fall in the range of 0.6 to 8.0 g C m^{-2} d^{-1} (Singh & Gupta, 1977; Nakayama, 1990). Mean respiration rates and annual fluxes of CO_2 from the soil are difficult to estimate, because soil respiration is highly variable in time and space. Rochette et al. (1991) calculated that 30 to 190 measurements were required to estimate soil respiration (within 10% of the mean at $P = 0.05$) under 1 ha of wheat in Ontario. At the same site, a burst of respiration immediately after a rainfall suggested that short periods of intense soil respiration may account for significant portions of the annual flux of CO_2 from soil.

Management influences soil respiration by altering soil temperature and moisture, root mass and activity, and inputs of decomposable organic matter. Several investigators have explored the influence of agronomic or silvicultural practices on soil respiration, but the treatment effects usually were insignificant or minor (Table 7–3). In an Ontario study, soil respiration was 25% lower under barley (*Hordeum vulgare* L.) than under bare fallow, and the difference was attributed to lower soil temperature and moisture under barley (Rochette et al., 1992). Hendrix et al. (1988) reported that CO_2 fluxes were slightly greater in no-tillage plots than in conventionally tilled plots, despite the expectation that tillage would stimulate soil respiration. Respiration in forest soils seems insensitive to tree removal and residue management (Edwards & Ross-Todd, 1983), site fertility and species (Jurik et al., 1991), and landscape position (Hanson et al., 1993) (Table 7–3).

Estimates of annual fluxes of CO_2 from soils depend on the assumptions used to scale up daily or hourly measurements of respiration at specific points. The annual fluxes ranged from 800 to 11000 kg C ha^{-1} yr^{-1} for several studies of agricultural and forest soils (Table 7–3). The smaller fluxes usually were calculated from respiration during the growing season of agricultural crops, without accounting for CO_2 released during the period from late fall to early spring. Jurik et al. (1991) assumed that the respiration rate measured in the late fall and early spring (1.1 g C m^{-2} d^{-1}) represented the winter rate (Table 7–3). Recent evidence confirmed that soils emit significant amounts of CO_2 (0.1–0.7 g C m^{-2} d^{-1}) during the snow-covered period (Sommerfield et al., 1993).

In Situ Nitrogen Mineralization

Net N mineralization in field soils is assessed directly from the accumulation of inorganic N in situ, provided losses to plant uptake, leaching and denitrification are prevented. The original and simplest method to prevent loss of inorganic N consists of enclosing soil samples in polyethylene bags, burying the bags in the field, and analyzing the inorganic N accumulated after a specified incubation period (Eno, 1960). Modifications to the "buried bag" method include:

Table 7–3. Soil respiration under contrasting managment regimes, including typical measurements of short–term fluxes into field-placed chambers and estimates of annual CO_2 fluxes.

Soil, texture, location, method & reference	Management or landscape position	Respiration	Annul flux
		g C m^{-2} d^{-1}	kg C ha^{-1} yr^{-1}
Haplaquolls, clay loam (ON) Portable IRGA† (Rochette et al., 1992)	Barley‡ Bare fallow	3.89 5.15	2720 3590
Haplaquolls, clay loam (ON) Alkali traps (Kowalenko et al., 1978)	Bare fallow, control Bare fallow, fertilized	0.61 0.52	1000 880
Rhodudults, sandy loam (GA) Alkali traps (Hendrix et al., 1988)	Conventional tillage No tillage	1.36–13.6 1.36–13.6	10100 11700
Haplorthods, sand (Germany) Alkali traps (Beyer, 1991)	Spruce forest‡ Ecological farming Conventional farming	0.80 1.80 1.52	na§ na
Paleudults, loam (TN) Alkali traps (Edwards & Ross-Todd, 1983)	Oak clear-cut, no residues‡ Oak clear-cut, residues left Oak forest	0.2–1.5 0.2–2.0 0.2–2.3	4920 4830 5290
Haplorthods, sand (MI) Alkali traps (Jurik et al., 1991)	High fertility aspen‡ Low fertility aspen	2.18 2.10	4640–6140 4640–6140
Paleudults, loam (TN) Portable IRGA (Hanson et al., 1993)	Oak forest valleys‡ Oak forest NE slopes Oak forest SW slopes Oak forest ridges	0.83–5.91 0.83–5.91 0.83–5.91 0.83–5.91	7400 8200 8500 9300

†IRGA = infrared gas analyzer.
‡Barley = *Hordeum vulgare*, Spruce = *Picea* spp., Oak = *Quercus* spp, Aspen = *Populus tremuloides*
§na = not available.

incubation of intact cores to curtail the effects of sample disturbance (Nadelhoffer et al., 1984), isolation of soil cores in tubes with closed tops to avoid disturbance and facilitate sampling (Raison et al., 1987; Adams et al., 1989), and placement of resin bags on the tops and bottoms of soil cores to attain more realistic moisture conditions (Di Stefano & Gholz, 1986). Accumulation of inorganic N on ion exchange resins that are buried in the field also indicates N availability, but interpretation is difficult because mineralization often is confounded with the transfer of inorganic N from soil to resin.

The rates of in situ N mineralization summarized by Binkley and Hart (1989) typically ranged from 20 to 70 kg ha^{-1} yr^{-1}, but rates in old conifer forests sometimes approached zero, while those in the Tropics exceeded 125 kg ha^{-1} yr^{-1}. In a mixed conifer forest the rates of in situ N mineralization were 0.8, 8.0 and 13.7 mg N m^{-2} d^{-1} for the control, regenerating and clearcut forests, but mineralization rates were much greater when soils were incubated under ideal conditions in the laboratory (Frazer et al., 1990). In a series of old agricultural fields, the amount of N mineralized in situ increased from 44kg ha^{-1} in a soil that had been abandoned for 16 yr to 65 kg ha^{-1} yr^{-1} for a native oak (*Quercus* genus) savanna, but the proportion of total soil N in the mineralizable fraction showed the opposite trend (Pastor et al., 1987).

Soil Microbial Biomass

Soil microbial biomass is the living part of soil organic matter, excluding plant roots. Soil microorganisms are critical as a source and sink of biologically mediated nutrients (especially C, N, P and S), and as the driving force behind soil nutrient transformations (i.e., a living catalyst). Several methods have been used to estimate the pool of organic matter in soil microbes, including plate counts, direct microscopy, indicator compounds [e.g., adenosine triphosphate (ATP), ergosterol], substrate-induced respiration, and fumigation-incubation or fumigation-extraction (Jenkinson, 1988; Parkinson & Coleman, 1991). The widely adopted fumigation method fails to distinguish between active and resting microbes, or between fungi, bacteria and protozoa, and sometimes between microbial and root biomass. Nevertheless, soil microbial biomass estimated by the fumigation method is one of the simpler and more reliable indices of actively cycling soil organic matter.

Microbial biomass C accounts for 2 to 5% of total soil C, and the quantities generally increase with soil C in the order: arable < forest < grassland (Adams & Laughlin, 1981; Smith & Paul, 1990). Powlson et al. (1987) suggested that microbial biomass served as an early warning indicator of soil C losses or gains. Soils that are plowed routinely contain less microbial biomass than soils under no-tillage, reduced tillage, or perennial crops (Smith & Paul, 1990; McGill et al., 1986). Some authors contend that the proportions of soil C present as microbial biomass, rather than the absolute amounts of biomass C, are better indicators of management-induced changes (Carter, 1991; Sparling, 1992). Inverse correlations between proportions of soil C present as microbial biomass and total amounts of soil C have been cited as evidence that microbial biomass is limited by C availability in soils with relatively low total C contents (Wardle, 1992).

The influence of management on microbial biomass was assessed for surface soils at two sites in Ontario (Ellert, 1992, unpublished). Microbial biomass in field moist soils was estimated by the chloroform fumigation-direct extraction method ($k_c = 0.35$) according to the method outlined by Voroney et al. (1993). Both the absolute amounts of microbial biomass C and the proportions of soil C present as microbial biomass decreased in the order of forest > perennial grasses > maize fields (Fig. 7–5). The lower proportions of soil C in the microbial biomass of soils under intensive maize production suggested that soil C may be lost preferentially from the microbial biomass. Previous studies indicated that C and N in the microbial biomass are closely related to the amounts in the mineralizable fraction (Paul & Voroney, 1984; Myrold, 1987). Close relationships between mineralizable C and microbial biomass C may indicate that both variables are dictated by the same set of soil properties, but it is unlikely that mineralizable C is derived entirely from the microbial biomass.

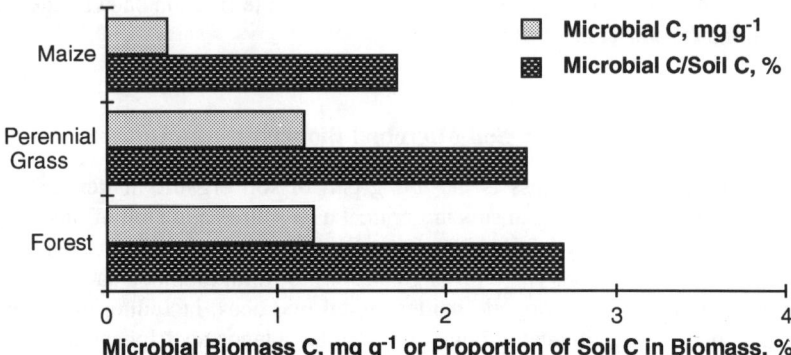

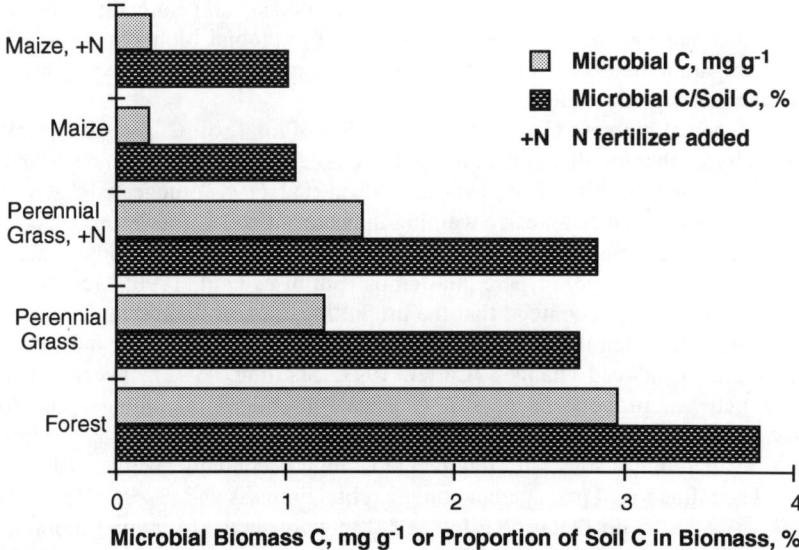

Fig. 7–5. Amount of soil microbial biomass C and proportion of soil C present as microbial biomass in surface soils under contrasting management at (a) Plainfield and (b) Woodslee, Ontario (Ellert, 1992, unpublished).

PHYSICAL FRACTIONATION TO ISOLATE ACTIVE ORGANIC MATTER

The physical separation of whole soils by size and density yields fractions that contain organic matter with different turnover times (Christensen, 1992). Most fractions contain some mineral particles, but the sizes or densities of the fractions can be adjusted to exclude mineral components. Sometimes the distribution of soil organic matter among aggregate size classes is analyzed to expose the influence of organo-mineral interactions on nutrient cycling (Oades, 1989). More intricate fractionation schemes consist of two or more separations according to size and density. Visual inspection usually is required to separate plant roots and rhizomes from detrital material in the organic fractions. Organic matter in the liquid phase can be isolated from the solid components of soil, but unaltered soil solution, especially that in small soil pores, is difficult to obtain. Consequently, additional water or dilute electrolytes usually are used to extract soil organic matter, even though extractable organic matter is not equivalent to organic matter in unaltered soil solutions (Ellert, 1990).

Size and Density Fractions

Physical fractionation of soils according to size and density provides information on the extent to which plant and animal residues have been processed by the decomposer community, and the degree of organo-mineral complexation. Soils are suspended in various liquids and disrupted by shaking or sonifying the suspensions to break the soil down into reproducible size or density fractions. The extent of disruption has a major influence on organic matter distribution among isolated fractions (Christensen, 1992).

Light Fraction

The light fraction (LF) is isolated from soils by flotation on dense liquids (Gregorich & Ellert, 1993). Light fraction largely consists of recognizable plant residues that have not been highly processed by decomposers in the soil, but also may contain spores, seeds, animal remains, charcoal and mineral particles attached to organic fragments. The densities of solutions used to isolate LF typically range from 1.5 to 2.0 g cm^{-3}. Since the densities of most soil minerals exceed 2.0 g cm^{-3}, flotation on dense solutions yields light fractions dominated by organic material with densities of <2.0 g cm^{-3}.

Cultivated soils generally contain 0.1 to 3% of the soil mass in the LF, but forests and grasslands with perennial roots often contain >3% of the soil mass in the LF. Management practices that alter the balance between soil C inputs and outputs also influence the proportions of soil mass and C in the LF (Table 7–4). Forests and grasslands contain greater amounts of LF than arable soils, but the LF in arable soils may be more decomposable (Cambardella & Elliott, 1992). In agricultural soils, LF is greater under perennial forages or continuous cropping than under frequent bare fallow (Janzen et al., 1992). Light fraction is enriched in C and N, to the extent that 0.5 to 14% of the soil mass accounts for 5 to 40% of the soil C (Table 7–4). Consequently, LF is the site of intense decomposer activity that dictates organic matter turnover. Kanazawa and Filip (1986) reported >30% of the microbial counts and >60% of the enzyme activities were associated with organic particles (density <1.0 g cm^{-3}) that accounted for only 2.2% of the soil mass.

Macro-organic Matter

Macro-organic matter (macroOM) consists of relatively large fragments of non-humified plant residues that are isolated from soils by sieving. The size of macroOM varies among investigators, but macroOM in mineral soils often is

Table 7–4. Proportion of soil mass and C in the LF of soils under contrasting management regimes.

Fraction, soil, location, & reference	Management	Soil mass	Soil C
		% in LF	
<1.8 g cm^{-3}†	Forest	1.38	8.8
Hapludalfs (Plainfield, ON)	Pasture	0.91	5.6
(Ellert, 1992, unpublished)	Cultivated	0.42	5.4
<1.6 g cm^{-3}	Conifer (Douglas fir)‡	6.4	36
Haplumbrept (Wind River, WA)	Alder & conifer‡	13.6	42
(Sollins et al., 1984)			
<2.0 g cm^{-3}	Continuous pasture	2.81	28.5
Rhodoxeralf (Australia)	Continuous wheat	1.27	20.6
(Greenland & Ford, 1964)	Wheat/fallow	0.78	25.3

†LF was determined by the method in Gregorich and Ellert (1993).
‡Douglas-fir = *Pseudotsuga menziesii* (Mirbel) Franco, Alder = *Alnus* spp.

equated with organic matter in the sand-sized fraction (0.05–2.00 mm) (Gregorich & Ellert, 1993). Sometimes coarse fragments (>2 mm) are included in the macroOM to obtain information on the breakdown of coarse plant residues (Allmaras et al., 1988). Methods used to isolate roots and weed seeds from soil are similar, and macroOM contains roots, seeds and detrital material. Under perennial grasses, roots may account for 70 to 80% of the macroOM, and the remainder consists of stem bases and shoot residues (Garwood et al., 1972). A comparison of organic matter in cropped and fallowed soils indicated that 2 yr of bare fallow preferentially decreased organic C, N and carbohydrates in the sand-sized fraction, but only had a small effect on the silt- or clay-sized fractions (Angers & Mehuys, 1990).

Both LF and macroOM contain fresh plant residues, but it is uncertain whether organic matter in the light and sand-sized fraction is equivalent. Comparisons between light and sand-sized fractions from 20 Ontario soils with contrasting texture, genesis and management suggested that the two fractions differ in several respects (Table 7–5). The mean percentage of soil mass in the sand-sized fractions was 25 times greater than in the LF. High ash contents and low C enrichments in the sand-sized fractions reflects the prevalence of mineral grains (Table 7–5). To examine the composition more closely, macroOM may be separated from the mineral grains in the sand-sized fraction by dry panning or flotation (Cambardella & Elliott, 1992). MacroOM and LF both contain noncomplexed organic matter that decomposes quickly, but the fractions have wide C/N ratios and tend to immobilize N (Greenland & Ford, 1964; Sollins et al., 1984; Christensen, 1992).

Root Assessments

Root investigations usually involve physical separation of roots from soil using sieving, washing and flotation. The measurement of root mass and density traditionally has been studied from the standpoint of plant nutrition, rather than the cycling of C, N, P and S through soil organic matter. Forest ecologists, how-

Table 7–5. Comparison between the mean compositions of sand-sized fractions and light fractions in 20 contrasting soils (Ellert, 1992, unpublished).†

Variable	Sand-sized		Light	
	Mean	95% CI‡	Mean	95% CI
Soil mass in fraction (%)	33.10	23–43	1.31	0.0–2.7
Soil C in fraction (%)	20.81	16–25	7.58	4.2–11
C enrichment§	0.73	0.58–0.87	11.42	8.0–15
Ash concentration (%)	93.63	91–96	42.41	37–46
C/N of fraction	19.28	17–21	26.44	24–29

†The sand-sized fraction was 0.05 to 2.00 mm, and the LF was <1.8 g cm^{-3}, both fractions were determined by the methods in Gregorich and Ellert (1993).
‡95% confidence intervals.
§Ratio of C concentrations in fraction/whole soil.

ever, have explored the involvement of fine roots (<1 to <5 mm, depending on the study) in the turnover of soil organic matter. The amount of C allocated to fine roots ranged from 25 to 820 -g m^{-2} yr^{-1} for 43 forests worldwide (Nadelhoffer & Raich, 1992). Fine roots likely are an important component of the actively cycling fraction of soil organic matter, because death and replacement take place simultaneously (Persson, 1983). The contribution of fine roots to organic matter cycling is poorly understood and data are scarce, because fine roots are difficult to study and even more difficult to distinguish from other fractions of soil organic matter with short turnover times.

Soluble Organic Matter

Nutrients that are immediately available to the soil microbial biomass and to plants are situated in the soil solution. The soil solution is viewed as a bottleneck through which organic matter in the solid phase must pass before it is converted to CO_2, CH_4, or a source of reducing power for denitrification. In contrast, organic matter in solid phases is expected to be less susceptible to decomposition. Most studies of solution focus on abiotic processes, such as sorption/desorption or precipitation/dissolution, and less emphasis has been placed on the cycling of organic matter through the liquid phase of the soil.

The influence of management on the amounts of soluble organic matter in soils likely reflects differences in soil climate, water fluxes, and the quantity, composition and placement of plant residues. Concentrations of C in solutions from Cryoboralf soils in Saskatchewan decreased in the order: aspen forest > recently cleared forest > wheat/fallow field (Fig. 7–6). Proportions of total soil C contained in the solution ranged from 0.04 to 0.39%. The proportion of soil C in solution was greatest in the forest E horizon, because organic solutes were translocated from the overlying litter layer and C was depleted from the solid phase during pedogenesis. Some authors reported that soil disturbance caused by tree harvesting increased solute concentrations in drainage water from clear-cut watersheds (Hart et al., 1981; Sollins & McCorison, 1981). In contrast, decreased exports of soluble C from disturbed forests were attributed to reduced litterfall and slower litter decay (Tate & Meyer, 1983) or changes in hydrologic pathways (Moore & Jackson, 1989).

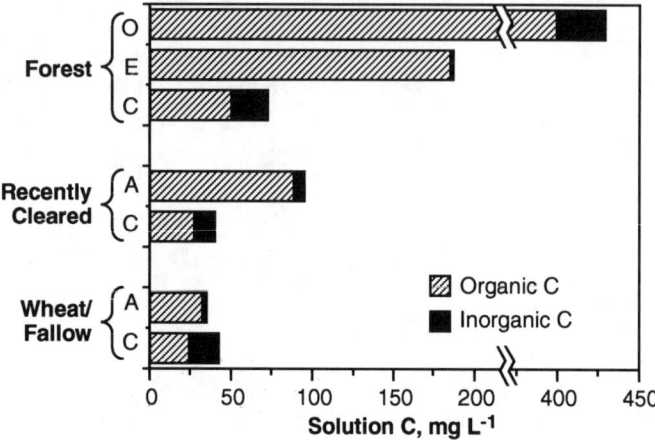

Fig. 7–6. Mean concentrations of organic and inorganic C in soil solutions isolated from forest O layers and the A, E and C horizons of Cryoboralf soils under contrasting management regimes (Ellert, 1990).

Conceptual models often envisage depolymerization and solubilization of organic matter as a prerequisite to mineralization, but data on the turnover of organic matter in soil solution are scarce. Data from laboratory studies on the relationship between soluble organic matter and C mineralization are inconclusive (Table 7–6). The CO_2 evolution from soils in western Tokyo was closely related to soluble C, but the turnover time of soluble C was <3 h (Seto & Yanagiya, 1983). In contrast, C mineralization in old field soils was related to water-extractable C only for early incubation times, thereafter C mineralization decreased while water-extractable C remained constant (Cook & Allan, 1992). If bioavailable C is taken up immediately by soil microorganisms (normally in a state of starvation), it may be expected that only the less decomposable organic matter will persist in solutions that have been isolated from soils. Zak et al.

Table 7–6. Relationships between dissolved organic C (DOC) and CO_2 evolution from soils incubated in the laboratory.

DOC method, & reference	Relationships
Soil solution DOC (Seto & Yanagiya, 1983)	Strong: CO_2 rate = 0.298 $DOC^{1.09}$, r^2 = 0.95
Water-extractactable DOC (Davidson et al., 1987)	Dry soil: CO_2 flux = 8 + 4.0 DOC, r^2 = 0.95 Wet soil: CO_2 flux = 135 + 11 DOC, r^2 = 0.70
Water-extractable DOC (Cook & Allen, 1992)	Weak: DOC constant as CO_2 decreased
Water-extractable DOC (Linn & Doran, 1984)	Weak: DOC and CO_2 not correlated
K_2SO_4-extractable DOC (Walters & Joergensen, 1991)	Weak: DOC and CO_2 not correlated

(1990) observed that water-extractable C and microbial biomass C were highly correlated in soils from 14 old agricultural fields that had been abandoned. Since the extractable C was about 20% of microbial biomass C, they suggested that extractable C must turn over rapidly if it was immediately decomposable.

SUMMARY AND CONCLUSIONS

Several fractions of soil organic matter that are more decomposable than whole soil organic matter have been identified. The mineralizable fraction perhaps is one of the most appealing measures of active organic matter, because soil microorganisms determine what is "actively cycling" in a bioassay. Elementary simulation models that use soil temperature and moisture as driving variables are useful to illustrate the dynamics of actively cycling fractions under field conditions. Field incubations to estimate residue decay, in situ N mineralization and soil respiration realistically indicate how organic matter transformations vary in time and space. Soil respiration reflects the intimate association of primary productivity and soil biological processes. Soil microbial biomass and undecomposed plant residues with recognizable cell structures have shorter turnover times and are more sensitive to management than whole soil organic matter. Microbial biomass serves as a source, sink and catalyst, whereas fresh residues in the LF or macroOM are sources of energy for the turnover of active organic matter. A portion of organic matter in soil solutions and extracts is immediately decomposable, but microbial uptake likely prevents the accumulation of metabolizable solutes.

The management-induced changes in the quantity of whole soil organic matter are difficult to ascertain, because soil variability hampers the detection of small changes in relatively large pools of organic matter. The potential advantage of measuring actively cycling organic matter to detect management impacts is based on the elimination of analytical noise from the resistant fractions. Since actively cycling organic matter usually is correlated with total soil organic matter, management impacts may be clarified by calculating the percentage of soil organic C or N in the actively cycling fraction.

The proportions of soil C and N in the actively cycling fractions often are influenced by management, but management impacts depend on original site conditions. When forest or grassland soils with thick Ah horizons are cleared and tilled to produce annual crops, the proportions of actively cycling C decline. Proportions may increase, however, when trees are cleared from sandy or acid soils, and lime or fertilizer is added to produce perennial forages. The impacts of agronomic practices on actively cycling organic matter are more clear than are the impacts of silvicultural practices. Proportions of total C in actively cycling fractions typically range from 1 to 20% for mineralizable C, 1 to 5% for microbial biomass C, and 3 to 50% for LF and macroOM.

Field measurements of litter decay, soil respiration and in situ N mineralization assess transformation rates rather than pool sizes. Rates of in situ transformations may be equally or more sensitive to management-induced changes in soil environmental conditions than to changes in the amounts of actively cycling organic matter. Estimates of annual CO_2 fluxes from soils and annual losses of plant litter frequently are compared with estimates of annual phytomass production. Since actively cycling organic matter represents the balance between C inputs and outputs, the relationship between actively cycling organic matter and annual phytomass production warrants more attention. Further research on the dynamics of actively cycling organic matter in soils under contrasting management regimes is required to clarify the factors that determine the balance between soil C inputs and outputs.

REFERENCES

Adams, T.McM., and R.J. Laughlin. 1981. The effects of agronomy on the carbon and nitrogen contained in the soil biomass. J. Agric. Sci. (Cambridge) 97:319–327.

Adams, M.A., P.J. Polglase, P.M. Attiwill, and C.J. Weston. 1989. *In situ* studies of nitrogen mineralization and uptake in forest soils; some comments on methodology. Soil Biol. Biochem. 21:423–429.

Allmaras, R.R., J.L. Pikul Jr., J.M. Kraft, and D.E. Wilkins. 1988. A method for measuring incorporated crop residue and associated soil properties. Soil Sci. Soc. Am. J. 52:1128–1133.

Anderson, J.P.E. 1982. Soil respiration. p. 831–871. *In* A.L. Page et al. (ed.) Methods of soil analysis. Part 2. 2nd ed. Agron. Monogr. 9. ASA and CSSA, Madison, WI.

Angers, D.A., and G.R. Mehuys. 1990. Barley and alfalfa cropping effects on carbohydrate contents of a clay soil and its size fractions. Soil Biol. Biochem. 22:285–288.

Beyer, L. 1991. Intersite characterization and variability of soil respiration in different arable and forest soils. Biol. Fertil. Soils 12:122–126.

Binkley, D., and S.C. Hart. 1989. The components of nitrogen availability assessments in forest soils. Adv. Soil Sci. 10:57–112.

Blair, J.M., and D.A. Crossley, Jr. 1988. Litter decomposition, nitrogen dynamics and litter microarthropods in a southern appalachian hardwood forest 8 years following clearcutting. J. Appl. Ecol. 25:683–698.

Cambardella, C.A., and E.T. Elliott. 1992. Particulate soil organic matter changes across a grassland cultivation sequence. Soil Sci. Soc. Am. J. 56:777–783.

Campbell, C.A., B.H. Ellert, and Y.W. Jame. 1993. Nitrogen mineralization potential in soils. p. 341–349. *In* M.R. Carter (ed.) Soil sampling and methods of analysis. Lewis Publ., Boca Raton, FL.

Carter, M.R. 1991. The influence of tillage on the proportion of organic carbon and nitrogen in the microbial biomass of medium-textured soils in a humid climate. Biol. Fertil. Soils. 11:135–139.

Christensen, B.T. 1992. Physical fractionation of soil organic matter in primary particle size and density fractions. Adv. Soil Sci. 20:1–90.

Cook, B.D., and D.L. Allan. 1992. Dissolved organic carbon in old field soils: Total amounts as a measure of available resources for soil mineralization. Soil Biol. Biochem. 24:585–594.

Davidson, E.A., L.F. Galloway, and M.K. Strand. 1987. Assessing available carbon: Comparison of techniques across selected forest soils. Commun. Soil Sci. Plant Anal. 18:45–64.

de Jong, E., and E.A. Paul. 1979. Aeration, respiration, and atmosphere. p. 10–21. *In* R.W. Fairbridge and C.W. Finkl (ed.) Encyclopaedia of soil science. Dowden, Hutchinson and Ross, Inc., Stroudsburg, PA.

Díaz-Raviña, M., M.J. Acea, and T. Carballas. 1993. Microbial biomass and C and N mineralization in forest soils. Bioresour. Technol. 43:161–167.

Di Stefano, J.F., and H.L. Gholz. 1986. A proposed use of ion exchange resins to measure nitrogen mineralization and nitrification in intact soil cores. Commun. Soil Sci. Plant Anal. 17:989–998.

Dormaar, J.F. 1991. Decomposition as a process in natural grasslands. p. 121–136. *In* R.T. Coupland (ed.) Ecosystems of the world. 8A. Natural grasslands—introduction and western hemisphere. Elsevier, New York.

Douglas, C.L. Jr., R.R. Allmaras, P.E. Rasmussen, R.E. Ramig, and N.C. Roager Jr. 1980. Wheat straw composition and placement effects on decomposition in dryland agriculture of the Pacific Northwest. Soil Sci. Soc. Am. J. 44:833–837.

Edwards, N.T., and B.M. Ross-Todd. 1983. Soil carbon dynamics in a mixed deciduous forest following clear-cutting with and without residue removal. Soil Sci. Soc. Am. J. 47:1014–1021.

Ellert, B.H. 1990. Kinetics of nitrogen and sulfur cycling in Gray Luvisol soils. Ph.D. diss. Univ. Saskatchewan, Saskatoon, Canada (Diss. Abstr. NN6821).

Eno, C.F. 1960. Nitrate production in the field by incubating the soil in polyethylene bags. Soil Sci. Soc. Am. Proc. 24:277–279.

Frazer, D.W., J.G. McColl, and R.F. Powers. 1990. Soil nitrogen mineralization in a clearcutting chronosequence in a northern California conifer forest. Soil Sci. Soc. Am. J. 54:1145–1152.

Garwood, E.A., C.R. Clement, and T.E. Williams. 1972. Leys and soil organic matter. III. The accumulation of macro-organic matter in the soil under different swards. J. Agric. Sci. (Cambridge) 78:333–341.

Greenland, D.J., and G.W. Ford. 1964. Separation of partially humified organic materials from soils by ultrasonic dispersion. p. 137–147. In Trans. Int. Cong. Soil Sci. 2, 8. Bucharest, Romania.

Gregorich, E.G., and B.H. Ellert. 1993. Light fraction and macroorganic matter in mineral soils. p. 397–407. In M.R. Carter (ed.) Soil sampling and methods of analysis. Lewis Publ., Boca Raton, FL.

Gregorich, E.G., M.R. Carter, D.A. Angers, C.M. Monreal, and B.H. Ellert. 1994. Towards a minimum data set to assess soil organic matter quality in agricultural soils. Can. J. Soil Sci. 74:367–385

Hanson, P.J., S.D. Wullschleger, S.A. Bohlman, and D.E. Todd. 1993. Seasonal and topographic patterns of forests floor CO_2 efflux from an upland oak forest. Tree Physiol. 13:1–15.

Hart, G.E., N.V. DeByle, and R.W. Hennes. 1981. Slash treatment after clearcutting lodgepole pine affects nutrients in soil water. J. For. 79:446–450.

Hart, S.C., M.K. Firestone, and E.A. Paul. 1992. Decomposition and nutrient dynamics of ponderosa pine needles in a Mediterranean type climate. Can. J. For. Res. 22:306–314.

Hendrickson, O.Q., and J.B. Robinson. 1984. Effects of roots and litter on mineralization processes in forest soil. Plant Soil 80:391–405.

Hendrix, P.F., H. Chun-Ru, and P.M. Groffman. 1988. Soil respiration in conventional and no-tillage agroecosystems under different winter cover crop rotations. Soil Tillage Res. 12:135–148.

Holling, C.S. 1986. The resilience of terrestrial ecosystems: Local surprise and global change. p. 292–317. In W.C. Clark and R.E. Munn (ed.) Sustainable development of the biosphere. Int. Inst. Appl. Systems Analysis, Laxenburg, Austria and Cambridge Univ. Press, Cambridge, England.

Janzen, H.H., C.A. Campbell, S.A. Brandt, G.P. Lafond, and L. Townley-Smith. 1992. Light-fraction organic matter in soils from long-term crop rotations. Soil Sci. Soc. Am. J. 56:1799–1806.

Jenkinson, D.S. 1988. Determination of microbial biomass carbon and nitrogen in soil. p. 368–386. In J.R. Wilson (ed.) Advances in nitrogen cycling in agricultural ecosystems. CAB Int., Wallingford, England.

Jurik, T.W., G.M. Briggs, and D.M. Gates. 1991. Soil respiration of five aspen stands in northern lower Michigan. Am. Midl. Nat. 126:68–75.

Kanazawa, S., and Filip, Z. 1986. Distribution of microorganisms, total biomass, and enzyme activities in different particles of brown soil. Microb. Ecol. 12:205–215.

Kowalenko, C.G., K.C. Ivarson, and D.R. Cameron. 1978. Effect of moisture content, temperature and nitrogen fertilization on carbon dioxide evolution from field soils. Soil Biol. Biochem. 10:417–423.

Linn, D.M., and J.W. Doran. 1984. Effect of water-filled pore space on carbon dioxide and nitrous oxide production in tilled and nontilled soils. Soil Sci. Soc. Am. J. 48:1267–1272.

McGill, W.B., K.R. Cannon, J.A. Robertson, and F.D. Cook. 1986. Dynamics of soil microbial biomass and water-soluble organic C in Breton L after 50 years of cropping to two rotations. Can. J. Soil Sci. 66:1–19.

Moore, T.R, and R.J. Jackson. 1989. Dynamics of dissolved organic carbon in forested and disturbed catchments, Westland, New Zealand. 2. Larry River. Water Resour. Res. 25:1331–1339.

Myrold, D.D. 1987. Relationship between microbial biomass nitrogen and a nitrogen availability index. Soil Sci. Soc. Am. J. 51:1047–1049.

Nadelhoffer, K.J., and J.W. Raich. 1992. Fine root production estimates and belowground carbon allocation in forest ecosystem comparisons. Ecology 73:1139–1147.

Nadelhoffer, K. J., J.D. Aber, and J. M. Melillo. 1984. Seasonal patterns of ammonium and nitrate uptake in nine temperate forest ecosystems. Plant Soil 80:321–335.

Nakayama, F.S. 1990. Soil respiration. Remote Sens. Rev. 5:311–321.

Oades, J.M. 1989. An introduction to organic matter in mineral soils. p. 89–159. In J.B. Dixon and S.B. Weed (ed.) Minerals in soil environments. 2nd. SSSA, Book Ser. 1. SSSA, Madison, WI.

Parkinson, D., and D.C. Coleman. 1991. Microbial communities, activity and biomass. Agric. Ecosyst. Environ. 34:3–33.

Pastor, J.A., M.A. Stillwell, and D. Tilman. 1987. Nitrogen mineralization and nitrification in four Minnesota old fields. Oecologia 71:481–485.

Paul, E.A. 1989. Soils as components and controllers of ecosystem processes. p. 353–374. In P.J. Grubb and J.B. Whittaker (ed.) Towards a more exact ecology. Blackwell Sci. Publ., Oxford, England.

Paul, E.A., and R.P. Voroney. 1984. Field interpretation of microbial biomass activity measurements. p. 509–514. *In* M.J. Klug and C.A. Reddy (ed.) Current perspectives in microbial ecology. Proc. Int. Symp. Microbial Ecol., Lansing, MI. 7–12 Aug. 1983. American Society for Microbiology, Washington, DC.

Perry, D.A., J.G. Borchers, S.L. Borchers, and M.P. Amaranthus. 1990. Species migrations and ecosystem stability during climate change: The belowground connection. Conserv. Biol. 4:266–274.

Persson, H. 1983. The distribution and productivity of fine roots in boreal forests. Plant Soil 71:87–101.

Polglase, P.J., E.J. Jokela, and N.B. Comerford. 1992. Nitrogen and phosphorus release from decomposing needles of southern pine plantations. Soil Sci. Soc. Am. J. 56:914–920.

Powlson, D.S., P.C. Brookes, and B.T. Christensen. 1987. Measurement of soil microbial biomass provides an early indication of changes in total soil organic matter due to straw incorporation. Soil Biol. Biochem. 19:159–164.

Raison, R.J., M.J. Connell, and P.K. Khanna. 1987. Methodology for studying fluxes of mineral-N *in situ*. Soil Biol. Biochem. 19:521–530.

Rochette, P., R.L. Desjardins, and E. Pattey. 1991. Spatial and temporal variability of soil respiration in agricultural fields. Can. J. Soil Sci. 71:189–196.

Rochette, P., R.L. Desjardins, E.G. Gregorich, E. Pattey, and R. Lessard. 1992. Soil respiration in barley (*Hordeum vulgare* L.) and fallow fields. Can. J. Soil Sci. 72:591–603.

Seto, M., and K. Yanagiya. 1983. Rate of CO_2 evolution from soil in relation to temperature and amount of dissolved organic carbon. Jpn. J. Ecol. 33:199–205.

Singh, J.S., and S.R. Gupta. 1977. Plant decomposition and soil respiration in terrestrial ecosystems. Bot. Rev. 43:449–528.

Smith, J.L., and E.A. Paul. 1990. The significance of soil microbial biomass estimates. p. 357–396. *In* Jean-Marc Bollag and G. Stotzky (ed.) Soil biochemistry. Vol. 6. Marcel Dekker, New York.

Sollins, P., and F.M. McCorison. 1981. Nitrogen and carbon solution chemistry of an old growth coniferous forest watershed before and after cutting. Water Resour. Res. 17:1409–1418.

Sollins, P., G. Spycher, and C.A. Glassman. 1984. Net nitrogen mineralization from light- and heavy-fraction forest soil organic matter. Soil Biol. Biochem. 16:31–37.

Sommerfield, R.A., R.A. Mosier, and R.C. Musselman. 1993. CO_2, CH_4 and N_2O flux through a Wyoming snowpack and implications for global budgets. Nature (London) 361:140–142.

Sparling, G.P. 1992. Ratio of microbial biomass carbon to soil organic carbon as a sensitive indicator of changes in soil organic matter. Aust. J. Soil Res. 30:195–207.

Sparrow, S.D., E.B. Sparrow, and V.L. Cochran. 1992. Decomposition in forest and fallow subarctic soil. Biol. Fertil. Soils. 14:253–259.

Tate, C.M., and J.L. Meyer. 1983. The influence of hydrologic conditions and successional state on dissolved organic carbon export from forested watersheds. Ecology 64:25–32.

Voroney, R.P., J.P. Winter, and R.P. Bayert. 1993. Soil microbial biomass C and N. p. 277–286. *In* M.R. Carter (ed.) Soil sampling and methods of analysis. Lewis Publ., Boca Raton, FL.

Wardle, D.A. 1992. A comparative assessment of factors which influence microbial biomass carbon and nitrogen levels in soil. Biol. Rev. 67:321–358.

Wieder, R.K., and G.E. Lang. 1982. A critique of the analytical methods used in examining decomposition data obtained from litter bags. Ecology 63:1636–1642.

Williams, S.T. 1985. Oligotrophy in soil: Fact or fiction? p. 81–110. *In* M. Fletcher and G.D. Floodgate (ed.) Bacteria in their natural environments. Soc. Gen. Microbiol. Spec. Publ. 16. Acad. Press, London.

Wolters, V., and R. G. Joergensen. 1991. Microbial carbon turnover in Beech forest soils at different stages of acidification. Soil Biol. Biochem. 23:897–902.

Zak, D.R., D.R. Grigal, S. Gleeson, and D. Tilman. 1990. Carbon and nitrogen cycling during old-field succession: Constraints on plant and microbial biomass. Biogeochemistry 11:111–129.

8 Long-Term Changes in Organic Matter in Soils Receiving Applications of Municipal Biosolids

Robert B. Harrison, Charles L. Henry, Dale W. Cole, and Dongsen Xue

University of Washington
Seattle, Washington

Municipal biosolids (sewage sludge) have been widely applied to forest land for several decades and have been shown in many circumstances to be an excellent soil amendment that enhances forest productivity (Gaynor & Halstead, 1976; Chaney & Giordano, 1977; Beckett et al., 1977; McCaslin & O'Connor, 1982). The benefits of municipal biosolids application include improving soil chemical and physical properties (Epstein et al., 1976; Zasoski, 1981). Though short-term benefits of municipal biosolids application have been observed in many studies (King et al., 1986), considerations of the long-term fate of organic matter and nutrient constituents and the potential for a permanent increase in nutrients and organic matter contents of soils have received much less research attention (Chang et al., 1984, 1986; Jokela et al., 1990; Zasoski, 1983).

There are several factors affecting the fate of biosolids-applied organic matter and nutrients, including:

1. The form of the applied nutrient.
2. Physical, chemical and biological retention or loss mechanisms in the soil, and the effects of external factors, such as temperature, soil aeration, drainage, etc.
3. Productivity response of the vegetation to the nutrient input and integration of nutrients into the nutrient cycle.
4. Resultant accumulations or losses of a nutrient due to enhancement or decrease of another input mechanism (e.g., N fixation).

Speculation on the long-term fate of organic matter and nutrients applied as municipal biosolids include the following (Fig. 8–1):

1. Increases in organic matter or nutrients above the level of application due to enhancement of the productive capacity and/or internal retention mechanisms.

Copyright © 1995 Soil Science Society of America, 677 S. Segoe Rd., Madison, WI 53711, USA.
Carbon Forms and Functions in Forest Soils.

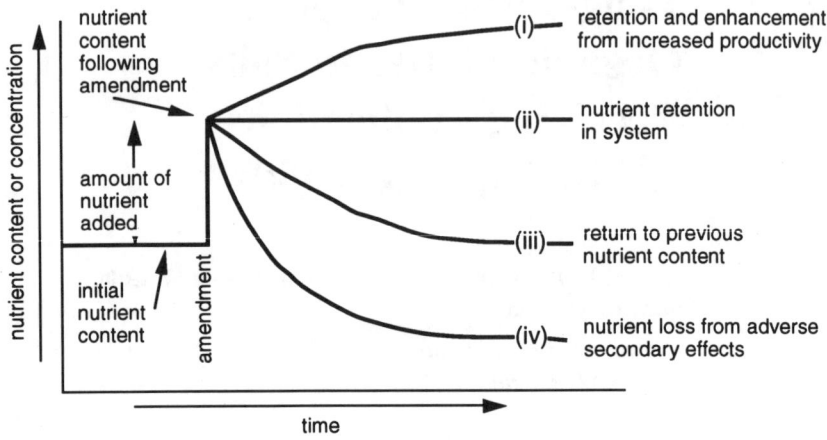

Fig. 8–1. Potential paths of change over time following nutrient amendment.

2. A permanent increase in organic matter or nutrient content at the amended level (Cole et al., 1982).
3. A gradual loss of added organic matter or nutrients to original nutrient level.
4. Loss of organic matter or nutrients beyond that added due to secondary effects of the original amendment, such as increased cation leaching due to nitrification (Van Miegroet & Cole, 1984).

Little is known about which path organic matter and nutrients added through municipal biosolids application would take. Many researchers have speculated that the application of municipal biosolids will result in a long-term retention of nutrients, since municipal biosolids contain all necessary nutrients in the approximate proportions necessary for plant growth, and the forest nutrient cycle will conserve nutrients through increased growth and cycling (Cole et al., 1982). However, this has not been demonstrated through research results.

Since one of the potential long-term advantages of municipal biosolids applications is an increase in site quality through permanently higher organic matter and nutrients (Henry, 1990), questions of long-term retention should be answered. In addition, the potential for decreasing inputs of greenhouse gases to the atmosphere by avoiding incineration (which produces CO_2) or landfilling (which produces methane) of municipal biosolids also is of considerable interest, as well as the potential for increased uptake and storage of C in soil.

The objectives of this study were:

1. To determine the differences in soil concentrations of C, N, P, Ca, Mg, K, pH and CEC in soils from two forest sites of contrasting mineralogy and management.
2. To evaluate the potential for movement of C and nutrients into the soil profile.

3. To estimate the total amounts of C, N and P remaining in the soil by a mass-balance approach.

METHODS AND MATERIALS

One application experiment was initiated in 1975 at the University of Washington's Pack Forest, located about 100 km south of Seattle, and another in 1981 at Pilchuck Tree Farm located about 100 km north of Seattle. Pack Forest has an average annual rainfall of 114 cm and a mean annual temperature of 9°C, while Mt. Pilchuck Tree Farm has an average annual rainfall of 95 cm and a mean annual temperature of 5°C. The soil on the Pack site is an extremely coarse-textured outwash soil (Everett series—Loamy-skeletal, mixed, mesic Dystric Xeropsamment; U.S.SCS, 1979) while the soil at Pilchuck Tree Farm is a sand textured outwash soil of the Indianaola soil series (mixed, mesic, Dystric Xeropsamments). Vegetation prior to the initiation of the study at Pack Forest was approximately 80-yr-old second growth Douglas-fir [*Pseudotsuga menziesii* (Mirbel) Franco] with an understory of salal (Gaultheria shallon Pursh), while at Pilchuck Tree Farm a second growth stand of approximately 60-yr-old Douglas-fir with an understory of salal was harvested. Both sites were clearcut harvested and cleared prior to the initiation of the studies.

At the Pack site approximately 500 Mg ha^{-1} and at the Pilchuck Tree Farm site approximately 300 Mg ha^{-1} of municipal biosolids from the Municipality of Metropolitan Seattle (Metro) were spread and disked into the surface soil layers to a depth of approximately 27 cm for the Pack Forest site and 15 cm at Pilchuck. Controls on actual application rates were poor, and it is estimated that the actual application rates range from 400 to 600 Mg ha^{-1} for the Pack and 200 to 350 Mg ha^{-1} for the Pilchuck Tree Farm sites. An analysis of Metro biosolids in 1981 showed that the C, N, P, Ca, Mg and K concentrations were 183.4, 26.2, 18.0, 26.4, 0.035 and 15.6 g kg^{-1}, respectively (Table 8–1). However, this analysis is only an approximation of the municipal biosolids applied, since no samples were taken from the applied biosolids and the actual concentrations would be expected to vary on such a large application.

Sites at Pack Forest were planted in separate, unreplicated blocks to either Lombardy poplar (*Populus nigra* var. italica Muenchh.), Douglas-fir or ponderosa pine (*Pinus ponderosa* Dougl.). Adjacent, unamended blocks also were planted to these three species. Sites at Pilchuck Tree Farm were planted to Douglas fir or grand fir [*Abies grandis* (Dougl.) Forbes] in separate, replicate blocks. At Pilchuck Tree Farm, a total of 63 soil samples (9 profiles with 7 depth increments for each profile) were collected from six biosolids-treated blocks and three untreated adjacent control blocks in November 1989. Soil samples were taken by depth since the subsurface showed little horizonation. No forest floor was present at the Pilchuck site. Soil was sampled uniformly from depths of 0 to 15, 15 to 30, 30 to 45, 45 to 60, 60 to 90, 90 to 120 and 120 to 140 cm. At the Pack Forest site, a total of 96 soil samples (12 profiles with 8 depth increments for each profile) were collected from three biosolids-treated and three adjacent untreated control (two profiles each) sites in November 1989. An average 3-cm thick forest floor

Table 8–1. Total elemental analysis of biosolids.

Element	Concentration
	mg kg^{-1}
C	183 400
N	26 200
P	18 000
Ca	26 400
K	15 600
Fe	27 100
Na	8 800
Mg	350
Zn	2 000
Cu	1 320
Cd	64
Ni	123
Pb	840

was present at Pack Forest. Soil was sampled uniformly from depths of 3 to 0, 0 to 7, 7 to 17, 17 to 27, 27 to 45, 45 to 75, 75 to 135 and 135 to 185 cm.

Because we thought that soil might vary in C, N or P concentrations with depth and because we wished to construct a per area estimate of content, soil depth increments were split according to Table 8–2 and 8–3. Soil samples were taken throughout the depth indicated in an effort to proportionally sample the full depth increment by collecting equal thickness of soil throughout the depth increment. Samples were air dried at 22°C and screened to <2mm before physical and chemical analysis. Fragments of organic matter >2mm were dried, ground and mixed back into the <2-mm fraction. Bulk density was estimated by excavation for the Pack Forest site and by a bulk density corer for the Pilchuck Tree Farm site.

Soil pH was measured using a 1:2 soil/water ratio; organic C content was determined by dry combustion (LECO model 176-100C Determinator St. Joseph, MI). Total N was analyzed utilizing a Li_2SO_4-H_2SO_4 modified Kjeldahl digest method (Parkinson & Allen, 1975). Total P, Ca, Mg and K were analyzed utilizing HNO_3-H_2O_2-HCl digestion (USEPA, 1986) with final determination by inductively coupled plasma emission spectroscopy (Thermo Jarrel Ash ICAP

Table 8–2. Summary sampling depths and physical analysis of Pack Forest soils.

Soil horizon	Depth	Bulk density		Fraction passing 2-mm sieve	
		Control	Treated	Control	Treated
	cm	Mg m^{-3}		%	
O	3– 0				
Ap	0– 7	1.06	0.81	0.40	0.62
Ap	7– 17	1.22	0.85	0.47	0.66
Ap	17– 27	1.53	1.52	0.31	0.30
Bs	27– 45	1.60	1.60	0.25	0.25
B	45– 75	1.70	1.70	0.24	0.23
Bc	75–135	1.70	1.70	0.24	0.24
C	135–185	1.70	1.70	0.24	0.24

Table 8–3. Summary sampling depths and physical analysis of Pilchuck soils.

Soil horizon	Depth	Bulk density		Fraction passing 2-mm sieve	
		Control	Treated	Control	Treated
	cm	Mg m^{-3}		%	
Ap	0– 15	1.25	1.24	0.98	0.96
Bs1	15– 30	1.42	1.39	0.98	0.99
Bs2	30– 45	1.51	1.53	0.96	0.93
Bs3	45– 60	1.53	1.45	0.96	1.00
BC	60– 90	1.55	1.61	1.00	0.97
C1	90–122	1.65	1.67	0.97	0.99
C2	122–152	1.68	1.74	0.95	0.98

61E, Thermo Jarrel Ash, Franklin, MA). Soil CEC was determined utilizing cation displacement with unbuffered 1.0 M NH$_4$Cl, ethanol leaching and displacement of NH$_4^+$ with 1.0 M KCl (Johnson et al., 1982).

Significance of differences in means were tested using a Tukey Honest Significant Difference test (SYSTAT, 1990).

RESULTS AND DISCUSSION

Effect of Amendment on Physical Properties

Except for the forest floor (Pack Forest only) and the Ap horizons, there were no distinct physical differences in soil horizonation between the biosolids-amended and unamended soils. Table 8–2 shows the average depths of the soil O (3–0 cm), Ap (0–27 cm), and subsurface soil for the Pack Forest soils. There was almost no difference in the physical properties of the soil with depth from 27 to 185 cm. The forest floors were different in the biosolids-amended compared to the unamended soil, with a great deal more highly humified material (Oa) and fewer fine roots. The forest floor ranged from 2 to 4 cm in depth, but its thickness was not related to treatment. The upper surface mineral soil also showed more fine roots in the unamended compared to the biosolids-amended soil in the Pack site.

Table 8–3 shows the average depths of the soil sampled for the Pilchuck Tree Farm site. There were very few differences in the physical properties of the soil with depth below 15 cm. No differences in proliferation of fine roots between the biosolids-amended and unamended soils were observed throughout the sampled soil profile.

Both soils had coarse soil textures; however, the Pack Forest soil had a much higher concentration of >2-mm rock fragments, particularly in the subsurface soil. For instance, the depths below 27 cm were 75% rock fragments >2 mm in the Pack Forest soil, while the Pilchuck Tree Farm had nearly all soil material pass through a 2-mm sieve. The Pack Forest soil was derived from andesitic glacial outwash from Mt. Rainier area, while the Pilchuck Tree Farm soil was derived from a granitic sandy outwash.

Effect of Amendment on Soil Properties

The original intent of the municipal biosolids application was to supplement the low organic matter and nutrient concentrations of the coarse-textured outwash soil, and this was clearly accomplished. Large increases in C, N and P concentrations were restricted to the top 27 cm of soil, which approximates the incorporation layer of the disk used in this study.

For the Pack Forest site, samples from treated sites had significantly higher mean total C concentration (139 vs. 67 g kg^{-1} in 0–7 cm soil), while the Pilchuck Tree Farm soils had almost identical total C contents (38 vs. 40 g kg^{-1} in 0–15 cm soil) in biosolids-amended vs. unamended soils (Fig. 8–2). There was no significant difference in C content at depths below 20 cm.

Biosolids-amended samples had a significantly higher total N concentration for both Pack Forest (12 vs. 3.4 g kg^{-1} in 0–7 cm soil) and for the Pilchuck Tree Farm (3.8 vs. 2.4 g kg^{-1} in 0–15 cm soil) in biosolids-amended vs. unamended soils, respectively (Fig. 8–3). There was no significant difference in N content at depths below 40 cm. Thus, the C/N ratio of the 0 to 7 cm soil at Pack Forest was reduced from 20:1 to 12:1 by application of municipal biosolids, a condition that should indicate relative sufficiency of N in the soil (Edmonds & Hsiang, 1987). The C/N ratio of the Pilchuck Tree Farm 0 to 15 cm soil was reduced from 10:1 to 7:1 with municipal biosolids application. It is not known why the initial C/N ratio at Pilchuck Tree Farm was much lower than at Pack Forest.

For both sites, biosolids-amended samples had significantly higher total P concentrations: 14 vs. 2.2 g kg^{-1} in 0 to 7 cm soil at Pack Forest and 3.0 vs. 1.2 g kg^{-1} in 0 to 15 cm soil at Pilchuck Tree Farm in biosolids-amended vs.

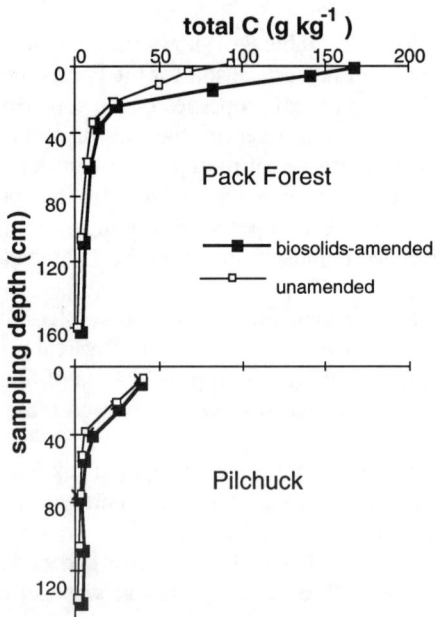

Fig. 8–2. Total C concentration vs. depth for Pack Forest and Pilchuck soils.

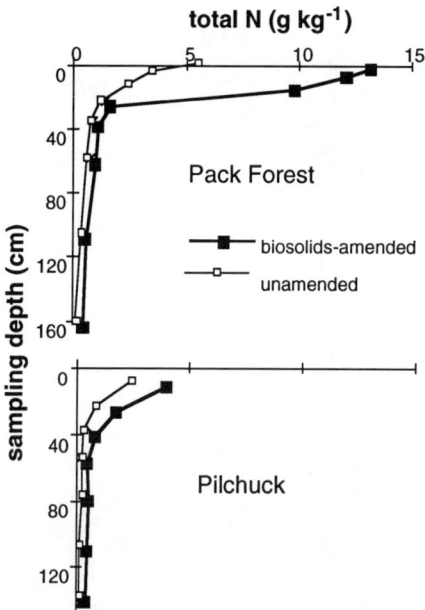

Fig. 8–3. Total N concentration vs. depth for Pack Forest and Pilchuck soils.

unamended soils (Fig. 8–4). There was no significant difference in P content at depths below 20 cm.

For the Pack Forest site, biosolids-amended samples had a significantly higher total Ca concentration (12.8 vs. 6.2 g kg^{-1} in 0–7 cm soil). The Pilchuck Tree Farm soils had no significant difference in total Ca concentrations (approximately 5.2 g kg^{-1} in 0–15 cm soil) in biosolids-amended vs. unamended soils (Fig. 8–5). The Pack Forest soil had significantly higher Ca concentrations in the deep soil profile at 160 cm in the unamended vs. amended soil, while the Pilchuck Tree farm soil showed little difference in the biosolids-treated soil throughout the sampled soil depth.

For the Pack Forest site, there was no significant difference in total K concentration (2.1 vs. 2.5 g kg^{-1} in 0–7 cm soil). The Pilchuck Tree Farm soils also had no significant difference in total K concentration (2.2 vs. 3.0 g kg^{-1} in 0–15 cm soil) in biosolids-amended vs. unamended soils (Fig. 8–6). The Pack Forest site had slightly but not significantly higher K concentrations in the deep soil profile below 40 cm in the unamended vs. amended soil, while the Pilchuck Tree farm soil had slightly but not significantly lower K concentrations in the biosolids-treated soil throughout the sampled soil profile, indicating the possibility of K leaching through the sampled soil profile in the amended sites.

For the Pack Forest site, biosolids-amended samples had a significantly higher total Mg concentration (1.9 vs. 1.5 g kg^{-1} in 0–7 cm soil). The Pilchuck Tree Farm showed no significant differences in total Mg contents (0.12 vs. 0.12 g kg^{-1} in 0–15 cm soil) in biosolids-amended vs. unamended soils (Fig. 8–7).

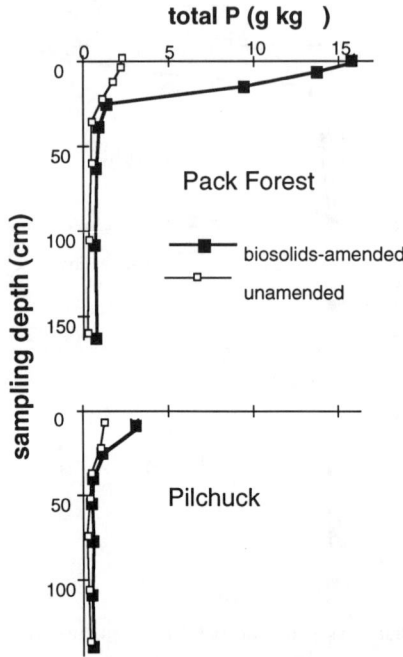

Fig. 8–4. Total P concentration vs. depth for Pack Forest and Pilchuck soils.

Pilchuck Tree Farm soils had lower total Mg throughout the soil profile than the Pack Forest Soils, reflecting the silicic nature of the parent material at Pilchuck vs. the andesite of Pack Forest. There is a Mg deficiency seen in some of the trees at Mt. Pilchuck Tree Farm (Harrison et al., 1994), and not at Pack Forest.

For the Pack Forest soil, the pH ranged from 0.4 to 1.0 pH units lower in biosolids-amended vs. unamended soil throughout the sampled soil profile. For the Pilchuck Tree Farm soil, the pH ranged from 0.2 to 0.9 pH units lower in biosolids-amended vs. unamended soil, with the largest differences in the surface soil (Fig. 8–8). There were significant differences in both the Pack Forest (5.1 vs. 6.1 in 0–7 cm soil for biosolids-amended vs. unamended soils) and Pilchuck Tree Farm (4.4 vs. 5.3 in 0–15 cm soil for biosolids-amended vs. unamended soils). There also were significant pH differences in Pack Forest soil throughout the sampled soil profile, but only above 40 cm at Pilchuck.

For the Pack Forest soil, soil CEC was significantly higher in the surface soil (30 vs. 18 $cmol_c kg^{-1}$ in 0–7 cm soil), but there were no significant differences below 50 cm. For the Pilchuck Tree Farm site, soil CEC was significantly lower in the surface soil of the biosolids-amended vs. unamended soil (5.9 vs. 10.0 $cmol_c kg^{-1}$ in 0–15 cm soil), but there were no significant differences below 60 cm (Fig. 8–9).

The results of base cation analysis indicate that cation leaching may be occurring in these soils. This is probably related to the nitrification of ammonium plus mineralizeable organic N added with the municipal biosolids to nitrate. The associated production of H^+ displaces cations from the soil CEC, and the high

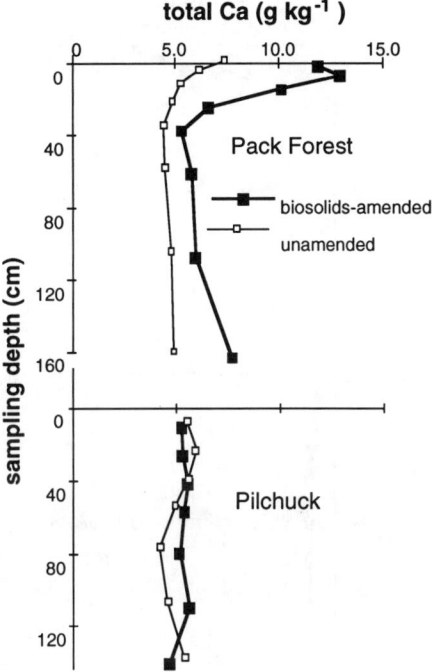

Fig. 8–5. Total Ca concentration vs. depth for Pack Forest and Pilchuck soils.

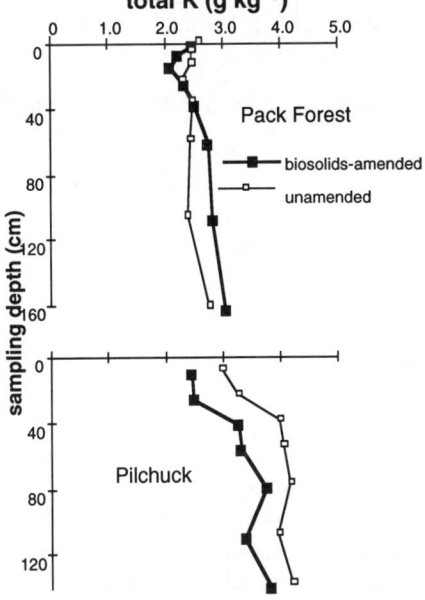

Fig. 8–6. Total K concentration vs. depth for Pack Forest and Pilchuck soils.

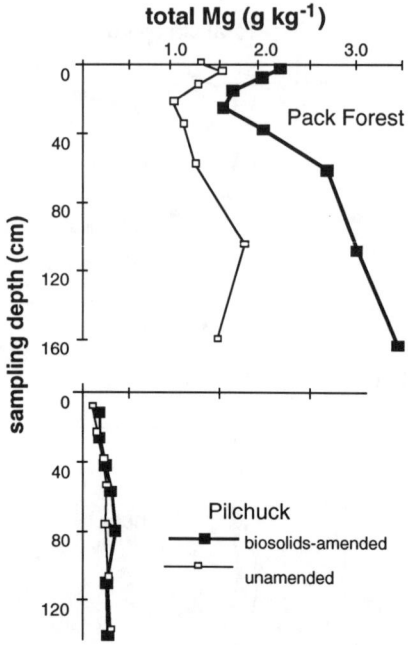

Fig. 8–7. Total Mg concentration vs. depth for Pack Forest and Pilchuck soils.

mobility of the excess NO_3^- generated by nitrification mobilizes Ca^{2+}, Mg^{2+}, or K^+ (Van Miegroet et al., 1990). However it is thought that the decrease in pH seen deep in soil profiles wouldn't be due to the penetration of H^+ into the soil profile, but the "salt effect" where a lower pH is measured in soils when a neutral salt is added and the ionic strength is increased (Reuss & Johnson, 1986).

Effect of Amendment on per Area Nutrient Contents

Estimates of total per hectare soil contents of C, N, and P, were made using the total analysis, bulk density and coarse fragment analysis data (Table 8–2 and 8–3). For the Pack Forest site, total C was significantly higher (140 vs. 78 Mg ha^{-1} for biosolids-amended vs. unamended soils) in the 0 to 27 cm soil of the biosolids-amended vs. unamended soil (Table 8–4), while at the Pilchuck Tree Farm site, total C was almost identical in the 0- to 30-cm profile (124 vs. 123 Mg ha^{-1} for biosolids-amended vs. unamended soils). Approximately 92 Mg C ha^{-1} was added at the Pack Forest and 55 Mg C ha^{-1} was added at the Pilchuck Tree Farm sites. Thus, approximately 67% of the original C added can still be seen as a higher C level at the Pack Forest site, while there is no sign of significantly higher C content due to the municipal biosolids application at Pilchuck Tree Farm. Though there was an additional 30 Mg C ha^{-1} in 27- to 185-cm depth of the Pack Forest soil, and an additional 48 to 53 Mg C ha^{-1} in the 30- to 135-cm depth in the Pilchuck soil, there was no significant difference between municipal biosolids amended and unamended soils or subsoils.

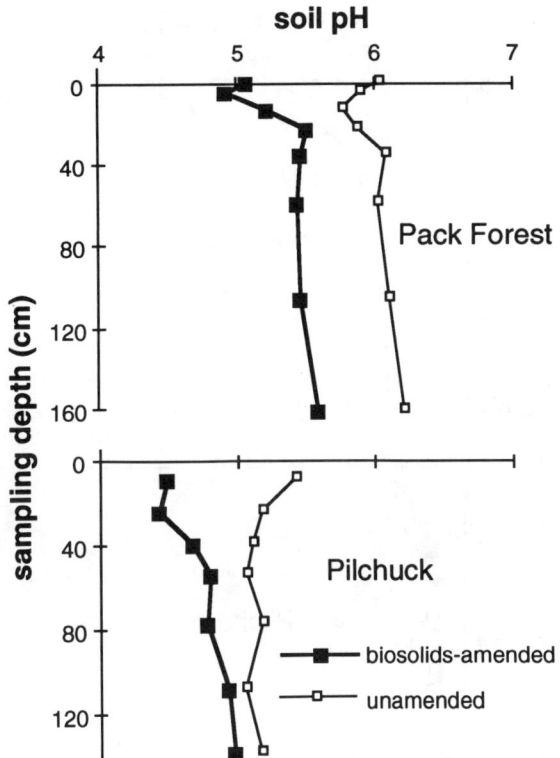

Fig. 8–8. Soil pH vs. depth for Pack Forest and Pilchuck soils.

At the Pack Forest site, total N was three times (significantly) higher (13.0 vs. 4.1 Mg ha^{-1} for biosolids-amended vs. unamended soils) in the 0 to 27 cm soil (Table 8–5). At the Pilchuck Tree Farm site, total N also was significantly greater (10.0 vs. 6.7 Mg ha^{-1} for biosolids-amended vs. unamended soils) in the 0- to 30-cm soil profile. Approximately 13.1 Mg N ha^{-1} was added at the Pack Forest and 7.9 Mg N ha^{-1} was added at the Pilchuck Tree Farm sites. Thus, approximately 68% of the original N added can still be seen as a higher N level at the Pack Forest site, while approximately 43% of the original N added can still be seen as a higher N level due to the municipal biosolids application at Pilchuck Tree Farm. In addition, a nearly 20% increase in recovered N was seen when the 30- to 135-cm layer was evaluated at Pilchuck Tree Farm, increasing the N recovery to 67%.

For the Pack Forest site, total P was 5.3 times higher (12.5 vs. 2.4 Mg ha^{-1} for biosolids-amended vs. unamended soils) in the 0 to 27 cm soil, while at the Pilchuck Tree Farm site, total P was 1.7 times greater (7.6 vs. 4.5 Mg ha^{-1} for biosolids-amended vs. unamended soils) in the 0- to 30-cm soil profile (Table 8–6). Approximately 9.0 Mg P ha^{-1} was added at the Pack Forest and 5.4 Mg P ha^{-1} was added at the Pilchuck Tree Farm sites. Thus, approximately 111% of the original P estimated to have been added can still be seen as a higher P level at the

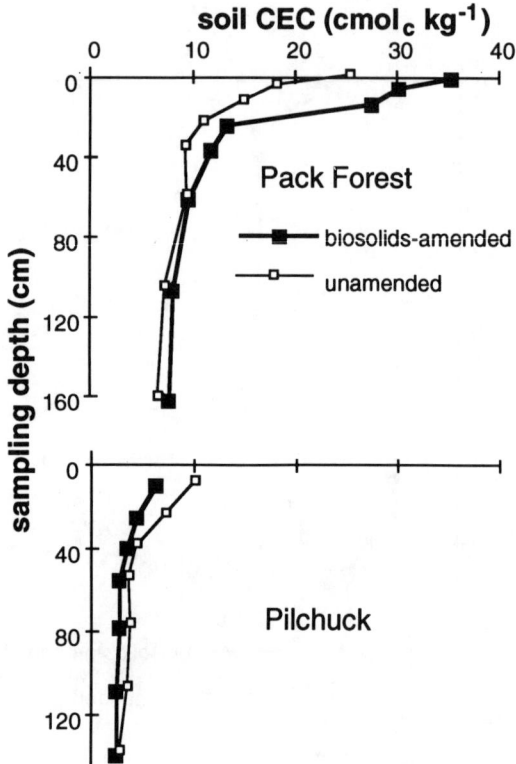

Fig. 8–9. Soil CEC vs. depth for Pack Forest and Pilchuck soils.

Table 8–4. Estimates of whole-profile nutrient pools for C.

Site	Depth	Control soils	Treated soils	Difference	Applied	Recovery
			Total C			
	cm		——— Mg ha^{-1} ———			%
Pack Forest	0– 27	78	140	61	92	67
	0–185	105	171	67	92	73
Pilchuck Tree Farm	0– 30	123	124	1	55	2
	0–135	171	177	6	55	11

Pack Forest site, while approximately 57% of the original P added can still be seen as a higher P level due to the municipal biosolids application at Pilchuck Tree Farm. Some increase in P recovery can be seen when 27 to 185 cm soil at Pack and 30 to 135 cm soil at Pilchuck Tree Farm are added, indicating small amounts of P may have been moved below 30 cm.

Table 8–5. Estimates of whole-profile nutrient pools for N.

Site	Depth	Total N				
		Control soils	Treated soils	Difference	Applied	Recovery
	cm	Mg ha^{-1}				— % —
Pack Forest	0– 27	4.1	13.0	8.9	13.1	68
	0–185	6.6	15.9	9.3	13.1	71
Pilchuck Tree Farm	0– 30	6.7	10.0	3.3	7.9	43
	0–135	10.0	15.3	5.3	7.9	67

Table 8–6. Estimates of whole-profile nutrient pools for P.

Site	Depth	Total P				
		Control soils	Treated soils	Difference	Applied	Recovery
	cm	Mg ha^{-1}				— % —
Pack Forest	0– 27	2.4	12.5	10.0	9.0	111
	0–185	4.6	15.9	11.4	9.0	126
Pilchuck Tree Farm	0– 30	4.5	7.6	3.1	5.4	57
	0–135	11.4	14.8	3.4	5.4	63

Implications for Long-Term Retention

When the pools of organic matter and nutrients are evaluated long after municipal biosolids application at Pack Forest and Pilchuck Tree Farm, it can be seen that most of the results fall between a retention of nutrients and a return to the previous nutrient contents, with variability between sites and obvious differences in nutrients. The added P, for instance, would be expected to be more strongly retained than N (Harrison & Adams, 1987), and this was observed.

It is not apparent why almost 70% of the estimated C added to the Pack Forest soil is still present in the soil profile where there is no legacy of municipal biosolids C at Pilchuck. Though the C concentration of the <2-mm fraction at Pack forest is higher than at Pilchuck, the C content per volume of unamended whole soil (rocks included) at Pilchuck is greater than at Pack Forest. Thus, there is the possibility that additional C added at Pilchuck cannot be retained in the long term, while the soil at Pack Forest is capable of additional retention under the present conditions. The lower retention of N (which is dominantly biologically retained), at the Pilchuck site, would seem to support this. In addition, the fact that there was no weed control of the Pack Forest site, while there was a significant weed-control effort at Pilchuck, also may have had some effect.

Perhaps the most significant difference between the two sites is the extremely low levels of Mg at Pilchuck Tree Farm. Magnesium deficiency resulted in lower rates of plant growth at the Pilchuck site (Harrison et al., 1994). Since municipal biosolids are a very poor source of plant-available Mg (Table 8–1), the resulting lack of enhanced plant growth may have resulted in the inability of the site to retain the added organic matter and nutrients. The higher retention of N relative to C could be accounted for by the microbial cycling of N during the breakdown of organic matter (and resultant loss of C as CO_2) on the site. Further study of a variety of sites crossing a range of soil types and management regimes would be required before the specific mechanisms most responsible for C retention or loss following the application of large amounts of organic matter from municipal biosolids can be evaluated for any given site.

REFERENCES

Beckett, P.H.T., R.D. Davis, A.F. Milward, and P. Brindley. 1977. A comparison of the effect of different sewage biosolids on young barley. Plant Soil 48:129–141.

Chaney, R.L., and P.M. Giordano. 1977. Microelements are related to plant deficiencies and toxicities. p. 234–279. *In* L.F. Elliot and F.J. Stevenson (ed.) Soils for management of organic wastes and waters. SSSA, ASA, CSSA, Madison, WI.

Chang, A.C., A.L. Page, J.E. Warneke, and E. Grgurevic. 1984. Sequential extraction of soil heavy metals following a biosolids application. J. Environ. Qual. 13:33–38.

Chang, A.C., T.J. Logan, and A.L. Page. 1986. Trace element considerations of forest land applications of municipal biosolids. p. 85–99. *In* D.W. Cole et al. (ed.) The forest alternative of treatment and application of municipal and industrial wastes. Univ. Washington Press, Seattle, WA.

Cole, D.W., C.L. Henry, P. Schiess, and R.J. Zasoski. 1982. The role of forests in sludge and wastewater utilization programs. p. 125–146. *In* A.L. Page et al. (ed.) Utilization of municipal wastewater and sludge on land. Univ. California, Riverside, CA.

Edmonds, R.L., and T. Hsiang. 1987. Influence of forest floor and soil properties on response of Douglas-fir to urea. Soil Sci. Soc. Am. J. 57:1332–1337.

Epstein, E., J.M. Taylor, and R.L. Chaney. 1976. Effects of sewage biosolids and biosolids compost applied to soil on some soil physical and chemical properties. J. Environ. Qual. 5:422–426.

Gaynor, J.D., and R.L. Halstead. 1976. Chemical and plant extractability of metals and plant growth on soils amended with biosolids. Can. J. Soil Sci. 56:1–8.

Harrison, R.B., and F. Adams. 1987. Solubility characteristics of residual phosphate in a fertilized and limed Ultisol. Soil Sci. Soc. Am. J. 51:963–969.

Harrison, R.B., C.L. Henry, and D. Xue. 1994. Magnesium deficiency in Douglas-fir and Grand Fir growing on a sandy outwash soil amended with sewage sludge. Water Air Soil Pollut. 75:37–50.

Henry, C.L. 1990. Nitrogen dynamics of pulp and paper sludge amendment to forest soils. Ph.D. diss. Univ. Washington, Seattle (Diss. Abstr. 90-36762).

Johnson, D.W., D.W. Cole, F.W. Horng, H. Van Miergroet, and D.E. Todd. 1982. Chemical characteristics of two forested Ultisols and two forested Inceptisols relevant to anion production and mobility. ORNL/TM-7646. Oak Ridge Natl. Lab., Oak Ridge, TN.

Jokela, E.J., W.H. Smith, and S.R. Colbert. 1990. Growth and elemental content of slash pine 16 years after treatment with garbage composted with sewage sludge. J. Environ. Qual. 19:146–150.

King, R.C., T.M. Hinckley, and C.C. Grier. 1986. Growth response of forest trees to wastewater and sludge application. p. 209–220. *In* D.W. Cole (ed.) The forest alternative of treatment and application of municipal and industrial wastes. Univ. Washington Press, Seattle, WA.

McCaslin, B.D., and G.A. O'Connor. 1982. Potential fertilizer value of gamma irradiated sewage biosolids on calcareous soils. New Mexico Agric. Exp. Stn. Bull. 692.

Parkinson, J.A., and S.E. Allen. 1975. A wet oxidation procedure suitable for the determination of nitrogen and mineral nutrients in biological material. Commun. Soil Sci. Plant Anal. 6:1–11.

Reuss, J.O., and D.W. Johnson. 1986. Acid deposition and the acidification of soils and waters. Springer-Verlag, New York.

SYSTAT. 1990. Version 5.1. SYSTAT, Evanston, IL.

U.S. Soil Conservation Society. 1979. Soil survey of Pierce County Area, Washington. USDA-SCS, Reston, WA.

U.S. Environmental Protection Agency. 1986. Test methods for evaluating solid waste. USEPA Rep. SW-846. USEPA. Washington, DC.

Van Miegroet, H., and D.W. Cole. 1984. The impact of nitrification on soil acidification and cation leaching in a red alder ecosystem (*Alnus rubra, Pseudotsuga menziesii*, Washington). J. Environ. Qual. 13:586–590.

Van Miegroet, H., D.W. Cole, and P.S. Homann. 1990. The effect of alder forest cover and alder forest conversion on site fertility and productivity. p. 333–354. *In* S.P. Gessel et al. (ed) Sustained productivity of forest soils. Univ. British Columbia, Vancouver, BC.

Zasoski, R.J. 1981. Effects of biosolids on soil chemical properties. p. 45–48. *In* C.S. Bledsoe (ed.) Municipal biosolids application to Pacific Northwest forest lands. Inst. For. Resour., Univ. Washington, Seattle.

Zasoski, R.J. 1983. Fate of heavy metals contained in municipal sludge following forest application. p. 67–75. *In* C.L. Henry and D.W. Cole (ed.) Use of dewatered sludge as an amendment for forest growth. Vol. 4. Inst. For. Resour., Univ. Washington. Seattle.

9 Soil Carbon, Soil Formation, and Ecosystem Development

Keith Van Cleve

University of Alaska
Fairbanks, Alaska

Robert F. Powers

USDA-FS
Pacific Southwest Research Station
Reading, California

Soil, climate, topography, biota, and time are the principal state factors controlling the structure, health, productivity and evolution of forest ecosystems. The importance of soil in ecosystem structure and function depends largely on its organic carbon (C) compounds. Organic C is one of the key mediators of soil development. It reflects the biotic factor of Jenny (1941, 1980) in its influence on soil formation. Interactions of soil organic C with other physical, chemical, and biological environmental controls result in the unique features of soil in equilibrium with a given constellation of state factors. Soil organic matter is an energy source for soil organisms which, through their activity and interactions with mineral matter, impart the structure to soil that affects its stability and its capacity to provide water, air, and nutrients to plant roots. Organic C plays a controlling role in nutrient cycling. Consequently, the amount and kind of soil organic C both reflects and controls soil development and, ultimately, ecosystem productivity.

Physical, chemical and biological interactions on parent material, mediated by the environmental constraints imposed by state factors, are termed pedogenesis. As described by Simonson (1959), pedogenic processes involve organic and inorganic material additions to, and losses from, the soil; translocation of materials from one point to another; and transformations of organic and inorganic matter within the soil profile. Through physical and biological transformations, organic C compounds mediate soil formation processes. Chelation, eluviation, and illuviation lead to horizons of distinct chemical and physical properties and biotic activity. The result is a surface organic layer (O), a leached (eluvial) horizon (E), an accumulation horizon (illuvial or B) and a zone of material showing little or no evidence of pedogenic processes (C) that develop in concert with environmental control.

Copyright © 1995 Soil Science Society of America, 677 S. Segoe Rd., Madison, WI 53711, USA.
Carbon Forms and Functions in Forest Soils.

Here, we consider the significance of organic soil C to water and nutrient supply, soil development, and ecosystem function and productivity. We illustrate this through both chronosequence studies and designed experiments using the state factor approach. Emphasis is placed on forms and pool sizes of organic C at the ecosystem level across a range of climatic and management regimes. Critical areas are discussed and proposed for further research.

OCCURRENCE AND NATURE OF SOIL ORGANIC CARBON

Source and Distribution

Plant net primary production (NPP) is the principal source of soil organic C. Soil C content reflects the balance between inputs from NPP and losses during decomposition and erosion. Optimum conditions result in maximum rates of production, decomposition, synthesis, and formation of soil organic matter. In cold environments, such as the northern boreal forest of interior Alaska, accumulation may be large even though NPP is low because decomposition is inhibited through cold temperatures and/or waterlogged soil.

Important climatic controls of NPP include mean annual temperature and precipitation (Leith, 1975). The NPP of forest ecosystems also depends on leaf biomass (Webb et al., 1983) and stand leaf area index (Gholz, 1982). Root production in some coniferous forests may contribute up to 67% of NPP compared with aboveground production. Root turnover may be as important as aboveground litterfall as a source of soil organic C (Vogt et al., 1982). Proportional allocation of photosynthate to root growth is related inversely to soil fertility (Waring and Schlesinger, 1985). The importance of root growth and distribution to soil C flux is one of the least studied and understood controls of forest soil development.

The ecosystem total and compartmental distributions of organic matter and C show substantial differences among biomes. Between three and four times more total plant biomass (and C) is encountered in tropical forests and temperate rain forests than in the boreal forest (Table 9–1, Fig. 9–1). Net annual aboveground production of biomass (and fixation of C) is between seven and eight times greater in these biomes than in the boreal forest. Greatest unit area accumulations of C in mineral soil, between 149 and 170 Mg ha^{-1}, are found in boreal forest and tundra ecosystems (Table 9–1, Fig. 9–1). While the greatest ecosystem total unit area accumulation of C occurs in the northern boreal forest or taiga (347 Mg ha^{-1}), the greatest biome total (863 Mg ha^{-1}) occurs in the tropical biome because of its greater areal extent (Powers & Van Cleve, 1991).

Worldwide totals for soil organic C span nearly a 20-fold range from Vertisols (19 Pg), Inceptisols (352 Pg), and Histosols (357 Pg, Table 9–2; Eswaran et al., 1993). Entisols, Alfisols, Oxisols, Aridisols, and Ultisols fall into a group of moderately high C storage (148–105 Pg). Andisols, Mollisols, and Spodisols range from 78 to 71 Pg of C stored.

Large accumulation of tree biomass and high annual production reflect favorable environmental controls of primary production in temperate and tropical ecosystems compared with boreal ecosystems. Greater accumulation of

Table 9–1. Distribution of organic matter and C in examples of major terrestrial ecosystems.†

Ecosystem type	Biomass‡	Annual net§ production	Mineral soil C
		Mg ha^{-1}	
Boreal forest	120	2	149
Temperate deciduous forest	200	11	139
Temperate coniferous forest	200	10	110
Temperate rain forest	500	15	78
Tropical¶ forest	325	17	104
Temperate grassland	3	3	192

†From Aber and Melillo, 1991; Schelsinger, 1977.
‡Total plant biomass.
§Aboveground production.
¶Average for tropical seasonal and tropical rain forests.

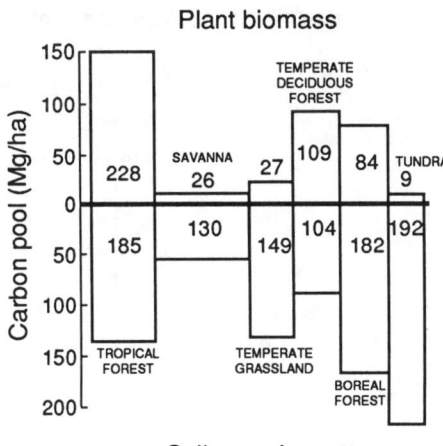

Fig. 9–1. Carbon pools in plant biomass and soil organic matter (including litter) in major ecosystem types. The width of each histogram is proportional to the area-extent of each biome. The figures within the histograms indicate total C storage in petagrams (Pg) in that compartment worldwide (modified from Anderson, 1991).

Table 9–2. Mass of organic C in soil orders of the world (from Eswaran et al., 1993).

Soil order	Organic C
	Pg
Histosols	357
Andisols	78
Spodosols	71
Oxisols	119
Vertisol	19
Aridisols	110
Ultisols	105
Mollisols	72
Alfilsols	127
Inceptisols	352
Entisols	148

mineral soil C in boreal ecosystems shows the influence of cold soil in reducing the rate and extent of organic matter decomposition by soil microbes. Typically, soil organic C distribution follows a *J*-shaped curve, with concentrations highest near the surface and declining with depth (Fig. 9–2).

Forms of Soil Carbon

The pool sizes of C differ within and among biomes but the composition of organic matter varies little among soil orders (Oades et al., 1989). Humic acids from six soils, including Mollisols, Alfisols, Spodosols, and Entisols, showed similar elemental (C, H, O, N) composition and spectra when subjected to cross polarization/magic angle spinning (CPMAS) ^{13}C-NMR (nuclear magnetic reso-

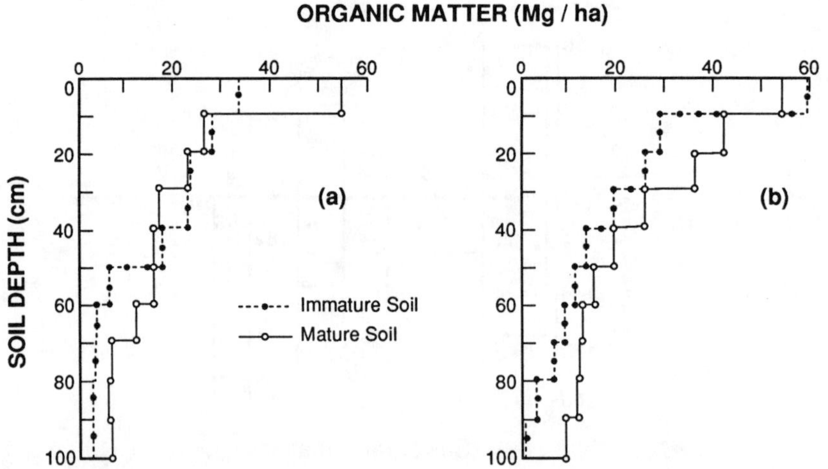

Fig. 9–2. Organic matter content to 1-m depth for mineral soils typical of northern California and southwest Oregon. Solid lines are for mature soils (Typic Haploxerults and Ultic Haploxeralfs), broken lines for immature soils (Dystric Xerochrepts and Umbric Vitrandepts). Organic matter contents are shown for (*a*) soft sediments of the Klamath Mountains, (*b*) volcanic soils of the Cascades (from Powers, 1989).

nance) spectral analysis (Malcolm, 1990). Fulvic acid fractions from these same samples showed varying elemental composition and spectra. Qualitatively, the NMR spectra of humic acids extracted from forest and agricultural soils do not show substantial differences (Preston, 1991). While humic and fulvic acids may have similar C and H contents, they can differ significantly in N and total acidity (Tan, 1986). The N concentrations of humic acids from selected temperate and tropical soils are nearly twice those for fulvic acids. The total acidity of fulvic acid is about twice that of humic acid.

Organic matter in some variable-charge soils is richer in aliphatic and carboxyl-C than organic matter in other soils (Fig. 9–3). Moreover, Andisols may have a higher ratio of hexoses to pentoses than other soils (Oades et al., 1989). These materials may be stabilized by adsorption to variable-charged surfaces and trapped in fine pores (Oades et al., 1989). Amounts of fulvic and humic acids vary substantially among A horizons of selected temperate region soils (Table 9–3; Tan, 1986). Large amounts of these constituents are found in Mollisols and Alfisols. Humic acids are more abundant in Mollisols, and fulvic acids are more abundant in Ultisols. Parent material, climate, and flora undoubtedly play major roles in determining soil organic matter composition. Although the concentrations of these humified acids are greater than that of nonhumified acids, the significance of their roles in soil formation depends on vegetation type and amounts of organic acids produced therein.

Organic C in soil exists in a wide variety of chemical forms that is reflected in the rate of compound metabolism by soil microbes. Simple carbohydrates (glucose, sucrose), early products of photosynthesis, are readily metabolized. Starch,

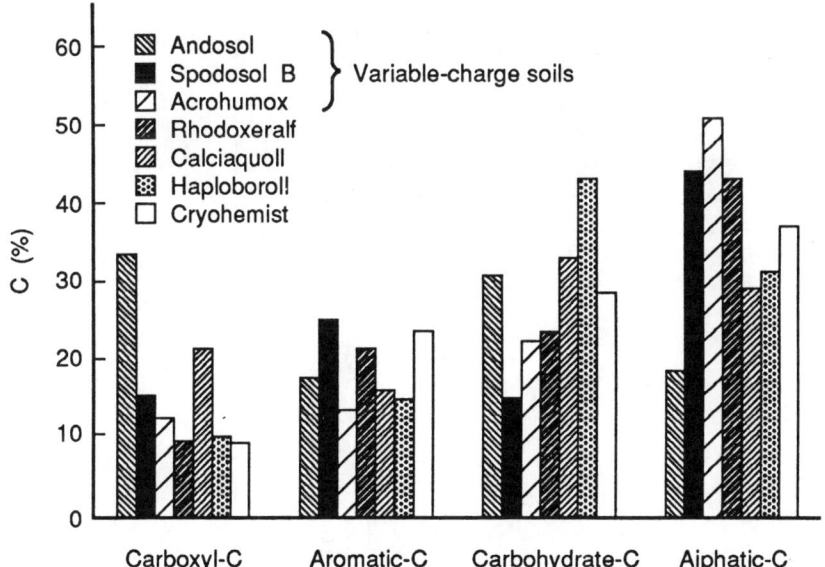

Fig. 9–3. Distribution of C in chemical groups as shown by solid state ^{13}C-NMR spectroscopy of selected soils (from Oades et al., 1989).

Table 9–3. Humic (HA) and fulvic acid (FA) contents in A horizons of some temperate region soils (Tan, 1986).

Soil	Na$_2$P$_2$O$_7$ extraction			NaOH extraction (after Na$_2$P$_2$O$_7$)			Total FA + HA
	FA	HA	FA/HA	FA	HA	FA/HA	
	g 100 kg^{-1}			g 100 kg^{-1}			g 100 kg^{-1}
Dubuque (Alfisol)	592.0	432.0	1.4	288.0	42.4	6.7	1354
Grenada (Alfisol)	588.0	128.0	4.6	192.0	96.0	2.0	1004
Marshall (Mollisol)	244.0	444.0	0.6	156.0	176.0	0.9	1020
Tama (Mollisol)	288.0	780.0	0.4	156.0	288.0	0.6	1512
Cecil (Ultisol)	484.0	80.0	6.1	132.0	84.0	1.6	780
Orangeberg (Ultisol)	516.0	80.0	6.5	80.0	76.0	1.1	752
Burleson (Vertisol)	160.0	376.0	0.4	—	80.0	—	616
Houston (Vertisol) Black	192.0	236.0	0.8	20.0	68.0	0.3	516

a storage carbohydrate polymer of glucose, is decomposed at a slower rate than simple sugars because of its polymeric (amylos and amylopectin) nature. The low concentration of simple sugars and starch in plant litter reflects their ease of metabolism.

Cellulose, a glucose polymer with 1–4 linkage between adjacent molecules, is the most abundant constituent in plant residues. It is the principal component of plant cell walls and is associated with hemicellulose and lignin in cell wall structure. Cellulose is of moderate quality as a source of energy for soil microbes. Cellulase, an extracellular enzyme complex produced by a variety of bacteria and fungi, eliminates the crystalline structure and depolymerizes cellulose to cellobiose. The latter disaccharide is hydrolyzed by cellobiase to glucose. Glucose can be absorbed by decomposer organisms or can enter the soil C pool (Paul & Clark, 1989; Stevenson, 1986). The rates of cellulose and hemicellulose decay are similar (Aber & Melillo, 1991). The hemicellulose pectin is a component of the middle lamella of plant cell walls.

Polyphenols are composed of several phenolic acid units and include classes of compounds of particular importance to soil formation and ecosystem behavior. Tannins are polyphenols of high molecular weight (between 500 and 3000) which with the complex and amorphous lignin, comprise two categories of soil organic compounds of substantial significance. Interactions of tannins with plant proteins reduce the decomposability of the N-containing compounds. In contrast, lignin presents a variable and complex structure that cannot be precisely defined. It is formed as an encrusting material on the cellulose-hemicellulose matrix of plant cell walls. Because of the low metabolic energy yield to soil microbes, lignin is slowly decomposed and constitutes a long-lived component in soil organic matter dynamics (Zeikus, 1981).

Hydrocarbons, including fats and waxes, are long-chain C compounds. They tend to be hydrophobic. Their composition and molecular size result in a decomposition rate between that of carbohydrates and that of phenolics (Aber & Melillo, 1991). Proteins are an important additional source of organic C and N. These compounds and their constituent amino acids are energy-rich and may be readily metabolized by soil microbes. Interaction with polyphenols and lignins (lignoprotein complexes) may dramatically increase protein resistance to decay. Sequential extraction (proximate analysis) of plant tissue with nonpolar (dichloromethane), polar (hot water), and strong acid (72% H_2SO_4) reagents removes, respectively, fats and waxes, soluble phenolics and carbohydrates, and cellulose and hemicellulose. Lignin remains after the extraction is completed.

On the average, soil organic matter contains 10 to 20% carbohydrates (primarily microbial origin), 20% N-containing materials (amino sugars, amino acids), 10 to 20% aliphatic fatty acids, and alkanes. The remaining material is aromatic C (Paul & Clark, 1989). Analysis of soil organic matter requires separation of the organic constituents from the mineral soil matrix. Extraction is with NaOH or $Na_2P_2O_7$. The insoluble organic fraction is humin. Acidification of the extract yields a precipitate, humic acid, while fulvic acid remains in solution (Paul & Clark, 1989; Stevenson, 1986).

CARBON AND PEDOGENESIS

Soil organic matter is composed of living plant and animal tissue, detrital remains from these sources, exudates from plant root systems, products of microbial synthesis, and leachates from above- and belowground organic sources. A variety of organic substances, variously modified by microbial processing, are termed nonhumified substances (Tan, 1986; Stevenson, 1982). Under vegetative cover, these materials include carbohydrates, amino acids, protein, lipids, lignin, nucleic acids, pigments, and a number of organic acids (Table 9–4). These materials are continually introduced into soil. The concentrations are low, reflecting a relatively short turnover time because they are energy sources for soil microorganisms.

A variety of roles have been attributed to organic matter in soil genesis. Organic acids have been shown to be important weathering agents of primary and secondary minerals (Huang & Keller, 1970). Conversely, organic acids have been identified as catalysts in the neogenesis of secondary clay minerals. Organic matter has been demonstrated to be an important aggregating agent of soil structure, particularly in A horizons. Conversely, organic acids are partially responsible for the flocculation and transport of clay minerals in the genesis of argillic horizons.

Several papers discuss the role of leaf extracts on deflocculating kaolinite and montmorillonite. Deflocculation in the A, E, and upper B horizons is a necessary precursor to the eluvial/illuvial mechanisms responsible for development of argillic horizons. *Soil Taxonomy* (Soil Survey Staff, 1975) states that argillic horizons are found primarily under deciduous forest ecosystems. Early research (Bloomfield, 1953; Rozanov, 1961 as cited by Grim, 1968, p. 384; Thorp et al., 1959) documents the importance of organic matter in the detachment and transport of clay micelles. The paper by Thorp et al. (1959) is a classic in pedological literature and is the basis for the genetic hypothesis of argillic horizon formation.

More recent work discussing both the deflocculation and aggregation of clay micelles by organic acids includes that of Caillier and Visser (1988) and Visser and Caillier (1988). The former paper shows that after 10 mo of percolating peat extract through columns of sand and various clay species (kaolinite, vermiculite, illite, and montmorillonite) discernible structural aggregates had formed. Structure morphology was related to clay species. The latter paper showed that humic acid concentrations of 40 mg L^{-1} were more effective than sodium hexametaphosphate or Calgon in dispersing clay. The dispersing ability of the humic acid was confined to a narrow concentration range.

Hem and Lind (1974) used quercetin, an organic compound, to catalyze kaolinite precipitation from a solution of Al and Si. Douglas (1982) reported that organic residues and pH's <6.7 are necessary to form beidellite. The organic acids were necessary to produce monomeric Al so that Al would substitute for Si in the tetrahedral layers. Wada (1980) reported that the presence of active humus in volcanic ash parent materials inhibits formation of allophane and imogolite.

Altschuler et al. (1963) described the conversion of montmorillonite to other secondary minerals (kaolinite, hematite, goethite) by groundwater enriched with organic matter. Organic acids increased the rate of weathering and affected composition of the weathering products. Schwertmann and Fischer (1973) examined

Table 9–4. Examples of nonhumified organic acids and inorganic acids present in soil (Tan, 1986; Robert & Berthelin, 1986).

Source of acid	Acid	Formula
Secreted by organisms in soil, in equilibrium with soil solution		
Rhizosphere	CO_2 (H_2CO_3), HNO_3, amino acids (aspartic, glutamic),	$HOOCCH_2CH(NH_2)COOH$ $C_{76}H_{52}O_{46}$
	Phenolic acids (tannic acid, gallic acid), aliphatic acids (oxalic, malic, citric, tartaric acids)	$HOOCCOOH$
Microorganisms	Aliphatic acids (oxalic, citric, formic, gluconic, lactic, malic, tartaric acids)	$HOC(CH_2COOH)_2COOH$
	Lichen acids (oxalic, orsellinic acid)	$HOOCCOOH$
In different environmental conditions		
Aerobic conditions	Phenolic acids (gallic, vanillic, hydroxybenzoic acids)	$C_7H_6O_5$
Anaerobic conditions	Aliphatic acids (acetic, butyric, formic, fumaric, succinic, lactic acid)	$CH_3CH_2CH_2COOH$
Under litter or plant conopies	Aliphatic acids (oxalic, citric, formic)	$HCOOH$
	Phenolic acids (gallic, vanillic)	C_8H_8O4

C-rich Fe deposits resulting from groundwater movement through soils. They concluded that the amorphous nature of the Fe deposits resulted from the presence of organic acids, which favored goethite production over hematite and "retarded" crystallization and aging by inclusion of organic compounds in the structure of the precipitate.

Kodama and Schnitzer (1977) showed that the goethite and hematite precipitated from aqueous suspension is affected by pH and concentrations of fulvic acid (FA). Fulvic acid concentrations of 0.5 g L^{-1} favored crystallization of hematite, but FA concentrations of 5.0 g L^{-1} or greater inhibited crystallization. This work confirmed earlier observations that low concentrations of organic acids were required for hematite formation in leaching environments (Schwertmann, 1971).

Soil minerals are weathered by direct attack from protons and by chelation of polyvalent cations in the crystal lattice structure. Mineral susceptibility to weathering depends on a number of factors including hardness, molecular structure, and bonding energies. The structural bonds that require the most energy for formation are the most resistant to weathering (Tan, 1986). Of substantial importance is the nature of the climate and vegetation in promoting conditions that maintain production of substantial quantities of these acid-weathering agents (Ugolini & Sletten, 1991). The ease of mineral weathering by organic compounds is: biotite > K-fledspars > muscovite (Tan, 1986).

Organic acids can be separated into two groups based on their functional activity. The acidic characteristics of the nonhumified group are attributed only to the presence of –COOH groups. The humified group has acidic character due to the presence of –COOH and phenolic –OH groups (Tan, 1986). The former group includes lower molecular weight compounds such as formic, acetic, and oxalic acids (Table 9–4). These chemicals may display complexing abilities, but their main effect is through acidity. A number have dissociation constants similar to those of strong mineral acids (Tan, 1986). Humic, fulvic, and other more complex humified acids comprise the second group. The presence of –COOH and phenolic –OH groups in their structure enables them to exert acidic and interactive effects. Interactive effects include electrostatic attraction, complex formation or chelation, and water bridging (Fig. 9–4; Tan, 1986). Because of the greater concentration of these materials compared to nonhumified organic acids, they may play a greater role in mineral weathering.

Humic acid (HA) and FA promote the dissolution of primary and secondary minerals in soil through acidic effects and the formation of organometallic chelates. Acidic effects may be manifested in secondary mineral weathering through increased mineral surface acidity, caused by proton dissociation from acids, with spontaneous mineral collapse. The H^+ may penetrate the crystal lattice replacing Al^{+3}, which is then preferentially adsorbed on clay surfaces. The exchangeable Al^{+3} generates protons. Mineral decomposition by proton attack is perpetuated (Tan, 1986). In general, octahedral cations are released more readily than tetrahedral cations.

In the Cascade Range of California, olivine and hypersthene andesite parent materials are subjected to moderate to intensive leaching by acidic solutions percolating through the soil profile (Hendricks & Whittig, 1968). Weathering of the parent materials to saprolite followed a similar progression. Early stages of

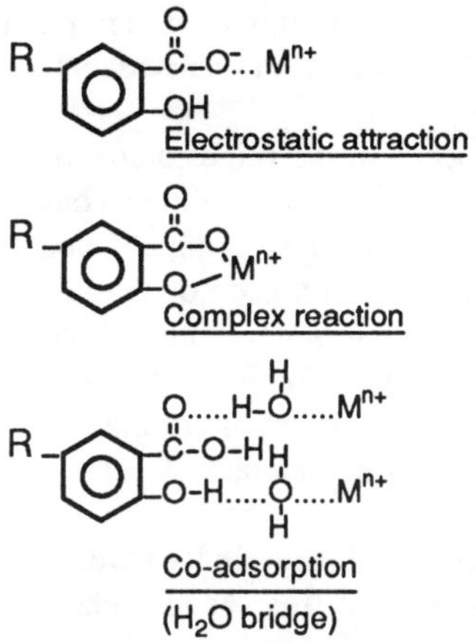

Fig. 9–4. Three types of interaction between HA and a metal cation. R = the remainder of the humic acid molecule, M^{n+} = cation with charge n^+ (Tan, 1986).

weathering of both rocks showed that relatively large phenocrysts of hypersthene were the most resistant. A decreasing order of resistance was found for plagioclase phenocrysts, fine-grained mafic minerals, olivine, and glassy matrix material. Quartz was relatively stable in the earliest stages of weathering, but it decreased rapidly as weathering progressed. The authors found that free Fe oxides and clay increased with weathering. Amorphous clay dominated in early stages but kaolinite increased relative to the amorphous material later in weathering.

Malcolm et al. (1969) demonstrated the formation of montmorillonite from 2:1 to 2:2 clay mineral in surface soil horizons under the influence of organic acids and pH below 4.5. The authors speculated that the organic acids and pH combined to protonate and solubilize Al-polymers in the clay interlayers. Organic matter was believed to be the catalyst and accelerator of this process, which theoretically could proceed in an acid environment without organic matter. Similar soils with low pH but lacking the organic matter did not contain montmorillonite in the surface horizons, but contained the 2:1 to 2:2 intergrade.

Organic acids similar to HA and FA were shown to selectively remove Si, Al, Fe, Ca, Mg and K from primary minerals, which were etched by the weathering process (Huang & Keller, 1970; Boyle & Voigt, 1973). Organic acids were shown to be catalysts in the recombination of Si in solution with Al to form kaolinite as a precipitate of the weathering solution.

Initial weathering of rock surfaces involves a combination of wetting–drying, freeze–thaw cycles, and biological/biochemical processes. Wetting and drying of lichen thalli and physical detachment may occur of surface minerals that adhere to the undersurface polymer (mucillagenous) layer of the lichen (Robert & Berthelin, 1986). Quartz and feldspars may be preferentially removed from granitic surfaces by this process. Oxalic acid also influences chemical weathering. A wide range of other lichen acids have been isolated and identified which may play roles in soil mineral weathering (Asahina & Shibata, 1971).

Lichen thalli have been observed to grow around minerals. The hyphae can physically separate biotite packets, which are subsequently dissolved by oxalic acid (Robert & Berthelin, 1986).

These activities also occur in the rhizospheres of plant root systems, where mycorrhizal hyphae penetrate and are in direct contact with mineral surfaces. Root exudates also may play a role in soil mineral weathering. Inoue et al. (1993) showed that mugineic acid, a root exudate of Fe-deficient oat (*Avena sativa* L.) and rice (*Oryza sativa* L.) grown in water culture, is an effective weathering agent of synthetically produced goethite, hematite, and lepidocrocite. Mechanisms of ligand exchange and nucleophilic substitution may control Fe dissolution. Thus, physical and chemical weathering interact in early stages of rock and mineral weathering.

Microbial activity substantially influences weathering processes by control of organic matter composition and flux. Their metabolism controls environmental redox potential, uptake, oxidation and reduction of mineral elements and ions (SO_4^{-2}, NO_3^-), decomposition of residues, and synthesis and recycling of low and high molecular weight organic chemicals that are involved in primary and secondary mineral weathering (Fig. 9–5; Robert & Berthelin, 1986). Through recycling of organics and organometallic complexes as energy sources, soil microbes are involved in the deposition of selected mineral elements. Microbial biosynthesis of high molecular weight organic compounds (HA and FA) and related organometallic complex formation is associated with development of surface coatings on soil particles and formation of soil aggregates (McKeague et al., 1986).

Mineral element deficiencies can promote organic acid production by microbes, thereby maintaining solubilization of mineral surfaces. Anaerobic conditions result in reduced biodegradation of organic chemicals, increasing the longevity of acids involved in mineral weathering (Robert & Berthelin, 1986). In most systems, a combination of acidic effects and complex formation involving both types of organic acids undoubtedly contributes to mineral transformation and dissolution (Stevenson, 1982; Tan, 1986; Robert & Berthelin, 1986; McKeague et al., 1986).

Role of Carbon in Horizonation

The action of organic C in soil development is through the influence of the acid, organometallic complexing characteristics of compounds produced by plants and microorganisms and modified by microbial activity. The nature and rates of reactions of these compounds in soil development depend on the soil-forming factors (Jenny, 1980). In the extreme examples of Spodosols and Oxisols, different mechanisms are responsible for differentiation of horizons. Warm, humid environments promote rapid decomposition of soil organic matter. Carbon acid weathering is important in Oxisol formation (Stevenson, 1982; Senstius, 1958). Cool, humid temperate climates support incomplete oxidation of organic matter and production of organic acids that favor Spodosol formation (Fig. 9–6). The extent of organic matter oxidation is a key factor for both soil

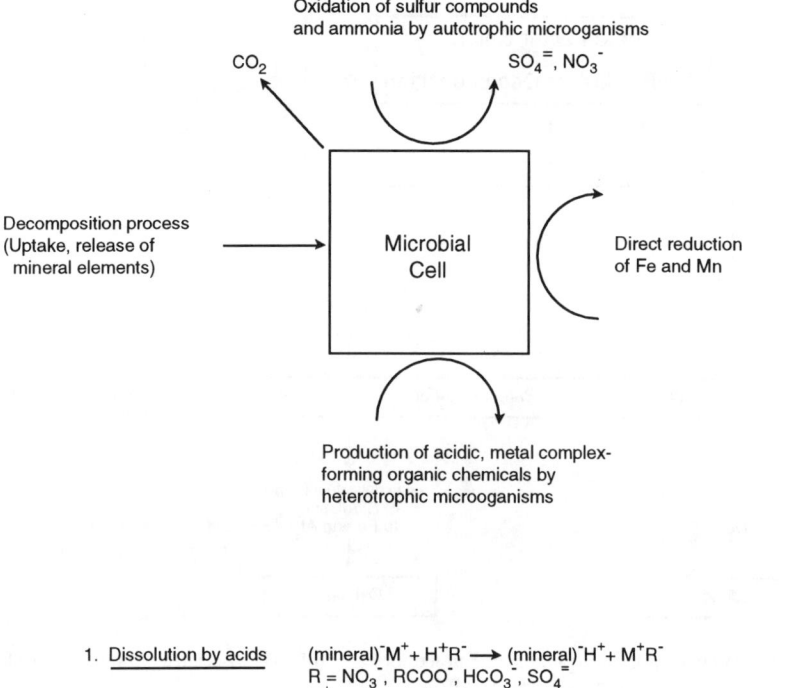

Fig. 9–5. Mechanisms of microbial weathering of minerals (modified from Robert & Berthelin, 1986).

orders. The state factor regime of Oxisols and Spodosols supports contrasting mechanisms of soil development.

Soil formation and horizon differentiation result from weathering and translocation processes in the presence of living or dead organic matter (Simonson, 1959; Ugolini & Sletten, 1991). The biota of each ecosystem, reflected primarily in plant community composition, determines the initial chemical composition of organic detritus. Subsequently, through microbial activity, the composition of organic materials that participate in soil development is determined. The state factors of soil formation ultimately control the chemical composition, strength and distribution of organic acids that mediate this process (Ugolini & Sletten, 1991; McKeague et al., 1986).

In *Abies amabalis* ecosystems of the central Cascades of Washington state, Spodosol development is explained by the action of organic and inorganic acids

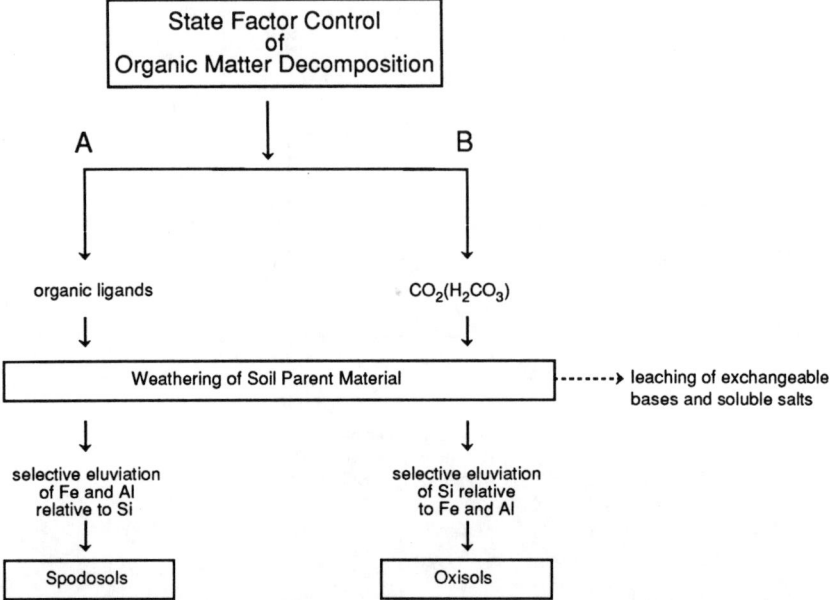

Fig. 9–6. Potential role of organic matter in the formation of soils associated with two soil orders (modified from Stevenson & Ardakani, 1972).

(proton donors; Ugolini & Sletten, 1991). Organic acids formed from incomplete decomposition or from root exudates are active weathering agents in the upper portion of the profile (O, E, and Bhs horizons). Carbonic acid plays a major role in the lower profile (Bs, BC, and C horizons; Ugolini & Sletten, 1991). Analysis of soil solutions has shown a variety of low molecular weight organic acids, including oxalic and citric acid. These acids are potential weathering agents. Organometallic complexes of FA with metals including Fe and Al are formed and translocated to the B horizon, where the metal load reaches a critical level (McKeague et al., 1986; Tan 1986). Root decomposition products probably are involved.

Relatively rapid decomposition (loss of CO_2 and H_2O) occurs in base-rich Mollisols of warm, temperate regions. The remaining organic products of microbial activity, including HA's and humin, form sorption complexes and contribute to development and maintenance of soil structure (McKeague et al., 1986; Oades, 1988). Translocation of organic matter in these base-rich soils probably is minor. Much of the organic matter in B horizons probably comes from root decomposition in situ.

Cation bridging among colloidal organic matter, polyvalent cations and clay is an important mechanism for soil aggregate formation (Fig. 9–7; Oades, 1988; Tate, 1987). Key ions are Ca^{2+} and Mg^{2+} in neutral to alkaline soil and Al^{3+} and Fe^{3+} in acid and ferallitic soils. The process occurs between clay particles, the ions (Ca^{2+}), and negatively charged organic materials (such as polyuronides and resistant cell wall debris from bacteria) which are polycarboxylic and can be

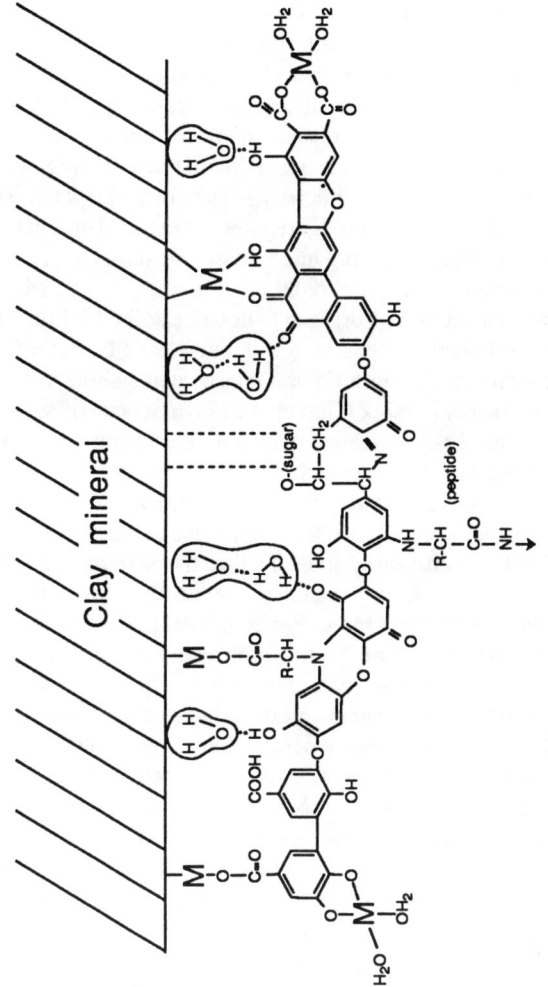

Fig. 9–7. Schematic of clay-humate complex in soil (from Stevenson, 1985).

equated with humic materials (Oades, 1988). Microaggregates of 2 to 20 μm form through this mechanism (Tisdall & Oades, 1982). Macroaggregates (20–250 μm) are stabilized by entanglement of particles within the aggregates by roots and fungal hyphae. Calcium plays a key role in this process. Low Ca concentration increases solubility of organic colloids and their mineralization. Addition of Ca has the long-term effect of stabilizing organic matter by cation bridging (Oades, 1988; Anderson, 1992). Allophane soils are organic-rich through oxide-organic colloid interactions that stabilize organic C. Turnover times for C may be greater than in adjacent nonallophanic soils (Oades, 1988). The mechanisms of stabilization are not clear. Stable macroaggregation is controlled by root growth and fungal activity, which in turn depend on resource management. Stable aggregation rapidly disappears from exploited soil. Elliott (1986) found that more organic matter was associated with macroaggregates in grassland soil and that it was more labile than microaggregate organic matter. He suggested that the organic matter binding microaggregates into macroaggregates is the main source of nutrients released when organic matter is lost on cultivation.

Complex formation between colloidal organic matter and Ca is an important factor in the depth distribution of organic C in some soils. Mobile humic materials (FA's) may be relatively recent in age in surface soils, but with downward migration, complex formation with Ca and consequent stabilization, they remain in soil as a more permanent C sink (Goh et al., 1976; Schoenau & Bettany, 1987). Calcium humates showed the greatest age in a chernozem soil (1410+/–95 yr) compared with mobile humic fractions (785+/–50 yr; Campbell et al., 1967). The various humic fractions were from 3 to 11 times older in the chernozem than a podzol. The age difference reflects the higher base status of the chernozem. Results of ^{13}C NMR spectroscopy indicate that the stability of residual soil C results from physical protection mechanisms rather than inherent differences in recalcitrance of the organic structures (Preston, 1991).

Depth distributions of organic matter in forest and grassland soils are influenced by leaching and by the base element contents (Schoenau & Bettany, 1987; Ugolini & Sletten, 1991). Fulvic acids produced in surface organic layers may be translocated to the B and C horizons where higher Ca contents precipitate the colloidal organic materials. Fulvic acid extracted from the B and C horizons of forest and grassland soils had higher N concentrations than humic and FA's extracted from surface horizons. Leaching of nutrient-rich organic acids may be a valid explanation for narrowing of C/N ratios with depth (Schoenau & Bettany, 1987). Analysis of ^{13}C NMR spectra of HA's from a forest soil indicate that lignin may be depleted in the Ae horizon, enriched in the Bm horizon, and reflect characteristics of decomposed lignin in the Bc horizon (Preston, 1991). Aromatic fragments of lignin decomposition may be incorporated into structures that no longer retain the spectral signature of lignin.

Integrative measures of the status of soil organic matter have been developed that separate light and heavy fractions on the basis of their specific gravities in standard solutions (Sollins et al., 1984). These authors found that light fraction organic matter was composed mainly of partially decayed root fragments and other plant and microbial parts. The heavy fraction was adsorbed or deposited on mineral surfaces or protected from decomposition within soil microaggregates (Sollins et al., 1984). Wider C/N ratios of the light fraction were associated with lower net N mineralization than the heavy fraction. In these soils, whole soil net N mineralization was less than that of the heavy fraction because of N immobilization by the light fraction.

STATE FACTORS AS PROCESS CONTROLS

Soil classification units, such as soil orders, differ in their organic C properties and this helps in depicting general trends. However, soil classification systems reflect modal tendencies. In that respect, they are not very helpful in understanding the dynamic nature of soil C relative to soil and ecosystem development. We believe that the most effective way of viewing the interrelationships between organic C and soil development is through an ecosystem perspective. We do this using the state factor approach of Jenny (1941, 1980).

State factors are the general environmental properties of a region that determine the nature of its ecosystems (Jenny, 1980; Van Cleve et al., 1991). The state factors condition the phenomena that directly affect ecosystem structure and function, they control the cycling of C and nutrients in ecosystems. The ecosystem controls operate at both macro- and microscales and include *climate*; the elevation, slope, and aspect components of *topography*; *soil* including the physical and chemical properties; *biotic* properties which include both micro- and macroorganisms; and successional phenomena which reflect the *time* and course of ecosystem development (Fig. 9–8; Van Cleve et al., 1991).

Climate

Climate has a commanding effect on soil C storage through its influence on primary production and decomposition. Post et al. (1982), analyzing 2,700 soil profiles from throughout the world, classified soil organic C storage into world life zones according to potential evapotranspiration, precipitation, and biotemperature. Soil C mass generally increased directly with precipitation and inversely with temperature. Boreal, desert, and tropical forest soils had similarly

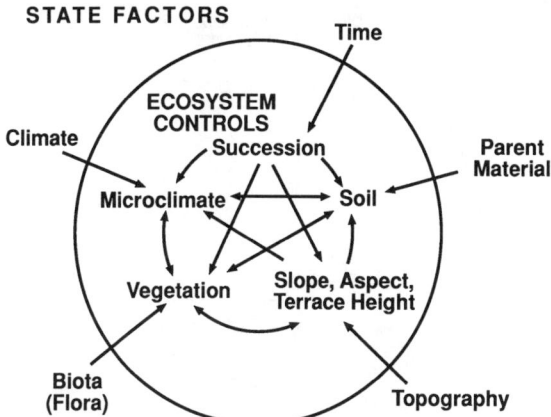

Fig. 9–8. State factors and ecosystem controls of C and element cycling (Van Cleve et al., 1991).

low C storage values (8–10 kg C m^{-2}). The highest values (24–30 kg C m^{-2}) occurred in soils of polar wet tundra and boreal rain forests. Where potential evapotranspiration was balanced by precipitation, soil C density averaged about 10 kg m^{-2}.

Regional climate plays a dominant role in controlling the pool size and flux of soil C in primary and secondary successional ecosystems (Fig. 9–8 and 9–9). The control is manifested through the influences of temperature and precipitation on the magnitude of annual NPP, biomass recycling (aboveground and belowground detrital accumulation), and organic matter decomposition (respectively, Fig. 9–10, 9–11, and 9–12). Temperature and moisture (precipitation) interact through evapotranspiration to influence decomposition and loss of soil C on a regional basis (Meentemeyer & Berg, 1986; Meentemeyer, 1978). Climate plays a key role in the longevity of organometallic chelates that may be involved in mineral weathering, element transport, and soil development. In warmer regions, these organic materials may not long survive in the soil because of rapid decomposition. Consequently, in cool temperate regions, climate favors the plant communities and decomposition processes that result in podzolization (Ugolini & Sletten, 1991). In the humid mesophytic forests, hydrologic gradients as affected by stratigraphy and relief are more important in affecting distribution of species, forest productivity, and organic matter production (Carmean, 1975).

The magnitude of C flux to the soil changes dramatically with latitude. Average net annual biomass production declines by at least threefold (15 Mg ha^{-1} to less than 5 Mg ha^{-1}) as the mean annual temperature declines from 15°C to –3°C (Fairbanks, Alaska, Fig. 9–10). Carbon recycling in litterfall also increases by about two- to three-fold from the latitude of the northern boreal forest of interior Alaska (64°45'N) to the latitude of Cascade Douglas fir (*Pseudotsuga menziesii*) forests in Oregon (44°14'N) and Eastern deciduous forests at Coweeta

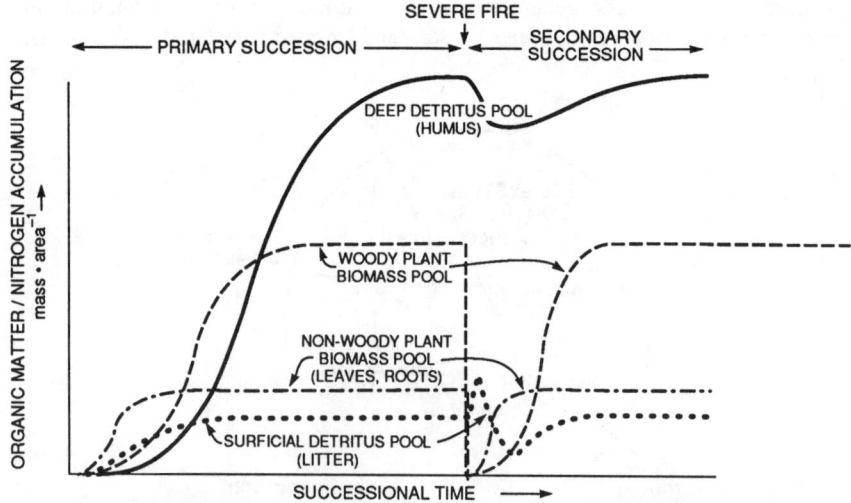

Fig. 9–9. Generalized patterns for changes in organic matter and N pools during primary and secondary succession (from Reiners, 1981).

SOIL C, FORMATION & ECOSYSTEM DEVELOPMENT

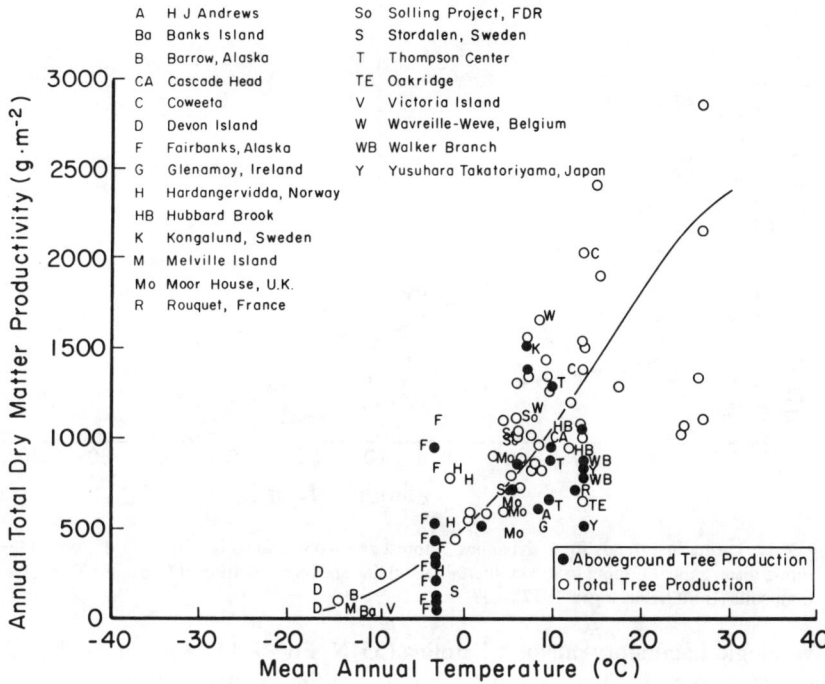

Fig. 9–10. Annual dry matter production in relation to mean annual temperature (from Van Cleve et al., 1983; Van Cleve & Alexander, 1981; Cole & Rapp, 1981, p. 341–409; Leith, 1975; unlabeled points are from Leith, 1975).

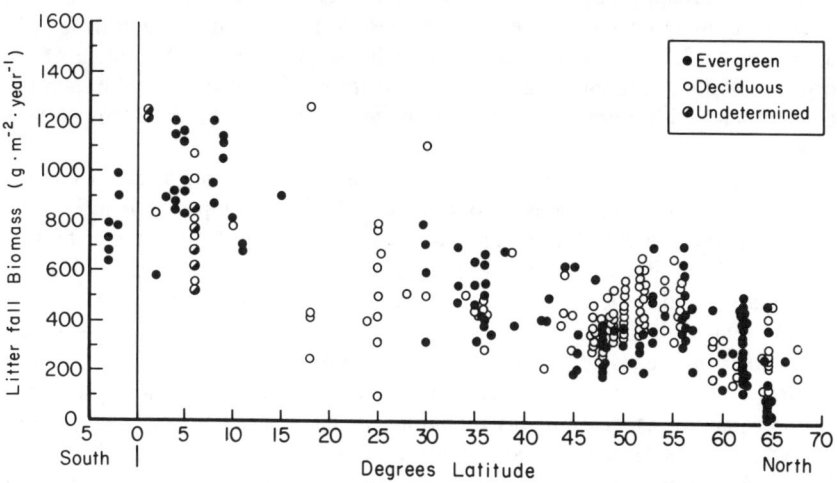

Fig. 9–11. Forest litterfall biomass in relation to latitude (from Van Cleve et al., 1983).

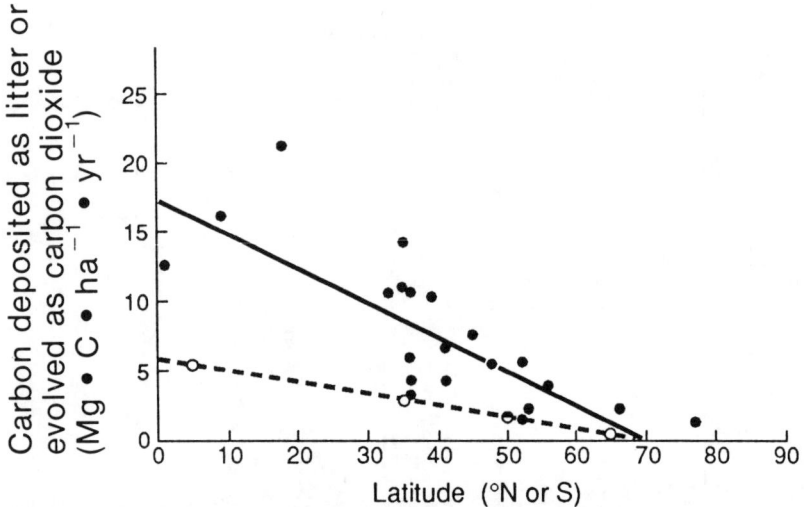

Fig. 9–12. Latitudinal trends for C dynamics in forest and woodland soils of the world. Dashed line shows mean annual C input to soil in litterfall; solid line shows the portion of C loss as CO_2 evolved from soils (from Schlesinger, 1977).

Hydrologic Laboratory in North Carolina (35°N; Fig. 9–11). Carbon loss by soil respiration increases from about 1.6 Mg ha^{-1} at the latitude of Fairbanks to 8.8 Mg ha^{-1} at Coweeta (Fig. 9–12).

Within climatic regions, microclimate interacts with other ecosystem controls to mediate soil C flux. Along an elevation transect in the west-central Sierra Nevada of California, the C/N ratio of the 0- to 20-cm soil depth increases with elevation from approximately 11 at 305 m to nearly 20 at 2400 m (Jenny, 1980). This is a consequence of a three-fold increase in the C concentration from 0.8% at 300 m, to 2.2% at 2400 m compared with a lag in rise in N content of soil organic matter. Organic matter decomposition also declines with summer drought, and with increasing elevation and accompanying lowered temperatures (Powers, 1990).

Soil temperature in the circumpolar boreal forest is a major factor controlling soil C flux. Interactions among solar radiation, canopy interception, soil moisture, and the forest floor are important determinants of the soil thermal regime with significant consequences for soil organic matter (Bonan, 1992). Soil temperature and moisture in the northern boreal forest of interior Alaska are closely related to the accumulation of soil C (Fig. 9–13). Mature black spruce (*Picea mariana*) ecosystems with low productivity developing on cold, wet, permafrost soils accumulate between 40 and 80 Mg soil C ha^{-1}. More productive mature quaking aspen (*Populus tremuloides*), paper birch (*Betula papyrifera*), and white spruce (*P. glauca*) ecosystems accumulate between 27 and 46 Mg soil C ha^{-1}. Although warmer, drier sites (and soils) support greater annual net primary production, they also support higher rates of soil organic matter decomposition resulting in smaller accumulations of soil C. The microclimate also interacts with vegetation and topographic controls of ecosystem processes to mediate

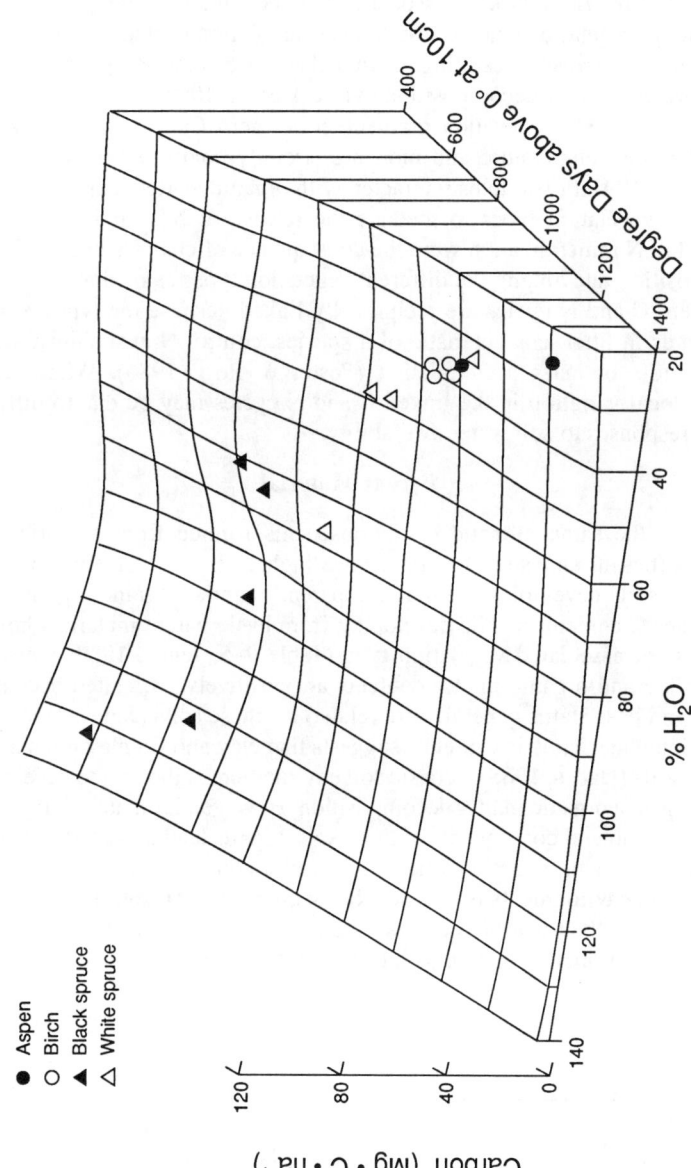

Fig. 9–13. Response relating temperature sum (degree days > 0°C for period May 20 to September 10 at 10-cm depth below surface of forest floor) and average annual moisture content of surface 10 cm of mineral soil to organic C content of 30-cm soil profiles from mature site types in northern boreal forest of interior Alaska.

the observed accumulation of organic soil C (Fig. 9–8 and 9–13). A regional analysis of U.S. grassland soils showed that organic C content increased with precipitation and clay content but decreased with temperature (Burke et al., 1989).

Carbon accumulation within primary successional soils of the Tanana River floodplain of interior Alaska is associated with declining soil temperatures. Here, successional vegetation change results in accumulation of thick forest floors that insulate the mineral soil, resulting in gradual cooling of the soil profile and reduction in organic matter decomposition (Viereck et al., 1993).

Along a 2000-m elevation gradient in northern California, Powers (1990) found that soil temperature and moisture strongly controlled mineralization of soil organic N. Both variables interacted with organic matter substrate chemistry across six vegetation types to mediate the release of NH_4^+ and NO_3^-. Changes observed in N mineralization were the consequence of changes in organic matter decomposition rate among the different vegetation types as mediated by climate. Ecosystem C and N cycles are reciprocally linked because the types of organic compounds in litter, characteristic of a species, control N availability, which in turn controls biomass accumulation (Pastor & Post, 1986). Within climatic regions, local variation in the linked C and N cycles may be due to differential species responses to soil water availability.

Parent Material

Under the same climatic conditions, soils formed from base-rich parent materials (basalt, andesite) generally have higher clay and organic matter contents than soils developing from acidic parent materials (granite, granodiorite). The higher C content in soils developing from basic parent materials holds true regardless of associated vegetation type (Table 9–5, Jenny, 1980; Jenny et al., 1968). Clay and organic matter contents are positively correlated because clay content and base status generally are related for these soils (Jenny, 1980; Oades, 1988). Some mechanistic evidence suggests that clay and Ca play roles in retaining C in soils (Oades, 1988). A basic soil environment is thought to accelerate the initial stages in organic matter decomposition. However, intimate mixing of detritus with the mineral components of these soils retards later stages of organic matter decay, increasing retention time and the soil C content. Initial litter breakdown in acidic soil environments is retarded. Subsequent mineralization proceeds fairly quickly to CO_2 because of the lack of stabilizing mechanisms (Oades, 1988). Stabilization of organic matter with clay in base-rich soils is the controlling fac-

Table 9–5. Mean values for selected properties of soils developing from acid igneous and basic igneous rocks (Jenny, 1980).*

Soil property	Acid igneous soils	Basic igneous soils
Clay (%)	12	21
Slit (%)	21	33
Sand (%)	58	35
C (%)	1.7	2.8
N (%)	0.07	0.12
Sum of exchangable ions (cmol kg^{-1})	5.3	10.9

*Differences between properties on acid and basic igneous rock significant at $\alpha = 0.01$.

tor. Calcium bridging of organic detritus to clays to form aggregates which are physically, chemically, and biologically stable is one mechanism proposed to account for these conditions (Oades, 1988).

Soils developing from metavolcanic rocks in southwestern Oregon showed higher contents of fine fraction particle sizes and generally higher concentrations of C and N in the various particle-size fractions compared with soils developing on quartz diorite (Borchers & Perry, 1992). Higher silt and clay content of the metavolcanic soils, their greater aggregation, and associated protection of organic C and N were considered the factors responsible for higher N mineralization in the fine soil fractions. Larger quantities of fine-textured soil, interacting with organic C to form aggregates, act as a slow release mechanism that sustains organic matter mineralization and N availability (Borchers & Perry, 1992).

Soils with appreciable quantities of volcanic ash (Andisols and Andepts) often have A horizons which are particularly rich in organic matter. In part, this is because many volcanic soils occur at higher elevations or latitudes where temperatures reduce organic matter decomposition, but organic complexes with allophane and other alumina compounds also are responsible (Tan, 1986).

Topography

Topography strongly controls the distribution of mass and energy across the landscape as well as soil aeration and water supply (Daniels & Hammer, 1992; Ugolini & Sletten, 1991; Jenny, 1980). Slope, aspect, and elevation interact with microclimate, parent material, and vegetation to influence rates of biochemical reactions, biomass production, organic matter decomposition and, consequently, soil C accumulation (Fig. 9–8). Daniels and Hammer (1992) recognize topography, parent material origin, and time as determinants of site geomorphology. Site stratigraphy is a component of parent materials.

Across the 12-state North Central Region of the USA, mineral soil C content (to 1-m depth) ranged from 20 Mg ha^{-1} in the Badlands of North Dakota to 210 Mg ha^{-1} in poorly drained soils of eastern North Dakota. The C contents of Alfisols and Spodosols, for the same depth interval in this region, ranged from 23 to 156 Mg ha^{-1} (Franzmeier et al., 1985). Soil C contents were controlled in this region by interactions of parent material, vegetation, and topography.

Along an elevation transect (2700–3400 m) in New Mexico, forest type played a key role in regulating seasonal and yearly variations in N mineralization and nitrification (Gosz & White, 1986). Aspect and slope were held relatively constant. The elevation range was not sufficient to exert major control over organic matter mineralization or the supply of NH_4^+. The chemistry of organic matter produced by the respective vegetation types was considered to be an additional control of C and N mineralization in these ecosystems.

Aspect and slope position had little influence on soil organic N (or C) mineralization in upland pin oak (*Quercus ellipsoidalis*) ecosystems of east-central Minnesota (Zak et al., 1991). Lower slope positions showed significantly lower nitrification than mid- or upper-slope positions. Time since disturbance was more important than the subtle differences in topography in controlling N cycling in this part of Minnesota. Ecosystems occurring in different landscape units (topographic positions) with characteristic soils and species composition will exhibit concomitant patterns of soil C and N turnover (Zak et al., 1986). The elevation transect of Powers (1990) illustrates the dominance of topography (elevation) and its interaction with microclimate and vegetation type-specific substrate chemistry to control soil C and N mineralization. When soil moisture was held constant, N mineralization rate per unit of soil organic C approximately doubled with each 10°C rise in soil temperature. But when soil moisture was allowed to vary naturally, mineralization rates at low elevations were reduced despite warm soil temperatures because of soil aridity.

Under the chaparral shrubs *Cercocarpus betuloides* and *Q. turbinella* in Arizona, more organic C accumulates on north-facing slopes than on south aspects (Klemmedson & Weinhold, 1992). Southerly aspects support more open stands of vegetation with larger areas of exposed soil between shrubs. Surface runoff and erosion result in shallower soils and lower C contents than north aspects.

Topography plays a dramatic role in modifying the regional climate of the northern boreal forest and hence the accumulation of mineral soil organic C in interior Alaska (Fig. 9–14; Van Cleve et al., 1991). Elevation ranges from 120 m on the floodplain of the Tanana River to 470 m on adjacent ridgetops. At this latitude (64°51'N), north and south aspects receive 70 and 94%, respectively, of the radiation received on a horizontal surface (Seifert, 1981). Consequently, Histosols, developing on north aspects, are cold, wet, and underlaid by permafrost. Although net primary production is lowest of all sites, large accumulations of organic C occur in these soils because their cold temperatures reduce the rate of microbial processing of detritus (Fig. 9–13 and 9–14). Evapotranspiration is markedly reduced compared with south aspects. Alfisols, developing on south aspects, are warmer, drier, and free of permafrost. Net primary production is higher on these sites, but smaller accumulations of organic C result because warmer soil temperatures support greater rates of microbial activity (Fig. 9–13 and 9–14). Histosols also occur in cold, wet lowland areas underlaid by permafrost. These locations are removed from the active portions of the river floodplains.

Biota

The plant's substrate environment bears long-term testimony to organism occupation (Jenny, 1980). We emphasize the importance of vegetation, vertebrate and invertebrate animal populations. The human influences are considered with respect to the status of soil organic C and soil development.

Vegetation

Vegetation is a major control of the status of soil organic C despite interactions with microclimate, soil, successional processes, slope, aspect, and elevation (Fig. 9–8). Vegetation directly influences soil C accumulation and soil development through above- and belowground NPP (Zak et al., 1990). The chemistry of these organic materials, including leachates and root exudates, plays a substantial role in (i) controlling the rates of microbial processing of the detritus, (ii) the nature of materials synthesized by soil microbial activity, and (iii) the roles of low

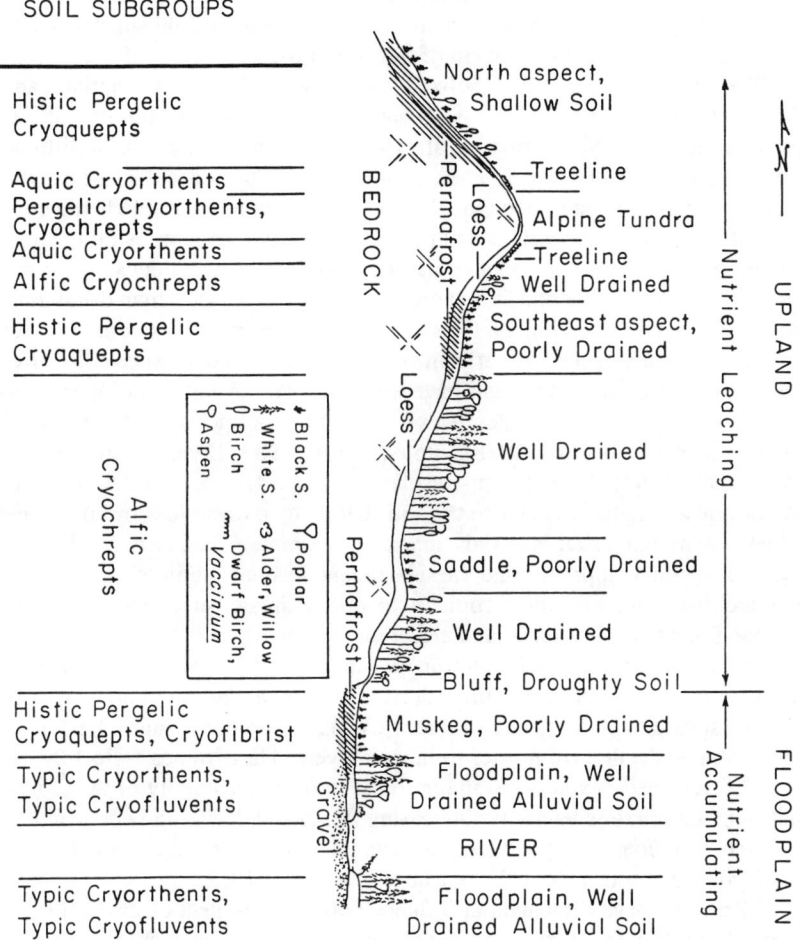

Fig. 9–14. Cross-section of the Yukon-Tanana Uplands in the vicinity of the Bonanza Creek Experimental Forest, 20 km southwest of Fairbanks, Alaska. Topography plays a major role in determining the microclimate and forest type distribution across this subarctic landscape and consequently the role of organic C in soil development.

and high molecular organic acids in soil development (Jenny, 1980; Stevenson, 1982; Tan, 1986; Robert & Berthelin, 1986). Detrital chemistry directly controls, through its influence on organic matter mineralization rates, plant nutrient supply (Flanagan & Van Cleve, 1983; Pastor et al., 1987). Consequently, this feedback mediates adequate or deficient element supplies for net primary production and the status of soil organic C.

The chemistry of forest litter, at the time of deposition on or within the soil profile, largely determines the rates of mass loss (Paul & Clark, 1989). The portions of readily metabolizable (low molecular weight carbohydrates, cellulose) and recalcitrant (lignin) organic constituents are important controls (Berg, 1986;

McClaugherty & Berg, 1987; Melillo et al., 1989). The early, rapid phases of decomposition reflect the metabolisms of low molecular weight substances. Later stages reflect the slow metabolism of lignocellulose.

The portions of easily metabolized to recalcitrant C vary substantially among plant parts and species. Woody parts have the highest concentrations of decay-resistant materials (Nadelhoffer et al., 1983; Harmon et al., 1990; Melillo et al., 1984; Berg, 1986). Edmonds et al. (1986) found that red alder (*Alnus rubra*) wood decomposed faster than Douglas fir wood. Simple sugar and hemicellulose concentrations were important controls of alder wood decomposition. Moisture and temperature were dominant controls of Douglas fir wood decay. Tree needle decomposition rates are one to two orders of magnitude faster than decay rates of woody materials, depending on the size of the woody substrate (Edmonds, 1991).

Plant species occurring early in primary and secondary successional communities often grow rapidly and have high nutrient contents and lower concentrations of decay-resistant chemicals—conditions promoting rapid turnover of organic C (Pastor & Post, 1988; Vitousek et al, 1989; Bryant et al., 1991; Van Cleve et al., 1993). Plant chemistry plays an important part in determining the rate of organic matter transfer to the soil. Ultimately, regardless of initial chemical composition, detrital materials approach a common chemistry (Melillo et al., 1989). The rate of approach and the subsequent decomposition of humified materials are determined by interactions with microclimate, successional processes, soil conditions, slope, aspect, and elevation (Fig. 9–8).

In mature forest types in interior Alaska, decomposition rates of each of the genetic horizons of the forest floor decline from quaking aspen and paper birch to white spruce to black spruce (Table 9–6). Within each vegetation type, decomposition rates decline from litter to humus layers. These trends reflect the influences of tree species and depth in the forest floor on the quality of the organic C that supports microbial activity: increasing lignin and declining N concentrations from the deciduous hardwood species to white and black spruce and with increasing depth in the forest floor (Flanagan & Van Cleve, 1983).

Plants can have a substantial influence on the C content of mineral soil in a relatively short time. Maximum rates of soil C accumulation occur during the first 15 yr of vegetation succession on the Tanana River floodplain of interior Alaska (Van Cleve et al., 1993). In the southeastern USA, establishing loblolly pine (*Pinus taeda*) with herbaceous and hardwood species in factorial combination led

Table 9–6. Mean laboratory respiration rates (microliters $O_2 \cdot g^{-1} \cdot h^{-1} \pm SD$) for forest floor materials collected in spring from selected sites in interior Alaska.†

Vegetation type	Layer of the forest floor		
	01 (litter)	021 (fermentation)	022 (humus)
Aspen	123.1 ± 53.6	150.9 ± 11.3	68.7 ± 7.8
Birch	159.3 ± 31.4	107.7 ± 0.7	73.6 ± 1.9
White Spruce	94.2 ± 9.6	86.4 ± 22.5	59.3 ± 9.5
Black Spruce	45.8 ± 11.7	23.4 ± 5.0	10.9 ± 2.3

†Material was air dried, moistened to a standard water content (250% of dry weight), and incubated at 15°C (from Flanagan & Van Cleve, 1983).

to more rapid increases in organic N contents of surface mineral soils compared with treatments with pine alone (Wood et al., 1992). Soil C and N dynamics were studied in five broadleaf deciduous and four conifer forest types in the prairie-border region of southern Wisconsin (Univ. Wisconsin Arboretum, Madison; Nadelhoffer et al., 1983). All stands were at least 35 yr old with closed canopies, although dominant individuals on oak sites were 125 yr old. The soil organic matter and N pools were largest in red pine (*P. resinosa*) stands where rates of net organic C and N mineralization were lowest. White pine (*P. strobus*) stands were associated with highest net C and N mineralization rates and smallest C and N pools (Nadelhoffer et al., 1983).

Soils developing under Lake States jack pine (*P. banksiana*) consistently had the lowest C contents (10–30 Mg C ha^{-1}) whereas those under balsam fir (*A. balsamea*; Fig. 9–15) that had the highest C contents (30–50 Mg C ha^{-1}) (Grigal & Ohman, 1992). The authors found that precipitation, available water and soil clay

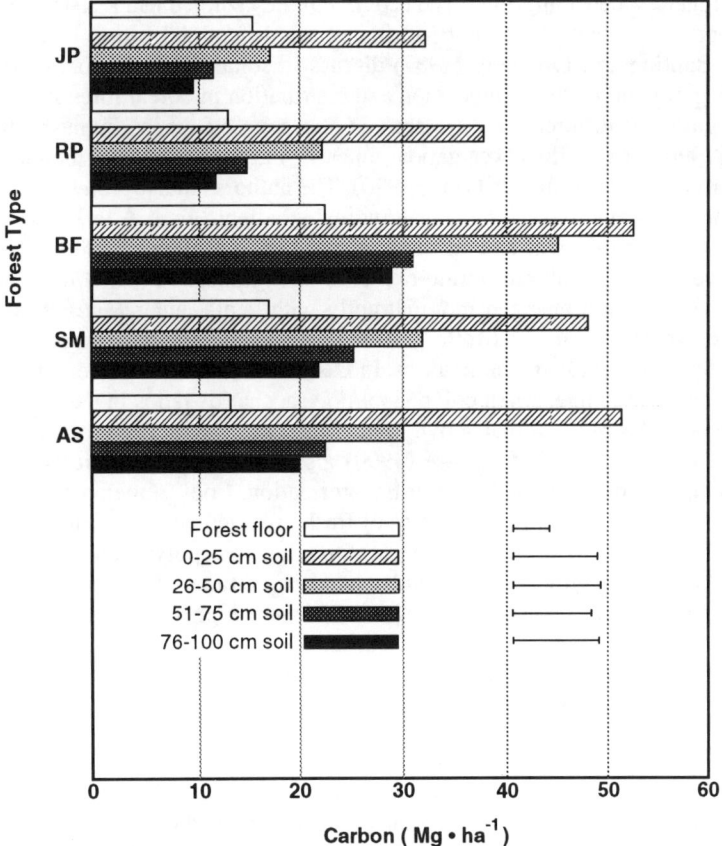

Fig. 9–15. Carbon storage in the forest floor and four, 25-cm depth layers of mineral soil in five forest types across the Lake States. Bays least significant differences indicated at bottom. Types are jack pine (JP), red pine (RP), balsam fir (BF), sugar maple (SM), aspen (AS) (after Grigal & Ohman, 1992).

content were additional variables controlling the sizes of soil C pools. Organic C stabilization by clay probably is an important mechanistic control of soil C dynamics in these systems (Oades, 1988; Sollins et al., 1984).

Schuppli et al. (1988) compared the compositions of black and brown B horizons of Spodosols in the Hudson Bay-James Bay area. The percentages of extractable C, pyrophosphate-extractable Fe and Al, and the ratios of CHA/CFA (the ratio of HA to FA) and Ca/CFA ratios were not different among the two colors of B horizons. The authors noted that the darker B horizons were over calcareous subsoils, while the lighter-colored B horizons were in acid parent materials. Ratios of extractable-to-organic C, the amount of C in the HA and FA components, and the CHA/CFA ratios were similar to measured relationships in other Canadian Spodosols. The Oi, Oe, Oa, A, and E horizons had lower proportions of extractable C and higher CHA/CFA ratios than the B horizons. The authors also noted that lichen-rich O horizons had very low CHA/CFA ratios. Mokma and Vance (1989) also observed that darker B horizons were associated with higher Ca concentrations. The ecosystem they studied had a history of periodic burns, to which the authors attributed the differential cation levels in soil pools. Bunting and Lundberg (1987) discuss in some detail the role of fires in affecting organic matter composition and distribution in boreal forests.

Significant differences were seen in the composition of humus extracted from A horizons of three vegetation zones in the Sierra Nevada mountains of northern Mexico (Drijber & Lowe, 1990). The authors studied *Abies, Pinus,* and grassland communities along an altitudinal gradient. Total C and CHA/CFA ratios were not different among the communities, but higher E400/E600 ratios and lower levels of polysaccharide-rich FA's were associated with *Abies*. Organic matter composition changed more abruptly with depth under *Pinus* than under *Abies* or grass. In part, this reflects differences in the nature of organic matter inputs by different plant communities. In Denmark, *Quercus* invasion of *Calluna* heath sites led to increases in polyphenol/polysaccharide ratios in the FA fraction and higher rates of microbial activity (Neilsen et al., 1987). Therefore, the elevational study of Drijber and Lowe (1990) also reflects temperature controls of microbially mediated organic matter transformation. Cool temperatures at higher elevations reduce decomposition rates of fresh organic matter and rates of depletion of such readily metabolized compounds as polysaccharides. Thus, E400/E600 ratios will be lower under harsh conditions that reduce microbial activity. Kahn (1969) showed that HA levels were higher in soils under a 5-yr rotation of small grain and legumes than under a wheat (*Triticum aestivum* L.) fallow system. Manure additions increased the HA content, while N, P, K, and S fertilizer additions widened the CHA/CFA ratio.

Soil Organisms

Plant production is the primary source of fixed C and the main source of C to the soil system that drives ecosystem processes and pedogenesis. But to a large degree, the transport of fixed C into the soil depends on soil organisms. Hole (1981) discussed 12 activities conducted by exopedonic (living outside the soil) and endopedonic (living within the soil) animals that affect soil development. They include mounding, mixing, forming voids, back-filling voids, forming and destroying peds, regulating soil erosion, regulating movement of water and air in soil, regulating plant litter, regulating nutrient cycling, regulating biota, and producing special constituents. Tate (1987) listed eight biologically mediated processes by which this occurs:

1. Catabolism of colloidal soil organic matter.
2. Modification of soil pH.
3. Synthesis of chelators.
4. Alteration of the soil redox potential.
5. Oxidation or reduction of soil cations and anions.
6. Synthesis of polysaccharides.
7. Physical shredding of organic debris.
8. Production of cell or mycelial biomass.

The aerobic depolymerization of polysaccharides produces monomers. Simple oxidation of monomers reduces free O_2, at least in localized soil microsites. The corresponding change in soil redox potential can change cation solubility and mobility. Soil acidity is generated by the complete mineralization of carbohydrates to CO_2 and H_2O, and conversion to weak carbonic acid. Organic intermediates, such as acetic acid, produced under reducing conditions also contribute acidity. Under reducing conditions, cations such as Fe and Mn become electron acceptors, and their solubilities are increased. Chelation by organic acids is an important weathering mechanism (Stevenson, 1982). Chelation also increases cation mobility in the soil solution from surface horizons to an illuvial horizon. An added effect of chelation is reduced toxicity from ionic metals and improved bioavailability of nutrients. Chelates of Al are not toxic to plants at concentrations known to be highly toxic for inorganic Al (Hue et al., 1986). Oxalate, common in most forest soils, is readily adsorbed at Al-oxide surfaces, displacing P (Goldberg & Sposito, 1985) and increasing its availability in the soil solution (Fox & Comerford, 1992).

Microbial polysaccharides form and stabilize soil aggregates by cementing mineral particles via H-bonding and coordination with polyvalent cations. In the absence of organic coatings, soil particles disperse readily in water. Polysaccharides are easily degraded by microorganisms, which explains why inputs of fresh organic matter must be sustained to maintain stable soil aggregates (Stevenson, 1982). Although they are ephemeral, there are important distinctions between polysaccharides and similar organic materials in their rates of metabolism by microbes and in the duration of their effects on aggregate stability (Fig. 9–16). Energy-rich materials like glucose are easily metabolized. The glucose effect on granulation is rapid and strong, but glucose is degraded quickly and the effect does not persist. In contrast, cellulose has a delayed but more lasting effect (Oades, 1984). Humic acids also are thought to favor aggregate stability through cation bridging (Tate, 1987; Oades, 1988).

Fungal mycelia also promote soil aggregate stability by encompassing aggregates in a physical network of hyphae (Tisdall & Oades, 1982). Maintaining this network requires a supply of suitable energy substrate for fungi, and losses of a key plant community may have serious consequences for aggregate stability. Perry et al. (1987) found that soils in unreforested clearcuts with herbaceous vegetation had lower hyphal activity and fewer large aggregates and macropores than soils beneath adjacent uncut forests.

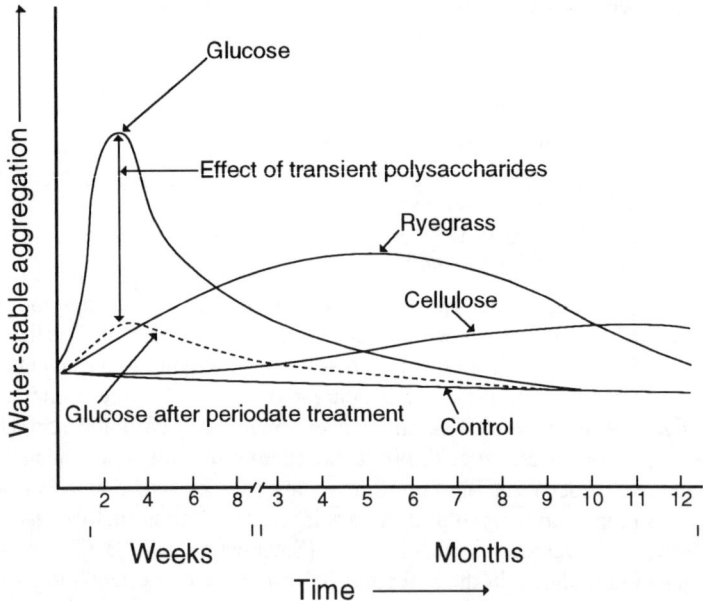

Fig. 9–16. Alteration of water stable aggregates in soil receiving organic amendments (from Tisdall & Oades, 1982).

Organic inputs and corresponding microbial activity do not always promote pedogenesis. Buildup of microbial biomass can severely restrict water movement through the soil profile by occluding pores at the soil surface (Frankenberger & Troeh, 1982), thereby retarding plant community and soil development. This problem is thought to be confined to feedlots and sewage disposal sites (Tate, 1987), but also could occur in soils treated with large amounts of sewage sludge.

Soil fauna provide linkages between fresh organic matter from plants and soil microorganisms (Anderson, 1988). They accomplish this through channeling activities and modification of substrates in grazing (examples being shredding and intermingling of shredded particles with soil materials during digestive processes and subsequent depositing of feces). Earthworm casts, fecal material, and organic matter-rich nests of soil macrofauna can produce microsites of high N mineralization because deposited materials have lower C/N ratios than the original organic matter (Gosz, 1984). Until recently, soil faunal activity has been thought insignificant with respect to soil processes and ecosystem function. However, important linkages have been identified between soil fauna, humus forms, and ecosystem behavior (Petersen & Luxton, 1982; Shaw et al., 1991). Estimates of biomass for the major classes of soil fauna vary between 2.4 and 8.0 g m^{-2} for coniferous forests and deciduous forests with mull humus, respectively (Shaw et al., 1991).

Certain soil fauna affect rates of gas and water exchange between the atmosphere and soil. Earthworms such as *Lumbricus* spp. create vertical, feces-lined tunnels in the soil, promoting the entry of gravitational water and facilitating the movement of capillary water. In an experiment in Nigeria, Aina (1984) found that soil macropore volume and water infiltration rates decreased rapidly when earthworms were excluded from forested and cultivated sites. After 8 mo, infiltration rates were nearly four times greater where earthworm activity was unrestricted.

Because of their mobility, soil meso- and macrofauna have a major role in transporting organic surface detritus to mineral soil horizons. For example, earthworm activity in a temperate hardwood forest was judged to have transported nearly a year's worth of litterfall belowground (Nielson & Hole, 1964). Carbon utilization efficiency of such organisms is low, compared with that of soil bacteria or fungi (Shaw et al., 1991), meaning that much of their ingestate is egested as feces to become substrate for other decomposers. Crossley (1977) considers this to be a major contribution of arthropods to forest soil fertility. Soil meso- and macrofauna also aid in soil aggregation and organic matter stabilization by passing soil and organic materials through their digestive tracts. In greenhouse tests, mean size, weight and water stability of soil aggregates was increased markedly by earthworm activity (Kladivko et al., 1986). Elsewhere, Shaw and Pawluk (1986) found that organic matter was stabilized by kaolinite in fecal material.

The significance of interactions between environment, vegetation, and soil biota in the development of forest humus forms has been recognized since the late nineteenth century (Muller, 1978), and an extensive classification system anchored in this concept has been proposed for forested North America (Green et al., 1993). However, the dynamic nature of soil faunal activity on pedogenesis has been demonstrated most graphically through the San Dimas lysimeter studies in southern California (Graham & Wood, 1991). After 41 yr under *Q. dumosa* or *Ceanothus crassifolia*, earthworm activity was largely responsible for incorporation of organic detritus into A horizons and increased clay content of this zone compared with underlying horizons. Earthworm activity was minimal under *Adenostoma fasciculatum* and absent under *Pinus coulteri*. During this period a thin A horizon developed under the former vegetation type, and clay translocation occurred under pine. Plant species-dependent soil faunal activity played a major role in soil development and incorporation of organic matter into an A horizon. In the absence of this pedoturbation, argillic horizons may form in several decades (Graham & Wood, 1991).

Vertebrates Other than Man

Browsing by small and large vertebrate animals such as snowshoe hare (*Lepus americanus*) and moose (*Alces alces*) has substantial effects on the nutrient content of litter (Irons et al., 1991). Senescent paper birch leaves collected from branches that recently regrew following browsing had higher N concentration than leaves from unbrowsed plants. Browsing of paper birch winter twigs by moose and hare has been observed to increase leaf N concentration and decrease the phenolic (Tannin) concentration of next season's green leaves (Bryant et al., 1991). Severe insect leaf defoliation, on the other hand, has been observed to cause the opposite effect—decreased green leaf N concentration and increased concentration of phenolics (tannins). Winter twig browsing by moose and hare removes twigs and buds competing for nutrients, decreasing competition for growth-limiting nutrients among the remaining buds (Bryant et al., 1991). Carbon reserves are utilized to support growth by leaves developing from these buds. Severe, repeated early season insect defoliation, as occurs in interior Alaskan birch (*Betula* genus) and aspen (*Populus* genus), results in fine root mortality, decline in nutrient absorption, and with no bud death, reduced leaf nutrient concentration (Bryant et al., 1991). If leaf carbohydrate supply is not limiting, this material is utilized in production of C-based secondary metabolites such as phenolics, whose concentrations rise in the tissue. The consequence of these effects for lignin has yet to be determined, but is hypothesized to be the same. The recalcitrance of leaves to decomposition is conditioned by these forms of herbivory, with as yet unknown consequences for the rate of transfer or organic C to the soil pool. Moreover, the long-term effects of browsing on soil C transformations may be an important feedback control of element supply for plant growth. Recent experiments indicate that leaf litter from moose-browsed willow (*Salix* spp.) decomposes more rapidly than litter from the same willow species inside large mammal exclosures (M. Wagener, personal communication).

Humankind

The importance of the human component of the biota to forest ecosystem structure and function is a central feature of the resource management today. The capacity of soil to support plant growth and maintain ecosystem integrity centers to a large degree around its organic matter content. Human activities in resource development have substantially affected soil organic matter in many cases and apparently have had relatively slight impact in other cases. Forest harvesting generally appears to have relatively slight impact on soil organic C reserves unless it is followed by intense burning or cultivation of the soil (Johnson, 1992). Harvesting on steep slopes may cause excessive soil erosion. Fertilization and encouragement of N-fixing species may increase soil C (Johnson, 1992). The impact of disturbance on soil organic C reserves will depend on the extent of initial disturbance, local climate, rate of site revegetation and soil physical and chemical characteristics. A regional analysis of cultivated and rangeland soils found that organic C losses due to cultivation increased with precipitation (Burke et al., 1989). Relative losses of C were lowest in clay textured soils.

Golden Gate Park in San Francisco is distinguished by extensive woods and forests and fairly productive soils (Amundson & Tremback, 1989). Before 1870, the area was an unstabilized dune landscape in the coastal fog climate. From 1870 to 1880 workers at the Park added horse manure, seeded barley (*Hordeum vulgare* L.) and lupine (Lupinus genus), and planted *Pinus radiata, Cupressus macrocarpa*, and *Eucalyptus* spp. The subsequent 100 yr have seen the undisturbed growth of trees and soil genesis (Amundson & Tremback, 1989). With respect to time, Amundson and Jenny (1991) describe the native dune ecosystem as Segment 1; the approximately decade-long human intervention involving dune fertilization and artificial revegetation as Segment 2; and the following century-long period of tree growth and soil development, with minimal human impact in many areas of the Park, as Segment 3. Approximately 8 to 16 times more organic

C accumulated (50 Mg ha^{-1} compared with 400–800 Mg ha^{-1}) in the vegetated sites compared with undisturbed dune locations a few kilometers south of the Park (Amundson & Tremback, 1989). The three segments of ecosystem state define three consecutive chronosequences all having the same climate and topography and nearly the same parent material but each differing in the biotic factor due to human intervention that introduced plants and fertilizer to the dunes (Amundson & Jenny, 1991).

Time

The influence of time is considered from the standpoint of the successional change in vegetation and soil characteristics. Primary succession occurs after disturbance that leaves little or no trace of the pre-existing ecosystem. The most common examples follow volcanic eruptions, glacial recession, and sand dune movement. Fluvial processes on river floodplains also result in primary successions, but the initial soil may have a residue of organic C and nutrients reflecting soil conditions at the upstream source of alluvium. Early stages of primary succession are associated with harsh environments, and soil and plant communities develop slowly (Vitousek et al., 1989).

Secondary succession represents situations where disturbance removes vegetation, the soil remains, and plant regrowth begins immediately (Vitousek et al., 1989). Familiar examples include forest harvesting, fire, windthrow and old-field succession. Some residual soil organic C remains following perturbation. In some cases dormant seedbanks may contribute to establishment of the new plant community.

A generalized example of the course of organic matter accumulation in ecosystems (including soil organic matter or C) is summarized in Fig. 9–4. Four examples of C accumulation are summarized for soils associated with primary successional ecosystems representing tropical montane forest developing on volcanic flow/tephra on the island of Hawaii; forest developing on the mud flows/volcanic soils at Mt. Shasta, California (Xeropsamments, Xerorthents); coastal forests developing on a glacial moraine at Glacier Bay, Alaska (Spodosol); and subarctic forests developing on fluvial sediments in the northern boreal forest of interior Alaska (Cryofluvents). Within each sequence the state factors, with the exception of time, were relatively constant.

Carbon accumulation at the Hawaii sites ranged between 65 and 75 Mg ha^{-1} for soils developing in tephra (at 4000 yr) and pahoehoe (at 3500 yr) volcanic materials, respectively (Table 9–7). Rates of net annual accumulation for the first 1 to 2 centuries were approximately twice as rapid for the ash sequence (121 kg ha^{-1} compared with 61 kg ha^{-1} for the pahoehoe). Ash deposits are a substantially more favorable environment for soil formation in general, probably reflecting the greater surface area of volcanic ash in support of mineral weathering (Vitousek et al., 1983).

The moraine sequence at Glacier Bay encompasses a substantially shorter time interval (250 yr) than the Hawaiian sequence (Ugolini, 1968). Up to 39 Mg C ha^{-1} accumulated during the 2.5 centuries of soil development. During the first 55 yr C accumulated at an average net annual rate of 125 kg ha^{-1} and at the rate of 153 kg ha^{-1} for the 250-yr period (Table 9–7).

Table 9–7. Average C accumlated in selected chronosequences.

Location/vegetation type	Parent material	Age of surface	Depth of profile	Organic C
		(yr)	(cm)	Mg ha^{-1}
Hawaii, montane	Pahoehoe	126	1.5	7.7
rain forest		1500	11.0	52.8
(Vitousek et al.,		2500	16.0	75.8
1983)		3500	20.0	75.1
Hawaii, montane	Tephra	191	10.0	21.6
rain forest (Vitousek		191	10.0	24.4
et al., 1983)		1000	10.0	22.7
		4000	10.0	65.4
Alaska, Glacier	Glacial	5	25.0	0.9
Bay Spodosol	moraine	24	25.0	3.4
(Ugolini, 1968)		41	25.0	10.6
		55	25.0	10.1
		250	25.0	38.3

At Mt. Shasta, total soil organic C increased from about 8 Mg ha^{-1} at 55 yr (Flow A) to more than 70 Mg ha^{-1} at an approximate mudflow age of more than 1200 yr (Flow E) (Sollins et al., 1983; Fig. 9–17). The light and heavy fractions of organic C generally increased with flow age to between 33 and 38 Mg ha^{-1}, respectively, for the 0- to 30-cm depth. At 0 to 10 cm, the light fraction contributed 63 and 57%, respectively, to the organic C pool of the D and E flows, and 45 and 44%, respectively to the A and B flow C pools. At 10- to 30-cm depth, the heavy fraction comprised the bulk of soil C, contributing between 54 and 72% to total organic C (Sollins et al., 1983). Differences in the accumulations of these fractions indicate functional differences in the two pools. The mass and composition of light fraction C correlated with short-term changes in ecosystem biolog-

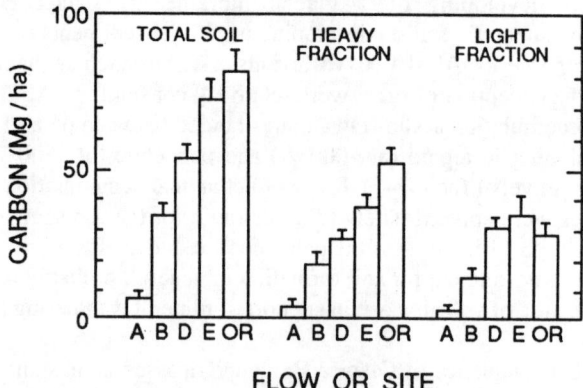

Fig. 9–17. Carbon accumulation in light and heavy fractions in 0- to 30-cm soil profiles at the Mt. Shasta mudflows (from Sollins et al., 1983).

ical activity, while heavy fraction C correlated with long-term processes including mineral weathering and horizon development.

The fluvial soils of the Tanana River floodplain contain an initial level of organic C near 10 Mg ha^{-1} which increases to approximately 60 Mg ha^{-1} at 200 yr (Fig. 9–18). For a 60-cm profile depth, the maximum net annual rate of C accumulation occurs during the 5- to 25-yr period of plant succession dominated by *Alnus tenuifolia*, with a maximum of 1550 kg ha^{-1} at about 15 yr. The net annual rate of accumulation during the white spruce period of succession (150 yr) is 8 kg ha^{-1}. A maximum rate of approximately 775 kg C ha^{-1} would accrue in a 30-cm profile at 15 yr. This amount is 5 to 6 times greater than the annual C accumulations estimated for the Glacier Bay sequence (30-cm profiles) and more than twice the rate estimated for the Hawaiian sequence (10-cm profiles). These accumulations assume a linear relation between net annual C accumulation and soil profile depth.

In a secondary successional context, the extent to which soil C reserves are depleted and the rate of their recovery to predisturbance levels depend on the type and extent of disturbance and influence of environmental controls over net primary production and organic matter decomposition (Vitousek & Walker, 1987; Allen, 1985). Most studies show relatively small change (about 10%) or little change in the soil C pool, respectively, with forest harvesting alone or low-intensity prescribed fire (Alban & Perala, 1992; Johnson et al., 1991, Binkley et al., 1992; Johnson, 1992). Intense wildfire and prescribed fire may result in

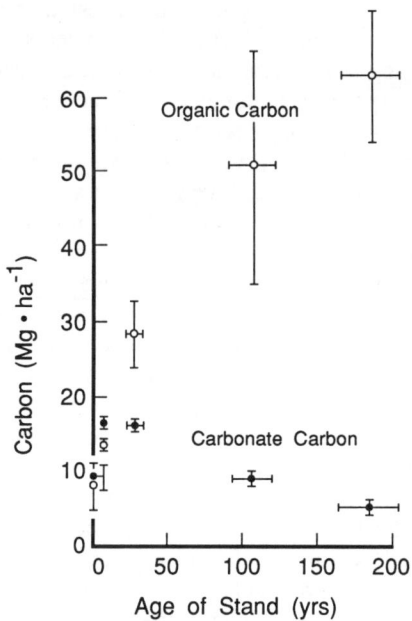

Fig. 9–18. Soil C accumulation in fluvial soils of the Tanana River floodplain (from Van Cleve et al., 1993).

substantial losses of soil C, with most of the loss occurring from the forest floor or surface layers of mineral soil (Dyrness et al., 1989).

In most cases clearing and cultivation result in up to 50% loss of soil C (Johnson, 1992). Differences in soil response to clearing are related to the ecosystem controls (microclimate, soil physical and chemical properties, vegetation, Fig. 7–8) that affect net primary production and decomposition processes (Allen, 1985; Zak et al., 1990). Zak et al. (1990) found a recovery of soil C of approximately 20 Mg ha^{-1} in a 60-yr-old field in east-central Minnesota. Although cultivation may lead to substantial loss of soil C, it does not seem to produce dramatic changes in C chemistry. Spectral analyses of HA's from forest and agricultural soil are qualitatively similar (Preston, 1991).

SYNTHESIS, INTEGRATION, AND DIRECTION

Modeling and Global Change

Combining process modeling and global change considerations seems appropriate, particularly when one considers the implications of global warming on organic C reserves of the higher latitude biomes (Table 9–1). Process modeling provides a means of synthesizing empirical and theoretical information from ecosystems (Agren et al., 1991). Much of our current thinking with regard to the effects of global climate change on ecosystem structure and function results from projections made by models. Model predictions of ecosystem response represent the best estimates that current knowledge, as synthesized by the models, can supply. Models can be employed to test hypotheses based on results of field experimental manipulations. In turn, these results can be tested both by field monitoring of ecosystem behavior and through critical experiments to test hypotheses in key ecosystems. Modeling can generate leads for future research directions.

Probably no single approach to simulation of the structure and function of the forest will precisely predict forest ecosystem responses to changing environmental controls (Shugart & Smith, 1992), but important insights can be gained. For example, two models of the boreal forest have provided substantial insight to the interaction of physical and biological controls on ecosystem structure and function and the potential effects of climate change on associated forest soils. LINKAGES (Pastor & Post, 1986) is a population-level model that bases net primary production and soil C dynamics on actual evapotranspiration (AET) and detrital lignin/N ratios. The biophysical, population-level model of Bonan (Bonan, 1990; Bonan & Van Cleve, 1991) employs the C/N ratios of detrital material as the indices of organic matter quality and ties decomposition to soil temperature and moisture regimes. It considers detrital chemistry and soil C relations through moss-organic matter accumulation and interactions with soil temperature and element supply for tree growth. Neither of these models, however, directly considers the effects of C in soil-forming processes.

Earlier field research revealed the reciprocal connection between the C and N cycles in the succession of birch and aspen to spruce forests (Flanagan & Van Cleve, 1983; Van Cleve et al., 1983). Net primary production, and consequently the flux of C to and within the soil, is limited by N availability. The latter is partially determined by the chemistry of C compounds in the forest floor. The reciprocal connection is related to the succession of tree species. Simulations with LINKAGES demonstrated the internal logic of the hypothesis and its consistency with the existing body of empirical information (Pastor et al., 1987). The consequences of these interactions for soil C accumulation are illustrated in Fig. 9–14. On warm, permafrost-free south aspects in interior Alaska, soils supporting birch and aspen accumulate less C than soils supporting white spruce.

The impact of detrital chemistry on organic matter decomposition rates among these forest types was illustrated in Table 9–6. Forest types occupying cool wet, permafrost soils (black spruce) produce soil organic detritus higher in lignin and lower in N than types occupying warmer, drier locations (white spruce, paper birch, quaking aspen). Consequently, decomposition rates are lower in black spruce soils. Forest floor horizons representing more advanced stages of decomposition (O21 and O22 horizons) also display the effects of increasing organic matter recalcitrance with declining decomposition rate (Table 9–6). Among tree species, differences in tissue chemistry also are seen in the general decline in N concentration, increase in lignin content, and widening of the lignin/N ratio from earlier successional birch, aspen and poplar to late-successional white spruce (Table 9–8). These trends hold for fresh litter from these forest types. The lignin/N ratio widens from 26 to 36. These relationships support conclusions of the LINKAGES simulations concerning the reciprocal connection of C and N cycles (Pastor et al., 1987). In the boreal forest of interior Alaska, development of a moss ground cover in later stages of forest development further contributes to the increase in thickness of the highly effective layer of decay-resistant forest floor organic matter. Decay-resistant supplies of recalcitrant organic matter accumulate on and within cooler soil profiles, creating a negative feedback for C mineralization and element supplies for tree growth that reinforce production of decay-resistant detritus. This limits the ability of organic C to interact with the mineral soil in weathering and soil development. A similar trend was noted for soil C in relation to site temperature and precipitation in New Zealand tussock grassland soils (Tate, 1992). However, interactions between Al and organic matter in formerly forested, acidic soils may stabilize and substantially increase organic C storage time.

Table 9–8. Average (±SE) chemical composition for green leaf tissue of plant species in early, mid- and late stages of secondary succession (uplands) in interior Alaskan forest types.

Species	Age†	%N	Lignin	% Lignin/% N
	yr	—————— % ——————		
Salix spp.	6	1.92 ± 0.05	7.11 ± 0.22	3.4 ± 0.2
Quaking aspen	6	2.04 ± 0.05	8.33 ± 0.39	4.2 ± 0.3
Paper birch	6	2.04 ± 0.05	6.30 ± 0.22	3.1 ± 0.1
Quaking aspen	75	2.17 ± 0.04	7.99 ± 0.15	3.7 ± 0.1
Paper birch	75	1.90 ± 0.05	9.42 ± 0.16	5.1 ± 0.1
Balsam poplar	75	1.76 ± 0.04	6.82 ± 0.11	3.9 ± 0.1
White spruce	75	1.04 ± 0.02	13.15 ± 0.21	12.7 ± 0.3
White spruce	200	1.07 ± 0.02	13.00 ± 0.17	12.3 ± 0.2

†Data on age of shrub and tree species provided by Dr. Leslie Viereck.

The consequences of global warming for forest type distribution in the northern boreal forest of Alaska have not been clearly evaluated using LINKAGES. In eastern North America on the current boreal/northern hardwood forest border, simulations indicate substantial potential type changes. With a doubling of CO_2, on soils where water availability did not decrease mixed spruce/fir/northern hardwood types were replaced by more productive northern hardwoods (Pastor & Post, 1988). The latter were more productive because of greater intrinsic growth rates and the warmer climate. Moreover, the higher N and lower lignin content of litter stimulated soil organic matter decomposition and N supply, amplifying the effect of warming on productivity. The simulations show increased net primary production and organic matter flux to the soil, but also increased soil organic matter decomposition. Increased soil moisture deficit was associated with forest type change to pine/oak, where higher lignin and lower N content of detrital material compensates for warmer temperatures with reduced organic matter decomposition and little or no change or local decrease in N availability (Pastor & Post, 1988).

With 1°, 3°, and 5°C atmospheric warming, currently cold, permafrost north aspect sites in interior Alaska may be converted from black spruce to mixed hardwoods and spruce (Bonan et al., 1990). However, without increased precipitation to offset greater evapotranspiration as a consequence of warming these sites, the thick organic layers will become drier, their thermal conductivity will decline, and depth of thaw will decrease despite warming temperatures (Bonan, 1992). Consequently, organic matter decomposition rate would not increase because soil temperature would remain cool. The impact of increased fire frequency in response to generally drier climatic conditions and drier forest floors could counteract these effects with decreased albedo, warmer mineral soil, and increased soil CO_2 production. The net result of a generally warmer soil on these sites would be more extensive soil profile development. Histic Pergelic Cryaquepts eventually would be replaced by Histic Cryaquepts as permafrost thaws on the warmest sites. Increased fire frequency and development of paper birch forests would be associated with improved base status of surface mineral soils and the occurrence of Typic Cryochrepts (>35% base saturation). Increased warming and precipitation would favor mixed white spruce and paper birch forests, improved ecosystem productivity, greater soil leaching and the development of Dystric Cryochrepts (<35% base saturation). Decomposition of organic matter would increase in response to changing forest types and improved organic matter chemistry. The soil would not be so much of a sink for C as under permafrost conditions (Fig. 9–13).

On south aspects these simulations indicate that at 3° and 5°C warming, none of the tree species currently encountered would grow. Increased evapotranspiration associated with warming resulted in excessive soil moisture stress and conversion of these upland forest sites to steppe. A 1°C temperature increase with no increase in precipitation predicts aspen-dominated forests. The 1°C temperature increase combined with greater precipitation to offset higher evapotranspiration leads to a mixed paper birch-white spruce forest (Bonan et al., 1990). Soil C dynamics should change little under these conditions.

Trace gas emission from northern latitude soils is an important concern in the global C budget. Aerobic soils act as methane sinks while waterlogged soils are sources of methane in arctic Alaska (Whalen & Reeburgh, 1990). The same relationships were encountered between well-drained upland and floodplain soils and floodplain soils with periodic high water tables and anaerobic conditions that occur early in primary succession (Whalen et al., 1991). Within the constraints noted earlier, climate warming may be associated with increased thawing of large areas of currently frozen Histosols with substantial contributions of CO_2 and methane to the atmosphere. How these ecosystems respond to warming will, to a large degree, depend on the precipitation regime.

On a global scale, modeling projections for the loss of the boreal forest indicate a 5°C cooling in January and July north of 60°N (Bonan et al., 1992). Increased circumpolar ground surface albedo, as a consequence of deforestation, interacting with oceanic weather circulation patterns resulted in the declining air temperatures.

Model validation for current levels of distribution of biomass, primary production, and decomposition for generalized examples of major forest types show close agreement with field measurements (Bonan, 1992; Bonan & Van Cleve, 1991; Pastor & Post, 1988). However, these models (Pastor & Post, 1986; Bonan, 1990) do not yet include mechanistic considerations of mineral weathering and soil horizon differentiation from the standpoint of the role of organic C. Moreover, although much information is available and some work completed (Cole & Rapp, 1981, p. 341–409; Vogt, 1991; Grigal & Ohman, 1992), C balances for each of the major forest types have yet to be synthesized for this northwestern part of the boreal region and for examples of forest types across the vast eastern extent of the biome.

Some estimates have been made concerning the potential change in tree species composition as a consequence of climate warming (Pastor & Post, 1988; Bonan et al., 1990). The ability of tree, shrub and herbaceous species to adapt to new environmental regimes is closely linked to soil C dynamics. The ranges of plant organic matter chemistries and their interactions with climate, decomposition, and element supply for plant growth are important controls of soil C flux and soil development.

Challenges to Research

In the milieu of climate change, a vital requirement for understanding the role of C dynamics in soil development is for experimental work that links above- and belowground plant and microbial processes (Anderson, 1992; Cropper & Gholz, 1993). A concerted effort should be made to understand how nutritional requirements and C flux for above- and belowground plant parts changes with forest development. Linking these processes to soil organic matter chemistry as a control of decomposition processes and element supply for tree growth largely determines the flux of C in soil and its role in soil development (Tate, 1992). For example, simulations of a 5°C air warming resulted in increased soil organic matter decomposition and N mineralization (Bonan & Van Cleve, 1991). Increased N mineralization promoted tree growth which offset increased C loss from decomposition. In permafrost black spruce tree growth increases continued for the 25 yr of simulation. The simulations emphasize the interaction between NPP and decomposition processes that control N supply for tree growth and, consequently, C flux in the soil (Pastor & Post, 1988; Bonan & Van Cleve, 1991; Anderson, 1992).

Is soil organic C comprised of larger fractions of recalcitrant constituents in warm temperate region forest soils than in cold soils? Eventually soil C should approach a common chemistry (Melillo et al., 1989). How these constituents may vary among forest types in general and across the mosaic of types encountered in the boreal forest in particular depends on the state factor control of ecosystem development (Preston, 1991). An assessment of the range in organic matter chemistry would substantially assist in understanding response of soil C pools to climate warming.

The effect of drainage on C dynamics of bogs and permafrost landscape in northern forests may have substantial effects for site productivity. Thirteen to 19 yr after drainage of peatlands in Alberta, growth of black spruce increased by up to 766% over that in undrained locations (Dang & Lieffers, 1989). The response is attributed to improved aeration, decomposition of soil organic matter, and resulting increased element supply for tree growth. Artificial warming of permafrost soil stimulated organic matter decomposition, element supply for tree photosynthesis and growth (Van Cleve et al., 1990). However, the impact of drainage may involve soil movement on slopes and increased runoff and loss of nutrients over the surface of the remaining frozen soil, especially on permafrost sites.

Experimental manipulations of soil through transplanting forest floors among extremes of site quality within forest types and among forest types (Powers, 1987, 1990) can provide substantial insight to physical and biological controls of decomposition and C flux in soils of varying developmental stages. Establishing standard soil tesseras among forest types could help to evaluate the rate of establishment of unique, type-specific signatures for soil C.

Evaluation of controls of soil C dynamics should be scaled from the stand level, to landscapes, to a regional basis. Excellent examples of "scaling up" exist for a number of ecosystem processes directly related to the status of soil C (Jenny et al., 1968; Jenny, 1980; Zak et al., 1986; Whalen & Reeburgh, 1990; Powers, 1990; Preston, 1991; Schlesinger, 1991; Bonan et al., 1992; Grigal & Ohman, 1992; Alban & Perala, 1992). Local, regional and latitudinal transect studies are a promising means for understanding how state factors interact in their control of net primary production, organic matter decomposition, soil C dynamics, pedogenesis, and ecosystem development.

REFERENCES

Aber, J.D., and J.M. Melillo. 1991. Terrestrial ecosystems. Saunders College Publ., Philadelphia, PA.

Agren, G.I., R.E. McMurtrie, W.J. Parton, J. Pastor, and H.H. Shugart. 1991. State-of-the-art of models of production-decomposition linkages in conifer and grassland ecosystems. Ecol. Appl. 1:118–138.

Aina, P.O. 1984. Contribution of earthworms to porosity and water infiltration in a tropical soil under forest and long-term cultivation. Pedobiologia 26:131–136.

Alban, D.H., and D.A. Perala. 1992. Carbon storage in Lake States aspen ecosystems. Can. J. For. Res. 22:1107–1110.

Allen, J.C. 1985. Soil response to forest clearing in the United States and the tropics: Geological and biological factors. Biotropica 17:15–27.

Altschuler, Z.S., E.J. Dwornik, and H. Kramer. 1963. Transformations of montmorillonite to kaolinite during weathering. Science (Washington, DC) 141:148–152.

Amundson, R.G., and H. Jenny. 1991. The place of humans in the state factor theory of ecosystems and their soils. Soil Sci. 151:99–109.

Amundson, R.G., and B. Tremback. 1989. Soil development on stabilized dunes in Golden Gate Park, San Francisco. Soil Sci. Soc. Am. J. 53:1798–1806.

Anderson, J.M. 1988. Spatiotemporal effects of invertebrates on soil processes. Biol. Fert. Soil 6:216–227.

Anderson, J.M. 1991. The effects of climate change on decomposition processes in grassland and coniferous forests. Ecol. Applic. 1:326–347.

Anderson, J.M. 1992. Responses of soils to climate change. Adv. Ecol. Res. 22:163–210.

Asahina, Y., and S. Shibata. 1971. Chemistry of lichen substances. Asher and Co., Amsterdam, the Netherlands.

Berg, B. 1986. Nutrient release from litter and humus in coniferous forest soils—a mini review. Scand. J. For. Res. 1:359–369.

Binkley, D., D. Richter, M.B. David, and B. Caldwell. 1992. Soil chemistry in a loblolly/longleaf pine forest with interval burning. Ecol. Appl. 2:157–164.

Bloomfield, C.A. 1953. Study of podzolization. I. The mobilization of iron and aluminum by Scots Pine needles. J. Soil Sci. 4:5–23.

Bonan, G.B. 1990. Carbon and nitrogen cycling in North American boreal forests. Biogeochemistry 10:1–28.

Bonan, G.B. 1992. Soil temperature as an ecological factor in boreal forests. p. 126–143. *In* H.H. Shugart et al. (ed.) A systems analysis of the global boreal forest. Cambridge Univ. Press, London.

Bonan, G.B., D. Pollard, and S.L. Thompson. 1992. Effects of boreal forest vegetation on global climate. Nature (London) 359:716–718.

Bonan, G.B., H.H. Shugart, and D.L. Urban. 1990. The sensitivity of some high-latitude boreal forests to climatic parameters. Clim. Change 16:9–29.

Bonan, G.B., and K. Van Cleve. 1991. Soil temperature, nitrogen, mineralization, and carbon source-sink relationships in boreal forests. Can. J. For. Res. 22:629–639.

Borchers, J.G., and D.A. Perry. 1992. The influence of soil texture and aggregation on carbon and nitrogen dynamics in southwest Oregon forests and clearcuts. Can. J. For. Res. 22:298–305.

Boyle, J.R., and G.K. Voigt. 1973. Biological weathering of silicate minerals. Implications for tree nutrition and soil genesis. Plant Soil 38:191–201.

Bryant, J.P., K. Danell, F. Provenza, P.B. Reichardt, T.A. Clausen, and R.A. Werner. 1991. Effects of mammal browsing on the chemistry of deciduous woody plants. p. 135–154. *In* Tallamy, D.W. and M.J. Raupp (ed.) Phytochemical induction by herbivores. John Wiley & Sons, New York.

Bunting, B.T., and J. Lundberg. 1987. The humus profile—concept class and reality. Geoderma 40:17–36.

Burke, I.C., C.M. Yonker, W.J. Parton, C.V. Cole, K. Flach, and D.S. Schimel. 1989. Texture, climate, and cultivation effects on soil organic matter content in U.S. grassland soils. Soil Sci. Soc. Am. J. 53:800–805.

Caillier, M., and S.A. Visser. 1988. Observations on the dispersion and aggregation of clays by humic substances. II. Short-term effects of humus-rich peat water on clay aggregation. Geoderma 43:1–9.

Campbell, C.A., E.A. Paul, D.A. Rennie, and K.J. McCallum. 1967. Applicability of the carbon dating method to soil humus studies. Soil Sci. 104:217–224.

Carmean, W.H. 1975. Forest site quality evaluation in the United States. Advan. Agron. 27:209–258.

Cole, D.W., and M. Rapp. 1981. Elemental cycling in forest ecosystems. Dynamic properties of forest ecosystems. Cambridge Univ. Press, London.

Cropper, W.P., Jr., and H.L. Gholz. 1993. Simulation of the carbon dynamics of a Florida slash pine plantation. Ecol. Model. 66:231–249.

Crossley, D.A., Jr. 1977. The role of terrestrial saprophagous arthropods in forest soils: Current status of concepts (with 1 figure). p. 49–56. *In* W.J. Mattson (ed.) The role of arthropods in forest ecosystems. Springer-Verlag, New York.

Dang, Q.L., and V.J. Lieffers. 1989. Assessment of patterns of response of tree ring growth of black spruce following peatland drainage. Can. J. For. Res. 19:924–929.

Daniels, R.B., and R.D. Hammer. 1992. Soil geomorphology. John Wiley & Sons, New York.

Douglas, L.A. 1982. Smectites in acid soils. p. 635–640. *In* H. Van Olphen and F. Veniale (ed.) Proc. Int. Clay Conf. 1981. Elsevier, New York.

Drijber, R.A., and L.E. Lowe. 1990. Nature of humus in Andosols under differing vegetation in the Sierra Nevada, Mexico. Geoderma 47:221–231.

Dyrness, C.T., K. Van Cleve, and J.D. Levison. 1989. The effect of wildfire on soil chemistry in four forest types in interior Alaska. Can. J. For. Res. 19:1389–1396.

Edmonds, R.L. 1991. Organic matter decomposition in western United States forests. p. 118–128. *In* A.E. Harvey and L.F. Neuenschwander (ed.) Gen. Tech. Rep. INT-280. USDA-FS, Intermountain Res. Stn., Ogden, UT.

Edmonds, R.L., D.J. Vogt, D.H. Sandberg, and C.H. Driver. 1986. Decomposition of Douglas-fir and red alder wood in clear-cuttings. Can. J. For. Res. 16:822–831.

Elliott, E.T. 1986. Aggregate structure and carbon, nitrogen, and phosphorus in native and cultivated soils. Soil Sci. Soc. Am. J. 50:627–633.

Eswaran, H., E. Van Den Berg, and P. Reich. 1993. Organic carbon in soils of the world. Soil Sci. Soc. Am. J. 57:192–194.

Flanagan, P.W., and K. Van Cleve. 1983. Nutrient cycling in relation to decomposition and organic matter quality in taiga ecosystems. Can. J. For. Res. 13:795–817.

Fox, T.R., and N.B. Comerford. 1992. Influence of oxalate loading on phosphorus and aluminum solubility in Spodosols. Soil Sci. Soc. Am. J. 56:290–294.

Frankenberger, W.T., Jr., and F.R. Troeh. 1982. Bacterial utilization of simple alcohols and their influence on saturated hydraulic conductivity. Soil Sci. Soc. Am. J. 46:535–538.

Franzmeier, D.P., G.D. Lemme, and R.J. Miles. 1985. Organic carbon in soils of North Central United States. Soil Sci. Soc. Am. J. 49:702–708.

Gholz, H. 1982. Environmental limits on aboveground net primary production, leaf area, and biomass in vegetation zones of the Pacific Northwest. Ecology 63:469–481.

Goh, K.M., J.D. Stout, and T.A. Rafter. 1976. Radiocarbon enrichment of soil organic matter fractions in New Zealand soils. Soil Sci. 123:385–391.

Goldberg, S., and G. Sposito. 1985. On the mechanism of specific phosphate adsorption by hydroxylated mineral surfaces: A review. Commun. Soil Sci. Plant Anal. 16:801–821.

Gosz, J.R. 1984. Biological factors influencing nutrient supply in forests soils. p. 119–146. *In* G.D. Bowen and E.K.S. Nambiar (ed.) Nutrition of plantation forests. Acad. Press, New York.

Gosz, J.R., and C.S. White. 1986. Seasonal and annual variation in nitrogen mineralization and nitrification along an elevational gradient in New Mexico. Biogeochemistry 2:281–297.

Graham, R.C., and H.B. Wood. 1991. Morphologic development and clay redistribution in lysimeter soils under chaparral and pine. Soil Sci. Soc. Am. J. 55:1638–1646.

Green, R.N., R.L. Trowbridge, and K. Klinka. 1993. Toward a taxonomic classification of humus forms. For. Sci. Monogr. 29. Soc. Am. For., Bethesda, MD.

Grigal, D.F., and L.F. Ohman. 1992. Carbon storage in upland forests of the Lake States. Soil Sci. Soc. Am. J. 56:935–943.

Grim, R.E. 1968. Clay mineralogy. 2nd ed. McGraw-Hill Book Co., New York.

Harmon, M.E., G.A. Baker, G. Spycher, and S.E. Greene. 1990. Leaf-litter decomposition in the *Picea/Tsuga* forests of Olympic National Park, Washington, U.S.A. For. Ecol. Man. 31:55–66.

Hem, J.D., and C.D. Lind. 1974. Kaolinite synthesis at 25°C. Science (Washington, DC) 184:1171–1173.

Hendricks, D.M., and L.D. Whittig. 1968. Andesite weathering. I. Mineralogical transformations from andesite to saprolite. J. Soil Sci. 19:135–146.

Hole, F.D. 1981. Effects of animals on soils. Geoderma 25:75–112.

Huang, P.M., and W.D. Keller. 1970. Dissolution of rock-forming silicate minerals in organic acids: simulated first-stage weathering of fresh mineral surfaces. Am. Miner. 55:2076–2094.

Hue, N.V., G.R. Craddock, and F. Adams. 1986. Effect of organic acids on aluminum toxicity in subsoils. Soil Sci. Soc. Am. J. 50:28–34.

Inoue, K., S. Hiradata, and S. Takagi. 1993. Interaction of mugeneic acid with synthetically produced iron oxides. Soil Sci. Soc. Am. J. 57:1254–1260.

Irons III, J.G., J.P. Bryant, and M.W. Oswood. 1991. Effects of moose browsing on decomposition rates of birch leaf litter in a subarctic stream. Can. J. Fish. Aquat. Sci. 48:442–444.

Jenny, H. 1941. Factors of soil formation. McGraw-Hill, New York.

Jenny, H. 1980. The soil resource. Origin and behavior. Springer-Verlag, New York.

Jenny, H., A.E. Salem, and J.R. Wallis. 1968. Interplay of soil organic matter and soil fertility with state factors and soil properties. p. 1–37. *In* Study week on organic matter and soil fertility, Rome, Italy. 22–27 April. North Holland Publ. Co., Amsterdam.

Johnson, C.E., A.H. Johnson, T.G. Huntington, and T.G. Siccama. 1991. Whole-tree clear-cutting effects on soil horizons and organic-matter pools. Soil Sci. Soc. Am. J. 55:497–502.

Johnson, D.W. 1992. Effects of forest management on soil carbon storage. Water Air Soil Pollut. 64:83–120.

Kahn, S.U. 1969. Humic acid fraction of a gray wooded soil as influenced by cropping systems and fertilizers. Geoderma 3:247–254.

Kladivko, E.J., A.D. Mackay, and J.D. Bradford. 1986. Earthworms as a factor in the reduction of soil crusting. Soil Sci. Soc. Am. J. 50:191–196.

Klemmedson, J.O., and B.J. Weinhold. 1992. Aspect and species influences on nitrogen and phosphorus accumulation in Arizona chaparral soil-plant systems. Arid Soil Res. Rehab. 6:105–116.

Kodama, H., and M. Schnitzer. 1977. Effect of fulvic acid on the crystallization of FE(III) oxides. Geoderma 19:279–292.

Leith, H. 1975. Modeling the primary production of the world. p. 237–262. In H. Leith and R.H. Whitaker (ed.) Primary productivity of the biosphere. Vol. 14. Springer-Verlag, New York.

Malcolm, R.L. 1990. Variations between humic substances isolated from soils, stream waters, and groundwaters as revealed by ^{13}C-NMR spectroscopy. p. 13–35. In P. MacCarthy et al. (ed.) Humic substances in soil and crop sciences: Selected readings. SSSA and ASA, Madison, WI.

Malcolm, R.L., W.D. Nettleton, and R.J. McCracken. 1969. Pedogenic formation of montmorillonite from a 2:1–2:2 intergrade clay mineral. Clays Clay Min. 16:405–414.

McClaugherty, C., and B. Berg. 1987. Cellulose, lignin and nitrogen concentrations as rate regulating factors in late stages of forest litter decomposition. Pedobiologia 30:101–112.

McKeague, J.A., M.V. Chesire, F. Andreux, and J. Berthelin. 1986. Organo-mineral complexes in relation to pedogenesis. p. 549–592. In P.M. Huang and M. Schnitzer (ed.) Interaction of soil minerals with natural organics and microbes. SSSA Spec. Publ. 17. SSSA, Madison, WI.

Meentemeyer, V. 1978. Macroclimate and lignin control of litter decomposition rates. Ecology 59:465–472.

Meentemeyer, V., and B. Berg. 1986. Regional variation in rate of mass loss of *Pinus sylvestris* needle litter in Swedish pine forests as influenced by climate and litter quality. Scand. J. For. Res. 1:167–180.

Melillo, J.M., J.D. Aber, A.E. Linkens, A. Ricca, B. Fry, and K.J. Nadelhoffer. 1989. Carbon and nitrogen dynamics along the decay continuum: Plant litter to soil organic matter. p. 53–62. In M. Clarholm and L. Bergstrom (ed.) Ecology of arable land. Kluwer Acad. Publ., New York.

Melillo, J.M., R.J. Naiman, J.D. Aber, and A.E. Linkens. 1984. Factors controlling mass loss and nitrogen dynamics of plant litter decaying in northern streams. Bull. Mar. Sci. 35:341–356.

Mokma, D.L., and G.F. Vance. 1989. Forest vegetation and origin of some spodic horizons, Michigan. Geoderma 43:311–324.

Muller, P.E. 1978. Studier over Skovjord. Tidsskr. Skogbrk 3:1–124.

Nadelhoffer, K.J., J.D. Aber, and J.M. Melillo. 1983. Leaf-litter production and soil organic matter dynamics along a nitrogen-availability gradient in Southern Wisconsin (U.S.A.). Can. J. For. Res. 13:12–21.

Nielsen, G., and F. Hole. 1964. Earthworms and the development of coprogeneous Al horizons in forest soils of Wisconsin. Soil Sci. Soc. Am. Proc. 28:426–430.

Nielsen, K.E., K. Dasgaard, and P. Nornberg. 1987. Effects on soils of an oak invasion of a Calluna heath, Denmark, II. Changes in organic matter and cellulose decomposition. Geoderma 412:97–106.

Oades, J.M. 1984. Soil organic matter and structural stability: Mechanisms and implications for management. Plant Soil 76:319–337.

Oades, J.M. 1988. The retention of organic matter in soils. Biogeochemistry 5:35–70.

Oades, J.M., G.P. Gillman, and G. Uehara. 1989. Interactions of soil organic matter and variable-charge clays. p. 69–95. In D.C. Coleman et al. (ed.) Dynamics of soil organic matter in tropical ecosystems. Dep. Agron. Soil Sci., Univ. Hawaii, Honolulu.

Pastor, J., R.H. Gardner, V.H. Dale, and W.M. Post. 1987. Successional changes in nitrogen availability as a potential factor contributing to spruce declines in boreal North America. Can. J. For. Res. 17:1394–1400.

Pastor, J., and W.M. Post. 1986. Influence of climate, soil moisture, and succession on forest carbon and nitrogen cycles. Biogeochemistry 2:3–27.

Pastor, J., and W.M. Post. 1988. Response of northern forests to CO_2-induced climate change. Nature (London) 334:55–58.

Paul, E.A., and F.E. Clark. 1989. Soil microbiology and biochemistry. Acad. Press, New York.

Perry, D.A., R. Molina, and M.P. Amaranthus. 1987. Mycorrhizae, mycorrhizospheres, and reforestation: Current knowledge and research needs. Can. J. For. Res. 17:929–940.

Petersen, H., and M. Luxton. 1982. A comparative analysis of soil fauna populations and their role in decomposition processes. Oikos 39:287–388.

Post, W.M., W.R. Emanuel, P.J. Zinke, and A.G. Stangengerger. 1982. Soil carbon pools and world life zones. Nature (London) 298:156–159.

Powers, R.F. 1987. Predicting growth responses to soil management practices: Avoiding "future shock" in research. p. 391–403. *In* L.L. Boersma et al. (ed.) Future developments in soil science research. SSSA, Madison, WI.

Powers, R.F. 1989. Maintaining long-term forest productivity in the Pacific Northwest: Defining the issues. p. 3–16. *In* D.A. Perry et al. (ed.) Maintaining the long-term productivity of Pacific Northwest forest ecosystems. Timber Press, Inc., Portland, OR.

Powers, R.F. 1990. Nitrogen mineralization along an altitudinal gradient: Interactions of soil temperature, moisture, and substrate quality. For. Ecol. Manage. 30:19–29.

Powers, R.F., and K. Van Cleve. 1991. Long-term ecological research in temperate and boreal forest ecosystems. Agron. J. 83:11–24.

Preston, C.M. 1991. Using NMR to characterize the development of soil organic matter with varying climate and vegetation. p. 27–36. *In* Stable isotopes in plant nutrition, soil fertility and environmental studies. IAEA, Vienna, Austria.

Reiners, W.A. 1981. Nitrogen cycling in relation to ecosystem succession. Ecol. Bull. (Stockholm) 33:507–528.

Robert, M., and J. Berthelin. 1986. Role of biological and biochemical factors in soil mineral weathering. p. 453–495. *In* P.M. Huang and M. Schnitzer (ed.) Interaction of soil minerals with natural organics and microbes. SSSA Spec. Publ. 17. SSSA, Madison, WI.

Rozanov, B.G. 1961. Aluminum and iron migration during soil formation in dense forests. Vestn. Mosk. Univ. (Ser. VI) Biol. Pochvoved. 16:67–78.

Schlesinger, W.H. 1977. Carbon balance in terrestrial detritus. Ann. Rev. Ecol. Syst. 8:51–81.

Schleslinger, W.H. 1991. Biogeochemistry. An analysis of global change. Acad. Press, New York.

Schoenau, J.J., and J.R. Bettany. 1987. Organic matter leaching as a component of carbon, nitrogen, phosphorus and sulfur cycles in a forest, grassland and gleyed soil. Soil Sci. Soc. Am. J. 51:646–651.

Schuppli, P.A., P. Protz, and J.A. McKeague. 1988. Extractable organic fractions of Podzolic and associated soils in the Hudson Bay-James Bay lowlands. Geoderma 41:263–274.

Schwertmann, U. 1971. Transformation of hematite to goethite in soils. Nature (London) 232:624–625.

Schwertmann, U., and W.R. Fischer. 1973. Natural "amorphous" ferric hydroxide. Geoderma 10:237–247.

Seifert, R.D. 1981. A solar design manual for Alaska. Bull. Inst. Water. Res. Vol. 1. Univ. Alaska, Fairbanks.

Senstius, M.W. 1958. Climax forms of rock weathering. Am. Sci. 46:355–367.

Shaw, C., and S. Pawluk. 1986. Faecal microbiology of *Octolasion tyrtaeum, Aporrectodea turgida,* and *Lumbricus terrestris* and its relation to the carbon budgets of three artificial soils. Pedobiologia 29:377–389.

Shaw, C.H., H. Lundkvist, A Moldenke, and J.R. Boyle. 1991. The relationships of soil fauna to long-term forest productivity in temperate and boreal ecosystems: Processes and research strategies. p. 39–77. *In* W.J. Dyck and C.A. Mees (ed.) Long-term field trials to assess environmental impacts of harvesting. IEA/BE T6/A6 Rep. no. 5. For. Res. Inst. Bull. 161. FRI, Rotorua, New Zealand.

Shugart, H.H., and T.M. Smith. 1992. Modelling boreal forest dynamics in response to environmental change. Unasylva 43:30–38.

Simonson, R.W. 1959. Outline of a generalized theory of soil genesis. Soil Sci. Soc. Am. Proc. 23:152–156.

Soil Survey Staff. 1975. Soil taxonomy: A basic system of USDA-SCS soil classification for making and interpreting soil surveys. USDA-SCS Agric. Handb. 436. U.S. Gov. Print. Office, Washington, DC.

Sollins, P., G. Spycher, and C.A. Glassman. 1984. Net nitrogen mineralization from light and heavy fraction forest soil organic matter. Soil Biol. Biochem. 16:31–37.

Sollins, P., G. Spycher, and C. Topik. 1983. Processes of soil organic matter accretion at a mudflow chronosequence, Mt. Shasta, California. Ecology 64:1273–1282.

Stevenson, F.J. 1982. Humus chemistry. Genesis, composition, reactions. John Wiley & Sons, New York.

Stevenson, F.J. 1985. Geochemistry of soil humic substances. p. 13–52. *In* G.R. Aiken et al. (ed.) Humic substances in soil, sediment, and water. John Wiley & Sons, New York.

Stevenson, F.J. 1986. Cycles of soil. carbon, nitrogen, phosphorus, sulfur, and micronutrients. John Wiley & Sons, New York.

Stevenson, F.J., and M.S. Ardakani. 1972. Organic matter reactions involving micronutrients in soils. p. 79–114. *In* J.J. Mortvedt et al. (ed.) Micronutrients in agriculture. SSSA, Madison, WI.

Tan, K.H. 1986. Degradation of soil minerals by organic acids. p. 1–27. *In* P.M. Huang and M. Schnitzer (ed.) Interaction of soil minerals with natural organics and microbes. SSSA Spec. Publ. 17. SSSA, Madison, WI.

Tate, R.L. III. 1987. Soil organic matter. Biological and ecological effects. John Wiley & Sons, New York.

Tate, K.R. 1992. Assessment, based on a climosequence of soils in tussock grasslands, of soil carbon storage and release in response to global warming. J. Soil Sci. 43:697–707.

Thorp, J., J.G. Cadey, and E.E. Gamble. 1959. Genesis of Miami silt loam. Soil Sci. Soc. Am. Proc. 23:156–161.

Tisdall, J.M., and J.M. Oades. 1982. Organic matter and water-stable aggregates in soils. J. Soil Sci. 33:141–163.

Ugolini, F.C. 1968. Soil development and alder invasion in a recently deglaciated area of Glacier Bay, Alaska. p. 115–140. *In* J.M. Trappe et al. (ed.) Biology of Alder. USDA-FS, Portland OR.

Ugolini, F.C., and R.S. Sletten. 1991. The role of proton donors in pedogenesis as revealed by soil solution studies. Soil Sci. 151:59–75.

Van Cleve, K., and V. Alexander. 1981. Nitrogen cycling in tundra and boreal ecosystems. Ecol. Bull. (Stockholm) 33:375–404.

Van Cleve, K., C.T. Dyrness, G.M. Marion, and R. Erickson. 1993. Control of soil development on the Tanana River floodplain, interior Alaska. Can. J. For. Res. 23:941–955.

Van Cleve, K., L. Oliver, R. Schlentner, L.A. Viereck, and C.T. Dyrness. 1983. Productivity and nutrient cycling in taiga forest ecosystems. Can. J. For. Res. 13:747–766.

Van Cleve, K., F.S. Chapin III, C.T. Dyrness, and L.A. Viereck. 1991. Element cycling in taiga forests: State-factor control. BioScience 41:78–88.

Van Cleve, K., W.C. Oechel, and J.L. Hom. 1990. Response of black spruce (*Picea mariana*) ecosystems to soil temperature modification in interior Alaska. Can. J. For. Res. 20:1530–1535.

Viereck, L.A., K. Van Cleve, P. Adams, and R.L. Schlentner. 1993. Climate of the Tanana River floodplain near Fairbanks, Alaska. Can. J. For. Res. 23:914–922.

Visser, S.A., and M. Caillier. 1988. Observations on the dispersion and aggregation of clays by humic substances. I. Dispersive effects of humic acids. Geoderma 42:331–337.

Vitousek, P.M., P.A. Matson, and K. Van Cleve. 1989. Nitrogen availability and nitrification during succession: Primary, secondary and old-field seres. Plant Soil 115:229–239.

Vitousek, P.M., K. Van Cleve, N. Balakrishnan, and D. Mueller-Dombios. 1983. Soil development and nitrogen turnover in montane rainforest soils on Hawaii. Biotropica 15:268–274.

Vitousek, P.M., and L.R. Walker. 1987. Colonization, succession, and resource availability: Ecosystem-level interactions. p. 207–223. *In* A.J. Gray et al. (ed.) Colonization, succession, and stability. Blackwell Sci. Publ. Oxford, England.

Vogt, K.A. 1991. Carbon budgets of temperate forest ecosystems. Tree Physiol. 9:69–86.

Vogt, K.A., C.C. Grier, C.E. Meier, and R.L. Edmonds. 1982. Mycorrhizal role in net primary production and nutrient cycling in *Abies amabilis* ecosystems in western Washington. Ecology 63:370–380.

Wada, K. 1980. Mineralogical characteristics of Andosols. p. 87–107. *In* B.K.G. Theng (ed.) Soils with variable charge. New Zealand Soc. Soil Sci., Lower Hutt, New Zealand.

Waring, R.H., and W.H. Schlesinger. 1985. Forest ecosystems. Concepts and management. Acad. Press, New York.

Webb, W.L., W.K. Lauenroth, S.R. Szarek, and R.S. Kinerson. 1983. Primary production and abiotic controls in forests, grasslands, and desert ecosystems in the United States. Ecology 64:134–151.

Whalen, S.C, and W.S. Reeburgh. 1990. Consumption of atmospheric methane by tundra soils. Nature (London) 346:160–162.

Whalen, S.C., W.S. Reeburgh, and K.S. Kizer. 1991. Methane consumption and emission by taiga. Global Biogeochem. Cycles 5:261–273.

Wood, C.W., R.J. Mitchell, B.R. Zutter, and C.L. Lin. 1992. Loblolly pine plant community effects on soil carbon and nitrogen. Soil Sci. 154:410–419.

Zak, D.R., D.F. Grigal, S. Gleeson, and D. Tilman. 1990. Carbon and nitrogen cycling during old-field succession: Constraints on plant and microbial biomass. Biogeochemistry 11:111–129.

Zak, D.R., A. Hairston, and D.F. Grigal. 1991. Topographic influences on nitrogen cycling within an upland pin oak ecosystem. For. Sci. 37:45–53.

Zak, D.R., K.S. Pregitzer, and G.E. Host. 1986. Landscape variation in nitrogen mineralization and nitrification. Can. J. For. Res. 16:1258–1263.

Zeikus, J.G. 1981. Lignin metabolism and the carbon cycle. Polymer biosynthesis, biodegradation, and environmental recalcitrance. Adv. Microb. Ecol. 5:211–243.

10 Soil Organic Carbon in the Missouri Forest-Prairie Ecotone

R. David Hammer, Gray S. Henderson, and Ranjith Udawatta

University of Missouri
Colunbia, Missouri

Donna K. Brandt

USDA-SCS
Rose Lake Plant Materials Center
East Lansing, Michigan

The possible impacts of elevated atmospheric CO_2 concentration on global climate change have received attention in the both the popular media and among scientists. Increased atmospheric C often has been attributed to fossil fuel burning and to destruction of forests (Woodwell et al., 1983). Both activities are coincident with the Industrial Revolution, which began in earnest midway through the nineteenth century—about the same time as large-scale conversion of grasslands to agriculture in the American Midwest. Soil C volatized as a result of anthropogenic factors may be a contributor to altered global atmospheric C dynamics (Anderson, 1992).

However, the soil organic C pool as a possible source of atmospheric C has received less attention than fossil fuel consumption and forest harvest. Magnitudes, distributions, and dynamics of the soil C pool are not understood sufficiently to permit reliable modeling of the global C system under the projected influences of global climate change (Anderson, 1992; Schlesinger, 1990). Because of this lack of understanding, predictions of the fate of atmospheric change on soil organic C are more speculative than substantive (Anderson, 1992).

This manuscript will focus on the distribution of soil organic C in the Missouri forest–prairie ecotone. Our hypotheses are: (i) that soil organic C pools are larger and more spatially variable than previous studies suggest, and (ii) the majority of soil organic C is below surface horizons. If either of these hypotheses is true, two important ramifications exist. First, the soil organic C pool may be a more important contributor to increased atmospheric C concentrations than previously suspected. Second, conversion of atmospheric C to soil organic C through ecosystem management may be a more important method of countering rising atmospheric CO_2 concentrations than previously has been envisioned.

Copyright © 1995 Soil Science Society of America, 677 S. Segoe Rd., Madison, WI 53711, USA.
Carbon Forms and Functions in Forest Soils.

Our initial objectives were to: (i) review the relevant literature of soil organic C in temperate forest and prairie biomes and, (ii) examine and compare soil organic C pools within and among botanical communities in the Missouri forest-prairie ecotone. Results of our investigation caused us to examine commonly used soil sampling strategies and existing soil science paradigms. We then discussed both topics.

LITERATURE REVIEW

Existing Estimates of Soil Carbon Pools

Global Estimates

Global C pools in the soil, atmosphere and living biomass are estimated to be 2, 0.72 and 0.56×10^{12} t, respectively (Scharpenseel & Becker-Heidemann, 1990). The soil C pool is approximately 1.6 times the atmospheric and biomass pools combined. Birdsey (1992) also estimated the soil C pool as greater than the biomass pool, and recognized that the proportions of C in the compartments vary among regions and forest types. Generally, however, estimates of ecosystem C dynamics, including C pools and fluxes in soils are imprecise. For example, in the Harvard Forest, Wofsy et al. (1993) measured gross ecosystem C production (11.1 t C ha^{-1}/yr^{-1}) "notably larger" than previous estimates for temperate forest ecosystems. Tinker and Ineson (1990) estimated the soil C pool to be 1.5×10^{18} g, but place the "uncertainty range" as 1.2 to 1.8×10^{18} g, which is 40% of their estimated mean.

Investigations of C cycles should more carefully consider the soil C pool in terms of its magnitudes and patterns of distribution. When the abundance and distribution of soil organic C have been more precisely documented and related to causal factors and processes, the release of soil C as CO_2 to the atmosphere and the potential of the soil as a C sink can be understood more completely.

Regional Estimates

Franzmeier et al. (1985) estimated the soil organic C levels in the upper 1 m of Missouri soils to be 10 to 15 kg m^{-2} in the northwestern one-half of the state and 1 to 10 kg m^{-2} in the southeastern one-half. Franzmeier et al. (1985) based their estimates on weighted mean average organic C in soil association map units of the region. These map units were taken from the 1:2 500 000 soil map of the North Central Region (Technical Committee on Soil Survey, 1960). Neither the numbers of samples averaged per map unit nor the calculated ranges and standard deviations for the data were reported. Map unit compositions were estimated from a variety of sources including "pedologists' knowledge of the soils." The authors stated that in the north central region ". . . the range of precipitation and temperature within areas in which the other variables, especially parent material, are relatively constant is not great enough to produce definite trends in soil C content." We submit that trends exist, but the sampling scheme did not allow Franzmeier and colleagues to recognize them.

Scrivner and Cooper (1985) summarized soil organic C distribution in Missouri agricultural soils. Their data included 1000 soil profiles sampled to a depth of 1 m as part of the Missouri Cooperative Soil Survey. They also used

1100 samples collected from plow layers for fertility recommendations. Scrivner and Cooper stratified results by landscape positions within "major land systems" which they patterned after Major Land Resource Areas (MLRA's). An MLRA is a geographically associated land resource unit. Land resource units are geographic areas, usually several thousand acres in extent, that are characterized by a particular pattern of soils, climate, water resources and land uses (USDA, 1981). Scrivner and Cooper (1985) also stratified their results by soil classification. They grouped soils within landscape positions by Mollisols, by mollic subgroups of all soil orders examined, and by nonmollic subgroups of all soil orders.

In Missouri, where the climate is wetter from southeast to northwest, but warmer from northwest to southeast (Fig. 10–1), several general relationships were revealed by Scrivner and Cooper's summary. Surface soil organic C tended to increase in all landscape positions from southeast to northwest (Fig. 10–2 and 10–3). This confirms the well-documented relationship that soil organic C generally increases with increasing coolness (Anderson, 1992). The temperature gradient apparently is more important than the precipitation gradient in affecting regional soil organic C in Missouri. Second, upland Mollisols contained more soil organic C to a depth of 100 cm than mollic subgroups and nonmollic subgroups, respectively (Table 10–1). Bottomland Mollisols had less organic C in the upper

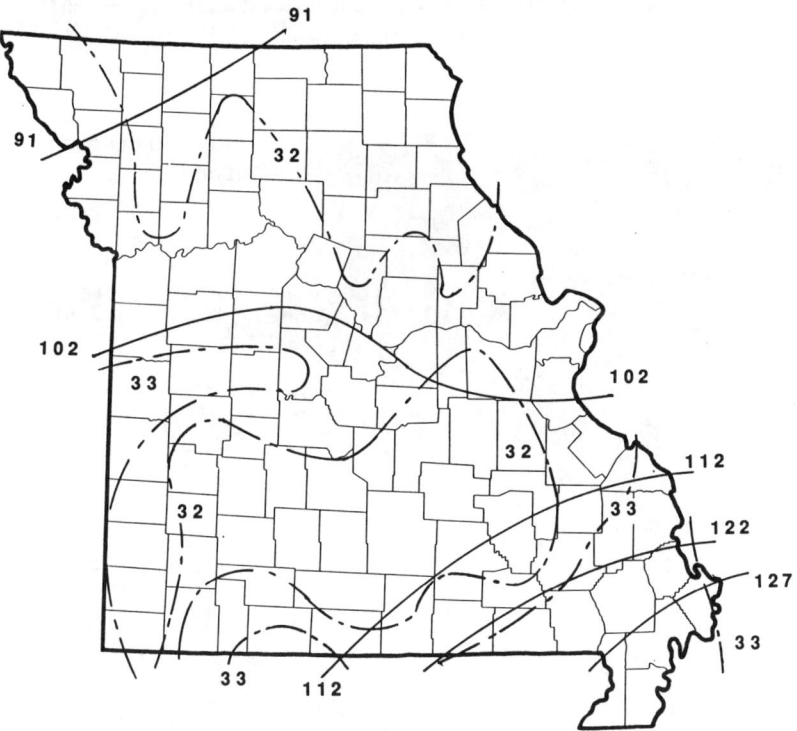

Fig. 10–1. Thirty-year average precipitation in centimeters (solid line) and mean maximum July temperatures in degrees Centigrade (broken line) in Missouri. Precipitation data were provided by Dr. Wayne Decker, State Climatologist. Temperature data were from Water Resources Information Center (1974).

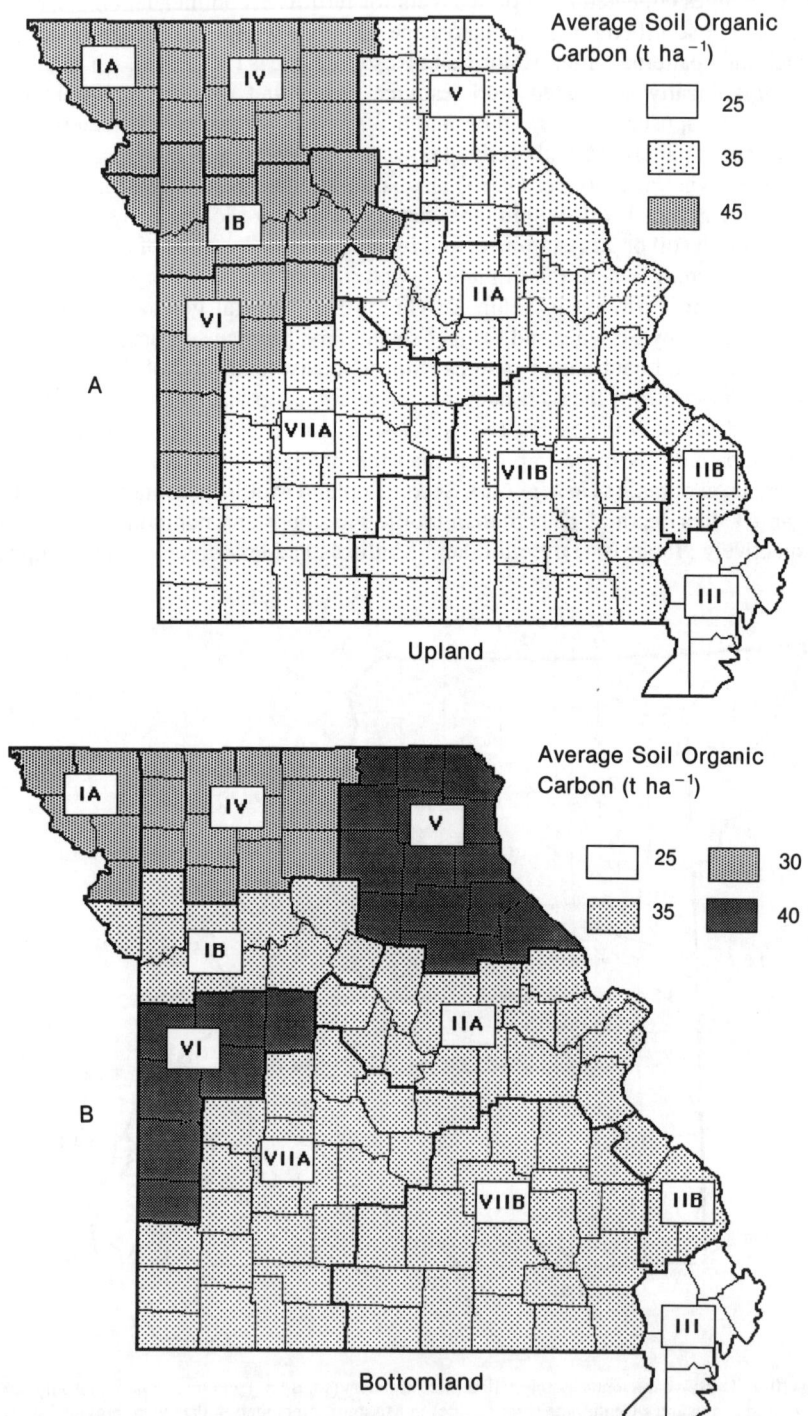

Fig. 10–2. Average amount of organic C within the upper 20 cm of upland (*A*) and bottomland (*B*) soils for major land systems in Missouri (Scrivner & Cooper, 1985).

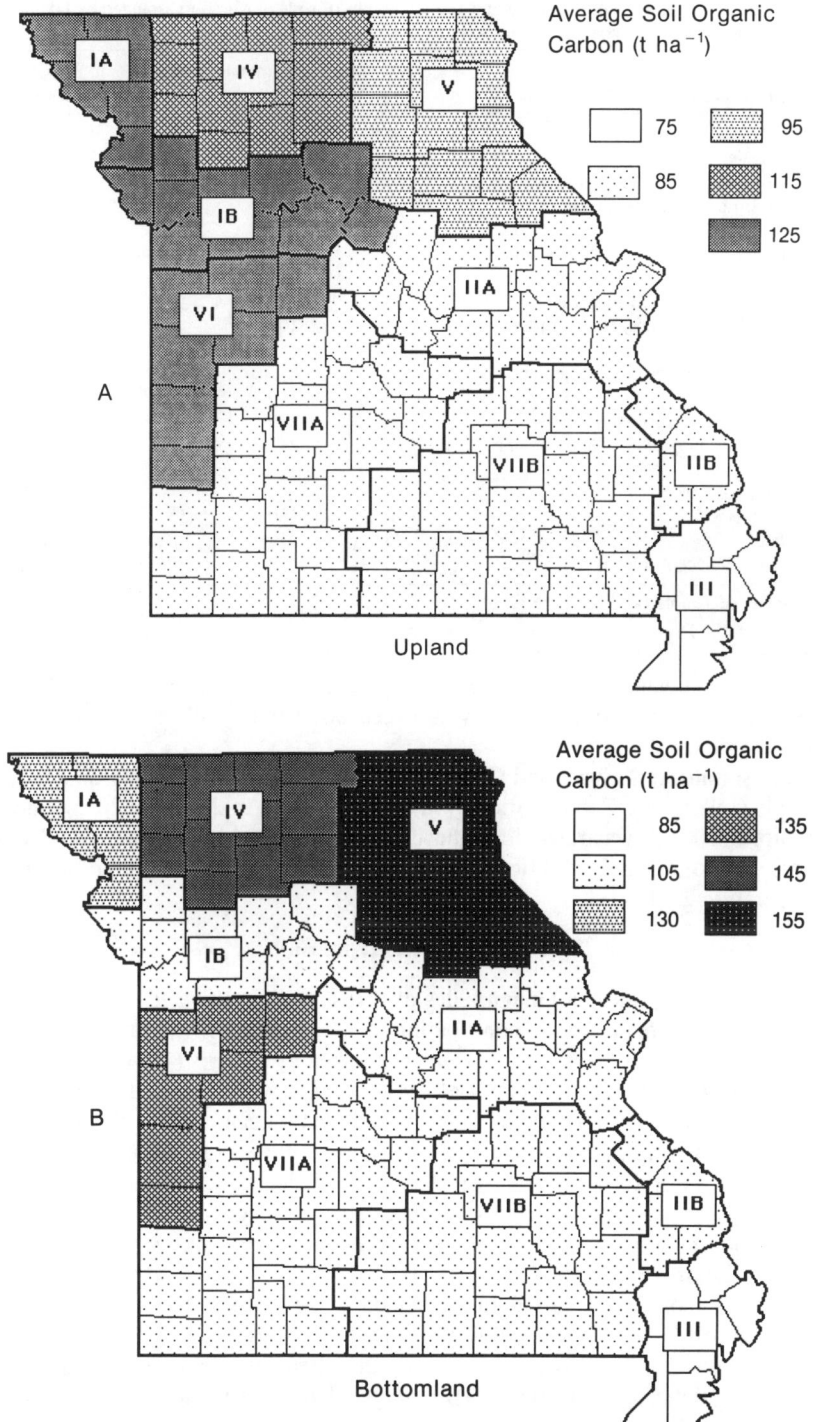

Fig. 10–3. Average amount of organic C within the upper 100 cm of upland (*A*) and bottomland (*B*) soils for major land systems in Missouri (Scrivner & Cooper, 1985).

Table 10-1. Organic C content in 20-and 100-cm depths of soils of Missouri arranged by taxonomic classes and landscape position (from Scrivner & Cooper, 1985).

Landscape position & soil taxonomy	Average amount of organic C in			
	upper 20 cm of soil		upper 100 cm of soil	
Uplands, summits only	t ha⁻¹	N†	t ha⁻¹	N†
Mollisols	44	86	131	82
Mollic subgroups, all soil orders	34	95	103	73
Nonmollic subgroups, all soil orders	30	22	88	20
Uplands, shoulders & backslopes only				
Mollisols	48	114	130	105
Mollic subgroups, all soil orders	36	76	100	71
Nonmollic subgroups, all soil orders	31	104	82	98
Uplands, all slope-profile components				
Mollisols	47	310	133	290
Mollic subgroups, all soil orders	36	209	103	181
Non mollic subgroups, all soil orders	31	241	81	224
Bottomlands				
Mollisols	40	180	137	174
Alfisols	25	79	82	78
Entisols & Inceptisols	28	108	98	103

†Number of sampled soil profiles.

20 cm than upland Mollisols, but had more organic C in the upper 100 cm. Finally, the subsoil organic C levels exceeded surface soil levels for all soil groups examined.

In summary, Scrivner and Cooper (1985) identified local and regional trends in total soil organic C distribution. Soil organic C both in the upper 20 cm and in the upper 100 cm was related to landscape position (local temperature and wetness), and to regional conditions related to parent materials and to the prevalent temperature gradient.

Carbon Losses from Soils

Quantifying rates and magnitudes of C loss from cultivated soil systems is necessary to estimate the soil contribution to increased atmospheric CO_2. General C loss rates from cultivated soils were established by Jenny's (1941) exponential decay function. Studies which confirmed Jenny's work were summarized by Stevenson (1982). These studies show that cultivation can result in the loss of 40 to 60% of the original soil organic C pool within 40 to 50 yr. Establishing more precise estimates for individual soil systems is difficult because local conditions including management practices, texture, soil drainage and parent materials affect both the amount of C a soil can retain and the amount lost due to anthropogenic factors (Stevenson, 1982).

Based upon data acquired during standard soil surveys, Mann (1985) postulated that soil C losses due to cultivation have been overestimated. She questioned how C losses from cultivated soils are partitioned into losses by leaching vs. losses by volatization. Mann (1985) compared depth distributions of organic C in cultivated and uncultivated soils of the same series. She reported that the C content of the total profile (for her study this was a depth of 100–150 cm) was 28% lower in cultivated vs. uncultivated Udalfs. Her comparison of subsoil C changes as a result of cultivation was inconclusive, although some cultivated soils showed larger subsurface C levels than the uncultivated soils with which they were compared.

Mann's review is important because it considered the soil below the plow layer and because she proposed illuviation into B horizons as an alternative pathway for C lost from A horizons through agricultural practices. Additionally, her attempts to draw conclusions from data collected from a variety of management practices across the parent material, age and climatic gradients of a 12 state area points to the necessity for carefully designed long-term research which can withstand rigorous statistical analysis.

Soil C can be redistributed with sediments in eroded landscapes (De Yong & Kachonoski, 1988). Redistribution complicates both estimates of pre-erosion soil C patterns and attempts to estimate volatization losses. The soil C system is open, so C is lost by leaching to groundwater as well organic matter oxidation and removal from the soil through erosion to perennial water bodies (Schimel et al., 1985).

Establishing Baseline Data from Undisturbed Sites

Estimating the contribution of soil C to atmospheric CO_2 levels poses a considerable challenge because determining current, pre-Industrial Revolution, and potential soil C pools is difficult. A large portion of the soil resource has been cultivated or otherwise altered, particularly in the Midwest. Forest and prairie communities on land not previously clearcut or cultivated are rare.

In Missouri, for example, less than 0.01% of the presettlement prairie remains (Schroeder, 1981). Native prairie remnants are small, fragmented, and usually occupy only portions of first-order watersheds. Establishing the patterns of botanical and soil diversity in uncultivated landscapes from the fragmented remnants is difficult. As Anderson (1972) pointed out, as scale of observation is changed in natural systems, interactions among processes often change and new processes are observed. Therefore, "more" is often "different," and modeling pattern and process in large systems from smaller templates often results in failure.

Remaining uncultivated soils may not represent the regional soil population. For example, the senior author has visited about 20 prairie remnants in southwest Missouri. Each of these remnants was at the highest elevation in the local landscape. Most sites had rocky sola and were on convex summits. They were the droughtiest sites in their respective landscapes and probably were left as prairie because they could not be converted profitably to agriculture.

TOTAL ORGANIC CARBON IN FOREST-PRAIRIE ECOTONE SOILS

Brandt (1993) compared soil properties under established native warm season grasses with properties of adjacent cultivated soils. She reviewed the literature in which cultivated and uncultivated forest and prairie soils were compared. Brandt concluded that previous investigations of effects of forest and prairie communities and cultivation on soils had limited usefulness. Her conclusion was based on four observations. The published studies generally suffered from: (i) shallow sampling (sampling seldom was conducted deeper than 15 cm); (ii) failure to describe important stratigraphic, geomorphic and pedologic parameters; (iii) lack of replication; and (iv) failure to obtain a complete suite of data-soil textures, including rock content, and bulk densities often were not acquired, for example, so C concentrations could not be converted to quantities.

These oversights in previous research limit our abilities to extrapolate from local to regional conditions. Additionally, incomplete or poorly designed studies do not contribute to our understanding of fundamental soil-forming processes. Stone's (1975) observation that past field investigations often have suffered from "... lack of rigor and burden of supposition..." is true for most comparisons of organic matter pools and fluxes in cultivated soils and their uncultivated analogues.

Extent and Location of the Forest-Prairie Ecotone in Missouri

Where was the forest-prairie ecotone and how extensive was it? The question cannot be answered on the basis of soil morphology alone. However, Brandt's (1993) literature review revealed that many investigators assumed that if they were sampling a Mollisol, they were investigating a soil formed exclusively under prairie. Mollisols are thought to form as a result of the annual die-back of grass roots in prairie ecosystems (Soil Survey Staff, 1975). It is generally conceded that most Mollisols had a grass vegetation at some time in their recent developmental history, but polygenesis may be typical of most Mollisols in continental North America (Fenton, 1983).

For example, on the basis of bog pollen and stratigraphic analysis, Walker (1966) concluded that forest was the dominant vegetation in Iowa until prairies became established approximately 8000 years before the present (YBP). Ruhe (1970, 1975) also has proposed climatic change and soil polygenesis during the Pleistocene. Geiss and Boggess (1968) suggested that grasslands expanded into north-central Illinois about 7600 to 4500 YPB. Therefore, soils on stable landscape positions in the region of the forest-prairie ecotone probably have formed under both forest and prairie and have been subjected to climatic change.

Brandt (1993) also reported that investigators commonly attributed soils with thin A horizons and argillic horizons, regardless of site erosion history, to formation under forest. The taxonomic paradigm of argillic horizon development in Alfisols and Ultisols is that clay illuviation into the Bt horizons results from periodicity of wetting and drying under a forest (usually deciduous) canopy (Thorp et al., 1959; Soil Survey Staff, 1975).

Thorp and Smith (1949) proposed that clay removal from the soil surface is a necessary precursor for melanization (development of a mollic epipedon). A related hypotheses is that calcareous parent material must be leached of calcium carbonates before significant concentrations of organic matter can accumulate. As carbonates are leached, clay particles previously cemented in the carbonates are released for subsequent eluvial/illuvial processes (Cady, 1960).

Soil Taxonomy (Soil Survey Staff, 1975) includes classes for Mollisols with subsurface cambic and argillic horizons and mollic classes for Alfisols with thick, dark, base-rich epipedons which don't meet all the criteria for Mollisols.

However, no clear consensus exists upon the pathways of origin of soils containing mollic epipedons and argillic horizons.

Unpublished data from the University of Missouri Soil Characterization Laboratory (SCL) indicate that 103 (of 428 soil series on soil survey legends in Missouri) soil series and taxadjuncts have both mollic and argillic diagnostic horizons (Table 10–2). These soils include Alfisols with mollic, aquollic, and udollic subgroups and Mollisols with argillic horizons. Several soils in this group are benchmark soils mapped on thousands of acres each. We estimate that 30 to 40% of Missouri's land area is mapped and classified with both argillic and mollic horizons.

Schroeder's (1981) map of presettlement prairie (Fig. 10–4) is based on historical survey data which noted the existence of forest and prairie at survey locations. Kucera (1961) modified Transeau's (1935) work using botanical records, soil and geologic materials, climate, and topography to produce a vegetation map of Missouri (not shown). Kucera's map shows prairie in the same regions as Schroeder's map, but Kucera has delineated a less extensive prairie.

The ecotone probably shifted in location throughout the Pleistocene as climate changed. The possibility exists that forest and prairie species coexisted in dissected landscapes (W.A. Schroeder, personal communication). Within the ecotone, forest communities were favored on north and east aspects and along drainageways while prairies existed on summits and south-facing slopes (Transeau, 1935).

Table 10–2. Missouri soil subgroups with diagnostic horizons characteristic of forest and prairie vegetation (argillic and mollic horizons, respectively). A total of 428 soil series is on survey legends in Missouri.

Subgroup	Number of families	Number of series and taxadjuncts
Alfisols		
Aquollic Hapludalfs	3	12
Mollic Albaqualfs	2	6
Mollic Fragiudalfs	1	1
Mollic Hapludalfs	8	19
Mollic Ochraqualfs	3	3
Mollic Paleudalfs	2	2
Udollic Albaqualfs	1	1
Udollic Ochraqualfs	2	2
Mollisols		
Abruptic Argiaquolls	2	2
Abruptic Argiudolls	1	1
Aquic Argiudolls	4	15
Argiaquic Argialbolls	2	4
Lithic Argiabolls	1	7
Typic Argiabolls	4	10
Typic Argiudolls	8	17
Vertic Argiaquolls	1	1
Totals	46	103

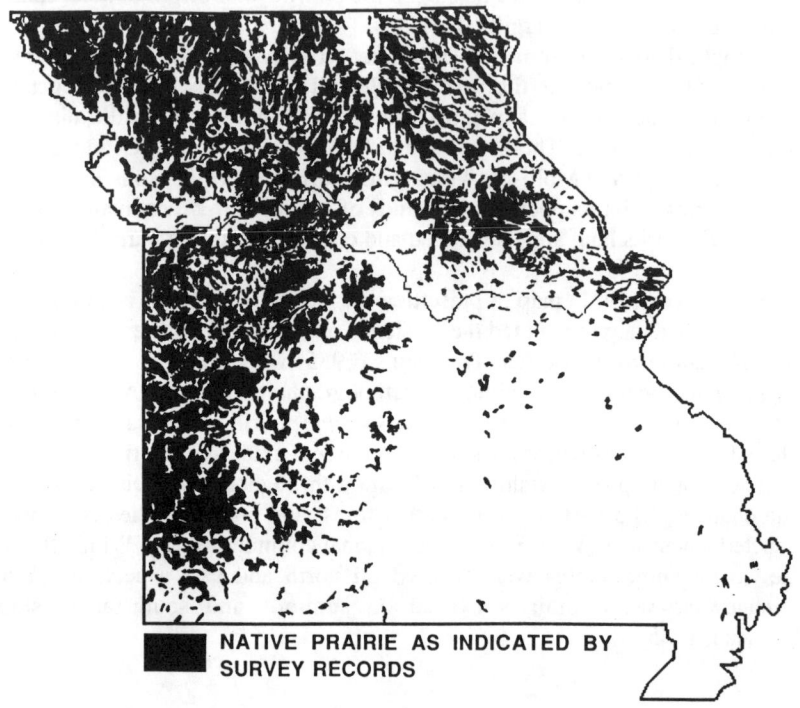

Fig. 10–4. Extent of presettlement prairie in Missouri (adapted from Schroeder, 1981, with author's consent).

Surface creep of A horizons and lateral subsurface clay illuviation in sloping landscapes are two geomorphic processes which could have created multiple diagnostic horizons within profiles regardless of site botanical history. For example, hillslope sediments from mollic horizons on summits and shoulders could have overthickened A horizons in Alfisols on footslopes and toeslopes. Prairies could have invaded previously forested sites as climate shifted. Alternatively, under certain conditions, perhaps argillic horizons can develop under prairie vegetation and mollic epipedons can develop under forest.

Arnold (1965) has pointed out that similar soils can develop from different genetic pathways. Our point is that caution must be exercised when attributing depth distribution patterns of soil organic C to site factors which have not existed for decades and which cannot be documented.

MATERIALS AND METHODS

Measuring Carbon Concentrations

Soil C concentrations were measured on air-dried samples in a Leco CR 12 combustion furnace (LECO Corp., St. Joseph, MI) in the University of Missouri

SCL. Our laboratory procedures require that every 10th sample be a reference standard whose known C value lies within 0.5% of the concentration of the sample being analyzed. All of our reference samples also have been analyzed by the Soil Conservation Service National Laboratory in Lincoln, Nebraska, to ensure precision and accuracy of our analyses. Several years of experience and analyses of over 15 000 soil samples have indicated that the furnace produces accurate results to within 0.05% on samples with minimum C content of 0.1%.

Most soils in Missouri do not have free carbonates in the solum. All soils in Missouri are leached by percolating water at least once a year in most years (Scrivner et al., 1973). Much of the Missouri landscape north of the Missouri River was covered by Kansan drift, the upper portion of which has been leached and oxidized. Loess of both Loveland (Illinosan) and Peorian (Wisconsinan) origins blankets much of the state. However even the deep loess deposits adjacent to the Missouri River are carbonate-free to depths of several meters (Krusekopf & Scrivner, 1962).

The SCL procedures require that all samples be visually inspected for free carbonates prior to C determinations. If free carbonates appear to be visible or if the parent material is suspected of having free carbonates, samples are treated with dilute HCl. If effervescence is observed, the sample is treated with 6% sulfurous acid to remove the free carbonates (USDA-SCS, 1982).

The soil C measured in the SCL is soil organic C. When we refer to "carbon" in the remainder of the manuscript, we mean soil organic C.

Calculation of Total Organic Carbon from Soil Profile Data

When possible, we used bulk density data to determine total C contents of soil horizons and profiles. Where bulk density data were not available, we used an estimate of 1.45 g cm^{-3} to convert C percentages to weight per unit area. This would tend to overestimate surface soil C and to underestimate subsoil C. All sampled soil profiles were air dried, weighed and passed through a 2-mm sieve. If rock fragments were present, they were weighed after being separated by sieving from the <2-mm fraction. Soil C quantities for each horizon were corrected for rock fragment content.

Sources of Soil Data

Most data were acquired from completed student theses. One thesis was in progress (Ranjith Udawatta, unpublished data). Data from Sanborn Field, the oldest continuously cultivated experimental field west of the Mississippi River, were acquired from the data base accumulated during the centennial sampling in 1988.

Some data were randomly selected from the SCL data base which was acquired during the accelerated cooperative soil survey. A list of laboratory sample numbers was produced from soils having both mollic and argillic horizons. Seven numbers were produced with a random number generator from a pocket calculator and the soils labeled with those numbers were selected. Data from those seven soil profiles were examined to determine the relative distributions of soil organic C in the sampling depths.

RESULTS

Deep Loess Soils under Forest and Prairie

Indorante (1990) used grid sampling to investigate relationships of soil properties to landscape position in deep loess deposits adjacent to the Missouri River. Two locations were used, Van Meter State Park in Saline County and Schnabel Woods in Boone County. The sites at Schnabel and Van Meter were similar in age, elevation, surface shape, aspect and relief.

Native soils and cultivated analogues were compared at both locations. Schnabel Woods had a mixed oak-hickory (*Quercus* sp., *Carya* sp.) overstory and has never been clear-cut. Soils were Menfro silt loams (fine-silty, mixed, mesic Typic Hapludalfs).

Van Meter State Park had early successional forest species with canopy dominants about 150 yr old on Knox soils (fine-silty, mixed, mesic Mollic Hapludalfs). Historical survey records indicated that the site was in prairie in 1817 (Indorante, 1990). The mollic epipedon is a remnant of prairie vegetation influence.

Most of the soil C at both sites was below the surface (Table 10–3). Surface horizon C as a percentage of measured profile C ranged from 22.1 to 45.6% in the native Schnabel soils (Menfro). The cultivated Menfro soils, which had been in a forage legume/cool season grass mixture for about 40 yr, had surface horizon C contents which were higher percentages of measured profile C than the

Table 10–3. Total organic carbon in cultivated and uncultivated Menfro (fine-silty, mixed, mesic Typic Hapludalfs) and Knox (fine-silty, mixed, mesic Mollic Hapludalfs) soils in Boone and Saline Counties, MO, respectively. These soils formed in deep loess adjacent to the Missouri River (from Indorante, 1990).

Landscape position	Total C	A horizon C	Sample depth
	—— t ha^{-1} ——		cm
Schnabel uncultivated			
Summit	109.5	50.0	185
Shoulder	101.0	25.3	160
Backslope	112.7	39.3	171
Schnabel cultivated			
Summit	122.7	52.3	180
Shoulder	106.2	40.4	180
Van Meter uncultivated			
Summit	166.9	60.5	183
Shoulder	128.0	62.7	191
Backslope	182.0	78.2	192
Van Meter cultivated			
Summit	99.4	34.9	170
Backslope	97.1	25.3	160
Summit	87.6	19.4	171

forested analogues. However, Menfro surface horizon C was less than 50% of the measured profile C under both forest and forage vegetation.

The Knox soil had less than 50% of its measured profile C in the surface, even with a mollic epipedon. The cultivated Knox soils had lower surface organic C concentrations than the forested analogues. Argillic horizons at or very near the soil surface in the cultivated Knox soils indicated that erosion had removed most of the epipedons.

The uncultivated Knox soils had higher total C contents (128–167 t ha^{-1}) than the forested Menfro soils (109.5–112.7 t ha^{-1}). Carbon in cultivated analogues ranged from 87 to 99 kg ha^{-1} in the Knox soil and from 106 to 123 t ha^{-1} in the Menfro soil. Total profile soil C was related to geomorphic surfaces at both native sites. The relative order of magnitude was backslope > summit > shoulder.

Taberville Native Prairie

The Taberville native prairie in southwest Missouri (St. Claire county) has never been cultivated and was mapped (Howard, 1987) as a Barco silt loam (coarse-loamy, mixed, thermic Mollic Hapludalfs). We opened 10 pits, and removed from each a monolithlike profile slice for detailed soil morphological descriptions. Detailed data were reported by Conway-Nelson (1991).

Carbon content ranged from 75 t ha^{-1} in the middle finger ridge to 155 ha^{-1} in the backslope (Table 10–4). In all profiles except the two on the summit (the "saddle" profile was a concave microsite on the summit), the percentage of C in the epipedon was more than 50% of the total C measured in the profiles.

This was the only landscape we studied in which quantities of surface horizon C exceeded quantities of measured subsoil C. However, only four pits (the saddle, summit, bottom, and noseslope) were dug to bedrock. The other profiles could not be opened below the argillic horizons because the pit sizes, dictated by Missouri Department of Conservation personnel, restricted deeper digging. In all of these profiles, the subsoil C concentrations were at least 0.2% at the base of the pit. A strong possibility exists that subsoil C would have exceeded the C in the epipedons had we been able to sample the entire solum.

Table 10–4. Organic C in 10 soil profiles from Taberville Prairie, an uncultivated prairie remnant in southwest Missouri (Conway-Nelson, 1991). The saddle, summit noseslope, and bottom profiles were dug to bedrock.

Landscape position	Total C	A horizon C	Sample depth
	t ha^{-1}		cm
Summit	146.5	60.8	151
Saddle	111.7	45.9	137
Backslope	155.0	101.4	171
Footslope	146.4	96.9	172
Bottom	140.7	91.4	82
Upper finger ridge	101.8	88.7	87
Middle finger ridge	74.8	48.8	71
Finger ridge backslope	126.1	97.8	127
Noseslope	104.9	88.6	94
Headslope	104.7	67.6	147

Carbon in the epipedons ranged from 45.9 to 101.4 t ha^{-1} with the highest levels in the backslope and footslope positions. These were concave surfaces which received water and hillslope sediments from higher positions in the landscape. The epipedons at these sites were the thickest in the prairie. Profiles with the lowest total and epipedon C were the headslope, followed by profiles on convex surfaces. The headslope profile had thick Bg and E horizons overlying the argillic horizon. The C content was low in the Bg and E horizons but increased in the argillic. We were able to sample only the upper 40 cm of the argillic horizon. If the C distribution in the Bt of the headslope profile is similar to other Bt horizons in the landscape, it is likely that substantial C is contained within the unsampled depths of this soil.

Distribution of soil C in this landscape was related to geomorphic and pedologic conditions affecting the temporal and spatial distributions of soil water. With the exception of the headslope, wetter portions of the landscape had higher C contents. These observations were similar to those of Aandahl (1948), who observed that soil N content in the A horizons of an uncultivated Iowa prairie was strongly related to soil microclimate as affected by topography.

Prism faces in the Taberville Prairie subsoils had C-rich coatings. This characteristic has been observed by other scientists investigating uncultivated soils (Thorpe et al., 1959; Buol & Hole, 1959; Heil & Buntley, 1965; Simonson, 1959). Conway-Nelson (1991) scraped some of these coatings and analyzed them for C content, which ranged from 2.6 to 3.1%. The senior author has observed that C coatings on prism surfaces are common in subsoils under native grasses, but commonly are absent under annual row crops. This is a topic worthy of intensive investigation. The implications are that agricultural practices might result in substantial subsoil C losses, which would represent a previously unmeasured contribution to atmospheric CO_2.

The Sanborn Field Continuous Cultivation Site

We examined centennial data acquired from five Sanborn Field plots (Table 10–5). A variety of cropping systems exist on the field. We chose three continuous corn (*Zea mays* L.) treatments, continuous wheat (*Triticum aestivum* L.), and a four-crop rotation of corn, soybean (*Glycine max* Mer.), wheat, and red clover (*Trifolium pratense* L.) with annual additions of cow manure. The continuous corn treatments have no amendments (Plot 17), full fertility according to soil test recommendations (Plot 6), and 13.6 t ha^{-1} yr^{-1} manure (Plot 18). The continuous wheat (Plot 10) and the four-crop rotation (Plot 34) receive the same amount of annual manure supplement as Plot 18.

The continuous corn treatments are on the summit of Sanborn Field, Plot 10 is on the shoulder and Plot 34 is on the footslope. Total relief in the field is less than 4 m, and the maximum slope is about 7% (Hammer & Brown, 1990). The entire field was mapped as Mexico silt loam (fine, montmorillonitic, mesic, Udollic Ochraqualfs) (Upchurch et al., 1985).

Plots 6, 17, and 18 contained less than 100 t ha^{-1} C to depths of 1.2 m. Carbon on the wheat and rotation cropping systems was 121 and 142 t ha^{-1},

Table 10–5. Soil organic C after 100 yr of continuous cultivation in selected plots within Sanborn Field, Columbia, Missouri. The entire field was mapped as Mexico silt loam (fine, montmorillontic, mesic Udollic Ochraqualfs) (data from Hammer & Brown, 1990).

Plot number†	Total C	Ap horizon C	Sample depth
	——— t ha^{-1} ———		cm
6	96.0	31.8	120
18	90.0	40.3	120
17±	81.0	48.3	120
10	120.8	54.7	120
34	141.8	89.9	120

†Plot treatments: 6, continuous corn + full fertility; 18, continuous corn + 13.6 t ha^{-1} yr^{-1} manure; 17, continuous corn; 10, continuous wheat + 13.6 t ha^{-1} yr^{-1} manure; and 34, corn-soybean-wheat-red clover + 13.6 t ha^{-1} yr^{-1} manure. ±Total organic C in core pulled to 312 cm = 180.3 t ha^{-1}.

respectively. We later returned to the field and pulled a core from Plot 17 to a depth of 3.12 m. An additional 100 t ha^{-1} soil C was in the profile between the depths of 1.2 m and 3.12 m. When these data are included, the surface soil C is only 26.8% of the total measured C. These data support our hypothesis that deep sampling is necessary to quantify the soil C pool in soil systems.

Plot 34 had the highest soil C total in the upper 1.2 m to 142 t ha^{-1}. Sixty-three percent of the measured profile C was in the A horizon of this plot. Deeper sampling will be necessary to quantify the C pool for this treatment.

The Sanborn Field data show that soil C reserves can be increased through a combination of fertility management and organic matter additions. They confirm the importance of deep sampling, and they reveal that most of the soil C is in the subsoil of cultivated soils.

Ozark Region Forest Soils

Ranjith Udawatta's Ph.D. dissertation in progress was the source of soil profile data from forested Ozark soils. Udawatta's sample sites were in the University Forest near Popular Bluff in southeast Missouri and from Pioneer Forest near Salem in south central Missouri. To acquire a subset of these data, the senior author asked the SCL database manager to provide a list of all of the soil data from the project. Data from individual profiles were on individual sheets with site descriptions at the top of the page. All of the 50 profiles were grouped by aspect into four subsets (north, south, east, and west) then two sheets from each subset were removed, without regard to any other characteristic, for profile C calculations.

Six of the selected profiles were skeletal, with rock fragment content ranging to 92%, and all were in the thermic soil temperature regime. Most of the soil C in seven of the eight soils was in the subsoils (Table 10–6). Total profile C ranged from 32.9 to 73.4 t ha^{-1}, and A horizon C as a percent of the total measured C ranged from 22.3 to 60.2%. Soils on north and east aspects had higher surface horizon C, but lower total profile C. Should this finding withstand more

Table 10–6. Measured soil organic C in randomly selected forested Ultisols from sourthern Missouri (data are from R. Udawatta's Ph.D. dissertation in progress).

Aspect	Total C	A horizon C	Sample depth
	t ha^{-1}		cm
W	51.5	11.5	202
W	73.4	24.4	170
N	48.8	22.5	170
N	34.5	8.9	165
E	32.9	19.8	164
E	62.4	23.0	198
S	47.7	12.1	195
S	70.1	28.5	195

rigorous scrutiny, it will have serious potential implications for long-standing paradigms about aspect effects on soil organic matter in temperate deciduous forests.

The two profiles with 70 t ha^{-1} C were close to the older Mississippi River floodplain in southeast Missouri, and had loess over residuum. High rock fragment was encountered only in the residuum.

Data from the Soil Characterization Laboratory

Data from the SCL data base indicate that most of the C in these profiles was in the subsoils, even though only the Bates profile was opened and sampled to bedrock. Total C in these profiles ranged from 65 to 113 t ha^{-1} (Table 10–7). The

Table 10–7. Total organic C in randomly selected profiles from north central Missouri soils. The Bates soil was the only profile dug to bedrock. These are unpublished data from University of Missouri SCL.

Subgroup (series)	Total C	A horizon C	Percentage C in A Horizon	Sample depth
	t ha$_{-1}$			cm
Aquic Hapludalf (Keswick)	112.9	54.0	47.8	119
Typic Argialboll (Edina)	125.6	41.0	32.5	191
Aquic Argaidoll (Adair)	116.5	63.4	54.4	152
Aquollic Hapludalf (Armstrong)	94.2	30.2	32.1	152
Typic Arguidoll (Bates)	73.2	24.5	33.5	94
Aquic Arguidoll (Lamoni)	133.2	62.9	47.2	152
Aquollic Hapludalf (Pershing)	64.8	28.7	44.4	94

A horizon C content ranged from 24.5 to 63.4 t ha^{-1}. The A horizon C contents were 32.5 to 54.4% of measured profile C. Soils in aquic subgroups had the highest A horizon C concentrations. Generally, Alfisols had lower whole profile C quantities than Mollisols.

These data while useful in showing that subsoil C levels exceeded surface horizon C in both Alfisols and Mollisols, have limited application for determining soil C reserves in mapping units, landscapes or regions. Their exact locations are unknown and we don't know how well these profiles represent soil variability within the mapping units from which they were taken.

DISCUSSION

General Considerations

Why have nine decades of study of soils in their natural settings not provided sufficient data to construct a usable database of soil C? Why do investigators sample only the upper few centimeters of soil profiles? For example, Hseih's (1992) treatise on "Pool Size and Mean Age of Stable Soil Organic Carbon in Cropland" is based on samples from the surface 20 cm, which does not include the complete A horizon of one sample plot.

Soil Science Paradigms

Perceptions of the soil have affected how the resource has been studied, characterized and managed (Cline, 1961; Hudson, 1992). We believe that the imprecise quantification of soil C to data is due in part to incomplete paradigms. The factors of soil formation (Jenny, 1941) and a conceptual model from a widely used text illustrate this point.

State Factor Topography

The state factor "topography" initially was defined as surface shape (Jenny, 1941). Stratigraphy, subsoil morphology (including horizons restrictive to vertical subsoil water movement), and microrelief were not discussed. A revision (Jenny, 1961) stated that topography is constant through time. The most recent treatise of the topic, in its discussion of state factor topography, used catenary studies in uniform parent materials as examples (Jenny, 1980). Clearly, this presentation of topography as a soil-forming factor is incomplete.

Landscapes and soils coevolve, so in many landscapes, similar soils are found on landforms with similar genetic histories (Daniels & Hammer, 1992). Soils can be expected to vary predictably in downslope transects. For example, Pregitzer et al. (1983) illustrated the effects of parent materials and stratigraphy in creating soil heterogeneity in a sloping forested landscape. Hammer et al. (1991a) have discussed other transect analyses of landscapes.

Yaalon's (1983) discussion of "feedback" soil properties—those which alter internal processes as soil development proceeds—is also germane. The influence of argillic horizons on vertical water movement in sloping landscapes is an excellent example. As development of the argillic horizons proceeds with clay illuviation by percolating water, the Bt horizons become increasingly less permeable.

Eventually, the argillic horizon becomes a restriction to vertical water movement. Seasonally perched water tables then can develop above the argillic horizon (Beasley, 1976).

Our purpose is not to disclaim the factors of soil formation, but to show the incompleteness of state factor topography when applied to the analysis of complex landscapes. Surface shape alone is not the sole determinant of the temporal and spatial distribution of soil water. Subsurface attributes including stratigraphy and horizons of low structural development or high bulk density are important in affecting subsoil water distribution. The ultimate fate of all conceptual models in science should be alteration or obsolescence as knowledge accumulates (Cline, 1961; Dijkerman, 1974). Fifty years of data accumulation and study have created a need for revision of the soil landscape paradigm.

The Forest-Prairie Ecotone

The common paradigm of the forest-prairie ecotone is one of gradual change between botanical communities, with concomitant gradual soil transitions (Fig. 10–5A). This model assumes uniform parent material and lack of relief. In reality, local topography, parent materials, stratigraphy and aspect through their effects on soil water distribution and evapotranspiration, influence the presence, abundance and productivity of forest and prairie species as well as the forms, concentrations and quantities of soil organic matter. The ecotone location has probably fluctuated with climatic change through the Pleistocene. Additionally, the ecotone probably was a complex mosaic also which changed in composition and structure across regional temperature and precipitation gradients, rather than being a uniformly changing vegetational belt between two distinct biomes.

A relatively simple conceptual model becomes more complex as stratigraphic and topographic diversity are introduced (Fig. 10–5B). Different parent materials develop different soils. Alfisols form under forest and Mollisols form under prairie. Stratigraphy affects spatial and temporal distribution of throughflow (subsurface water movement). Aspect differences cause different evapotranspiration rates in landscapes, generating differential biomass production, decay, distribution and retention. Site-specific differences in amounts of retained C result, as demonstrated previously.

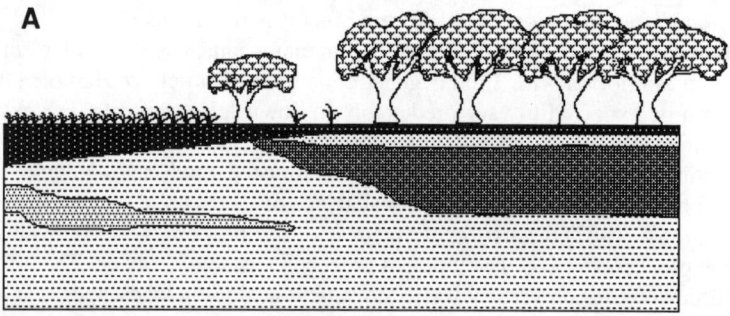

Fig. 10–5A. Conceptual models of the forest-prairie ecotone. Figure 5A assumes parent material and topographic uniformity (adapted from Buol et al., 1980).

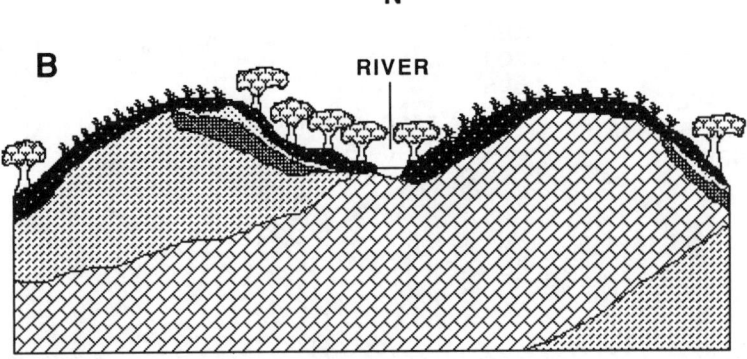

Fig. 10–5B is based on dissected sandstone and shale bedrock overlain by loess.

Conversion of the landscape to row crop agriculture causes erosion, removing some features, burying others under hillslope sediments, and changing local hydrology (Fig. 10–5C). Subsequent multiple use management of the landscape imposes a variety of new treatments, further increasing heterogeneity. Models of soil systems—genesis models, hydrological models, and land-use models—should include as components the stratigraphic and geomorphic features which affect the temporal and spatial distributions of soil water.

Soil Sampling and Statistical Analyses

Soil sampling strategies commonly suffer from several weaknesses: (i) failure to adequately describe site conditions and sampling methodology; (ii) failure to sample the genetic horizons which represent the alteration of the parent materials into the chemical, physical and biological arrangements representative of the soil; (ii) failure to design and sample with sufficient rigor to satisfy statistical requirements; and (iv) failure to recognize relationships among soils and the factors causing their development (geomorphology, land use, aspect, etc.).

One frequently reads in the "methods" sections of published papers that the authors selected for study "representative pedons" or "representative transects." The reader seldom is given an indication of the methods used to select the "representative" sampling units from the population. If field experience or investigator preference are the criterion as opposed to a statistical selection, the reader should be so advised.

Most landscapes are characterized by patterns of convexity and concavity which influence the temporal and spatial distributions of soil water, organic debris, and sediments, thus affecting pedogenesis in systematic ways. For example, McFee and Stone (1965) observed that forest floor variability was affected by soil surface microrelief on the Adams soil (sandy, mixed, frigid Typic Haplorthods). Cremeans and Kalisz (1968) found that forest soil variability was related to windthrow microrelief in Kentucky. Numerous studies (Hammer et al., 1991a) have shown the importance of landscape position on soil properties in

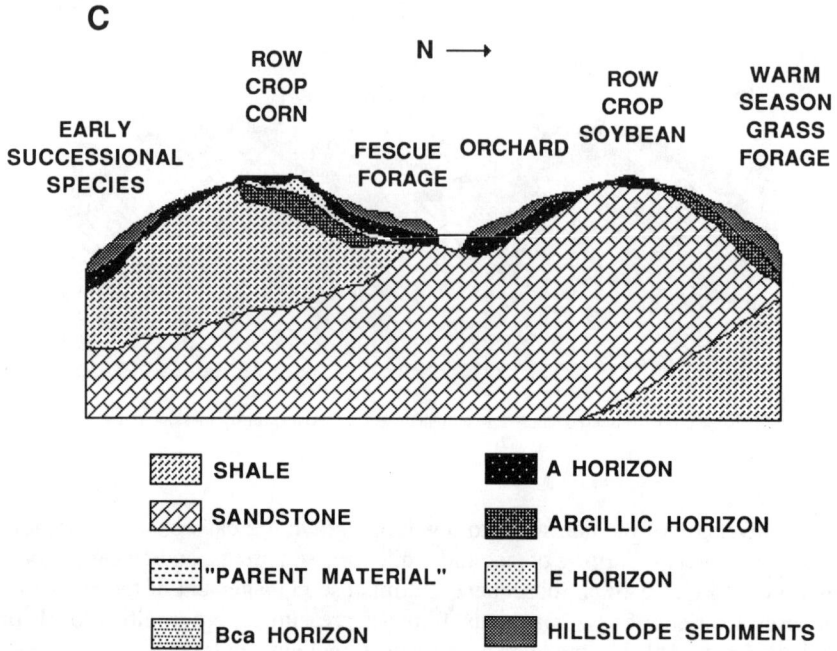

Fig. 10–5C is a model of the landscape in 4B after clearing of native vegetation and post-settlement erosion.

dissected or undulating topography. Seldom are surface shape or position in the landscape discussed together in the context of experimental design.

Boudeman (1989) studied forage legume yield as a function of landscape position in the Mexico soil. Figure 10–6 shows total profile C obtained from each of four transects in a 0.4-ha field in which relief was less than 3 m. Profile C values range from 76.4 to 157 t ha^{-1}. Without some prior knowledge of soil-landscape relationships, and without some existing data, how would one be able to determine where to place the "representative transect"? These data were acquired from a single map unit with little apparent surficial or stratigraphic variability.

Are Soil Properties Normally Distributed?

Parametric statistical procedures are based on the assumption that sample data are randomly acquired from normally distributed populations. Few studies have investigated the fundamental assumption of normality in populations of soil variables. Young et al. (1993) acquired 123 pedons from the Eudora soil (coarse-silty, mixed, mesic Fluventic Hapludolls). Neither A horizon thickness (Fig. 10–7A) nor C content (Fig. 10–7B) were normally distributed. The distribution of C did not mirror that of A horizon thickness. Distributions of A horizon textures and measured subsoil properties also were non-normally distributed.

Young et al. (1993) also compared confidence intervals produced using parametric and nonparametric statistical methods. For many of the measured soil

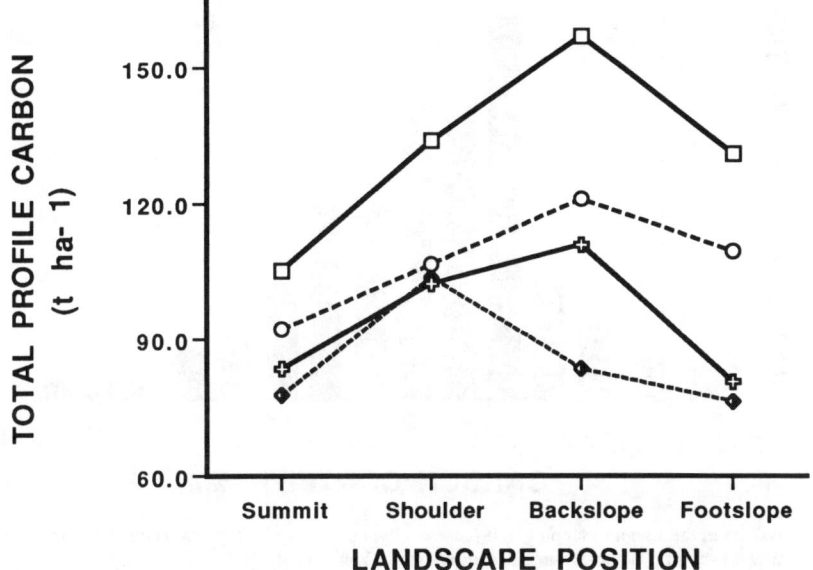

Fig. 10–6. Organic C to depths of 1.2 m in four parallel transects of a Mexico silt loam (fine, montmorillonitic, mesic Udollic Ochraqualfs) near Columbia, Missouri. Sampling intervals are approximately 15 m (from Boudeman, 1989).

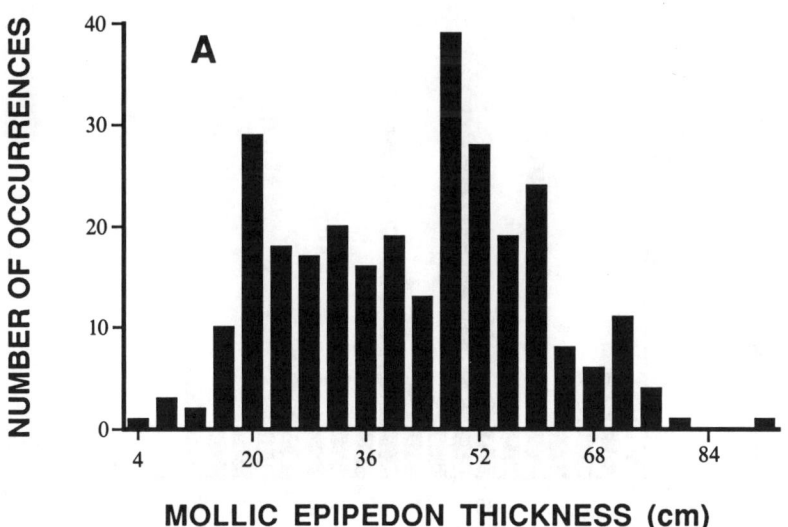

Fig. 10–7(A). Histograms of mollic epipedon thicknesses (A) and A horizon organic C.

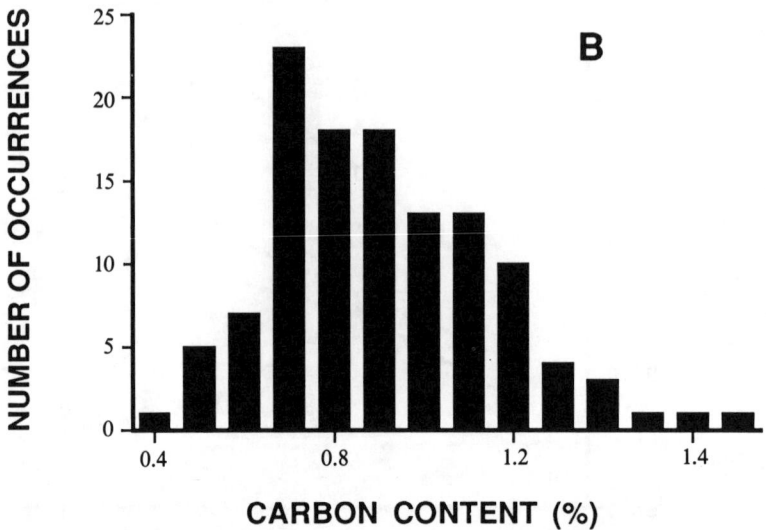

Fig. 10–7(B) in the Eudora mapping unit (coarse-silty, mixed, mesic fluventic Hapludolls) in Boone County, Missouri. Data are from 120 samples (from Young et al., 1993).

properties, they found significant differences between confidence intervals produced by the two statistical methods. Additionally, many of the data transformed for nonparametric statistical analyses were difficult to interpret.

Data acquired for Young's Ph.D. dissertation in progress (Fig. 10–8) show that mollic epidedon thickness is not normally distributed in the Lindley mapping

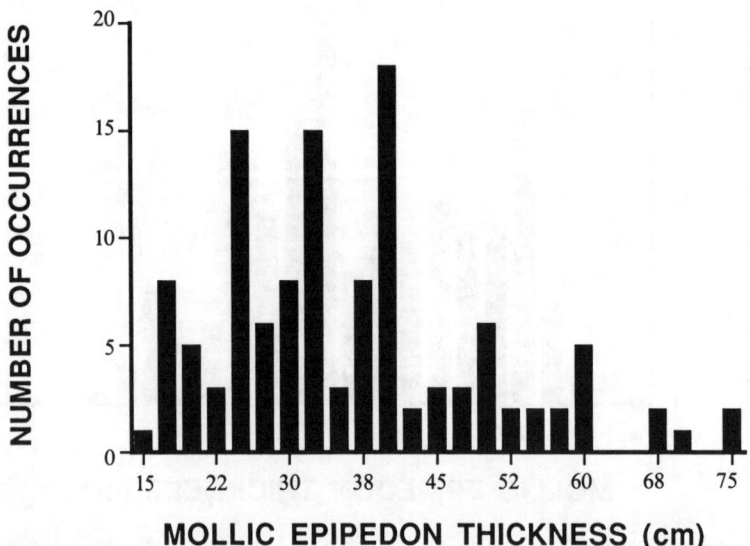

Fig. 10–8. Histogram of mollic epidedon thicknesses in the Lindley silt loam mapping unit (fine-loamy, mixed, mesic Mollic Hapludalfs) in Boone County, Missouri. Data are from 245 pedons taken from a 12-ha field (unpublished data from F.J. Young, Ph.D. dissertation in progress).

unit (fine-loamy, mixed, mesic Mollic Hapludalfs). This soil formed in loess over till in a gently undulating landscape in which sideslopes were about 4% and relief was about 8 m in a 40-ha field. In the same study, Young and Hammer (1993) tested the normality of 88 soil and chemical properties in each of three landscape positions (summit, shoulder, and backslope). Only 33% of the 264 soil properties were normally distributed.

Information on the normality of soil properties is limited, and questions remain about the proper statistical analyses required to analyze non-normal data. Several investigators believe that nonparametric statistical procedures will be required to analyze non-normal soil data (McIntyre & Tanner, 1959; Wagenet & Jurinak, 1978; Bell et al., 1992). Assumptions of normally distributed soil properties should be tested in other soil landscapes.

The "Dynamic" A Horizon

Shallow (incomplete) soil sampling is a problem which has received too little attention (Stone & Kalisz, 1991). One possible explanation is that foresters assume that the "dynamic" components of the soil ecosystem—the forest floor and A horizon—are the most important, and sample accordingly. However, Hammer et al. (1991b) showed that in forested Ultisol landscapes, most soil variance is in the subsoil. Another study (Hammer et al., 1987) compared spatial and temporal soil variability in forest soils. This study showed that landform was the significant cause of soil variance. Effects of time were not significant. These are two examples that landform and subsoil properties are important determinants of local soil variability. We suspect that the dynamic temporal nature of the A horizon in forest ecosystems has been overemphasized with respect to other factors causing soil variance.

The Importance of Soil Water

The importance of soil water for biomass production and organic matter retention in soils is well established (Anderson, 1992; Carmean, 1975). As previously discussed, factors affecting soil water are related to geomorphology, stratigraphy, and subsoil conditions including depth, stoniness, texture, structure, and bulk density.

Other considerations when assessing the relative distribution of soil water in landscapes include; location in the landscape with respect to other landforms (Daniels & Hammer, 1992), surface shapes of individual landforms (Huggett, 1975), and microrelief (Cremeans & Kalisz, 1988; Veneman et al., 1984). Soil organic matter and other soil chemical properties also are affected by aspect and slope position (Hammer et al., 1991a), proximity to large tree boles (Wolfe et al., 1987; Gersper & Holowaychuck, 1970; McFee & Stone, 1965; Zinke, 1962; Zinke & Crocker, 1962) and presence of fallen stems (McFee & Stone, 1965).

Aandahl's (1948) classic investigation of an Iowa native prairie showed the importance of surface shape and landscape position in causing differential organic matter production and soil C retention in the landscape. Variability would have appeared to be random if geomorphic surfaces were not considered in a systematic sampling scheme. Observed variance was related to factors affecting soil water distribution. Aandahl's study should have provided a conceptual framework for subsequent field soil sampling, but it has been largely ignored.

Temporal and spatial distributions of soil water are the driving forces for soil genesis and landscape evolution, but many investigators fail to consider the geomorphic and stratigraphic factors affecting soil water (Daniels & Hammer, 1992).

Soil Variability and Field Sampling

Wilding and Drees (1983) divided soil spatial variability into "systematic" and "random" components. Random variability was defined as that for which no cause was known. We suggest that the term "random" could be replaced by "unexplained." The difference in perspective is subtle but important. The former term implies that causal relationships may be undefinable. The latter assumes that order exists, but that we have yet to discern the patterns and causal processes.

Field studies commonly rely on random sampling or "representative transects" for data acquisition. Our previous discussion shows that much soil variability is systematically ordered by recognizable biological and physical attributes of the soil-vegetation continuum. We propose that field sampling would be more precise, less costly, and would produce results more easily transferrable to similar landscapes if soil sampling procedures were systematically planned in accordance with factors known to cause variability. Roberts et al. (1989) called this approach "hierarchical" sampling and have applied it successfully to regional studies.

A second consideration is sample size. Brandt's (1993) literature review revealed that lack of replication and low sample numbers was a recurring characteristic of field studies comparing either similar treatments, different soils, or different treatments on similar soils.

McFee and Stone (1965) reported that forest floor variability under yellow birch (*Betula alleghaniensis* Britton)-red spruce (*Picea rubens* Sarg.) was so great that about 50 samples were necessary to reduce the standard error to 10% of the mean forest floor weights in 0.04-ha plots. They observed similar variability in mineral horizon thicknesses.

Mollitor et al. (1980) determined that more than 1000 samples per plot would be necessary to establish K concentration levels with error within 10% of the sample mean on a northeastern flood plain. Plot size was not specified. In the Virginia Piedmont, Della-Bianca and Wells (1967) found that exchangeable Ca in the A2 horizon of the Cecil soil ranged from 2 to 1090 mg kg^{-1}. They concluded that the soil series concept was inadequate as a sampling entity for evaluating nutrient levels for forest management purposes.

Budgetary constraints usually limit the scope of field sampling. Clearly, however, a few "random" samples chosen without regard to factors causing soil variance are not likely to produce data which can be extrapolated to other systems or which will be upheld in confirmational studies.

The Mapping Unit as a Soil Sampling "Treatment"

The mapping unit is sometimes treated by investigators as an entity of sufficient purity that replication has minimal importance. Forest soil scientists who have studied site productivity are aware that numerous investigators have found the soil series as a mapping unit to be too heterogeneous to be useful as a measure of forest productivity (Farnsworth & Leaf, 1965; Jones, 1969; Shetron,

1972; Van Lear & Hosner, 1967). Grigal's (1984) discussion of the reasons for this situation is objective and thorough.

Some pedologists are concerned about mapping unit heterogeneity. Often, much important information about mapping units is not clearly expressed and delivered to the user (Arnold & Wilding, 1991). Three primary reasons exist for this situation. First, much of what is known by soil mappers is tacit information which is difficult to convey linguistically (Hudson, 1992). Second, no means exists by which all relevant information can be conveyed by maps produced at the scales commonly used to make soil surveys (Burrough, 1991). Finally, soil surveys traditionally have been designed for the nontechnical user, who has neither needed nor demanded quantitative information (Brown & Huddleston, 1991).

Numerous studies have quantified mapping unit composition. Soils in the previously cited Taberville, Missouri, native prairie were mapped as Barco (coarse-loamy, mixed, thermic Mollic Hapludalfs) (Howard, 1987). None of the 10 study pits classified as a Barco (Conway-Nelson, 1991). Several early studies showed that taxonomic purity of mapping units seldom exceeds 50% (Amos & Whiteside, 1975; McCormack & Wilding, 1969; Powell & Springer, 1965; Wilding et al., 1965). More recently, Young et al. (1993) sampled 120 soil cores from a single mapping unit in Boone County, Missouri. Only 25% of the profiles were the mapped soil series, and 28 taxonomic classes were represented (Fig. 10–9). Hartung et al. (1991) found only 26% of the mapped taxon in five transects of the Sharpsburg silty clay loam (fine, montmorillonitic, mesic Typic Argiudolls) on 0 to 2% slopes in Nebraska. The eroded phase of the soil on 6 to 12% slopes was 15% taxonomically pure, and the severely eroded phase on 6 to 12% slopes was 16% taxonomically pure (Hartung et al., 1991). Nordt et al. (1991) investigated four mapping units within a single survey area (Brazos

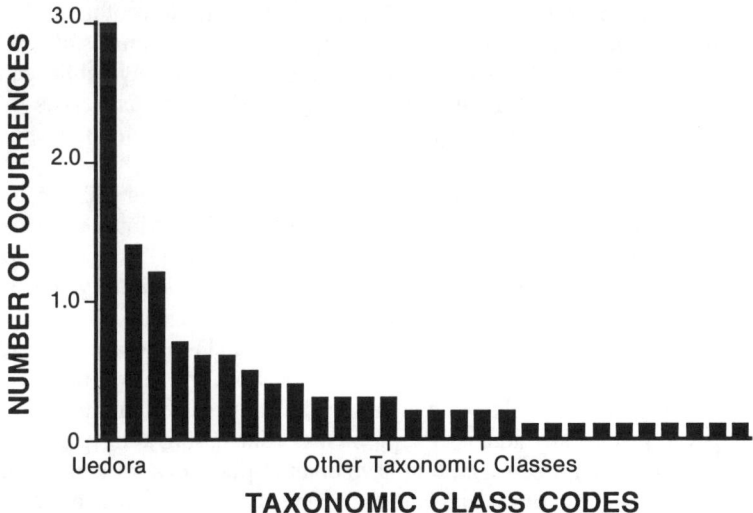

Fig. 10–9. Taxonomic classes in 120 samples of the Eudora silt loam mapping unit (coarse-silty, mixed, mesic Fluventic Hapludolls) in Boone County, Missouri. Taxonomic class "a" is the class representing the Eudora Soil (from Young et al., 1993).

County, TX) and found within-series taxonomic purities of 49, 45, 21 and 11%. Taxonomic variability among mapping units was not consistent in the survey area.

Nordt et al. (1991) and Wilding and Drees (1983) have cautioned soil survey users that taxonomic purity often is less than interpretive purity. However, Brown and Huddleston (1991) voice strong concerns that soil surveys contain qualitative rather than quantitative descriptions of soil variability. Several investigators have recommended that statistical data should be gathered in the course of soil survey activities and should be published as tables in soil surveys (Burrough, 1991; Brown & Huddleston, 1991; Young et al., 1991; Nordt et al., 1991; Lammers & Johnson, 1991; Upchurch & Edmonds, 1991; Brubaker & Hallmark, 1991).

For forest management and forest soil inventory purposes, these problems are further compounded by an apparent tendency of soil scientists to paint with a broader brush when mapping forest lands. Figure 10–9 illustrates this point. The two field sheets are facing pages from the St. Francois County, Missouri, soil survey (Brown, 1981). Both areas have the same parent materials and topographic relief. The difference is that Fig. 10–10A is from a portion of the Mark Twain National Forest and Fig. 10–10B is from agricultural land. The detail of mapping in the agricultural lands clearly is finer than in the forested area.

Users of soil survey information should be aware of the natural heterogeneity of soil survey mapping units. Hudson (1992) further cautions that the soil–landscape paradigm, which is the basis for soil surveys "... is nearly absent from soil science curricula and texts. Furthermore, there has been little substantive academic research dealing with soils as natural bodies on the landscape...."

SUMMARY AND CONCLUSIONS

Most of the soil organic C in the Missouri forest-prairie ecotone is in subsoils. This relationship is true even for many Mollisols. Our estimates of soil C in this part of the temperate forest-prairie ecotone exceed previously published estimates. Additionally, the distribution of soil organic C within landscapes is heterogeneous, but the heterogeneity is related to soil geomorphology and soil hydrology.

The implications are twofold. First, the high quantities of subsoil C in uncultivated soils probably invalidate existing models of soil C contributions to atmospheric CO_2. Second, the likelihood of accurately quantifying either soil C pools or the fluxes between soil C and other compartments in the soil C cycle is small until systematic, statistically rigorous inventories are conducted.

Rates of C loss or accumulation under different land use practices cannot be accurately measured or predicted until the pool sizes are quantified. Clearly, quantification of soil C pools and dynamics is a challenge which must be met if man hopes to realistically address the problems of increasing atmospheric C.

The results of our deep sampling have other implications for soil scientists, foresters, landscape ecologists and land use planners. Shallow soil sampling, coupled with sampling schemes lacking statistical rigor, will continue to yield incomplete and inaccurate perceptions of the soil resource. We as soil scientists

A. FORESTED

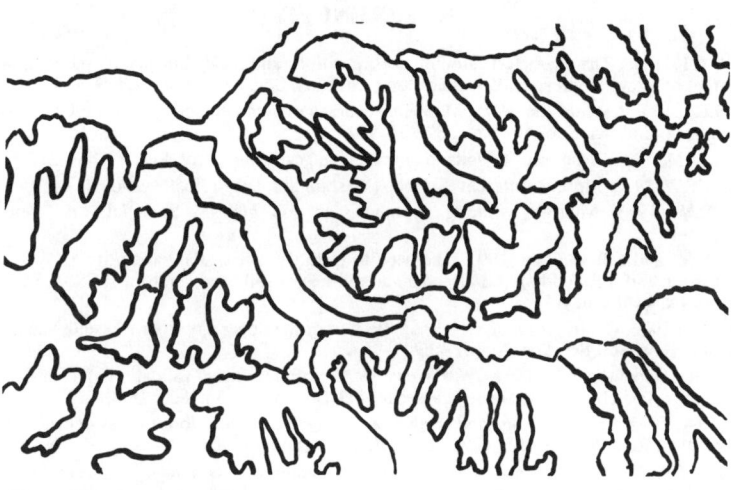

B. AGRICULTURAL

Fig. 10–10. Overlays of mapping units from forested (A) and agricultural (B) lands on adjacent terrain in the St. Francois County, Missouri, soil survey (from Brown, 1981).

should demonstrate the same respect with applications of statistical methods as we expect scientists from other disciplines to exercise when they study soils.

We do not expect that all biomes will have higher subsoil C quantities than the surface or forest floor. The boreal forests often have thick forest floors. However, subsoil organic C levels need careful quantification. McFee and Stone (1965) have shown that subsoil C quantities can be high, even in sandy parent materials.

Need exists for carefully planned, long-term investigations of soil C dynamics under a variety of management practices in agriculture, range management and forestry. The challenge will be great because of temporal and spatial soil heterogeneity. However, the challenge is important, and if met, should greatly enhance man's perception of the dynamics of the soil resource.

ACKNOWLEDGMENTS

Contribution no. 11,926 from the Missouri Agricultural Experiment Station. Partial funding for cited research was provided by Grants 86–1, 88–3, 88–4, and the continuing grant to the Soil Characterization Laboratory from the Missouri Department of Natural Resources. The Missouri Department of Conservation allowed access to the Taberville Prairie and provided partial funding for Ranjith Udawatta's research. We thank Dr. David Peterson for his stimulating discusison and encouragement. The manuscript was improved by the helpful comments of Dr. W.W. McFee and Dr. J.M. Kelly and by the suggestions of an anonymous reviewer.

REFERENCES

Aandahl, A.R. 1948. The characterization of slope positions and their influence on the total nitrogen content of a few virgin soils of western Iowa. Soil Sci. Soc. Am. Proc. 12:449–454.

Amos, D.F., and E.P. Whiteside. 1975. Mapping accuracy of a contemporary soil survey in an urbanizing area. Soil Sci. Soc. Am. Proc. 39:937–942.

Anderson, J.M. 1992. Responses of soils to climate change. Advan. Ecol. Res. 22:163–210.

Anderson, P.W. 1972. More is different. Science (Washington, DC) 177:393–396.

Arnold, R.W. 1965. Multiple working hypothesis in soil genesis. Soil Sci. Soc. Am. Proc. 29:717–724.

Arnold, R.W., and L.P. Wilding. 1991. The need to quantify soil spatial variability. p. 1–8. *In* M.J. Mausbach and L.P. Wilding (ed.) Spatial variabilities of soils and landforms. SSSA Spec. Publ. 28. SSSA, Madison, WI.

Beasley, R.S. 1976. Contribution of subsurface flow from the upper slopes of forested watersheds to channel flow. Soil Sci. Soc. Am. J. 40:944–957.

Bell, J.C., R.L. Connungham, and M.W. Havens. 1992. Calibration and validation of a soil-landscape model for predicting soil drainage class. Soil Sci. Soc. Am. J. 56:1860–1866.

Birdsey, R.A. 1992. Carbon storage and accumulation in United States forest ecosystems. USDA-FS. Gen. Tech. Rep. WO-59. U.S. Gov. Print. Office, Washington, DC.

Boudeman, J.W. 1989. Above- and below-ground production of three forage legumes related to soil properties and a soil-based productivity index. M.S. thesis, Univ. Missouri-Columbia.

Brandt, D.K. 1993. Soil properties under warm season grasses and continuous cultivation. M.S. thesis, Univ. Missouri-Columbia.

Brown, B.L. 1981. Soil survey of St. Francois County, MO. USDA-SCS and Missouri Agric. Exp. Stn. U.S. Gov. Print. Office, Washington, DC.

Brown, R.B., and J.H. Huddleston. 1991. Presentation of statistical data on map units to the user. p. 127–148. *In* M.J. Mausbach and L.P. Wilding (ed.) Spatial variabilities of soils and landforms. SSSA Spec. Publ. 28. SSSA, Madison, WI.

Brubaker, S.C., and C.T. Hallmark. 1991. A comparison of statistical methods for evaluating map unit composition. p. 73–88. *In* M.J. Mausbach and L.P. Wilding (ed.) SSSA Spatial variabilities of soils and landforms. Spec. Publ. 28. SSSA, Madison, WI.

Buol, S.W., and F.D. Hole. 1959. Some characteristics of clay skins on peds in the B horizon of a Gray-Brown Podzolic soil. Soil Sci. Soc. Am. Proc. 23:239–241.

Buol, S.W., F.D. Hole, and R.J. McCracken. 1980. Soil genesis and classification. 2nd ed. Iowa State Univ. Press, Ames, IA.

Burrough, P.A. 1991. Sampling designs for quantifying map unit composition. p. 89–126. *In* M.J. Mausbach and L.P. Wilding (ed.) Spatial variabilities of soils and landforms. SSSA Spec. Publ. 18. SSSA, Madison, WI.

Cady, J.G. 1960. Mineral occurrence in relation to soil profile differentiation. p. 418–424. *In* F.E. Bear (ed.) Trans. Congr. Soil Sci. 7, 4. Int. Soc. Soil Sci., Madison, WI.

Carmean, W.H. 1975. Forest site quality evaluation in the United States. Advan. Agron. 27:209–258.

Cline, M.G. 1961. The changing model of soil. Soil Sci. Soc. Am. Proc. 25:442–446.

Conway-Nelson, C. 1991. Soil variability associated with landscape position in a southwestern Missouri prairie. M.S. thesis, Univ. Missouri-Columbia.

Cremeans, D.W., and P.J. Kalisz. 1988. Distribution and characteristics of windthrow microtopography on the Cumberland Plateau region of Kentucky. Soil Sci. Soc. Am. J. 52:816–822.

Daniels, R.B., and R.D. Hammer. 1992. Soil geomorphology. John Wiley & Sons, New York.

Della-Bianca, L., and C.G. Wells. 1967. Some chemical properties of forest soils in the Virginia-Carolina Piedmont. USDA-FS Res. Pap. SE-28. U.S. Gov. Print. Office, Washington, DC.

De Yong, E., and R.G. Kachanoski. 1988. The importance of erosion in the carbon balance of prairie soils. Can. J. Soil Sci. 68:111–119.

Dijkerman, J.C. 1974. Pedology as a science: The role of data, models and theories in the study of natural soil systems. Geoderma 11:73–93.

Farnsworth, C.E., and A.L. Leaf. 1965. An approach to soil-site problems: Sugar maple-soil relations in New York. *In* C.T. Youngberg (ed.) Proc. Forest soil relationships in North America, 2nd North Am. For. Soils Conf., Corrallis, OR. Oregon State Univ. Press, Corvallis, OR.

Fenton, T.E. 1983. Mollisols p. 125–163. *In* L.P. Wilding et al. (ed.) Pedogenesis and soil taxonomy. Vol. 2. The soil orders, Elsevier Press, New York.

Franzmeier, D.P., G.D. Lemme, and R.J. Miles. 1985. Organic carbon in soils of north central United States. Soil Sci. Soc. Am. J. 49:702–708.

Geiss, J.W., and W.R. Boggess. 1968. The prairie peninsula: Its origin and significance in the vegetational history of Central Illinois. p. 89–95. *In* R.E. Bergstrom (ed.) Proc. Symp. Quaternary of Illinois, Champaign, IL. 12–13 February. College Agric. Spec. Publ. 14. Univ. Illinois, Urbana, IL.

Gersper, P.L., and N. Holowaychuk. 1970. Effects of stemflow on a Miami soil under a beech tree: II. Chemical properties. Soil Sci. Soc. Am. Proc. 34:786–794.

Grigal, D.F. 1984. Shortcoming of soil surveys for forest management. p. 148–164. *In* J.G. Bockheim (ed.) Proc. Symp. For. Land Classification: Experience, problems, perspectives, Madison, WI. 18–20 March. Dep. Soil Sci., Madison, WI.

Hammer, R.D., J.H. Astroth, G.S. Henderson, and F.J. Young. 1991a. Applications of geographic information systems and digital elevation models for soil survey and land use planning. p. 243–270. *In* M.J. Mausbach and L.P. Wilding (ed.) Spatial variabilities of soils and landforms. SSSA Spec. Publ. 28, SSSA, Madison, WI.

Hammer, R.D, and J.R. Brown. 1990. 100 years of cultivation on Sanborn Field—A pedologist's perspectives. p. 109–123. *In* Proc. Sanborn Field Cent., Columbia, MO. 27 June 1989. Univ. Missouri Agric. Exp. Stn. Spec. Publ. 415. Missouri Agric. Exp. Stn.

Hammer, R.D., J.W. Philpot. and J.M. Maatta. 1991b. Applying principal component analysis to soil-landscape research—quantifying the subjective. p. 90–105. *In* G.A. Milliken and J.R. Schwenke (ed.) Proc. 1990 Kansas State Univ. Conf. Applied Statistics Agriculture, Lawrence, KS. 29 Apr.–1 May 1990. Manhattan, KS.

Hammer, R.D., R.G. O'Brien, and R.J. Lewis. 1987. Temporal and spatial soil variability on three forested landtypes on the Mid-Cumberland plateau. Soil Sci. Soc. Am. J. 51:1320–1326.

Hartung, S.L., S.A. Scheinhost, and R.J. Ahrens. 1991. Scientific methodology of the national cooperative soil survey. p. 39–48. *In* M.J. Mausbach and L.P. Wilding (ed.) Spatial variabilities of soils and landforms. SSSA Spec. Publ. 28, SSSA, Madison, WI.

Heil, R.D., and G.J. Buntley. 1965. A comparison of the characteristics of the ped faces and ped interiors of the B horizon in a Chestnut soil. Soil Sci. Soc. Am. Proc. 29:583–587.

Howard, D.A. 1987. Soil survey of St. Clair County, MO. USDA-SCS and Missouri Agric. Exp. Stn. U.S. Gov. Print. Office, Washington, DC.

Hsieh, Y. 1992. Pool size and mean age of stable soil organic carbon in cropland. Soil Sci. Soc. Am. J. 56:460–464.

Hudson, B.D. 1992. The soil survey as paradigm-based science. Soil Sci. Soc. Am. J. 56:836–841.

Huggett, R.J. 1975. Soil landscape systems: A model of soil genesis. Geoderma 13:1–22.

Indorante, S.J. 1990. Comparison of soil properties on eroded and native deep loess soils. Ph.D. diss. Univ. Missouri-Columbia (Diss. Abstr. 81–26447).

Jenny, H.J. 1941. Factors of soil formation. McGraw-Hill, New York.
Jenny, H.J. 1961. Derivation of state factor equations of soils and ecosystems. Soil Sci. Soc. Am. Proc. 25:385–388.
Jenny, H.J. 1980. The soil resource. Springer-Verlag, New York.
Jones, J.R. 1969. Review and comparison of site evaluation methods. USDA-FS Res. Pap. RM–51. U.S. Gov. Print. Office, Washington, DC.
Krusekopf, H.H., and C.L. Scrivner. 1962. Soil survey of Boone County, MO. USDA-SCS. U.S. Gov. Print. Office, Washington, DC.
Kucera, C.L. 1961. The grasses of Missouri. Univ. Missouri Press, Columbia, MO.
Lammers, D.A., and M.G. Johnson. 1991. Soil mapping concepts for environmental assessment. p. 149–160. In M.J. Mausbach and L.P. Wilding (ed.) Spatial variabilities of soils and landforms. SSSA Spec. Publ. No. 28. SSSA, Madison, WI.
Mann, L.K. 1985. A regional comparison of carbon in cultivated and uncultivated alfisols and mollisols in the central United States. Geoderma 36:241–253.
McCormack, D.E., and L.P. Wilding. 1969. Variation of soil properties within mapping units of soils with contrasting substrata in northwestern Ohio. Soil Sci. Soc. Am. Proc. 33:587–593.
McFee, W.W., and E.L. Stone. 1965. Quantity, distribution, and variability of organic matter and nutrients in a forest Podzol in New York. Soil Sci. Soc. Am. Proc. 29:432–436.
McIntyre, D.S., and C.B. Tanner. 1959. Abnormally distributed soil physical measurements and nonparametric statistics. Soil Sci. 88:133–137.
Mollitor, A.V., A.L. Leaf, and L.A. Morris. 1980. Forest soil variability on a northeastern floodplain. Soil Sci. Soc. Am. J. 44:617–620.
Nordt, L.C., J.S. Jacob, and L.P. Wilding. 1991. Quantifying map unit composition for quality control in soil survey. p. 183–198. In M.J. Mausbach and L.P. Wilding (ed.) Spatial variabilities of soils and landforms. SSSA Spec. Publ. 28. SSSA, Madison, WI.
Powell, J.C., and M.E. Springer. 1965. Composition and precision of classification of several mapping units of the Appling, Cecil and Lloyd series in Walton County, Georgia. Soil Sci. Soc. Am. Proc. 29:454–458.
Pregitzer, K.S., B.V. Barnes, and G.D. Lemme. 1983. Relationship of topography to soils and vegetation in an Upper Michigan ecosystem. Soil Sci. Soc. Am. J. 47:117–123.
Roberts, T.L., J.R. Bettany, and J.W.B. Stewart. 1989. A hierarchical approach to the study of organic C, N, P, and S in western Canadian soils. Can. J. Soil Sci. 69:739–749.
Ruhe, R.V. 1970. Soils, paleosols and environment. p. 37–52. In W. Dort, Jr., and J.K. Jones, Jr. (ed.) Pleistocene and recent environments of the Central Great Plains. Univ. Press Kansas, Lawrence, KS.
Ruhe, R.V. 1975. Climate, geomorphology and fully developed slopes. Catena 2:309–320.
Scharpenseel, H.W., and P. Becker-Heidmann. 1990. Overview of the greenhouse effect global change syndrome; general outlook. p. 1–14. In H.W. Scharpenseel et al. (ed.) Soils on a warmer earth. Proc. Int. Workshop Effects Expected Climate Change on Soil Processes in the Tropics and Subtropics, Nairobi. 12–14 February, Elsevier, New York.
Schimel, D.S, D.C. Coleman, and K.A. Horton. 1985. Soil organic matter dynamics in paired rangeland and cropland toposequences in North Dakota. Geoderma 36:201–214.
Schlesinger, W.H. 1990. Evidence from chronosequence studies for a low carbon-storage potential of soils. Nature (London) 348:232–234.
Schroeder, W.A. 1981. Presettlement prairie of Missouri. Nat. Hist. Ser. no. 2. Missouri Dep. Conserv., Jefferson City, MO.
Scrivner, C.L., J.C. Baker, and D.R. Brees. 1973. Combined daily climatic data and dilute solution chemistry in studies of soil profile formation. Soil Sci. 115:213–223.
Scrivner, C.L., and D.T. Cooper. 1985. Organic carbon in soils of Missouri: A summary of accumulated research data. Res. Bull. 1055. Univ. Missouri, College Agric. Exp. Stn., Columbia, MO.
Shetron, S.G. 1972. Forest site productivity among soil taxonomic units in northern lower Michigan. Soil Sci. Soc. Am. Proc. 36:358–363.
Simonson, R.W. 1959. Outline of a generalized theory of soil genesis. Soil Sci. Soc. Am. Proc. 23:152–159.
Soil Survey Staff. 1975. Soil taxonomy: A basic system of soil classification for making and interpreting soil surveys. USDA-SCS Agric. Handb. 436. U.S. Gov. Print. Office, Washington, DC.
Stevenson, F.J. 1982. Origin and distribution of nitrogen in soils. p. 1–42. In F.J. Stevenson (ed.) Nitrogen in agricultural soils. Agron. Monogr. 22. ASA, CSSA, SSSA, Madison, WI.
Stone, E.L. 1975. Effects of species on nutrient cycles and soil change. Philos. Trans. R. Soc. London, B. 271:149–162.

Sonte, E.L., and P.J. Kalisz. 1991. On the maximum extent of tree roots. For. Ecol. Manag. 46:59–102.
Technical Committee on Soil Survey. 1960. Soils of the North Central Region of the United States. North Central Regional Publ. 76. Bull. 544. Univ. Wisconsin Agric. Exp. Stn., Madison, WI.
Thorp, J., and G.D. Smith. 1949. Higher categories of soil classification. Soil Sci. 67:117–126.
Thorp, J., J.G. Cady, and E.E. Gamble. 1959. Genesis of Miami silt loam. Soil Sci. Soc. Am. Proc. 23:156–161.
Tinker, P.B., and P. Ineson. 1990. Soil organic matter and biology in relation to climate change. p. 71–89. In H.W. Scharpenseel et al. (ed.) Soils on a warmer earth. Proc. Int. Workshop Effects Expected Climate Change Soil Processes in the Tropics and Sub-tropics, Nairobi. 12–14 February. Elsevier, New York.
Transeau, E.N. 1935. The prairie peninsula. Ecology 16:423–437.
Upchurch, D.R., and W.J. Edmonds. 1991. Statistical procedures for specific objectives. p. 49–72. In M.J. Mausbach and L.P. Wilding (ed.) Spatial variabilities of soils and landforms. SSSA Spec. Publ. 28. SSSA, Madison, WI.
Upchurch, W.J., R.J. Kinder, J.R. Brown, and G.H. Wagner. 1985. Sanborn Field: Historical perspective. Res. Bull. 1054. Univ. Missouri-Columbia Agric. Exp. Stn.
U.S. Department of Agriculture. 1981. Land resource regions and major land resource areas of the United States. USDA-SCS Agric. Handb. 296. U.S. Gov. Print. Office, Washington, DC.
U.S. Department of Agriculture. Soil Conservation Service. 1982. Procedures for collecting soil samples and methods of analysis for soil survey. Soil Surv. Invest. Rep. no. 1. USDA-SCS, Washington, DC.
Van Lear, D.H., and J.F. Hosner. 1967. Correlation of site index and soil mapping units poor for yellow-poplar in southwest Virginia. J. For. 65:22–24.
Veneman, P.L.M., P.V. Jacke, and S.M. Bodine. 1984. Soil formation as affected by pit and mound microrelief in Massachusetts, U.S.A. Geoderma 33:89–99.
Wagenet, R.J., and J.J. Jurinak. 1978. Spatial variability of soluble salt content in a Mancos Shale watershed. Soil Sci. 126:342–349.
Walker, P.H. 1966. Postglacial environments in relation to landscape and soils on the Cary drift, Iowa. Iowa Agric. Home Econ. Exp. Stn. Bull. 549.
Water Resources Information Center. 1974. Climates of the States. Vol. 2. Western states. Water Information Center, Manhasset Isle, Port Washington, NY.
Wilding, L.P., and L.R. Drees. 1983. Spatial variability and pedology. p. 83–116. In L.P. Wilding et al. (ed.) Pedogenesis and soil taxonomy. 1. Concepts and interactions. Elsevier Press, New York.
Wilding, L.P., R.B. Jones, and G.M. Schafer. 1965. Variation of soil morphological properties within Miami, Celina, and Crosby mapping units in west-central Ohio. Soil Sci. Soc. Am. Proc. 29:711–717.
Wofsy, S.C., M.L. Goulden, J.W. Munger, S.M. Fan, P.S. Bakwin, B.C. Daube, S.L. Bassow, and F.A. Bazzaz. 1993. Net exchange of CO_2 in a mid-latitude forest. Science (Washington, DC) 260:1314–1317.
Wolfe, M.H., J.M. Kelly, and J.D. Wolt. 1987. Soil pH and extractable sulfate-sulfur distribution as influenced by tree species and distance from the stem. Soil Sci. Soc. Am. J. 51:1042–1046.
Woodwell, G.M., J.E. Hobbie, R.A. Houghton, J.M. Melillo, B. Moore, B.J. Peterson, and G.R. Shaver. 1983. Global deforestation: Contribution to atmospheric carbon dioxide. Science (Washington, DC) 222:1081–1086.
Yaalon, D.H. 1983. Climate, time and soil development. p. 233–252. In L.P. Wilding et al. (ed.) Pedogenesis and soil taxonomy. 1. Concepts and interactions. Elsevier Press, New York.
Young, F.J., and R.D. Hammer. 1993. Relationships between soil properties and landscape positions on a small, loess-covered upland landscape in Missouri. p. 308. In Agronomy abstracts. ASA, Madison, WI.
Young, F.J., J.M. Maatta, and R.D. Hammer. 1991. Confidence intervals for important soil properties within mapping units. p. 213–219. In M.J. Mausbach and L.P. Wilding (ed.) Spatial variability and map units for soil surveys. SSSA Spec. Publ. 28. SSSA, Madison, WI.
Young, F.J., R.D. Hammer, and J.M. Maatta. 1993. Confidence intervals for soil properties based on differing statistical assumptions. In G.A. Milliken (ed.) Proc. Annu. Kansas State Univ. Conf. Applied Statistics in Agriculture, 4th, Lawrence, KS. 27–28 April. Manhattan, KS. Kansas State Univ., Manhattan, KS.
Zinke, P.J. 1962. The pattern of individual forest trees on soil properties. Ecology 43:130–133.
Zinke, P.J., and R.L. Crocker. 1962. The influence of giant sequoia on soil properties. For. Sci. 8:2–11.

11 Carbon Cycling in a Loblolly Pine Forest: Implications for the Missing Carbon Sink and for the Concept of Soil

D.D. Richter, D. Markewitz, J.K. Dunsomb, and P.R. Heine
Duke University
Durham, North Carolina

C.G. Wells
USDA-FS
Research Triangle, North Carolina

A. Stuanes
Norwegian Forest Research Institute
NLH, Ås, Norway

H.L. Allen and B. Urrego
North Carolina State University
Cary, North Carolina

K. Harrison
Columbia University
New York, New York

G. Bonani
ETH
Zurich, Switzerland

Recent environmental issues have renewed the long-standing interest in the forest C cycle, and many aspects of the cycle remain unresolved. Two environmental issues in particular have focused recent attention on the forest C cycle: the unbalanced global C cycle and regional acidic air pollution.

Copyright © 1995 Soil Science Society of America, 677 S. Segoe Rd., Madison, WI 53711, USA.
Carbon Forms and Functions in Forest Soils.

Even after several decades of intensive research, we are unable to balance the global C cycle (Sundquist, 1993). Forests mediate this global imbalance, but the impact of forest growth and disturbance on the global C cycle remains uncertain (Ewel et al., 1986; Mattson & Swank, 1989; Castelle & Galloway, 1990). There is little question that forests dominate the terrestrial biosphere's C cycle, although the role of the forest is complex (Schlesinger, 1991). Forests occupy about one-third of the world's land area and they store most of the terrestrial biosphere's organic C. Forests exchange enormous amounts of C during annual cycles of photosynthesis and respiration, and forest C dynamics have been greatly altered by human activities. Large areas of forest have been cleared over the last several millennia, but in the last two centuries we have cleared, burned, planted, and stimulated secondary succession on a very large fraction of the earth's forest land. These large-scale disturbances have perturbed regional and global C cycles (Post & Mann, 1990; Post et al., 1992).

In the southeastern USA forest, for example, about 250 yr of clearing and conversion to mixed land uses have reduced C in regional ecosystems by about 25 Gton of C, for an estimated annual release of 0.13 Gt (Delcourt & Harris, 1980). In recent decades, however, regrowth of secondary forests in the southeast are estimated to have resulted in an average regional sink of 0.07 Gton of C per year. Even in this well-studied forest of the southeastern USA, there is great uncertainty with C fluxes. Understanding whether regional forests are net C sources or sinks in relation to the global cycle can only be improved by better quantification of C dynamics in specific forests. Long-term studies of above- and belowground forest C are particularly needed (Sanchez et al., 1985).

The second environmental issue that has renewed research interest in the forest C cycle is that of acidic air pollution. Recent investigations of acid deposition effects on forests have given us a fuller appreciation for acidity generated by biotic sources in forest ecosystems, most specifically from two C-based systems: the organic acid and the carbonic acid systems. These acids are associated with important belowground processes in terrestrial ecosystems, including soil genesis, mineral weathering, soil leaching, and nutrient availability. Much remains to be learned about the effects of these C-based acids on soil-plant and soil-water systems, even though they have been topics of active soils research dating from the nineteenth century.

The overall objective of this study was to examine the C cycle in a pine (*Pinus* spp.) forest with special attention devoted to poorly characterized components and processes. One specific objective was to estimate rates of C accumulation above- and belowground during the first three decades of the development of an old-field loblolly pine (*P. taeda* L.) forest in the Piedmont of South Carolina. A second specific objective was to evaluate interactions of the C cycle with the acid chemistry of the soil profile down to bedrock at 6 m.

The chapter is organized in four parts:

1. The current C cycle, in which C capital and important C processes in the 34-yr-old forest are evaluated in the forest–soil system whose lower boundary is defined by the soil at bedrock at a minimum of 6-m depth.
2. The three-decade C sink, in which C accumulation is estimated in biomass, forest floor, and mineral soil from eight permanent plots during the

first three decades of pine stand development following long-term row cropping.
3. Carbon cycling and soil acidification, in which soil acidity is characterized to a depth of 6 m, and the major acid sources associated with hydrologic leaching are evaluated, including sulfate, bicarbonate, and dissolved organic carbon (DOC) leaching.
4. The C cycle and the concept of soil, in which soil C dynamics in gas, liquid, and solid phases are used to critically examine traditional concepts of soil, especially in relation to the C horizon.

METHODS

Calhoun Experimental Forest Research Area

The research area [studied in related investigations by Metz & Douglas, (1959); Wells & Jorgensen (1975); DeBell et al. (1989); Binkley et al. (1989); Buford (1990); and Richter et al. (1994)] is located in the South Carolina Piedmont on the Calhoun Experimental Forest of the Sumter National Forest in Union County, about 100 km northwest of Columbia. Annual precipitation averages about 1170 mm (1950–1987) and temperature about 16°C.

The study site is on two broad interfluves which have <3% slopes that are covered mainly by Udults (Soil Survey Staff, 1975) of the Appling series soils (clayey, kaolinitic, thermic Typic Kanhapludults) which have well-developed kandic horizons (Soil Survey Staff. 1992; Moormann, 1985). The profiles extend into deep saprolite, at least about 6 m thick above granite-gneiss bedrock. Detailed soil chemical and physical properties of these profiles are given in Richter et al. (1994). The A horizon is weakly expressed due to coarse textures and long-term cultivation, but appears to be reaccumulating under the thick pine O horizon. Thick eluvial E horizons with sandy loam textures overlie clayey kandic (low CEC) B horizons dominated by kaolinite. The B horizons' CEC (at soil pH) averages about 7.4 cmol$_c$/kg of clay. The upper C horizon is about 1.5 m from the soil surface, and is highly acidic with some layers having less than 10% base saturation.

From the late 1700s through the early twentieth century, soils such as these supported crops of cotton (*Gossypium hirsutum* L.), corn (*Zea mays* L.), and wheat (*Triticum aestivum* L.). Before the U.S. Civil War, a shifting cultivation was practiced; when crop productivity faltered, soil fertility was often regenerated by forest fallow (Trimble, 1974). Secondary forests were subsequently recleared by slaves. Following the war, agricultural practices included more continuous cropping of fields and increased use of lime, fertilizers, and soil conservation practices. In 1954, the last crop of cotton on the research site was followed by a 2-yr fallow, after which loblolly pine was planted in the winter of 1956–1957.

Carbon Contents in Biomass, Forest Floor, and Mineral Soil

The long-term ecosystem study site at the Calhoun Forest, has eight permanent plots used for biogeochemical analysis (Richter et al., 1994; Urrego, 1993; Binkley et al., 1989). These eight plots, four plots of 380 m^2 and four plots of

595 m² in area, were established in 1962, 5 yr after the site was planted with loblolly pine. Most of the ecosystem measurements reported in this paper are from these permanent plots, including the three-decade C accumulations in biomass, forest floor, and mineral soil, and C dynamics in throughfall, forest floor leachates, soil solutions, and soil atmospheres down 6-m depth.

Tree biomass was estimated from measurements of all trees on the eight permanent plots. Diameter and height have been measured on five occasions during stand development. A combination of loblolly pine allometric equations were used to estimate C in above- and below-ground biomass during the course of forest development. Up through age 25, equations of Shelton et al. (1984) were used. After age 25, equations of Van Lear et al. (1984) were employed. Site specific equations of Urrego (1993) based on a sample of 10 trees confirmed the applicability of the equations of Van Lear et al. (1984) for this site. Root estimates were entirely derived from Shelton et al. (1984), which are to our knowledge the best allometric root estimates for the species. Carbon was estimated as 50% of biomass.

Forest floor mass was estimated in the spring of 1992 from samples in each of the eight permanent plots. Four or five forest floor samples were taken randomly in each of the eight plots, each sample being 0.13 m² in area. An additional 1-m radius area was established around each forest floor sample to collect large woody debris. Carbon content was taken as 50% of ash-free mass.

Litterfall was estimated with 40 samplers, five per plot, each of which were 0.7 m² in area. Samples were divided into foliage and nonfoliage fractions. As with biomass estimations, C was taken at 50% of oven-dry mass.

Soil profiles were examined, first by excavating four large soil pits in 1990 (about 2 by 3 m with 2-m depth) and second by augering with a 10-cm diam. bucket auger. In each soil pit, six intact, volumetric samples (each >0.3 L) were taken from each genetic horizon to 1.65-m depth to estimate bulk density, stone volume, fine root mass, soil chemistry, and soil mineralogy. Fine root biomass (<2 mm) was estimated with the volumetric samples using methods outlined in Richter et al. (1990). Below the 2-m depth, auger samples of the CB and C horizons were collected from 0.5-m layers of saprolite down to 6 m.

Soil samples were air dried and sieved through a 2-mm screen. Subsamples were pulverized and analyzed for C with a Perkin Elmer CHNS analyzer. Soils also were analyzed for pH (water and 0.01 M $CaCl_2$), exchangeable cations (Ca, Mg, K, Na) by extracting with 1 M NH_4-acetate (pH 7), and exchangeable and total acidity at pH 8.2 (Richter, 1986; Binkley et al., 1989; Richter et al., 1994).

In order to estimate soil C changes, similar field and laboratory methods were used to collect and analyze soils in both 1962 and 1990. In both years, four composite soil samples representing four depths were collected from each of the eight permanent plots (32 composited samples per collection). In both 1962 and 1990, at least 20 individual punch-tube samples were collected in each plot to compose the composited samples from each of the four soil layers: 0 to 0.075, 0.075 to 0.15, 0.15 to 0.35, and 0.35 to 0.6 m. All soil samples collected from 1962 are archived. Soil C was determined in 1992 on all archived soils.

Two supplementary studies evaluated accumulation rates of C in coarse-textured soils of pine ecosystems typical of old fields in the southeastern USA.

First, to further examine the reaccumulation of soil C by pine forests, a set of soils was carefully selected in the vicinity of the research area in Union County, South Carolina, from areas that currently support uneven-aged hardwood forest ($n = 5$), 30- to 50-yr loblolly pine forest ($n = 5$), and row crops ($n = 2$). Careful consideration was given to selecting soils derived from acid igneous materials that were not greatly eroded, had no evidence of fire, were on nearly level upper landscape positions, and were close to an Appling series concept. The hardwood stands were previously used for forest pasture and woodlots, and were probably never intensively cultivated. At each site, soil samples were collected from five locations within a 100-m² plot with a 0.06-m diam. bulk density corer. Samples were divided into 0- to 0.075-, 0.075- to 0.15-, and 0.15- to 0.30-m depths. Samples were dried and sieved for both bulk density and C determination as previously described.

Second, to further describe turnover of soil C, we examined the rate at which the forest incorporated CO_2 into soil organic C by determining how rapidly ^{14}C from aboveground thermonuclear explosions was incorporated into the soil. Specifically, $^{14}C/^{12}C$ ratios were measured in superficial soil samples (0–0.075m) collected and archived from the eight permanent plots in 1962, 1968, 1977, 1990. Due to extremely high sample analysis cost for ^{14}C analysis, an additional sample compositing scheme was used. In each of these four collections, the composite samples from the eight permanent plots described previously were combined and pulverized for analysis of C isotopes. Carbon was oxidized to CO_2 and reduced to form a graphite target for accelerator mass spectrometry. Turnover rate of soil C was evaluated by the response of ^{14}C in surface soils to the doubling of ^{14}C in the atmosphere caused by thermonuclear bomb testing (Cambell et al., 1967; Martel & Paul, 1974). A model proposed by Harrison et al. (1993) was used to estimate passive and active soil C fractions in this surficial soil layer.

Carbon in Above- and Belowground Solutions and Soil Atmospheres

Fluxes of dissolved organic and inorganic carbon (DOC and DIC) were estimated in 1992–1993, from collections of wet-only precipitation, bulk throughfall, forest floor leachate, and soil solution, all of which were collected continuously from mid-January 1992 to mid-January 1993. Solutions were collected from field collection bottles every 2 wk. Solutions of precipitation were collected in a forest opening using a battery operated a wet-only rainfall collector. On each of the eight permanent forest plots, three bulk throughfall bottles and funnels collected canopy throughfall (Richter & Lindberg, 1988), tension-free lysimeters sampled soil water that drained through Oe horizons and through 0.15-m mineral soil layers (upper E horizons), and Teflon lysimeters under vacuum collected solutions at 0.6- and 1.75-m soil depths. Soil solutions also were collected from 6 m using tension-free piezometers that were specifically designed to collect soil water in contact with saprolite at these depths.

Two methods were used to measure P_{CO_2} in soil atmospheres in four of the eight permanent plots: at 0.15-m depths, the coarse-textured soil permitted a 1-L/min pump to draw soil air directly from the soil atmosphere through a

0.5-cm diam. Teflon straw into a portable infrared gas analyzer. Deeper in the profile, soil diffusivity was much lower than in surface soils. Polyvinyl chloride (PVC) reservoirs approximately 300 mL in volume had to be installed belowground to equilibrate with soil atmospheres at one of four depths. A stopper was fixed to the top of each PVC reservoir and two tubes (long enough to extend aboveground) were connected through the stopper to the atmosphere reservoir. Every 2 wk the portable gas analyzer was connected to the two tubes (as input and output for the analyzer's pump) to circulate air through the reservoir and tubes and thereby measure CO_2 in soil air. The PVC reservoirs were installed at 0.6-, 1.75-, 4-, and 6-m depths and measurements were made every 2 wk in 1992–1993.

Solutions were analyzed for alkalinity by autotitration to a fixed endpoint of pH 5.0, and for DOC by an infrared gas analyzer following filtration through 0.4-μm filters. Solutions also were carefully analyzed for all other major ions (H^+, Ca^{2+}, Mg^{2+}, K^+, Na^+, NH_4^+, SO_4^{2-}, NO_3^-, and Cl^-), to evaluate the anionic charge of the DOC in solutions. Metal cations were determined with flame emission and atomic absorption spectrophotometry, NH_4^+ by autoanalytic colorimetry, H^+ by glass electrode (following 12-min of equilibration with lab atmosphere for all solutions taken from 0.6 m or deeper), and strong acid anions by ion chromatography. Analysis benefitted from comparisons of measured and estimated specific conductance and the frequent use of standards in sample streams. Concentrations of total dissolved inorganic C or DIC [C_T in Stumm & Morgan (1981)] were estimated from the measurements of soil atmosphere P_{CO_2} and Henry's Equation (temperature corrected to 15°C)

$$DIC = (10^{-1.35})(P_{CO_2})\{1/(1+[K_1/(H^+)]+[(K_1K_2)/(H^+)^2]$$

Dissolved organic and inorganic C fluxes were estimated from the products of volume-weighted C concentrations and estimated hydrologic fluxes during the 1992–1993 collection period. For the aboveground fluxes, products were determined from measured volumes and observed elemental concentrations. Soil leaching, however, was estimated from the product of volume-weighted DOC or DIC concentrations in the 1992–1993 collections, and drainage estimates obtained from hydrologic simulation with the PROSPER model (Vose & Swank, 1992; Gnau, 1992; Richter et al., 1994).

RESULTS AND DISCUSSION

Results and discussion are organized in four parts: current contents and cycling of forest C, the forest as a three-decade C sink, C cycling and soil acidity, and the C cycle and the concept of soil.

Forest Carbon Cycle at 34 Years

During the 34 yr, biomass accumulated 140.6 Mg/ha of C (Table 11–1), or about 52% of the ecosystem total. Mean annual accumulation of C averaged 4.13 Mg/ha in biomass over the three decades, whereas over the last 10- and 5-yr

Table 11–1. Carbon contents in the loblolly pine ecosystem at 34 yr in the Calhoun Experimental Forest, South Carolina

Component	C content	
	— Mg ha^{-1} —	
Aboveground biomass	123.6	
Foliage		2.9
Live branch		14.2
Dead branch		6.8
Stemwood		90.4
Stembark		9.3
Belowground biomass	17.0	
Taproot		10.7
Lateral root		3.4
Fine root		2.9
Forest floor	34.9	
O		32.8
Large wood		2.1
Mineral soil	96.0	
0–3 m		83.7
3–6 m		12.4
Total biomass	140.6	
Total ecosystem	271.5	

periods, C accumulation rates averaged 2.33 and 1.30 Mg ha^{-1} yr^{-1} (Table 11–2). At 34 yr, the pine stand is still actively aggrading biomass C, with annual C contained in net primary productivity estimated at 3.78 Mg/ha in the most recent 5-yr period. Such rates are similar to other studies of biomass accumulation by this species (Wells & Jorgensen, 1975; Delcourt & Harris, 1980; Schiffman & Johnson, 1991; Johnson & Lindberg, 1992).

Forest floor in this ecosystem has accumulated about 34.9 Mg/ha of C in 34 yr, equivalent to 13% of the ecosystem total or 26% of the total C stored in the entire 6-m soil profile (Table 11–1). This is a relatively large amount for southern pine forest floor, and without fire the forest floor may be accumulating additional C. Other southern pine stands with such large forest floors are reported in the literature (Heyward & Barnett, 1936; Johnson & Lindberg, 1992). Without burning, it is uncertain how much C may eventually accumulate in the forest floor on this upland, well-drained site.

Six meters of mineral soil contained about 96.0 Mg/ha or about 35% of the ecosystem total. About 80% of soil C storage is in the upper-half of the 6-m profile, 83.7 Mg/ha (Table 11–1). Although soil C is concentrated in upper horizons, root biomass-C was even more concentrated in surficial layers (Fig.11–1). Ratios of root-C to soil organic matter-C decreased from 61.4% in the 0- to 0.35-m layer to 8.5% in the 0.6- to 1.65-m layer of subsoil. This pattern illustrates the large contribution of root C to total belowground C in surficial layers of soil (Richter et al., 1990).

Table 11–2. Annual C fluxes at 34 yr in the loblolly pine ecosystem in the Calhoun Experimental Forest, South Carolina, estimated over 5- and 10-yr periods.

C flux	1982–1992	1987–1992
	kg ha^{-1} yr^{-1}	
Annual precipitation input†		
Dissolved organic carbon (DOC)	—	26.8
Dissolved inorganic carbon (DIC)	—	2.7
Annual increments		
Foliage	–21	34
Branch	352	220
Stem	1893	1094
Root	102	–46
Annual transfers		
Litterfall‡	—	2475
Annual hydrologic output†		
Soil leaching of DOC	—	3.7
Soil leaching of DIC	—	59.5
Net primary productivity	4801	3777
Net ecosystem productivity (NEP)§	3410	2386

†1992–1993.
‡1991–1992.
§NEP includes estimates of mean annual increment of organic C in mineral soils and forest floor (34–yr means).

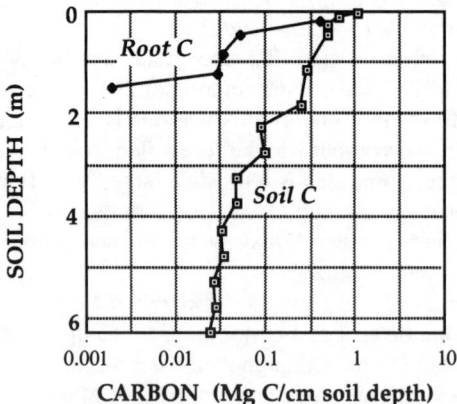

Fig. 11–1. Root and soil C distributions within the soil profile in the loblolly pine ecosystem in the Calhoun Experimental Forest, South Carolina.

Annual rainfall supplied 26.8 kg/ha of DOC, about 10-fold higher than DIC (Table 11–2). On the other hand, annual hydrologic removal of C estimated from the 6-m depth, was mainly as DIC, 59.5 kg/ha, compared with only 3.7 kg/ha as DOC (Table 11–2).

The Growing Forest as a Three-Decade Carbon Sink

During the 34-yr period, ecosystem C storage increased by a total of 175.5 Mg/ha, averaging 5.16 Mg ha^{-1} yr^{-1}. Nearly all (>98%) of this C accumulated in plant biomass and forest floor, with <2% accumulated in surficial 0 to 0.6 m mineral soil (Fig. 11–2). Annual rates of C accumulation in the redeveloping A horizon averaged about 0.07 Mg/ha, comparable to accumulation rates during primary succession (Schlesinger, 1990, 1991).

These relatively low rates of C accumulation in soil under the 34-yr pine forest were substantiated by a study of surface soils under row crops, 30- to 50-yr old pine stands, and old, uneven-aged hardwood stands all in the vicinity of the Calhoun study site (Dunscomb, 1992). Soil C concentrations and contents were similar in pine and row-cropped soils even in surficial 0.075-m layers (Fig. 11–3). The surface soils under pine and row crops averaged 13.0 Mg/ha less soil C than soils under hardwoods (0–0.3 m). These latter C differences were mainly expressed in the most surficial layers sampled, 0- to 0.075-m layers; no significant differences were observed in soils from the three systems that were sampled from 0.15 to 0.3 m (Fig. 11–3). Thus, in both the three-decade study and in the chronosequence study, C in the coarse-textured mineral soil appears to reaccumulate slowly under pine. Carbon accumulation may well occur more prominently later in secondary succession, especially during the later, transitional stage of forest succession when the mor-like pine forest floor shifts to a hardwood-dominated mull (Green et al., 1993).

Such relatively low C accumulation rates in surficial horizons indicate that nearly all of mineral soil C inputs (from fine root mortality, root exudates, and

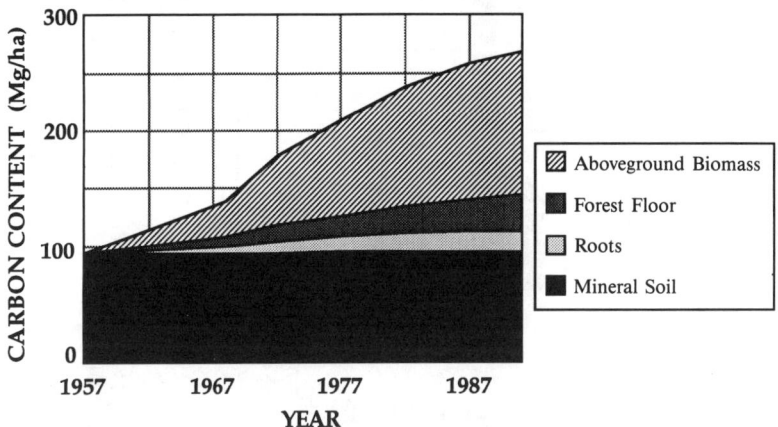

Fig. 11–2. Accumulation of C over 28 yr in a loblolly pine ecosystem in the Calhoun Experimental Forest, South Carolina.

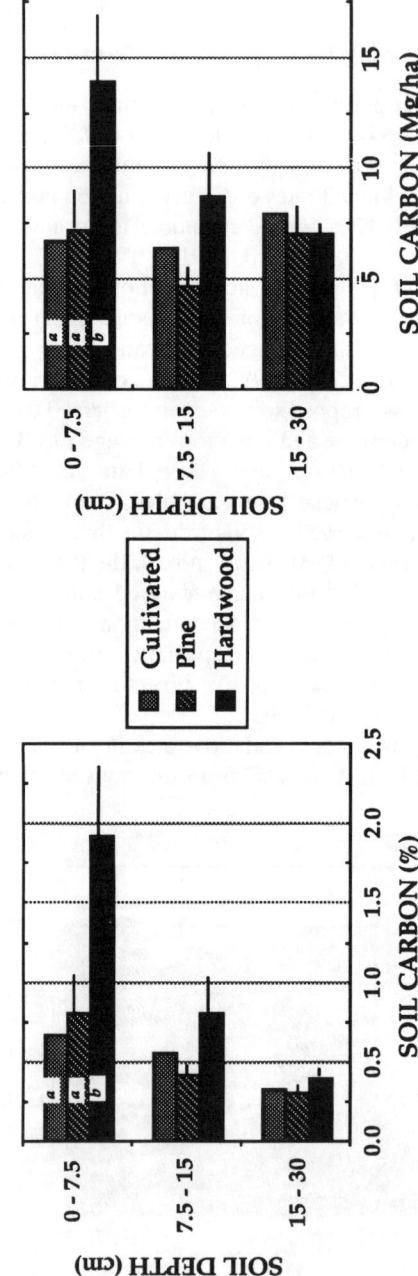

Fig. 11–3. Concentrations and contents of organic C in acid igneous, coarse-textured Udults under row crop, loblolly pine, and hardwood in Union County, South Carolina.

forest floor leachates) are decomposed with little accumulating as humus even over three decades. Soil C inputs are relatively high, perhaps as high as 3 Mg/ha/annum (Kinerson et al., 1977); thus, the accumulation of only about 2 Mg/ha of C over the three-decade life of the forest illustrates how rapid turnover and decomposition must be. Certainly the coarse sandy loam texture of the surface horizon promotes rapid decomposition, a contention supported by a recent study of the fine root decomposition (Ruark, 1993).

Rapid turnover of soil C in surface soils also was substantiated by a study of ^{14}C contents in soil C in the surficial 0.075-m soil layers. Radiocarbon measurements and C accumulation rates provided enough information to estimate turnover times of C in the surficial soil layers. We used a model recently described by Harrison et al. (1993) to evaluate proportions of active and passive soil C fractions, based on the changing $^{14}C/^{12}C$ ratios between 1962 and 1990. A subsoil sample from 0.6- to 1.1-m depth was measured to be about 77% modern C. In other words, the $^{14}C/^{12}C$ of the subsoil was 77% of the 14/12 ratio of the C standard of modern wood that dates from 1850. By taking this C in the deep profile to be entirely passive C (and subject only to decomposition loss and radioactive decay), radiocarbon measurements indicate that the average age of the passive fraction is about 2300 yr. The soil C turnover time was then estimated in the surficial 0.075-m layer by a curve-fitting technique described in Harrison et al. (1993), which used radiocarbon data of surface soils from 1962, 1968, 1977, and 1990 (Fig. 11–4). The rapid increase in soil ^{14}C demonstrates that bomb ^{14}C was rapidly incorporated into the upper soil profile and thus emphasizes the short turnover times of the soil C in this upper soil layer. The curve-fitting exercise suggested that 65% of the soil C in this surficial layer was in the active fraction which has a 12-yr average turnover time.

In aggregate, these soil C data emphasize a relatively limited capacity for C storage in surface horizons and suggest that the potential C sink in the coarse surficial soils may be about 10 to 15 Mg/ha (Fig. 11–3). This perspective does not,

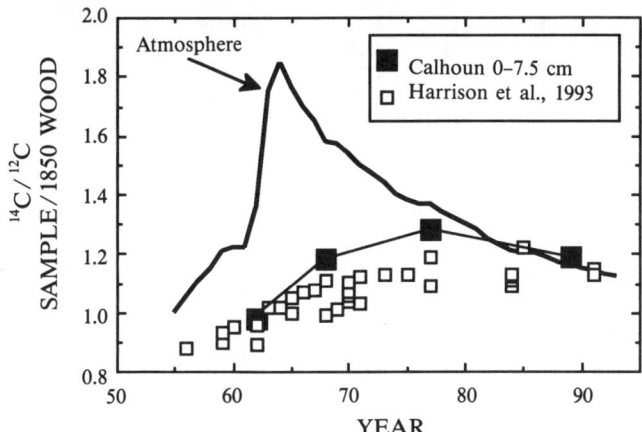

Fig. 11–4. Ratios of ^{14}C to ^{12}C in 0- to 0.075-m soils sampled between 1962 and 1990 within the loblolly pine ecosystem in the Calhoun Experimental Forest, South Carolina.

however, preclude the possibility of an additional C storage in subsoils. Subsoil C accumulation can not yet be estimated, if only because these changes can not easily be determined by resampling. Nonetheless, small increases of C concentrations in a large volume of subsoil may account for an additional C storage. One meter of subsoil, for example, that increases by only 0.05% C is equivalent to a storage of about 7.5 Mg/ha.

Carbon Cycling and Soil Acidification

The C cycle and its ability to generate acidity led us to consider it in the context of the content of acidity in the entire weathering profile of the Ultisol. Richter et al. (1994) describe the acidification process in the 34-yr of pine forest development and include all major processes that have acidified the surface 0.6 m of soil during reforestation. Of interest here, are acidification processes that are associated with soil leaching over pedologic time, and that are responsible for the acidity within the upper 6-m profile.

The Ultisol is strikingly acidic throughout all of the 6 m of material. Salt exchangeable acidity (KCl-acidity) in the 6-m totals about 2250 kmol/ha (estimated from data in Table 11–3). Soil pH is <4.2 in $CaCl_2$ and acid saturation (AS) is commonly greater than 90% of effective cation exchange capacity or ECEC, (Table 11–3). Two layers in the profile appear somewhat less acidic than the other horizons of the profile (Table 11–3): the upper B horizons, likely due to residual effects from agricultural liming (Richter et al., 1994); and the lower C horizon, apparently a weathering front immediately above bedrock (Calvert et al., 1980;

Table 11–3. Soil physical and chemical properties of four profiles under the loblolly pine ecosystem in Calhoun Experimental Forest, South Carolina.

Depth	Clay	Soil C	pH†	Exchangeable Ca‡	AS§	ECEC¶
m	g kg^{-1}	g kg^{-1}		cmol kg^{-1}		cmol kg^{-1}
0–0.075	100	6.99	4.21	0.060	0.911	1.24
0.075–0.15	140	4.45	4.21	0.048	0.892	0.83
0.15–0.35	180	3.23	4.21	0.227	0.653	1.01
0.35–0.60	393	3.45	4.20	0.995	0.480	2.79
0.60–1.10	627	2.40	4.08	0.687	0.740	4.61
1.10–1.35	519	2.40	3.91	0.029	0.962	5.29
1.35–1.65	545	2.40	3.88	0.062	0.952	4.82
1.65–2.0	461	1.34	3.98	0.108	0.913	2.99
2.0–2.5	377	0.70	4.01	0.031	0.945	3.28
2.5–3.0	285	0.76	4.00	0.041	0.945	3.61
3.0–3.5	234	0.38	3.91	0.062	0.946	3.51
3.5–4.0	153	0.38	3.93	0.109	0.926	3.23
4.0–4.5	115	0.25	3.93	0.187	0.879	2.73
4.5–5.0	84	0.27	4.04	0.381	0.780	2.64
5.0–5.5	85	0.21	4.14	0.442	0.704	2.23
5.5–6.0	74	0.22	4.20	0.632	0.643	2.41

†Soil pH in 0.01 M $CaCl_2$.
‡Salt exchangeable Ca.
§Salt exchangeable KCl acidity as a proportion of effective cation exchange capacity (ECEC).
¶Effective cation exchange capacity (sum exchangeable actions plus KCl acidity).

Stolt et al., 1992). Even this deep material has been strongly acidified, however, as AS averages >60% in the deepest 0.5 m of material (Table 11–3).

In the 6-m Ultisol profile, the C chemistry of the soil atmosphere and soil solution affects considerable acidification, largely from organic acids in surficial horizons and from carbonic acid in subsoils. Dissolved organic carbon, derived largely from forest canopies and forest floors, leaches through the A and E horizons into the uppermost B horizons where it is adsorbed or decomposed. Average DOC decreased from 23.8 to 1.5 mg/L in soil solutions at 0.15 and 0.6 m depth, respectively (Fig. 11–5). Due to this pine-derived DOC, these surficial solutions are acidic with a mean pH of <4.5 in O horizon leachate. However, this DOC and its effects are mainly limited to surficial layers of soil (Fig. 11–5), i.e., mainly to the upper 0.6 m soil. Similarly, sulfate, which is largely derived from atmospheric sources, leached through surficial horizons but was effectively adsorbed by B horizons (Fig. 11–5). Strong adsorption in B horizons has been confirmed in the laboratory (D. Markewitz, unpublished data).

In contrast to DOC and sulfate, CO_2 and carbonic acid increased in significance with soil depth. In surface horizons, gaseous diffusion to the aboveground atmosphere is rapid due to coarse-textured, sandy loam A and E horizons, and P_{CO_2} remained below 0.75% within 0.6 m of the soil surface (Fig. 11–6). Due to low diffusivity in clayey B horizons and soil respiration that is active in subsoils, CO_2 readily exceeded 1.5% at >1.75 m (Fig. 11–6). At these partial pressures of CO_2, carbonic acid creates an intense weathering potential as well as low solution pH. Measured concentrations of P_{CO_2} and HCO_3 indicated that in situ pH of soil solutions averaged 4.9 to 5.4 between 0.6- and 6-m depth (Fig. 11–5, 11–6, and 11–7). These relatively low pH's were verified by measurements of low solution pH prior to degassing soil solution of CO_2. Total DIC [C_T of Stumm & Morgan (1981)] averaged >0.70 mmol/L (Fig. 11–7) at >1.75-m soil depth, and although solutions had low ionic strength, HCO_3 was a dominant anion in soil solutions (Fig. 11–5). Annual hydrologic flux of DIC from the soil profile was 59.5 kg ha^{-1} yr^{-1} or about 5000 M ha^{-1} yr^{-1} of DIC, a flux with great potential to acidify soils and weather minerals.

Carbon Cycling and the Concept of Soil

Traditional concepts of soil and weathering processes are strongly influenced by organic C dynamics and have not emphasized that the intense effects of CO_2 are biogenic in origin and influence material within and potentially below the depth of rooting. Textbook concepts for generations have portrayed the weathering profile as comprised of two systems, an upper system of soil, and a lower system of parent material (Lyon & Buckman, 1946; Cook & Ellis, 1987; Harpstead et al., 1988; Brady, 1990; Plaster, 1992). Soil is defined as the upper system that is biogeochemical, enriched in organic C, intensively rooted, and strongly influenced by biologic activity. This upper system is taken to be the zone of pedogenesis, and is called the *solum*, formally defined to include the O, A, E, and B horizons. Sometimes these horizons are referred to as "true soil" (Troeh & Thompson, 1993).

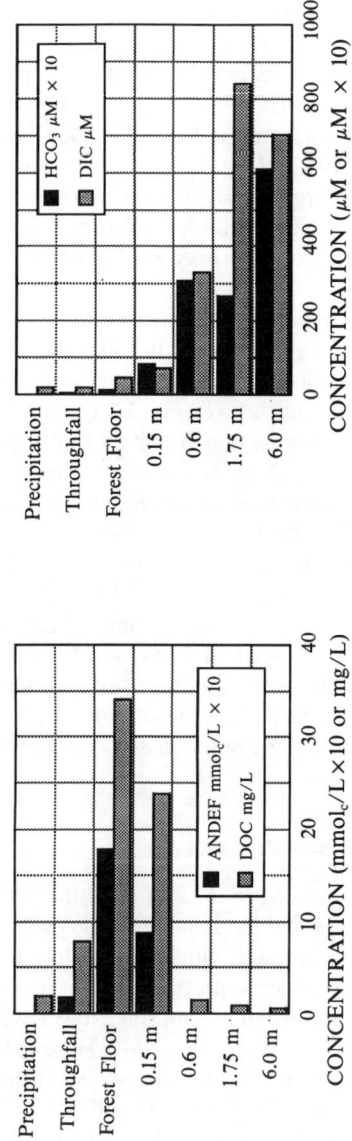
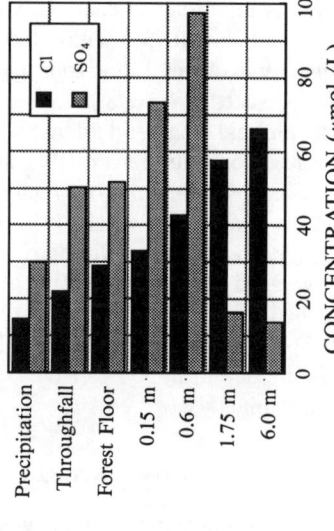

Fig. 11-5. Concentration of C constituents in leaching solutions of the loblolly pine forest.

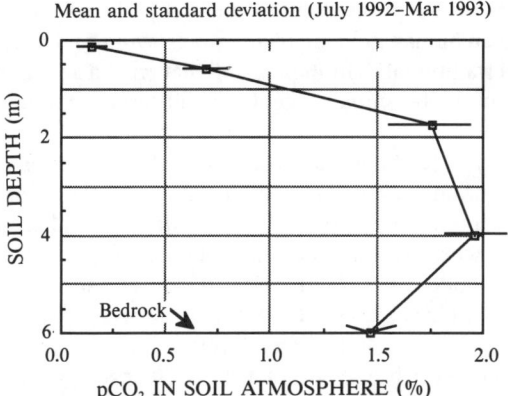

Fig. 11–6. Carbon dioxide concentrations in soil atmospheres through the full soil profile.

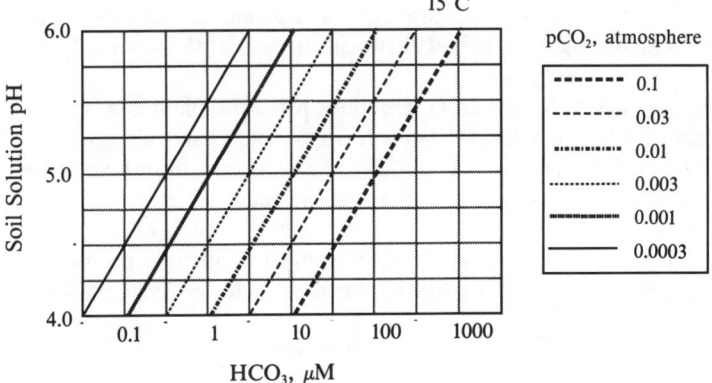

Fig. 11–7. Estimates of pH as a function of solution pCO_2 and HCO_3.

In contrast, the lower system, i.e., the parent material or C horizon, is typically conceived as a geological domain. The C horizon is said to be geological material that is "subjected to little soil development." In many soil, geology, and ecology books this material is described as "chemically" or "geochemically" weathered, which contrasts it with the biologically influenced material above. There are formal exceptions to the dichotomy of solum and parent material; for example, if an illuvial calcium carbonate accumulation occurs in the upper C horizons it is recognized as a pedogenic deposition in an otherwise geologic domain of the parent material (Soil Survey Staff, 1951, 1975, 1992).

This dichotomy of soil and parent material suggests that pedological perspectives of soil have been mixed with those that are more edaphological. In other words, perspectives of soil as natural bodies have not been considered apart from those that regard soil as a rooting medium. Soil (i.e., not parent material) is often defined by its ability to support plant life, yet such a concept is not entirely satisfactory as it makes soil specifically dependent on something other than soil or soil-forming process. As pointed out by Buol et al. (1989), lakes, oceans, glaciers, moss, and even human skin are media for growth of microbes and plants. We need to hold to a traditional definition of soil that emphasizes that soil is the interactive product of biota, climate, geologic substrate, topographic effects, and time.

We know very little about the development of soil through the long sweep of geologic time. During the evolution of forests and grasslands as communities (over the last several hundred million years), mineral weathering in the soil profile was strongly accelerated by increased biogenic P_{CO_2} in the soil atmospheres (Schwartzman & Volk, 1989, 1991). Forests and grasslands have altered many aspects of their edaphic substrates, but the increase in P_{CO_2} and impacts on weathering of all soil horizons including the C were certainly some of the most significant. In addition to these biotic effects, C horizon development is certainly influenced by topography, climate, time, and geologic substrata (e.g., Stolt et al., 1992). Since all five of the soil-forming factors control C horizon properties, C horizons must by definition be part of the soil. The C horizon may or may not be rooted and it may have low soil organic C and major deficiencies in plant nutrients, but it is subjected to biogeochemical processes of a distinctly pedogenic nature.

Soil and geologic substrata are thus part of continuum rather than a dichotomy of solum and parent material. On the other hand, although the C horizon is geologic material exposed to pedogenic processes, the effects of root penetration, bioturbation, and colloidal transport of material, are less intense than in the more biologically excited layers above. There is good reason, therefore, to first conceive of soil as the entire biogeochemical weathering profile, but secondarily to conceive of soil as having upper and lower systems (Fig. 11–8).

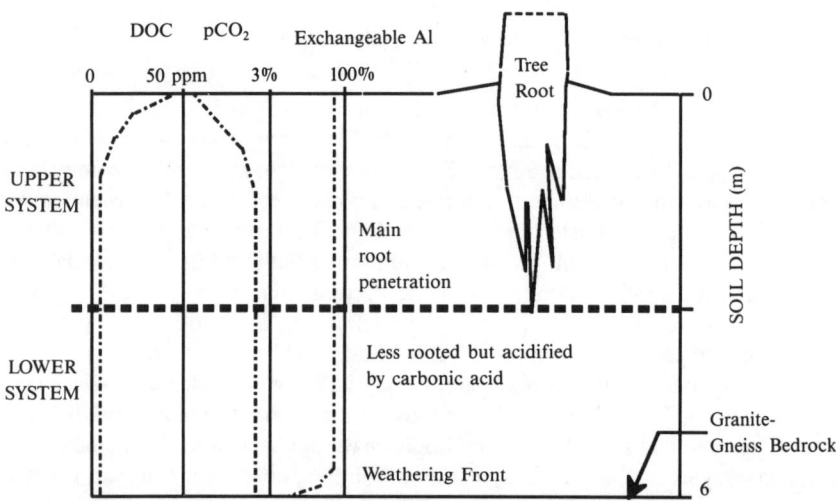

Fig. 11–8. Concept of an Uptisol of soil-ecosystem and the biogeochemical influence of the forest ecosystem on pedogenesis down to bedrock.

Recently, Brimhall et al. (1991) used a two-system concept to describe pedogenesis in a highly weathered, deep Oxisol. Carbon dynamics were important to Brimhall's concept of soil, although dissolved organic acids were credited as being the weathering agents of the lower system. No data were supplied by Brimhall et al. (1991) to support the hypothetical role of DOC in the lower system and, in fact, the dominant C weathering agent in lower B and C horizons of Ultisols and Oxisols would seem more likely to be carbonic acid rather than organic acids. More likely, Ultisols and Oxisols typically have relatively low concentrations of DOC and organic acids in their C horizon soil solutions (e.g., Fig. 11–5). Their atmospheres, however, are likely to have relatively high P_{CO_2} (e.g., Fig. 11–6) and DIC (Fig. 11–5).

CONCLUSIONS

Results of this study of the C cycle in a loblolly pine forest indicate how C cycling in three decades pine stand development has had a fundamental impact on the entire ecosystem. During the 34 yr of forest regrowth, ecosystem C increased by 175.5 Mg/ha, averaging 5.16 Mg ha^{-1} yr^{-1}. Nearly all (>98%) of this C accumulated in plant biomass and forest floor, with <2% of newly accumulated C stored in the surficial 0.6 m of mineral soil. These relatively low rates of C reaccumulation in surficial layers were attributed to rapid turnover of C in coarse surface soils. Low rates of C accumulation in these soils were supported by analyses of ^{14}C in soil organic matter and by a study of total C storage in similar local soils that have undergone different land use histories. The extent to which soil C has reaccumulated in this old-field soil at >0.6-m depth is not resolved.

While the forest reaccumulated relatively little C in surficial mineral soil, it exerted profound effects on soil chemistry throughout the 6-m profile. Dissolved organic carbon derived from forest canopies and forest floors, leached through the A and E horizons into BE horizons, where it was lost from solution presumably by adsorption and decomposition. Carbonic acid and HCO_3 derived from elevated CO_2 due to soil respiration were increasingly important to solutions at greater soil depth. In lower B and saprolite C horizons, soil CO_2 averaged >1.5% of soil air, and HCO_3 became a dominant anion in solutions, despite in situ pH ranging from 4.9 to 5.4. In fact, to explain the intense acidity of the material in the entire 6 m of profile, carbonic acidification appears to be the dominant process to have acidified the whole soil profile.

Finally, the analysis of the forest C cycle led to a critical evaluation of the concept of soil and pedogenesis. Textbook concepts of soil formally describe pedogenic processes as affecting the solum, the upper part of the weathering profile, the O, A, E, and B horizons. In contrast, C horizon parent material is said to be formed by chemical and geochemical weathering. From the perspective of the C cycle, however, the dichotomy of solum and parent material is artificial. In the loblolly pine system, for example, it is clear that the entire 6-m profile of O to C horizons is subjected to biogeochemical processes of a distinctly pedogenic nature. Definitions of soil and pedogenesis need to fully incorporate C horizons as part of the soil profile; to consider that soil often exists as a continuum with the underlying geologic substrata; and that the full soil profile has upper and lower systems, that in general take the place of the traditional solum and parent material. Soil C horizons are often far more voluminous than the surficial O to B horizons and as such are extremely important to a wide range of environmental problems, most especially water resource problems. Including C horizons within the biogeochemical domain of the soil will prove broadly useful for soil science in the coming century.

REFERENCES

Binkley, D., D. Valentine, C. Wells, and U. Valentine. 1989. An empirical analysis of the factors contributing to 20-yr decrease in soil pH in an old-field plantation of loblolly pine. Biochemistry 8:39–54.

Brady, N. 1990. The nature and properties of soils. 10th ed. The Macmillan Co., New York.

Brimhall, G.H., O.A. Chadwick, C.J. Lewis, W. Compston, I.S. Williams, K.J. Danti, W.E. Dietrich, M.E. Power, D. Hendricks, and J. Bratt. 1991. Deformational mass transport and invasive processes in soil evolution. Science (Washington, DC) 255:695–702.

Buford, M.A. 1990. Performance of four yield models for predicting stand dynamics of a 30-year-old loblolly pine (Pinus taeda L.) spacing study. For. Ecol. Manage. 46:23–38.

Buol, S.W., F.D. Hole, and R.J. McCracken. 1989. Soil genesis and classification. Iowa State Univ. Press, Ames.

Calvert, C.S., S.W. Buol, and S.B. Weed. 1980. Mineralogical characteristics and transformations of a vertical rock-saprolite-soil sequence in the North Carolina Piedmont: I. Profile morphology, chemical composition and mineralogy. Soil Sci. Soc. Am. J. 44:1096–1103.

Campbell, C.A., E.A. Paul, D.A. Rennie, and K.J. McCallum. 1967. Factors affecting the accuracy of the carbon dating method in soil humus studies. Soil Sci. 104:81–85.

Castelle, A.J., and J.N. Galloway. 1990. Carbon dioxide dynamics in acid forest soils in Shenandoah National Park, Virginia. Soil Sci. Soc. Am. J. 54:252–257.

Cook, R.B., and B.G. Ellis. 1987. Soil management: A world view of conservation and production. John Wiley & Sons, Inc., New York.

DeBell, D.S., W.R. Harms, and C.G. Whitesell. 1989. Stockability: A major factor in productivity differences between *Pinus taeda* plantations in Hawaii and the southern United States. For. Sci. 35:708–719.

Delcourt, H.R., and W.F. Harris. 1980. Carbon budget of the southern U.S. biota: Analysis of historical change in trend from source to sink. Science (Washington, DC) 210:321–323.

Dunscomb, J.K. 1992. Forest soil recovery after agriculture. Master's Project thesis, Duke Univ., Durham, NC.

Ewel, K.C., W.P. Cropper, Jr., and H. Gholz. 1986. Soil CO_2 evolution in Florida slash pine plantations. I. Changes through time. Can. J. For. Res. 17:325–329.

Gnau, C.B. 1992. Modeling the hydrologic cycle during 25 years of forest development. Master's Project thesis, Duke Univ., Durham, NC.

Green, R.N., R.L. Trowbridge, and K. Klinka. 1993. Towards a taxonomic classification of humus forms. For. Sci. 39:29.

Harpstead, M.I., F.D. Hole, and W.F. Bennett. 1988. Soil science simplified. Iowa State Univ. Press, Ames.

Harrison, K., W. Broecker, and G. Bonani. 1993. A strategy for estimating the impact of CO_2 fertilization on soil carbon storage. Glob. Biochem. Cycles 27:69–80.

Heyward, F., and R.M. Barnett. 1936. Field characteristics and partial chemical analyses of the humus layer of longleaf pine forest soils. Florida Agric. Exp. Stn. Bull. 302.

Johnson, D.W., and S.E. Lindberg. 1992. Atmospheric deposition and nutrient cycling in forest ecosystems. Springer-Verlag, New York.

Kinerson, R.S., C.W. Ralston, and C.G. Wells. 1977. Carbon cycling in a loblolly pine plantation. Oecologia 29:1–10.

Lyon, T.L., and H.O. Buckman. 1946. The nature and properties of soils. 4th ed. The Macmillan Co., New York.

Martel, Y.A., and E.A. Paul. 1974. Effects of cultivation on the organic matter of grassland soils as determined by fractionation and radiocarbon activity. Can. J. Soil Sci. 54:419–426.

Mattson, K.G., and W.T. Swank. 1989. Soil and detrital carbon dynamics following forest cutting in the southern Appalachians. Biol. Fert. Soils 7:247–253.

Metz, L. J., and J. E. Douglass. 1959. Soil moisture depletion under several Piedmont cover types. USDA Tech. Bull. 1207. U. S. Gov. Print. Office Washington, DC.

Moormann, F.R. 1985. Excerpts from the circular letters of ICOMLAC, International Committee from Low Acidity Clay Soils. Tech. Monogr. 8. Soil Manage. Support Serv., Washington, DC, and Honolulu, HI.

Plaster, E.J. 1992. Soil science and management. Delmar Publ. Inc., Albany, NY.

Post, W.M., and L.K. Mann. 1990. Changes in soil organic carbon and nitrogen as a result of cultivation. p. 401–406. *In* A.F. Bouman (ed.) Soil and the greenhouse effect. John Wiley & Sons Ltd., London.

Post, W.M., J. Pastor, A.W. King, and W.R. Emanuel. 1992. Aspects of the interaction between vegetation and soil under global change. Water Air Soil Pollut. 64:345–363.

Richter, D.D. 1986. Sources of acidity in some forested Udults. Soil Sci. Soc. Am. J. 50:1584–1589.

Richter, D.D., L.I. Babbar, M.A. Huston, and M. Jaeger. 1990. Effects of annual tillage on organic carbon in a fine-textured Udalf: The importance of fine root dynamics to soil carbon storage. Soil Sci. 149:78–83.

Richter, D.D., and S.E. Lindberg. 1988. Wet deposition estimates from long-term bulk and event wet-only samples of incident precipitation and throughfall. J. Environ. Qual. 17:619–622.

Richter, D.D., C.G. Wells, D. Markewitz, H.L. Allen, R. April, P.R. Heine, and B. Urrego. 1994. Soil chemical change during three decades in a loblolly pine (*Pinus taeda* L.) ecosystem. Ecology 75:1463–1473.

Ruark, G.A. 1993. Modeling soil temperature effects on in situ decomposition rates for fine roots of loblolly pine. For. Sci. 39:118–129.

Sanchez, P.A., C.A. Palm, L.T. Szott, and C.B. Davey. 1985. Tree improvers in the humid tropics? p. 331–362. *In* M.G.C. Cannell (ed.) Attributes of trees as crop plants. Inst. Terrestrial Ecol., Midlothian, Scotland.

Schiffman, P.M., and W. Johnson. 1991. Phytomass and detrital carbon storage during forest regrowth in southern United States Piedmont. Can. J. For. Res. 19:69–78.

Schlesinger, W.H. 1990. Evidence from the chronosequence studies for a low carbon-storage potential of soils. Nature (London) 348:232–234.

Schlesinger, W.H. 1991. Biogeochemistry. Acad. Press, New York.

Schwartzman, D.W., and T. Volk. 1989. Biotic enhancement weathering and the habitability of Earth. Nature (London) 340:457–460.

Schwartzman, D.W., and T. Volk. 1991. Biotic enhancement of weathering and surface temperatures on Earth since the origin of life. Palaeogeogr. Palaeoclimatol. Paleoecol. 90:357–371.

Shelton, M.G., L.E. Nelson, and G.L. Switzer. 1984. The weight, volume, and nutrient status of plantation-grown loblolly pine trees in the interior of flatwoods of Mississippi. Tech. Bull. 121. Mississippi State Agric. For. Exp. Stn, Starkville.

Soil Survey Staff. 1951. Soil survey manual. U.S. Gov. Print. Office, Washington, DC.

Soil Survey Staff. 1975. Soil Taxonomy. U.S. Gov. Print. Office, Washington, DC.

Soil Survey Staff. 1992. Keys to soil taxonomy. Cornell Univ., Ithaca, NY.

Stolt, M.H., J.C. Baker, and T.W. Simpson. 1992. Characterization and genesis of saprolite derived from gneissic rocks of Virginia. Soil Sci. Soc. Am. J. 56:531–539.

Stumm, W., and J. Morgan. 1981. Aquatic chemistry: An introduction emphasizing chemical equilibria in natural waters. John Wiley & Sons, Inc., New York.

Sundquist, E.T. 1993. The global carbon dioxide budget. Science (Washington, DC) 259:934–941.

Switzer, G.L., and M.G. Shelton. 1979. Successional development of the forest floor and soil surface on upland sites of the east gulf coastal plain. Ecology 60:1162–1171.

Trimble, S.W. 1974. Man-induced soil erosion on the southern Piedmont 1700–1970. Soil Water Conserv. Soc., Ankeny, IA.

Troeh, F.R., and L.M. Thompson. 1993. Soils and soil fertility. 5th ed. Oxford, London.

Urrego, J.B. 1993. Nutrient accumulation in biomass and forest floor of a 34-year old loblolly pine plantation. M.S. thesis. North Carolina State Univ., Raleigh.

Van Lear, D.H., J.B. Waide, and M.J. Teuke. 1984. Biomass and nutrient content of a 41-year-old loblolly pine (*Pinus taeda* L.) plantation on a poor site in South Carolina. For. Sci. 30:395–404.

Vose, J.M., and W.T. Swank. 1992. Water balances. p. 27–49. *In* D.W. Johnson and S.E. Lindberg (ed.) Atmospheric deposition and forest nutrient cycling. Springer-Verlag, New York.

Wells, C.G., and J.R. Jorgensen. 1975. Nutrient cycling in loblolly pine plantations, p. 137–158. *In* B. Bernier and C.H. Winget (ed.) Forest soils and forest land management. North American Forest Soils Conf., 4th, Québec, Canada. August 1973. Laval Univ. Press, Québec, Canada.

12 Toward a New Theory of Podzolization

Bryant A. Browne

University of Wisconsin
Stevens Point, Wisconsin

Spodosols (Buol et al., 1989) are a major class of soils that tend to develop under climatic, parent material and other conditions that promote formation of a dark-colored organic surface horizon (McKeague et al., 1983). This horizon serves as a source of downward migrating organic chelating agents that attack the mineral fabric releasing structural Al and Fe. Hence, in a well-developed Spodosol profile, the horizons express sharply differentiated patterns due to the transport and deposition of Al, Fe and organic matter (podzolization; Buol et al., 1989).

Studies of soil genesis are developing quantitative data on soil formation processes and their rate of change in response to environmental perturbatons and global change phenomena (Yaalon, 1990). Podzolization is of considerable interest with respect to environmental climate change for several reasons. First, Spodosols are prominent in climatic zones (e.g., subpolar Boreal Forests ecosystems) where assumed climate changes, especially temperature, will be sufficient to affect soil processes in detectable ways. Second, many attributes of Spodosols (e.g., spodic horizon) are assumed to develop on a short enough timescale for changes to be detectable. Third, processes operative in podzolization have counterparts or are operative under other pedological processes (Buol et al., 1989). Finally, podzolization theory has received considerable attention and may, therefore, provide an overall index of the quality of our theoretical constructs for assessing sudden change in soil formation processes.

Approaches for studying the dynamic response of soil-forming processes to new environmental conditions are often based on the assumption that solid phase soil attributes preserve a record of soil-forming processes. Thus, it is reasoned that future trends in soil-forming processes will be predictable from an analysis of measurable and observable soil attributes (e.g., spodic horizon). However, several problems underlie the use of soil attributes to interpret environmental change. First, several factors could be responsible for changes in soil attributes (Rosanov & Samoilova, 1990): climate change, geomorphological development of land surface, human impact, and self evolution of natural ecosystems. Second, the physical/chemical mechanisms responsible for many soil attributes have yet to be articulated. Third, measurability does not necessarily imply that changes in an

---

Copyright © 1995 Soil Science Society of America, 677 S. Segoe Rd., Madison, WI 53711, USA. Carbon Forms and Functions in Forest Soils.

attribute have relevance under specific conditions of investigation. And fourth, background variability in time and space may obscure important attribute response patterns.

Accordingly, a greater sensitivity to current soil-forming processes has been sought after. One approach has been to focus on the soil solution as the primary and dynamically responsive vehicle for the redistribution of mass and energy in the soil profile. For example, Ugolini and coworkers (1987, 1988) have applied the analysis and interpretation of soil solutions collected beneath the genetic horizons of Spodosols to formulate podzolization theory (Fig. 12–1). Using this type of approach, current transport and deposition patterns can be inferred from the

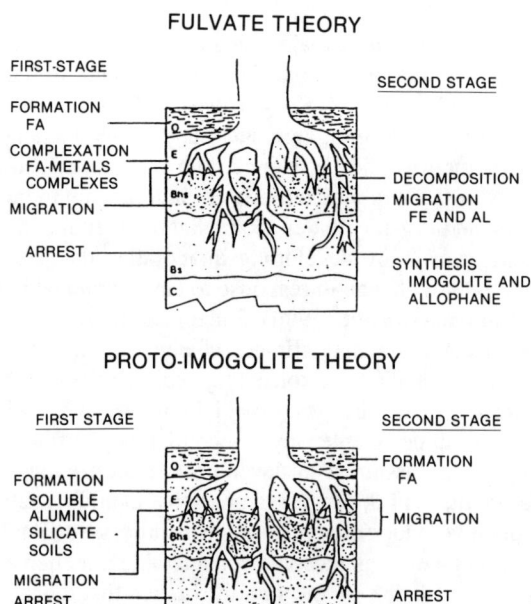

Fig. 12–1. Summary of three current theories of podzolization (modified from Ugolini & Dahlgren, 1987).

spatial attributes (concentration vs. depth profiles) of the soil solution (Fig. 12–2). However, the attributes of the soil solution alone cannot reveal the full story because many of the processes controlling the expression of these attributes have yet to be articulated. Hence, the elementary processes (i.e., physical/chemical reaction mechanisms) responsible for transformation/redistribution of mass and energy in the soil profile also must be explored.

This paper provides an analysis of podzolization from an elementary process level, and advances toward a revised theory of podzolization. The analysis is predicated on new information about the solution phase reactions of Al with three primary soil reactants (organic acids, water, and silica). It is therefore focused on the processes controlling the transport and deposition of Al in the Spodosol profile. Information to construct a similar analysis of Fe pedogenesis in Spodosols remains to be developed. The elementary process view of Al pedogenesis described herein provides a step toward simulation models of soil response to new environmental conditions.

CURRENT THEORIES OF PODZOLIZATION

Gross Morphology of the Soil Profile

A "typical" Spodosol profile (Fig. 12–1) will be used as a platform for discussion of pedogenic process. The typical pedon has the following features: An extensive accumulation of incompletely combusted organic matter occurs in the surface horizon (O horizon). This overlies a leached E horizon, residually enriched with Si and depleted in major structural cations (Fe and Al) and nutrient cations. This eluvial horizon overlies two darker colored illuvial horizons (B*hs* and B*s*) enriched in amorphous Fe and Al. The B*hs* horizon contains an illuvial accumulation of organic matter in combination with Al and Fe. The B*s* horizon contains a greater illuvial accumulation of sesquioxides of Fe and Al. Allophane and imogolite are assumed to be present here. The B*s* overlies the C horizon.

Attributes of the Soil Solution

Figure 12–2 presents the results of a lysimetry study by Ugolini and Dahlgren (1987). It illustrates the major spatial trends in attributes of the soil solution in our "typical" Spodosol profile. Three spatial trends are of particular interest for subsequent discussion: (i) pH increases with depth, (ii) Al decreases with depth, and (iii) dissolved organic matter decreases with depth.

Pedogenic Theory

Podzolization is a collection of processes controlling the transport and deposition of Al and Fe within the soil profile (Duchaufour, 1982). Buol et al. (1980, p. 93) define podzolization as "The chemical migration of aluminum and iron and/or organic matter, resulting in the concentration of silica in the layer eluviated." Three current theories of podzolization (Fig. 12–1) have been summarized and discussed recently by Ugolini and Dahlgren (1987). They differ largely in treatment of the formation of the B*hs* horizon, and the synthesis of

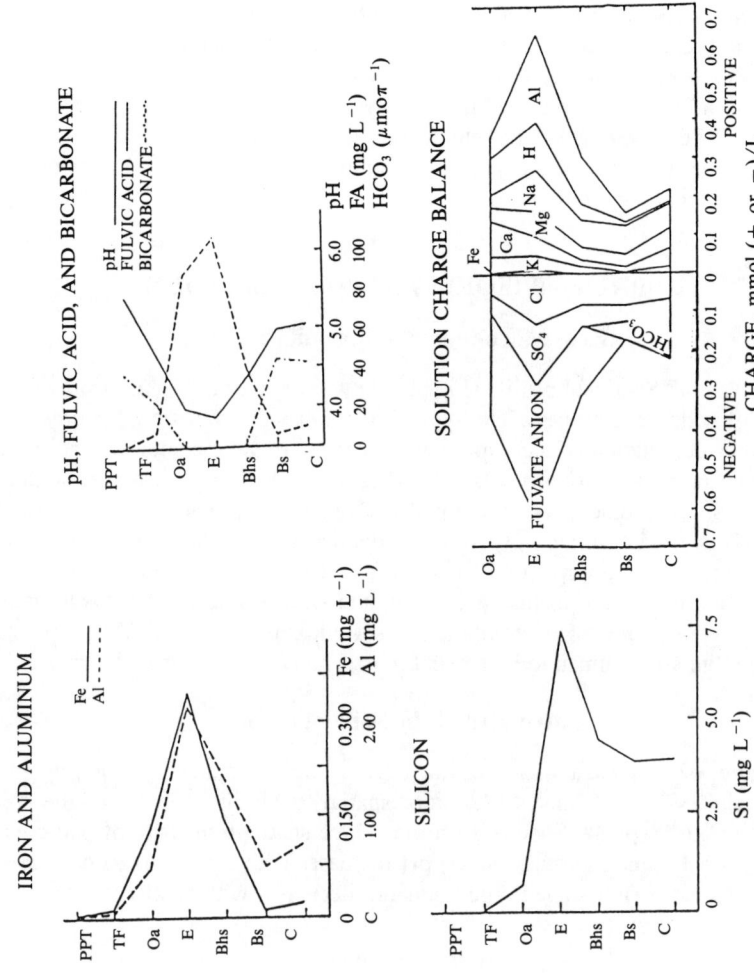

Fig. 12-2. Spatial patterns in the soil solution chemistry of a "typical" Spodosol profile (modified from Ugolini & Dahlgen, 1987).

imogolite/allophane in the Bs horizon. Each is presented briefly to provide a summary of current thinking.

Fulvate Theory

The fulvate theory is a two-stage theory. In the first stage organic acids attack the mineral fabric in the E horizon. Soluble Al and Fe chelates form and migrate through the E and humus rich Bhs horizon. The concentrations of organic acids, Al and Fe decrease across the Bhs as selective retention of organic material causes increasing saturation of Al and Fe complexation sites of soluble chelates and a corresponding decrease in the solubility of Al and Fe chelates. At the top of the Bs horizon a critical metal to fulvic acid ratio is reached, causing nearly complete arrest of the migration of the fulvic acid and Al and Fe chelates. In the second stage, microbial decarboxylation of fulvic acids causes the liberation of Al and Fe. Migrating as free metals, they precipitate as trihydroxides in the Bs as pH increases. Imogolite/allophane forms when Si released as a product of weathering in the upper profile combines with the amorphous Al and Fe.

Proto-Imogolite Theory

Farmer and coworkers (1979, 1980, 1982; Anderson et al., 1982) revised the fulvate theory in light of the occurrence of imogolite in sesquioxide enriched Bs horizons of Spodosols. In the first stage, they suggest that soluble Al and Si combine in the E horizon to form a positively charged 2:1 Al/Si colloid. After migrating through the E and Bhs horizons, the colloid is arrested in the Bs horizon due to decreased solubility with increased pH and the availability of negatively charged sorption sites on mineral surfaces. In the second stage, fulvic acid migrates through the E horizon to form the Bhs horizon.

Fulvate/Bicarbonate Theory

Ugolini and Dahlgren (1987) identified weaknesses in both the traditional "fulvate theory" and the "proto-imogolite" theory of podzolization. Casting out features of both, they presented a revised theory which integrates the spatial features of soil solution chemistry. They objected to the second stage of the fulvate theory based on reported rates of C turnover in the Bhs horizon (1500–4000 yr; DeConinck, 1980). Hence, they reasoned that the release of Fe and Al from the Bhs as organics decompose is too slow to account for the synthesis of imogolite/allophane and the accumulation of other amorphous Al and Fe products in the Bs horizon. With respect to the protoimogolite theory, they suggested that the solution composition in the E horizon does not provide the thermodynamic conditions to support the formation of the protoimogolite colloid.

The alternative theory proposed by Ugolini and Dahlgren (1987) has two "stages" which may occur sequentially or simultaneously. The first stage involves in situ formation of imogolite/allophane in the Bs horizon by a carbonic acid weathering scheme. High P_{CO_2} levels from soil respiration processes promote carbonic acid formation. The carbonic acid attacks mineral surfaces forming an amorphous Al-rich residue in the Bs. Synthesis of imogolite/allophane results as soluble Si, presumably released elsewhere in the soil profile, combines with Al.

The second stage of the theory is similar to the first stage of the "fulvate theory." Fulvic acid forms Al and Fe chelates in the E and Bhs horizons. These migrate through the E and Bhs horizons and are arrested at the top of the Bs as they interact with amorphous Al-rich residue of the Bs horizon.

Some conceptual difficulties also arise in applying the carbonic acid/organic acid weathering scheme posited in the fulvate/bicarbonate theory of podzolization. Under a simultaneous fulvate/carbonate weathering scheme, it is hard to explain the incorporation of illuvial Si into the amorphous Al-rich material in the Bs horizon to form imogolite/allophane, since Si would have to be simultaneously leached from the Bs to form the Al-rich residue. Moreover, the time scale of weathering under carbonic acid may be substantially longer than under organic acids (Schwartzman & Volk, 1989) (since these serve as chelating agents as well as a proton source). Thus, the eluvial processes (from the fulvate dominated soil horizons) perhaps happen at a faster rate than can be accommodated by an incipient Bs horizon. Under a sequential fulvate/carbonate weathering scheme, it remains unclear whether carbonic acid weathering occurs fast enough to account for formation of the Bs in the early or first stage of profile development. Moreover, left unexplained are the specific or general conditions that motivate the carbonic acid weathering regime to eventually shift to an organic acid weathering regime.

SOIL REACTANTS AND SOIL GENESIS: IMPLICATIONS FOR PODZOLIZATION THEORY

Soil reactants as they are defined here are path controlling soluble intermediaries between primary and secondary mineral forms. These intermediaries and their reactions with Al have received little attention, but they have important roles in the redistribution of mass and energy in the soil profile, and are key to understanding soil genesis on a real time basis. In the following discussion each of three soluble soil reactants (organic acids, water, and silica) are considered individually. The implications of the individual discussions are then aggregated into a theory of podzolization.

Organic Acids

Several studies have been performed in recent years to characterize the binding of Al with natural organic ligands (e.g., Pott et al., 1985; Young & Bache, 1985; Backes & Tipping, 1987; Plankey & Patterson, 1987; Tipping et al., 1991). The pedogenic implications of Al-organic interactions in soil solution include weathering of soil minerals, control over primary and secondary mineral equilibria in soils, transport and deposition of Al in the soil profile, formation of secondary minerals (Siffert, 1962; Linares & Huertes, 1971; Huang, 1991), and control of soil pH (Bloom et al., 1979; Hargrove & Thomas, 1982). Here, the focus of discussion will be on recent developments in three areas which have implications for podzolization theory and fundamental pedogenic process: a pH-dependent model of Al-organic binding, polynucleation of Al-organics, and pH buffering by Al-organic complexes.

Recently, Browne and Driscoll (1993) in a series of studies of Al-fulvates, formulated a pH dependent model of Al-fulvate binding. On a theoretical and experimental basis, the intensity of binding between Al and fulvic acid was shown to be a simple polynomial function (second- to fourth-order) of the reciprocal of H⁺ activity. This pattern of binding was attributed to two mechanisms incorporated in the following reaction scheme for the ith type complexing site (L) for mononuclear Al

$$Al^{3+} + H_qL_i + rH_2O = Al(OH)_rL_i^{(3-q-r)} + (q + r)H^+ \quad [1]$$

where q and r are stoichiometric coefficients with integer values of 0, 1 or 2 for the ith type complexing site of the fulvic acid. Here, protons may be displaced/released from the fulvic acid in the process of binding with Al, or protons may be released from a water in the coordination sphere of Al to produce a hydroxy-Al fulvate complex. Both mechanisms shift the reaction equilibrium toward formation of Al-organic complexes as pH increases. Thus, in acidic natural waters, organic ligands become increasingly competitive with other Al binding ligands as pH increases (Fig. 12–3 and 12–4). Consequently, Al-organic complexes tend to dominate the species distribution, and therefore the transport, of Al even as the dissolved organic C declines with depth in Spodosol profiles.

Hence, the transport and deposition patterns of Al in Spodosol profiles are strongly tied to the physical-chemical properties of organic Al complexes as the traditional fulvate theory of podzolization tells us. Additional insight related to

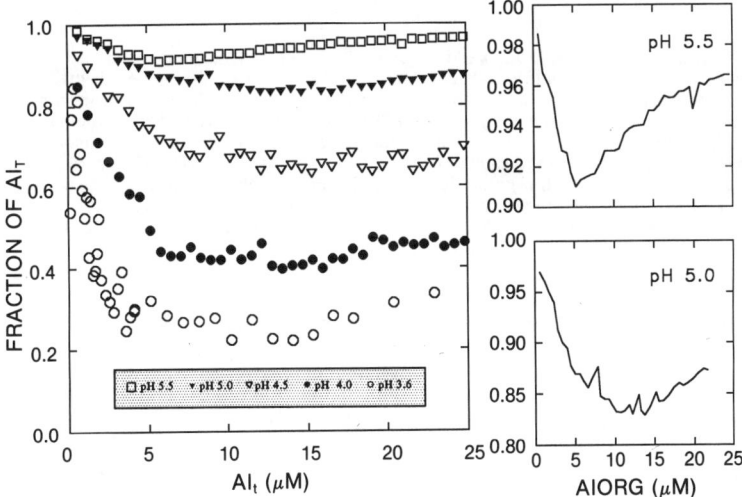

Fig. 12–3. pH-dependent binding of Al by 10 mg/L Suwannee River fulvic acid (from Browne & Driscoll, 1993). The fraction of Al bound by the fulvic acid is shown a function of total dissolved Al (Al$_t$) and organically complexed Al (AlORG). The Al$_t$ vs. AlORG plots highlight evidence of polynucleation at pH 5.0 and 5.5. The apparent increase in the bound fraction of Al$_t$ signals a shift in the stoichiometry coefficient of Al (p) from mononuclear ($p = 1$) to polynuclear ($p > 1$) as Al increases. Explanations for the stoichiometric shift include formation of polynuclear Al fulvates or formation of polynuclear hydroxy Al ions (Browne & Driscoll, 1993).

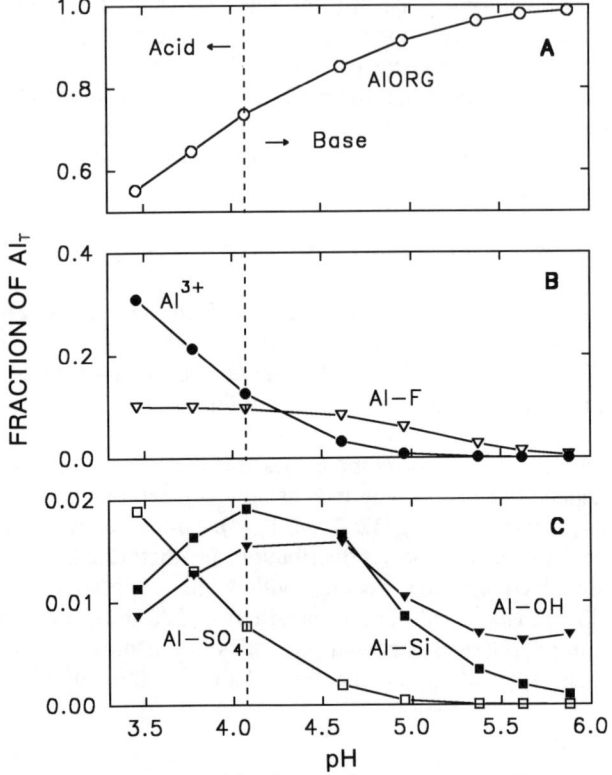

Fig. 12–4. Effect of pH on the formation of organic Al complexes. Due to the impermeable till underlying shallow Spodosols (Typic haplorthods and fragiorthods), the stream chemistry of first-order forested watersheds within the Hubbard Brook Valley in the White Mountains of New Hampshire strongly reflects the influence of soil processes (Browne, 1988; Hedin et al., 1990). Vertical patterns in the attributes of soil solutions are expressed horizontally in first-order stream channels in the sense of a soil catena (Browne, 1988; see Fig. 12–5). This stream sample was collected at the 2550-ft (770-m) elevation in the Cascade Brook watershed in November 1983. A titration was performed to evaluate the influence of pH on the species distribution of Al (after Browne, 1988). Sample chemistry was as follows: pH 4.07, 21.5 μM Al, 1.38 μM dissolved organic C, 72μM SO_4, 2.2 μM fluoride, and 82 μM Si.

the transport of Al under podzolization comes if the pH-dependent charge of complexed Al is considered. By Eq. [1], the charge of complexed Al will likely range from 1+ to 2+ at low pH values in the upper soil profile (before hydrolysis reactions result in the formation of hydroxy Al organic complexes). Until hydrolysis depletes this charge, an anion must escort cationic organic Al, as it migrates downward in the profile, to maintain charge balance (Fig. 12–2). The properties of this independent anion will necessarily influence transport of Al. Deposition of large charge balancing organic anions in the Bhs horizon will restrict further downward movement of cationic Al organic complexes. However, small anions (e.g., NO_3^-, Cl^- and SO_4^{2-}) with limited sorption affinity for soil surfaces will facilitate their downward transport. These observations suggest that the Bhs/Bs

deposition boundary may be a sensitive function of anion mobility. Browne (1988) has suggested that the elevated levels of Al in low-order stream systems in podzolic forested watersheds of the northeastern USA reflects a downward shift in the Bhs/Bs deposition boundary in response to the presence of SO_4^{2-} in acidic deposition (Fig. 12–5).

The binding of metal ions and natural organic ligands (e.g., fulvic acid) has largely been treated as a 1:1 metal/ligand stoichiometry (Dzombak et al., 1986; Pott et al., 1985). This assumption appears to hold well for Al when the fraction of binding sites occupied by Al is low and the solution is at or below pH 4.5 (e.g., A and E horizons of Spodosols). However, as binding site coverage and pH increases, there is a tendency for polynucleation of Al. In studying Al-fulvates, Browne and Driscoll (1993) observed a shift in stoichiometric coefficient (p) of Al from $p = 1$ to $p > 1$ above pH 4.5 (Fig. 12–3) and attributed it to two mechanisms: (i) formation of polynuclear hydroxy Al fulvates (Al–FA), and (ii) facilitated formation of polynuclear hydroxy Al ions.

Assuming a maximum stoichiometric coefficient of $p = 2$ and allowing for dihydroxy bridge bonding between Al ions, two types of reactions can describe the equilibrium status of solutions behaving according to the first mechanism alone

$$Al^{3+} + FA = Al - FA \qquad [2]$$

$$Al - FA + Al(OH)_2^+ = Al_2(OH)_2 - FA \qquad [3]$$

where charge and proton displacement reactions have been omitted for simplicity. Inclusion of the dihydroxy Al here affords the mechanism for hydroxy bridge formation, resulting in chainlike structures with at least two Al per chain.

The second mechanism identified by Browne and Driscoll (1993) is an extension of the first. However, the fulvic acid is now visualized as a template for the formation of polynuclear hydroxy Al ions.

$$Al_2(OH)_2 - FA = Al_2(OH)_2^{4+} + FA \qquad [4]$$

Supporting Eq. [4] is the polynucleation behavior of Al in the presence of phosphate. Browne (1988) was able to isolate facilitated formation of polynuclear hydroxy Al ions from the formation of polynuclear hydroxy Al phosphates. In this set of experiments, Al was in stoichiometric excess to phosphate, such that phosphate Al complexes alone could not account for the majority (>90%) of the polynuclear Al in solution. In the absence of phosphate no polynucleation was observed.

Based on these considerations, Browne and Driscoll (1993) suggest that fulvic acid and other natural ligands (e.g., phosphate) reduce the activation energy necessary for hydroxy bridge bonding between Al ions

$$pAl^{3+} + qH_2O = Al_p(OH)_q^{(3p-q)} + qH^+ \qquad [5]$$

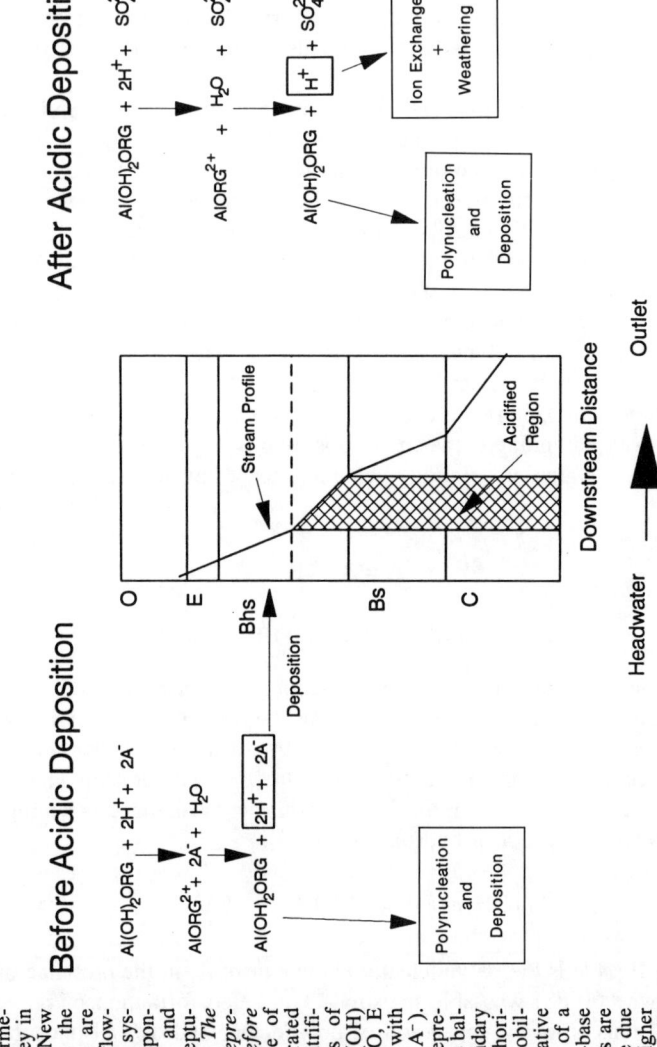

Fig. 12–5. Catena concept of surface water acidification suggested by Browne (1988) for forested watersheds with shallow Spodosols overlying impermeable till (e.g., Hubbard Brook Valley in the White Mountains of New Hampshire). Vertical trends in the attributes of the soil solution are expressed horizontally along the flow-path of first-order streams in these systems (Browne, 1988). The correspondence between the soil solution and stream chemistry is depicted conceptually by the stream profile line. *The dashed horizon boundary line represents the Bhs/Bs boundary before acidic deposition.* In the absence of acidic deposition or soil generated strong inorganic acids (e.g., via nitrification), the acid-base reactions of organic Al complexes [e.g., Al(OH) ORG^{2+}] in the upper soil profile (O, E horizons) reflect an interaction with organic acids (HA = H$^+$ + A$^-$). Uncomplexed organic anions are represented by A$^-$ to depict their charge balance function. The deposition boundary of organic Al complexes (Bhs/Bs horizons) is therefore limited by the mobility of these large nonconservative anions. However, in the presence of a source of strong acidity, the acid-base reactions of organic Al complexes are expressed deeper in the soil profile due to (i) the presence of a small higher mobility anion (SO$_4^{2-}$), and (ii) additional protons. These factors promote a downward shift in the polynucleation and deposition zone of organic Al complexes, and a corresponding expression of soil solution acidity further downstream. According to this catena concept, the acidification of surface waters would be viewed as a symptom of the perturbation of pedogenesis in Spodosols. This suggests that podzolization may be highly responsive to environmental change.

and thereby promote the attainment of equilibrium in solutions metastable with respect to polynuclear hydroxy Al ions. This observation has a number of implications regarding the role of organic C in pedogenesis. First, it helps explain the formation of amorphous Al minerals in B horizons (e.g., the Bs horizon in Spodosols). Where pH rises and dissolved organic carbon (DOC) drops, there occurs a transition from mononuclear hydroxy-Al-organic, to polynuclear hydroxy-Al organic, to polynuclear hydroxy Al ions. Polynuclear hydroxy Al ions have a higher affinity than mononuclear Al (Jackson, 1963) for surfaces, where they will coalesce further into solid phases (e.g., hydroxy Al interlayers, Barnhisel & Bertsch, 1989) or coatings that provide aggregate stability (Hsu, 1989). Second, the polynuclear hydroxy Al ions themselves may combine with Si in solution or on surfaces to form secondary Al-silicate (see discussion of Si below). They can therefore be viewed as structural subunits for clay formation, and would perhaps determine the initial Al and OH and H_2O stoichiometries of the mineral precursor. Hence, the stoichiometry of the polynuclear hydroxy Al ion formed under the influence of soil organic acids may be a determining factor in the composition of short-range ordered Al-silicate and crystalline Al-Si in the soil profile (e.g., in the formation of imogolite and allophane materials in the Bs horizons of Spodosols).

The important role of Al-organic matter (AlORG) in the pH buffering in soils has been known descriptively for some time (Bloom et al., 1979; Hargrove & Thomas, 1982). Surprisingly, little is known quantitatively, however, in order to simulate this buffering system or predict soil responses to changes in environmental conditions. This can be attributed in large part to our inability to distinguish AlORG and organic acid buffering systems. Most quantitative work has therefore been performed by acid/base titration of synthetic systems (e.g., fulvic acids, with and without Al).

Recently, Browne and Glaser (Glaser, 1993) were able to distinguish organic acid and AlORG pH buffering mechanisms in natural water samples. They confirmed the multiprotic model of organic Al acidity of Browne and Driscoll (1993) through a set of titrations in the presence and absence of Fl^-. Fluoride was used to strip Al from organic ligands, masking the pH buffering of AlORG, and thereby isolate the pH buffering of organic acids alone. Figure 12–6 illustrates the relative contributions of strong and weak acids (organic acids, and AlORG) in an acidic natural water sample in equilibrium with atmospheric CO_2. The AlORG behaved as a weak diprotic acid with effective pK_a's of approximately 4, and accounted for the majority of weak acid neutralizing capacity and pH buffering at pH values below approximately 5.5.

The acid-base characteristics of AlORG complexes are important on at least three pedogenic levels. First, this weak acid system in conjunction with organic acids provides an important component of buffering intensity in the soil solution and a source of mobile protons (base neutralizing capacity) for weathering/exchange reactions in the soil profile. Second, through its influence on soil solution pH, AlORG exerts influence on the possible secondary mineral phases which could form by neoformation or transformation reactions in the soil profile. (Until it is depleted from soil solution by precipitation/sorption reactions or the soil solution pH drops well below or rises well above its pK_a, AlORG will tend to

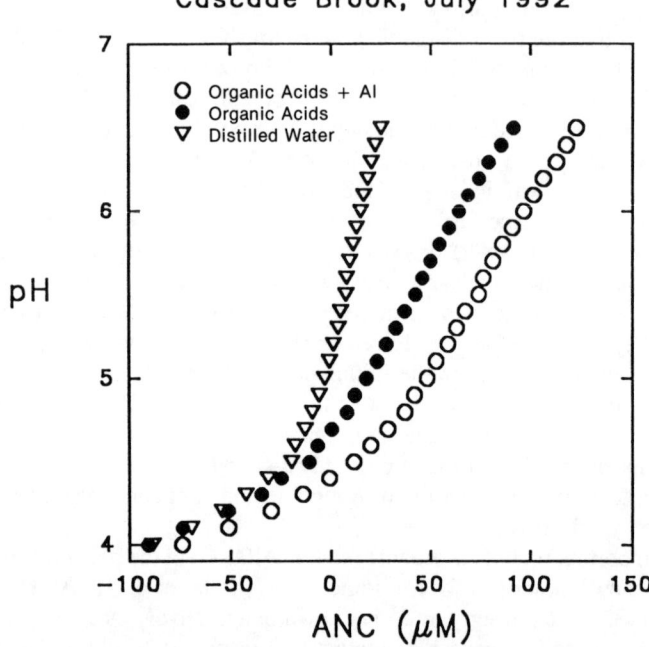

Fig. 12–6. Distribution of acid neutralizing capacity between organic acids and AlORG (Al organic complexes; Glaser, 1993). The titration curve for distilled water represents the titration of strong acid (perchloric acid) with strong base (NaOH) in the presence of atmospheric CO_2. This curve is shown as a reference curve for the strong base titration of a water sample collected in July 1992 from Cascade Brook (see Fig. 12–4 legend) in the Hubbard Brook Valley. The AlORG complexes provide a major portion of the buffering intensity and capacity in the range of pH 4 to 5. The composition of the sample was as follows: ambient pH 4.36, 18 μM Al, 1040 μM dissolved organic C, 56.7 μM SO_4, 1.1 μM F, 94 μM Si).

poise soil solution pH within a narrow range.) Third, the proton buffering reactions of AlORG will influence the depth within the soil profile at which (i) polynucleation and deposition of inorganic Al occurs, (ii) pH buffering of the soil solution shifts to the carbonate system (as soil solution pH rises with depth in the soil profile, the base neutralizing capacity of AlORG must be depleted before this can occur), and (iii) carbonic acid mineral weathering becomes significant.

As a major component of the acid/base neutralizing capacity of soil solutions in Spodosols, the AlORG buffering system can be expected to be sensitive and rapidly responsive to new environmental conditions. For example, if an environmental perturbation were to temporarily stimulate nitrification within the upper soil profile, we could expect a downward shift in the polynucleation/deposition zone as the H^+ generated by nitrification is consumed by mobile hydroxy Al organic complexes and transported to the Bs horizon. This imputed responsiveness suggests that Spodosols are particularly labile to environmental conditions. Browne (1988) hypothesized that surface water acidification in podzolic watersheds reflects a downward shift in the B*hs*/Bs boundary (Fig. 12–5).

Water (Polynuclear Hydroxy Aluminum Ions)

The hydrolysis of Al has received considerable attention in recent years. Many reaction schemes and products have been proposed (Hsu, 1989; Bertsch, 1989; Parker & Bertsch, 1992a). However, a consensus has yet to emerge about the composition and structure of the predominant polynuclear species.

Much attention of late has focused on the tridecameric polycation (Parker & Bertsch, 1992a,b). This species has been observed in laboratory studies using concentrated Al solutions which are not representative of dilute natural solutions. However, recent detection of the tridecamer by ^{27}Al-NMR (nuclear magnetic resonace) in an acid forest soil from Vermont suggests it may be environmentally relevant under certain conditions (Hunter & Ross, 1991). The pedological significance of the tridecamer remains unclear, however, because its size and structure are not consistent with secondary Al-Si soil mineralogy.

Less attention of late has focused on low-order polynuclear ions despite their potential to serve as structural subunits for secondary mineral formation in soils. This is perhaps because their thermochemical stability has been assumed to be too small to account for a significant fraction of Al in natural solutions. However, Browne and co-workers (Browne, 1988; Labiosa, 1993) have recently shown (i) that the dinuclear Al, $Al_2(OH)_2^{4+}$ (Fig. 12–7), forms appreciably in dilute Al solutions representative of natural waters, (ii) that other (as yet undetermined) low-order species (e.g., trinuclear) may form under similar conditions if sufficient activation energy is available (Browne, 1988; Labiosa, 1993), and (iii) that ligands (e.g., fulvic acid and $H_2PO_4^-$) prevalent in natural waters catalyze the formation of such low-order polynuclear species from metastable solutions (Browne, 1988; Browne & Driscoll, 1993).

The formation and reactions of low-order polynuclear hydroxy Al ions have several pedogenic implications including:

1. Formation of low-order polynuclear Al ions will influence the set of stable mineral phases in soils by depressing the equilibrium concentration of the central cation (Al^{3+}) in soil solution. Low-order polynuclear Al ions perhaps serve as precursors (alone or in combination with Si) of the secondary minerals formed in soil.
2. The reaction paths of low-order polynuclear species may have significant roles in the formation of soil minerals (e.g., imogolite and allophane formation, Browne & Driscoll, 1992).
3. The reactions of low-order polynuclear hydroxy Al ions may be a central feature of the formation of the Bs horizon in Spodosols. Polynucleation reaction paths stimulated by fulvic acid may promote the deposition of Al in the Bs horizon with simultaneous or subsequent formation of allophane and imogolite.

Silica

Despite the abundance of Al silicate (Al-Si) minerals in soils and geological formations, little attention has been accorded to the solution phase interactions of

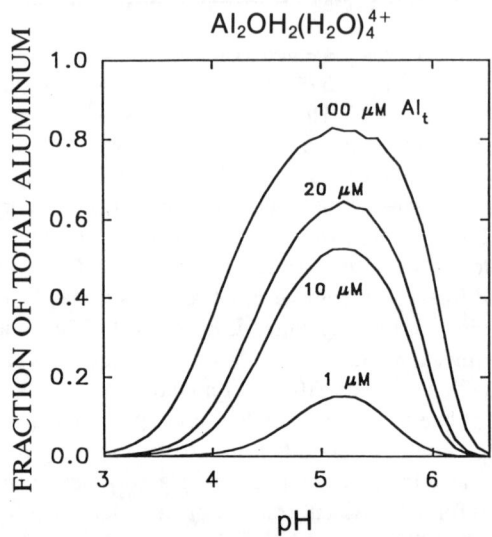

Fig. 12–7. Formation of dinuclear hydroxy Al. Theoretical formation curves for dinuclear Al at 0.001 M ionic strength. The formation curves were generated using a stability constant for $Al_2OH_2^{4+}$ based on data reported by Browne (1988) in a study of the influence of phosphate on polynucleation reactions. The value of the stability constant used in these calculations ($1.2 \times 10^{-4} M$) is two to three orders of magnitude stronger than reported previously. Confirmation of the stoichiometry and stability of this dinuclear species has recently been obtained (Labiosa, 1993).

Si with Al. However, recent studies have suggested the existence of soluble (Browne & Driscoll, 1992) and colloidal Al-Silicates (Farmer et al., 1979; Farmer & Fraser, 1982; Wada & Wada, 1980, 1981) and have stirred considerable interest in soil chemistry, geochemistry and pedology. Soluble Al-Si may be important intermediates in the weathering reactions of primary and secondary minerals, and may serve as templates or structural subunits in secondary mineral formation (Browne & Driscoll, 1992). Stoichiometric and thermochemical description have yet to be attached to colloidal Al-Si, frequently referred to as hydroxy aluminosilicate ions (HAS), but they have received attention as possible precursors of short-range ordered Al silicates such as allophane and imogolite in Andosols (Wada, 1989, 1980; Ugolini et al., 1988) and Spodosols (Ross & Kodama, 1979; Farmer et al., 1980, 1982; Ugolini et al., 1988; Ugolini & Dahlgren, 1989).

Browne and Driscoll (1992) studied the stoichiometry and stability of low-order (mono- and dinuclear) soluble Al-Si using fluorescence spectrometry. They reported the formation of one mononuclear (1:1:0 Al/Si/OH) and two dinuclear (2:2/r and 2:1/r) complexes. Figure 12–8a illustrates formation curves for the mononuclear complex $\{Al[OSi(OH)_3]^{2+}\}$ for a range of Si concentrations in natural waters. In the absence of competing ligands in acidic solutions (Fig. 12–8a), $Al[OSi(OH)_3]^{2+}$ may be a prominent fraction of soluble Al. However, at low Si concentrations in acidic natural waters (Fig. 12–8b), organic ligands and F have a large competitive impact on the fraction of mononuclear Al present as $\{Al[OSi(OH)_3]^{2+}\}$.

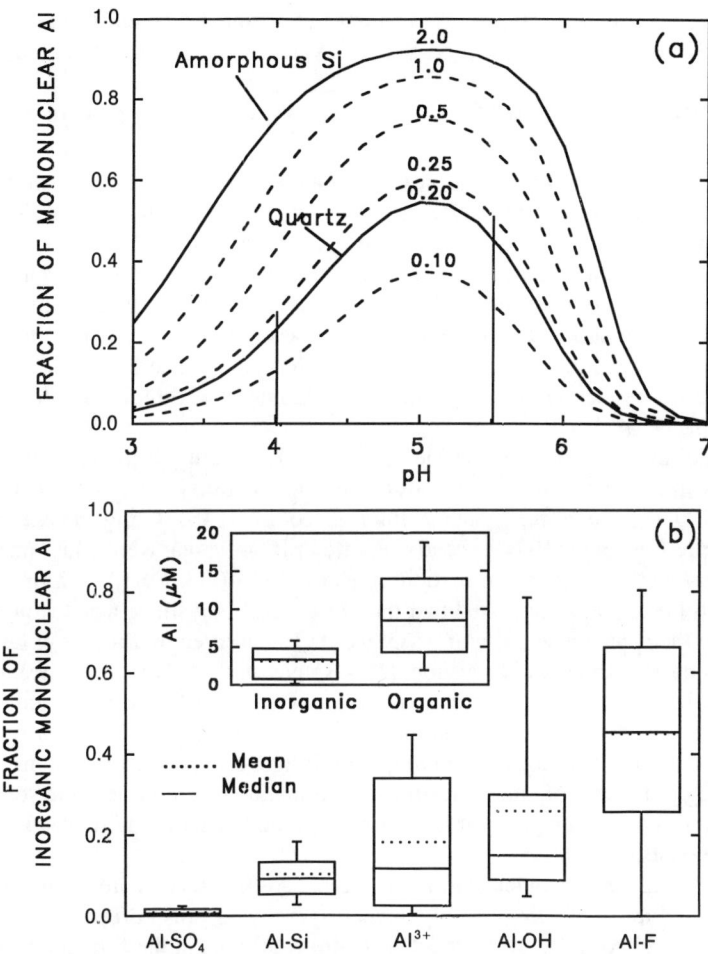

Fig. 12–8. From Browne and Driscoll (1992) (A) Theoretical formation curves for $Al[OSi(OH)_3^{2+}]$ in dilute Si solutions. The value of $[H_4SiO_4]$ for each curve is given in units of millimoles per liter. The area delimited by the vertical lines represents the range of experimental conditions employed by Browne and Driscoll (1992). Upper and lower limits of complex formation in natural waters are shown on the basis of the solubility limits of quartz and amorphous Si, respectively. The presence of other Al binding ligands in natural waters will reduce the formation of the complex. (B) Species distribution ($n = 216$) of inorganic mononuclear Al in dilute acidic natural waters (Browne, 1988). Box plots represent the interquartile range, error bars represent the 10th and 95th percentiles. Inset illustrates the distribution of Al between inorganic and organic mononuclear Al. The overall mean ±standard deviation values for pH and concentrations were as follows: pH = 5.0 ± 0.6, Al_t = 13 ± 8 μM, H_4SiO_4 = 82 ± 21 μM, dissolved organic C = 470 ± 319 μM, Fl⁻ = 3 ± 2 μM, and sulfate = 57 ± 20 μM. Farmer & Lumsdom (1994) recently reported a lower stability for mononuclear Al-Si than reflected above.

Browne and Driscoll (1992) assumed a stoichiometric coefficient of two for hydroxide in the two dinuclear Al-Si species ($Al_2OH_2Si_q$) identified in their study

$$2Al^{3+} + qH_4SiO_4 + 2H_2O = Al_2(OH)_2[OSi(OH)_3]_q^{(4-q)} + (2 + q)H^+ \qquad [6]$$

They estimated conditional stability constants (K_{pqr}, where p, q, and r are the stoichiometric coefficients for Al, Si, and H_2O, respectively) for each species at pH 5.5. The values of $K_{2,2,2}$ and $K_{2,1,2}$ were estimated to be 1.9×10^{-7} and 7.7×10^{-6}, respectively. Based on these formation constants, both species could become significant fractions of soluble Al near pH 5 at typical concentrations of Si within soil solutions. As such they would be significant as potential templates or structural subunits for secondary mineral formation. The 2:1 Al/Si species is noteworthy for its consistency with the stoichiometry of protoimogolite reported by Farmer et al. (1979).

Additional studies of soluble Al-Si and HAS are warranted. Conflicting reports now exist in the literature on the stability and significance of $Al(OSi(OH)_3)^{2+}$ and the general importance of HAS. Using potentiometry, Farmer & Lumsdon (1994) recently reported pH shifts that were about one-tenth those predicted from the formation constant for $Al(OSi(OH)_3)^{2+}$ of Browne & Driscoll (1992), and suggested that the presence of polysilicic acids would have inflated the apparent stability of $Al(OSi(OH)_3)^{2+}$ observed by Browne & Driscoll. On this basis, Farmer and Lumsdom (1994) have suggested that the concentration of soluble Al-Si can be neglected in many natural waters. Similarly, recent reports have suggested that HAS are not stable at Al^{3+} and Si concentrations typically found in natural solutions containing natural ligands. [Huang (1991) and others have suggested that organic acids prevent formation of protoimogolite sol which Farmer and coworkers suggest is important in the transport and deposition of Al in Spodosols.]

Advances in our understanding of soluble Al-Si and HAS are confounded by our limited understanding of polynuclear hydroxy Al ions (Browne & Driscoll, 1992). The importance of polynuclear hydroxy Al ions in the formation of soluble and colloidal Al-Si becomes explicit when Eq. [6] is recast in terms of dinuclear hydroxy Al ions

$$Al_2(OH)_2^{4+} + qH_4SiO_4 = Al_2(OH)_2(OSiO_3)_q^{(4-q)} + (q)H^+ \qquad [7]$$

Equation [7] reveals that there may be a strong linkage between the Al-Si precursors of short-range ordered (e.g., imogolite and allophane) and crystalline (kaolinite) aluminosilicates and the pedogenic processes that promote formation of polynuclear hydroxy Al ions in soils. Factors that lead to the achievement of the activation energy necessary to attain equilibrium with polynuclear hydroxy Al ions remain poorly understood (Bertsch, 1989). These factors likely exert similar constraints on spontaneous formation of polynuclear hydroxy-Al silicates, and will therefore tend to confound the development of accurate thermodynamic and stoichiometric data. This consideration suggests that if we are to map the reaction paths of secondary mineral formation, and understand mechanisms involved in the

A NEW THEORY OF PODZOLIZATION

formation of spodic horizons, we will need to focus on the chemistry of low-order hydroxy Al ions as fundamental structural units of mineral formation reactions.

There are several potential pedogenic implications related to the formation of soluble Al-Si and HAS.

1. Formation of soluble Al-Si and HAS may affect mineral equilibria. Failure to account for the formation of these species may lead to incorrect calculations regarding the stable mineral phase for given soil conditions. For example, Browne and Driscoll (1992) showed that solubility equilibrium controlled by kaolinite could appear to reflect gibbsite mineral equilibrium if no adjustment is made for soluble Al-Si species. Hence, our understanding of pedogenically derived mineral phases may be limited by our understanding of the thermochemical and kinetic characteristics of soluble Al-Si and HAS species.

2. Some forms of soluble Al-Si and HAS may reflect the formation of activated complexes on the surface of minerals (e.g., potassium feldspars, Fig. 12–9) during hydrolytic reaction in soil solutions (Aagaard & Helgeson, 1982). These mineral transformation products (as opposed to neoformation species) may include the actual activated complexes, metastable species intermediate in the decomposition of activated complexes, or the stable progeny of activated complexes. Nonstoichiometric release of Al and Si to solution may therefore emerge, depending on the stoichiometry of the Al-Si activate complex and decomposition products thereof. (For example, a 2:1 Al/Si complex released to or formed in solution could result in an apparent stoichiometric excess of Al over Si in the solution phase and a layer enriched in Si on the surface of a primary soil

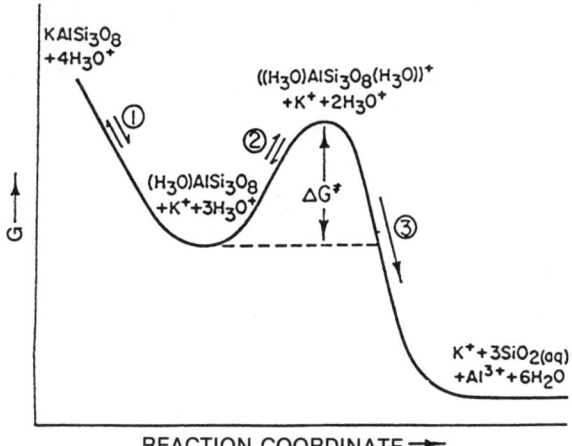

Fig 12–9. Transition state theory for feldspar hydrolysis (from Aagaard & Helgeson, 1982). Transition state theory posits that the formation of an activated complex is a rate-limiting step in feldspar hydrolysis. The activated complex and its daughter complexes may have important roles in secondary mineral formation.

mineral [Browne & Driscoll, 1992].) This type or reaction has implications for understanding the mechanism(s) and spatial features of congruent and incongruent weathering reactions in soils.

3. Soluble Al-Si and HAS released from mineral surfaces or formed in solution may serve as structural subunits or templates for the formation of secondary minerals (e.g., imogolite, allophane, kaolinite, smectite), affecting the nature of secondary minerals formed in soils.

4. Soluble Al-Si and HAS are comprised of hydroxy Al silicate structures. Polynuclear forms of Al-Si species are therefore linked to the formation of polynuclear hydroxy Al ions. Thus, the reaction paths of polynuclear hydroxy Al ions in soil solution may (i) influence the nature of soluble Al-Si and HAS in soil solution that potentially serve as precursors for secondary mineral formation, and (ii) therefore be significant arbiters of secondary mineral formation.

TOWARD A NEW THEORY OF PODZOLIZATION

The foregoing discussions have established that soluble soil reactants can be important intermediaries of pedogenic expression. With respect to podzolization, the polynucleation reactions of Al are central to the deposition of eluviated Al, and the subsequent formation of amorphous and crystalline secondary minerals within the spodic horizons. Organic acids, water (OH) and Si each have significant roles in these reactions. In combination with the important role of organic Al complexes in buffering soil solution pH, the polynucleation reactions of Al represent a rapidly responsive system to new environmental conditions. Hence, shifts in Al pedogenesis due to environmental change (e.g., climate change) may occur without a detectable corresponding change in the attributes of the soil profile during limited term (e.g., human lifetime) observation. Further research on the paths and products of low-order polynucleation reactions and the overall pedogenic role of soil reactants is important for developing process-oriented simulation models of the podzolization process.

A proposed theory of podzolization is summarized in Fig. 12–10 as the more or less simultaneous formation of the B*hs* and B*s* horizons. Organic complexes of Al migrate downward from the O and E horizons. The rising pH (Fig. 12–2) maintains the majority of soluble Al in complexation with organic matter, despite losses of organic matter from solution. As the occupation of organic Al binding sites increase, a critical pH value is reached within the B*hs* at which polynucleation is activated and polynuclear hydroxy Al fulvates are formed (Eq. [3]). Dissociation of the polynuclear species from the organic ligands then establishes the presence of inorganic polynuclear hydroxy Al ions in solution (Eq. [4]). Further rise in pH at the B*hs*/B*s* boundary shifts the majority of polynuclear Al into the inorganic form, causing a swift depletion from solution. Silica reacts with (i) the soluble polynuclear species within the B*hs* and B*s* solutions or (ii) the coatings of deposited polynuclear Al in the B*s* to establish allophane/imogolite in the B*s* horizon. The latter reaction likely occurs on a longer timescale than the solution phase reactions.

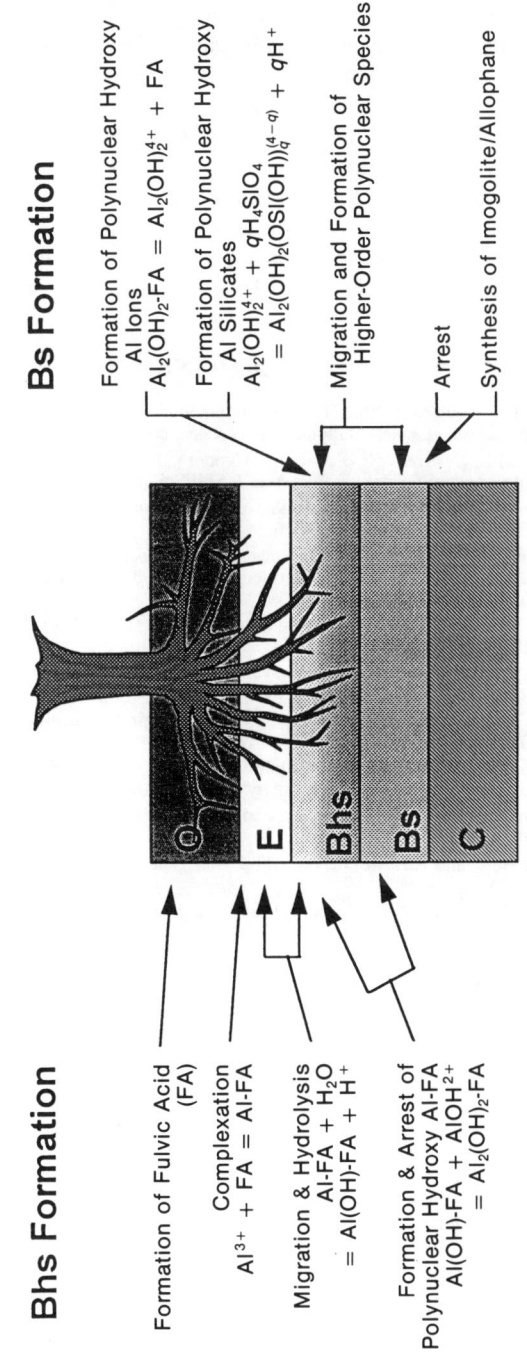

Fig. 12-10. Revised podzolization theory suggesting simultaneous formation of the *Bhs* and Bs horizons. Charge and proton displacement reactions have been omitted to maintain simplicity. (See Fig. 12–5 for explicit consideration of the influence of charge and proton displacement reactions on the deposition of Al.) Aluminum organic complexes are represented here by Al-FA (Al fulvates) in order to be consistent with the depiction of podzolization theories by Ugolini and Dahlgren (1987) in Fig. 12–1.

SOIL REACTANTS, SOIL GENESIS AND GLOBAL CHANGE

The elementary process analysis developed in this paper suggests that fundamental insights into soil genesis can be gathered by focusing on the intermediary soluble soil reactants in weathering and mineral formation processes. This type of approach is essential if we are to develop process-oriented simulation models capable of predicting the response of soil properties to new environmental conditions (e.g., changing temperature and precipitation). However, much remains to be done if we are to achieve a full description of the role of soil reactants in the redistribution of mass and energy within the soil profile. In this regard, the reaction paths and products of low-order polynuclear hydroxy Al ions, and their role in Al-silicate chemistry deserve particular attention.

REFERENCES

Aagaard, P., and H.C. Helgeson. 1982. The thermodynamic and kinetic constraints on reaction rates among minerals and aqueous solutions: I. Theoretical considerations. Am. J. Sci. 282:237–285.

Anderson, H.A., M.L. Berrow, V.C. Farmer, A. Hepburn, J.D. Russell, and A.D. Walker. 1982. A reassessment of podzol formation processes. J. Soil Sci. 33:125–136.

Backes, C.A., and E. Tipping. 1987. Aluminum complexation by an aquatic humic fraction under acidic conditions. Water Res. 21:211–216.

Barnhisel, R.J., and P.M. Bertsch. 1989. Chlorites and hydroxy-interlayered vermiculite and smectite. p. 729–788. In J.B. Dixon and S.B. Weed (ed.) Minerals in soil environments. 2nd ed. Soil Sci. Soc. Am. Book Ser. 1; SSSA, Madison, WI.

Bertsch, P.M. 1989. Aqueous polynuclear aluminum species. p. 87–115. In G. Sposito (ed.) The environmental chemistry of aluminum. CRC Press; Boca Raton, FL.

Bloom, P.R., M.B. McBride, and R.M. Weaver. 1979. Aluminum organic matter in acid soils: Buffering and soil solution aluminum activity. Soil Sci. Soc. Am. J. 43:488–493.

Browne, B.A. 1988. Transformations of Al in waters draining podzolic forest soils. Ph.D. diss. Syracuse Univ. Syracuse, NY (Diss. Abstr. 89-01823).

Browne, B.A., and C.T. Driscoll. 1992. Soluble aluminum silicates: Stoichiometry, stability and implications for environmental geochemistry. Science (Washington, DC) 256:1667–1670.

Browne, B.A., and C.T. Driscoll. 1993. pH-dependent binding of aluminum by a fulvic acid. Environ. Sci. Technol. 27:915–922.

Buol, S.W., F.D. Hole, and R.J. McCracken. 1980. Soil genesis and classification. Iowa State Univ. Press, Ames, IA.

Buol, S.W., F.D. Hole, and R.J. McCracken. 1989. Soil genesis and classification. Iowa State Univ. Press, Ames, IA.

De Coninck, F. 1980. Major mechansims in formation of spodic horizons. Geoderma 24:101–128.

Duchaufour, P. 1982. Pedology: Pedogenesis and classification. George Allen & Unwin, London.

Dzombak, D.A., W. Fish, F. Morel. 1986. Metal-humate interactions. 1. Discrete ligand and continuous distribution models. Environ. Sci. Technol. 20:669–675.

Farmer, V.C., and A.R. Fraser. 1982. Chemical and colloidal stability of sols in the Al_2O_3-Fe_2O_3-SiO_2-H_2O system: Their role in podzolization. J. Soil Sci. 33:737–742.

Farmer, V.C., A.R. Fraser, and J.M. Tait. 1979. Characterization of the chemical structures of natural and synthetic aluminosilicate gels and sols by infrared spectroscopy. Geochim. Cosmochim. Acta 43:1417–1420.

Farmer, V.C. and D.G. Lumsdom. 1994. An assessment of complex formation between aluminum and silicic acid in acidic solutions. Geochim. Cosmochim. Acta 58:3331–3334.

Farmer, V.C., J.D. Russell, M.L. Berrow. 1980. Imogolite and protoimogolite allophane in spodic horizons: Evidence for a mobile aluminum silicate complex in podzol formation. J. Soil Sci. 31:673–684.

Glaser, C. 1993. A technique to measure the ANC of aluminum organic complexes in natural waters. Master's Project, Duke Univ., Durham, NC.

Hargrove, W.L., and G.W. Thomas. 1982. Titration properties of Al-organic matter. Soil Sci. 134:216–225.

Hedin, L.O., G.E. Likens, K.M. Postal, and C.T. Driscoll. 1990. A field experiment to test whether organic acids buffer acidic deposition. Nature (London) 345:798–800.

Hsu, P.H. 1989. Aluminum oxides and oxyhydroxides. p. 331–378. *In* J.B. Dixon and S.B. Weed (ed.) Minerals in soil environments. 2nd ed. SSSA Book Ser. 1. SSSA, Madison, WI.

Huang, P.M. 1991. Ionic factors affecting the formation of short-range ordered aluminosilicates. Soil Sci. Soc. Am. J. 55:1172–1180.

Hunter, D., and D.S. Ross. 1991. Evidence for a phytotoxic hydroxy-aluminum polymer in organic soil horizons. Science (Washington, DC) 251:1056–1058.

Jackson, M.L. 1963. Aluminum bonding in soils: A unifying principle in soil science. Soil Sci. Soc. Am. Proc. 27:1–10.

Labiosa, W.B. 1993. Chemical speciation of dissolved aqueous Al(III) species in well-defined solutions by ^{27}Al nuclear magnetic resonance spectroscopy. M.S. thesis, Duke Univ., Durham, NC.

Linares, J., and F. Huertas. 1971. Kaolinite: Synthesis at room temperature. Science (Washington, DC) 171:896–897.

McKeague, J.A., F. DeConinck, and D.P. Franzmeier. 1983. Spodosols. p. 217–252. *In* L.P. Wilding et al. (ed.) Pedogenesis and soil taxonomy: II. The soil orders. Elsevier, Amsterdam.

Parker, P.R., and P.M. Bertsch. 1992a. Identification and quantification of the "Al_{13}" tridecameric polycation using ferron. Environ. Sci. Technol. 26:908–914.

Parker, P.R., and P.M. Bertsch. 1992b. Formation of the "Al_{13}" tridecameric polycation under diverse synthesis conditions. Environ. Sci. Technol. 26:914–921.

Plankey, B.J., and H.H. Patterson. 1987. Kinetics of Aluminum-fulvic acid complexation in natural waters. Environ. Sci. Technol. 21:595–601.

Pott, D.B., J.J. Alberts, A.W. Elzerman. 1985. The influence of pH on the binding capacity and conditional stability constants of aluminum and naturally occurring organic matter. Chem. Geol. 48:293–304.

Rosanov, B.G., and E.M. Samoilova. 1990. Soils of the subboreal region on a warmer earth. p. 185–190. *In* H.W. Scharpenseel et al. (ed.) Soils on warmer earth. Elsevier, New York.

Ross, G.J., and H. Kodama. 1979. Evidence for imogolite in Canadian soils. Clays Clay Miner. 27:297–300.

Schwartzman, D.W., and T. Volk. 1989. Biotic enhancement of weathering and the habitability of the Earth. Nature (London) 340:457–460.

Siffert, B. 1962. Some reactions of silica in solution: Formation of clay. Rep. Geol. Map Serv. Alsace-Lorraine no. 21. (In French.)

Tipping, E., C. Woof, M.A. Hurley. 1991. Humic substances in acid surface waters: Modeling aluminum binding, contribution of ionic charge-balance, and control of pH. Water Res. 25:425–435.

Ugolini, F.C., and R.A. Dahlgren. 1987. The mechanism of podzolization as revealed through soil solution studies. p. 195–203. *In* D. Righi and A. Chauvel (ed.) Podzols et podzolization. AFES and INRA, Paris.

Ugolini, F.C., and R. Dahlgren. 1989. Aluminum fractionation of soil solutions from unperturbed and tephra-treated spodosols, Cascade Range, Washington USA. Soil Sci. Soc. Am. J. 53:559–566.

Ugolini, F.C., R. Dahlgren, S. Shoji, and T. Ito. 1988. An example of andisolization and podzolization as revealed by soil solution studies, southern Hakkoda, northeast Japan. Soil Sci. 145:111–125.

Wada, K. 1989. Allophane and imogolite. p. 1051–1088. *In* J.B. Dixon and S.B. Weed (ed.) Minerals in soil environments. 2nd ed. SSSA Book Ser. 1. SSSA, Madison, WI.

Wada, K. 1980. Mineralogical characteristics of Andisols. p. 87–107. *In* B.K.G. Theng (ed.) Soils with variable charge. N.Z. Soc. Soil Sci., Palmerston North, NZ.

Wada, S., and K. Wada. 1980. Formation, composition and structure of hydroxy-aluminosilicate sols. J. Soil Sci. 31:457–467.

Wada, S., and K. Wada. 1981. Reactions betwen aluminate ions and orthosilicic acid in dilute, alkaline and neutral solutions. Soil Sci. 132:267–273.

Yaalon, D. 1990. The relevance of soils and paleosols in interpreting past and ongoing climate changes. Paleogeogr. Paleoclimatol. Paleoecol. 82:63–64.

Young, S.D., and B.W. Bache. 1985. Aluminum-organic complexation: Formation constants and a speciation model for the soil solution. J. Soil Sci. 36:261–269.

13 A Model of Soil Organic Matter and its Function in Temperate Forest Soil Development

Robert C. Santore, Charles T. Driscoll, and Michael Aloi
Syracuse University
Syracuse, New York

There is considerable need to simulate transformations involving organic carbon (C) in forest soils. There is a large and dynamic reservoir of organic C associated with soils that is likely to be an important part of the global C cycle (Schlesinger, 1990). Soil C plays an important role in determining the physical and chemical properties of soils. Transformations with the C cycle also have interactions with the biogeochemical cycles of other major elements. To facilitate discussion about processes involving soil organic C it is essential to define specific organic C pools and fractions. Soil organic matter (SOM) is a solid phase soil pool composed of organic C and other associated elements and is derived from detrital litter inputs and/or from sorption of dissolved organic solutes. Typically SOM is measured as loss on ignition. Soil organic carbon (SOC) is functionally identical to SOM, however this pool is expressed on a per C basis. Dissolved organic carbon (DOC) is soluble C which passes through a glass fiber filter.

Several terrestrial processes control SOC pools including: (i) production of plant biomass and litter, (ii) release of dead biomass, (iii) mineralization of SOC and CO_2 production, (iv) decomposition and production of DOC, and (v) sorption on soil surfaces. However, modeling efforts to date have concentrated on a small number of these processes, and typically ignore abiotic soil processes and linkages of the C cycle with other biogeochemical cycles (N, S, P, H, Ca, Al, Si). Probably, the most prominent of the soil C models is the Century model. This model was developed to predict transfers of organic C between labile and nonlabile soil pools (Parton et al., 1988, 1993). While models such as Century seem adequate in simulating stocks and transfers of SOC, they were not developed to evaluate the function of soil organic matter (SOM) nor do they rigorously consider linkages of SOC or DOC with other biogeochemical cycles. Several models have been developed to simulate element cycling in forest ecosystems and the response of these ecosystems to disturbances like atmospheric deposition (Gherini et al., 1985; Cosby et al., 1985; Nikolaidis et al., 1988; Furrer et al.,

---

Copyright © 1995 Soil Science Society of America, 677 S. Segoe Rd., Madison, WI 53711, USA. Carbon Forms and Functions in Forest Soils.

1990). These biogeochemical models generally focus on abiotic process (e.g., cation exchange, sulfate adsorption, mineral weathering, and dissolution and aqueous speciation of Al) in an effort to predict the concentrations of major solutes in drainage waters, pH and acid neutralizing capacity (ANC). For the most part, biogeochemical models have not considered transformations involving organic matter. Important exceptions are the ILWAS model (Gherini et al., 1985), the NuCM model (Liu et al., 1991) and STEADYQL (Furrer et al., 1990). These models include reactions involving organic matter, such as mineralization of SOM and associated elements, and the production and chemistry of organic acids in drainage water, within the framework of simulating biogeochemical processes. Despite the comprehensive nature of these models they were not developed with simulating SOC pools as a primary goal.

Most biogeochemical models have distinct limitations because they do not address the role of naturally occurring organic matter in the function of complex ecosystems, including soil chemical processes. Pools and transfers of organic matter are intimately coupled with the biogeochemistry of forest ecosystems. Large and important pools of many major (e.g., N, Ca, Mg, K, S) and trace (e.g., P, Cu, Zn) elements are associated with living and dead biomass (Aber & Melillo, 1991; Schlesinger, 1991). Mineralization of litter and SOM are prominent transfers in the cycles of these elements.

The production of organic acids in the canopy and forest floor, organic acid mediated dissolution of mineral soil in upper soil horizons and subsequent deposition of organic compounds in the lower mineral soil, are important components of soil development (Ugolini et al., 1977). Biogeochemical models generally have not considered these processes. This deficiency may be due to a lack of quantitative experimental data to allow for process formulation. For example, there is considerable information on the stoichiometry and rates of litter mineralization over the time scale of months to 1 to 2 yr (Aber & Melillo, 1991). However, there is very little quantitative information on long-term mineralization rates of SOM, the stoichiometry of these reactions, whether by-products of these reactions occur as CO_2 or organic acids, or which factors regulate mineralization rates (e.g., temperature, moisture, chemical characteristics; Oades, 1988).

We also know very little about rates of DOC mineralization and factors which influence these rates. Qualls and Haines (1992) investigated rates of DOC mineralization in forest soils finding that rates of microbial oxidation of DOC in soil water were slow. They and others (e.g., McDowell & Wood, 1984) have suggested that soil water concentrations of DOC are regulated by soil adsorption reactions. The adsorption of DOC and subsequent mineralization by microbial activity are the processes that regulate SOC pools.

Efforts to model soil development have been thwarted by a poor quantitative understanding of the mechanisms by which organic C is immobilized in the mineral soil. The removal of DOC from solution has been represented by a surface adsorption reaction (McDowell & Wood, 1984; Nodvin et al., 1986; Dahlgren & Marrett, 1991; Moore et al., 1992). However, DOC adsorption isotherms never reach the adsorption capacity of the soil and are usually depicted as a partition coefficient. The long-term accumulation of SOC is not considered by surface adsorption. Organic C deposition may occur through the process of complexation of organic acids with Al and/or Fe during transport through the soil profile. As the DOC to metal (Al and Fe) ratio of the complexes increases, these solutes decrease in solubility and are precipitated in the lower mineral soil (Ugolini et al., 1977; David & Driscoll, 1984; McDowell & Wood, 1984). Unfortunately, there is little information in the literature to allow for representation of this process in simulation models.

It is well established that DOC released from the forest canopy, litter, or SOM has H^+ and metal binding characteristics associated with functional groups (Stevenson, 1982). Organic acids may be important in regulating the acid-base chemistry and Al speciation in drainage waters (Driscoll et al., 1989; Driscoll & Schecher, 1990). Considerable effort has been made to model these reactions with good success (Oliver et al., 1983; Perdue et al., 1984; Perdue, 1990; Driscoll et al., 1994). However, these processes have generally not been included in biogeochemical models. Ignoring organic acids likely contributes to the failure of biogeochemical models to accurately simulate the pH buffering of waters draining forest ecosystems (Sullivan et al., 1992; Sullivan et al., 1994, unpublished data).

Process-level research has demonstrated that rates of mineral weathering are accelerated markedly by organic acids (Zepp & Wolfe, 1987; Schnoor, 1990; Stumm & Wieland, 1990). However, expressions used to represent mineral weathering rates in biogeochemical models are usually simple zero-order formulations with H^+ dependence (Eary et al., 1989). The role of organic acids in mineral weathering has not been incorporated into biogeochemical models.

Functional groups associated with SOM are important in the surface properties of soil (Stevenson, 1982; Sposito, 1984). For many forest soils, much of the cation exchange capacity (CEC) is associated with SOM (Kalisz & Stone, 1980; Federer & Hornbeck, 1985; Ross et al., 1991). The production and loss of CEC from the soil exchange complex is undoubtedly closely coupled with deposition of SOC and mineralization/leaching of DOC, which in turn, should affect the nutrient status of the forest stand. Again, this important process linkage has been ignored in biogeochemical modeling efforts.

Biogeochemical simulation models are important research tools for the study and management of forest ecosystems. Biogeochemical models can be developed as research tools to investigate important disturbances such as clear-cutting, land-use change, climatic events or air pollution. When confidence is attained in model performance, they can be used as management tools by resource managers to assess disturbance issues. Because of the importance of SOC and DOC in forest ecosystems, it is critical to consider the pools and transfers of organic matter as well as the linkages with other biogeochemical processes (e.g., weathering, cation exchange, soil solution chemistry) in the development of simulation models.

In this study, we developed a framework to simulate the function of naturally occurring organic matter in soil and solutions. Unfortunately, we have a limited knowledge base to justify process formulations. Nevertheless, we present some field observations to support process formulations and parameter values. Finally, we conducted a hypothetical simulation using the model. The simulation was made, not only to develop insight into the biogeochemistry of forest ecosystems, but also to illustrate deficiencies in model formulation and suggest the need for additional quantitative research.

MODEL FORMULATION

System Description

To simulate the acid-base chemistry of a mineral soil profile, a soil column model was established. This model was calibrated using field data available from the Hubbard Brook Experimental Forest (HBEF), New Hampshire (Likens et al., 1977). The physical characteristics used for model simulation are summarized in Table 13–1. Hydrologic flow through the profile is assumed to be one-dimensional plug flow, with some loss due to evaporation and transpiration, which was partitioned according to the distribution of fine roots (Driscoll et al., 1985). The soil profile is split into 20 layers in order to allow differential rates and fluxes to exist with depth.

We used a simplified approach. Rather than simulating all major elements, we selected H^+, organic C (A_T), inorganic C, Al, Ca, SO_4^{2-}, NO_3^-, Si and soil surface cations and anion binding sites, as components in the soil system. We believe that the predominant acid-base characteristics of soil acidification are represented by reactions involving these components. The primary minerals used in these simulations were plagioclase and hornblende, which are important minerals supplying Ca, Al, and H_4SiO_4 to drainage waters at the study site. Calcium and Al participate in exchange reactions, while Al also can form aqueous complexes with OH^-, SO_4^{2-}, Si, and organic ligands (A^{3-}). Hydrolysis reactions also can result in precipitation of gibbsite [$Al(OH)_3$]. Sulfate is largely supplied by atmospheric deposition to the site (Likens et al., 1977) and is an input variable to the model. Sulfate participates in soil adsorption through surface complexation reactions with soil surfaces and forms aqueous complexes with Al. Naturally occurring organic acids (A_T), represented as DOC, originate from throughfall leachate or by mineralization of the SOM in the forest floor.

Table 13–1. Initial soil conditions and values of driving parameters.

Catchment area (ha)	1
Soil depth (m)	0.5
Soil bulk density (kg/L)	0.6
Soil organic C (g/kg)	0
Precipitation inputs (mm)	1175
Evaporation and transpiration loss (mm)	275
Drainage (mm)	900
Plagioclase (% by weight)	10.5
Hornblende (% by weight)	10.5
Gibbsite (% by weight)	0
Concentrations of solutes entering top mineral soil profile (µmol/L)	
H	219
Ca	33
Al	30
SO_4	40
DOC	1500
DIC	200
NO_3	70
H_4SiO_4	100

Dissolved organic C is supplied as an input variable based on observations from the study site (Fig. 13–1). The organic acids can either bind with H^+ and Al in solution, be sorbed onto soil surfaces, or mineralized to CO_2 by microbial action. Dissolved inorganic C (C_T) originates from the mineralization of DOC (A_T) to CO_2, and in solution dissociates to HCO_3^- and CO_3^{2-} (Table 13–2).

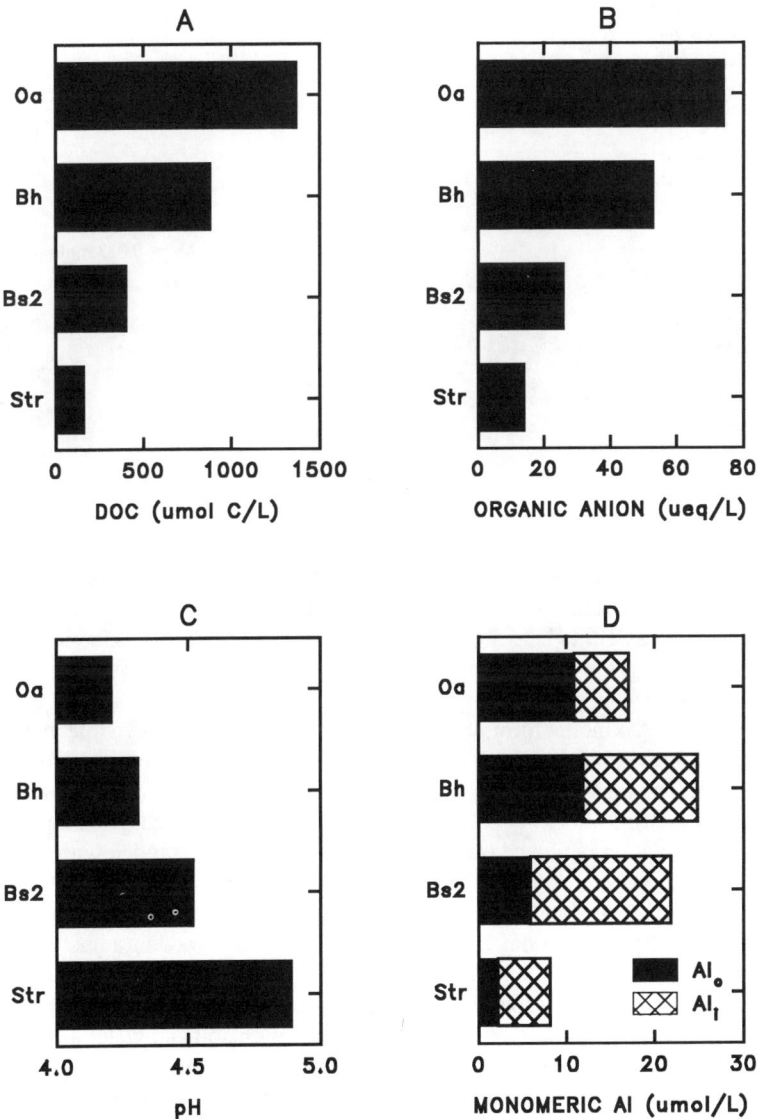

Fig. 13–1. Mean values of dissolved organic carbon (DOC concentration, µmol C/L; A), organic anion (A^{n-} equivalence/L; B), pH (C) and inorganic monomeric Al (Ali) and organic monomeric Al (Ali, Al_o concentrations µmol/L; D) in Oa, Bh and Bs2 horizon solutions and streamwater at the Hubbard Brook Experimental Forest, New Hampshire.

Table 13–2. Fast reactions used in this analysis and associated equilibrium constants (A^{3-} represents organic anion).

Aqueous reactions		log K
$H^+ + OH^-$	$= H_2O$	14.00
$CO_3^{2-} + H^+$	$= HCO_3^-$	10.25
$CO_3^{2-} + 2H^+$	$= H_2CO_3$	16.55
H_4SiO_4	$= H_3SiO_4^- + H^+$	–0.79
$Al^{3+} + H_2O$	$= Al(OH)^{2+} + H^+$	–4.99
$Al^{3+} + 2H_2O$	$= Al(OH)_2^+ + 2H^+$	–9.30
$Al^{3+} + 4H_2O$	$= Al(OH)_4^- + 4H^+$	–15.00
$Al^{3+} + SO_4^{2-}$	$= Al(SO_4)^+$	3.02
$Al^{3+} + H_4SiO_4$	$= H^+ + Al(H_3SiO_4)^{2+}$	–9.46
$A^{3-} + H^+$	$= HA^{2-}$	6.48
$A^{3-} + 2H^+$	$= H_2A^-$	11.69
$A^{3-} + 3H^+$	$= H_3A$	13.71
$Al^{3+} + A^{3-}$	$= Al(A)$	7.89
$Al^{3+} + A^{3-} + H^+$	$= Al(H)A^+$	12.86
$(A)_T$	$= M_D \times DOC$	$M_D = 0.023$ mol/mol C
Surface reactions		
$>MOH^0 + H^+$	$= >MOH_2^+$	4.28
$>MOH^0$	$= >MO^- + H^+$	–9.43
$>MOH^0 + SO_4^{2-} + H^+$	$= >MOH_2\text{-}SO_4^-$	6.24
$>MOH^0 + 3H^+ + A^{3-}$	$= >MOH\text{-}H_4A$	20.46
$>MOH^0 + Al^{3+} + A^{3-}$	$= >MOH\text{-}AlA$	12.05
$2X_3Al + 3Ca^{2+}$	$= 3X_2Ca + 2Al^{3+}$	0.41
Solid phase reactions		
$Al^{3+} + 3H_2O$	$= Al(OH)_{3(s)} + 3H^+$	8.11

The Soil Acidification/Soil Development (SASD) model was designed to coordinate the various processes included in these simulations. Processes were characterized as either "fast" or "slow" depending on reaction kinetics and the time scale of a given simulation (Furrer et al., 1990). Fast processes are those that can be described as equilibrium reactions, while slow processes require a kinetic formulation. A kinetic formulation also is required for irreversible reactions. Subroutines within the SASD model calculate the contribution from slow processes. The information from fast reactions is supplied by the CHESS model (Santore & Driscoll, 1994). The CHESS model was designed to be called as a subroutine and can be linked into models such as the SASD model, without modification.

Chemical fluxes and transformations were simulated separately by layer to allow the soil profile to develop. The model time step was an adjustable parameter, however for these simulations a timestep of 100 yr was used. During each time step fast and slow processes were simulated as follows: (i) incoming water and associated solutes entered a layer and mixed with soil and water already present in that layer, (ii) water loss due to evapotranspiration was removed, (iii) a call to the CHESS model was made to obtain information required for development of rate expressions for slow processes, (iv) slow processes were simulated as the resulting set of simultaneous ordinary differential equations. The net effect of the slow processes on the mass balance of chemical components within the layer was calculated, (v) a final call to CHESS completed the simulation of fast processes, and (vi) water was drained off to field capacity carrying solutes out of this soil layer and into the next. This sequence was repeated until all layers were simulated for all time steps. The simulations were run for 15 000 yr to represent development since deposition of a uniform glacial till.

As we have indicated, the focus of this modeling effort was to examine interactions between DOC and SOC and other biogeochemical cycles. Many processes important to soil development were omitted or simplified. In these simulations we did not consider: (i) variations in temperature and rainfall; (ii) vegetation growth, succession, and nutrient uptake; (iii) direct mineralization of SOC; (iv) clay translocation; (v) erosion or physical disturbances to soil; (vi) incongruent weathering reactions; and (vii) secondary clay formation.

Dissolved Organic Carbon

Due to the complex nature of DOC, it is difficult to incorporate an accurate representation of the number and types of functional groups attributed to naturally occurring organic solutes. Nevertheless, DOC has been shown to participate in acid-base and metal complexation reactions (Perdue et al., 1984; Driscoll et al., 1994). We chose to use a triprotic organic analog that has been used to model interactions between DOC and Al in surface waters (Schecher & Driscoll, 1995; Driscoll et al., 1994), such that

$$A_T = M_D \cdot DOC \qquad [1]$$

where:

A_T is the total concentration of organic functional groups (mol sites/L),
M_D is the site density of naturally occurring organic solutes (mol sites/mol C), and
DOC is the concentration of dissolved organic carbon (mol C/L).

There are several advantages to this approach. It relates the concentration of organic ligands to a measurable value (DOC). The effectiveness of the organic analog in describing interactions with H^+ and Al has been demonstrated in surface waters (Driscoll & Schecher, 1990). The organic analog can be extended to include surface adsorption and the rate dependence weathering reactions on organic acids. Finally, the triprotic organic analog can be easily incorporated into an equilibrium model.

Soil Carbon

The model considered the formation of SOC as the result of two types of processes. Dissolved organic C was allowed to accumulate on soil surfaces by a reversible adsorption mechanism. In the model, this adsorbed C pool was defined as a labile soil C pool (LSC) corresponding to pyrophosphate extractable C. Once adsorbed, the LSC can be converted to a nonlabile soil C pool (NSC) by a slow, irreversible reaction. The sum of these two C pools (LSC + NSC) constitutes the total pool of SOC.

The reaction mechanism for the adsorption of LSC was chosen to favor adsorption of neutral species of the triprotic organic analog.

$$3H^+ + A^{3-} + >MOH \Leftrightarrow >MOH-H_3A$$

$$Al^{3+} + A^{3-} + >MOH \Leftrightarrow >MOH-AlA \quad [2]$$

where

>*MOH* represents a surface adsorption site, and
>*MOH–H$_3$A* and >*MOH–AlA* represent surface adsorbed species.

This choice of mechanism results in enhanced adsorption of DOC at low pH, and desorption at higher pH (Davis & Gloor, 1981; Tipping & Woof, 1991). The LSC pool was defined as the sum of these two species (Eq. [3]).

The transformation of LSC to NSC occurs by a slow, irreversible reaction that is controlled by LSC concentrations.

$$LSC = [>MOH-H_3A] + [>MOH-AlA] \quad [3]$$

$$LSC \rightarrow NSC \quad [4]$$

$$R_{NSC} = k_{LSC}[LSC] \quad [5]$$

where

R_{NSC} is the rate of formation of NSC (mol/yr), and
k_{LSC} is the first-order rate constant (1/yr).

As the pool of SOC accumulates, it contributes sites for cation exchange reactions. It has been observed that a correlation exists between SOM/SOC and CEC in northern temperate forest soils (Kalisz & Stone, 1980; Ross et al., 1991). This empirical relationship can be used to quantify the linkage between CEC and SOC.

$$CEC = SOM \cdot M_s + b \quad [6]$$

$$SOM = 2 \cdot SOC \quad [7]$$

where

the units of SOM and SOC are (g/kg soil) and the units of CEC are (cmol$_c$/Kg soil),
M_s is the site density of SOM (mol reactive sites/mol C), and
b is the intercept obtained from the empirical relationship.

Chemical Equilibrium Calculations

The set of reactions defining the equilibrium chemistry (i.e., the "fast" reactions; Table 13–2) were solved using the CHESS model (Santore & Driscoll, 1995). The CHESS model uses a matrix approach to chemical equilibrium calculations based on the tableau notation (Morel, 1983). Input to the model included a list of reactions including reaction stoichiometry, thermodynamic equilibrium constants and associated assumptions (such as method of chemical activity corrections) that were to be used in the chemical equilibrium calculations. A complete description of the chemical system included physical properties (such as volume of water, amount of soil, temperature), and the chemical composition of solutes entering the mineral soil. Model output included the equilibrium concentrations of all species defined in the reaction list.

Chemical Kinetic Calculations

Rate equations were expressed as functions of the soil and solution composition. This approach was developed to make use of empirical rate laws that describe slow reactions, such as mineral weathering, as having a fractional order dependence on chemical factors, such as H$^+$ concentration (Sverdrup, 1990; Stumm & Wieland, 1990). The rate was expressed as

$$R_p = k_{p,0} + k_{p,H}[H]^{\eta_{p,H}} + k_{p,A}[nA^{n-}]^{\eta_{p,A}} \qquad [8]$$

where $k_{p,H}$ and $k_{p,A}$ are rate constants for mineral p associated with H and nA^{n-} concentration,

$\eta_{p,H}$ and $\eta_{p,A}$ are the order of the reaction with respect to H and nA^{n-} concentration, and

nA^{n-} is the equivalent charge concentration of organic anion in solution (eq/L).

Furthermore, as the mass of these minerals decreased due to weathering, we assumed that the observed weathering rates decreased due to changes in total surface area of mineral pools. These changes were considered by adjusting the rate constants downward as mineral mass was lost to reflect decreases in mineral surface area.

The set of equations R_p (Eq [4], [5], and [8]; Table 13–3) were treated as a set of coupled ordinary differential equations. They were integrated over each time step of the simulation by a fourth-order Runge-Kutta method (Press et al., 1986). An adaptive step size was implemented to control integration errors.

Model Calibration

Estimation of Constants for Triprotic Organic Acid Analog

The thermodynamic equilibrium constants and site density (M_D) parameters for the organic acid analog were optimized for streamwater and soil solutions

Table 13–3. The rate expressions and stoichiometry of slow reactions used in this analysis.

Stoichometry
Hornblende + 13.33H$^+$ = 1.65Ca^{2+} + 1.87Al^{3+} + 7.08H$_4$SiO$_4$
Plagioclase + 4.9H$^+$ = 0.22Ca^{2+} + 1.21Al^{3+} + 2.77H$_4$SiO$_4$
DOC (and associated N and S) = H$_2$CO$_3$ + 0.05NO$_3^-$ + 0.005SO$_4^{2-}$ + 0.06H$^+$

Weathering rate expressions
$R_{weathering} = k_0 + k_H [H^+]^{\eta_H} + k_A [nA^{n-}]^{\eta_A}$

	k_0 (mol/yr)	k_H (1/yr)	η_H	k_A (1/yr)	η_A
Hornblende	8.35E-2	10.02E-0	0.7	3.4E-1	1
Plagioclase	8.35E-3	10.02E-1	0.5	2.4E-1	1

from the Oa, Bh, and Bs soil horizons at Hubbard Brook (Fig. 13–1b, d, and 13–2). The parameters were obtained by a nonlinear regression program based on the Levenberg-Marquardt algorithm (Press et al., 1986). The program is designed to find the best set of parameters (**a**) by minimizing the weighted sum of squares of the discrepancy between measured and model predicted values of organic monomeric Al (Al_o) (Chi-squared, Eq. [9]).

$$X^2(a) = \sum_{i=1}^{N} \left[\frac{Al_{o_i} - f_i(a)}{\sigma_i} \right]^2 \quad [9]$$

where

a is the set of adjustable parameters [**a** = ($M_D, K_{H_3A}, K_{H_2A}, K_{HA^2}, K_{AlA}, K_{AlHA^+}$)],

Al_{o_i} is a measured value of organic monomeric Al (Al_o),

$f_i(a)$ is a value calculated from CHESS model output given the parameters **a** ($f_i(a) = [AlA]_i + [AlHA^+]_i$), and

an σ_i is the uncertainty in the measured value for Al_{o_i}.

Rather than require uncertainty values for each observation as an input to the program, these values were calculated based on the analytical precision ($E_{rel} = 0.05$) and detection limit ($E_{abs} = 0.5$ µmol/L) associated with measurement of Al_o.

$$\sigma_i = E_{abs} + E_{rel} \cdot Al_o \quad [10]$$

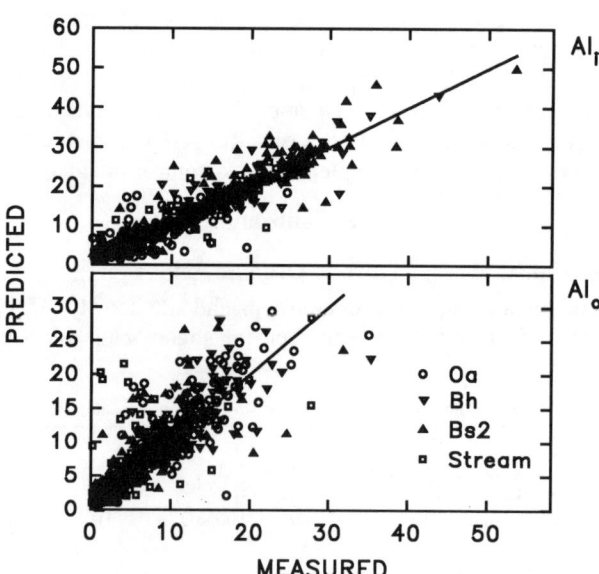

Fig. 13–2. A comparison of measured and predicted values of inorganic monomeric Al (Ali, µmol/L) and organic monomeric Al (Al_o, µmol/L) concentrations based on the calibration of a triprotic organic analog model (Table 13–1).

Estimation of Dissolved Organic Carbon Adsorption Constants

The adsorption constants for DOC complexes were estimated in a similar manner to the H^+ and Al binding constants of the triprotic organic acid analog. For DOC adsorption, the prediction from CHESS given a set of adsorption constants and average soil solution chemistry was compared to measured labile soil C for HBEF soil horizons (Fig. 13–3). The adsorption constants were adjusted to minimize the weighted sum of squares of the discrepancy between these two values.

Estimation of Rate Constants

Controlled studies measuring rates of mineral dissolution have shown a strong dependence on surface coordination, particularly with regard to binding of H^+ (Sparks, 1988; Stumm & Wieland, 1990) and concentrations of organic acids (Zepp & Wolfe, 1987; Sparks, 1988). Other authors have noted that dissolution rates increase at low pH (Sverdrup, 1990). Weathering rates measured in the laboratory, however, are often several orders of magnitude higher than those measured in the field (Schnoor, 1990). On the other hand, weathering rates measured from field studies cannot supply the detailed information required to formulate H^+ and organic acid rate constants. In the development of the SASD model, we chose to compromise between laboratory and field rate information. The rate expressions used to calculate weathering rates were based on those developed from laboratory studies, but the rate constants were chosen to keep the annual supply of cations from weathering close to values observed from field studies. When

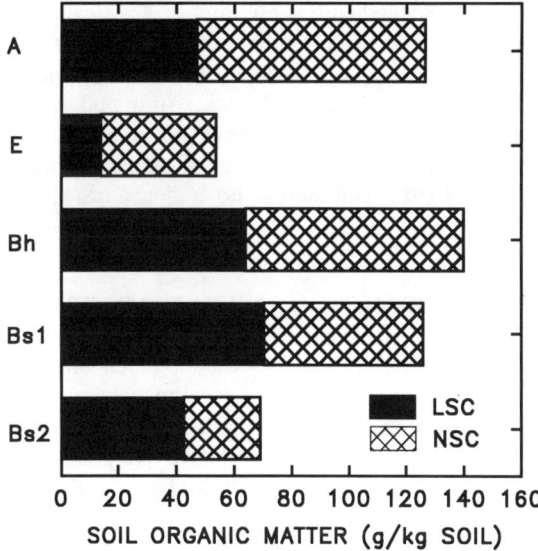

Fig. 13–3. Concentrations of SOM in soil horizons at HBEF Watershed 5. The LSC pool is defined as reversibly adsorbed soil C and is measured by pyrophosphate extraction. The total soil C (sum of LSC and NSC) was measured using a pyrolytic C analyzer (Huntington et al., 1988). The NSC fraction was estimated as the difference between total SOC and pyrophosphate extractable C.

adjusting the rate constants all values were multiplied by a constant term but the relative values between constants are the same as those reported by Sverdrup (1990). These values are summarized in Table 13–3.

Values for DOC mineralization were not as readily available in the literature. The rate constants for DOC mineralization were obtained by calibration to give DOC concentrations similar to those in HBEF soil solutions (Fig. 13–1). A value of $k_{min} = 0.1$ (1/yr) was used. Similarly the value of rate constant for NSC accumulation was found to be $k_{NSC} = 1 \times 10^{-3}$ (1/yr).

SITE DESCRIPTION

We have used observations from the HBEF in the White Mountains of central New Hampshire to develop process formulations and parameter values in this modeling effort. The HBEF is a 300-ha forested area, with a series of experimental watersheds (Likens et al., 1977). Soils at the site are acidic, well-drained Spodosols (Haplorthods and Fragiorthods), with a well-drained organic layer (3–15 cm). They are underlain with variable depths of glacial till and siliceous bedrock (Littleton formation; schist) (Johnson et al., 1981). Soils are shallow at high elevations and increase in depth with decreasing elevation (Lawrence et al., 1986).

Climate at the HBEF is cool-temperate, humid-continental with mean July and January temperatures of 19°C and –9°C, respectively (at 450-m elevation). Mean annual precipitation for Watershed 6 is approximately 1390 mm (SD = 188), with 25 to 33% of the total occurring as snow (Federer et al., 1990). Mean annual streamflow from Watershed 6 is 869 mm (SD = 175).

Northern hardwood vegetation dominates most of the HBEF, consisting of American beech (*Fagus grandifolia* Ehrh.), yellow birch (*Betula alleghaniensis* Britt.) and sugar maple (*Acer saccharum* Marsh.), from 500 to 730 m. Coniferous vegetation, consisting primarily of red spruce (*Picea rubens* Sarg.) and balsam fir [*Abies balsamea* (L.) Mill] dominates elevations above 730 m. The HBEF was logged between 1909 to 1917, and there is no evidence of recent fire (Whittaker et al., 1974).

PROCESS INFORMATION AND PARAMETER VALUES

Information available from monitoring the biogeochemical reference watershed, Watershed 6, at the HBEF was used to develop and parameterize the model. These data included information on soil solution and stream chemistry, and soil chemical parameters. Soil solutions were collected using tension-free lysimeters (installed in fall 1983) beneath the Oa and Bh horizons, and within the lower Bs horizon, adjacent to Watershed 6 at two elevations in the deciduous forest (600 and 730 m; Driscoll et al., 1988). Streamwater was collected at the base of the watershed (540 m) just above the stream gauging station. Soil solutions and streamwater were collected at monthly intervals (January 1984–December 1991). Solutions were monitored for all major solutes, including DOC, pH, inorganic monomeric Al (Ali) and organic monomeric Al (Al_o; Lawrence et al., 1986).

Concentrations of organic anions ($nA^{n-} = [H_2A^-] + 2[HA^{2-}] + 3[A^{3-}]$) were not measured directly, but estimated as the difference between the sum of measured cation concentrations and the sum of measured anion concentrations (Driscoll et al., 1989, 1994). Using standard methods for the analysis of dilute waters, Schecher and Driscoll (1988) indicated that the uncertainty in estimates of nA^{n-} concentrations by discrepancy in charge balance is ± 2 µeq/L. A detailed description of soil water and streamwater collections is provided in Driscoll et al. (1988) and Likens et al. (1994).

Mean values of soil solution and stream chemistry showed that DOC concentrations were high in soil waters draining the Oa horizon (1370 µmol C/L) and declined through the mineral soil (Fig. 13–1a). Stream concentrations draining the northern hardwood stand exhibited low concentrations of DOC (mean value 160 µmol C/L). This pattern in DOC was similar to conditions reported at Hubbard Brook (Driscoll et al., 1985; McDowell & Likens, 1988) and elsewhere (Dawson et al., 1978; David & Driscoll, 1984; Cronan & Aiken, 1985). Coincident with these patterns in DOC were high concentrations of nA^{n-} in Oa horizon leachate (74 µeq/L) and low pH values (4.21). As waters drained through the mineral soil to the stream, concentration of nA^{n-} declined and pH values increased (Fig. 13–1).

The pH and DOC concentrations are important master variables regulating the concentration and speciation of Al. Because of low pH conditions, drainage waters at HBEF exhibited elevated concentrations of Al (Fig. 13–1d). Mean concentrations of Al were 17 µmol/L in Oa horizon solutions, which largely occurred as the Al_o fraction (11 µmol/L). Concentrations of Al peaked in the Bh horizons (25 µmol/L), and showed higher concentrations of both Al_i and Al_o than in Oa horizon leachate. Concentrations of Al_o declined markedly in Bs2 soilwater and streamwater, coincident with decreases in DOC and nA^{n-}. Inorganic monomeric Al concentrations were highest in the Bs2 horizon (16 µmol/L). The Al_i fraction was the major form of monomeric Al in mineral soil solutions and streamwaters.

Measured values of pH, DOC, Al_i and Al_o, and estimated values of nA^{n-} for each of the Oa, Bh and Bs2 soil solutions were used with CHESS to fit the field observations to H$^+$ and Al binding constants using a triprotic organic analog. The equilibrium constants and the values of the site density (mol sites/mol C; M_D) obtained from this analysis are summarized in Table 13–4. There was generally good agreement between measured Al_i and Al_o values with values predicted from the calibrated triprotic organic acid analog, although some scatter from the 1:1 line was evident (Fig. 13–2). The values obtained for H$^+$ and Al equilibrium constants were similar throughout the soil profile. There was a systematic increase in estimates for the M_D parameter from Oa to Bh to Bs2 horizon solutions. Increases in M_D with depth could result from several possible mechanisms. Organic solutes with lower functional group density may be preferentially deposited in the soil. This hypothesis is supported by observations that hydrophobic organic acids (i.e., those with lower densities of organic functional groups) are preferentially retained (Vance & David, 1989). Alternatively, increases in M_D could be due to creation of new reactive sites in the DOC due to partial microbial oxidation.

Soils at HBEF were sampled from 60 quantitative 0.5-m² pits, randomly located in Watershed 5 adjacent to Watershed 6. A description of the soil

Table 13-4. Summary of total site density (M_D) and log values of the H⁺ and Al equilibrium constants with the triprotic organic acid analog. The parameter values were determined using a nonlinear optimization routine with the CHESS model and Oa, Bh and Bs horizon leachate and steamwater samples from the Hubbard Brook Experimental Forest, New Hampshire. The overall fit of model parameters is given by values of the correlation coefficient (r) between measured and predicted Alo values. The SASD model used parameter values obtained from an optimization performed on all samples simultaneously.

Solution	M_D	Log K_{HOrg}	Log K_{H_2Org}	Log K_{H_3Org}	Log K_{AlOrg}	Log K_{AlHOrg}	r
	mol/mol C						
Oa	0.014	6.178	11.769	13.179	8.235	13.361	0.7673
Bh	0.024	6.500	11.684	13.691	7.905	12.884	0.7261
Bs2	0.044	6.788	13.027	14.960	8.682	13.167	0.7015
Stream	0.028	6.705	11.852	13.695	7.766	13.005	0.7419
All samples	0.023	6.484	11.691	13.713	7.888	12.857	0.7947

sampling procedure is provided by Huntington et al. (1989) and the analysis for soil physical and chemical parameters is given by Johnson et al. (1991a,b). As observed in other studies, a correlation was evident between CEC and soil organic matter in both organic and mineral soil horizons at HBEF (Fig. 13–4). The linear relationships between CEC and SOM are characterized by intercepts near the origin and may suggest that most of the CEC of the soil is associated with SOM. Our values of CEC/SOM (1.8 and 3.9 $cmol_c$/kg SOM for the Oa and mineral soil samples, respectively; Fig. 13–4) are somewhat lower than other values reported for the Northeast (2.9 $cmol_c$/kg SOM for New Hampshire organic horizons and 5.2 $cmol_c$/kg SOM for organic horizons in the Adirondacks in New York, Kalisz & Stone 1980; 2.9 $cmol_c$/kg SOM for Vermont soil, all horizons, Ross et al., 1991).

Based on estimates of 50% of the SOM occurring as SOC (Huntington et al., 1989), the site density of SOC (M_s) was 0.036 mol sites/mol C for the organic horizon and 0.078 mol sites/mol C for the mineral soil. A comparison of these values with the site density values obtained for DOC (M_D = 0.014 mol sites/mol C in the Oa horizon, M_D = 0.044 mol sites/mol C in the Bs2 horizon; Table 13–4), suggest that SOC has greater values of functional group site density than DOC. The increase in observed reactive site densities in both the DOC and SOC of mineral soil horizons compared to organic soil horizons would suggest that preferential retention of hydrophobic DOC cannot explain all of the observed trends in the site density. However, the alternative explanation that the increase in site

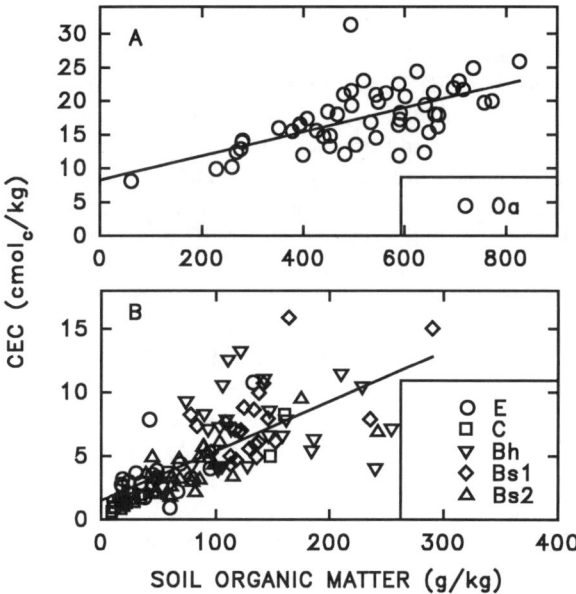

Fig. 13–4. Empirical relationships between soil organic matter (SOM) in grams SOM per kilogram of soil (g/kg) and cation exchange capacity (CEC) in centimoles of charge per kilogram of soil ($cmol_c$/kg) in organic soil horizons (A) and mineral soil horizons (B) at Hubbard Brook. The lines on the figure represent linear regressions of CEC versus SOM (CEC = 8.27 + 0.0179*SOM; r^2 = 0.39 for organic soil horizons; and CEC = 1.5 + 0.039 *SOM; r^2 = 0.52 for mineral soil horizons).

density is due to the creation of new reactive sites by partial microbial oxidation is consistent with observed site density increases for both the DOC and SOC pools. Recently, Qualls and Haines (1992) have proposed that a combination of adsorption followed by biodegradation is responsible for observed patterns of increased labile C fraction in lower soil horizons. It is possible that a similar mechanism is responsible for the increases in site density in DOC and SOC with depth at HBEF.

Additional analyses on the soil pit samples included total and pyrophosphate extractable C content (Fig. 13–3). These data were used in the development of a DOC sorption model. The pyrophosphate extractable C was assumed to be reversibly adsorbed on soil surfaces and is referred to as a LSC pool. The quantity of SOC that was not removed by pyrophosphate extraction is assumed to be irreversibly bound to the soil and is referred to as NSC. The NSC was estimated as the difference between total SOC and phosphate extractable C.

RESULTS AND DISCUSSION

Simulation using the SASD model, calibrated to the HBEF, revealed marked changes in the chemistry of soil and soil waters for the 15 000 yr since glaciation of the region (Fig. 13–5 to 13–7). Patterns of DOC concentrations were constant over time, but declined with depth from about 1400 µmol C/L at the surface of the mineral soil to 300 µmol C/L of the bottom of the profile (Fig. 13–5). The decreases in DOC concentrations in model simulations were due to mineralization reactions and the formation of SOC.

Soil organic matter was largely deposited in the upper mineral soil, due to the functional dependence of DOC adsorption on concentrations of DOC, H^+, and Al (Fig. 13–5). In model simulations there was a rapid initial accumulation of SOM due to sorption reactions and formation of a LSC pool. Later in the simulation, some of the LSC pool was lost due to export of DOC complexed with aqueous Al (Fig. 13–6). After this initial accumulation and loss there was a slow steady increase in SOM due to the formation of a NSC pool. Patterns of rapid initial accumulation (~3 000 yr) followed by prolonged periods of slow accumulation (~10 000 yr) have been observed in chronosequence studies of recently deglaciated areas (Schlesinger, 1990). Although the overall concentrations of SOM in the soil are similar to those found at HBEF (Fig. 13–5), the absence of a spodic horizon in the model output suggests that the relative values for the adsorption constants of the H^+ saturated and Al saturated LSC forms may need to be adjusted. The adsorption constants were obtained from fitting with observed SOM profiles, but there is no information available on the metals associated with SOM at HBEF. The lack of information on specific forms of SOM and the importance of this information in model simulations underscores the need for additional work in this area.

Because of their empirical linkage in the SASD model, patterns of CEC were similar to SOM (Fig. 13–5). Values of CEC increased with time, coincident with the formation of SOM. Soil CEC was highest in the upper mineral soil and diminished with depth, this vertical pattern in CEC became more marked during the later years of simulation.

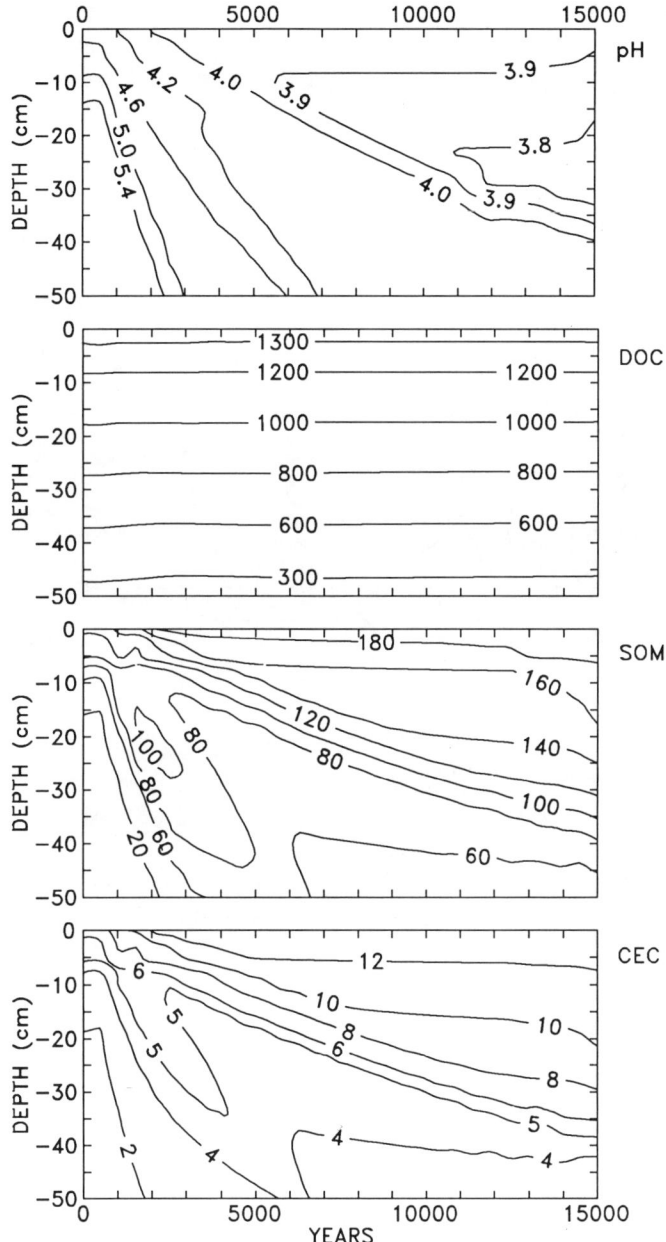

Fig. 13–5. Simulations of changes in soil water pH, dissolved organic C concentrations (DOC μmol C/L), soil organic matter concentration (SOM, g OM/kg), and soil cation exchange capacity (CEC, cmol$_c$/kg) with depth in a temperate mineral soil horizon over 15 000 yr. Concentrations are expressed as isopleths.

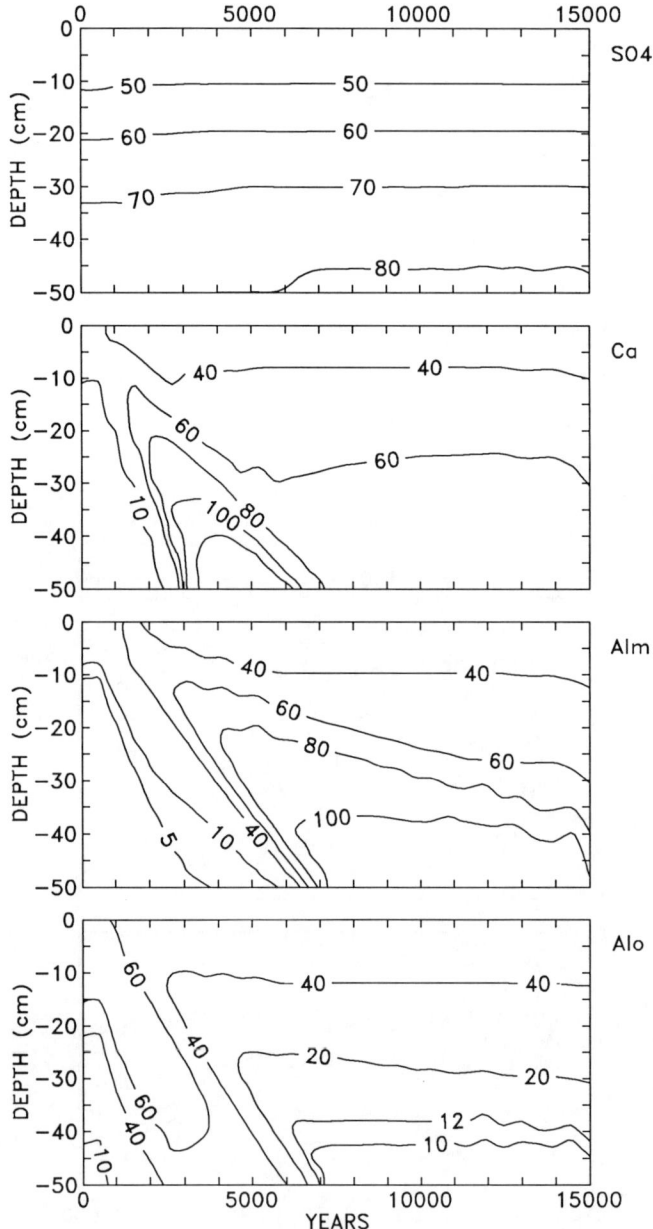

Fig. 13–6. Simulations of changes in soil water concentrations of SO_4^{2-} (µmol/L), Ca^{2+} (µmol/L), monomeric Al (Al_m, µmol/L), and percentage of monomeric Al that occurs as organic monomeric Al (Al_o%) with depth in a temperate mineral soil horizon over 15 000 yr. Concentrations are expressed as isopleths.

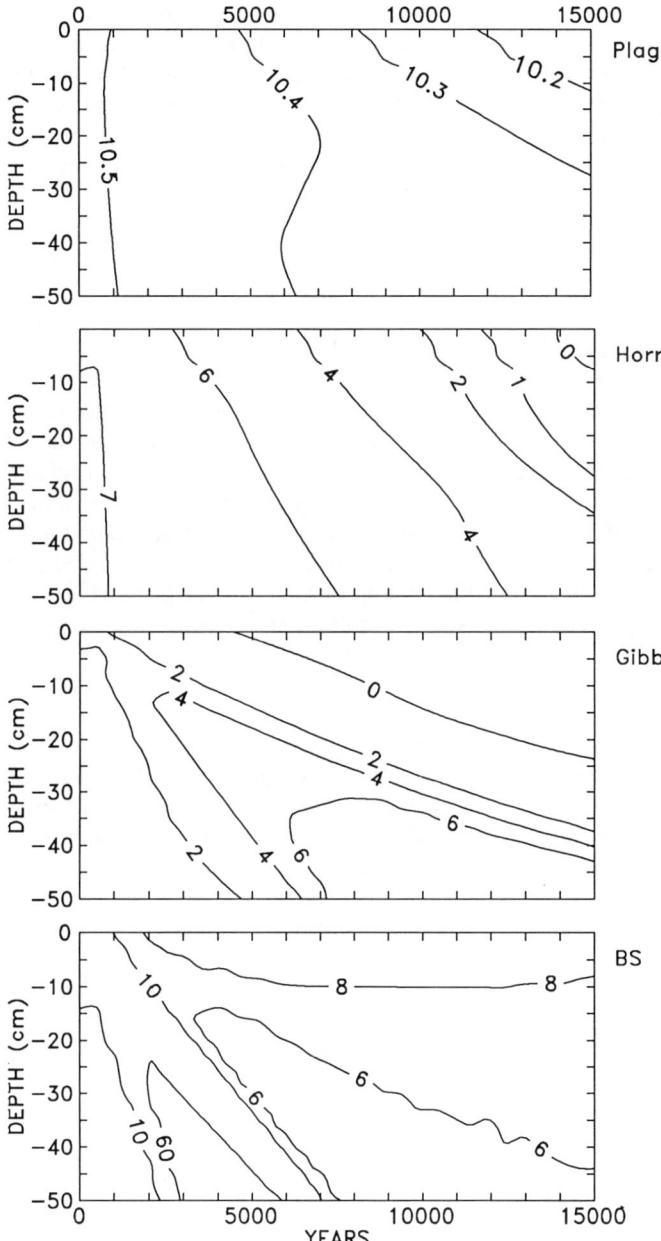

Fig. 13–7. Simulations of changes in concentrations of plagioclase (*Plag.* %wt), hornblende (*Horn.* %wt), gibbsite (*Gibb.* %wt), and percent base saturation (% *BS*) in soil with depth in a temperate mineral soil horizon over 15 000 yr. Concentrations are expressed as isopleths.

The initial conditions of the simulation specified that primary minerals (hornblende, plagioclase) were distributed uniformly throughout the soil profile (Table 13–1). Model calculations showed an initial, rapid loss of mineral mass resulting in a period of high pH (Fig. 13–5), high Ca concentrations (Fig. 13–6), and high base saturation (Fig. 13–7). Most of these effects were due to the weathering of hornblende, which weathered at a faster rate than plagioclase (Table 13–3). Weathering rates were highest in the upper soil due to the high concentrations of H^+ and nA^{n-} and the concentration dependence of these solutes to the overall expression for weathering rate (Table 13–3). As concentrations of DOC declined through the profile, H^+-dependent component of weathering dominated the overall rate expression at intermediate depths. At the lowest soil depths, H^+ neutralization resulted in diminished sensitivity of the weathering rate to H^+, so mineral depletion was largely controlled by the base rate expression.

By the end of the simulation the heavy mineral, hornblende, occurred at lower concentrations due to its faster weathering rate than plagioclase (Sverdrup, 1990). Nevertheless, hornblende is Ca rich and an important source of basic cations in glaciated soils of the region (April & Newton, 1985; Table 13–3). As weathering of hornblende began to slow in the upper soil (about 8000 yr), the associated release of Ca^{2+} and percentage base saturation (*%BS*) of the soil decreased, resulting in acidification of the soil and soil water (Fig. 13–5, 13–6).

Initially, simulated pH values in soil water were lowest in the upper soil horizons and increased with soil depth. This pattern was largely determined by the release of Ca^{2+} and neutralization of H^+ from weathering. After the initial period of rapid mineral weathering the production of H^+ by mineralization of DOC exceeded the rate of H^+ neutralization by weathering resulting in decreases in soil water pH. The mineralization of DOC was assumed to result in the production of H_2SO_4 and HNO_3 (Table 13–3) and provided an important input of strong acids to the soil profile (Fig. 13–6). The production of strong acids from mineralization reactions was responsible for the minimum in solution pH at middle depths of the soil profile (Fig. 13–5). Mineralization reactions of DOC appeared to be a critical process regulating the acid-base status of base depleted soils. A better understanding of this process is an important research issue.

Concentrations of SO_4^{2-} are mainly determined by atmospheric inputs. Adsorption on soil surfaces plays a minor role in regulating soil solution SO_4^{2-} concentrations. However, adsorption of SO_4^{2-} does result in slightly lower initial solution SO_4^{2-} concentrations, especially at the deepest layers.

Aluminum was supplied to the mineral soil by the input solution (from Oa horizon leachate; Table 13–1) and as a byproduct of primary mineral weathering (Table 13–3). Inputs of Al initially resulted in conditions of oversaturation with respect to the solubility of gibbsite $[Al(OH)_3]$ and precipitation of this secondary mineral. Pools of gibbsite were zero initially but this solid phase accumulated during the simulation (Fig. 13–7). Deposition of gibbsite was rapid during the first 5 000 yr of the simulation when weathering rates were high. As the rate of supply of Al from weathering reactions slowed and the soil began to acidify, soil gibbsite pools that previously had accumulated dissolved and elevated concentrations of Al were exported from the profile in drainage water. This pattern was especially apparent in the upper soil layers where gibbsite was entirely depleted after about 7 000 yr. The low pH and high concentrations of nA^{n-} in these upper horizon soil solutions readily dissolved previously precipitated gibbsite pools. Aluminum concentrations in the later years of the simulation were higher than expected for HBEF soil solutions due to the low pH and the arbitrary choice of gibbsite dissolution as a "fast" process regulating the solubility of solution Al.

SUGGESTIONS FOR FUTURE RESEARCH

This modeling exercise demonstrates the potential linkages of DOC and SOM with biogeochemical cycles. Unfortunately, there is limited quantitative information available to allow for model formulations and to parameterize the model. As a result, the SASD model is a research tool which helps demonstrate uncertainty in knowledge and can be used to direct future research activities. We have identified a series of research needs to improve the formulations of processes represented in the SASD model.

1. Information is needed to help develop a quantitative theory of the mechanism of DOC sorption to mineral soil. In particular the strength of various metal complexes and their relative solubilities needs to be addressed.
2. There is very little information available on the mechanism, stoichiometry and rates of DOC and SOM mineralization in soil and environmental factors which influence these rates. Dissolved organic C mineralization in these simulations was an important source of strong acids to the mineral soil. These results are based on the assumption that mineralization of DOC produces CO_2 with no organic acid intermediates. The exact pathway, however, is not likely to be this simple.
3. Experimental data are needed to quantify the dependence of mineral weathering rates on naturally occurring organic solutes. Most of the work in this area has been done using simple organic acids. It is not clear how these results can be extrapolated to field conditions and natural organic solutes.
4. There is a limited data on stability constants for organic C complexation with other metallic cations (e.g., Ca^{2+}, Fe^{3+}). Complexes with H^+ and Al in these simulations show important interactions that could be extended to these other metals.
5. Linkages between deposition of SOC and development of cation binding sites need to be developed and possible interference with adsorption of anions (e.g., SO_4^{2-}) needs to be considered.
6. A better understanding of the role of plant and microbial processes in element cycling is needed.

ACKNOWLEDGMENTS

This work is a contribution to the Hubbard Brook Ecosystem Study and was supported with funds from the National Science Foundation, the USEPA, and the USDA through the USDA-FS. Although the research described in this article has been funded in part by the USEPA, it has not been subjected to the Agency's review and therefore does not necessarily reflect the views of the Agency, and no official endorsement should be inferred.

REFERENCES

Aber, J.D., and J.M. Melillo. 1991. Terrestrial ecosystems. Saunders College Publ., Orlando, FL.

April, R., and R. Newton. 1985. Influence of geology on lake acidification in the ILWAS watersheds. Water Air Soil Pollut. 26:373–386.

Cosby, B.J., G.M. Hornberger, J.N. Galloway, and R.F. Wright. 1985. Modeling the effects of acid deposition: Assessment of a lumped parameter model of soil and streamwater chemistry. Water Resour. Res. 21:51–63.

Cronan, C.S., and G.R. Aiken. 1985. Chemistry and transport of soluble humic substances in forested watersheds of the Adirondack Park, New York. Geochim. Cosmochim. Acta 49:1697–1705.

Dahlgren, R.A., and D.J. Marrett. 1991. Organic carbon sorption in arctic and subalpine Spodosol B horizons. Soil Sci. Soc. Am. J. 55:1382–1390.

David, M.B., and C.T. Driscoll. 1984. Aluminum speciation and equilibria in soil solutions of a Haplorthod in the Adirondack Mountains (New York, U.S.A.). Geoderma 33:297–318.

Davis, J.A, and R. Gloor. 1981. Adsorption of dissolved organics in lake water by aluminum oxide. Effect of molecular weight. Environ. Sci. Technol. 15:1223–1229.

Dawson, H.J., F.C. Ugolini, B.F. Hrutfiord, and J. Zachara. 1978. Role of soluble organics in the soil processes of a podzol, central Cascades, Washington. Soil Sci. 126:290–296.

Driscoll, C.T., R.D. Fuller, and W.D. Schecher. 1989. The role of organic acids in the acidification of surface waters in the eastern U.S. Water Air Soil Pollut. 43:21–40.

Driscoll, C.T., R.D. Fuller and D.M. Simone. 1988. Longitudinal variations in trace metal concentrations in a northern forested ecosystems. J. Environ. Qual. 17:101–107.

Driscoll, C.T., M.D. Lehtinen, and T.J. Sullivan. 1994. Modeling the acid–base chemistry of organic solutes in Adirondack, NY, lakes. Water Resour. Res. 30:297–306.

Driscoll, C.T., and W.D. Schecher. 1990. The chemistry of aluminum in the environment. Environ. Geochem. Health 12: 28–49.

Driscoll, C.T., N. van Breemen, and J. Mulder. 1985. Aluminum chemistry in a forested Spodosol. Soil Sci. Soc. Am. J. 49:437–444.

Eary, L.E., E.A. Jenne, L.W. Vail, and D.C. Girvin. 1989. Numerical models for predicting watershed acidification. Arch. Environ. Contam. Toxicol. 18:29–53.

Federer, C.A., L.D. Flynn, C.W. Martin, J.W. Hornbeck, and R.S. Pierce. 1990. Thirty years of hydrometeorologic data at the Hubbard Brook Experimental Forest, New Hampshire. USDA–FS, General Techn. Rep. NE-141. USDA–FS, Radnor, PA.

Federer, C.A., and J.W. Hornbeck. 1985. The buffer capacity of forest soils in New England. Water Air Soil Pollut. 26:163–173.

Furrer, G., P. Sollins, and J.C. Westall. 1990. The study of soil chemistry through quasi–steady–state models: II. Acidity of soil solution. Geochim. Cosmochim. Acta 54:2363–2374.

Gherini, S.A., L. Mok, R.J.M. Hudson, G.F. Davis, C.W. Chen, and R.A. Goldstein. 1985. The ILWAS model: Formulation and application. Water Air Soil Pollut. 26:425–459.

Huntington, T.G., C.E. Johnson, A.H. Johnson, T.G. Siccama, and D.F. Ryan. 1989. Carbon, organic matter and bulk density relationships in a forested Spodosol. Soil Sci. 148:380–386.

Johnson, C.E., A.H. Johnson, T.G. Huntington, and T.G. Siccama. 1991a. Whole–tree clear–cutting effects on exchangeable cations and soil acidity. Soil Sci. Soc. Am. J. 55:502–508.

Johnson, C.E., A.H. Johnson, and T.G. Siccama. 1991b. Whole–tree clear–cutting effects on soil horizons and organic–matter pools. Soil Sci. Soc. Am. J. 55:497–502.

Johnson, N.M., C.T. Driscoll, J.S. Eaton, G.E. Likens and W.H. McDowell. 1981. "Acid rain," dissolved aluminum and chemical weathering at the Hubbard Brook Experimental Forest, New Hampshire. Geochim. Cosmochim. Acta 45:1421–1437.

Kalisz, P.J., and E.L. Stone. 1980. Cation exchange capacity of acid forest humus layers. Soil Sci. Soc. Am. J. 44:407–413.

Lawrence, G.B., R.D. Fuller, and C.T. Driscoll. 1986. Spatial relationships of aluminum chemistry in the streams of the Hubbard Brook Experimental Forest, New Hampshire. Biogeochemistry 2:115–135.

Likens, G.E., F.H. Bormann, R.S. Pierce, J.S. Eaton, and N.M. Johnson. 1977. Biogeochemistry of a forested ecosystem. Springer–Verlag, New York.

Likens, G.E., C.T. Driscoll, D.C. Buso, T.G. Siccama, C.E. Johnson, G.M. Lovett, D.F. Ryan, T. Fahey, and W.A. Reiners, 1994. The biogeochemistry of potassium at Hubbard Brook. Biogeochemistry 25:1–65.

Liu, S., R. Munson, D. Johnson, S. Gherini, K. Summers, R. Hudson, K. Wilkinson, and L. Pitelka. 1991. Applications of a nutrient cycling model (NuCM) to a northern mixed hardwood and a southern coniferous forest. Tree Physiol. 9:173–184.

McDowell, W.H., and G.E. Likens. 1988. Origin, composition, and flux of dissolved organic carbon in the Hubbard Brook valley. Ecol. Monogr. 58:177–195.

McDowell, W.H., and T. Wood. 1984. Podzolization: Soil processes control dissolved organic carbon concentrations in stream water. Soil Sci. 137:23–32.

Moore, T.R., W. Desouza, and J.F. Koprivnjak. 1992. Controls on the sorption of dissolved organic–carbon by soils. Soil Sci. 154:120–129.

Morel, F.M.M. 1983. Principles of aquatic chemistry. John Wiley & Sons, New York.

Nikolaidis, N.P., H. Rajaram, J.L. Schnoor, and K.P. Georgakakos. 1988. A generalized soft water acidification model. Water Resour. Res. 24:1983–1996.

Nodvin, S.C., C.T. Driscoll, and G.E. Likens. 1986. Simple partitioning of anions and dissolved organic carbon in a forest soil. Soil Sci. 142:27–35.

Oades, J.M. 1988. The retention of organic matter in soils. Biogeochemistry 5:35–70.

Oliver, B.G., E.M. Thurman, and R.L. Malcolm. 1983. The contribution of humic substances to the acidity of colored natural waters. Geochim. Cosmochim. Acta 47:2031–2035.

Parton, W.J., J.M.O. Scurlock, D.S. Ojima, T.G. Gilmanov, R.J. Scholes, D.S. Schimel, T. Kirchner, J–C. Menaut, T. Seastedt, E. Garcia Moya, A Kamnalrut, and J.L. Kinyamario. 1993. Observations and modeling of biomass and soil organic matter dynamics for the grassland biome worldwide. Global Biochem. Cycles (In press.)

Parton, W.J., J.W.B. Stewart, and C.V. Cole. 1988. Dynamics of C, N, P and S in grassland soils: A model. Biogeochemistry 5:109–131.

Perdue, E.M. 1990. Modeling the acid–base chemistry of organic acids in laboratory experiments and fresh waters. p. 111–126. *In* E.M. Perdue and E.T. Gjessing (ed.) Organic acids in aquatic ecosystems. Dahlem Konferenzen. Berlin, Germany. Life Sci. Res. Rep. 48. John Wiley & Sons, New York.

Perdue, E.M., J.H. Rueter, and R.J. Parrish. 1984. A statistical model of proton binding by humus. Geochim. Cosmochim. Acta 48:1257–1263.

Press, W.H., B.P. Flannery, S.A. Teukolsky, and W.T. Vetterling. 1986. Numerical recipes. Cambridge Univ. Press, New York.

Qualls, R.G., and B.L. Haines. 1992. Biodegradability of dissolved organic matter in forest throughfall, soil solution, and stream water. Soil Sci. Soc. Am. J. 578–586.

Ross, D.S., R.J. Bartlett, and F.R. Magdoff. 1991. Exchangeable cations and the pH–independent distribution of cation exchange capacities in Spodosols of a forested watershed. p. 81–92. *In* R.J. Wright (ed.) Plant–soil interactions at low pH. Kluwer Acad. Publ., the Netherlands.

Santore, R.C., and C.T. Driscoll. 1995. The CHESS model for calculating chemical equilibria in soils and solutions. p. 357–376. *In* R. Loeppert et al. (ed.) Chemical equilibrium and reaction models. SSSA, Madison, WI.

Schecher, W.D., and C.T. Driscoll. 1988. An evaluation of equilibrium calculations within acidification models: The effect of uncertainty in measured chemical components. Water Resour. Res. 24:533–540.

Schecher, W.D., and C.T. Driscoll. 1995. ALCHEMI: A chemical equilibrium model to assess the acid–base chemistry and speciation of aluminum in dilute solutions. p. 325–356. *In* R. Loeppert et al. (ed.) Chemical equilibrium and reaction models. SSSA and ASA spec. Publ. 42., Madison, WI.

Schlesinger, W.H. 1990. Evidence from chronosequence studies for a low carbon–storage potential of soils. Nature (London) 348:232–234.

Schlesinger, W.H. 1991. Biogeochemistry: An analysis of global change. Acad. Press, San Diego, CA.

Schnoor, J.L. 1990. Kinetics of chemical weathering: A comparison of laboratory and field weathering rates. p. 475–504. *In* W. Stumm (ed.) Aquatic chemical kinetics. John Wiley & Sons, New York.

Sparks, D.L. 1988. Kinetics of soil chemical processes. Acad. Press, San Diego, CA.

Sposito, G. 1984. The surface chemistry of soils. Oxford Univ. Press, New York.

Stevenson, F.J. 1982. Humus chemistry. John Wiley & Sons, New York.

Stumm, W., and E. Wieland. 1990. Dissolution of oxide and silicate minerals: rates depend on surface speciation. p. 367–400. *In* W. Stumm (ed.) Aquatic chemical kinetics. John Wiley & Sons, New York.

Sullivan, T.J., R.S. Turner, D.F. Charles, B.F. Cumming, J.P. Smol, C.L. Schofield, C.T. Driscoll, B.J. Cosby, H.J. Birks, A.J. Uutala, J.C. Kingston, S.S. Dixit, J.A. Bernert, P.F. Ryan, and D.R. Marmorek. 1992. Use of historical assessment for evaluation of process–based model projections of future environmental change: Lake acidification in the Adirondack Mountains, New York, USA. Environ. Pollut. 77:253–262.

Sverdrup, H.U. 1990. The kinetics of base cation release due to chemical weathering. Lund Univ. Press, Lund, Sweden.

Tipping, E., and C. Woof. 1991. The distribution of humic substances between the solid and aqueous phases of acid organic soils; a description based on humic heterogeneity and charge–dependent sorption equilibria. J. Soil Sci. 42:437–448.

Ugolini, F.C., R. Minden, H. Dawson, and J. Zachara. 1977. An example of soil processes in the Abies amabilis zone of Central Cascades, Washington. Soil Sci. 124:291–302.

Vance, G.F., and M.B. David. 1989. Effect of acid treatment on dissolved organic carbon retention by a spodic horizon. Soil Sci. Soc. Am. J. 53:1242–1247.

Whittaker, R.H., F.H. Bormann, G.E. Likens, and T.G. Siccama. 1974. The Hubbard Brook Ecosystem Study: Forest biomass and production. Ecol. Monogr. 44:233–254.

Zepp, R.G., and N.L. Wolfe. 1987. Abiotic transformations of organic chemicals at the particle–water interface. p. 423–455. *In* W. Stumm (ed.) Aquatic surface chemistry: Chemical processes at the particle–water interface. John Wiley & Sons, New York.

14 Role of Carbon in the Cycling of Other Nutrients in Forested Ecosystems

Dale W. Johnson

Desert Research Institute,
University of Nevada
Reno, Nevada

With current concerns over increasing atmospheric CO_2 levels and consequent effects upon climate change, the topic of C cycling in forests is very timely. In particular, there is great interest in the effects of land management practices on terrestrial C balances. A cursory review of the estimates of C cycling on a global scale reveal why this is so: C fixation by terrestrial photosynthesis (120×10^{15} g) and C inputs to the soil via detritus (60×10^{15} g) are an order of magnitude greater than either fossil fuel emissions (approximately 6×10^{15} g) or atmospheric CO_2 increase (approximately 3.4×10^{15} g) (Post et al., 1990). Of particular interest is the "missing sink" of C, which arises from the difference in CO_2 release by fossil fuels and the annual CO_2 increase in the atmosphere (Lugo, 1992). There is speculation and some evidence that temperate forests of the northern hemisphere are the site of this "missing sink" (Tans et al., 1990; Kauppi et al., 1992; Sedjo, 1992).

Forest soil scientists have long been concerned with the cycling of C and its interactions with other nutrients in forest ecosystems because of the basic issues of fertility and productivity. Conceptual models of nutrient cycling in forests were borne from an effort to understand how nutrients affect productivity (i.e., C cycling) following fertilization (Cole & Gessel, 1965; Curlin, 1970; Cole et al., 1968; Miller et al., 1979), and how biomass production and harvesting affect nutrient budgets (Rennie, 1955; Ovington, 1962; Wheetman & Webber, 1972). Thus, a review of the role of C in the cycling of other nutrients can be as encompassing as a review of forest nutrient cycling itself, a task well beyond the scope of this paper. Instead, this review will be concerned with the role of C in nutrient cycling processes, drawing examples from individual studies.

In reviewing the role of C in the cycling of other nutrients, it is convenient to refer to Switzer and Nelson's (1972) definitions of cycles within forest nutrient cycles: (i) the geochemical cycle, which encompasses inputs and outputs from the ecosystem; (ii) the biogeochemical cycle, which encompasses soil–plant relationships; and (iii) the biochemical cycle, which encompasses the internal

---

Copyright © 1995 Soil Science Society of America, 677 S. Segoe Rd., Madison, WI 53711, USA.
Carbon Forms and Functions in Forest Soils.

translocation of nutrients within the vegetation. This paper will deal with the role of C in cycling nutrients through geochemical and biogeochemical cycles; although C clearly plays a role in biochemical cycling as well, a discussion of those processes is beyond the scope of this paper and is left to a more in-depth physiological treatment of the subject (e.g., Bugbee & Monje, 1992; Zelich, 1992; Grodzinski, 1992).

At first glance, it would seem that C plays only a minor role in the geochemical cycles, which are usually dominated by hydrologic processes. However, as will be shown, both organic and inorganic forms of C play a major role in the geochemical transport of some nutrients from forest ecosystems via leaching. Organic C plays a major role in the cycling of nutrients within the biogeochemical cycle (uptake, litterfall, root turnover), yet at the same time, nutrients can play a major role in the biogeochemical cycling of C as well. The rate at which nutrients cycle within biogeochemical cycles is dependent on primary productivity and the concentrations of nutrients in plant tissues. Both of these factors are in turn dependent on soil nutrient status, age and species composition of the site in question.

Carbon can have significant indirect effects on primary productivity and the cycling of other nutrients via soils. Soil organic matter has a major influence on site productivity because of its effects on the physical (bulk density, water holding capacity), biological (microbial populations) and chemical (cation exchange capacity) properties of soils.

BIOGEOCHEMICAL CYCLING OF CARBON AND NUTRIENTS

Carbon plays a major role in the biogeochemical cycling of nutrients in the solid phase, that is, by uptake, litterfall and root turnover. The fluxes and accumulation of nutrients associated with biomass in biogeochemical cycles are a function of biomass and tissue nutrient concentrations. Biomass is a function of site productivity, age, and species. Mean annual temperature is an important factor in determining biomass and biomass fluxes (Olson, 1963; Cole & Rapp, 1981), and it will be used as an index for comparative purposes in this chapter. Live tissue nutrient concentrations vary with species, age, growth rate, and site nutrition (Marion, 1979; Cole & Rapp, 1981), and detrital tissue concentrations vary with these factors as well as translocation prior to senescence and the stage of decay. Thus, the linkage between C and nutrients in biogeochemical cycles is strong, but not stoichiometrically exact. In the following sections, the degree of interdependence of organic matter (live and dead) and nutrients in forest ecosystems is explored, using data from the Integrated Forest Study (IFS) (Johnson & Lindberg, 1991) for the purpose of illustration. Table 14–1 gives the location and a brief description of the IFS sites.

Vegetation Uptake and Accumulation

By way of background for a discussion of vegetation uptake and accumulation, it is convenient to refer to the conceptual model of C/nutrient interactions posed by Shaver et al. (1992) for Arctic ecosystems of Alaska. The model is outlined in Fig. 14–1, using C/N ratios in both the Arctic and temperate forest

Table 14–1. A summary of geographic and climatic information for the field research sites participating in the IFS (from Johnson & Lindberg, 1991).

Site code	Forest type	Location	Latitude, longitude	Elevation	Mean annual precipitation†	
				m	cm	% as snow
CH‡	Southern hardwoods	Coweeta Hydrologic Laboratory, NC	35°03'N 83°27'W	725	138	<3
CP	White pine (*Pinus monticola* Douglas ex D. Don)	Coweeta Hydrologic Laboratory, NC	35°03'N 83°27'W	725	144	<3
DF	Douglas–fir	Thompson Forest, WA	47°23'N 121°56'W	220	114	<5
DL	Loblolly pine	Duke Forest, NC	36°12'N 79°17'W	213	113	<2
FL‡	Fir (*Abies* spp.) hemlock (*Tsuga* spp.)	Findley Lake, WA	47°04'N 121°25'W	1130	270	80
FS	Slash pine (*Pinus elliottii* Engelm.)	Gainesville, FL	29°44'N 82°30'W	38	112	0
GL	Loblolly pine	B. F. Grant Forest, GA	33°22'N 83°27'W	145	87	<1
HF	Northern hardwoods	Huntington Forest, NY	43°59'N 74°14'W	530	97	47
LP	Loblolly pine	Oak Ridge, TN	35°54'N 84°20'W	300	114	3
MS	Red spruce (*Picea rabens Sargi*)	Howland, ME	45°10'N 68°40'W	65	79	38

(continued on next page)

Table 14–1 (continued).

Site code	Forest type	Location	Latitude, longitude	Elevation m	Mean annual precipitation† cm	% as snow
NS	Norway spruce [*Picea abies* (L.) Karsten]	Nordmoen, Norway	60°16'N 11°06'E	200	107	29
RA	Red alder (*Alnus rubra* Bong.)	Thompson Forest, WA	47°23'N 121°56'W	220	114	<5
SB‡	American Beech (*Fagus grandifolia* Ehrh.)	Great Smokey Mountains National Park	35°37'N 83°26'W	1600	151	10
SS‡	Red spruce	Great Smokey Mountains National Park	35°34'N 83°29'W	1800	151	10
ST	Red spruce	Great Smokey Mountains National Park	35°34'N 83°28'W	1740	203	10
TL	Northern hardwoods	Turkey Lakes, Ontario	47°03'N 84°25'W	350	121	27
WF	Red spruce, balsam fir [*Abies balsamea* (L.) Miller] white birch (*Betula papyrifera* Marshall)	Whiteface Mt., NY	44°22'N 73°54'W	1000	115	19

† During the study period.
‡ Bulk deposition only.

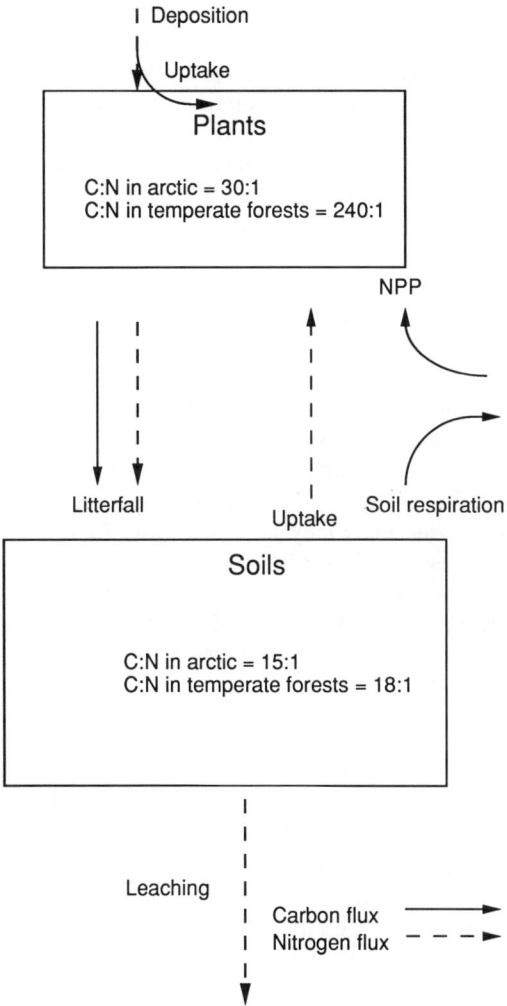

Fig. 14–1. A conceptual model of C/N interactions in terrestrial ecosystems [modified from Shaver et al. (1992), adding data from the IFS sites (Johnson & Lindberg, 1991)].

ecosystems (the IFS data) by way of illustration. The model has only two components—plants and soils. Shaver et al. (1992) reported typical C/N ratios of 30:1 for plants and 15:1 for soils in Arctic ecosystems, and speculated that ratios of 200:1 were more typical for temperate forests. The IFS data support this speculation (average vegetation C/N ratio is 237) showing that vegetation components of these two ecosystem types are very different with regard to N concentrations. As noted by Shaver et al. (1992), however, the basic concepts are the same even though the numbers differ substantially. It is interesting to note that soil C/N ratios do not differ substantially.

In the model, C enters the plants via net primary production, or photosynthesis minus respiration, enters the soil via litterfall (foliar and root), and leaves via soil heterotrophic respiration. Shaver et al. (1992) show N entering the soil directly via atmospheric deposition, fixation, and lateral groundwater transport. This model is not entirely appropriate for the temperate forests of the IFS in that it does not allow for foliar N uptake, which is known to occur in several of the IFS sites (Lovett, 1991). Thus, the model has been modified somewhat to show N entering the ecosystem via both plants and soils. This modification, while significant for some purposes, does not substantially alter the nature of the discussion to follow.

Shaver et al. (1992) note three major constraints on the C budget depicted by this model: (i) limits on the variation in C/N ratios in both pools and fluxes, (ii) controls on the distribution of N between plants and soils, and (iii) controls on N inputs and outputs. The first two constraints place limits on the size of C pools and fluxes for given amounts of N in these pools and fluxes, and the third constraint determines the amount of N in the ecosystem. The authors view N budgets and C/N ratios as the key variables through which all other ecosystem perturbations act. For instance, they note that these Arctic ecosystems are not responsive to either elevated CO_2 or light on a long-term basis, a consequence of the fact that both of these factors affect only photosynthesis and not the N (or P) cycles. In short, ecosystem productivity and C cycling can only be significantly affected on a long-term basis through actions on nutrient cycles.

The model of Shaver et al. (1992) is a convenient starting point for a discussion of nutrient distribution between vegetation and soils; the question now becomes, what are the error bars associated with the numbers in Fig. 14–1, both between ecosystems and between nutrients? The magnitudes of the variations in ecosystem biomass and nutrient contents among the IFS sites (Johnson & Lindberg, 1991) are shown in Fig. 14–2 to 14–6. In Fig. 14–2, soil organic matter and N are included because most soil N is incorporated into organic matter. In Fig. 14–3 to 14–6, soils are omitted because of the importance of inorganic pools for P, S, K, Ca, and Mg and the lack of information as to soil organic pools of these nutrients. In each case, the sites are ordered according to organic matter content. Table 14–2 gives the correlation coefficients for regressions of biomass vs. nutrient content for this data set.

It can easily be seen that the patterns in nutrient content deviate significantly from those in biomass in several respects. One particularly striking deviation between biomass and nutrient content is in vegetation S content (Table 14–2). The lack of correlation between S content and organic matter can be attributed to differences in S deposition. In particular, the Findley Lake (FL) site has low vegetation S content relative to the other sites even though its biomass is over twice as great. This difference reflects the influence of air pollution in the form of S deposition; the FL site is relatively pristine, with the lowest precipitation sulfate concentrations and the lowest total S deposition rates of the IFS site, despite its high precipitation rate (Lindberg, 1991).

As has been noted many times before (Duvigneaud & Denaeyer-DeSmet, 1970; Cole & Rapp, 1981, Shaver et al., 1992), the soil constitutes a much greater

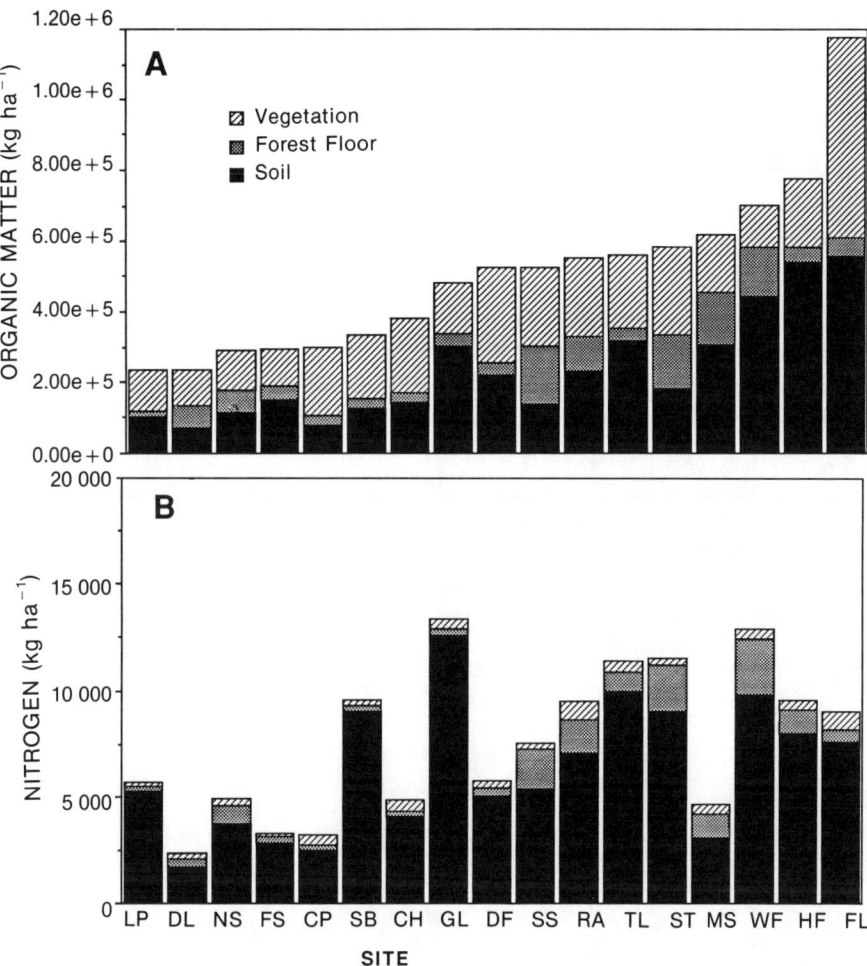

Fig. 14–2. Ecosystem organic matter (A) and N (B) contents in the IFS study (data from Johnson & Lindberg, 1991; see Table 14–1 for legend).

reservoir of nutrients than organic matter. Among the IFS sites, the soil contains 44% of the total ecosystem organic matter, on average, but 81% of the N (Fig.14–2). The soil contains even greater percentages of other nutrients, a large fraction of which are in inorganic form. Thus, the soil exerts a disproportionately large effect on ecosystem nutrient content.

Neither soil nor ecosystem N content corresponds well to ecosystem organic matter content (Fig. 14–2). There is a significant degree of scatter in the plot of soil organic matter vs. N content (Fig. 14–3A), and the correlation coefficient is not particularly high (Table 14–2), indicating a substantial amount of variation in N concentration. Anderson (1992) notes that although soils in cooler climates

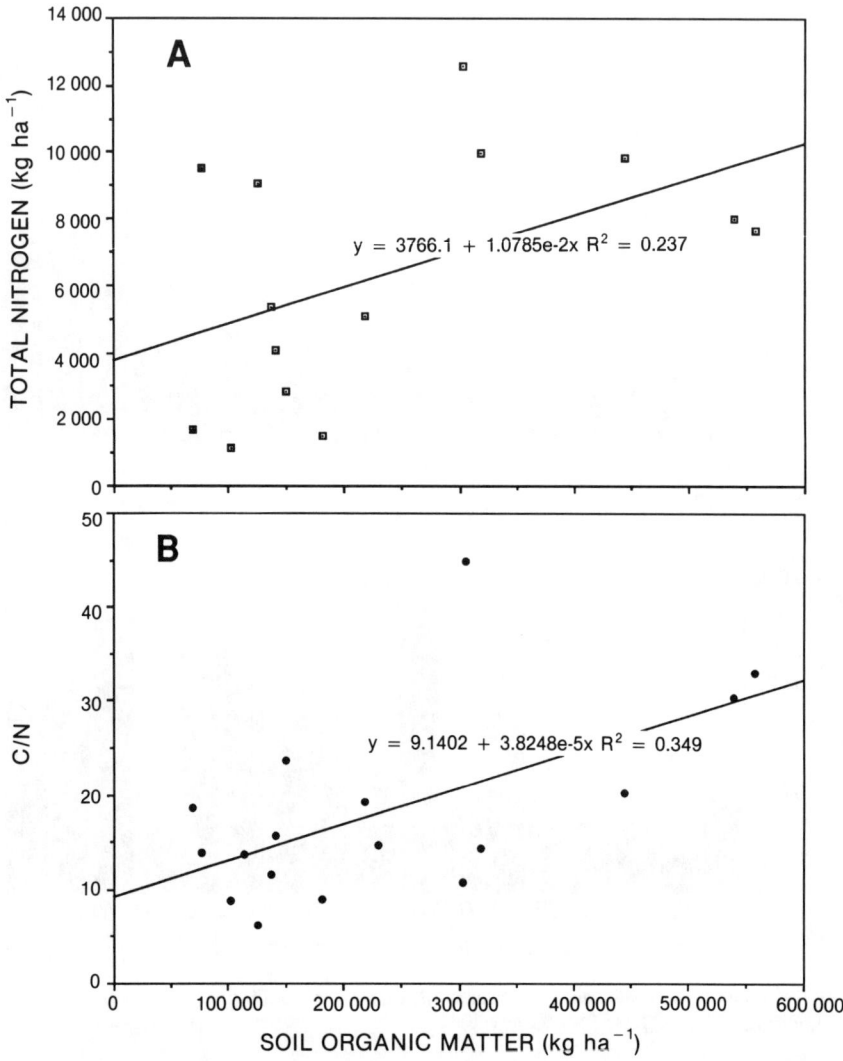

Fig. 14–3. Soil N vs. soil organic matter (*A*) and soil C/N ratio vs. soil organic matter (*B*) in the IFS study (data from Johnson & Lindberg, 1991; see Table 14–1 for legend).

generally have greater soil organic matter than soils in warmer climates, soil organic matter in cooler climates is generally in a less advanced stage of decomposition than soil organic matter in warmer climates. The same principles may apply to wetter and drier climates, but neither this relationship nor the interactions between temperature and moisture were explicitly discussed by Anderson (1992) with regard to effects on the nature of soil organic matter. In any event, one might expect a general upward trend in C/N ratio and soil organic matter if Anderson's (1992) hypothesis holds true. There was a statistically significant but

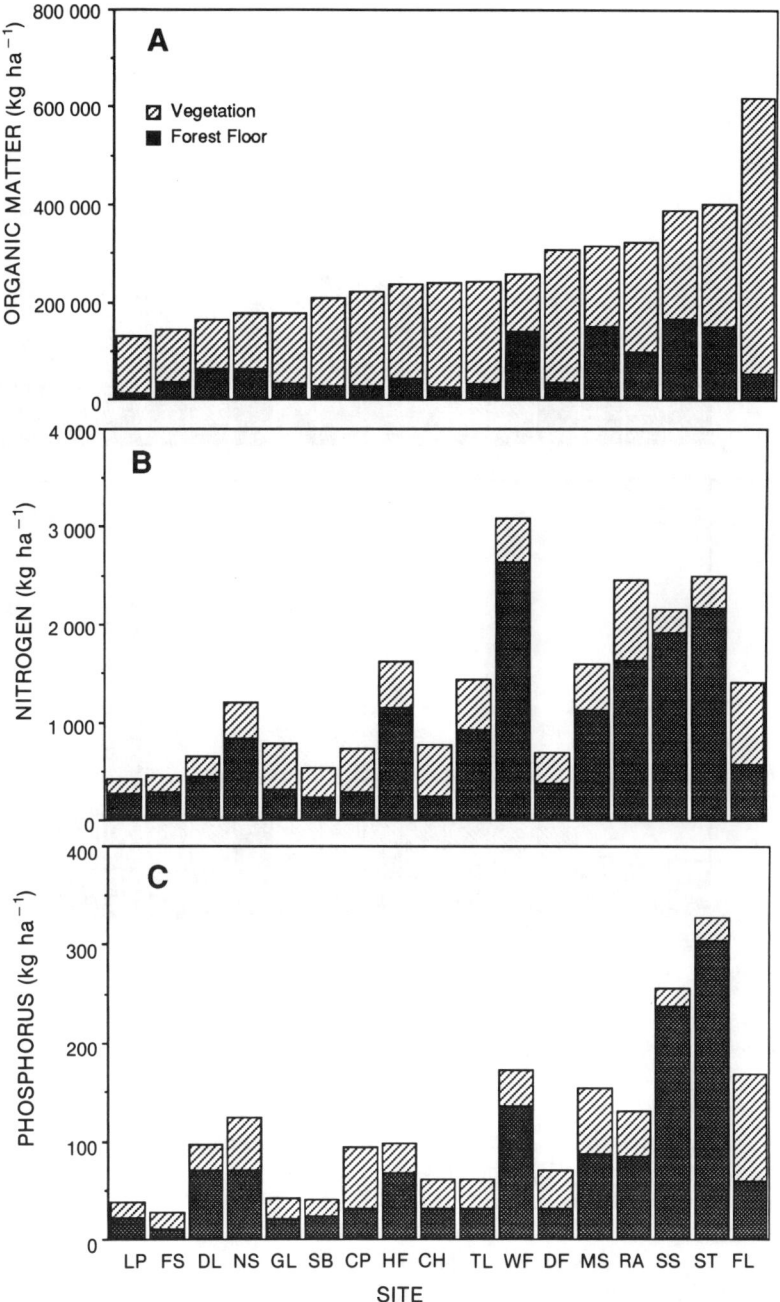

Fig. 14-4. Vegetation and forest floor contents of organic matter, N, and P (data from Johnson & Lindberg 1991; see Table 14-1 for legend).

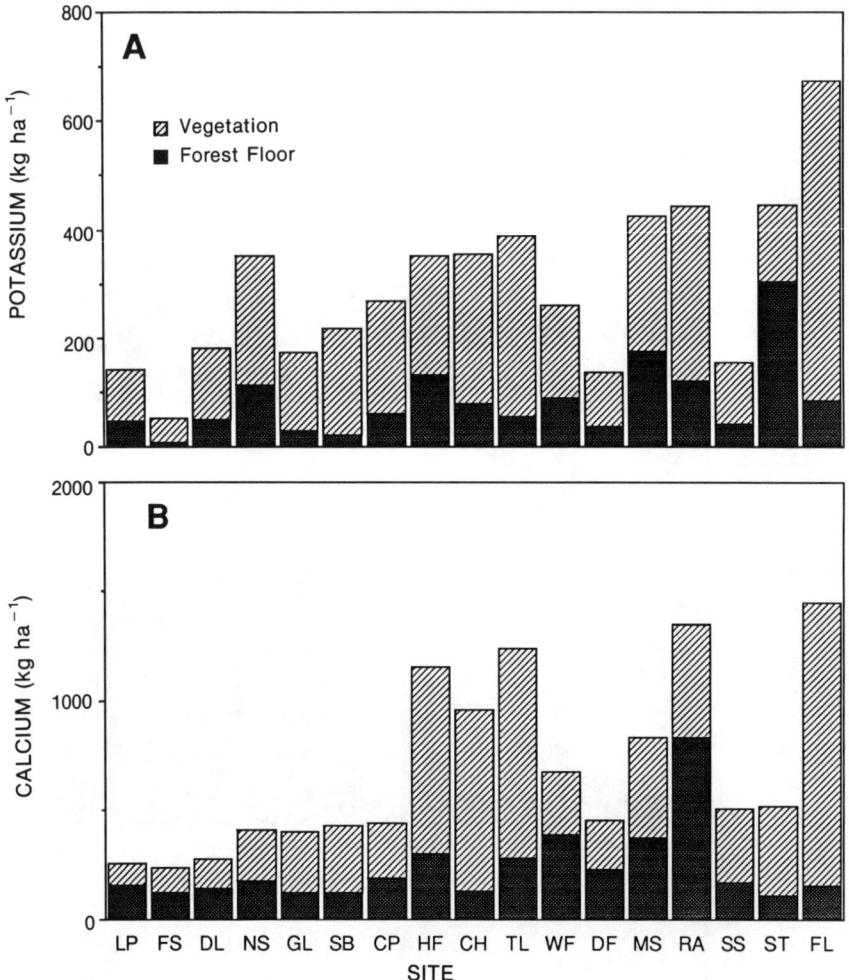

Fig. 14–5. Vegetation and forest floor contents of K and C (data from Johnson & Lindberg, 1991; see Table 14–1 for legend).

weak positive correlation between soil C/N ratio and organic matter content in the IFS data (Fig. 14–3B). The only other soil chemical parameter that significantly correlated to soil organic matter is total Ca (Table 14–2). This may relate to the effects of Ca on soil organic matter stabilization, as noted above (Oades, 1988). However, it should be noted that there is no relationship between exchangeable Ca and soil organic matter ($r^2 = 0.071$) among the IFS sites.

When only the nonsoil reservoirs of organic matter and nutrients are considered, it is clear that the forest floor constitutes a disproportionately important reservoir of both N and P, and, to a lesser extent, S, Mg, and Ca (Fig. 14–4 to 14–6). This phenomenon is due to the concentration of most nutrients in the

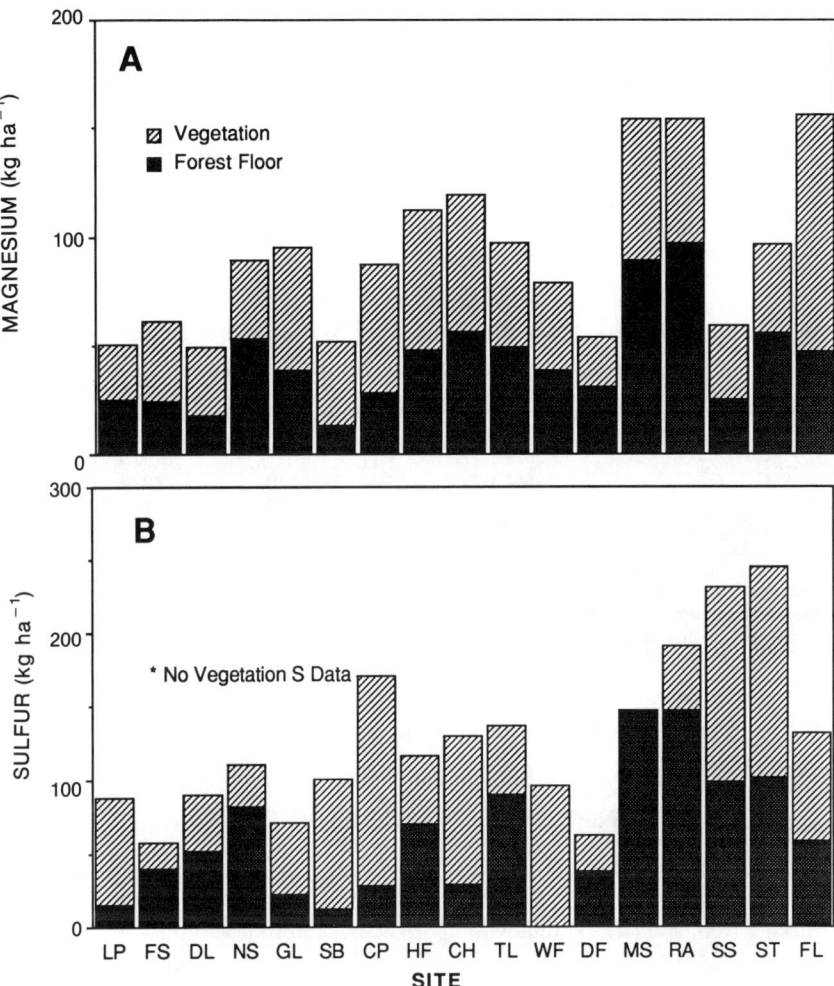

Fig. 14–6. Vegetation and forest floor contents of Mg and S (data from Johnson & Lindberg, 1991; see Table 14–1 for legend).

forest floor during decomposition, as indicated by the C/nutrient ratios (Table 14–3). The concentration effect is greatest for N and P (6-fold), less for S (4-fold), Ca, and Mg (approximately 2-fold) and zero for K. The greater concentration effects for N and P reflect greater conservation of these nutrients relative to others, and the lack of a concentration effect for K also reflects the lesser importance of organic matter and greater importance of hydrology in the K cycle as compared to the other nutrient cycles. In the case of N, concentrations increase an average of twofold further from the forest floor to the soil, reflecting further conservation of N in the soil. (C/nutrient ratios in soil are also given for P, S, Ca, K, and Mg for the sake of completeness, but are not particularly meaningful

Table 14–2. Correlation coefficients (r^2) of nutrient contents with organic matter in aboveground vegetation, forest floor, soils, and litterfall of the IFS sites.

	Vegetation	Forest floor	Soil	Litterfall
N	0.54	0.74	0.29	0.57
S	0.03	0.47	0.01	0.62
P	0.47	0.84	0.01	0.81
Ca	0.54	0.09	0.82	0.46
K	0.57	0.35	0.00	0.73
Mg	0.51	0.15	0.00	0.80

Table 14–3. Carbon nutrient ratios in vegetation, forest floor, and soils of the IFS sites (data from Johnson & Lindberg, 1991).

Carbon/nutrient	Vegetation	Forest floor	Soil
C/N	237 ± 24	40 ± 4	18 ± 2
C/P	2828 ± 322	511 ± 72	(45 ± 9)†
C/S	1681 ± 290	484 ± 50	(97 ± 17)†
C/K	517 ± 73	524 ± 124	(6 ± 3)†
C/Ca	269 ± 33	161 ± 37	(12 ± 5)†
C/Mg	1968 ± 247	821 ± 171	(5 ± 1)†

†Carbon to nutrient ratios for P, S, K, Ca, and Mg are presented for the sake of completeness, but may not be meaningful in that an unknown and probably large proportion of the latter nutrients are contained in minerals and other inorganic forms.

because of the fact that soil mineral pools often constitute the largest reservoirs of these nutrients.)

Nutrient Return to the Forest Floor

Two aspects of nutrient return to the forest floor must be considered when evaluating the role of C in nutrient cycling: (i) variations in litter nutrient concentration, and (ii) the relative importance of litterfall vs. foliar leaching and throughfall return. With regard to the former, deciduous species generally have higher litter concentrations, as is the case with live tissue (Cole & Rapp, 1981). Translocation can cause significant variations in the concentrations of most nutrients in litter (the exception is Ca, which is not translocated from senescing leaves). Turner (1977) showed that N translocation in Douglas fir is strongly affected by soil N status. He reduced soil N availability by adding carbohydrate to the forest floor on the one hand and increased soil N availability on the other hand by fertilization with urea. With increasing N stress in the carbohydrate plots, translocation increased, causing decreased litter N concentrations and increased foliage loss to senescence and litterfall. The reverse occurred when N stress was decreased by fertilization. Nambiar and Fife (1991) found that N concentrations in both senesced foliage and roots was proportional to live tissue concentrations in radiata pine (*Pinus radiata*), indicating that tree N status was indeed reflected in litter quality. They found evidence of significant translocation in foliar tissue (senesced tissues had lower concentrations than live tissues), but not in roots.

In most forest ecosystems, litterfall (and therefore organic C flux) is the most important medium for the transport of N, P, Ca, and Mg to the forest floor (Duvigneaud & Denaeyer-DeSmet, 1970; Cole et al., 1968; Cole & Rapp, 1981). Potassium and S are exceptional—in many cases, throughfall and/or foliar leaching of these nutrients exceed transport via litterfall. In the case of K, this is a natural consequence of the fact that K is primarily in inorganic forms in plants and that the K cycle is dominated by hydrologic processes. In the case of S, atmospheric deposition may exceed litterfall in polluted areas of the world (Johnson, 1984).

The IFS data sets conform to these general observations in most cases, but there were some significant exceptions (Fig. 14–7 to 14–10). In red spruce (*Picea rubens* Sorg.) forests of the Great Smoky Mountains (Sites SS and ST), foliar leaching plus deposition exceeded litterfall for N, Mg and Ca as well as K and S. Thus, the return of all but P to the forest floor was dominated by hydrologic processes at these sites. (Foliar leaching is not shown for N or S because there was apparent foliar uptake of these nutrients in several sites.) The importance of hydrologic as opposed to biomass fluxes at these sites was due to the high rainfall rates, low productivity, and high pollutant inputs.

Decomposition

During decomposition, there is a major change in the nature of nutrient transfers from solid (litter) to liquid (soil solution) phase. This phase change allows nutrients returned to the forest floor to become available once again for plant and microbial uptake, and thus decomposition exerts a major control over nutrient availability and productivity in forest ecosystems. A full review of decomposition and the factors affecting it are beyond the scope of this paper. The reader is referred to Paul and Clark (1989) for a comprehensive treatment. In this section, some of the major factors affecting decomposition, forest floor, and soil organic matter accumulation are discussed, and the connections between organic matter and nutrient contents are explored using the IFS data set.

Olson's (1963) classic paper provides a useful background for a discussion of decomposition. He writes,

$$\frac{dX}{dt} = L - kX \qquad [1]$$

where X = quantity of litter, t = time, L = litterfall, and k = decomposition constant (typically determined from litterbag studies).

At steady-state

$$\frac{dX}{dt} = 0 \text{ (by definition) and } L = -kX \qquad [2]$$

Using Eq. [2] in combination with estimates of litterfall and decomposition constants (k's), it is a simple matter to calculate steady-state detrital pools. Given knowledge of the effect of climate, species, or nutrient status on k, it is a simple

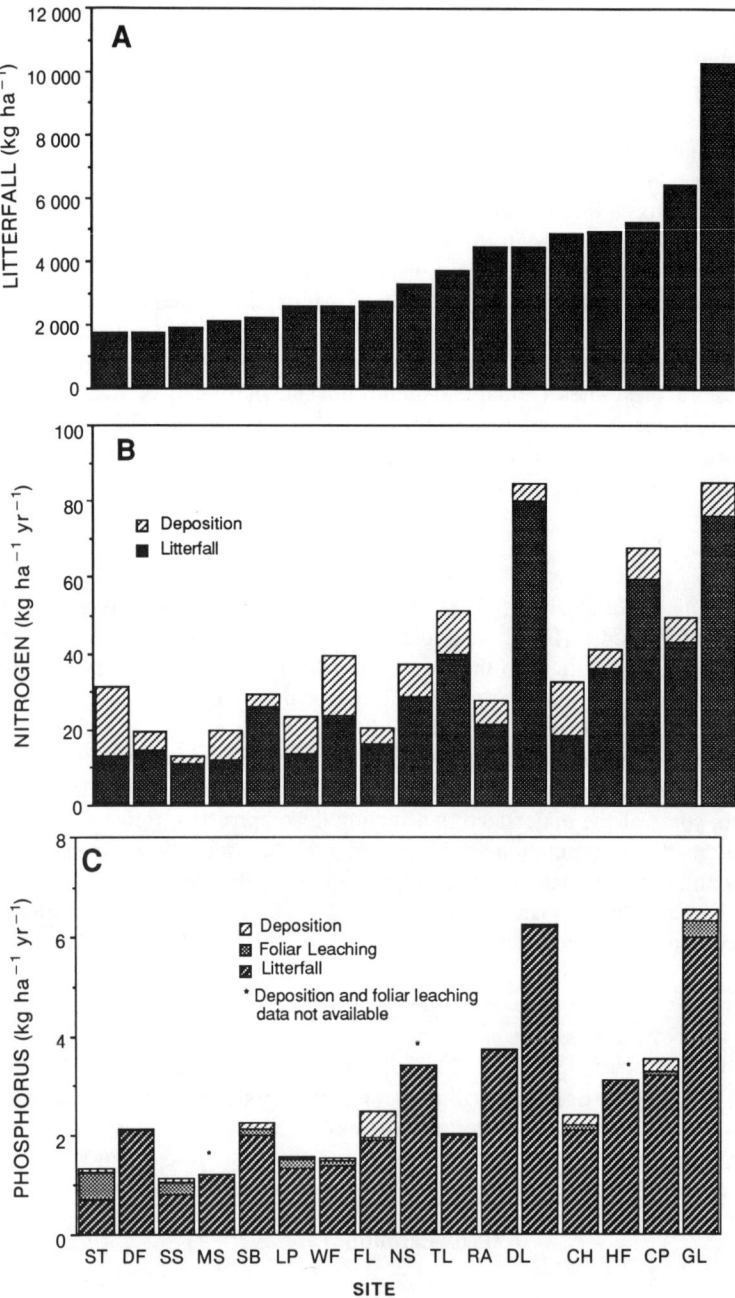

Fig. 14–7. Litterfall, atmospheric deposition and foliar leaching fluxes of organic matter, N and P in the IFS sites (data from Johnson & Lindberg, 1991). Foliar leaching data are not given in cases where net foliar uptake occurred (see Table 14–1 for legend).

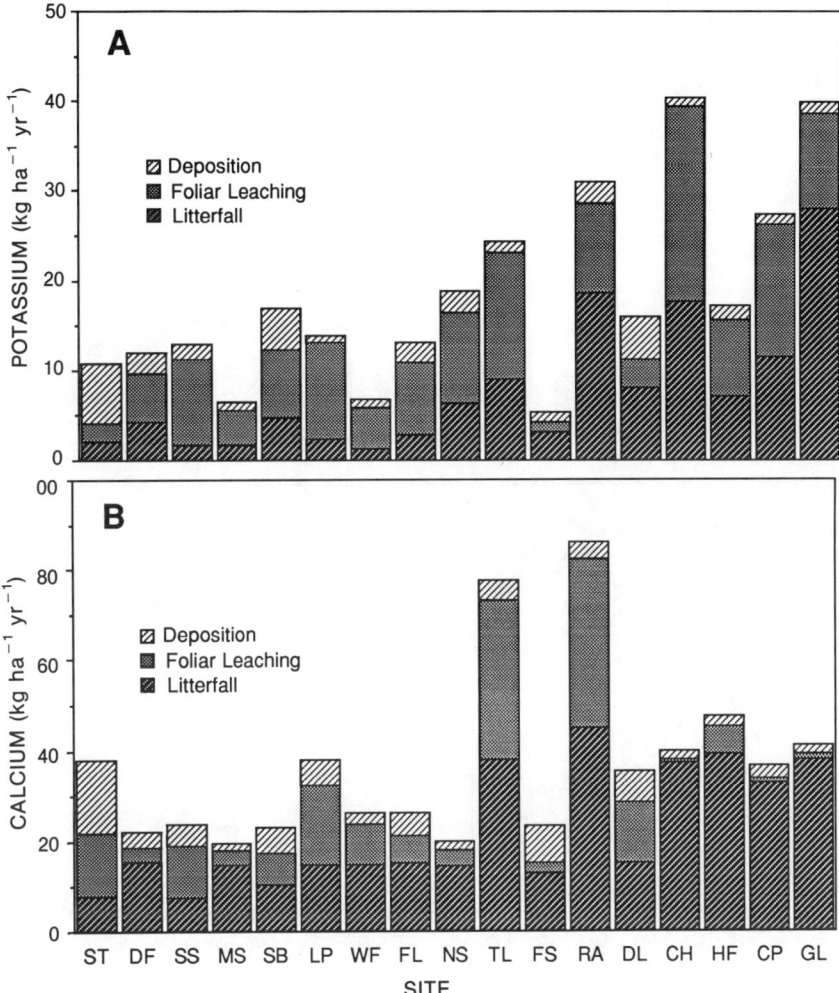

Fig. 14–8. Litterfall, atmospheric deposition and foliar leaching fluxes of K and C in the IFS sites (data from Johnson & Lindberg, 1991; see Table 14–1 for legend).

matter to calculate the effects of these variables on steady-state detrital pools. The calculation for nonsteady state estimates is only slightly more complicated and requires no additional information

Rearranging and integrating Eq. [1] yields

$$\ln (L/k - X) = -kt - \text{constant} \qquad [3]$$

Assuming $X = 0$ at $t = 0$, the constant equals $-\ln (L/k)$, which in turn allows the expression of X as function of L, k and t

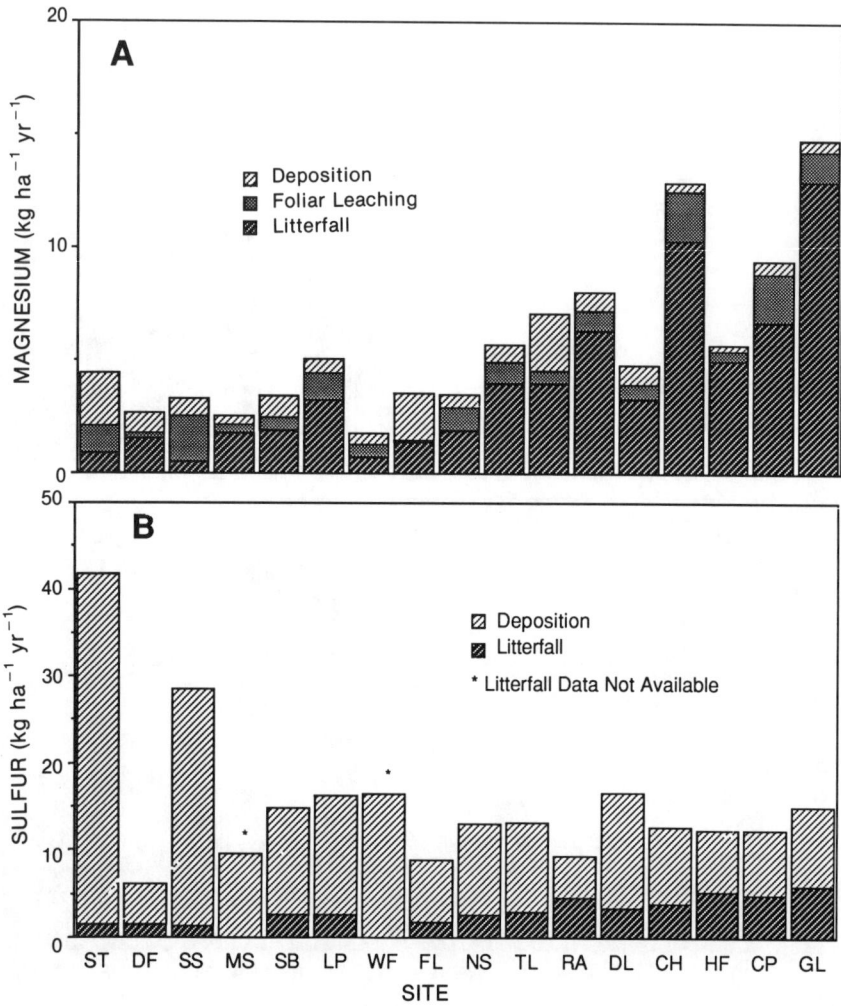

Fig. 14–9. Litterfall, atmospheric deposition and foliar leaching fluxes of Mg and S in the IFS sites (data from Johnson & Lindberg, 1991). Foliar leaching data are not given in cases where net foliar uptake occurred (see Table 14–1 for legend).

$$X = (L/k)(1 - e^{-kt}) \qquad [4]$$

Olson's (1963) simple decay model is useful in obtaining a first cut at predicting times to quasi steady-state and final forest floor biomass, especially when combined with the added information on lignin, tannins, and N effects on decomposition rates (k) (Meetemeyer, 1978; Melillo et al., 1982; Gallardo & Merino, 1992). However, there are questions as to whether it can be applied to mineral soil C responses to climate change. First, from a purely technical perspective, litterbag-derived k values overestimate C losses because weight loss represents

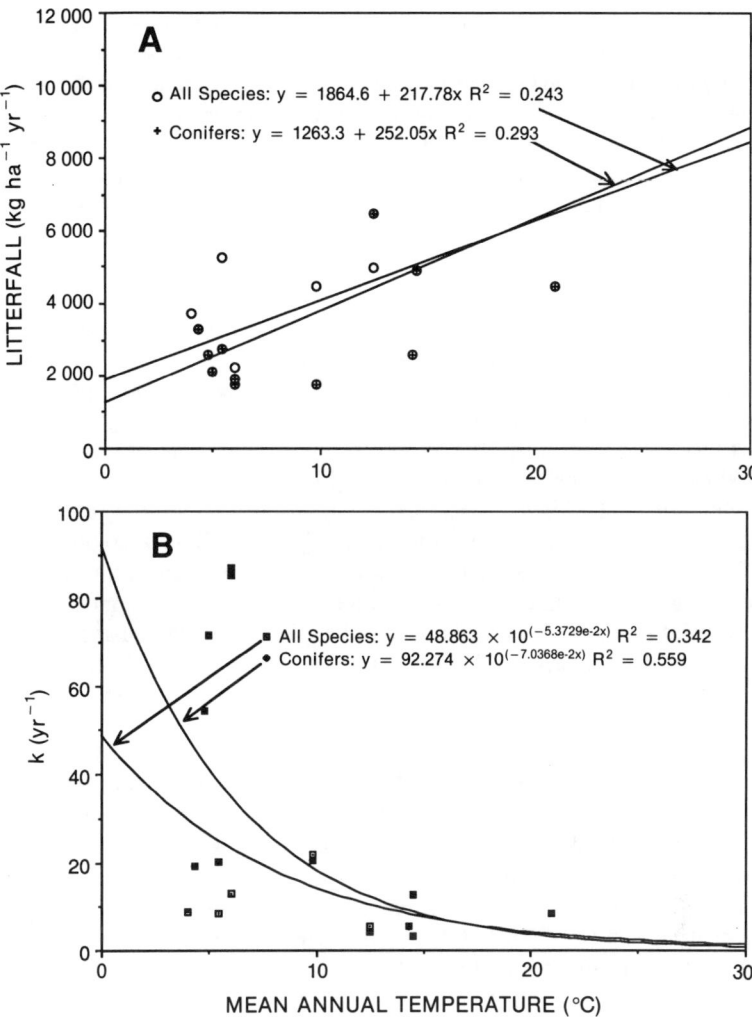

Fig. 14–10. Litterfall vs. mean annual temperature (A) and forest floor turnover times (forest floor content/litterfall) vs. mean annual temperature (B) in the IFS sites (data from Johnson & Lindberg, 1991; see Table 14–1 for legend).

not only organic matter decomposed to CO_2, but also the conversion of coarse organic matter to fine and soluble organic matter, which, in turn becomes soil organic matter. These nongaseous losses of organic matter from the forest floor may be negligible relative to CO_2 evolution in terms of overall litter decay, but not with regard to the long-term accumulation of soil organic matter.

Second, soil organic matter accumulation is strongly affected by binding with clays, polyvalent cations and inorganic reactions with N as well as temperature and moisture. Intercalation of organic matter in smectite clays and carbohydrate adsorption onto clay surfaces can result in substantial soil organic matter stabilization (Oades, 1988; Anderson, 1992). Oades (1988) observes that the often-noted positive correlation between clay and organic matter content in soils may not represent a true cause–effect relationship because clay content is usually highly correlated with other factors that influence soil organic matter accumulation such as polyvalent cations. Polyvalent cations stabilize soil organic matter by a number of mechanisms, including electrostatic cation bridging of organic colloids (Ca and Mg in alkaline and circumneutral soils; Al in acidic soils) and specific adsorption onto Fe and Al hydrous oxide surfaces (acidic soils) (Oades, 1988).

In addition to polyvalent cations, the effects of nonbiological reactions between N and organic matter on soil organic matter stabilization deserve mention here. Nonbiological condensation reactions of phenols with ammonium are important in the production of humus (Mortland & Wolcott, 1965; Paul & Clark, 1989; Johnson, 1992b). These reactions are enhanced by high pH (because NH_3 is the reactive form of N) and high NH_3 concentrations. The reaction can occur slowly at pH's below neutrality, also (Mortland & Wolcott, 1965), but the importance of nonbiological reactions in forest soil or litter N retention under ambient pH conditions is unclear. Nommik (1970) found little nonbiological NH_4^+ retention in acid Norway spruce [P. abies (L.) Karsten] humus unless pH was raised to neutrality or higher. On the other hand, Axelsson and Berg (1988) found that chemical fixation of ammonia occurred at pH 4 on decomposing Scots pine (*Pinus sylvestris*) litter, and Schimel and Firestone (1989) found that nonbiological reactions accounted for 20% of the N retained in an acid (pH 4.3–4.5) forest soil.

Despite the importance of nonclimatic factors in the accumulation of soil organic matter, there is some empirical evidence that climate does have a significant influence. Post et al. (1982) found that soil organic matter increased with increasing precipitation, decreasing temperature (at constant precipitation), and decreasing evapotranspiration to precipitation ratio. High C, N, and C/N ratios observed in wet tundra regions were interpreted as resulting from slow decomposition. Anderson (1992) noted that detrital turnover becomes more superficial in the profile and divorced from soil organic matter as soils become colder. Anderson (1992) also observed that litter quality (i.e., N, lignin, or other factors that affect the rate at which litter decomposes in a given microclimatic condition) has a greater influence on soil organic matter in cold climates, where there is little stabilization of soil organic matter (by reactions with clays and polyvalent cations), and soil organic matter is generally younger. Presumably the reverse is true as soils warm.

The IFS data illustrate a general trend of increasing litterfall and decomposition constants and decreasing forest floor and soil organic matter accumulation with increasing mean annual temperature, as Olson (1963) and Post et al. (1982) would predict (Fig. 14–11). The correlation coefficients are not high; however, there is considerable scatter due to differences in species and site conditions. Consideration of coniferous forests alone does not alter the general patterns in either litterfall (total or foliage) or soil organic matter substantially, but it does alter the patterns in k values and the forest floor accumulation (Fig. 14–11).

With regard to the decomposition constants (k), there also may be some error due to the assumptions involved in the calculations. In the absence of litterbag

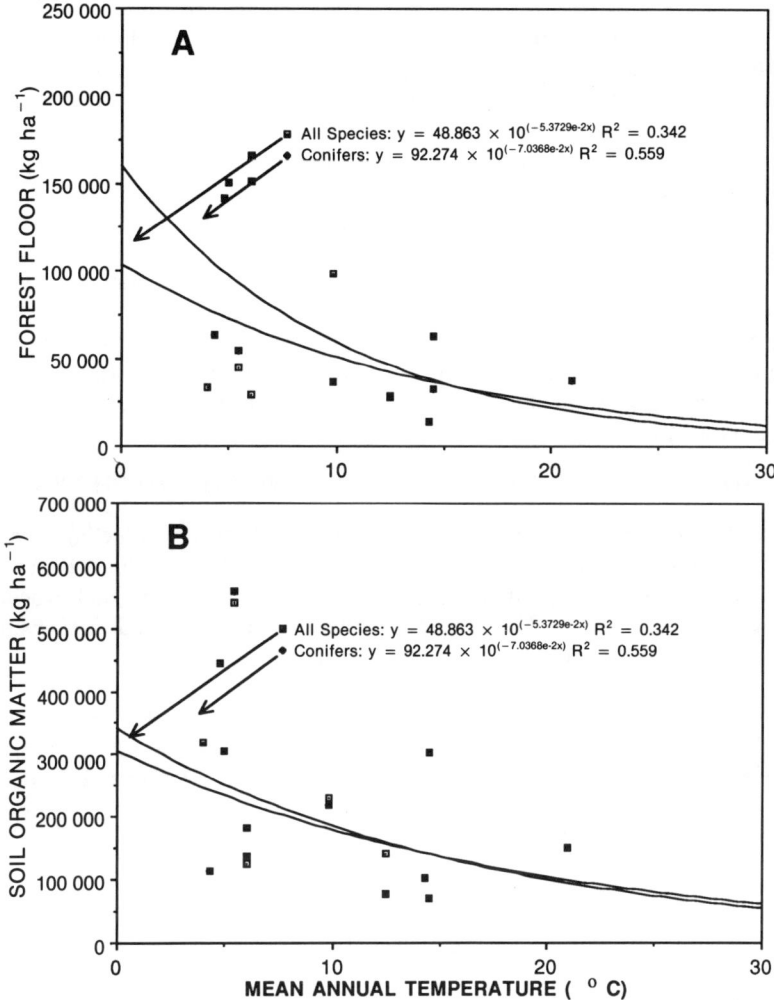

Fig. 14–11. Forest floor (A) and soil organic matter (B) vs. mean annual temperature in the IFS sites (data from Johnson & Lindberg, 1991; see Table 14–1 for legend).

decomposition values, the k values were calculated simply as the ratio of forest floor mass to litterfall (total) inputs, which implicitly assumes a steady-state condition. The assumption of steady-state forest floor may be valid for older, undisturbed sites [e.g., Smokies spruce (*Picea* spp.) and beech (*Engus grandifolia* Ehrh.), Findley Lake, Whiteface], or sites with rapid litter turnover rates [e.g., Coweeta hardwood, Huntington Forest, the loblolly pine (*Pinus taeda* L.) sites]. In other cases, however, steady-state may not have been achieved at the time that these studies were done, and the k values would be overestimates.

Collectively, these patterns suggest that (i) litterfall, decomposition rates, and forest floor accumulation follow the patterns with respect to mean annual temperature (MAT) predicted by Olson (1963); (ii) soil organic matter accumulation follows the pattern with respect to MAT predicted by Post et al. (1982); and (iii) forest floor accumulation is affected by species whereas soil organic matter accumulation is not.

The correlation coefficients in Table 14–2 suggest that forest floor accumulation significantly affects the accumulation of N, P, S, and K in the IFS sites. The forest floor is a major storage pool for N and P in the colder sites, often exceeding vegetation content (Fig. 14–4). The poor correlations between forest floor organic matter, Ca, and Mg contents are due to very low litter Ca and Mg concentrations in some northern/high-elevation coniferous forest sites (Smokies spruce, Maine, and Norway). Although forest floor biomass is relatively high at these sites, forest floor Ca and Mg contents are low (Fig. 14–5 and 14–6). This may be due to species effects (*Picea* spp. in all cases), to the very low exchangeable Ca and Mg pools at these sites, or both. It is noteworthy that one of the Smokies spruce sites has shown a slight growth response to Ca fertilization (Van Miegroet et al., 1993).

GEOCHEMICAL CYCLING OF CARBON AND NUTRIENTS

Carbon can affect the geochemical cycling of nutrients by affecting nutrient concentrations in soil solution, which can in turn affect leaching losses. In unpolluted ecosystems, carbonic and organic acids dominate soil leaching (Johnson et al., 1977, 1983), and both of these acids are linked to C cycling processes in general and decomposition in particular.

In temperate soils that are not extremely acidic, increased partial pressure of CO_2 (pCO_2) in the soil atmosphere will stimulate increased carbonic acid leaching (McColl & Cole, 1968; Johnson et al., 1977). This can be shown from the following equations describing the dissolution of CO_2 in solution and the dissociation of carbonic acid

$$[H_2CO_3] = (K_h)(pCO_2) \qquad [5]$$

$$\frac{[HCO_3^-][H^+]}{[H_2CO_3]} = K_1 \qquad [6]$$

where K_h = Henry's Law constant and K_1 = first dissociation constant for carbonic acid. Combining and solving for $[HCO_3^-]$ yields

$$[HCO_3^-] = \frac{(K_1)(K_h)(pCO_2)}{[H^+]} \qquad [7]$$

Thus, bicarbonate concentration (and, therefore, cation leaching associated with bicarbonate) is a function pCO_2 in the soil atmosphere and soil solution pH (Johnson, 1975). McColl and Cole (1968) artificially increased PCO_2 in soil columns, and demonstrated increased carbonic acid leaching. Johnson et al. (1977) found that elevated pCO_2 due to greater biological activity in soils at a tropical site (La Selva, Costa Rica) caused greater carbonic acid leaching rates than in more northern forest soils with lower pCO_2's.

Soil pCO_2 is a function of soil respiration (both autotrophic [root] and heterotrophic), the diffusivity of CO_2 in the soil (D), and depth in the soil. This can be seen from the equation used to calculate CO_2 flux (assuming steady-state conditions; Wesseling, 1962; deJong & Schappert, 1972; Amundsen & Davidson, 1990)

$$q = D\frac{dC}{dz} \qquad [8]$$

where C = soil CO_2 concentration (g cm^{-3}), z = depth (cm), and D = diffusion coefficient (cm^2 s^{-1}). Wesseling (1962) provides an equation for the diffusion coefficient (D) as a function of the diffusion coefficient of CO_2 in air (D_a) and effective soil porosity (p)

$$D = (D_a)(0.9p - 0.1) \qquad [9]$$

By combining Eq. [8] and [9], integrating, and rearranging we can obtain CO_2 concentration at depth z

$$C_z = \frac{(q)(z)}{(D_a)(p - 0.1)} + C_a \qquad [10]$$

where C_z = CO_2 concentration at depth z and C_a = CO_2 concentration in ambient air.

Using Wesseling's (1962) equations for root respiration with depth, pCO_2 profiles were calculated for different fluxes and p values typical of forest soils (Fig. 14–12). These comparisons show that, for values selected, soil pCO_2 profiles were about equally sensitive to flux and p. Thus, one would expect that pCO_2 and the potential for carbonic acid leaching would increase as CO_2 flux increases and soil porosity decreases, e.g., in warm, clay-rich soils. The degree to which this potential is realized is dependent on soil solution pH, which is in turn dependent on base saturation and the concentrations of strong mineral acid anions (NO_3^-, SO_4^{2-}, and Cl^-) or organic acids (Reuss & Johnson, 1986; Johnson et al., 1986).

Soil solution bicarbonate concentrations increase with mean annual temperature among the IFS sites, with the notable exception of the Florida slash pine site (FS in Fig. 14–13). The Florida site soil is a Spodosol (Uletic Haplaquad), implying that organic acids dominate soil leaching and genesis processes at that site. Thus, the lowering of soil solution pH by organic acids at that site negates the potential for high rates of carbonic acid leaching. Similarly, many of the sites with low mean annual temperature have Spodosols or heavily podzolized soils, and thus the potential for carbonic acid leaching (which would presumably be quite low) is negated by low soil solution pH. Spodosols are characteristic of cold-region soils and in warm humid and tropical soils with quartz-rich sand parent material and fluctuating groundwater tables, such as the Florida site. Where Spodosols are present, natural leaching and soil genesis processes in the upper soil horizons (i.e., above the Spodic horizon) are dominated by organic acids. Organic ligands and their chelated cations (principally Fe and Al) precipi-

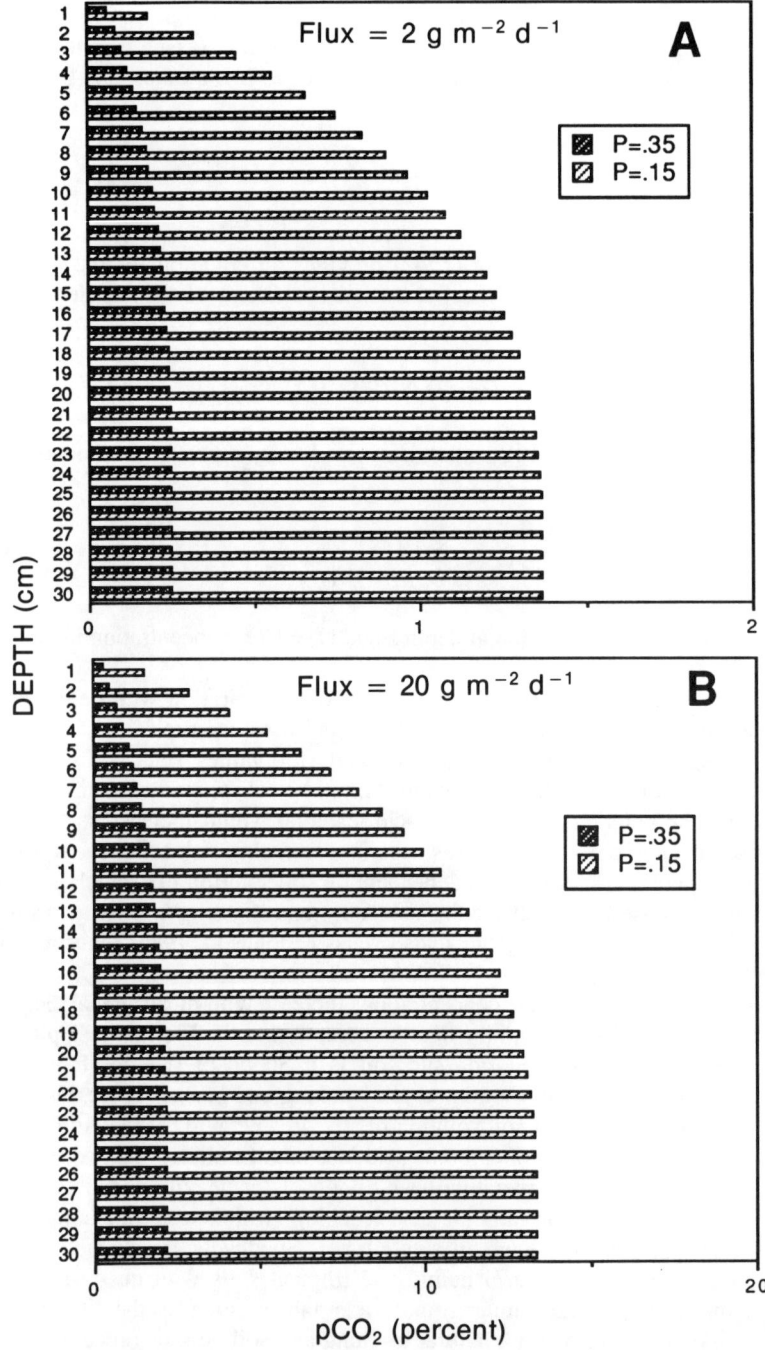

Fig. 14–12. Calculated soil pCO_2 profiles using the model of Wesseling (1962) for CO_2 fluxes and effective porosities typical of soils. (A) Constant flux of 2 g m^{-2} d^{-1} with effective porosities of 0.15 and 0.35, (B) Constant flux of 20 g m^{-2} d^{-1} with effective porosities of 0.15 and 0.35.

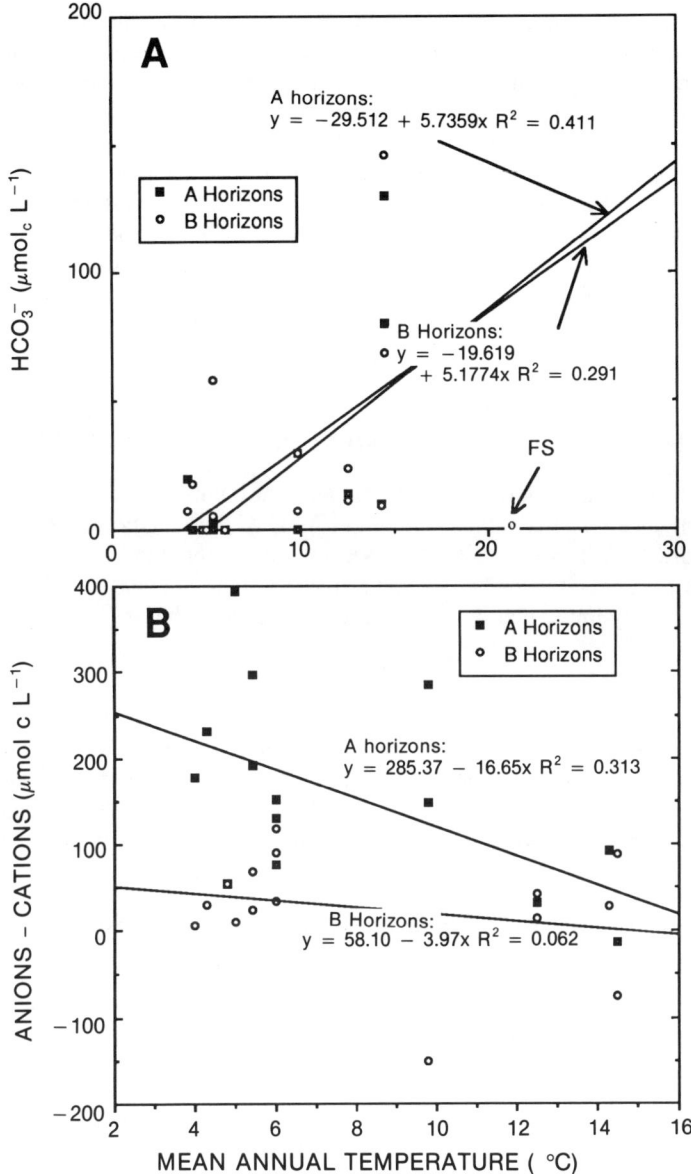

Fig. 14–13. Soil solution bicarbonate (A) and anion deficit (B) vs. mean annual temperature in the IFS sites (data from Johnson & Lindberg, 1991; see Table 14–1 for legend).

tate in the Spodic horizon, allowing pH rise and dissociation of carbonic acid below the Spodic horizon (Johnson et al., 1977; Ugolini et al., 1977).

Where quantitative data on organic anion concentrations are lacking (which is usually the case), the anion deficit (inorganic cations − inorganic anions, in $\mu mol_c\ L^{-1}$) can be used as a surrogate. The anion deficit of A horizons decreases with mean annual temperature among the IFS sites (Fig. 14–13B), as would be expected to result from a decrease in the importance of organic acids with mean annual temperature. There is no pattern in B horizons, and none would be expected given the normal precipitation of organic acids in the spodic horizons.

Thus, C plays a major role in the transport of cations via leaching in natural systems; that is, by organic acids in Spodosols and carbonic acid in other soils. There are exceptions to this general rule, however, where strong mineral acids may dominate soil cation leaching processes. Nitric acid is known to dominate soil leaching in some naturally N-rich sites (either by fixation or high N availability; Van Miegroet & Cole, 1984; Foster, 1985), and in sites with high rates of atmospheric N deposition (e.g., Van Breemen et al., 1982; Johnson et al., 1991). Similarly, SO_4^{2-} is known to dominate soil leaching processes in sites with high rates of atmospheric S deposition (e.g., Cronan et al., 1978; Johnson et al., 1985; Matzner & Prenzel, 1992).

EFFECTS OF CARBON AND NUTRIENT AMENDMENTS

Significant alterations of both C and nutrient inputs to forest ecosystems are occurring as a result of human activities. Some of these activities are deliberate (e.g., forest fertilization) and some are not (e.g., pollutant inputs). No value judgment is implied here, the fact that a given nutrient input is not deliberate does not necessarily imply that its consequences are undesirable.

Effects of Carbon Inputs

Carbon can be added to forest ecosystems in the form of organic matter, in which case its primary effect is on the soil, and in the form of CO_2, in which case its primary effect is on the vegetation. In either case, both the soil and vegetation are ultimately affected by C additions. Much has been written about each of these subjects, and a full review is beyond the scope of this paper. What follows is a brief summary of both established and hypothesized effects.

Effects of Organic Matter Additions to the Soil

Additions of organic matter to soils can affect the cycling of other nutrients through its effects on the physical (bulk density, water holding capacity), biological (microbial populations) and chemical (cation exchange capacity) properties of soils. In some instances, these effects can be beneficial to plant growth and therefore cause increased rates of nutrient cycling. With respect to N, however, the effects can be negative, at least temporarily. Heterotrophs are effective competitors for available N, and heterotrophic demand for N is greatly stimulated by adding labile organic C substrates (Paul & Clark, 1989). Turner (1977) demonstrated that addition of carbohydrates to a forest soil in Washington caused increased N deficiency in Douglas fir (*Pseudotsuga menziesii*) trees, presumably by stimulating heterotrophic competition for N. Johnson and Edwards (1979) found that addition of carbohydrate substrate to a forest soil caused an immediate reduction in NO_3^- leaching and net nitrification during laboratory incubation of soil from a yellow poplar (*Liriodentron tulipifera*) forest in Tennessee.

Additions of nonlabile organic matter to soils may result in the immobilization of N through nonbiological processes, as noted earlier (Wollum & Davey, 1975; Paul & Clark, 1989). The well-documented inhibitory effect of lignin on N mineralization is due in part to the formation of stable nitrogenous compounds from lignin by-products, reducing N availability to decomposer organisms (Berg et al., 1984; Axelsson & Berg, 1988; Johnson, 1992a). Axelsson and Berg (1988) found significant amounts of chemical fixation of ammonia during the early stages of decomposition in Scots pine litter. They also noted in inverse relationship between N concentration and N fixation capacity; thus, litter at later stages of decay fixed less N because of its higher N concentration.

Effects of Elevated Carbon Dioxide

Studies to date suggest that the N cycle may be affected by elevated CO_2 in a number of ways. First and perhaps foremost, there are good reasons to suspect that N fixation rates will increase (and perhaps have already increased) with increasing atmospheric CO_2. Elevated CO_2 usually causes increased carbohydrate allocation to roots (Norby et al., 1986a; Rogers et al., 1992), and it is logical to assume that N fixation will be favored by increasing atmospheric CO_2 levels (Norby, 1987). This hypothesis has been borne out by studies on agricultural crops (see review by Norby, 1987a,b), but, to date, only one study on forest tree species has been conducted. Norby (1987) tested the effects of ambient (350 $\mu L\ L^{-1}$) and elevated (700 $\mu L\ L^{-1}$) CO_2 on growth and N fixation in seedlings of one forest legume (*Robinia pseudoacacia*) and two actinorhizal species (*Alnus glutinosa* and *Elaeagnus angustifolia*). He found that N fixation was increased substantially due to increases in nodule mass (but not due to increases in fixation rate per unit of nodule mass). These results suggest two important, related hypotheses: (i) N-fixing species will be favored (and perhaps have already been favored) by increasing atmospheric CO_2, and (ii) N-fixation rates (including those by herbaceous understory species) will increase as atmospheric CO_2 levels increase.

Several investigators have noted reduced tissue N concentrations in tree seedlings exposed to elevated CO_2 (Norby et al., 1986a,b; Campagna & Margolis, 1989; Brown, 1991; Johnson & Henderson, 1994). This effect could obviously allow greater biomass production even under N limitation, should it also occur in mature forests. However, there are questions as to whether tree internal reserves of N are sufficient to sustain this increased N "use efficiency" over longer periods (Norby et al., 1986a). There has been speculation that reduced foliar N concentrations will cause both litter quality and decomposition rates to be reduced (Strain, 1985; Norby et al., 1986b). Such decreases in litter (foliar or root) quality may create or exacerbate N deficiency, perhaps causing further deterioration of litter quality and creating a feedback loop that ultimately must result in a decline in primary productivity. Results to date have not supported the latter hypothesis, however. Norby et al. (1986b) concluded that the effects of CO_2 fumigation on the chemical composition of senesced leaves from white oak (*Quercus alba*) seedlings were too small to significantly affect litter decomposition rates. Furthermore, Couteaux et al. (1991) demonstrated that initial indices of litter quality such as C/N or lignin/N ratio are not accurate predictors of decomposition rates over the long term. Our initial studies of the effects of CO_2 on the decomposition of aspen (*Populus* spp.) leaves also have shown that decomposition and N mineralization are very dynamic processes and that the effects of treatment are very transient (Johnson & Henderson, 1994).

There is some evidence that elevated CO_2 can affect soil C and N mineralization through rhizosphere effects. Körner and Arnone (1992) found a reduction in soil C and increases in soil respiration and nitrate leaching in an artificial tropical ecosystem subjected to elevated CO_2. They attributed these responses to increased soil organic matter decomposition in the rhizosphere. Similarly, Zak et al. (1993) found increased microbial biomass, C and N mineralization in the rhizosphere soils of *Populus grandidentata* seedlings subjected to elevated CO_2.

The mechanisms and potential feedbacks by which increased atmospheric CO_2 could affect nutrient cycling are very complex, making predictions very difficult, especially in view of the lack of information on an ecosystem scale. Studies to date have been limited to seedlings, and effects of CO_2 on nutrient cycling in mature forest ecosystems must be deduced from small-scale process-level studies. Thus, the net ecosystem C balance and the nature of nutrient cycling with elevated CO_2 is very difficult to predict with the current paucity of information.

Effects of Nutrient Additions

Nutrient additions can affect the cycling of C in two basic ways: (i) by causing increases in primary productivity, and (ii) by causing chemical stabilization (or perhaps destabilization) of soil organic matter. Obviously, increases in primary production will occur only in those cases where the nutrient being added is growth-limiting. The effects of fertilizer additions on the C cycling of nutrient-limited stands is obvious and, in many cases, long-lasting (Heilman & Gessel, 1963; Pritchett & Comerford, 1982; Turner & Lambert, 1986). In addition to causing changes in primary productivity, the addition of certain nutrients can cause changes in the stability of soil organic matter. The potential importance of nonbiological condensation reactions between ammonia and soil organic matter has already been noted. These reactions are enhanced by high pH and high NH_3 concentrations, both of which occur following urea fertilization, causing substantial nonbiological fixation of fertilizer N (Foster et al., 1985). In the case of Ca and other polyvalent cations, cation bridging of organic colloids causes condensation and stabilization of organic matter, as noted above (Oades, 1988). Because Ca is rarely limiting to tree growth, the positive effects of liming on soil C noted by Gilmore (1980) and Jenkinson et al. (1991) are likely due to these reactions rather than a direct effect on plant primary productivity. Several studies also have reported that liming caused decreases in forest floor and soil organic matter because of its stimulatory effect on microbiological activity (Marschner & Wilczynski, 1991).

There are some fundamental differences in the nature of forest ecosystem response to N vs. other nutrients that directly relate to the unique nature of C and N cycles. Nitrogen does not accumulate in soils in inorganic forms for prolonged periods; thus, fertilizer N is usually rapidly consumed by heterotrophs, plants, nonbiological reactions with soils, or nitrifiers (and leached as nitrate). Prolonged growth responses to N fertilization are therefore due primarily to internal retention and translocation within the tree itself, or simply to the advancement of the stage of stand development (Miller et al., 1979; Miller, 1981). In contrast, fertilization with Mg, P, or K often results in a long-term enrichment of the soil with inorganic, exchangeable reserves of these nutrients, allowing growth responses to carry from one rotation to another in some cases (Turner & Lambert,

1986). Thus, the efficiency of fertilization with N is usually much less than that of fertilization with other nutrients because of the very fact that N and C cycles are so closely coupled.

SUMMARY AND CONCLUSIONS

Although organic C and other nutrients are coupled in the biogeochemical cycle, the correlation is not stoichiometrically exact—there are significant variations due to differences in species, nutrient status, and air pollution inputs. There is a general trend of increasing litterfall, increasing decomposition rate, decreasing forest floor, and decreasing soil organic matter with increasing mean annual temperature which affects the cycling of nutrients. Carbon also plays a major role in the geochemical cycling of base cations via carbonic and organic acid leaching. Carbonic acid leaching is greatest in warmer, clay-rich soils where the diffusivity of CO_2 is low and root and microbial respiration are high. Organic acid leaching is greatest in cool or waterlogged coarse-textured soils where decomposition and adsorption of organics to soil surfaces are low. Carbon additions as organic matter to the soil can affect nutrient cycles by affecting soil physical properties, cation exchange capacity, and N availability. Carbon addition as CO_2 to the vegetation can affect nutrient cycles by altering primary productivity, nutrient uptake, and decomposition rates. Nutrient additions can affect the cycling of C by causing increases in primary productivity and by causing changes in the chemical stability of soil organic matter.

In concluding this review of the role of C in the cycling of nutrients in temperate forests, it is appropriate to revisit the hypotheses and conclusions posed by Shaver et al. (1992) in their similar review of Arctic ecosystems. In essence, they argue that C cycling in these systems can be thought of as being almost entirely dependent on N and P cycling rather than the reverse. They contend that effects of environmental variables such as light, CO_2, and even temperature on these ecosystems are manifested primarily through the cycles of N and P. If environmental variables do not affect N and P cycles, they do not materially affect C cycles or the basic functioning of the ecosystem. In support of this contention, they document many studies showing that nutrients strongly affect the productivity of Arctic ecosystems, and that the effects of manipulating other environmental variables (light, CO_2, temperature) can be explained on the basis of how they affect N and P cycles.

Although there have been considerably more studies in temperate forests than in Arctic ecosystems, we lack a sufficiently comprehensive data base to either support or refute the conceptual model posed by Shaver et al. (1992). Reasons for this include the greater diversity of ecosystems we must deal with and the logistical difficulties of conducting whole-ecosystem manipulative studies of temperature and CO_2. Nonetheless, this model seems to have considerable merit as a means of evaluating temperate forest response to environmental changes such as increasing CO_2, temperature, and precipitation. We know from numerous fertilizer studies that additions of limiting nutrients have major effects on forest C cycles even though water, temperature, and light, or CO_2 also may be constraining growth to various degrees. Ecosystem-level experiments involving manipulations of non-nutritional environmental variables and careful monitoring of nutrient cycling processes are needed in order to adequately test the hypothesis of Shaver et al. (1992) in temperate forest ecosystems.

ACKNOWLEDGMENTS

Research supported by the Electric Power Research Institute (RP3041-02) and its Nevada Agricultural Experiment Station, University of Nevada, Reno.

REFERENCES

Anderson, J.M. 1992. Responses of soils to climate change. Adv. Ecol. Res. 22:163–210.

Amundson, R.G., and E.A. Davidson. 1990. Carbon dioxide and nitrogenous gases in the soil atmosphere. J. Geochem. Explor. 38:13–41.

Axelsson, G., and B. Berg. 1988. Fixation of ammonia (^{15}N) to Pinus sylvestris needle litter in different stages of decomposition. Scand. J. For. Res. 3:273–279.

Berg, B., G. Ekbohm, and C. McClaugherty. 1984. Lignin and holocellulose relations during decomposition in a Scots pine forest. Can J. Bot. 62:2540–2550.

Brown, K.R. 1991. Carbon dioxide enrichment accelerates the decline in nutrient status and relative growth rate of *Populus tremuloides* Michx. seedlings. Tree Physiol. 8:161–173.

Bugbee, B., and O. Monje. 1992. The limits of crop productivity. Bioscience 42:494–502.

Campagna, M.A., and H.A. Margolis. 1989. Influence of short-term atmospheric CO_2 enrichment on growth, allocation patterns, and biochemistry of black spruce seedlings at different stages of development. Can. J. For. Res. 19:773–782.

Cole, D.W., and S.P. Gessel. 1965. Movement of elements through forest soil as influenced by tree removal and fertilizer additions. p. 95–104. *In* C.T. Youngberg (ed.) Forest soil relationships in North America. Oregon State Univ. Press, Corvallis, OR.

Cole, D.W., S.P. Gessel, and S.F. Dice. 1968. Distribution and cycling of nitrogen, phosphorus, potassium, and calcium in a second-growth Douglas-fir forest. p. 197–213. *In* H.E. Young (ed.) Primary production and mineral cycling in natural ecosystems. Univ. Maine Press, Orono, ME.

Cole, D.W, and M. Rapp. 1981. Elemental cycling in forest ecosystems. p. 341–409. *In* D.E. Reichle (ed.) Dynamic properties of forest ecosystems. Cambridge Univ. Press, London.

Couteaux, M.-M., M. Mousseau, M.-L. Celkerier, and P. Bottner. 1991. Increased atmospheric CO_2 and litter quality decomposition of sweet chestnut litter with animal food webs of different complexities. Oikos 61:54–64.

Cronan, C.W., W.A. Reiners, R.L. Reynolds, and G.E. Lang. 1978. Forest floor leaching: Contributions from mineral, organic, and carbonic acids in New Hampshire subalpine forests. Science (Washington, DC) 200:309–311.

Curlin, J.W. 1970. Nutrient cycling as a factor in site productivity and forest fertilization. p. 313–326. *In* C.T. Youngberg and C.R. Davey (ed.) Tree growth and forest soils. Oregon State Univ. Press, Corvallis, OR.

deJong, E., and H.J.V. Schappert. 1972. Calculation of soil respiration and activity from CO_2 profiles in the soil. Soil Sci. 119:328–333.

Duvigneaud, P., and S. Denaeyer-DeSmet. 1970. Biological cycling of minerals in temperate deciduous forests. p. 199–255. *In* D.E. Reichle (ed.) Analysis of forest ecosystems. Springer-Verlag, New York.

Foster, N.W. 1985. Acid precipitation and soil solution chemistry within a maple-birch forest in Canada. For. Ecol. Manage. 12:215–231.

Foster, N.W., E.G. Beauchamp, and C.T. Corke. 1985. Immobilization of ^{15}N-labelled urea in a Jack pine forest floor. Soil Sci. Soc. Am. J. 49:448–452.

Gallardo, A., and J. Merino. 1992. Nitrogen immobilization in leaf litter at two Mediterranean ecosystems of SW Spain. Biogeochemistry 15:213–228.

Gilmore, A.R. 1980. Changes in a reforested soil associated with tree species and time. IV. Soil organic content and pH in pine plantations after 24 years. For. Res. Rep. no. 80–3. Univ. Illinois Agric. Exp. Stn.

Grodzinski, B. 1992. Plant nutrition and growth regulation by CO_2 enrichment. Bioscience 42:517–525.

Heilman, P.E., and S.P. Gessel. 1963. Nitrogen requirements and the biological cycling of nitrogen in Douglas-fir stands in relationship to the effects of nitrogen fertilization. Plant Soil 18:386–402.

Jenkinson, D.S., D.E. Adams, and A. Wild. 1991. Model estimates of CO_2 emissions from soil in response to global warming. Nature (London) 351:304–306.

Johnson, D.W. 1975. Processes of elemental transfer in some tropical, temperate, alpine, and northern forest soils: Factors influencing the availability and mobility of major leaching agents. Ph.D. diss. Univ. Washington, Seattle (Diss. Abstr. 75–37/02-B:542).

Johnson, D.W. 1984. Sulfur cycling in forests. Biogeochemistry 1:29–44.

Johnson, D.W. 1992a. Nitrogen retention in forest soils. J. Environ. Qual. 21:1–12.

Johnson, D.W. 1992b. Effects of forest management on soil carbon storage. Water Air Soil Pollut. 64:83–120.

Johnson, D.W., D.W. Cole, S.P. Gessel, M.J. Singer, and R.V. Minden. 1977. Carbonic acid leaching in a tropical, temperate, subalpine, and northern forest soil. Arct. Alp. Res. 9:329–343.

Johnson, D.W., D.W. Cole, H. Van Miegroet, and F.W. Horng. 1986. Factors affecting anion movement and retention in four forest soils. Soil Sci. Soc. Am. J. 50:776–783.

Johnson, D.W., and N.T. Edwards. 1979. Effects of stem girdling on biogeochemical cycles in a mixed deciduous forest in eastern Tennessee. II. Soil nitrogen mineralization and nitrification rates. Oecologia 40:259–271.

Johnson, D.W., and P. Henderson. 1994. Effects of forest management and elevated carbon dioxide on soil carbon storage. Adv. Soil Sci. (In press.)

Johnson, D.W., and S.E. Lindberg (ed.). 1991. Atmospheric deposition and forest nutrient cycling: A synthesis of the integrated forest study. Ecol. Ser. 91. Springer-Verlag, New York.

Johnson, D.W., D.D. Richter, G.M. Lovett, and S.E. Lindberg. 1985. The effects of atmospheric deposition on potassium, calcium, and magnesium cycling in two deciduous forests. Can. J. For. Res. 15:773–782.

Johnson, D.W., H. Van Miegroet, D.W. Cole, and D.D. Richter. 1983. Contributions of acid deposition and natural processes to cation leaching from forest soils: A review. J. Air Pollut. Control Assoc. 33:1036–1041.

Johnson, D.W., H. Van Miegroet, S.E. Lindberg, R.B. Harrison, and D.E. Todd. 1991. Nutrient cycling in red spruce forests of the Great Smoky Mountains. Can. J. For. Res. 21:769–787.

Kauppi, P.E., K. Mielikäinen, and K. Kuusela. 1992. Biomass and carbon budget of European forests, 1971 to 1990. Science (Washington, DC) 256:70–74.

Körner, C., and J.A. Arnone. 1992. Responses to elevated carbon dioxide in artificial tropical ecosystems. Science (Washington, DC) 257:1672–1675.

Lindberg, S.E. 1991. Atmospheric deposition and canopy interactions of sulfur. p. 74–89. In D.W. Johnson and S.E. Lindberg (ed.) Atmospheric deposition and forest nutrient cycling: A synthesis of the integrated forest study. Ecol. Ser. 91. Springer-Verlag, New York.

Lovett, G.M. 1991. Atmospheric deposition and canopy interactions of nitrogen. p. 152–165. In D.W. Johnson and S.E. Lindberg (ed.) Atmospheric deposition and forest nutrient cycling: A synthesis of the integrated forest study. Ecol. Ser. 91. Springer-Verlag, New York.

Lugo, A.E. 1992. The search for carbon sinks in the tropics. Water Air Soil Pollut. 64:3–9.

Marschner, B., and A.W. Wilczynski. 1991. The effect of liming on quantity and chemical composition of soil organic matter in a pine forest in Berlin, Germany. Plant Soil 137:229–236.

Marion, G.K. 1979. Biomass and nutrient removal in long-rotation stands. p. 98–110. In A.L. Leaf (ed.) Impact of intensive harvesting on forest nutrient cycling. State Univ. New York, Syracuse.

Matzner, E., and J. Prenzel. 1992. Acid deposition in the German Solling area: Effects on soil solution chemistry and Al mobilization. Water Air Soil Pollut. 61:221–234.

McColl, J.G., and D.W. Cole. 1968. A mechanism of cation transport in a forest soil. Northwest Sci. 42:132–140.

Meetenmeyer, V. 1978. Macroclimate and lignin control of litter decomposition rates. Ecology 59:465–472.

Melillo, J.M., J. Aber, and J.F. Muratore. 1982. Nitrogen and lignin control of hardwood leaf litter decomposition dynamics. Ecology 63:621–626.

Miller, H.G. 1981. Forest fertilization: Some guiding concepts. Forestry 54:157–167.

Miller, H.G., J.M. Cooper, J.D. Miller, and O.J.L. Pauline. 1979. Nutrient cycles in pine and their adaption to poor soils. Can. J. For. Res. 9:19–26.

Mortland, M.M., and A.R. Wolcott. 1965. Sorption of inorganic nitrogen compounds by soil materials. p. 150–197. In W.V. Bartholomew and F.E. Clark (ed.) Soil nitrogen. Agron. Monogr. 10. ASA, Madison, Wisconsin.

Nambiar, E.K.S., and D.N. Fife. 1991. Nutrient translocation in temperate conifers. Tree Physiol. 9:185–207.

Nommik, H. 1970. Non-exchangeable binding of ammonium and amino nitrogen by Norway spruce raw humus. Plant Soil 33:581–595.

Norby, R.J. 1987. Nodulation and nitrogenase activity in nitrogen-fixing woody plants stimulated by CO_2 enrichment of the atmosphere. Physiol. Plant. 71:77–82.

Norby, R.J., E.G. O'Neill, and R.J. Luxmoore. 1986a. Effects of atmospheric CO_2 enrichment on the growth and mineral nutrition of *Quercus alba* seedlings in nutrient-poor soil. Plant Physiol. 82:83–89.

Norby, R.J., J. Pastor, and J.M. Melillo. 1986b. Carbon-nitrogen interactions in CO_2-enriched white oak: physiological and long-term perspectives. Tree Physiol. 2:233–241.

Oades, J.M. 1988. The retention of organic matter in soils. Biogeochemistry 5:35–70.

Olson, J.S. 1963. Energy storage and the balance of producers and decomposers in ecological systems. Ecology 44:322–331.

Ovington, J.D. 1962. Quantitative ecology and the woodland ecosystem concept. Adv. Ecol. Res. 1:103–192.

Paul, E.A., and F.E. Clark. 1989. Soil microbiology and biochemistry. Acad. Press, New York.

Post, W.M., W.R. Emannel, P.J. Zinke, and A.G. Stangenberger. 1982. Soil carbon pools and world life zones. Nature (London) 298:156–159.

Post, W.M., T-H Peng, W.R. Emmannel, A.W. King, V.H. Dale, and D.L. DeAngelis. 1990. The global carbon cycle. Am. Sci. 78:310–326.

Pritchett, W.L., and N.B. Comerford. 1982. Long-term response to phosphorus fertilization on selected southeastern coastal plain soils. Soil Sci. Soc. Am. J. 46:640–644.

Rennie, P.J. 1955. The uptake of nutrients by mature forest growth. Plant Soil 7:49–95.

Reuss, J.O., and D.W. Johnson. 1986. Acid deposition and the acidification of soil and water. Ecol. Stud. no. 59. Springer-Verlag, New York.

Rogers, H.H., C.M. Peterson, J.N. McCrimmon, and J.D. Cure. 1992. Response of plant roots to elevated atmospheric carbon dioxide. Plant Cell Environ. 15:749–752.

Sedjo, R.A. 1992. Temperate forest ecosystems in the global carbon cycle. Ambio 21:274–277.

Schimel, J.P., and M.K. Firestone. 1989. Inorganic N incorporation by coniferous forest floor material. Soil Biol. Biochem. 21:41–46.

Shaver, G.R., W.D. Billings, F.S. Chapin III, A.E. Giblin, K.J. Nadelhoffer, W.C. Oechel, and E.B. Rastetter. 1992. Global change and the carbon balance of acitic ecosystems. Bioscience 42:433–441.

Strain, BR. 1985. Physiological and ecological controls on carbon sequestering in terrestrial ecosystems. Biogeochemistry 1:219–232.

Switzer, G.L., and L.E. Nelson. 1972. Nutrient accumulation and cycling in loblolly pine (*Pinus taeda* L.) plantation ecosystems: The first twenty years. Soil Sci. Soc. Am. Proc. 36:143–147.

Tans, P.P., I.Y. Fung, and T. Takahashi. 1990. Observational constraints on the global atmospheric CO_2 budget. Science (Washington, DC) 247:1431–1438.

Turner, J. 1977. Effect of nitrogen availability on nitrogen cycling in a Douglas-fir stand. For. Sci. 23:307–316.

Turner, J., and M.J. Lambert. 1986. Fate of applied nutrients in a long-term superphosphate trial in *Pinus radiata*. Plant Soil 93:373–382.

Ugolini, F.C., R. Minden, H. Dawson, and J. Zachara. 1977. An example of soil processes in the Abies amabilis zone of Central Cascades, Washington. Soil Sci. 124:291–302.

Van Breemen, N., P.A. Burrough, E.J. Velthorst, H.F. van Dobben, T. de Witt, T.B. Ridder, and H.F.R. Reijuders. 1982. Soil acidification from atmospheric ammonium sulphate in forest canopy throughfall. Nature (London) 299:548–550.

Van Miegroet, H., and D.W. Cole. 1984. The impact of nitrification on soil acidification and cation leaching in a red alder ecosystem. J. Environ. Qual. 13:586–590.

Van Miegroet, H., D.W. Johnson, and D.E. Todd. 1993. Foliar responses of red spruce seedlings to fertilization with Ca and Mg in the Great Smoky Mountains National Park. Can. J. For. Res. 23:89–95.

Weetman, G.L., and B. Webber. 1972. The influence of wood harvesting on the nutrient status of two spruce stands. Can. J. For. Res. 2:351–369.

Wesseling, J. 1962. Some solutions of the steady state diffusion of carbon dioxide through soils. Neth. J. Agric. Sci. 10:109–117.

Wollum, A.G., and C.B. Davey. 1975. Nitrogen accumulation, transformation, and transport in forest soils. p. 67–106. *In* B. Bernier and C. Winget (ed.) Forest soils and land management. Les presses de l' Universite Laval, Quebec, Canada.

Zak, D.R., K.S. Pregitzer, P.S. Curtis, J.A. Teeri, R. Fogel, and D.L. Randlett. 1993. Elevated atmospheric CO_2 and feedback between carbon and nitrogen cycles. Plant Soil 151:105–117.

Zelich, I. 1992. Control of productivity by regulation of photorespiration. Bioscience 42:510–516.

15 Carbon Controls on Spodosol Nitrogen, Sulfur, and Phosphorus Cycling

Mark B. David

University of Illinois
Urbana, Illinois

George F. Vance and Anna J. Krzyszowska

University of Wyoming
Laramie, Wyoming

The movement, accumulation, and transformations of organic C, N, S, and P are poorly understood in Spodosols and soils developed by similar formation processes. The spodic horizon is rich in these organic forms, but their accumulation (from leaching of solubilized products from the forest floor) or degradation have not been examined sufficiently. Recent studies have evaluated the movement of dissolved organic carbon (DOC) from the forest floor and its sorption in spodic horizons (Vance & David, 1989, 1991a; David & Zech, 1990; Guggenberger & Zech, 1992). However, the role of DOC leaching and accumulation in relation to N, S, and P cycles is virtually unknown in Spodosols. Controls on C solubilization and leaching from the forest floor (organic horizon) have not been established. Because of the importance of C, N, S, and P in nutrient cycles of northern hardwood and conifer forests found on Spodosols, a better understanding of their biogeochemical cycles is needed, with particular emphasis on the dynamics of the organic forms.

Soil leachates play an important role in the nutrient chemistry of forested ecosystems (Fahey & Yavitt, 1988; McDowell & Likens, 1988). Leachates from the forest floor comprise a variety of inorganic and organic nutrients that are transported to subsurface horizons, groundwaters, and/or surface waters. The transport of these inorganic and organic constituents contribute to pedogenic and biological processes (Candler & Van Cleve, 1982; Candler, 1985; Qualls et al., 1991) and can influence surface water quality, especially where terrestrial inputs are significant (Ertel et al., 1986; Driscoll et al., 1989; Howell, 1989; Cronan, 1990). Qualls and Haines (1991) determined that over 90 and 66%, respectively, of dissolved N and P leaching from a deciduous forest soil were in organic forms.

Copyright © 1995 Soil Science Society of America, 677 S. Segoe Rd., Madison, WI 53711, USA. Carbon Forms and Functions in Forest Soils.

The cycling of nutrients in organic forms in forested ecosystems is an area of research that requires further examination. Current organic nutrient cycling concepts have evolved in part from past research dealing with humic substances (i.e., humic and fulvic acid composition and reactivity; Stevenson, 1982; Parsons, 1988). Recently, DOC from soils and natural waters has been examined using adsorption chromatography to separate DOC into six general fractions; hydrophobic acids, bases and neutrals, and hydrophilic acids, bases and neutrals (Leenheer, 1981; David et al., 1989; Vance & David, 1989, 1991b; Qualls & Haines, 1991; Qualls et al., 1991). Following this approach, individual organic fractions can be isolated and used for further examination of the nutrient status of the migrating constituents.

OBJECTIVES

In this paper we examine the mechanisms and processes involved in the biogeochemistry of organic C, N, S, and P in forest ecosystems. We emphasize the accumulation, transformation, and translocation of these elements through the role of soil C. Our approach considers the movement of soluble organic constituents as they are released or retained in the forest floor and mineral soil. This discussion focuses on hardwood and coniferous forests that have developed on Spodosols in cool temperate to boreal climates. The objectives of this paper are to: (i) examine the translocation and chemistry of organic C, N, S, and P in forest floor leachates and mineral soil solutions; (ii) discuss the solid and solution chemistry of organic matter found in the forest floor and spodic horizons; (iii) examine the components of the C cycle, including C solubilization and translocation from the forest floor, and accumulation of C in the spodic horizon; and (iv) discuss a conceptual biogeochemical model that considers the processes and mechanisms of soluble organic matter involved in C, N, S, and P nutrient cycling.

ORGANIC MATTER POOLS IN SPODOSOLS

Within the last decade many studies have evaluated organic matter pools in Spodosols either in individual studies (Huntington Forest, New York, David et al., 1987) or large, integrated projects covering a wide range of soils (Integrated Forest Study, Johnson & Lindberg, 1992). These results have clearly shown that both the forest floor and spodic horizons (Bh and Bhs horizons) have large pools of organic matter, and therefore accumulations of organic forms of C, N, S, and P. Typical patterns for C, N, S, and P in Spodosols include high concentrations in the O horizons, low levels in E horizons, then high concentrations in Bh and Bhs horizons, followed by declining concentrations with depth (Fig. 15–1). Generally, the forms of C and N found in Spodosols are composed of organic constituents, with S and P forms consisting of greater inorganic concentrations with depth. Adsorbed sulfate is highest in Bhs and Bs horizons, leading to accumulations of inorganic S concentrations within the spodic horizon. Phosphorus follows a different pattern, with inorganic concentrations increasing with soil depth, reaching nearly 100% of soil P in the C horizon. The highest concentrations of C, N, S,

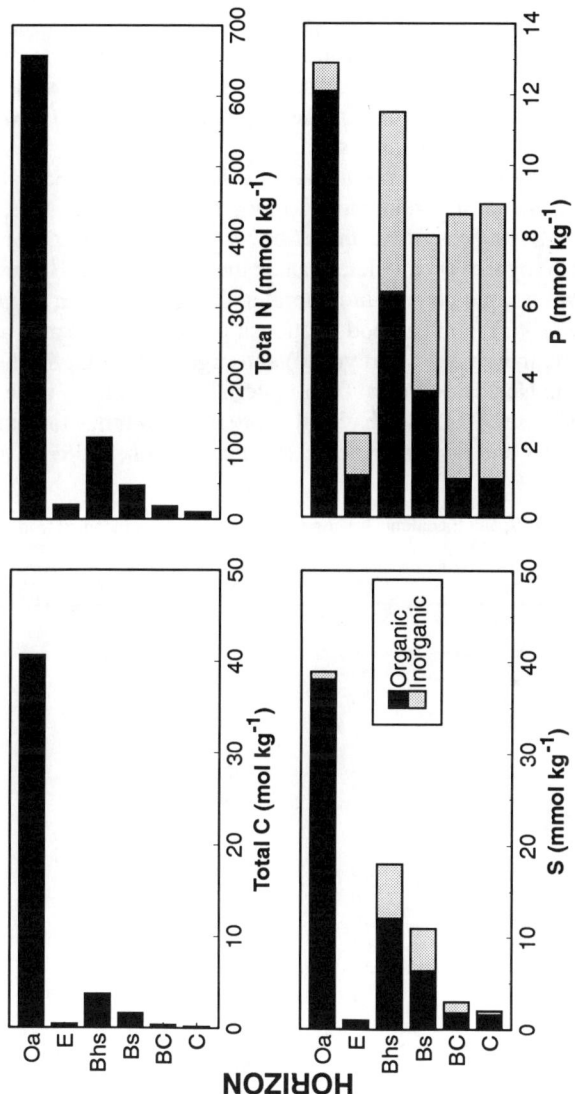

Fig. 15–1. Mean total C, N, S, and P and organic S and P concentrations in four profiles sampled in a spruce *(Picea* spp.)-fir *(Abies* spp.) forest at Howland, Maine (M.B. David, G.F. Vance, & A.J. Krzyszowska, unpublished data).

and P are always found in the forest floor, which also tends to have the highest root activity. However, on a content basis, the greatest amounts of these nutrients are often found in the mineral soil, with the B horizons dominating (Table 15–1). This pattern of nutrient content is due primarily to the mass of soil present in these horizons. Where thick forest floors have built up under some forest types, and where the >2-mm mineral fractions are dominant, the forest floor may be the largest pool of some organic nutrients (e.g., C and N at the Howland site in Maine). McFee and Stone (1965) measured forest floor and mineral soil C contents of 114 and 92 Mg C ha^{-1}, respectively, in forest Spodosols from the Adirondack Mountains in New York, suggesting similar pools of C exist in surface and subsurface horizons of some soils.

It is clear from these concentration patterns that the forest floor, along with Bh and Bhs horizons, are important sinks for organic forms of C, N, S, and P in Spodosols. In most forest ecosystems these horizons undergo slow decomposition and generally are thought to accumulate organic matter over time. However, some studies have shown a rapid change in mineral soil organic matter. For example, McFee and Stone (1965) found a good relationship between estimated age of the forest floor (approximately 90–>300 yr old) and organic C in the Bh horizon of Spodosol profiles in New York. Forest floor materials were thought to be removed by fire, and the increase of 0.06 Mg ha^{-1} yr^{-1} in organic C was due to both increasing horizon thickness and percentage of C (McFee & Stone, 1965). Using their

Table 15–1. Total C, N, S, and P contents by horizon in four Spodsols (adapted from Johnson & Lindberg, 1992).

Horizon	Depth	C	N	S	P
	cm	Mg ha^{-1}		kg ha^{-1}	
Huntington Forest, New York (hardwood)					
FF		22	1152	70	68
A	0–4	34	1424	108	173
E	4–8	7	340	49	148
Bh	8–12	18	853	96	306
Bhs	12–20	43	1522	209	893
Bs	20–58	167	3846	717	5151
Findley Lake, Washington (conifer)					
FF		27	575	59	60
E	0–12	17	790	83	133
Bhs	12–17	20	690	79	145
Bs	17–41	179	4090	786	1486
BC	41–71	64	2050	454	1152
Howland Forest, Maine (conifer)					
FF		74	1121	148	87
E	0–8	8	209	25	400
B	8–13	10	283	76	510
B	13–28	40	961	424	2818
B	28–52	20	484	246	2871
Turkey Lakes, Ontario (hardwood)					
FF		16	927	90	32
E	0–4	10	645	43	47
Bhs1	4–7	15	1013	103	103
Bhs2	7–15	19	1802	197	244
Bs	15–57	116	6520	753	1427

data, the average Bh horizon with 24 Mg C ha^{-1} would form in 377 yr, assuming no decomposition or other inputs. They hypothesized that the Bh horizon is reduced after fire and then slowly accumulates. Huntington and Ryan (1990) found no change in mineral soil C and N pools 3 yr after whole-tree-harvesting at Hubbard Brook, New Hampshire, whereas Mroz et al. (1985) found a large loss of mineral soil N after a similar harvest. Cruickshank and Cruickshank (1981) reported that the C in organic-rich B horizons of podsols from northern Ireland to be quite old (e.g., >1000 yr), suggesting great stability in deposited organic matter. Additional studies such as these are needed to determine whether mineral horizon organic matter responds rapidly to changes in forest floor C content.

Although much is known about the general organic C trends in Spodosols, few studies have examined C, N, S, and P organic forms in detail. Most have measured the total concentrations of these elements and/or extractable inorganic forms such as sulfate and phosphate. In the following discussion, examples are given which illustrate our understanding of the forms of C, N, S, and P that exist within soil organic matter constituents, particularly that which resides in the forest floor and spodic horizons.

Carbon

Spodosols generally have two distinct zones of C accumulation, one at the surface, the forest floor, and the other the spodic horizon (e.g., Bh and Bhs horizons). The forest floor is comprised of C-based materials such as plant and root litter, partially decomposed plant and microbial products, and humic substances. Figure 15–2 illustrates the interacting processes that are responsible for the breakdown and decomposition of organic materials, and the resulting forest floor layers. The forest floor has traditionally been separated into three layers based on the degree of organic matter decomposition and includes the: Oi layer, the least decomposed and is comprised of recognizable plant litter (e.g., needles and twigs), Oe layer, partially decomposed litter, and Oa layer, completely decomposed. Rates of C turnover in these layers are governed by the site-specific environment (i.e., soil moisture, temperature, and pH), resource quality (litter and residue composition), and microbial activity (Anderson et al., 1989).

Carbon accumulation in spodic horizons is due to the retention of DOC leached from the forest floor, by in situ decomposition of root litter and microbial products, and from the transport of organic matter by soil animals. Organic matter in spodic horizons consists, in part, of C compounds that can be identified as products of known structures. Kögel-Knabner (1992) suggested that greater than 40% of the mineral horizon C is composed of biomacromolecules which can be traced back to plant or microbial origins, the remaining C is comprised of humic substances that are typically separated into humic and fulvic acids.

A commonly used method of determining the general components of the C pools that exist with depth in Spodosols has been the extraction and quantification of the humic substances, i.e., humic acid, fulvic acid, and humin (Table 15–2) (Stevenson, 1982; David et al., 1990). Typically, the forest floor is dominated by humin because of the large amounts of intact or partially decomposed organic material. The percentage of organic matter comprised of fulvic acids

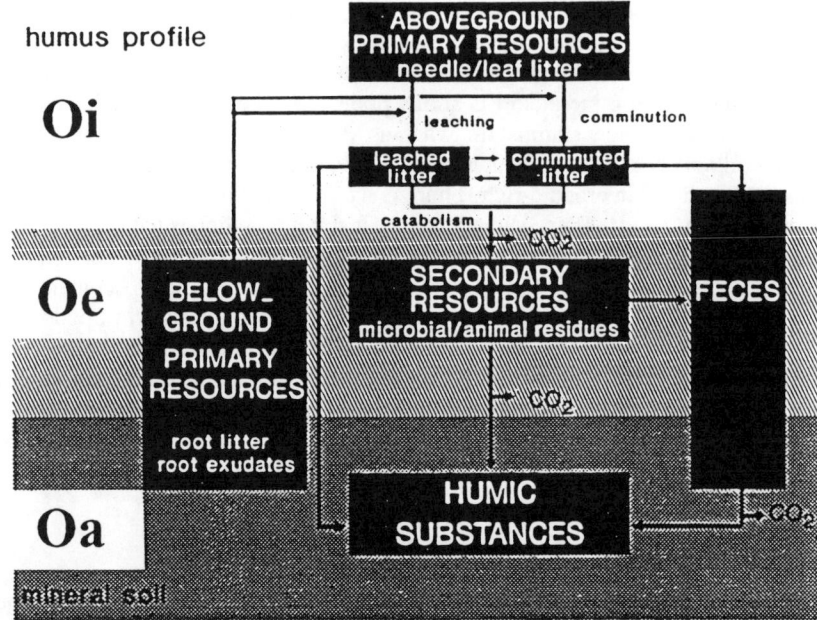

Fig. 15–2. Diagram depicting the processes involved in the transformation of primary and secondary C-based resource materials that comprise the forest floor (adapted from Kögel-Knabner, 1992).

Table 15–2. Quantification of total C and humic substances from 18 soil profiles of the Bear Brook Watershed in Maine. Values represent the average and one standard deviation except for the humin which was obtained by difference (data from David et al., 1990).

Horizon	Total C	Humic acid	Fulvic acid	Humin
		g kg^{-1}		
O	383 ± 57	84 ± 13	35 ± 6	263
E	19 ± 9	6 ± 2	4 ± 2	9
B				
0–2 cm	42 ± 12	14 ± 5	15 ± 5	13
2–7 cm	37 ± 11	12 ± 5	15 ± 5	10
7+ cm†	27 ± 6	9 ± 3	12 ± 2	6
B/C‡	13 ± 4	3 ± 1	8 ± 2	2
C	7 ± 4	2 ± 1	4 ± 2	1

†Represents the material from 7 cm below the B down to 25 cm below the surface of the mineral soil.
‡Represents the material from 25 cm below the surface of the mineral soil to 5 cm above the bottom of the pit.

increases with depth; fulvic acids are usually the dominant humic substance within the lower mineral soil horizons.

Studies of specific C compounds in Spodosols generally focus on certain types of organic compounds such as aromatic or aliphatic low-molecular-mass compounds (Vance et al., 1985; 1986; Fox & Comerford, 1990). However, these types of organic compounds are often rapidly metabolized, thus their existence is in general short unless they are protected by formation of metal complexes.

Carbon compounds in soil solutions have been characterized by DOC fractionation which has indicated the dominant organic compounds present in soil leachates are hydrophobic and hydrophilic acids. Methods used in characterizing humic substances also have been applied to these DOC fractions. More recently, ^{13}C nuclear magnetic resonance (NMR), Fourier Transform Infrared (FTIR) spectroscopy, and cupric oxide oxidation techniques have been used to further assist in the identification of the C-based compounds.

Nitrogen

Little information is available concerning Spodosol organic N compounds. Although many studies have evaluated N mineralization rates, both actual and potential, few have examined the specific N compounds present. The fractionation of soil N has generally been examined using acid hydrolysis, with amino acids and amino sugars the main identifiable organic N compounds (Stevenson, 1982; Schnitzer, 1985). Stevenson (1982) indicated that one-third to one-half of the organic N in most soils can be found in these types of known compounds. The acid hydrolysate of a typical soil may contain the following N constituents: acid insoluble N (20–35%), amino acid N (30–45%), amino sugar N (5–10%), and unknown N (10–20%). Swoden (1956) found that as a percentage of total N amino acids in three Canadian podzol profiles ranged from 22.5 to 41.8% and 19.9 to 24.9% in A and B horizons, respectively. Clearly, additional studies using older methods (e.g., chemical extractions) along with modern techniques (e.g., NMR, ^{15}N dilution) are needed to examine specific compounds for this important nutrient (Schnitzer, 1985).

Sulfur

Sulfur has been extensively studied due to its importance as a major component of acidic deposition, but little is known about the specific organic S compounds that exist in soils. David et al. (1984) and Mitchell et al. (1992a) summarized our current knowledge of S biogeochemistry, with a particular emphasis on ecosystems containing Spodosols. In most studies involving Spodosols, organic S (C-bonded and ester sulfate S) dominates the soil S pool, with smaller amounts of the S pool comprised of sulfate or other inorganic S forms (Mitchell et al., 1992a,b). Few specific organic S compounds, other than S-containing amino acids, have been identified, primarily because most analytical techniques focus on classes of compounds. Soil S concentrations in a boreal conifer forest at Lake Laflamme, Quebec, were dominated by C-bonded S, with smaller amounts of ester sulfate and sulfate (Fig. 15–3). Ester sulfate is formed microbially in the forest floor, and perhaps B horizons, from the decomposition of C-bonded S or microbial utilization of sulfate (David et al., 1987). A dominant process that retains soil sulfate is sorption by the mineral horizons, which in high sulfate deposition areas can result in large inorganic S pools (Mitchell et al., 1992b).

Phosphorus

Little is known about the chemistry of organic P in soil, however, some studies have reported that as much as 50 to 70% of organic P in soils could be accounted for by known P compounds (Anderson, 1980; Cole & Heil, 1981).

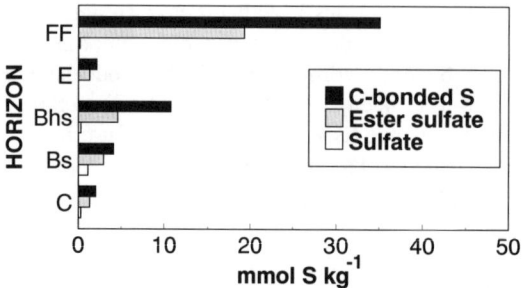

Fig. 15–3. Distribution of organic S constituents in a boreal conifer forest at Lake Laflamme, Quebec (adapted from Houle & Carignan, 1992).

Mechanisms for organic P transformations are not well documented and most available data are from indirect techniques (McGill & Christie, 1983; Smeck, 1985). Organic P in soil occurs primarily as ester linkages to inositol phosphates (2–50% of organic P), with smaller quantities occurring as phospholipids (1–5% of organic P) and nucleic acids (0.2–2.5% of organic P) (McGill & Cole, 1981; Stevenson, 1982). Small amounts (trace levels) of P may be present as phosphoproteins and metabolic phosphates (Stevenson, 1982). Nucleic acids and phospholipids are readily hydrolyzed and returned to soluble and labile pools, whereas inositol phosphates are less readily hydrolyzed and tend to accumulate in soil (Smeck, 1985). Inositol phosphates (esters of hexahydroxybenzene) can become stabilized through formation of insoluble complexes with Al, Fe, and Ca and other organic substances (Stevenson, 1982).

PODZOLIZATION PROCESS

The podzolization process redistributes organic matter in soil profiles from the forest floor to the mineral soil. Dissolved organic C concentrations are greatest in forest floor leachates as chemical and microbial decomposition and solubilization of surface litter and organic matter is a dominant process that controls DOC levels. Concentrations of DOC leaching from the forest floor generally are greater in conifer as compared to hardwood ecosystems (Cronan, 1990). Formation of complexes with Fe and Al enhances DOC deposition in mineral horizons through sorption and precipitation processes. The movement of organic N, S, and P is integral to leaching and deposition processes. With increasing soil depth, DOC concentrations decrease rapidly reaching as little as 100 to 200 μmol L^{-1} in the lower B horizons (McDowell & Likens, 1988; Cronan, 1990). Organic N, S, and P are therefore removed from the forest floor and deposited in the mineral horizons with C, part of the primary mechanism of Spodosol formation.

DECOMPOSITION/MICROBIAL ACTIVITY

Directly linked to the podzolization process is the decomposition of plant materials, including lignin, in the forest floor. As this process progresses, an Oa horizon composed of humus is formed of decomposed plant components and microbial biomass, with CO_2 released during the decomposition process. There is repeated recycling of the microbial biomass C, N, P, and S, with polymerization into high molecular weight compounds (Stevenson, 1986). As the organic matter in the Oa horizon undergoes further decomposition, some of the C (and other elements) is converted into water soluble forms which are then leached during precipitation events. Controls on the amount of organic matter leached from the forest floor include temperature, moisture, ionic strength, pH, and frequency of precipitation events (David et al., 1989; Vance & David, 1989, 1991a; Grieve, 1990, 1991; Cronan et al., 1992). The first two factors affect microbial activity, which is needed to produce water soluble compounds. Ionic strength and pH affect the chemical solubilization of the decomposed materials. Frequency of precipitation is important in determining the soluble organic matter at any given time because microbial activity can only produce so much material per unit time (Grieve, 1990). If two large rainfall events occur over a short time period, soluble organic matter may not be available for leaching by the second event.

ORGANIC CHEMISTRY OF SPODOSOLS

Soil Solutions

Extensive studies have been conducted on DOC in forest and aquatic ecosystems in Maine, particularly with reference to DOC composition and reactivity (David et al., 1989, 1992; David & Vance, 1991; Vance & David, 1991a,b). One of the forest sites studied is a northern hardwood ecosystem in eastern Maine (Bear Brook Watershed), where work has focused on effects of acidic deposition in the solubilization and movement of natural organic acids. These studies have shown (i) the importance of the forest floor in generating DOC that is resistant to decomposition and (ii) mineral soil sorption of solubilized C (Vance & David, 1992).

At the Bear Brook Watershed site, samples of precipitation, stemflow, throughfall, lysimeter soil solutions from Oa and Bs horizons, and stream solutions were analyzed for total DOC and fractionated into five DOC classes (Fig. 15–4). Results of these studies indicated there were low levels of DOC in precipitation and throughfall, higher levels of DOC in stemflow, large releases of DOC from the forest floor, removal of DOC in the B horizon, and stream DOC concentrations that were similar to lower B horizon samples. For most samples the dominant fractions were hydrophobic and hydrophilic acids, although large amount of neutrals also were detected in the stemflow and throughfall samples. A general pattern of greater proportions of hydrophobic acids to hydrophilic acids in the forest floor, with the reverse in the B horizons and stream water also were observed (Fig. 15–4). This pattern is thought to be due to selective adsorption of hydrophobic acids compared to hydrophilic acids in the B horizons, although all are sorbed to a large degree (Cronan & Aiken, 1985; Vance & David, 1989; Cronan, 1990).

Concentrations of DOC in forest floor leachates can vary significantly over time. Levels found in leachates from the Bear Brook hardwood forest floor averaged 4835 µmol L^{-1} and ranged from 2228 to 7193 mmol L^{-1} (Vance &

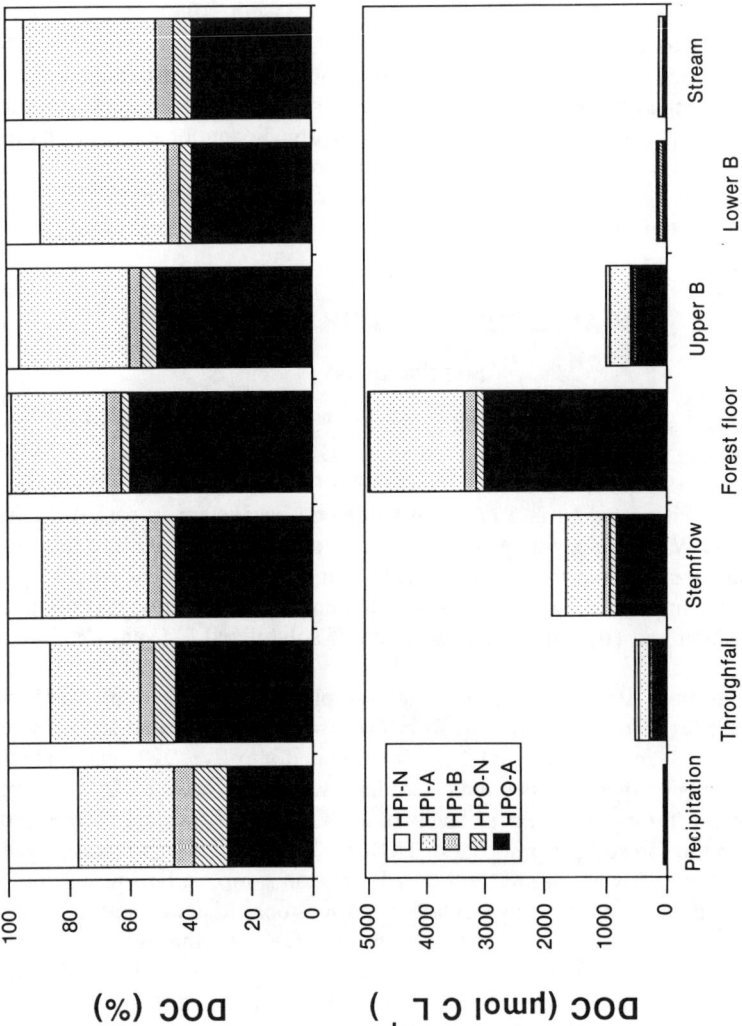

Fig. 15–4. Dissolved organic C concentrations and fractions in various strata of a northern hardwood forest at Bear Brook Watershed, Maine. Fractions are as follows: HPO-A, hydrophobic acids; HPO-N, hydrophobic neutrals; HPI-B, hydrophilic bases; HPI-A, hydrophilic acids; and HPI-N, hydrophilic neutrals (adapted from Vance & David, 1991a; David et al., 1992; M.B. David & G.F. Vance, unpublished data).

David, 1991a). These leachates were comprised on average of organic acids (i.e., hydrophobic and hydrophilic acids) totaling 92% of the DOC, with hydrophobic acids comprising the larger fraction at 60%. The remaining DOC fractions (bases plus neutrals) reached a maximum of 14% of the DOC, hydrophilic bases were most often dominant followed by hydrophobic neutrals. Although the DOC concentrations leached from this forest floor were quite variable, B horizon lysimeter solutions and stream waters draining the hardwood ecosystem were low and generally consistent; 250 to 500 µmol L^{-1} and less than 150 µmol L^{-1} for the B horizon and stream water solutions, respectively. Solutions from a New Hampshire watershed dominated with Spodosols also had relatively low DOC levels in the B horizons and stream waters (McDowell & Wood, 1984).

The organic chemistry of soil solutions during snowmelt was examined by Yavitt and Fahey (1985) in a lodgepole pine (*Pinus centorta* Douglas ex Loudon) forest of Wyoming. The DOC concentrations in forest floor leachates decreased with increasing water flux during the spring snowmelt period. This was attributed to buildup of decomposition products during the winter until they were leached by melting snow. Most of the dissolved organics (85%) were acidic, with the majority of the remaining DOC as neutrals, similar to the findings of Vance and David (1991b). The C/N ratio of snowmelt leachates increased from 25:1 to >75:1 during the snowmelt period, indicating a change in the solubilized dissolved organic matter. Most of the organic compounds were then retained by the mineral soil, again similar to studies on Spodosols and Ultisols in the eastern USA.

Differences in the elemental composition of forest floor solid and dissolved fractions reflect variation in their chemical structure. Hydrophobic acids are similar in elemental composition to soil and aquatic fulvic acids (Stevenson, 1982; Thurman, 1985; Vance & David, 1991b). Concentrations of dissolved N and S have been found to be comparable between the two acid fractions (hydrophobic and hydrophilic) of forest floor leachates; however, C/N and C/S ratios indicate different controls on their solubility (Vance & David, 1991b). Less N is leached from the forest floor in comparison to C while a greater amount of S is leached. Both N and S in forest mineral horizons are typically comprised of organic forms that make up >90% of the total N and S, respectively, which suggests that large amounts of organic N and S are translocated in soil solutions (David et al., 1982; Stevenson, 1982). Studies by Schoenau and Bettany (1987) support the premise that N, S, and P are translocated in soils as soluble, nutrient-rich organic matter.

Qualls and Haines (1991) and Qualls et al. (1991) examined fluxes of dissolved organic nutrients and C in a deciduous ecosystem at Coweeta, North Carolina. Although their work was not conducted in forest ecosystems containing Spodosols, their studies include some of the first detailed results on organic nutrient fluxes in soil solutions. They evaluated the importance of dissolved organic matter in the movement of N and P from the canopy and the forest floor into the mineral soil. Leaching of dissolved organic matter was important for translocation of N and to a smaller degree for P (Qualls et al., 1991). Soluble inorganic N and P compounds were primarily removed by immobilization in the forest floor. After fractionating dissolved organic matter they found that most of the dissolved organic N leaching from the forest floor was in the form of hydrophobic and hydrophilic acids, and not as free proteins or amino acids (Qualls & Haines, 1991; Qualls et al., 1991). They also observed that most of the dissolved organic P migrated in the acid fractions. The N and P in the organic molecules seemed to have little effect on the behavior of dissolved organic substances, whose movement was dominated by the carboxylic and phenolic functional groups present. Qualls et al. (1991) also pointed out how the movement of particulate organic N and P is important to nutrient cycling in forested ecosystems.

In laboratory incubation studies, Qualls (1989) found that dissolved organic matter can be relatively resistent to degradation, especially that leaching from the forest floor. Adsorption of DOC by mineral soils was believed to be responsible for the decline in dissolved organic matter with soil depth. This would suggest that material that is leached from the forest floor is relatively stable and first removed by adsorption, and then is accumulated or slowly decomposed over time.

David et al. (1987) and Mitchell et al. (1989) found substantial concentrations of organic S in throughfall and forest floor leachates in New York and New Hampshire Spodosols, respectively. Most of the organic S (>72%) leaching from the forest floor was in an ester sulfate form (David et al., 1987). It also was found in this study that organic S was not detectable after solutions had passed through the B horizon. This supports the hypothesis that ester sulfate is formed in the forest floor by microbial activity and is leached and deposited in B horizons (David et al., 1987). Ester sulfate forms rapidly in soils from the Huntington Forest, New York, particularly in the forest floor (David & Mitchell, 1987; Schindler et al., 1986). In a range of Integrated Forest Study sites, Homann et al. (1990) also found large amounts of organic S in throughfall and soil solutions. Their study also indicated that S cycling budgets within forests can be altered by including organic S in solution in addition to sulfate alone, because internal transfers and outputs may be greater. Mitchell et al. (1989) estimated that the accumulation of organic S in the mineral soil from forest floor leachates could account for the organic S accumulated over a period of approximately 2500 yr. This demonstrates the long-term importance of this mode of organic S formation in mineral soils, which suggests that soil storage of sulfate deposition (from acidic deposition) in organic forms may take decades to accumulate.

Although organic S in many Spodosols is mainly derived from atmospheric deposition, P is generally supplied from parent materials. Generally, accumulation of organic and other forms of P in mineral soils is largely a result of pedogenesis (Walker & Syers, 1976). However, few studies have examined the movement of particulate or dissolved organic P, especially as it relates to accumulation in the mineral soil. Stewart and Tiessen (1987) presented a conceptual model that included the transformations of soil organic P. They concluded that microbial uptake and subsequent release and redistribution of P were the dominant processes in the soil organic P cycle. Controlling factors of soil organic P transformations were the solution P concentration and the activities of soil biomass. Most of these results were based on short-term laboratory studies, which, unfortunately, were not field tested.

Wood (1980), Wood et al. (1984), and Yanai (1991, 1992) concluded that solution P levels in forest floor and A horizons (if present) of a hardwood forest were controlled principally through biological controls (i.e., microbial mineralization, tree uptake). The underlying B horizons, however, showed strong chemical control (i.e., inorganic sorption) because of high concentrations of amorphous Fe and Al. Wood (1980) concluded that the P cycle was dominated by recirculation of organic P pools between the forest vegetation and the forest floor.

Yavitt and Fahey (1985) found that P appeared to be the most susceptible to deep leaching in organic forms as compared to N and S, and least susceptible to leaching in the inorganic form. Qualls and Haines (1991) determined that concentrations of dissolved organic P declined with soil depth at a greater rate than DOC or dissolved organic N (Table 15–3). In their study, most dissolved organic P occurred in the hydrophilic acid, hydrophobic acid, and hydrophilic neutral fractions. The ester phosphate functional group is negatively charged at neutral pH and hydrophilic ester phosphates occurred in the hydrophilic acid fraction (Qualls et al., 1991; Qualls & Haines, 1991). Model phosphate compounds such as glucose-phosphate (ester phosphate) and phytic acid (inositol P) are approximately 89 and 95% hydrophilic acids, respectively (Qualls & Haines, 1991; M.B. David, unpublished data). The hydrophilic fraction as a carrier of P is therefore a more important component than is the hydrophobic fraction in lower A and B horizons (Qualls & Haines, 1991).

Soil Organic Solid Phases

Several new techniques and instrumentation have become available for characterizing solid phase organic constituents. Thus far only C has been analyzed by many of the state-of-the-art methods due to the amount of C in forest floor and spodic horizon samples generally being sufficient to conduct the analysis. This is not always the case for organic N, S, and P contents, which are typically two to four orders of magnitude lower than C. However, with concentration and refinement of these new techniques, advancements in the area of solid phase characterization of organic nutrients is expected to provide many new and exciting results.

Great advances have taken place in the area of spectroscopic measurements for the characterization of soil organic compounds. Analysis of organic compounds by FTIR and NMR techniques have provided information that has changed our view on organic matter composition. Infrared spectroscopy has been used to confirm the carboxylic, ester, ketone, and hydroxyl nature of soil and aquatic organic substance, and is capable of quantifying ester and alcohol groups

Table 15–3. Dissolved organic C/dissolved organic P ratios, by weight, for the inital sample and fractions of dissolved organic matter (forested watershed in the Appalachian Mountains of North Carolina; adapted from Qualls & Haines, 1991).

Horizon	Initial		Hydrophobic acids		Hydrophilic acids	
	August	December	August	December	August	December
Oa	2700	3340	3900	7700	1400	2700
A	1700	3820	4000	2000	900	2800
B	403	1140	700	8400	250	340
C	330	820	404	2200	234	272

(Bloom & Leenheer, 1989). The ^{13}C NMR spectra of soil organic matter indicate these materials are not of a predominant aromatic nature, as was earlier thinking, but are rather highly aliphatic.

Humic and fulvic acids extracted from forest floor and mineral horizons have been analyzed by ^{13}C NMR (Fig. 15–5). From studies using NMR techniques the nature of the C-based materials present in forest soils has been elucidated. Carbon in forest litter layer is comprised of approximately 55% O-alkyl C (mainly polysaccharides, i.e., cellulose and hemicellulose), 20% aromatic C (primarily lignins and tannins), 5% carboxyl C, and 20% alkyl C (largely proteins and lipids) (Kögel-Knabner, 1992). Figure 15–5 indicates C-based materials in humic acids extracted from forest floor (Oi, Oe, and Oa) and mineral soil change with depth. With increasing depth there is an increase in the alkyl (0–50 ppm) and carboxyl (160–220 ppm) C with a corresponding decrease in the amount of O-alkyl (50–110 ppm) C.

Vance and David (1991a) isolated hydrophobic and hydrophilic acid fractions from forest floor leachates and examined their chemical composition by several analytical methods. With FTIR they identified several distinct differences between the two fractions (Fig. 15–6). Both fractions contained a significant amount of carboxylic (1720 cm^{-1}) and hydroxyl (broad band with maximum adsorption at 3420 cm^{-1}) functional groups; however, hydrophobic acid had additional low molecular mass aromatic acids (1500 cm^{-1}) and hydrophilic acids contained greater amounts of carbohydrate material (1085 and 955 cm^{-1} bands).

Dissolved organic C in forest floor leachates is adsorbed by spodic horizons as solutions move deeper in the soil profile. Preliminary results based on ^{13}C NMR analysis of the solid phase organic C in spodic horizons and forest floor DOC fractions suggest the C accumulated or adsorbed by the mineral horizons is

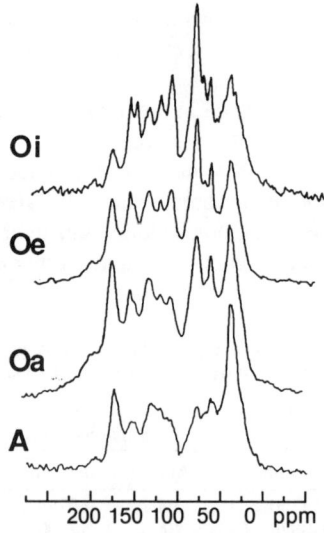

Fig. 15–5. The ^{13}C nuclear magnetic resonance spectra of humic acids extracted from the forest floor layers and a mineral horizon of a German soil (adapted from Kögel-Knabner, 1992).

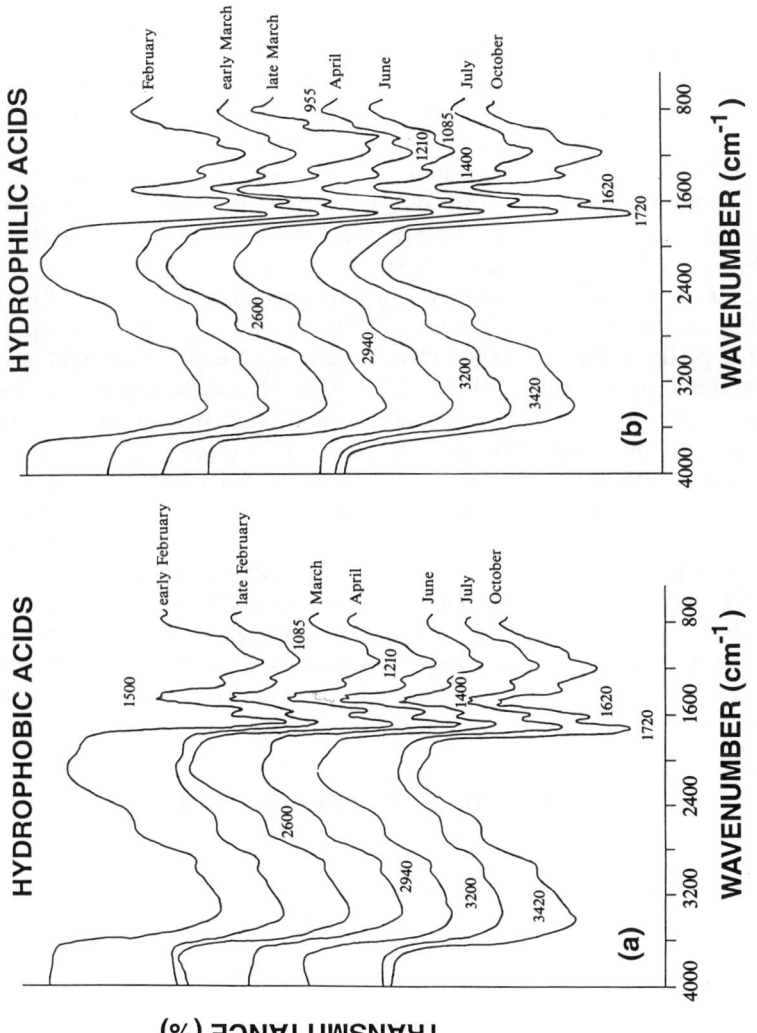

Fig. 15–6. Fourier Transform Infrared spectroscopy spectra of hydrophobic (*a*) and hydrophilic acids (*b*) isolated from forest floor leachates at different times of the year (adapted from Vance & David, 1991a).

similar in character as the leachate DOC (G.F. Vance & M.B. David, unpublished data).

Use of NMR for the evaluation of organic N, S, and P would enhance our understanding of the formation, solubilization and accumulation of these elements in forested ecosystems. Bishop et al. (1992) used ^{31}P NMR to study the mineralization of organic P compounds in soils by enzymatic processes. Addition of phytase enzymes to soils resulted in concentrations of phytic acid, which is thought to be the primary form of organic P in soils, to decrease with a concomitant increase in orthophosphate. Bishop et al. (1992) found the ^{31}P NMR technique to be useful for studying mineralization, however, they also suggested soils must have sufficiently high levels of organic P in order to use this method for examining P transformations. Other studies also have shown that ^{31}P NMR can be used to detect organophosphorus compounds in soils. Unless labeled ^{15}N-labeled or tracer studies are performed, ^{15}N NMR is generally not sensitive enough to detect the natural low levels of this isotope.

Information derived from studies on the structural characterization of forest floor organic matter using advanced analytical techniques (e.g., FTIR, CuO oxidation, ^{13}C NMR, and pyrolysis-gas chromatography) has provided additional clues into the nature of the organic materials involved, and the transformations that occur, in the process of C production, solubilization and accumulation in forested ecosystems. Kögel-Knabner (1992) proposed a theoretical model (Fig. 15–7) for the decomposition and humification pathways that take place in the formation of soil organic matter. Lignin structural units are transformed through the loss of oxygen-containing functional groups, with the recalcitrant lignin molecules remaining primarily intact. Much of the carbohydrate and protein materials are readily decomposed and microbially resynthesized. The cutin/suberin and other aliphatic biomacromolecules are transformed through pathways that are not well known. In addition, our knowledge of the pathways that contribute to the synthesis of N, S, and P containing organic matter is limited, and further research is certainly needed in this area.

INTERACTIONS AMONG CARBON, NITROGEN, SULPHUR, AND PHOSPHORUS

Chemical, physical, biological, and hydrological processes control the solubilization of organic C, N, S and P, as well as the movement of these constituents within forested ecosystems (Fahey et al., 1985; Mahendrappa, 1989; Vance & David, 1991b). Organic acids are constantly being produced and released into solution within the forest floor. These organic acids, along with the organic forms of N, S, and P, are transported to subsurface horizons and into stream waters. The amount of DOC and organic acidity leaving the forest floor is generally high compared to that exiting the B horizon (Cronan & Aiken, 1985; McDowell & Likens, 1988). This would suggest adsorption, precipitation and degradation mechanisms must play a significant role in the movement of organic C, N, S, and P. Much of these forms are removed from solution when forest floor leachates percolate through mineral subsurface horizons (McDowell & Wood, 1984; Cronan & Aiken, 1985; Cronan, 1990).

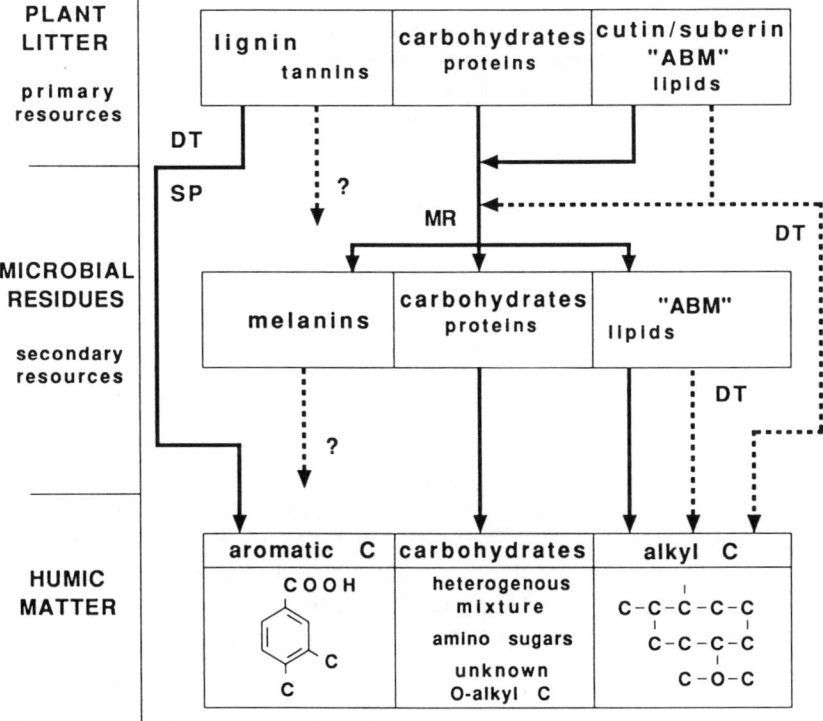

DT = direct transformation
MR = microbial resynthesis   "ABM" = aliphatic biomacromolecules
SP = selective preservation

Fig. 15–7. Proposed pathways for the decomposition and transformations of organic compounds during the different stages of humification that determines the composition of soil organic matter. Solid lines are examples of known pathways, dashed lines are suggested pathways, and question marks represent further research is need (adapted from Kögel-Knabner, 1992).

Many attempts have been made to examine elemental ratios to better understand the linkages among C, N, S, and P in forest soils. Generally, over various soil and forest types few clear patterns emerge (Homann & Harrison, 1992). A few important patterns are presented below for Spodosols, however.

Soil S appears to be closely linked to the overall soil organic matter cycling of C and N. Typically, ratios of total N to C-bonded S do not change greatly with soil depth, whereas total N/ester sulfate ratios decrease sharply (Fig. 15–8). This suggests greater ester sulfate leaching compared to C-bonded S, with higher amounts of ester sulfate deposited into deep B and even C horizons in Spodosols. Carbon-bonded S is stabilized and released (or mineralized) in conjunction with soil C, and is leached from the forest floor and deposited in mineral soils along with C. Ester sulfate is formed and solubilized in the forest floor, but apparently leaches deeper in the mineral soil (David et al., 1987).

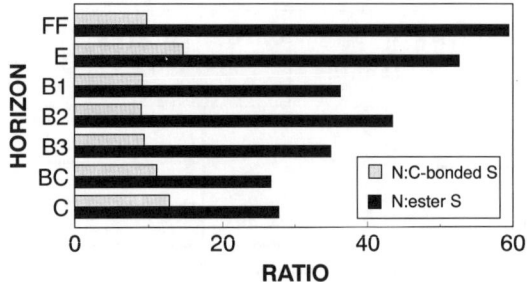

Fig. 15–8. Mean ratios of soil total N to C-bonded S and ester sulfate by soil depth ($n = 18$) from hardwood plots at Bear Brook Watershed, Maine. Depths are as follows: B1, 0 to 2 cm; B2, 2 to 7 cm; B3, 7 cm into B to 25 cm below the surface of the mineral soil; B/C, from 25 cm below the surface of the mineral soil to 5 cm above the bottom of the pit (calculated from David et al., 1990).

Phosphorus leaching and accumulation is similar to ester sulfate in some Spodosols. A decrease in the total N/organic P ratio is often found, suggesting greater P leaching with depth compared to N (Fig. 15–9). The mechanism for this may be the leaching of ester P components within the hydrophilic forms which are weakly sorbed by mineral horizons. Once deposited, protection of the ester P components may occur by way of sorption to sesquioxides (Stewart & Tiessen, 1987).

Schoenau and Bettany (1987) observed that soil organic C/N, C/P, and C/S ratios narrowed with depth in a deciduous forest soil (Table 15–4). A similar pattern also was found in the fulvic acid fraction in this soil, which when combined with decreasing humic acid to fulvic acids ratios with increasing soil depth, is consistent with the dominant forms of organic matter translocated in percolating water (Schoenau & Bettany, 1987). They suggested that these patterns demonstrate that organic matter translocated in this soil is richer in N, P, and S than the fractions not solubilized and leached. Organic P (compared to C, N, and S) was the element found to be the most susceptible to deep leaching.

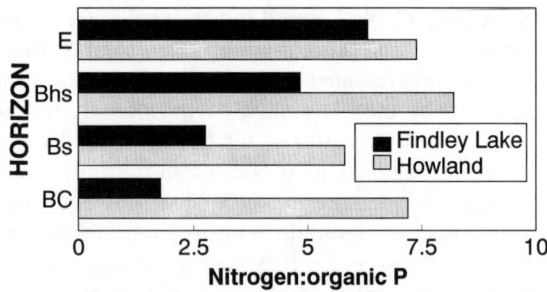

Fig. 15–9. Ratio of soil total N to organic P at Findley Lake, Washington and Howland, Maine, conifer forests. Findley Lake data from Homann and Harrison, 1992; Howland data from A.J. Krzyszowska, M.B. David and G.F. Vance (unpublished data).

Table 15–4. Organic C/N, C/P, C/S ratios from a Canadian forest soil in three stope positions (Typic Cryoboralf; data from Schoenau & Bettany, 1987).

Horizon	Slope position	C/N	C/P	C/S
L–H	Upper	17.5	286	231
Ae		14.1	87	152
Bt		11.3	53	100
Ck		5.8	18	17
L–H	Middle	16.6	249	240
Ae		13.2	71	133
Bt		11.3	36	80
Ck		7.2	29	20
L–H	Lower	15.8	256	215
Ae		12.6	49	127
Bt		10.3	31	69
Cca		7.9	38	41

BIOGEOCHEMISTRY OF SPODOSOLS

Conceptual Cycle of Carbon, Nitrogen, Sulfur, and Phosphorus

McGill and Cole (1981) presented a conceptual model that compares the processes/mechanisms involved in the cycling of organic C, N, S, and P in soils through soil organic matter. Two mineralization mechanisms were proposed: *biological mineralization* for C, N, and C-bonded S; and *biochemical mineralization* for organic P and sulfate esters. They defined biological mineralization as a microbial process that releases N and S as the organic matter is oxidized for energy. Biochemical mineralization releases P and S by extracellular enzymatic processes. This conceptual model provides a framework for the following discussion of C, N, S, and P cycles.

Our studies, and those of others, have shown that large amounts of DOC and organic nutrients are solubilized and leached from the forest floor. We have discussed our current understanding of organic chemistry of these compounds, both in solution and in the solid phase. Carbon and N appear to be stabilized together in the forest floor, with solubilization and leaching controlled primarily by microbial activity. The use of NMR and other results (see earlier discussion) have provided valuable information on the mechanism of C decomposition. Sulfur cycling has been thought to be controlled both by the biological mineralization in conjunction with C and N, and by biochemical mineralization; however, there is little evidence in Spodosols for biochemical control of S. The McGill and Cole (1981) model would predict an increase in the size of the ester sulfate pool with increased sulfate availability, such as from acidic deposition inputs. This has not been found in numerous laboratory and field experiments, over short to long (several years) time spans (e.g., David et al., 1990, 1991; Mitchell et al., 1994). Also, Houle and Carignan (1992) found no relationship between forest floor organic S speciation and sulfate inputs in a summary of 10 sites. They concluded that microbial S immobilization of S in soils of forested catchments was not related to S inputs. Furthermore, they indicated that S transformations likely follow overall soil organic matter transformations, and controls due to microbial requirements for S are probably not important.

According to the McGill and Cole (1981) model, organic P is mineralized through biochemical mineralization in response to microbial enzymes that are synthesized when low amounts of inorganic P are available in the soil (similar to S). They concluded that: (i) mineralization of organic P can occur independently of C mineralization, (ii) organic P content is a function of the relative rates of biomass production and the hydrolysis of phosphate esters, and (iii) hydrolysis of phosphate esters is repressed by high levels of labile P and induced by low levels of labile P.

Stability of soil organic P is due more to the phosphate group than to that of the C pools. Consistent with this are observations that P association with dissolved organic matter increases rather than decreases the mobility of P in soils, and half or more of the P in soil solution is organic. The phosphate group in organic P molecules can be very accessible for adsorption to, or reaction with, soil inorganic material. Organic P compounds are known to form metal-organic P complexes, much the same as inorganic P (Anderson et al., 1974). Specifically, reactions with Al and Fe tend to control P availability in Spodosols (Ballard & Fiskell, 1974).

Northern Spodosol Conceptual Cycle

By adapting the McGill and Cole (1981) model, a conceptual diagram of soil C, N, S, and P cycling can be described, as shown in Fig. 15–10. Forest floor organic matter decomposition regulates C, N, and S cycling, with P cycling regulated both by decomposition and microbial need. Overall microbial activity regulates DOC and N, S, and P release from the forest floor. Both ester S and P forms are released, along with DOC, organic N (C–N), and C-bonded S (C–S). These organic forms are leached to B horizons where they are stabilized as complexes or precipitates with Fe and Al. Ester sulfate and P move into deeper mineral horizons than C, N, and C-bonded S because they are found in hydrophilic acids and neutrals, which are weakly adsorbed. Mineralization of all elements in the forest floor is controlled by C metabolism, with much slower decomposition in the mineral soil.

IMPLICATIONS

Importance of Organic Cycles

The biogeochemistry of C, N, S, and P in forest ecosystems is of major importance in maintaining tree health and productivity. However, our insight into these elemental cycles is limited with respect to much of the organic chemistry that is involved, as discussed in this paper. The accumulation of above- and belowground litter to the forest floor and mineral soil, solubilization of organics from the canopy and the forest floor, and formation of a spodic horizon are dominant aspects of C, N, S, and P biogeochemistry. We need a better understanding of the controls in which organic forms of these elements have on the

ORGANIC ELEMENTS IN FOREST ECOSYSTEMS

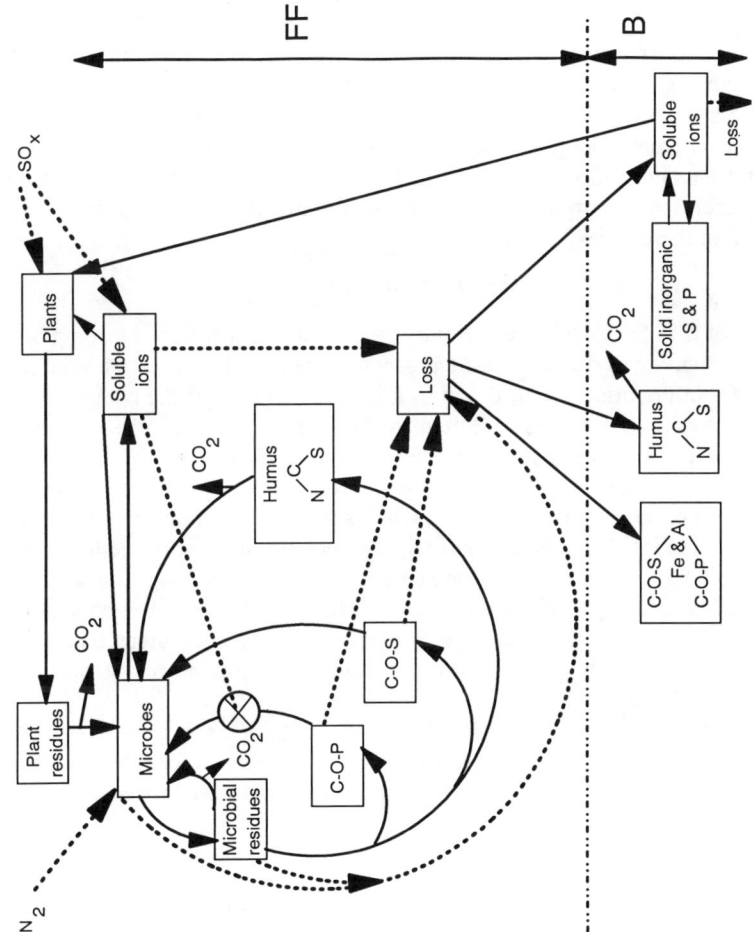

Fig. 15–10. Conceptual model of C, N, S, and P cycling in northern Spodosols (adapted from McGill & Cole, 1981).

biogeochemical processes that affect their accumulation, decomposition, and uptake by forest species. Because forests found on Spodosol-type soils are a major forest resource in the USA, particularly in the northeastern and northwestern states and in Canada, a better understanding of the biogeochemistry of C, N, S, and P would be of immediate benefit in managing these resources.

Additional information is needed on the role of C in N, S, and P cycling for several reasons. First, to enhance our management of our forest ecosystems, this type of information would be useful in determining basic site quality, including how management might alter the organic chemistry of C, N, S, and P. Second, the long-term effects of atmospheric pollutants (e.g., acidic deposition, trace metals, ozone) are not well understood for forest ecosystems. Research in this area would be applicable in predicting how these stresses might alter the cycling of C, N, S, and P. Third, predicted long-term climate change could severely impact the formation and decomposition of organic matter in forest soils. Studies would be extremely useful in determining what these impacts might be on the cycling of C, N, S, and P in these types of ecosystems.

Major Unknowns

Forms of C in Spodosols are currently being identified, which is leading to a better understanding of decomposition processes. There is, however, the need for much greater information for N, P, and S. It is not at all clear if the controls of S and P organic cycling are really different as compared to C and N. Also, sources of spodic horizon organic matter have not been well quantified. We have suggested that leaching of DOC from the forest floor can account for much of the spodic horizon organic matter. However, what is the role of inputs from fine root turnover and exudates? Are these inputs quickly decomposed and not added to the long-term soil organic matter pool or are they incorporated into long-term organic C, N, S, and P storage as well? How resistant to change or how stable are the large pools of organic materials found in B horizons? These are all areas that need additional studies to fully understand the complexity of forest ecosystems containing Spodosol soils.

Clearly, the organic chemistry of C, N, S, and P is complex, controlled by both biotic and abiotic processes. Current research is addressing this area, with new and powerful techniques that should increase our knowledge of these processes.

REFERENCES

Anderson, G. 1980. Assessing organic phosphorus in soils. p. 411–431. *In* F.E. Khasawneh et al. (ed.) The role of phosphorus in agriculture. ASA, CSSA, and SSSA, Madison, WI.

Anderson, G., E.G. Williams, and J.O. Moir. 1974. A comparison of the sorption of inorganic orthophosphate and inositol hexaphosphate by six acid soils. J. Soil Sci. 25:51–62.

Anderson, J.M., P.W. Flanagan, E. Caswell, D.C. Coleman, E. Cuevas, D.W. Freckman, J.A. Jones, P. Lavelle and P. Vitousek. 1989. Biological processes regulating organic matter dynamics in tropical soils. p. 97–123. *In* D.C. Coleman et al. (ed.) Dynamics of soil organic matter in tropical ecosystems. University of Hawaii Press, Honolulu, HI.

Ballard, R., and J.G.A. Fiskell. 1974. Phosphorus retention in coastal plain forest soils: I. Relationship to soil properties. Soil Sci. Soc. Am. Proc. 38:250–254.

Bishop, M.L., R.W.K. Lee, and A.C. Chang. 1992. Enzymatic mineralization of soil organic phosphorus. p. 234. *In* Agronomy abstracts. ASA, Madison, Wi.

Bloom, P.R., and J.A. Leenheer. 1989. Vibrational electronic and high energy spectroscopic methods for characterizing humic substances. p. 409–446. *In* M.H.B. Hayes et al. (ed.) Humic substances. II. John Wiley & Sons, Chichester, England.

Candler, R. 1985. Water extractable material from the B horizon of a contiguous aspen and birch forest in interior Alaska. Org. Geochem. 8:127.

Candler, R., and K. Van Cleve. 1982. A comparison of aqueous extracts from the B horizon of a birch and aspen forest in interior Alaska. Soil Sci. 134:176–180.

Cole, C.V., and R.D. Heil. 1981. Phosphorus effects on terrestrial nitrogen cycling. Ecol. Bull. 33:363–374.

Cronan, C.S. 1990. Patterns of organic acid transport from forested watersheds to aquatic ecosystems. p. 245–260. *In* E.M. Perdue and E.T. Gjessing (ed.) Organic acids in aquatic ecosystems. Wiley-Interscience Publ., New York.

Cronan, C.S., and G.R. Aiken. 1985. Chemistry and transport of soluble humic substances in forested watersheds of the Adirondack Park, NY. Geochim. Cosmochim. Acta 49:1697–1705.

Cronan, C.S., S. Lakshman, and H.H. Patterson. 1992. Effects of disturbance and soil amendments on dissolved organic carbon and organic acidity in red pine forest floor. J. Environ. Qual. 21:457–463.

Cruickshank, J.G., and M.M. Cruickshank. 1981. The development of humus-iron podsol profiles, linked by radiocarbon dating and pollen analysis to vegetation history. Oikos 36:238–253.

David, M.B., W.J. Fasth, and G.F. Vance. 1991. Forest soil response to acid and salt additions of sulfate: I. Sulfur constituents and net retention. Soil Sci. 151:136–145.

David, M.B., R.D. Fuller, I.J. Fernandez, M.J. Mitchell, L.E. Rustad, G.F. Vance, A.C. Stam, and S.C. Nodvin. 1990. Spodosol variability and assessment of response to acidic deposition. Soil Sci. Soc. Am. J. 54:541–548.

David, M.B., and M.J. Mitchell. 1987. Transformations of organic and inorganic sulfur: Importance to sulfate flux in an Adirondack forest soil. J. Air Pollut. Control Assoc. 46:847–852.

David, M.B., M.J. Mitchell, and J.P. Nakas. 1982. Organic and inorganic sulfur constituents of a forest soil and their relationship to microbial activity. Soil Sci. Soc. Am. J. 46:847–852.

David, M.B., M.J. Mitchell, and S.C. Schindler. 1984. Dynamics of organic and inorganic sulfur constituents in hardwood forest soils. p. 221–245. *In* E.L. Stone (ed.) Forest soils and treatment impacts. Proc. North American Forest Soils Conf., 6th, Knoxville, TN. 19–23 June 1983. Univ. Tennessee, Knoxville.

David, M.B., T.J. Scott, and M.J. Mitchell. 1987. Importance of biological processes in the sulfur budget of a northern hardwood ecosystem. Biol. Fert. Soils 5:258–264.

David, M.B., and G.F. Vance. 1991. Chemical character and origin of organic acids in streams and seepage lakes of central Maine. Biogeochemistry 12:17–41.

David, M.B., G.F. Vance, and J.S. Kahl. 1992. Chemistry of dissolved organic carbon and organic acids in two streams draining forested watersheds. Water Resour. Res. 28:389–396.

David, M.B., G.V. Vance, J.M. Rissing, and F.J. Stevenson. 1989. Organic carbon fractions in extracts of O and B horizons from a New England Spodosol: Effects of acid treatment. J. Environ. Qual. 18:212–217.

David, M.B., and W. Zech. 1990. Adsorption of dissolved organic carbon and sulfate by acid forest soils in the Fichtelgebirge, FRG. Z. Pflanzenernähr. Bodenk. 153:379–384.

Driscoll, C.T., R.D. Fuller, and W.D. Schecher. 1989. The role of organic acids in the acidification of surface waters in the eastern U.S. Water Air Soil Pollut. 43:21–40.

Ertel, J.R., J.I. Hedges, A.H. Devol, J.E. Richey, and M. de N.G. Ribeiro. 1986. Dissolved humic substances of the Amazon River system. Limnol. Oceanogr. 31:739–754.

Fahey, T.J., and J.B. Yavitt. 1988. Soil solution chemistry in lodgepole pine (*Pinus contorta* sp. latifolia) ecosystems, southeastern Wyoming, USA. Biogeochemistry 6:91–118.

Fahey, T.J., J.B. Yavitt, A.E. Blum, and J.I. Drever. 1985. Controls of soil-solution chemistry in lodgepole pine forest ecosystems, Wyoming. p. 473–484. *In* D.E. Caldwell et al. (ed.) Planetary ecology. Van Nostrand Reinhold Press, New York.

Fox, T.R., and N.R. Comerford. 1990. Low-molecular-weight organic acids in selected forest soils of the southeastern USA. Soil Sci. Soc. Am. J. 54:1139–1144.

Grieve, I.C. 1990. Variations in chemical composition of the soil solution over a four-year period at an upland site in southwest Scotland. Geoderma 46:351–362.

Grieve, I.C. 1991. A model of dissolved organic carbon concentrations in soil and stream waters. Hydrol. Processes 5:301–307.

Guggenberger, G., and W. Zech. 1992. Retention of dissolved organic carbon and sulfate in aggregated acid forest soils. J. Environ. Qual. 21:643–653.

Homann, P.S., and R.B. Harrison. 1992. Relationships among N, P, and S in temperate forest ecosystems. p. 215–232. *In* D.W. Johnson and S.E. Lindberg (ed.) Atmospheric deposition and forest nutrient cycling. Ecol. Stud. 91. Springer-Verlag, New York.

Homann, P.S., M.J. Mitchell, H. Van Miegroet, and D.W. Cole. 1990. Organic sulfur in throughfall, stem flow, and soil solutions from temperate forests. Can. J. For. Res. 20:1535–1539.

Houle, D., and R. Carignan. 1992. Sulfur speciation and distribution in soils and aboveground biomass of a boreal coniferous forest. Biogeochemistry 16:63–82.

Howell, G.D. 1989. Seasonal patterns of mineral and organic acidification in two streams in southwestern Nova Scotia. Water Air Soil Pollut. 46:165–175.

Huntington, T.G., and D.F. Ryan. 1990. Whole-tree-harvesting effects on soil nitrogen and carbon. For. Ecol. Manage. 31:193–204.

Johnson, D.W., and S.E. Lindberg (ed.). 1992. Atmospheric deposition and forest nutrient cycling. Ecol. Stud. 91. Springer-Verlag, New York.

Kögel-Knabner, I. 1992. Forest soil organic matter: Structure and formation. Bayreuther Bodenkundliche Berichte, Univ. Bayreuth, Germany.

Leenheer, J.A. 1981. Comprehensive approach to preparative isolation and fractionation of dissolved organic carbon from natural waters and wastewaters. Environ. Sci. Technol. 15:578–587.

Mahendrappa, M.K. 1989. Impacts of forests on water chemistry. Water Air Soil Pollut. 46:61–72.

McDowell, W.H., and G.E. Likens. 1988. Origin, composition, and flux of dissolved organic carbon in the Hubbard Brook Valley. Ecol. Monogr. 58:177–195.

McDowell, W.H., and T. Wood. 1984. Podzolization: Soil processes control dissolved organic carbon concentrations in stream water. Soil Sci. 137:23–32.

McFee, W.W., and E.L. Stone. 1965. Quantity, distribution, and variability of organic matter and nutrients in a forest podzol in New York. Soil Sci. Soc. Am. Proc. 29:432–436.

McGill, W.B., and E.K. Christie. 1983. Biogeochemical aspects of nutrient cycle interactions in soils and organisms. p. 271–301. *In* B. Bolin and R.B. Cook (ed.) The major biogeochemical cycles and their interactions. John Wiley & Sons, New York.

McGill, W.B., and C.V. Cole. 1981. Comparative aspects of cycling of organic C, N, S, and P through soil organic matter. Geoderma 26:267–286.

Mitchell, M.J., M.B. David, I.J. Fernandez, R.D. Fuller, K. Nadelhoffer, L.E. Rustad, and A.C. Stam. 1994. Response of buried mineral soil bags to experimental acidification of forest systems. Soil Sci. Soc. Am. J. 58:556–563.

Mitchell, M.J., M.B. David, and R.B. Harrison. 1992a. Sulphur dynamics of forest ecosystems. p. 215–254. *In* R.W. Howarth et al. (ed.) Sulphur cycling on the continents: Wetlands, terrestrial ecosystems, and associated water bodies. John Wiley & Sons, Chichester, England.

Mitchell, M.J., C.T. Driscoll, R.D. Fuller, M.B. David, and G.E. Likens. 1989. Effect of whole-tree harvesting on the sulfur dynamics of a forest soil. Soil Sci. Soc. Am. J. 53:933–940.

Mitchell, M.J., R.B. Harrison, J.W. Fitzgerald, D.W. Johnson, S.E. Lindberg, Y. Zhang, and A. Autry. 1992b. Sulfur distribution and cycling in forest ecosystems. p. 90–140. *In* D.W. Johnson and S.E. Lindberg (ed.) Atmospheric deposition and forest nutrient cycling. Ecol. Stud. 91. Springer-Verlag, New York.

Mroz, G.D., M.F. Jurgensen, and D.J. Frederick. 1985. Soil nutrient changes following whole tree harvesting on three northern hardwood sites. Soil Sci. Soc. Am. J. 49:1552–1557.

Parsons, J.W. 1988. Isolation of humic substances from soils and sediments. p. 3–14. *In* F.H. Frimmel and R.F. Christman (ed.) Humic substances and their role in the environment. Wiley Interscience Publ., New York.

Qualls, R.G. 1989. Geochemical and biological properties of dissolved organic matter in the soil and stream of a deciduous forest ecosystem: Their influence on retention of N and P. Ph.D. diss. Univ. of Georgia, Athens (Diss. Abstr. 90-03448).

Qualls, R.G., and B.L. Haines. 1991. Geochemistry of dissolved organic nutrients in water percolating through a forest ecosystem. Soil Sci. Soc. Am. J. 55:1112–1123.

Qualls, R.G., B.L. Haines, and W.T. Swank. 1991. Fluxes of dissolved organic nutrients and humic substances in a deciduous forest. Ecology 72:254–266.

Schindler, S.C., M.J. Mitchell, T.J. Scott, R.D. Fuller, and C.T. Driscoll. 1986. Incorporation of ^{35}S-sulfate into inorganic and organic constituents of two forest soils. Soil Sci. Soc. Am. J. 50:457–462.

Schnitzer, M. 1985. Nature of nitrogen in humic substances. p. 303–325. *In* G.R. Aiken et al. (ed.) Humic substances in soil, sediment, and water. Wiley-Interscience, New York.

Schoenau, J.J., and J.R. Bettany. 1987. Organic matter leaching as a component of carbon, nitrogen, phosphorus, and sulfur cycles in a forest, grassland, and gleyed soil. Soil Sci. Soc. Am. J. 51:646–651.

Smeck, N.E. 1985. Phosphorus dynamics in soils and landscapes. Geoderma 36:185–199.
Stevenson, F.J. 1982. Humus chemistry: Genesis, composition, reactions. John Wiley & Sons, New York.
Stevenson, F.J. 1986. Cycles of soil carbon, nitrogen, phosphorus, sulfur, micronutrients. John Wiley and Sons, New York.
Stewart, J.W.B., and H. Tiessen. 1987. Dynamics of soil organic phosphorus. Biogeochemistry 4:41–60.
Swoden, F.J. 1956. Distribution of amino acids in selected horizons of soil profiles. Soil Sci. 82:491–496.
Thurman, E.M. 1985. Organic geochemistry of natural waters. Martinus Nijhoff/dr. W. Junk, Dordrecht, the Netherlands.
Vance, G.F., S.A. Boyd, and D.L. Mokma. 1985. Extraction of phenolic compounds from a Spodosol profile: An evaluation of three extractants. Soil Sci. 140:412–420.
Vance, G.F., and M.B. David. 1989. Effect of acid treatment on dissolved organic carbon retention by a spodic horizon. Soil Sci. Soc. Am. J. 53:1242–1247.
Vance, G.F., and M.B. David. 1991a. Chemical characteristics and acidity of soluble organic substances from a northern hardwood forest floor. Geochim. Cosmochim. Acta 55:3611–3625.
Vance, G.F., and M.B. David. 1991b. Forest soil response to acid and salt additions of sulfate: III. Solubilization and composition of dissolved organic carbon. Soil Sci. 151:297–305.
Vance, G.F., and M.B. David. 1992. Adsorption of dissolved organic carbon and sulfate by Spodosol mineral horizons. Soil Sci. 154:136–144.
Vance, G.F., D.L. Mokma, and S.A. Boyd. 1986. Phenolic compounds in soils of hydrosequences and developmental sequences of Spodosols. Soil Sci. Soc. Am. J. 50:992–996.
Walker, T.W., and J.K. Syers. 1976. The fate of phosphorus during pedogenesis. Geoderma 15:1–19.
Wood, T.E. 1980. Biological and chemical control of phosphorus cycling in a northern hardwood forest. Ph.D. diss. Yale Univ., New Haven (Diss. Abstr. 80-26440).
Wood, T., F.H. Bormann, and G.K. Voigt. 1984. Phosphorus cycling in a northern hardwood forest: biological and chemical control. Science (Washington, DC) 223:391–393.
Yanai, R.D. 1991. Soil solution phosphorus dynamics in a whole-tree-harvested northern hardwood forest. Soil Sci. Soc. Am. J. 55:1746–1752.
Yanai, R.D. 1992. Phosphorus budget of a 70-year-old northern hardwood forest. Biogeochemistry 17:1–22.
Yavitt, J.B., and T.J. Fahey. 1985. Organic chemistry of the soil solution during snowmelt leaching in *Pinus contorta* ecosystems, Wyoming, p. 485–496. *In* D.E. Caldwel et al. (ed.) Planetary ecology. Van Nostrand Reinhold Press, New York.

16 Carbon and Nitrogen Cycling within Mid- and Late-Rotation Jack Pine

Neil W. Foster, Ian K. Morrison, Paul W. Hazlett, and Gary D. Hogan
Canadian Forest Service
Sault Ste. Marie, Ontario, Canada

Maria I. Salerno
Convenio Ministerio de Asuntos Agrarios
Universidad Nacional de La Plata
La Plata, Argentina

Numerous studies of nutrient accumulation patterns in various aged forests, largely in temperate regions, have been published for pine (*Pinus* spp.) (e.g., Ovington, 1959a; Switzer & Nelson, 1972; Gholz et al., 1985) and other species (see Cole & Rapp, 1981). Results from these studies have led to generalizations regarding C accumulation and nutrient cycling in late rotation closed-canopy forests. In general, slow growth of older forests on infertile sites has been attributed in part to changing patterns of nutrient cycling with stand age.

It has been proposed that: (i) older forests become N deficient due to a slowdown in N cycling resulting from N sequestration in the forest floor; (ii) there is a shift with age in N uptake from the mineral soil to uptake from the forest floor; and (iii) as stands age, particularly on less fertile sites, more of their requirement for N is met by internal retranslocation than by uptake from the soil.

Our objective is to assess the applicability of these concepts to boreal jack pine (*P. banksiana* Lamb.) ecosystems by re-evaluating C and N pools and fluxes in two stands, of different starting N capital, that were initially examined in the early 1970s (Foster & Morrison, 1976; Morrison & Foster, 1977).

A frequently used method of assessing organic and nutrient accumulation over time is to compare a chronosequence of forests, i.e., forests that are different in age but similar otherwise (e.g., composition, stocking, soil, elevation, slope). In selecting a chronosequence, however, there is the possibility of overlooking factors that regulate productivity. Consequently, phytomass and nutrient differences observed may not be a function of age alone. For example, site characteristics that influence rooting volume, such as soil depth and coarse fragment

---

Copyright © 1995 Soil Science Society of America, 677 S. Segoe Rd., Madison, WI 53711, USA. Carbon Forms and Functions in Forest Soils.

content, can strongly influence jack pine site index (Schmidt & Carmean, 1988). Instead, we determined site nutrient contents sequentially, within a single stand, over more than a 20-yr period.

STUDY AREAS

Two sites were included in the study. One is the Dupuis site (47°38' N lat, 83°15' W long) located in the Sudbury District of Ontario. The average length of growing season, based on a 5.5°C index, is 161 d, roughly May through September inclusive (Chapman, 1953). Mean annual precipitation (Chapleau) is 834 mm, mean evapotranspiration is 480 mm. The growing season receives 53% of the precipitation. In profile, the soil is a Typic Dystrochrept (Orthic Dystric Brunisol) (Canada Soil Survey Committee, 1978), developed in approximately 30 cm of silt loam over loamy sand. The site index 50 of the fire-origin natural jack pine forest is 16.5 m. Between age 45 and 68, stand basal area and dry matter increased from 31 to 34 m^2 ha^{-1} and 127 to 167 t ha^{-1} and stocking decreased from 2926 to 1478 trees per hectare.

The second is the Wells site (46°25' N lat, 83°23' W long) located in Algoma District, Ontario. Average length of growing season is 175 d, roughly from late April to mid-October. Mean annual precipitation is 860 mm, mean evapotranspiration is 533 mm. The soil, developed in a coarse to medium sand, is classed as a Typic Cryorthod (Orthic Humo-Ferric Podzol). The site index 50 of the fire origin, jack pine-dominated forest, with a minor component of red pine (*P. resinosa* Aiton), is 19 m. Between age 35 and 56, stand basal area and dry matter increased from 27 to 35 m^2ha^{-1} and 117 to 175 t ha^{-1}, respectively, and stocking decreased from 1952 to 1174 trees per hectare.

METHODS

Tree phytomass was determined as follows: first, allometric equations relating each tree component (foliage, fruit, live branch, dead branch, stem wood, stem bark, total aboveground dry weight, and total belowground dry weight) to diameter and age were developed from a data set derived from a sampling of 108 trees ranging in age from 30 to 100 yr (though dominated by trees aged from 50–70 yr). These equations were used in conjunction with 1972, 1982, and 1993 stand inventories of four permanent sample plots for the Wells site, and 1970, 1975, 1980, 1985, and 1993 inventories of five permanent sample plots for the Dupuis site, to determine stand level phytomass by components over time.

Phytomass was converted to the C equivalent using component specific conversion factors (0.48–0.49). Nitrogen content was calculated by multiplying component weight by N concentration as determined at different points in time. In both stands, the intermediate values were interpolated from N values measured in the early 1970s and either the late 1980s (Wells) or the early 1990s (Dupuis). Quality control was provided through the reanalysis of reference samples. Standing dead and fallen dead dry matter were estimated from recorded standing dead trees in 1992 and tree mortality records in 10 permanent sample plots in the Dupuis stand.

Uptake of N by the trees was calculated in the following manner: $U = A - P + T + S + L$ where A, P, T, S, and L represent the amount of N in annual accumulation in perennial organs, precipitation, throughfall, stemflow, and litterfall, respectively (Morrison & Foster, 1974). The nutrient requirement of the trees was the sum of annual N accumulation in perennial tissues and N in current foliage production. Translocation was the difference between requirement and uptake.

Replicated soil samples were collected from four plots (0.04 ha) in each stand. Samples to determine fine root and fine earth mass were obtained in 1992 by volumetric extraction of soil in 10-cm increments to a depth of 60 cm, of all material from six pits in each stand. Nitrogen and C contents in mineral soil were determined by multiplying fine earth weights by concentrations measured in different years. Forest floor weights, however, were assessed each time that N and C contents were evaluated.

Litterfall was determined, from a minimum of 10 locations, in a single plot per stand according to procedures described in Foster (1974). Litter chemistry at Dupuis is based on chemical analysis of all samples in 1993. Nitrogen was determined by means of a semimicro-Kjeldahl procedure or by Hewlett-Packard 180B CNH Analyzer (conversion to Kjeldahl × 1.1). Carbon was determined by loss on ignition for fresh and decomposing litter and by Walkley-Black wet combustion procedure for soil.

RESULTS AND DISCUSSION

Organic Matter and Nitrogen Distribution Patterns

Stand

More than one-half of the ecosystem C in the 68-yr-old Dupuis pine stand and the 56-yr-old Wells stand was contained in the tree layer (Fig. 16–1). Both in amount and in proportion in the tree layer, these stands are more like coniferous forests of the temperate, as opposed to the boreal region (e.g., Cole & Rapp, 1981). Age-adjusted, a greater phytomass was associated with the coarser textured, more N-rich Wells site. At the common age of 45 yr, for example, the Wells stand contained 15% more C and 42% more N than the Dupuis forest (unpublished data). Previous work demonstrated that phytomass production in jack pine stands was positively related to the amount of N in the soil (Fig. 16–2).

Ecosystem

Overall the aboveground phytomass and N contents of the Ontario stands exceeded published values for jack pine forests (Fig. 16–3). Aboveground N contents in late rotation jack pine were similar to those reported for younger, though at the same stage of maturity, southern pines (Fig. 16–4). In both regions between 20 to 30% of the total ecosystem N was contained in living and decomposing plant materials on top of the mineral soil. However, decomposition rates are likely faster in the southern USA, so less N was apparently distributed in the forest floor and more N in the trees in southern pine ecosystems.

The two Ontario jack pine forests differed considerably in the distribution of C in the ecosystem (Fig. 16–1). Approximately 50% of the C in the Dupuis soil

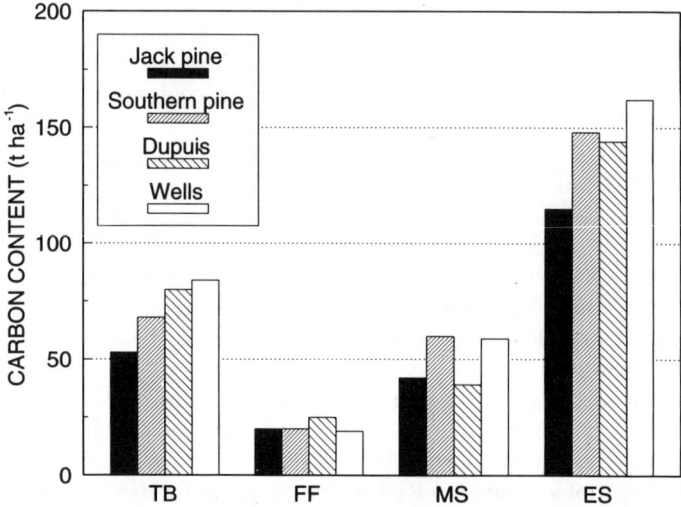

Fig. 16–1. Published C contents in tree phytomass (TB), forest floor (FF), mineral soil (MS) and the ecosystem (ES) for southern pines (mean age 26 yr, $n = 6$), jack pine (mean age 52 yr, $n = 8$) and 56-yr-old pine at Wells and 68-yr-old pine at Dupuis (Binkley & Johnson, 1992; Foster & Morrison, 1976; Gholz & Fisher, 1982; Green & Grigal, 1980; Johnson & Todd, 1987; MacLean & Wein, 1977; Morrison & Foster, 1974; 1977; Perala & Alban, 1982; Weetman & Algar, 1983; Wells & Jorgensen, 1975).

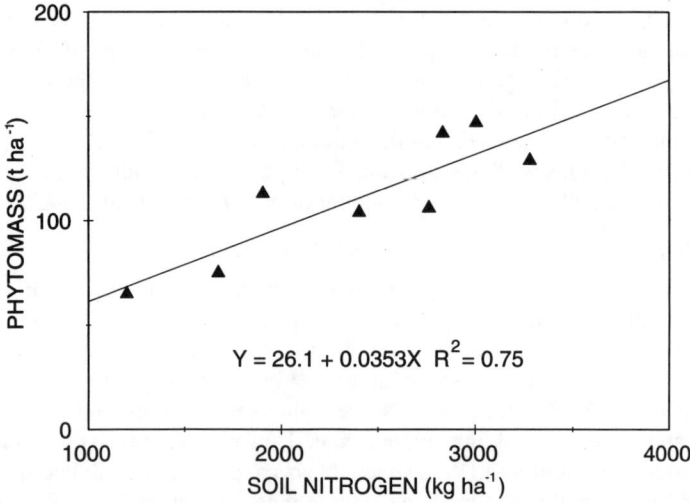

Fig. 16–2. The relationship between phytomass accumulation in jack pine and soil N contents (redrawn from Foster & Morrison., 1989).

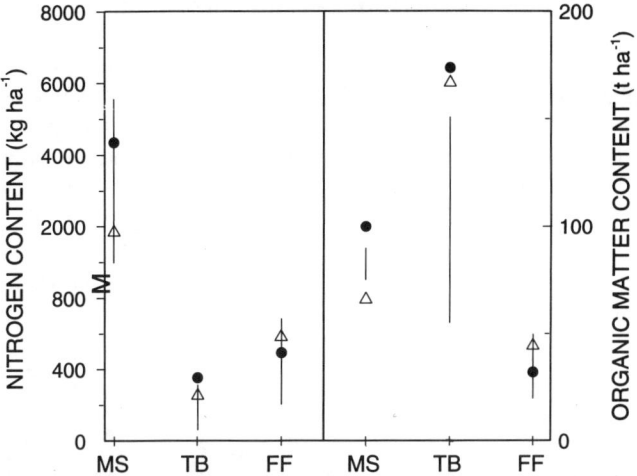

Fig. 16–3. Range of published N and organic matter contents in mineral soil (MS), aboveground tree phytomass (TB), and forest floor (FF) for jack pine ecosystems showing the position of the Wells (•) and Dupuis (Δ) forests at age 56 and 68, respectively (citations listed in Fig. 16–1).

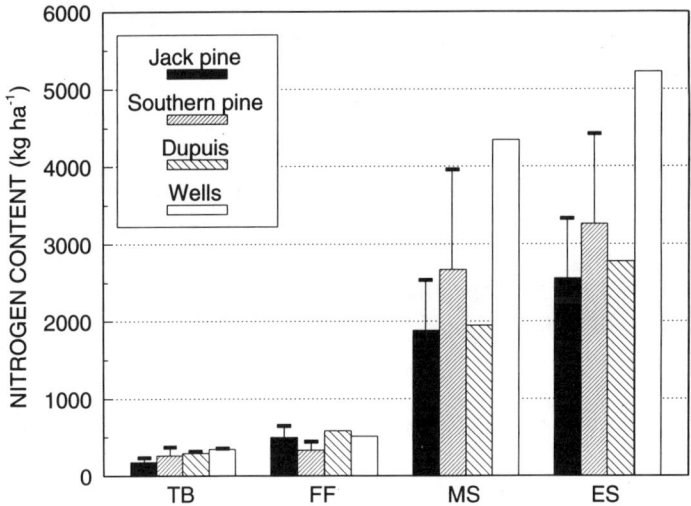

Fig. 16–4. Nitrogen contents in tree phytomass (TB), forest floor (FF), mineral soil (MS) and the ecosystem (ES) for jack pine (mean age 52 yr, $n = 8$); southern pine (mean age 28, $n = 9$) and 56-yr-old pine at Wells and 68-yr-old pine at Dupuis (southern pine from Binkley & Johnson, 1992; jack pine citations listed in Fig. 16–1, means plus one standard deviation).

was contained in the forest floor, 25% in the Wells soil. The Wells mineral soil contained twice as much N as the Dupuis soil, to a depth of 60 cm. Forest floor N contents were ~10 and 30% of the mineral soil N reserves in the Wells and Dupuis sites, respectively. Forest floor and mineral soil organic matter and N amounts were within the range of published values for jack pine sites (Fig. 16–4).

Accumulation of Nitrogen and Carbon with Age

Stand

Both closed canopy pine forests increased their C and N contents between mid- and late rotation (Fig. 16–5, 16–6). Nitrogen accumulated somewhat more rapidly than C. Increases in foliar N concentrations accounted for a significant proportion of the N accretion at Wells; but at Dupuis, stemwood was the most important (Table 16–1). Foliar mass stabilized at age 55 in the Dupuis forest, but

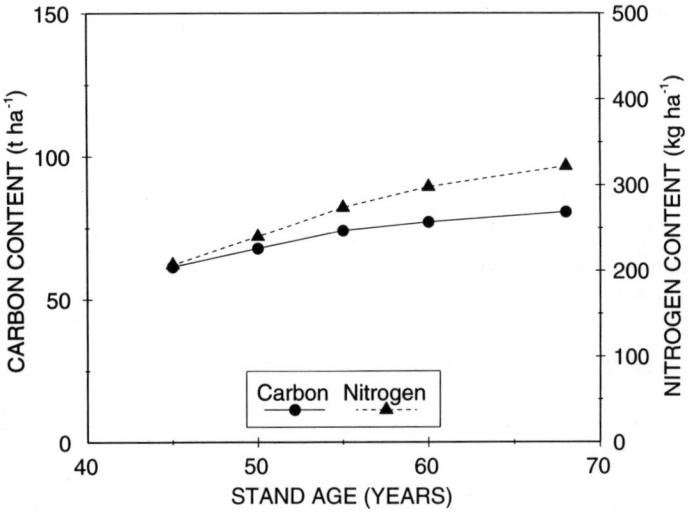

Fig. 16–5. Age effects on C and N contents in the Dupuis jack pine stand.

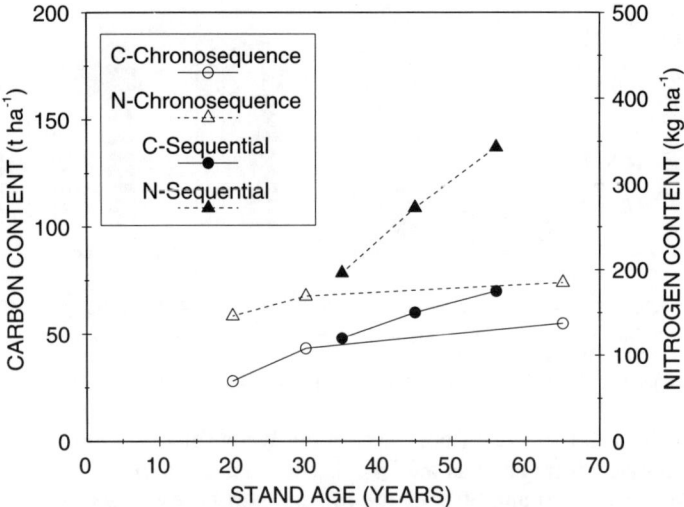

Fig. 16–6. Age effects on C and N contents in the Wells jack pine stand (chronosequence adapted from Morrison, 1973).

Table 16–1. Nitrogen in two mid-to late-rotation jack pine stands in Ontario.

	Wells		Dupuis	
	35 yr	56 yr	45 yr	68 yr
	kg ha^{-1}			
Foliage	72 ± 2.6†	118 ± 5.0	61 ± 7.8	91 ± 8.2
Fruit	1 ± 0.1	3 ± 0.3	1 ± 0.2	2 ± 0.3
Live branches	34 ± 1.1	100 ± 4.3	43 ± 5.8	57 ± 5.2
Dead branches	10 ± 0.5	23 ± 1.1	10 ± 1.1	13 ± 1.0
Stem wood	60 ± 3.7	76 ± 3.9	56 ± 5.3	105 ± 9.0
Stem bark	20 ± 1.4	22 ± 1.3	21 ± 1.9	24 ± 2.0
Stump & roots	18 ± 0.6	26 ± 1.0	15 ± 1.8	29 ± 3.0
Total trees	214 ± 9	369 ± 16	207 ± 23	322 ± 28

†Plus or minus one standard deviation.

not foliar N. Needle retention, computed as in Gessel and Turner (1976), in both stands was ~3.0 needle classes. Foliar phytomass increased with site quality as reported for other species (Turner, 1981).

Carbon accretion rates for jack pine were lower than those for temperate radiata, loblolly, slash, *P. elliottii* Engelm. Corsican, *P. nigra* var. *marrtima* (Ait.) Melv. and Scots pines *P. sylvestnis* L. (Table 16–2). Rates of N accumulation, however, were within the range reported for these pine species. Increases in foliage production and foliar N contents accounted for a significant proportion of the changes observed in both jack pine forests (Table 16–1, 16–3). Warm-dry weather associated with the 1980s (Foster et al., 1992) may have contributed to enhanced decomposition and N release from soil and increased dry-matter production during this period.

Sequential Versus Chronosequence

Phytomass and N assimilation beyond age 35 in the Wells forest were greatly in excess of that contained in an adjacent 65-yr-old jack pine stand on the same glaciofluvial deposit (Fig. 16–6). Late rotation C and N accretion determined by sequential sampling, therefore, were much greater than those determined by Foster and Morrison (1976) using the chronosequence approach (Table 16–2). The 65-yr-old jack pine forest, therefore, likely contained, 30 yr ago, less phytomass and N than the 35-yr stand that was examined. We conclude that the 65-yr-old forest was not representative, with respect to phytomass and nutrient content, of an older 35-yr-old stand. The measured accumulation patterns indicate that the chronosequence approach was a poor choice for determining production and development of pine forests at this location. Differences in pine productivity at this site may relate to fire intensity influences on levels of pine stocking and postburn levels of soil nutrients and/or climatic changes over time.

Forest Floor

Forests in northern latitudes are generally characterized by low production and decomposition rates with significant accumulation of organic matter and nutrients within the forest floor. In 1971 the two midrotation jack pine stands had

Table 16–2. Average annual organic matter and N accumulation in aboveground pine trees and forest floor (beyond crown closure).

Species	Initial age	Growth period	Method†	Organic matter		N		Reference
				Trees	Humus	Trees	Humus	
	yr	yr		t ha⁻¹ yr⁻¹		kg ha⁻¹ yr⁻¹		
Radiata pine	12	23	2	7.6	—	2.6	5.8	Will, 1964
Loblolly pine	5	15	3	5.7	0.9	10.1	7.3	Switzer & Nelson, 1972
Slash pine	10	24	2	4.2	1.1	3.7	12.5	Gholz & Fisher, 1982; Gholz et al., 1985
Corsican pine	18	30	2	2.9	—	3.5	—	Wright & Will, 1958
Scots pine	17	38	2	2.7	0.7	4.0	5.7	Ovington, 1957, 1959a,b
Jack pine	35	21	1	2.2	0.2	7.0	5.8	This study (Wells stand)
Scots pine	18	46	2	1.4	—	1.6	—	Wright & Will, 1958
Jack pine	45	23	1	1.4	0.8	4.4	7.0	This study (Dupuis stand)
Jack pine	30	35	2	0.9	0.2	1.0	2.9	Foster & Morrison, 1976

† 1 = Sequential observations one stand; 2 = Chronosequence one location; 3 = Chronosequence, two or more locations, common site/soil properties.

Table 16-3. Carbon in two-mid to late-rotation jack pine stands in Ontario.

	Wells		Dupuis	
	35 yr	56 yr	45 yr	68 yr
	t ha^{-1}			
Foliage	3.3 ± 0.1†	4.5 ± 0.2	2.9 ± 0.3	3.5 ± 0.3
Fruit	0.2 ± 0.02	0.6 ± 0.1	0.1 ± 0.04	0.4 ± 0.05
Live branches	5.5 ± 0.2	8.0 ± 0.3	4.8 ± 0.6	6.0 ± 0.5
Dead branches	1.8 ± 0.09	2.7 ± 0.1	1.9 ± 0.2	2.5 ± 0.2
Stem wood	33.6 ± 2.1	50.0 ± 2.6	39.2 ± 3.7	51.0 ± 4.4
Stem bark	3.6 ± 0.2	4.7 ± 0.3	4.3 ± 0.4	4.9 ± 0.4
Stump & roots	8.2 ± 0.31	3.9 ± 0.6	8.1 ± 1.0	12.2 ± 1.1
Total trees	56.2 ± 2.9	84.1 ± 4.0	61.4 ± 6.3	80.5 ± 7.0

†Plus or minus one standard deviation.

substantial forest floors with an average depth of 8 cm and mass of 44 t ha^{-1}, and average C and N content of 26 t ha^{-1} and 400 kg ha^{-1}, respectively. Average forest floor N contents, obtained from jack pine sites in Canada and the USA, were 50% greater than those for southern pine of comparable size (Fig. 16–4). Forest floor organic matter and N contents for jack pine were only 33 and 70% of published values for boreal and temperate coniferous forests (Cole & Rapp, 1981).

The forest floor mass of 12 jack pine stands in New Brunswick, Canada, did not vary with age (MacLean & Wein, 1977). Others have reported that forest floor organic matter contents remained constant, i.e., beyond age 17 in Scots pine (Ovington, 1959b) or increased with age, i.e., in slash pine (Gholz & Fisher, 1982) and Douglas fir [*Pseudotsuga menziesii* (Mirbel) Franco] (Turner, 1981). The mass of the Oi (L) and Oe (F) layers at the Wells site was independent of age (Fig. 16–7). However, at the Dupuis site, the Oe mass increased to age 53 then stabilized (Fig. 16–8). The mass of the Oa (H) layer increased with age at both sites. Nitrogen and C concentrations, determined in 1969 and 1984 in Wells and 1971, 1985 and 1991 at Dupuis, for Oi, Oe, and Oa layers did not vary consistently with age. Differences in mass, therefore, produced a net accumulation of forest floor N and C with age (Fig. 16–9, Table 16–2).

Nutrient Cycle Changes with Age

Crown Litterfall

Forest floor organic matter accumulation varies with the rates of fauna and floral decomposition and inputs from litterfall and root detritus. The average annual litterfall for the Wells and Dupuis forests was 3.8 and 2.2 t ha^{-1} of organic matter (1.8 and 1.1 t C ha^{-1}). Average annual N deposition in litterfall was 25 and 15 kg ha^{-1}, respectively. Litterfall was average for pine forests (Bray & Gorham, 1964) at Wells and below average at the Dupuis site. Substantial year-to-year differences in litterfall were observed in both stands, but litterfall C and N inputs did not vary systematically with age (Fig. 16–10, 16–11). Foliage-dominated litterfall quantities were not influenced by the reduction in stem density of 40% between age 35 and 56 in one stand and 50% between age 45 and 68 in the other. Our results confirm those of chronosequence studies by many authors (Bray &

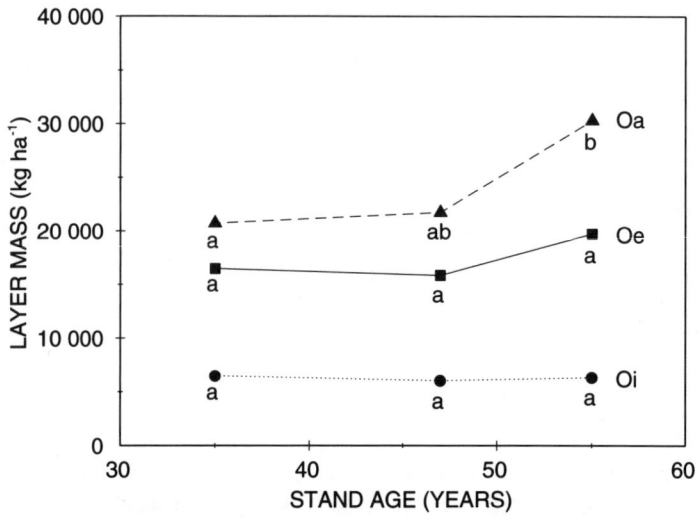

Fig. 16–7. Age effects on the mass of forest floor Oi, Oe, and Oa layers under Wells jack pine. Corresponding letters indicate no significant ($P = 0.05$) difference between values.

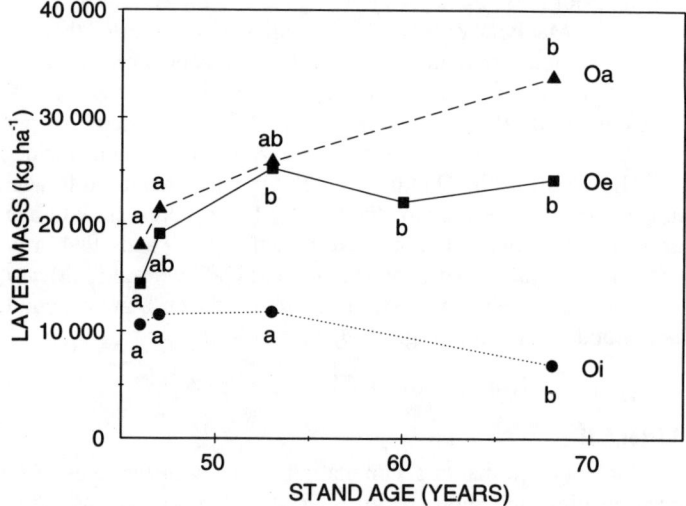

Fig. 16–8. Age effects on the mass of forest floor Oi, Oe, and Oa layers under Dupuis jack pine. Corresponding letters indicate no significant ($P = 0.05$) difference between values.

Gorham, 1964) that litter production in closed canopy forests does not change significantly with differences in stand age.

Stem Litterfall

Tree growth and mortality in the Dupuis stand were intensively examined over 23 yr (Morrison & Foster, 1990). This allowed us to examine the little-

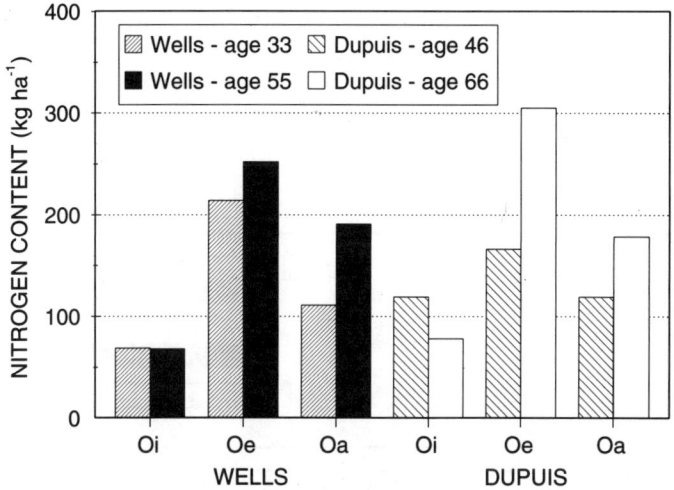

Fig. 16–9. Nitrogen contents in forest floor Oi, Oe, and Oa layers under jack pine.

studied pathway of nutrient cycling associated with tree mortality and stem litterfall. The considerable reduction in stocking associated with self-thinning in midrotation pine stands could add appreciable C to the forest floor. Fallen stem materials increased aboveground C transfer from the trees to soil by 40% (Fig. 16–12). The amount of N in freshly fallen boles, however, was equivalent to only 5% of that in foliage-dominated litterfall (Fig. 16–13). Stem litterfall has very wide C/N ratios of >500:1. Decomposition of this material will contribute to greater N immobilization in soil and a slowing of N cycling between the soil and tree.

Standing dead tree C and N pools increased and stem litter C and N decreased over time (Fig. 16–12, 16–13). In 1993 standing dead pine trees contained 5.5 t C ha^{-1} and 9.0 kg N ha^{-1} at the two sites. Almost all dead trees remained standing for 5 yr or longer, some were still standing after 23 yr. After wildfires or severe windstorms in mature pine forests fallen stems also will greatly enhance C additions to soil.

Litter Decomposition

Decomposition rates in closed canopy jack pine forests can be expected to fluctuate over time as microbial activity responds to fluctuations in soluble C, N, temperature and moisture (Foster et al., 1980a,b). The influence of age on decomposition has been examined in jack pine forests by determining organic matter and nutrient turnover rates from forest floor mass and litterfall. After fire on jack pine sites, forest floor turnover rates recovered to near equilibrium values of 23 yr for N and 13 yr for organics within 20 yr after disturbance (Weber, 1987). Slower decomposition in younger jack pine stands has been attributed to the presence of large amounts of slowly decomposing woody material of fire origin (MacLean & Wein, 1978). Nitrogen turnover rates at our sites (Table 16–4) were generally as low, or lower, than the values from other jack pine sites.

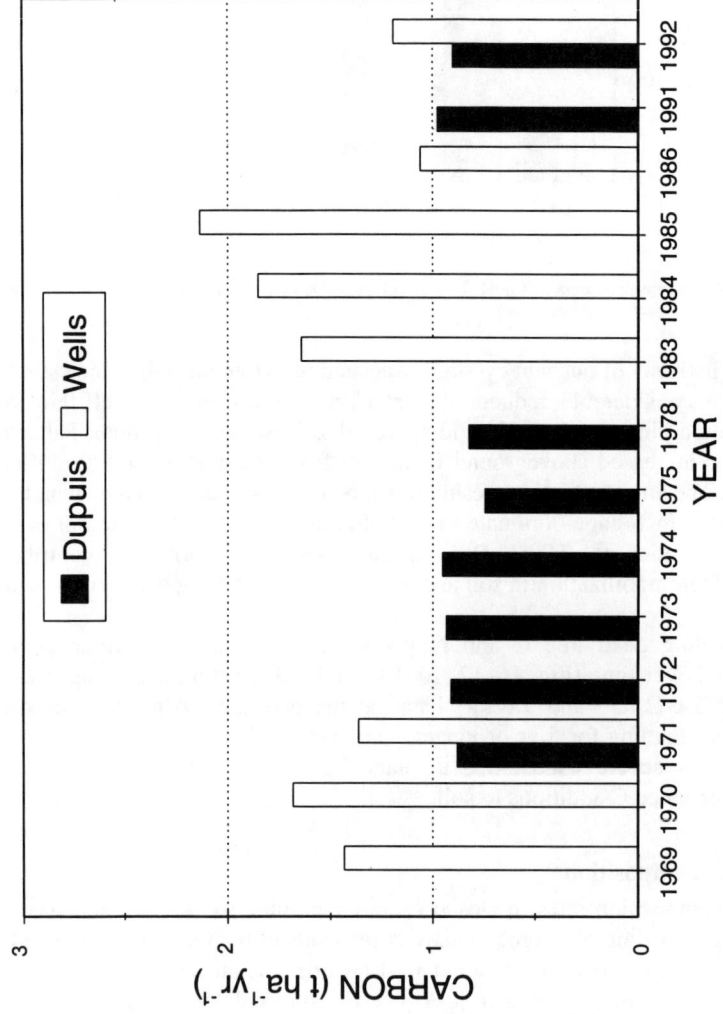

Fig. 16-10. Carbon content of jack pine litterfall.

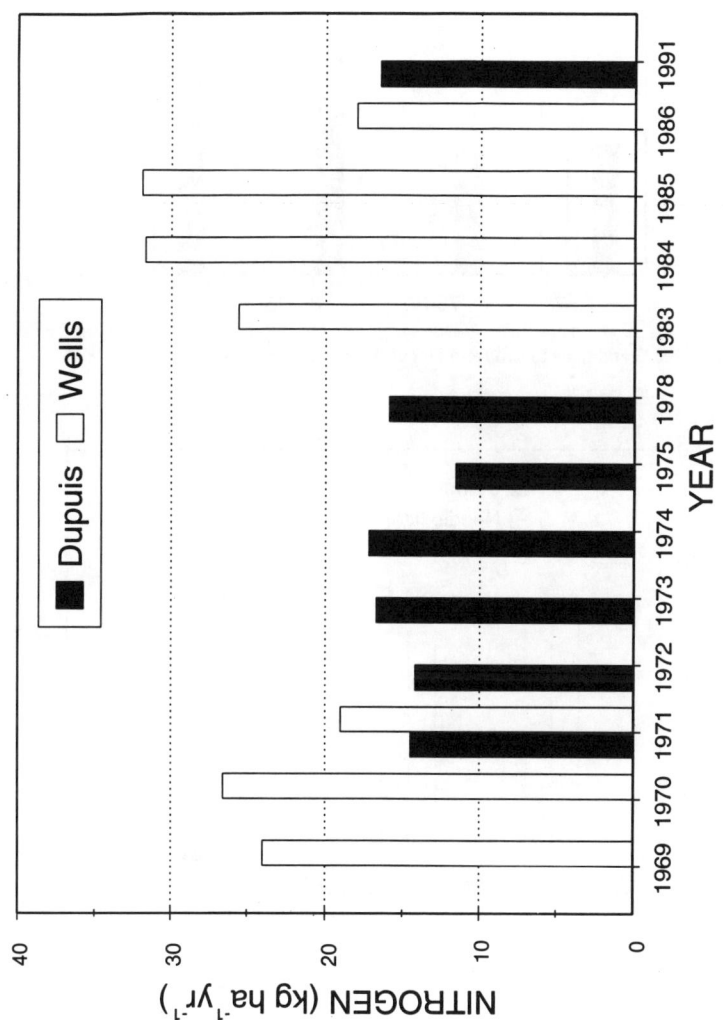

Fig. 16–11. Nitrogen content of jack pine litterfall.

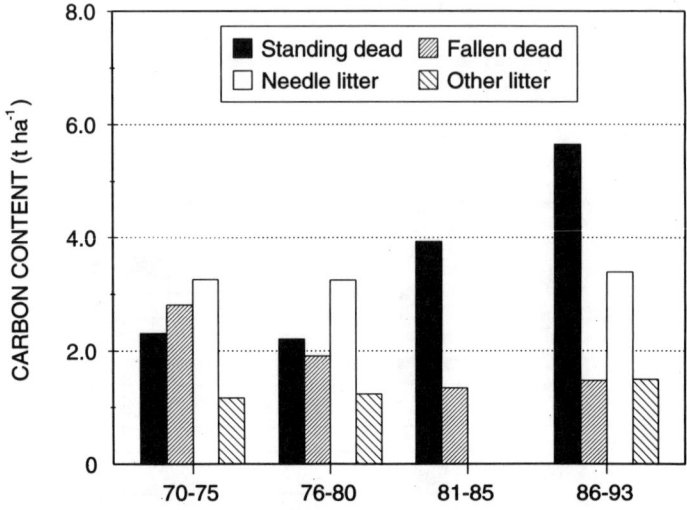

Fig. 16–12. Periodic C additions to soil from Dupuis jack pine age 45 to 68.

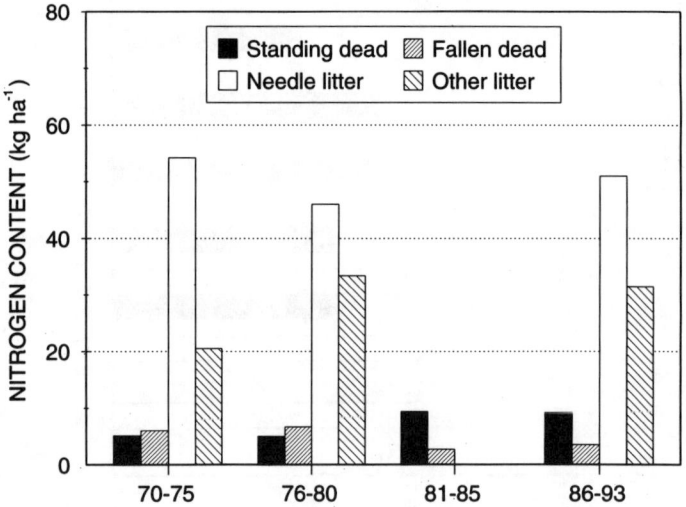

Fig. 16–13. Periodic N additions to soil from Dupuis jack pine age 45 to 68.

Mean residence time for N in forest floor (Oi + Oe layers) in the Dupuis forest was almost twice that at the Wells site (Table 16–4). The total forest floor (Oi + Oe + Oa) at Dupuis also accumulated C and N at a faster rate than at Wells (Table 16–2). This would only be possible with much higher decomposition rates at Wells since the Wells forest floor receives almost twice the C and N additions in litterfall (Fig. 16–10, 16–11). The residence time for N did not change between

Table 16–4. Variation with age in indices of N cycling for jack pine stands.

Age	Use efficiency†	Uptake‡	Requirement§	Potential uptake¶	Net uptake#	Forestfloor residence time††	Potential net mineralization‡‡
yr			——— kg ha⁻¹ ———			yr	mg N kg soil⁻¹ d⁻¹
			Wells				
35	509	26.6	30.7	18.6	12.8	12.0	—
47	458	30.4	44.2	18.3	12.5	11.9	10.0
56	425	27.0	44.3	—	—	—	9.1
			Dupuis				
45	576	—	27.3	9.0	—	—	—
50	547	16.3	26.2	12.7	5.7	19.0	—
55	523	—	28.4	—	—	—	—
60	502	—	30.8	—	—	—	4.7
68	484	18.6	31.3	13.0	6.0	—	5.9

† N use efficiency = kilograms of phytomass produced per kilogram of N uptake.
‡ Uptake = annual accumulation in perennial tissue – precipitation + throughfall + stemflow + litterfall.
§ Requirement = annual accumulation in perennial tissues and current foliage.
¶ Potential uptake = litterfall + throughfall + stemflow – forest floor leaching.
# Net uptake = potential uptake – accumulation in forest floor.
†† Residence time = forest floor pool size ÷ annual input litter.
‡‡ Potential net mineralization = 20 wk at constant 20°C, 33 k Pa, Oe layer.

age 35 and 47 yr in the Wells forest floor. Our results generally confirm results of various studies that stand age has little influence on decomposition and release of N in closed canopy pine forests (Lamb, 1976, Berg & Staff, 1980) and other conifers (Turner & Long, 1975; Edmonds, 1979).

Potential N mineralization in the Oe layer of the Wells forest floor was twice that at Dupuis (Table 16–4). On these nutrient poor outwash materials, there is an intense competition for N between microbes and trees, and net mineralization of N by microbes is slow. Fyles and McGill (1987), however, determined that N mineralization in jack pine forest floor increased with stand age. The trees and ground vegetation captured most of the mineralized N; little N was nitrified, and less than 1 kg N ha^{-1} yr^{-1} was leached beyond the zone of intensive rooting at our pine sites (P.W. Hazlett, unpublished data).

Nitrogen Uptake

One consequence of slow N release from soil organic matter would be an increase in the amount of phytomass produced by the trees per kilogram of N taken up. This is referred to as "N use efficiency" and should be highest when ecosystems are N deficient (Cole, 1981). Nitrogen use efficiency was higher in the Dupuis stand than in the Wells stand (Table 16–4). Nitrogen use efficiency in jack pine exceeds the average of 168 that was computed for coniferous forests by Cole (1981). The high N efficiency can be explained in part by greater translocation of N within the canopy than was observed by Cole.

To produce 15% more total phytomass, the Wells forest retained 40% more foliage with a 60% greater N content. The Wells soil contained more N throughout and exhibited greater potential N mineralization (P.W. Hazlett, unpublished data). Fertilization trials have confirmed that the growth of the Dupuis forest at age 45 was N limited (Morrison & Foster, 1990).

If N was being used more efficiently in older pine forests the ratio would widen. Since the ratio was observed to narrow as each of the forests aged, N availability appears to be increasing or, translocation increases. Since N uptake was calculated to increase at maturity (Table 16–4), the decrease in N use efficiency with age suggests that these forests are becoming less N deficient as they approach maturity, not more.

Older boreal coniferous forests are considered to become more N deficient due to a slowdown in N cycling between vegetation and soil (e.g., Williams, 1972; Miller, 1981). Several lines of evidence, produced from the current study, suggest that this is not the case at least in some mature jack pine forests. First, as the stands approached maturity they accumulated N faster than C. Second, N use efficiency decreased with age. Third, N was observed to accumulate in the forest floor at the same rate as C in the more N limited Dupuis forest. Last, in the period of lowest N uptake by the forest, potential rates of net N mineralization in the forest floor did not change.

Nitrogen Retranslocation

Our estimated N uptake values fell short of the requirement values (Table 16–4). Our results do not support the generalization that, on less fertile sites, more of the N required for current growth is derived from the biological cycle within the tree and less from uptake from soil. There was, however, a difference in the amount of N retranslocated from foliage prior to abscission in the two stands at age 45 or older (Wells = 15.6 kg ha^{-1} yr^{-1}, Dupuis = 11.3 kg ha^{-1} yr^{-1}). Retranslocation of N from foliage provided 35 to 40% of the annual N requirement of the trees (Wells age 35 excepted) (Table 16–4). Average N requirement of the forest was 40 and 29 kg ha^{-1} yr^{-1} at Wells and Dupuis, respectively. There was no evidence from either forest, however, that beyond age 45 substantially more of the N requirement was met by translocation from foliage as the stands increased in age (Table 16–4).

Nitrogen Supply

Uptake of N from soil was the principle source of N for jack pine growth. We have calculated that the potential N uptake from the forest floor averaged 18.4 and 12.8 kg ha^{-1} yr^{-1} at the Wells and Dupuis sites, respectively, which was equivalent to 46 and 44% of their overall N requirements (Table 16–4). Nitrogen uptake decreased and radial tree growth declined 30%, over 8 yr, with annual removal of the forest floor in a 50-yr-old jack pine stand (Weber et al., 1985). Our two maturing jack pine stands obtained 50 to 60% of their N uptake from the mineral soil (Table 16–4). Miller et al. (1979) estimated that 30% of N uptake by a 40-yr-old Corsican pine plantation was from mineral soil. Nearly all the N uptake in 60-yr-old Douglas fir (Cole, 1981) and 73% for 120-yr-old Scots pine (Staff & Berg, 1977), on the other hand, could be attributed to uptake from the forest floor.

There are several other lines of evidence demonstrating that mature jack pine forests are unlikely to rely on the forests floor for the majority of their N uptake. First, we could not determine any increase in potential N uptake from the forest floor with age (Table 16–4). Second, N uptake in closed canopy jack pine forests was only weakly correlated (R^2 = 0.28, n = 9) with forest floor N. In a fertilization experiment, jack pine foliar N concentrations were only weakly correlated with extractable NH_4^+ concentrations in the forest floor (Weetman & Algar, 1974). Third, jack pines continue to expand their rooting in the mineral soil with age (Strong & La Roi, 1983a).

Last, the opportunity for maturing jack pine trees to absorb water and nutrients from the forest floor may decrease with age due to increased competition from understory species. Vascular plants root prolifically in the forest floor and their phytomass increases as the overstory ages (MacLean & Wein, 1977). Smothering ground vegetation in a 40-yr-old jack pine stand increased tree growth and N uptake (Weetman & Algar, 1974). The natural pattern of jack pine succession to shallow-rooted black [*Picea mariana* (Miller) B.S.P.] or white spruce [*P. glauca* (Moench) Voss] has resulted in a subdominant spruce canopy with trees up to 10 m tall in the stands we examined. Horizontal roots of white spruce were reported to grow in litter and humus, whereas jack pine horizontal roots were in the mineral soil (Strong & La Roi, 1983b). We believe, though admittedly based on indirect arguments, that pine trees are unlikely to exhibit a shift in uptake of N from mineral soil to forest floor in mid- to late rotation.

Figure 16–14 illustrates the distribution of fine roots, reduced to C equivalent, through the soil profile at the two sites. The greatest concentration of fine roots in both stands was in the upper 10 cm of the mineral soil, as opposed to the

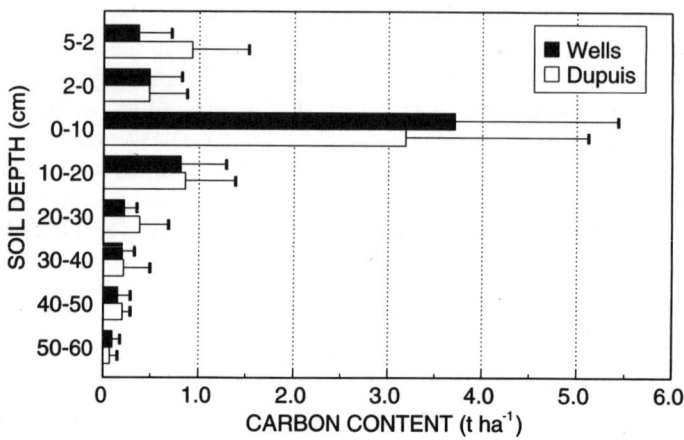

Fig. 16–14. Distribution of fine root C contents in two jack pine soils (means plus one standard deviation).

forest floor. The total fine root phytomass was 12.6 and 13.2 t ha^{-1} at the Wells and Dupuis sites, respectively. Pine aboveground phytomass was less at Dupuis than at Wells. The proportion of C allocated to fine root production, therefore, was greater at the less fertile Dupuis site than at Wells, in agreement with reports for other coniferous forests (e.g., Keyes & Grier, 1981). At the Wells site, with a coarser soil, fine root phytomass may have been increased to obtain water rather than nutrients.

Nitrogen Pools

Twenty years of nutrient cycling has resulted in a redistribution of 200 kg N ha^{-1} from mineral soil to living and detrital C pools. Litter and N inputs from the canopy and tree stems were generally stable through semimaturity; and N accumulation in the forest floor, therefore, paralleled or exceeded (Wells) the accumulation of C. Stability in nutrient accumulation in jack pine trees and soil after 40 to 45 yr (see also Wilde, 1964; MacLean & Wein, 1977) was not observed for N at our sites.

Forest Management Implications

Measured stand C and N contents were greater than those previously published for mature jack pine in Ontario (Foster & Morrison, 1976; Morrison & Foster, 1974). Full tree harvesting would remove up to 50% of the C and over 300 kg ha^{-1} of N from these ecosystems. The impact of these removals is largely unknown, but outwash pine sites typically contain low C concentrations of 1 to 3%. The forest floor layers of our two stands contained sufficient N to meet the expected nutrient requirements of the second-growth forest over a 20- to 30-yr period. Retaining humus layer N on site will minimize the requirement for second-growth forest to utilize reserves of N in the mineral soil. Within the maturing forest fallen tree stems produce an uneven ground surface with a pit- and mound microtopography with moss-overgrown logs. Much of the C associated with pine mortality would not be retained on site if midrotation pine forests were thinned. Thinning is a common practice in boreal pine forests in Scandinavia, but few jack pine forests in Canada are commercially thinned.

Light scarification techniques are preferable to those that would remove all or part of the N reserves in the forest floor (i.e., blading). Mixing of forest floor C with mineral soil may facilitate decomposition of plant tissues and ensure that young stands benefit from the nutrients contained within. Organic matter also will contribute to the development of soil structure and increase nutrient and moisture retention in these dry coarse-textured soils.

Natural high-intensity fires also are likely to cause a considerable loss of C and N from the canopy and forest floor of jack pine sites. A light prescribed burn, on the other hand, would raise the soil pH, increase decomposition and release of N without excessive loss of N to volatization (Gholz & Fisher, 1984).

Management practices that reduce forest floor C and N reserves are likely to exacerbate N deficiencies in pine soils similar to the Dupuis site, and possibly induce them in more fertile sites like Wells. We believe that, to stimulate pine growth, it would be more effective to improve N availability with fertilizers with midrotation applications to sites with low site quality and/or total ecosystem N reserves.

ACKNOWLEDGMENTS

We thank the many individuals who contributed to the project. Assistance with sample collection was provided by D. Ropke, H. Broderson, and G. Koteles; sample processing by L. Hawdon, K. Jones and J. Ralston; sample chemical analysis by J. Ramakers, D. Kurylo, and L. Irwin; data compilation by D. Glibota, and A. Boyonoski, and typing of tables and references by B.J. Rowlinson. The staff of the Ontario Ministry of Natural Resources in Chapleau and Blind River reserved and protected the study areas. The financial support from Natural Resources Canada (ENFOR program) and from the Canada-Ontario Northern Ontario Development Agreement (NODA) is gratefully recognized.

REFERENCES

Berg, B., and H. Staff. 1980. Decomposition rate and chemical changes of Scots pine needle litter. I. Influence of stand age. Ecol. Bull. 32:363–372.

Binkley, D., and D.W. Johnson. 1992. Southern Pines. p. 534–543. *In* D.W. Johnson and S.E. Lindberg (ed.) Atmospheric deposition and forest nutrient cycling. Ecol. Stud. 91. Springer Verlag, New York.

Bray, R.J., and E. Gorham. 1964. Litter production in forests around the world. Adv. Ecol. Res. 2:101–157.

Canada Soil Survey Committee. 1978. The Canadian system of soil classification. Canada Dep. Agric., Publ. no. 1646. Canada Dep. Agric. Ottawa, Ontario.

Chapman, L.J. 1953. The climate of northern Ontario. Can. J. Agric. Sci. 33:41–73.

Cole, D.W. 1981. Nitrogen uptake and translocation by forest ecosystems. Ecol. Bull. 33:219–232.

Cole, D.W., and M. Rapp. 1981. Elemental cycling in forest ecosystems. p. 341–407. *In* D.E. Reichle (ed.) Dynamic properties of forest ecosystems. Cambridge Univ. Press, Cambridge, England.

Edmonds, R.L. 1979. Decomposition and nutrient release in Douglas-fir needle litter in relation to stand development. Can. J. For. Res. 9:132–140.

Foster, N.W. 1974. Annual macroelement transfer from *Pinus banksiana* Lamb. forest to soil. Can. J. For. Res. 4:470–476.

Foster, N.W., and I.K. Morrison. 1976. Distribution and cycling of nutrients in a natural *Pinus banksiana* ecosystem. Ecology 57:110–120.

Foster, N.W., and I.K. Morrison. 1989. Effects of site preparation and full tree logging on nutrient cycling. p. 28–46. *In* P.M. Corbett (ed.) Aspects of site preparation biology and practice. Proc. Workshop Fort Frances, Ontario. 27–28 Sept. 1988. Tech. Workshop Rep. no. 2. Ontario Min. Natural Resour., For. Technol. Dev. Unit, Thunder Bay, Ontario, Canada.

Foster, N.W., E.G. Beauchamp, and C.T. Corke. 1980a. Microbial activity in a *Pinus banksiana* Lamb. forest floor amended with nitrogen and carbon. Can. J. Soil Sci. 60:199–209.

Foster, N.W., E.G. Beauchamp, and C.T. Corke. 1980b. The influence of soil moisture on urea hydrolysis and microbial respiration in jack pine humus. Can. J. Soil Sci. 60:675–684.

Foster, N.W., I.K. Morrison, Xiwei Yin, and P.A. Arp. 1992. Impact of soil water deficits in a mature sugar maple forest: stand biochemistry. Can. J. For. Res. 22:1753–1760.

Fyles, J.W., and W.B. McGill. 1987. Nitrogen mineralization in forest soil profiles from central Alberta. Can. J. For. Res. 17:242–249.

Gessel, S.P., and J. Turner. 1976. Litter production in western Washington Douglas-fir stands. Forestry 49:63–72.

Gholz, H.L., and R.F. Fisher. 1982. Organic matter production and distribution in slash pine (*Pinus elliottii*) plantations. Ecology 63:1827–1839.

Gholz, H.L., and R.F. Fisher. 1984. The limits to productivity: Fertilization and nutrient cycling in coastal plain slash pine forests. p. 105–120. *In* E.L. Stone (ed.) Proc. North Am. For. Soils Conf., 6th, Knoxville, TN. June 1983. Univ. Tennessee, Knoxville.

Gholz, H.L., R.F. Fisher, and W.L. Pritchett. 1985. Nutrient dynamics in slash pine plantation ecosystems. Ecology 66:647–659.

Green, D.C., and D.R. Grigal. 1980. Nutrient accumulations in jack pine stands on deep and shallow soils over bedrock. For. Sci. 26:325–333.

Johnson, D.W., and D.E. Todd. 1987. Nutrient export by leaching and whole-tree harvesting in a loblolly pine and mixed oak forest. Plant Soil 102:99–109.

Keyes, M.R., and C.C. Grier. 1981. Above- and below-ground net production in 40-year-old Douglas-fir stands on low and high productivity sites. Can. J. For. Res. 11:599–605.

Lamb, D. 1976. Decomposition of organic matter on the forest floor of Pinus radiata plantations. J. Soil Sci. 27:206–217.

MacLean, D.A., and R.W. Wein. 1977. Nutrient accumulation for postfire jack pine and hardwood succession patterns in New Brunswick. Can. J. For. Res. 7:562–578.

MacLean, D.A., and R.W. Wein. 1978. Litter production and forest floor nutrient dynamics in pine and hardwood stands of New Brunswick, Canada. Hol. Ecol. 1:1–15.

Miller, H.M. 1981. Nutrient cycles in forest plantations, their change with age and the consequence for fertilizer practice. p. 187–199. *In* N. Turvey (ed.) Australian forest nutrition workshop productivity in perpetuity. Canberra, Australia.

Miller, H.M., J.M. Cooper, J.D. Miller, and O.J.L. Pauline. 1979. Nutrient cycles in pine and their adaptation to poor soils. Can. J. For. Res. 9:19–26.

Morrison, I.K. 1973. Distribution of elements in aerial components of several natural jack pine stands in Northern Ontario. Can. J. For. Res. 3:170–179.

Morrison, I.K., and N.W. Foster. 1974. Ecological aspects of forest fertilization. p. 47–53. *In* F. Hegyi (ed.) Proc. Workshop on Forest Fertilization. 8–10 January. Ontario. For. Tech. Rep. no. 5, Canadian For. Serv., Sault Ste. Marie Ontario.

Morrison, I.K., and N.W. Foster. 1977. Fate of urea fertilizer added to boreal forest *Pinus banksiana* Lamb. stand. Soil Sci. Soc. Am. J. 41:441–448.

Morrison, I.K., and N.W. Foster. 1990. On fertilizing semimature jack pine stands in the boreal forest of central Canada. p. 416–431. *In* S.P. Gessel et al. (ed.) Proc. North Am. For. Soils Conf., 7th, Vancouver, Canada. Univ. British Columbia, Vancouver.

Ovington, J.D. 1957. Dry-matter production by *Pinus sylvestris* L. Ann. Bot. 21:288–314.

Ovington, J.D. 1959a. Mineral content of plantations of *Pinus sylvestris* L. Ann. Bot. 23:75–88.

Ovington, J.D. 1959b. The circulation of minerals in plantations of *Pinus sylvestris* L. Ann. Bot. 23:229–239.

Perala, D.A., and D.H. Alban. 1982. Phytomass, nutrient distribution and litterfall in *Populus*, *Pinus* and *Picea* stands on two different soils in Minnesota. Plant Soil 64:177–192.

Schmidt, M.G., and W.H. Carmean. 1988. Jack pine site quality in relation to soil and topography in north central Ontario. Can. J. For. Res. 18:297–305.

Staff, H., and B. Berg. 1977. Mobilization of plant nutrients in a Scots pine forest mor in central Sweden. Sil. Fenn. 11:210–217.

Strong, W.L., and G.H. La Roi. 1983a. Rooting depths and successional development of selected boreal forest communities. Can. J. For. Res. 13:577–588.

Strong, W.L., and G.H. La Roi. 1983b. Root-system morphology of common boreal forest trees in Alberta, Canada. Can. J. For. Res. 13:1164–1173.

Switzer, G.L., and L.E. Nelson. 1972. Nutrient accumulation and cycling in loblolly pine (*Pinus taeda* L.) plantation ecosystems: The first twenty years. Soil Sci. Soc. Am. Proc. 36:143–147.

Turner, J. 1981. Nutrient cycling in an age sequence of western Washington Douglas-fir stands. Ann. Bot. 48:159–169.

Turner, J., and J.N. Long. 1975. Accumulation of organic matter in a series of Douglas-fir stands. Can. J. For. Res. 5:681–690.

Weber, M.G. 1987. Decomposition, litterfall and forest floor nutrient dynamics in relation to fire in eastern Ontario jack pine ecosystems. Can. J. For. Res. 17:1496–1506.

Weber, M.G., I.R. Methven, and C.E. Van Wagner. 1985. The effect of forest floor manipulation on nitrogen status and tree growth in an eastern Ontario jack pine ecosystem. Can. J. For. Res. 15:313–318.

Weetman, G.F., and D. Algar. 1974. Jack pine nitrogen fertilization and nutrition studies: Three year results. Can. J. For. Res. 4:381–398.

Weetman, G.F., and D. Algar. 1983. Low-site class black spruce and jack pine nutrient removals after full-tree and tree-length logging. Can. J. For. Res. 13:1030–1036.

Wells, C.G., and J.R. Jorgensen. 1975. Nutrient cycling in loblolly pine plantations. p. 137–158. *In* B. Bernier and C.H. Winget (ed.) Proc. North Am. For. Soils Conf. 4th, Univ. Laval, Quebec City, Quebec, Canada.

Wilde, S.A. 1964. Changes in soil productivity induced by pine plantations. Soil Sci. 97:276–278.

Will, G.M. 1964. Dry matter production and nutrient uptake by *Pinus radiata* in New Zealand. Commonw. For. Rev. 43:57–70.

Williams. B.L. 1972. Nitrogen mineralization and organic matter decomposition in Scots pine humus. Forestry 45:177–188.

Wright, T.W., and G.M. Will. 1958. The nutrient content of Scots and Corsican pines growing on sand dunes. Forestry 31:13–25.

17 Carbon Chemistry and Nutrient Supply in Cedar–Hemlock and Hemlock–Amabilis Fir Forest Floors

Cindy E. Prescott and Gordon F. Weetman
University of British Columbia
Vancouver, British Columbia, Canada

Louise E. DeMontigny
Ministry of Forests
Victoria, British Columbia, Canada

Caroline M. Preston
Pacific Forestry Centre
Canadian Forest Service
Victoria, British Columbia, Canada

Rodney J. Keenan
Tropical Forestry Research Institute
Atherton, Queensland, Australia

The forests of northern Vancouver Island are primarily of two types: old-growth western red cedar (*Thuja plicata* Donn ex D. Don) and western hemlock [*Tsuga heterophylla* (Raf.) Sarg.], and second-growth western hemlock and amabilis fir [*Abies amabilis* (Dougl.) Forbes] originating from a windstorm in 1906. The two forest types occur adjacent to one another at similar topographic positions on similar parent materials. Following clearcutting and slashburning of the old-growth cedar–hemlock forests, young Sitka spruce [*Picea sitchensis* (Bong.) Carr.], western hemlock and western red cedar trees grow slowly and become chlorotic. In contrast, trees in adjacent cutovers of second-growth hemlock and amabilis fir forests grow well and do not show symptoms of nutrient deficiencies. Fertilization experiments have identified N and P deficiencies as the cause of the slow growth and chlorosis of trees in cedar–hemlock cutovers (Weetman et al., 1989a,b). The nutritional stress of the young trees can be at least partly attributed to the abundance of the ericaceous shrub, salal (*Gaultheria shallon* (Pursh.), which rapidly reoccupies cedar–hemlock cutovers after cutting and burning, and competes with young trees for nutrients (Messier & Kimmins, 1991). Findings of low nutrient availability in the residual humus in cedar–hemlock cutovers (Weetman et al., 1990) suggested that the nutrient supply problem in cedar–hemlock cutovers might originate in the forest floors of cedar–hemlock forests. We tested this hypothesis, by comparing nutrient availability in the forest floors of uncut old-growth cedar–hemlock and second-growth hemlock–amabilis forests. Our objectives were to determine if the low N and P availability in cedar–hemlock forest floors existed prior to clear-cutting, and to identify any differences in soil chemistry, litter decomposition rates or C chemistry of the forest floors that would explain the differences in nutrient supply between the two forest types.

STUDY AREA

The study area is in the wetter Coastal Western Hemlock biogeoclimatic zone (CWHb) (Pojar et al., 1987) between the towns of Port McNeill and Port Hardy, British Columbia (50°60' N lat, 127°35' W long). The climate at the Port Hardy airport, 15 km from the study area, is very wet maritime with an annual average precipitation of 1700 mm. Mean daily temperatures range from 3.0°C in January to 13.7°C in July. Topography is gently undulating, and elevations are all less than 300 m. Soils are well- to imperfectly-drained humo-ferric podzols (typic cryorthods) over periglacial and fluvio-glacial tills of igneous origin.

For the studies of nutrient availability and decomposition, three study sites were established within a 20-km^2 area in Block 4 of Tree Farm Licence 25 operated by Western Forest Products Ltd. Each study site had adjacent stands of uncut hemlock–amabilis and cedar–hemlock, with a transitional area of approximately 10 m between the two forest types. The cedar–hemlock forests were uneven-aged, and contain cedars more than 500 yr old, and hemlocks up to 400 yr. The average basal area of these stands was 108 m^2 ha^{-1} (84 cedar, 21 hemlock and 3 amabilis). The cedar–hemlock forests had relatively open canopies and a dense understory of salal. The hemlock–amabilis stands were even-aged, originating from a catastrophic windstorm in 1906. The average basal area of these stands was 77 m^2 ha^{-1} (63 hemlock and 14 amabilis). The hemlock–amabilis forests had dense canopies and small amounts of understory, primarily mosses and ferns. Measurements of soil chemistry, forest floor mass, nutrient availability and decomposition rates were conducted within one 50-by 50-m plot at each site. Samples for analysis of the C chemistry of the forest floors were taken from five additional cedar–hemlock and hemlock–amabilis forests.

METHODS

Soil Physical and Chemical Properties

At four randomly selected points in each of the three forests of each type, a soil pit was excavated to a depth of 1 m, or to an impenetrable layer. Bulk density samples were taken at intervals of 30 cm using a metal cylinder hammered into the side of the pit. The samples were air dried for 7 d, separated into coarse and fine fractions using a 2-mm sieve, and weighed. A subsample of the fine fraction was oven dried at 105°C for 24 h to determine the moisture content, and the weight of the fine fraction was adjusted accordingly. Bulk density was calculated by summing the weight of the coarse fraction and the adjusted weight of the fine fraction and dividing by the volume of the sampling cylinder (330 cm^3). The fine fraction from three samples from each depth in each plot were analyzed for silt and clay content using the hydrometer method of Gee and Bauder (1982), after removing the organic matter by digestion in sodium hypochlorite, and centrifuging. The sand fraction was determined after passing the soil solution through a 53-μ screen and drying the residue at 105°C for 24 h.

For chemical analyses, 10 samples of the 0- to 15-cm layer of the mineral soil were taken from each of the three forests of each type. Soil samples from deeper in the profile were taken from fine fractions of the three bulk density samples from each site. Samples were air dried for 14 d, and concentrations of N were measured with an RFA 300 autoanalyzer (Alpkem Corp., Wilsonville, OR) following Kjeldahl digestion (Bremner & Mulvaney, 1982). Available P was determined on the autoanalyzer following acid ammonium fluoride extraction (Ohlsen & Sommers, 1982). Concentrations of Fe and Al were determined on the spectrophotometer after extraction with sodium pyrophosphate (Bascomb, 1968).

Forest Floor Mass and Nutrient Content

Concentrations of nutrients in cedar–hemlock and hemlock–amabilis forest floors were estimated from five samples of each forest floor layer collected from each site in July 1990. Each of these samples was passed through a 2-cm mesh sieve to remove large pieces of wood, coarse roots, and shoots. Concentrations of total N and P were measured on the autoanalyzer, following sulphuric acid–hydrogen peroxide digestion of oven dried (70°C) samples (Parkinson & Allen, 1975). Total C was measured by the titration method of Walkley (1947). The pH was measured on a Model 750 pH meter (Allied Corp., Montreal, Quebec) from 5-g samples mixed with 20 mL of distilled water.

The mass of each forest floor layer in cedar–hemlock and hemlock–amabilis forest floors was estimated from eight 30-by 30-cm samples from each of the six plots (3 cedar–hemlock and 3 hemlock–amabilis). The samples were collected from random locations in each plot, avoiding coarse woody debris. The forest floor samples were sorted in the field into: L (litter), Lw (fine woody litter) F (fermentation), Fw (woody fermentation), Hs (humus), and Hw (woody humus). Live plants and large roots (>1-mm diam.) were removed, and the samples were dried at 70°C and weighed to estimate the mass of each forest floor layer at each site. The mass of coarse woody debris in each plot was estimated by surveying the volume and decay class of wood in each plot, and determining the density and nutrient content of samples of each decay class–species combination (Keenan et al., 1993). Values for the mass of each forest floor layer and woody debris were multiplied by the average concentration of each nutrient in that layer to estimate the mass of nutrients in the forest floors at each site.

Nutrient Availability

A 20-g fresh-weight portion of each field-moist sample (except Lw) collected in July 1990 was extracted with 50 mL of 2 M KCl (Keeney & Nelson, 1982) and analyzed for concentrations of NH_4–N and NO_3–N on the autoanalyzer. A second subsample was oven dried at 70°C to determine moisture content. The remainder of each sample was then air dried and extracted with Bray's solution (McKeague, 1978) and analyzed for extractable PO_4–P on the autoanalyzer.

Five-gram (dry weight equivalent) samples of each fresh forest floor sample collected in July 1990 were incubated in canning jars (Kerr wide mouth, 1 pint or 0.473 L) and opened to outside air for 30 min at weekly intervals. After 40 d, the contents of the jars were extracted with 50 mL of 2 M KCl and concentrations of extractable NH_4–N and NO_3–N were determined on the autoanalyzer. Differences between the amounts of KCl-extractable N before and after incubation were used to estimate the rates of net mineralization (if positive) or net immobilization (if negative), of N in each forest floor layer.

The capacity of the forest floors to provide available nutrients was measured by growing seedlings in material from the F layer at each site, which contained most of the fine roots and fungal mycelia. In January 1990, five samples of F layer material from each of the six sites were collected and kept refrigerated until April 1990. Four 60-g portions (dry weight equivalent) of each sample were mixed with perlite (4:1 by volume) and put into 15-cm diam. plastic pots. Ten seeds of either western red cedar, Sitka spruce, western hemlock or amabilis fir (locally collected by Western Forest Products Ltd.) were planted in each of the four pots per sample in April and May (1 species per week), and thinned to three seedlings in August. All 120 pots received a layer of granite grit to prevent moss growth, and daily watering for 1 yr. Roots and shoots from the three seedlings per pot were harvested in April and May 1991 (1 species per week), oven dried at 70°C and weighed separately. One bulked sample of shoot material from the five pots in each species × site combination was analyzed for total N and P as described earlier, root material was similarly bulked and analyzed. The mass of N and P in each pot was calculated to estimate the capacity of each forest floor to supply available nutrients to seedlings. The nutrient availability assessments have been described in more detail by Prescott et al. (1993).

Respiration and Litter Decomposition

Rates of CO_2 evolution from each layer of the forest floors in cedar–hemlock and hemlock–amabilis forests were measured during the 40-d aerobic incubation described above. Concentrations of CO_2 in the headspace gas in each jar were estimated at the end of the 40 d, 8 d after the last airing of the jars. Concentrations of CO_2 in each jar were measured by infrared gas analysis, and converted to estimate milligrams of CO_2 produced per gram of each forest floor layer during 8 d (Clegg et al., 1978).

The rate of decomposition of a standard litter substrate in cedar–hemlock and hemlock–amabilis forests was measured to compare the decomposition potentials of the two forest types. Thirty litterbags containing 1.0 g of dry lodgepole pine (*Pinus contorta* Douglas ex Loudon) needle litter from a single site were placed on the forest floor in each of the three cedar–hemlock and hemlock–amabilis plots in March 1990. Five bags were collected from each site every 6 mo for 3 yr, and the contents of each bag were dried and weighed.

The rates of decomposition of partially decomposed forest floor material were also measured in the three cedar–hemlock and hemlock–amabilis forests. Samples of the F layer of the forest floor in each of the six plots were collected in December 1989. Roots of diameter 1 mm or greater and wood were removed from the samples and discarded. Moist subsamples of about 1.0 g dry weight equivalent were placed into weighed fiberglass mesh bags (as above) and dried at 70°C and weighed. In March 1990, 15 bags of F material from each site were inserted horizontally into the forest floor at their site of origin. Five bags were harvested from each site annually for 3 yr. At each harvest, all live roots penetrating the bags were removed, and the F material remaining in the bags was dried and weighed.

Freshly fallen foliar litter was collected from three cedar–hemlock and three hemlock–amabilis forests in the fall of 1990 and dried at 70°C. Brown needles were separated from the litter and placed into fiberglass mesh bags with pore diameter of 1.5 mm. For hemlock–amabilis litter, 2.0 g of mixed hemlock and amabilis needle litter was placed in each bag. For cedar–hemlock litter, 1.0 g of mixed hemlock and amabilis needle litter and 1.0 g of cedar foliar litter were placed in each bag. Salal leaves showing signs of decay were collected in each cedar–hemlock forest, dried, weighed, remoistened, and put into litterbags. In March 1991, 25 bags of hemlock–amabilis litter were placed on the litter layer in each of the three hemlock–amabilis plots, and 25 bags of cedar–hemlock–amabilis litter and salal litter were placed in each of the three cedar–hemlock plots. Five bags of each type were collected from each plot every 4 mo for 1 yr and every 6 mo during the 2nd yr. At each harvest, the litter remaining in each bag was dried at 70°C and weighed.

Carbon Chemistry of the Forest Floors

Samples of forest floors for analyses of C chemistry were collected from five additional cedar–hemlock and hemlock–amabilis forests near the main study sites. At each site, one or two samples of the forest floor were collected and stored at 4°C until processed. Each sample was sieved through an 80-mm sieve to remove large roots and wood, then through a 40-mm sieve to break up wood and remove fine roots. From this collection, 4 to 6 samples of cedar–hemlock and hemlock–amabilis forest floors were air dried, ground to pass a 20-mesh sieve (850-µm) and stored in sealed plastic containers until analyzed. Samples of the L layer also were collected within cedar–hemlock and hemlock–amabilis plots, and composited into one cedar–hemlock and hemlock–amabilis sample. Salal litter was collected in cedar–hemlock plots and composited. Litter samples were air dried, ground and stored as above.

Concentrations of total and labile polysaccharides in each forest floor and litter sample were estimated using the phenol–sulphuric acid procedure of Dubois et al. (1956), following acid hydrolysis. For labile polysaccharides, hydrolysis was carried out by autoclaving for 1 h at 15 lb in^{-2} (103 kPa). For total polysaccharides, hydrolysis involved cold treatment with 0.5 M H_2SO_4, and subsequent autoclaving as for labile polysaccharides (Ivarson & Sowden, 1962; Cheshire, 1979). Cellulose was calculated as the difference between the total and labile polysaccharides.

Lipid concentrations were measured after shaking 5 g of each sample with 75 mL of 1:2 ethanol–benzene for 2 h and filtering. The leachate was evaporated in a fume hood, and the residue was weighed (Lowe, 1974).

Nuclear Magnetic Resonance Spectroscopy

One dry sample of each forest floor and litter type were packed into a bullet-type rotor that was placed in the probe of a M-100 nuclear magnetic resonance (NMR) spectrometer (Chemagnetics, In., Ft. Collins, CO) operating at 100 MHz for H_2. Solid-state ^{13}C NMR spectra with cross-polarization and magic-angle spinning at 3.5 KHz ("CPMAS" NMR spectra) were obtained as previously described by Hatcher (1987). Dipolar-dephased CPMAS spectra were generated by inserting a delay period of 40 to 100 µs without H decoupling between the cross-polarization and acquisition portions of the CPMAS pulse sequence. Chemical shifts are reported relative to tetramethylsilane (TMS) at 0 ppm.

Spectra were divided into chemical shift regions according to chemical types of C as follows: alkyl 0 to 50 ppm; methoxyl 50 to 60 ppm; O-alkyl 60 to 96 ppm; di-O-alkyl and aromatic 96 to 141 ppm; phenolic 141 to 159 ppm; carboxyl 159 to 185 ppm; and aldehyde and ketone, 185 to 210 ppm. Areas of the chemical shift regions were measured by cutting and weighing the paper, and were expressed as percentages of the total area (relative intensity). The proportions of lignin C and carbohydrate C (C_l/C_c) were then determined using the procedure of Preston et al. (1990). These procedures are described in more detail by deMontigny et al. (1993).

Statistical Analyses

Differences in the mean values for each parameter (except NMR analysis) were tested using two-way analysis of variance, assessing the effects of forest types and sites, and interactions between the two factors (Zar, 1974). Differences between the two forest types (cedar–hemlock and hemlock–amabilis) that were significant at $P < 0.05$ are noted. All analyses were done with SPSSx (Norusis, 1988) or SYSTAT (Wilkinson, 1990).

RESULTS

Soil Physical and Chemical Properties

Some physical properties of the mineral soils in cedar–hemlock and hemlock–amabilis forests are shown in Table 17–1. Bulk densities of soil in the upper 30 cm ranged from 0.58 to 0.94 g cm^{-3}, and those from 30 to 60 cm deep ranged from 0.58 to 0.82 g cm^{-3}. Sand content was 35 to 45%, silt 33 to 42% and clay 21 to 24%, allowing these soils to be classified as loams. There were no significant differences in any of these physical properties of soils between the two forest types.

Table 17-1. Physical properties of mineral soils under cedar–hemlock (CH) and hemlock–amabilis (HA) forests on northern Vancouver Island.

Type	Depth	Bulk density	Coarse fragment	Sand	Silt	Clay
	cm	g cm^{-3}	% mass		%	
CH	0–30	0.71 (0.06)†	34 (6.6)	41 (2.8)	37 (2.0)	22 (1.0)
	30–60	0.69 (0.12)	43 (4.2)	42 (2.9)	35 (0.4)	23 (2.7)
HA	0–30	0.71 (0.12)	43 (1.7)	39 (1.7)	38 (2.4)	23 (1.1)
	30–60‡	0.68	43 (0.8)	37	40	23

†Each value is the mean and standard errors (in parentheses) of three samples per site and three sites per forest type.
‡Average of two HA sites.

Table 17-2. Chemical properties of mineral soils under cedar–hemlock (CH) and hemlock–amabilis (HA) forests.

Type	Depth	Organic C	Total N	C/N	Available P	Fe	Al
	cm	——— mg g^{-1} ———			mg kg^{-1}	— mg g^{-1} —	
CH	0–15	141†	2.7*	52.2	4.34	9.0*	6.1
	15–30	50	1.3	38.5	1.96		
	30–60	34	1.2	28.3	2.06		
HA	0–15	152	3.8	40.0	2.71	12.1	8.7
	15–30	36	3.0	12.0	1.25		
	30–60	37	1.3	28.4	1.77		

†Values for the 0= to 15=cm depth are the mean of 10 samples per site, 3 sites per forest type. Values for the lower horizons are the mean of three samples per site, three sites per forest type.
*Asterisks indicate significant differences ($P<0.05$) between the two forest types based on Student's t-test.

Some chemical properties of the mineral soils in the two forest types are shown in Table 17–2. Concentrations of organic C, total N, and available P generally decreased with depth. Total N concentrations in the 0 to 15 cm and 15 to 30 cm layers were smaller in cedar–hemlock forests than in hemlock–amabilis forests. Organic C concentrations were similar in the two forest types, and C/N ratios were generally greater in cedar–hemlock forests. There were no significant differences in concentrations Al between the two forest types, but Fe concentrations were lower in cedar–hemlock forests.

Forest Floor Mass and Nutrient Content

Concentrations of N in each forest floor layer were lower in cedar–hemlock forests than in hemlock–amabilis forests (Table 17–3). Significantly lower P concentrations were measured in the L and Hw layers of cedar–hemlock forest floors. Carbon concentrations were similar in the two forest types. The pH of all nonwoody layers were significantly higher in cedar–hemlock forests. The average masses of forest floors and woody detritus in cedar–hemlock and hemlock–amabilis forests are shown in Table 17–4. The total detrital mass in cedar–hemlock forests (643 Mg ha^{-1}) was greater than in hemlock–amabilis

Table 17–3. Concentrations of N, P, and C, and pH of each forest floor layer from adjacent cedar–hemlock (CH) and hemlock–amabilis (HA) forests.

Layer†	Type	N	P	C	pH
		\- mg g^{-1} \-			
L	CH	6.7 (0.5)*‡	0.6 (0.1)*	543 (19)	4.75 (0.28)*
	HA	8.3 (0.5)	0.8 (0.1)	534 (23)	3.96 (0.11)
F	CH	9.0 (1.1)*	0.6 (0.1)	529 (28)	4.21 (0.31)*
	HA	9.9 (1.0)	0.7 (0.1)	531 (18)	3.45 (0.20)
Fw	CH	3.2 (1.6)	0.2 (0.1)	626 (35)	3.57 (0.15)
	HA	4.0 (2.1)	0.3 (0.1)	628 (45)	3.58 (0.02)
Hs	CH	9.7 (1.2)*	0.5 (0.1)	523 (17)	3.65 (0.24)*
	HA	11.7 (1.8)	0.6 (0.1)	517 (23)	3.19 (0.14)
Hw	CH	5.0 (1.5)*	0.2 (0.1)*	534 (43)	3.41 (0.23)
	HA	8.0 (3.3)	0.4 (0.1)	558 (36)	3.23 (0.14)

*Significant differences ($P < 0.05$) between the two forest types based on two-factor ANOVA.
†L = litter, F = fermentation, Hs = humus, w = woody.
‡The mean and standard deviations of 15 samples are presented.

forests (436 Mg ha^{-1}). The total amount of N in the forest floor and woody detritus were, however, only slightly greater in cedar–hemlock forests (2.2 Mg ha^{-1}) than in hemlock–amabilis forests (2.0 Mg ha^{-1}), as was the total amount of P (0.14 Mg ha^{-1} in cedar–hemlock and 0.11 Mg ha^{-1} in hemlock–amabilis). The mass of the H layer was greater in cedar–hemlock forest than in hemlock–amabilis forests, but the mass of the F layer was greater in hemlock–amabilis forests (Table 17–4).

Nutrient Availability

Concentrations of KCl-extractable N were significantly less in cedar–hemlock forests in all layers except Fw (Fig. 17–1a). Most of the N was present as NH_4–N, but some NO_3–N was detected in all forest floor layers. Concentrations of Bray-extractable PO_4–P were significantly lower in the L and Hw layers of cedar–hemlock forests (Fig. 17–1b). The amounts of N mineralized during the 40-d aerobic incubation were significantly smaller in the F, Fw and H layers of cedar–hemlock forest floors (Fig. 17–1c). Net immobilization of N occurred in L material from both forests. Almost all of the mineralized N was NH_4–N; the changes in NO_3–N concentrations during the 40-d incubation were between –2.6 and 1.2 ug g^{-1}. Seedlings of all four species grown in F material from cedar–hemlock forests produced significantly less biomass during the year than those grown in F material from the hemlock–amabilis forests (Fig. 17–1d). Shoot biomasses of all four species were significantly different between sites, root biomasses were significantly different for all species except hemlock. Concentrations of N and P in root and shoot material of each species did not differ between pots containing F material from cedar–hemlock and hemlock–amabilis forests. Thus, the total amounts of N and P taken up by seedlings (concentration × biomass) reflected the seedling biomass in each pot,

Table 17–4. Mass and nutrient content of forest floors and woody detritus in adjacent cedar–hemlock (CH) and hemlock–amabilis (HA) forests.

Layer	Type	Mass	N	P
		Mg ha^{-1}	kg ha^{-1}	kg ha^{-1}
L	CH	4.3	28.81	2.58
	HA	4.1	34.03	3.28
Lw	CH	1.7*	1.36	0.09
	HA	3.7	2.29	0.22
F	CH	18.1*	162.9	10.86
	HA	30.1	298.0	21.07
Fw	CH	29.7	95.04	5.94
	HA	25.9	101.0	7.7
Hs	CH	91.3*	885.6	45.7
	HA	43.0	507.4	21.5
Hw	CH	134.6	673.0	40.4
	HA	104.1	812.0	41.6
Forest floor	CH	113.7 (12.5)*	1077.3	59.1
(nonwoody)	HA	77.2 (9.9)	839.4	45.9
Forest floor	CH	166.1 (14.3)	769.4	46.4
(woody)	HA	133.7 (14.9)	915.3	49.5
Surface woody	CH	362.9 (28.8)*	331.9	36.2
	HA	225.5 (33.9)	294.7	22.9
Total	CH	642.5 (36.4)	2178.6	141.7
	HA	436.4 (39.4)	2049.4	118.3

*Significant ($P < 0.05$) differences in mass between the two forest types based on two-factor ANOVA.

and were significantly lower in pots containing forest floor material from cedar–hemlock forests than those from hemlock–amabilis forests.

Respiration and Litter Decomposition

Amounts of CO_2 respired during the 8-d incubation were significantly greater in the L and F layers of cedar–hemlock forest floors, and significantly lower in the Hw layer of cedar–hemlock forest floors (Fig. 17–2). Respiration rates declined with depth in the forest floor (i.e., L > F > Hs). Pine needle litter decomposed more rapidly in cedar–hemlock forests during the first 6 mo, but thereafter there were no differences in the weight of litter remaining in the two forests (Fig. 17–3a). In both forest types, pine needle litter lost more than 70% of its original weight during 3 yr. Forest floor F layer material decomposed significantly faster in cedar–hemlock forests, and lost just over 10% of its original weight during 3 yr (Fig. 17–3b). Of the endemic litter types, salal leaves decomposed most rapidly and cedar needle litter most slowly (Fig. 17–4). Mixtures of hemlock and fir needle litter from cedar–hemlock and hemlock–amabilis forests lost weight in an intermediate rate, and decomposed slightly faster in cedar–hemlock forests than in hemlock–amabilis forests (Fig. 17–4).

Carbon Chemistry of the Forest Floors

Concentrations of lipids were consistently higher in all forest floor layers in cedar–hemlock forests, but the differences were not significant (Fig. 17–5a).

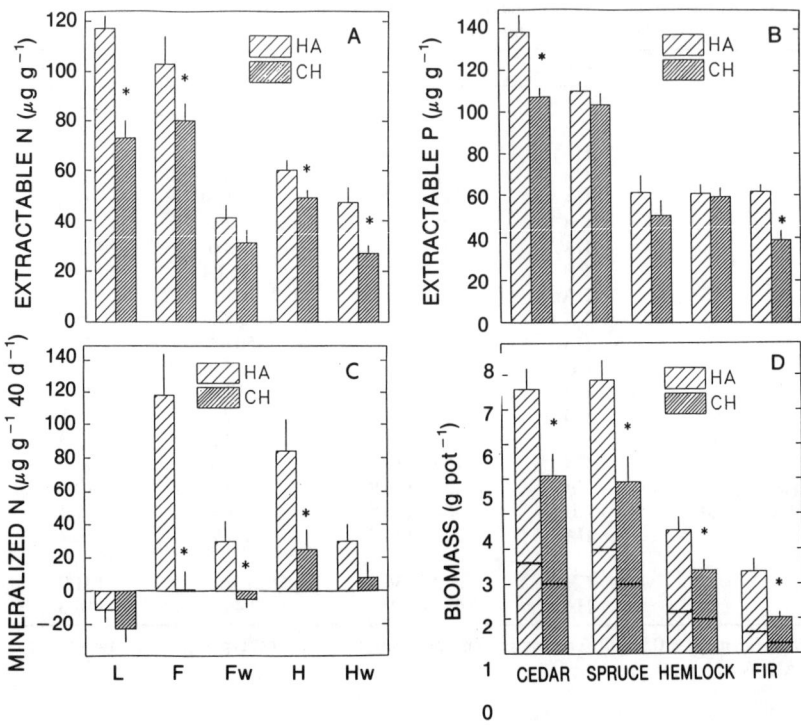

Fig. 17–1 Availability of nutrients in each forest floor layer of adjacent cedar–hemlock (CH) and hemlock–amabilis (HA) forests, measured as concentrations of KCl-extractable N (A) and Bray-extractable P (B), amounts of N mineralized during a 40-d aerobic incubation (C), and biomass production by seedlings of four species grown in F material from each forest type (D). Each value represents the mean (±SE) of 15 samples, asterisks indicate significant differences ($P \leq 0.05$) between the two forest types based on two-factor ANOVA. Abbreviations for forest floor layers as in Table 17–3. Horizontal lines separate values for shoots (above) and roots (below).

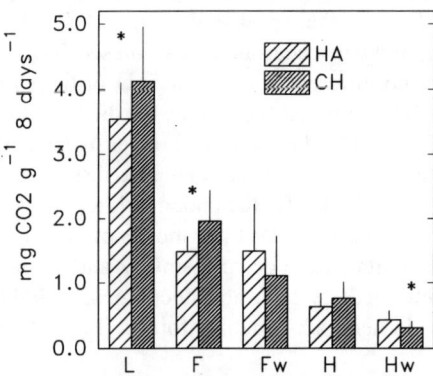

Fig. 17–2. Amount of CO_2 respired from each forest floor layer of adjacent cedar–hemlock (CH) and hemlock–amabilis (HA) forests during an 8-d laboratory incubation. Each value represents the mean (± SE) of 15 samples, asterisks indicate significant differences ($P \leq 0.05$) between the two forest types based on two-factor ANOVA. Abbreviations for forest floor layers as in Table 17–3.

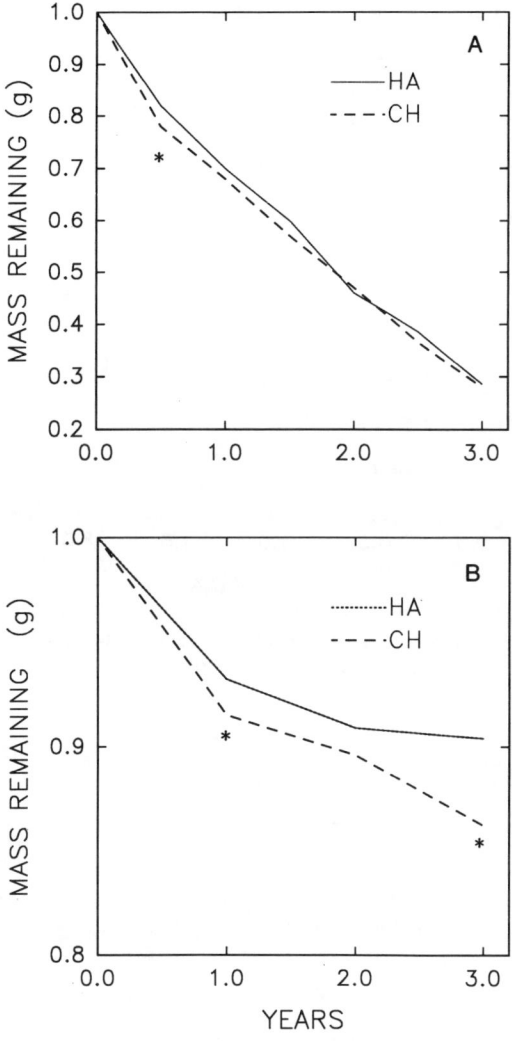

Fig. 17–3. Decomposition of lodgepole pine needles (A) and forest floor material (B) in adjacent cedar–hemlock (CH) and hemlock–amabilis (HA) forests. Each value represents the mean of five samples, asterisks indicate significant differences ($P \leq 0.05$) between the two forest types based on two-factor ANOVA.

Cedar–hemlock forests also had consistently higher total and labile polysaccharide concentrations than hemlock–amabilis sites, but the differences were significant only in the F layer (Fig. 17–5b,c). Cellulose concentrations, which were calculated as the difference between concentrations of total and labile polysaccharides, were similar in cedar–hemlock and hemlock–amabilis forests in all layers of the forest floor (Fig. 17–5d).

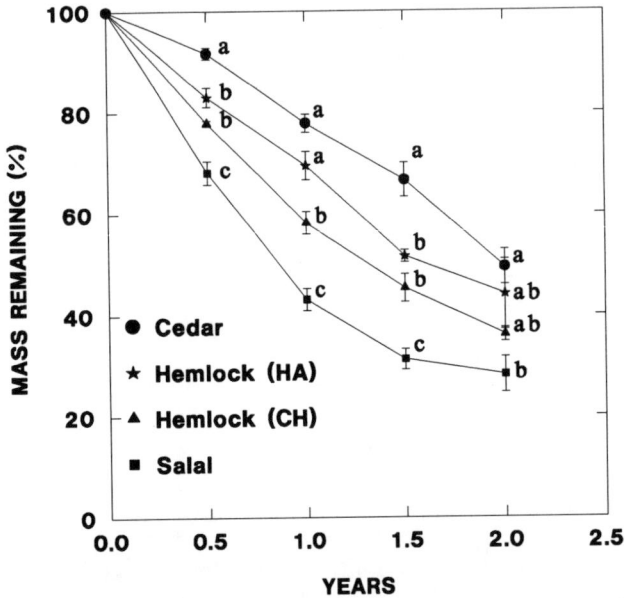

Fig. 17–4. Decomposition of foliar litter of hemlock and amabilis in HA forests and of hemlock and amabilis, cedar, and salal in CH forests. Each value represents the mean (±SE) of five samples.

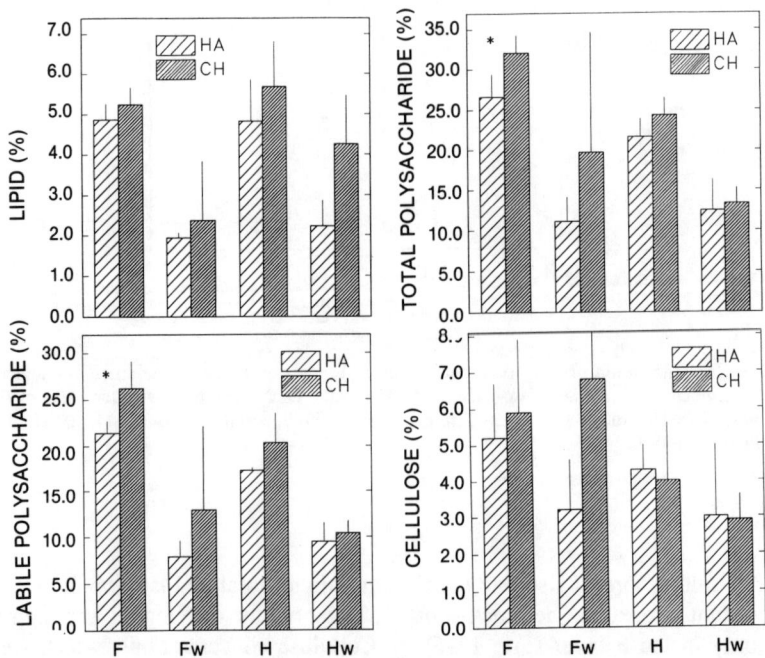

Fig. 17–5. Concentrations of lipids (A), polysaccharides (B,C) and cellulose (D) in each layer of the forest floor from cedar–hemlock (CH) and hemlock–amabilis (HA) forests. Each value represents the mean (±SD) of four samples, asterisks indicate significant differences ($P \leq 0.05$) between the two forest types based on oneway ANOVA. Abbreviations for forest floor layers are as described in Table 17–3.

Nuclear Magnetic Resonance Spectroscopy

Carbon-13 CPMAS NMR spectra of forest floor layers from cedar–hemlock and hemlock–amabilis forests are shown in Fig. 17–6. The general features of the spectra indicate that there is little to distinguish samples taken from cedar–hemlock and hemlock–amabilis forests. There is evidence of tannins in the spectra of the nonwoody layers. Tannins are difficult to identify in NMR spectra because the peaks occur in the same regions of those of lignin and carbohydrate (Morgan & Newman, 1986; Preston & Sayer, 1992). However, a peak due to condensed tannins occurs at 144 to 145 ppm, in a region that is usually clear in wood

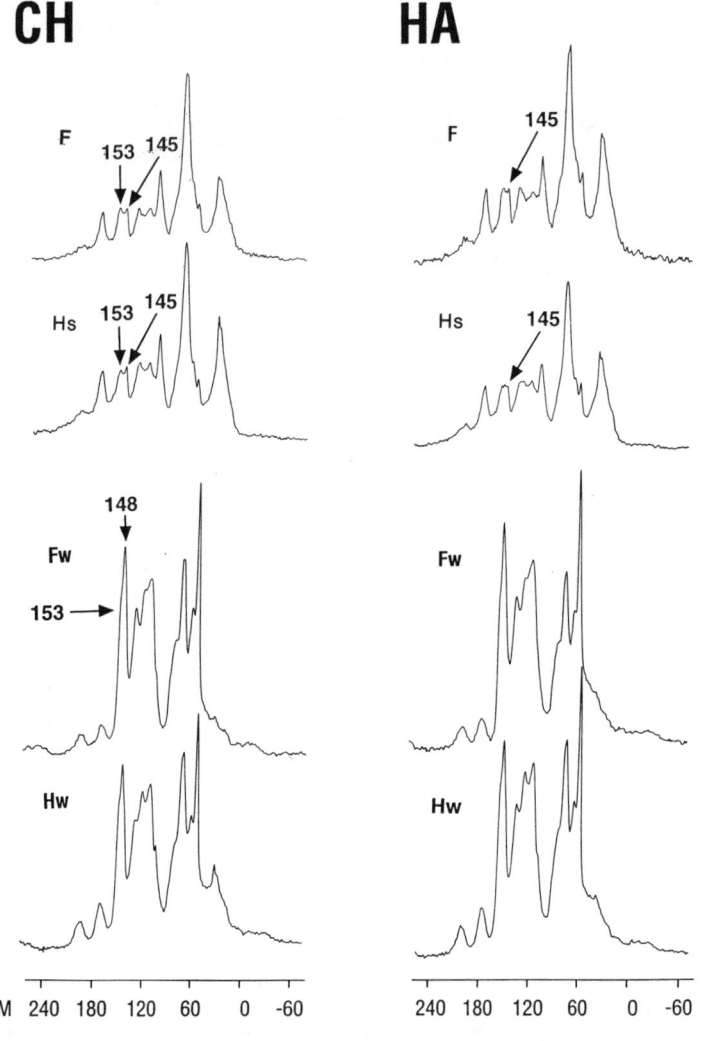

Fig. 17–6. Carbon-13 CPMAS NMR spectra of forest floor layers from cedar–hemlock (CH) and hemlock–amabilis (HA) forests showing tannin peaks at 144 to 145 ppm. Abbreviations for forest floor layers are described in Table 17–3.

and litter spectra. As can be seen in the expanded spectra (Fig. 17–7), the phenolic region of the Fw layer from both cedar–hemlock and hemlock–amabilis forests shows a well-defined peak at 148 ppm, with a slight shoulder at 153 ppm, typical of guaiacyl lignin C_3 and C_4. The F and H layers from cedar–hemlock forests show a broad signal presumably resulting from a combination of tannin, guaiacyl and syringyl phenolic inputs.

Dipolar-dephased spectra were used to examine features that may be masked in the normal CPMAS spectra. For the woody layers, the dipolar-dephased spectra in Fig. 17–8 show typical lignin peaks for methoxyl at 56 ppm, phenolic at 148 ppm with a shoulder at 153 ppm, and nonprotonated aromatic C (guaiacyl C_1) at 132 ppm, as well as weaker signals due to carboxyl (172 ppm) and carbonyl (195 ppm) C (Hatcher, 1987; Preston et al., 1990). Tannin signals are very weak or absent. The dipolar-dephased spectra of the nonwoody F and Hs layers from cedar–hemlock forests show the characteristic tannin peaks: a small but clearly defined peak at 145 ppm partially resolved from the main phenolic peak, as well as a strong peak at 108 ppm. The dipolar-dephased spectra of the F and Hs layers from the hemlock–amabilis forests lack a clearly defined peak at 145 ppm,

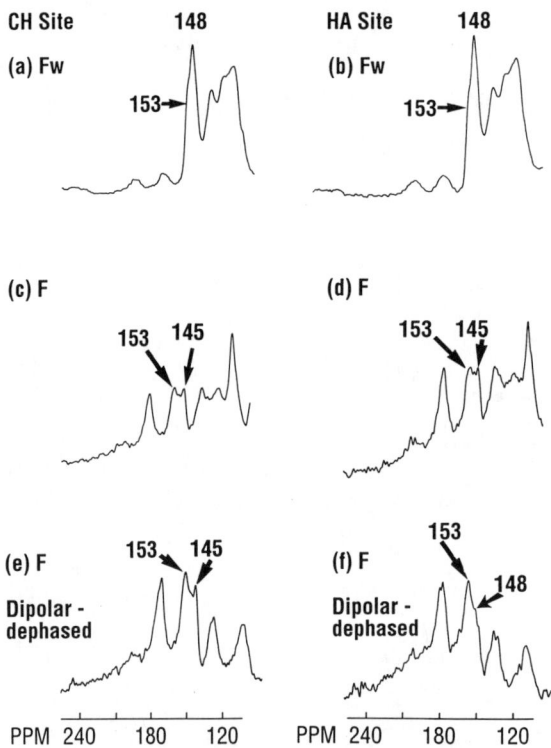

Fig. 17–7. Expanded ^{13}C CPMAS NMR spectra of the aromatic region for selected forest floor layers from cedar–hemlock (CH) and hemlock–amabilis (HA) forests. Abbreviations for forest floor layers are described in Table 17–3.

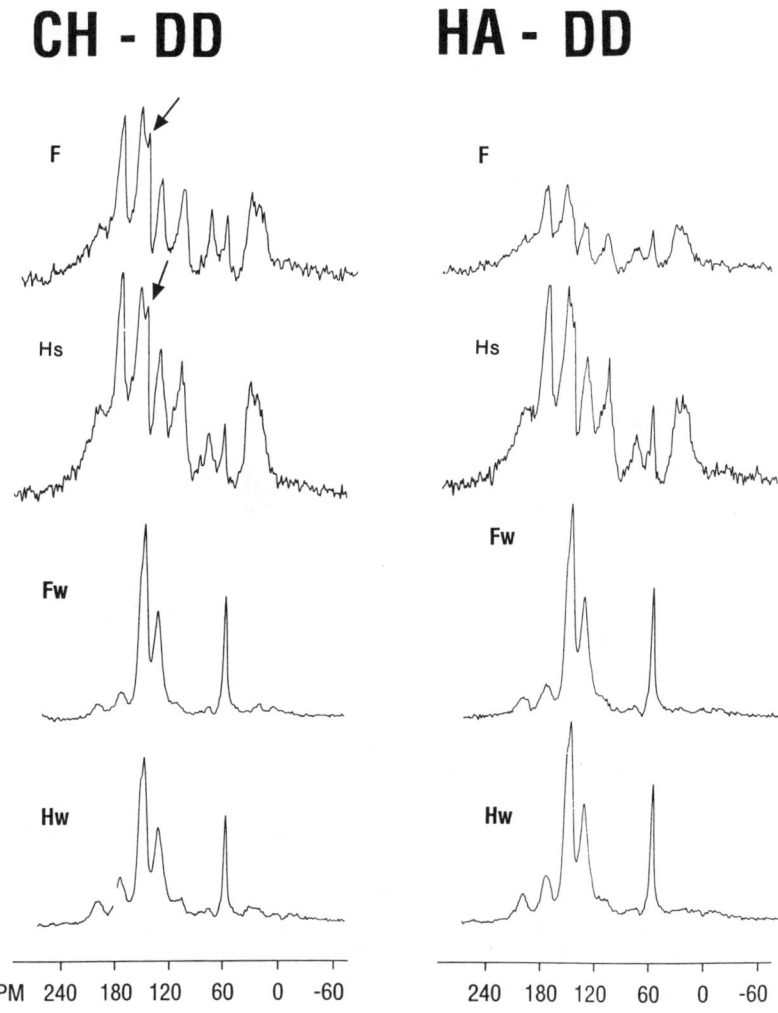

Fig. 17–8. Dipolar-dephased ^{13}C CPMAS NMR spectra of forest floor layers from cedar–hemlock (CH) and hemlock–amabilis (HA) forests showing tannin peaks at 144 to 145 ppm. Abbreviations for forest floor layers are described in Table 17–3.

although there is some intensity at 108 ppm. Together, these spectra suggest higher tannin concentrations in cedar–hemlock forest floors; however, the differences were small.

To trace the source of the tannins in cedar–hemlock forest floors, samples of the L layers, and leaf litter, roots and flowers of salal, were examined. Very low levels of tannins in L layer material from cedar–hemlock and hemlock–amabilis forests were indicated by poorly resolved shoulders in the phenolic region at 144 to 145 ppm (Fig. 17–9). Salal leaf litter, in contrast showed a well-defined peak at 144 ppm, and a large peak at 105, consistent with a sum of anomeric and

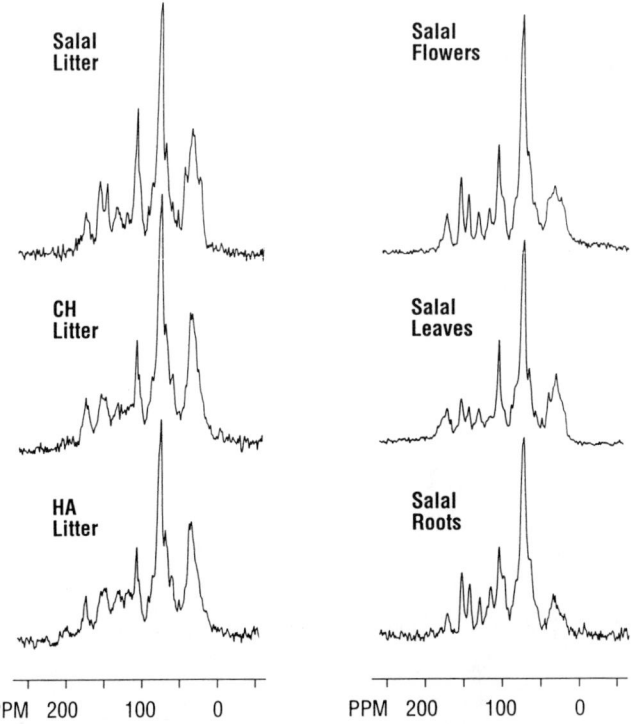

Fig. 17–9. Carbon-13 CPMAS NMR spectra of litter from cedar–hemlock (CH) and hemlock–amabilis (HA) forests, and of leaf litter, flowers, leaves and roots of salal.

tannin C. The spectra of salal flowers, leaves and roots all had strong tannin signals at 144 ppm.

The percentages of the total C in each forest floor layer that were lignin and carbohydrate are shown in Table 17–5. Lignin comprised 70 to 80% of the C in woody layers (Fw and Hw) and about 50% of the nonwoody layers (F and Hs). The percentage of C as lignin was greater in cedar–hemlock litter (44%) than in hemlock–amabilis litter (31%). Carbohydrate C was greater in hemlock–amabilis litter (42%) than in cedar–hemlock litter (25%) and declined with depth in the forest floor to about 14% in Hs layers of both forest types. Less than 5% of the C in woody layers (Fw and Hw) was in the form of carbohydrate. The ratio of carbohydrate to lignin monomer units (Cm/Lm) was greater in hemlock–amabilis litter than in cedar–hemlock litter, but there were no differences between forest types in the other forest floor layers. Values for Cm/Lm declined with depth in the forest floor and were least in woody material.

DISCUSSION

It is apparent from the consistently lower nutrient availability in the forest floors of old-growth cedar–hemlock forests that the nutritional stress encountered by trees in cedar–hemlock cutovers is at least partly a consequence of the low

Table 17–5. Lignin and carbohydrate C as percentage of total C, and ratios of carbohydrate to lignin monomer units (Cm/Lm) in each forest floor layer of cedar–hemlock (CH) and hemlock–amabilis (HA) forests.

Layer	Type	Lignin C	Carbohydrate C	Cm/Lm
		%		
L	CH	43.6	24.5	1.1
	HA	30.9	41.5	2.3
F	CH	48.1	21.1	0.7
	HA	48.7	17.2	0.6
Fw	CH	78.7	4.7	0.1
	HA	80.2	0	0
Hs	CH	52.8	13.3	0.4
	HA	50.0	14.4	0.5
Hw	CH	70.0	4.6	0.1
	HA	74.5	4.2	0.1

capacity of cedar–hemlock forest floors to provide nutrients for the regenerating forests. The causes of the lower nutrient availability in cedar–hemlock forest floors relative to adjacent hemlock–amabilis forests are less apparent. The physical and chemical properties of mineral soils in the two forest types were similar. There were smaller concentrations of N in mineral soil in cedar–hemlock forests, but because this N is mostly derived from the decomposition of litter, it is probably a consequence of the smaller N concentrations in cedar–hemlock forest floors, rather than a cause. The higher concentrations of pyrophosphate-extractable Fe in hemlock–amabilis forests could indicate more active weathering in these forests than in cedar–hemlock forests, or there may be differences in functional groups in organic ligands that cause more Fe to be in solution in hemlock–amabilis forests. deMontigny et al. (1993) found a higher ratio of vanillic acid to vanillin in forest floor material from hemlock–amabilis forests, and the acid would be a more effective ligand than the vanillin.

We expected to find evidence of slower rates of decomposition and microbial activity in cedar–hemlock forest floors, which would cause lower rates of N mineralization and lower nutrient availability. However, differences in rates of decomposition of similar substrates (pine needles or hemlock and fir needles) between the two forest types were small, and where they were significant, decomposition was usually faster in cedar–hemlock forests. Likewise, microbial activity, measured as CO_2 evolution from samples of the forest floors, was greater in the L and F layers of cedar–hemlock forests. These findings suggest that the decomposition "potential" of the two forest type are similar, and perhaps slightly greater in cedar–hemlock forests.

There is some evidence that characteristics of the litter in cedar–hemlock forests may contribute to slower mineralization of nutrients. Cedar foliar litter had lower N concentrations and higher lignin concentrations (Keenan et al., 1995, see Chapter 25), and lost mass more slowly than the other foliar litters. Other studies also have found that cedar foliar litter is lower in N (Daubenmire, 1953; Beaton et al., 1965), higher in lignin, and decomposes more slowly than other conifer foliar litters (Harmon et al., 1990). Cedar litter may be the cause of the greater proportions of lignin found in litter in cedar–hemlock forests compared with hemlock–amabilis forests, which may contribute to the accumulation of humus in cedar–hemlock forests. As Keenan et al. (1995, see Chapter 25) demonstrate through computer modeling, these characteristics of cedar litter may result in slower nutrient release and lower N availability in forest floors in cedar–hemlock forests.

The smaller amounts of N available in the forest floor in cedar–hemlock forests also might result from more N being immobilized therein, and less N circulating in litter. There were greater masses of forest floors in cedar–hemlock forests, although the total amounts of N in the forest floors in the two forest types were not very different. The mass of Hs in cedar–hemlock forests was greater than that in hemlock–amabilis forests, despite there being similar amounts of L and greater amounts of F in hemlock–amabilis forests. Similar results were reported by deMontigny (1992) in other forests of these types. This suggests that a greater proportion of the litter is converted to Hs in cedar–hemlock forests, rather than decomposing completely. This observation and the finding of consistently (albeit slightly) higher concentrations of polysaccharides and lipids in cedar–hemlock forest floors, also suggests that although the rates of mass loss may be similar in the two forest types, decomposition may be less "complete," causing more Hs to accumulate in cedar–hemlock forests. Results of the incubation experiment indicate that the N in Hs is not mineralized as rapidly as that in the L and F layers. This could result in more N being immobilized in cedar–hemlock forest floors than in hemlock–amabilis forest floors in which more of the N is in the L and F layers.

The NMR spectra suggested the presence of tannins in cedar–hemlock forest floors, apparently associated with salal. Tannins have been found to reduce the mineralization of organic matter by forming resistant complexes with proteins, by coating polysaccharides, and by inactivating enzymes important for decomposition (Benoit & Starkey, 1968; Zucker, 1983). The production of phenolic acids which "tan" proteins and enzymes, resulting in reduced mineralization of organic matter and accumulation of Hs, has been suggested as one of the causes of nutrient stress in plantations dominated by ericaceous shrubs (deMontigny & Weetman, 1990). Higher concentrations of some phenolic acids have been reported in cedar–hemlock forest floors (deMontigny, 1992), and these appeared to be associated with salal. Fyles and Fyles (1993) found that ground needle litter decomposed more slowly in the presence of ground salal litter, and suggested that an inhibitory substance may be produced during the decomposition of salal. The role of tannins in the immobilization of N in cedar–hemlock forest floors warrants further investigation.

The results of this study suggest that the nutritional difficulties encountered by trees growing in cedar–hemlock cutovers are at least partly the result of the low nutrient supplying capacity of forest floors in old-growth cedar–hemlock forests. This, in turn, is a consequence of low-quality organic matter in cedar–hemlock forests, which can be attributed to the presence of cedar and salal in these forests. Cedar foliar litter has high concentrations of lignin and low concentrations of N, which lead to slow rates of decomposition and N mineralization in the forest floor. Tannins, apparently associated with salal, also may interfere with decomposition and N mineralization in cedar–hemlock forest floors. After clear-cutting, the low N availability in cedar–hemlock forest floors may encourage growth of salal, which, like other ericaceous shrubs, is tolerant of low nutrient conditions. Competition from salal for the small amounts of nutrients mineralized from the remaining forest floors exacerbates the nutrient stress of trees, resulting in the observed decline in growth of trees in cedar–hemlock cutovers.

ACKNOWLEDGMENTS

This research was part of the Salal Cedar Hemlock Integrated Research Program (SCHIRP), and was supported by a Collaborative Research and Development Grant from the Natural Sciences and Engineering Council of Canada (NSERC), with Western Forest Products Ltd., MacMillan Bloedel Ltd., and Fletcher Challenge Canada Ltd. Western Forest Products Ltd. also provided field accommodation and seeds, and the MacMillan Bloedel Laboratory provided some of the chemical analyses.

REFERENCES

Bascomb, C.L. 1968. Distribution of pyrophosphate extractable iron and organic carbon in soils of various groups. J. Soil Sci. 19:251–268.
Beaton, J.D., A. Moss, I. MacRae, J.W. Konkin, W.P.T. McGhee, and R. Kosick. 1965. Observations on foliage nutrient content of several coniferous tree species in British Columbia. For. Chron. 41:222–236.
Benoit, R.E., and R.L. Starkey. 1968. Inhibition of decomposition of cellulose and some other carbohydrates by tannin. Soil Sci. 105:291–296.
Bremner, J.M., and C.S. Mulvaney. 1982. Nitrogen—total. p. 595–624. In A.L. Page et al. (ed.) Methods of soil analysis. Part 2. 2nd ed. Agron. Monogr. 9. ASA and SSSA, Madison, WI.
Cheshire, M.V. 1979. Nature and origin of carbohydrates in soils. Academic, New York.
Clegg, M.D., C.Y. Sullivan, and J.D. Eastin. 1978. A sensitive technique for the rapid measurement of carbon dioxide concentrations. Plant Physiol. 62:924–926.
Daubenmire, R. 1953. Nutrient content of leaf litter of trees in the northern Rocky Mountains. Ecology 34:786–793.
deMontigny, L.E. 1992. An investigation into the factors contributing to growth check of conifer regeneration on northern Vancouver Island. Ph.D. diss. Univ. British Columbia, Vancouver.
deMontigny, L.E., and G.F. Weetman. 1990. The effects of ericaceous plants on forest productivity. p. 83–90. In B.D. Titus et al. (ed.) The silvics and ecology of boreal spruces. For. Can. Information Rep. N-X-271.
deMontigny, L.E., C.M. Preston, P.G. Hatcher, and I. Kögel-Knaber. 1993. Comparison of humus horizons from two ecosystem phases on northern Vancouver Island using ^{13}C CPMAS NMR spectroscopy and CuO oxidation. Can. J. Soil Sci. 73:9–25.
Dubois, M., K.A. Gilles, J.K. Hamilton, P.A. Rebess, and F. Smith. 1956. Colorimetric method for determination of sugars and related substances. Anal. Chem. 28:350–356.
Fyles, J.W., and I.H. Fyles. 1993. Interaction of Douglas-fir with red alder and salal foliage litter during decomposition. Can. J. For. Res. 23:358–361.
Gee, G.W., and J.W. Bauder. 1982. Particle-size analysis. p. 383–411. In A. Klute (ed.) Methods of soil analysis. Part 1. 2nd ed. Agron. Monogr. 9. ASA and SSSA, Madison, WI.
Harmon, M.E., G.A. Baker, G. Spycher and S.E. Greene. 1990. Leaf-litter decomposition in the Picea/Tsuga forests of Olympic National Park, Washington, U.S.A. For. Ecol. Manage. 31:55–66.
Hatcher, P.G. 1987. Chemical structural studies of natural lignin by dipolar dephasing solid-state ^{13}C nuclear magnetic resonance. Org. Geochem. 11:31–39.
Ivarson, K.C., and F.J. Sowden. 1962. Methods for analysis of carbohydrate material in soil. 1. Colorimetric determination of uronic acids, hexoses and pentoses. Soil Sci. 94:245–250.
Keenan, R.J., J.P. Kimmins, and J. Pastor. 1995. Modeling carbon and nitrogen dynamics in western red cedar and western hemlock forests. p. 547–568. In W.W. McFee and J.M. Kelly (ed.) Carbon forms and functions in forest soils. SSSA, Madison, WI.

Keenan, R.J., C.E. Prescott, and J.P. Kimmins. 1993. Mass and nutrient content of woody debris and forest floor in western red cedar and western hemlock forests on northern Vancouver Island. Can. J. For. Res. 23:1052–1059.

Keeney, D.R., and D.W. Nelson. 1982. Nitrogen-inorganic forms. p. 643–698. *In* A.L. Page et al. (ed.) Methods of soil analysis. Part 2. 2nd ed. Agron. Monogr. 9. ASA and SSSA, Madison, WI.

Lowe, L.E. 1974. A sequential extraction procedure for studying the distribution of organic fractions in forest humus layers. Can. J. For. Res. 4:446–454.

McKeague, J.A. (ed.) 1978. Manual on soil sampling and methods of analysis. Can. Soc. Soil Sci., Ottawa, Canada.

Messier, C., and J.P. Kimmins. 1991. Above- and below-ground vegetation recovery in recently clearcut and burned sites dominated by *Gaultheria shallon* in coastal British Columbia. For. Ecol. Manage. 46:275–294.

Morgan, K.R., and R.H. Newman. 1986. Estimation of the tannin content of eucalypts and other hardwoods by carbon-13 nuclear magnetic resonance. Appita 40:450–454.

Norusis, M.J. 1988. SPSS/PC+ V2.0 base manual. SPSS, Chicago, IL.

Ohlsen, S.R., and L.W. Sommers. 1982. Phosphorus. p. 403–408. *In* A.L. Page et al. (ed.) Methods of soil analysis. Part 2. 2nd ed. Agron. Monogr. 9. ASA and SSSA, Madison, WI.

Parkinson, J.A., and S.E. Allen. 1975. A wet oxidation method for the determination of nitrogen and mineral nutrients in biological material. Commun. Soil Sci. Plant Anal. 6:1–11.

Pojar, J., K. Klinka, and D.V. Meidinger. 1987. Biogeoclimatic ecosystem classification in British Columbia. For. Ecol. Manage. 22:119–154.

Prescott, C.E., M.A. McDonald, and G.F. Weetman. 1993. Availability of N and P in the forest floors of adjacent stands of western red cedar–western hemlock and western hemlock–amabilis fir on northern Vancouver Island. Can. J. For. Res. 23:605–610.

Preston, C.M., and B.G. Sayer. 1992. What's in a nutshell: An investigation of structure by ^{13}C CPMAS NMR spectroscopy. J. Agric. Food Chem. 40:206–210.

Preston, C.M., P. Sollins, and B.G. Sayer. 1990. Changes in organic components of fallen logs in old-growth Douglas-fir forests monitored by ^{13}C nuclear magnetic resonance spectroscopy. Can. J. For. Res. 20:1382–1391.

Walkley, A. 1947. A critical examination of a rapid method for determining organic carbon in soils: Effect of variations in digestion conditions and of inorganic soil constituents. Soil Sci. 63:251–263.

Weetman, G.F., R. Fournier, J. Barker, E. Schnorbus-Panozzo, and A. Germain. 1989a. Foliar analysis and response of fertilized chlorotic Sitka spruce plantations on salal-dominated cedar-hemlock cutovers on Vancouver Island. Can. J. For. Res. 19:1501–1511.

Weetman, G.F., R. Fournier, J. Barker, and E. Schnorbus-Panozzo. 1989b. Foliar analysis and response of fertilized chlorotic western hemlock and western red cedar reproduction on salal-dominated cutovers on Vancouver Island. Can. J. For. Res. 19:1512–1520.

Weetman, G.F., R. Fournier, E. Schnorbus-Panozzo, and J. Barker. 1990. Post-burn nitrogen and phosphorus availability of deep humus soils in coastal British Columbia cedar/hemlock forests and the use of fertilization and salal eradication to restore productivity. p. 451–499. *In* S.P. Gessel et al. (ed.) Sustained productivity of forest soils. Proc. North Am. For. Soils Conf., 7th, Univ. British Columbia, Vancouver, Canada. Univ. British Columbia Faculty of For., Vancouver, Canada.

Wilkinson, L. 1990. SYSTAT: The system for statistics. SYSTAT, Evanston, IL.

Zar, J.H. 1974. Biostatistical analysis. Prentice-Hall, Toronto, Ontario, Canada.

Zucker, W.V. 1983. Tannins: Does structure determine function? Am. Nat. 121:335–365.

18 Belowground Responses to Atmospheric Carbon Dioxide in Forests

Richard J. Norby, E.G. O'Neill, and Stan D. Wullschleger

Oak Ridge National Laboratory
Oak Ridge, Tennessee

The increasing concentration of CO_2 in the atmosphere is probably the best documented and most fundamental change in the earth's environment since the advent of industrialization. As the primary substrate for photosynthesis, CO_2 has a most obvious role in plant growth and hence in all biological activity on the planet. Forests have a prominent role in the biological cycling of global C, and the extent to which trees and forest soils might store or release additional C as the atmospheric CO_2 concentration increases is a critical part of the analysis of the relationships between fossil fuel emissions, atmospheric CO_2, and the earth's climate system. The large size and long life of forest trees preclude the relatively simple experiments used to study the responses of crop plants to elevated atmospheric CO_2, and trees should be studied in the context of the forest ecosystem, including variable and often limiting supplies of other environmental resources. In particular, the importance of mineral nutrients to forest productivity mandates that we consider how nutrient availability modifies the response of trees to a CO_2-enriched atmosphere. Given the significance of the questions and the inherent limitations in experimental approaches to forest responses, our challenge is to make appropriate use of our understanding of tree physiological responses—particularly those that occur belowground—to explore the ramifications of increased atmospheric CO_2 on C and nutrient cycles in forests.

CARBON DIOXIDE AND LIMITING FACTORS

In 1981 Paul Kramer challenged the common assumption of many scientists that increasing atmospheric CO_2 concentrations would increase global photosynthesis and dry matter production (Kramer, 1981), thereby slowing down the effects of fossil fuel emissions of CO_2. While recognizing the importance of these assertions, he pointed out that they were based on the assumptions that photosynthesis is limited chiefly by CO_2 and that dry matter production is limited chiefly by photosynthesis. One critical problem he then highlighted was that of

---

Copyright © 1995 Soil Science Society of America, 677 S. Segoe Rd., Madison, WI 53711, USA.
Carbon Forms and Functions in Forest Soils.

limiting factors: in nature, photosynthesis and plant growth are more often limited by N or water deficiencies than by low CO_2. "Increasing the CO_2 concentration will have little effect if the stomata already are closed, the cell enlargement is inhibited by water stress, or the use of photosynthate is limited by lack of nitrogen" (Kramer, 1981).

The precept that limited nutrient availability in soil will restrict the capacity of trees to respond to increasing CO_2 is common in the literature on global change (Melillo et al., 1990; Bazzaz, 1990). These arguments are essentially invoking the so-called "Law of the Minimum" despite the oversimplification inherent in this doctrine. More than one substrate can be limiting at once, the supply of one resource can alter the supply of another, or at the population level, a host of environmental factors and interactions might limit growth at different places. Such considerations could certainly be relevant to interactions between CO_2 and nutrients. For example, increased C supply to plants in a CO_2-enriched atmosphere could stimulate the processes involved in nutrient turnover in soil or alter the capacity for roots to take up nutrients. On the other hand, the increased C supply could lead to sequestration of nutrients in wood or slower decomposition and nutrient release.

Several encompassing issues linking C and nutrient cycles are beginning to emerge. The three questions below provide a framework for analyzing the existing experimental data base on tree responses to elevated CO_2, as well as for designing future experiments.

1. Assuming that C cycling rates increase in elevated CO_2 (i.e., there is sustained enhancement of photosynthesis), does C cycling to belowground structures and processes increase nutrient cycling rates as well? If so, are there more nutrients available to the tree?
2. Is increased nutrient availability necessary for a forest system to sequester additional C from a CO_2-enriched atmosphere?
3. Does increased C sequestration reduce nutrient cycling rates and nutrient availability, providing a negative feedback on the primary response to elevated CO_2?

As will become clear, the existing data base on CO_2 responses of trees and forest processes is completely inadequate to answer these questions. Our first objective, then, must be to clarify the issues and the research priorities needed to address them.

A WORKING HYPOTHESIS

In his synthesis and challenge of the current understanding of CO_2 effects on plants, Kramer (1981) established a research agenda for the 1980s and beyond. Luxmoore (1981) responded to that challenge with an alternate hypothesis stating that increased photosynthesis in elevated concentrations of CO_2 would stimulate root and microbial processes, which could then increase potential nutrient and water supply to the plant. A shift in the nutrient pools in temperate forests from the soil to vegetation, he suggested with a bit of hyperbole, "could be the first step to another Carboniferous Age!" This hypothesis was elaborated by the

Microbial Effects panel at a 1982 conference, "Rising Atmospheric Carbon Dioxide and Plant Productivity: An International Conference" (Lamborg et al., 1983).

A research program at the Oak Ridge National Laboratory was initiated to consider the mechanisms whereby nutrient limitations to tree growth might be circumvented in elevated CO_2. The program was guided by an expanded version of Luxmoore's hypothesis (Fig. 18–1). In this scheme, the initial response of the plant to increased CO_2 is an increase in the rate of photosynthesis, a response that is very well documented (Eamus & Jarvis, 1989; Stitt, 1991). Increased photosynthesis is assumed to result in relatively more C allocated to root systems. A general principle of C allocation in plants is that the organs farthest from the source of C are those most responsive to changes in C supply (Waring & Schlesinger, 1985). With more C allocated to root systems, two general types of responses might be expected—increased root growth and stimulation of root processes. Those processes that might be responsive to C or energy flow include exudation of organic compounds into the rhizosphere, which can stimulate microbial organisms involved in nutrient transfers, symbiotic relationships (mycorrhizae and N_2 fixation), and membrane-bound nutrient uptake. An increase in fine root proliferation and soil exploration can increase nutrient availability directly, as well as increasing the total capacity of the root systems for symbiotic or rhizosphere processes. All of these possible

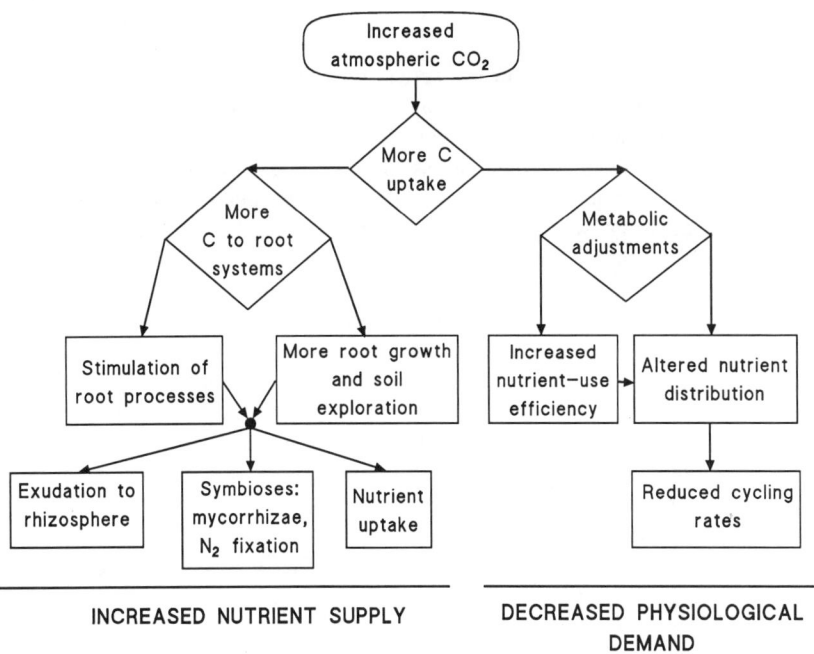

Fig. 18–1. Proposed mechanisms whereby nutrient limitations to tree growth might be circumvented in elevated CO_2.

responses to elevated CO_2 can be expected to increase nutrient supply to the plant, perhaps alleviating an existing or incipient deficiency.

There also are mechanisms of response to elevated CO_2 whereby nutrient demand might decrease (Fig. 18–1). For example, some plants that have become acclimated to elevated CO_2 maintain in their leaves lower amounts of the photosynthetic enzyme ribulosebiphosphate carboxylase/oxygenase (rubisco) (Stitt, 1991), which comprises a sizeable fraction of the N in a leaf. Photosynthesis is nonetheless higher because of the higher internal CO_2 concentration, but the amount of N required is less. Lower foliar requirements for N could make N available for other tissues and processes. The mechanisms of nutrient-use efficiency and nutrient distribution within trees are not specifically belowground responses and are not the focus of this paper, although they undoubtedly will have implications for C and nutrient cycles in soil (Norby et al., 1986b).

MECHANISTIC STUDIES ON CARBON DIOXIDE AND NUTRIENT INTERACTIONS

Experiments to investigate the responses of tree seedlings to CO_2 under variable nutrient regimes have been conducted with different objectives, species, and experimental designs. Not surprisingly, the results have been variable and ambiguous, even with regard to the seemingly straightforward question of whether elevated CO_2 can increase growth when nutrients are limiting. Nevertheless, experiments with potted tree seedlings in growth chambers have been useful for defining the larger-scale issues pertaining to C–nutrient interactions in a forest.

The overriding hypothesis for a series of experiments designed around Fig. 18–1 was: nutrient deficiency does not preclude a growth enhancement in tree species in response to CO_2 enrichment because stimulation of belowground processes will alleviate nutrient stress. The first test of this hypothesis was whether growth increases in elevated CO_2 even when one or more nutrients are deficient. Seedlings of white oak (*Quercus alba* L.) and yellow-poplar (*Liriodendron tulipifera* L.) were grown in nutrient-poor forest soil in pots in growth chambers containing ambient or twice-ambient CO_2 concentrations (Norby & O'Neill, 1989, 1991). A growth increase after the addition of fertilizer was taken as evidence that the unfertilized plants were nutrient deficient. In both species, the proportionate increase in dry mass attributable to CO_2 enrichment after 24 wk was somewhat *greater* in unfertilized seedlings than in fertilized ones (Fig. 18–2). Furthermore, the relative effect of CO_2 was similar in the two species even though the yellow-poplar was much more responsive to nutrient additions.

Experiments with other species also have shown that nutrient deficiency does not preclude growth responses to CO_2 enrichment (McLaughlin & Norby, 1991). Carbon dioxide enrichment increased dry mass of *Eucalyptus grandis* Hill ex Maid. seedlings at each rate of N and P fertilization, and the greatest relative increases occurred at the lowest level of P nutrition (Conroy et al., 1992). Two-year-old sweet chestnut (*Castanea sativa* Mill.) seedlings grown in infertile soil with and without fertilizer additions responded to CO_2 enrichment in much the same way as the white oak and yellow-poplar shown in Fig. 18–2 (El Kohen et

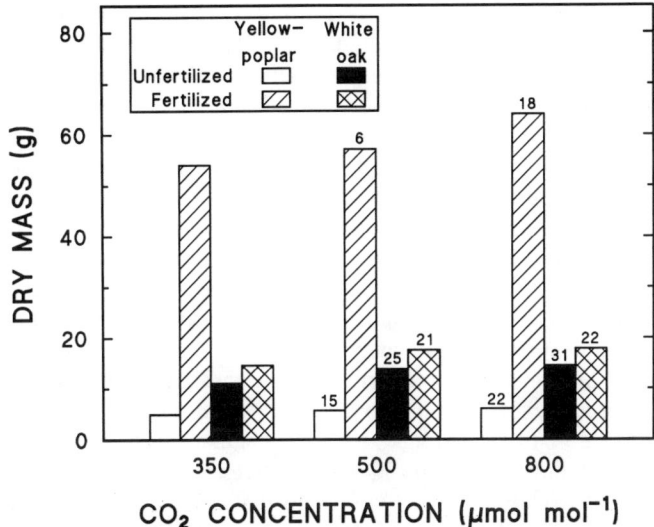

Fig. 18–2. Whole-plant dry mass of yellow-poplar and white oak seedlings grown for 24 wk in different concentrations of CO_2 in controlled environments. The seedlings were planted in nutrient-poor forest soil with or without added fertilizer. The numbers over bars represent the percentage increase in dry mass relative to plants in ambient CO_2 (data from Norby & O'Neill, 1989, 1991).

al., 1992). Fertilization more than doubled plant growth, confirming the nutrient deficient status of the unfertilized seedlings, but CO_2 increased total plant dry mass about 20% in both unfertilized and fertilized soil. Without suggesting a mechanism of response or addressing the critical issue of whether such responses could be sustained in a forest, these experiments have demonstrated that nutrient limitation does not necessarily preclude a response to elevated CO_2.

Other experiments, however, have pointed toward different conclusions. Conroy et al. (1986) showed that *Pinus radiata* D Don. (Monterey pine) seedlings did not increase in growth in elevated CO_2 if P was deficient. The P requirement apparently is higher in CO_2-enriched seedlings, but in low P soil, P uptake was increased by CO_2 enrichment, perhaps because of changes in mycorrhizal competition (Conroy et al., 1990). Results from the unusual environment of the arctic tundra (Oechel et al., 1991) have been used to support a general conclusion that soil nutrient limitations will significantly lessen the response of plants to elevated CO_2 (Mueller-Dombois, 1992). Bazzaz (1990) concluded that the effects of elevated CO_2 disappear under N and P limitation. Some of the experimental evidence used to support this conclusion may not be generally applicable. For example, Thomas et al. (1991) reported that the tropical legume tree *Gliricidia sepium* (Jacq.) Walp. increased in mass with CO_2 enrichment only when additional N was provided, but they pointed out that the unfertilized seedlings growing in an artificial medium with no fixed N were severely N deficient. Likewise, in a comparative study of *Pinus taeda* L. (loblolly pine) and *P. ponderosa* Laws (ponderosa pine) to CO_2 enrichment in interaction with soil fertility, seedlings responded to elevated CO_2 only when N and P were adequate (Strain & Thomas, 1992;

Johnson et al., 1994). However, the low fertility regimes were extremely N deficient—more so than would be expected in the field. The role of soil fertility in modifying growth responses to CO_2 enrichment should be viewed as a continuum rather than an on/off situation (Johnson et al., 1994).

These growth chamber studies of CO_2 and nutrient interactions are useful only if they suggest mechanisms of response that may help to predict the responses that will prevail in a forest. A general conclusion from many growth chamber studies is that belowground processes are especially sensitive to increased CO_2 in the atmosphere. The belowground processes for which there is some evidence of a CO_2 response include fine root production, symbiotic N_2 fixation, mycorrhizal establishment, exudation, and rhizosphere microbial processes. Because these processes are at the interface between C and nutrient cycles in soil, they may provide mechanisms for increased nutrient availability, thereby circumventing current nutrient limitations to growth (Fig. 18–1).

Root growth of tree seedlings in growth chamber CO_2 enrichment experiments is usually studied without regard to differences between fine and coarse roots, which may explain the variety of results that have been reported. The question of whether root-to-shoot ratio increases in elevated CO_2 has been quite difficult to answer, with increases, decreases, and no change in the ratio all reported in the literature (Eamus & Jarvis, 1989). A compilation of 224 observations on tree seedlings in elevated CO_2 (Wullschleger et al., 1995) indicated a log-transformed mean response of only a 6% increase in root-to-shoot ratio (Fig. 18–3). To extend these observations to larger trees in a forest, which have a

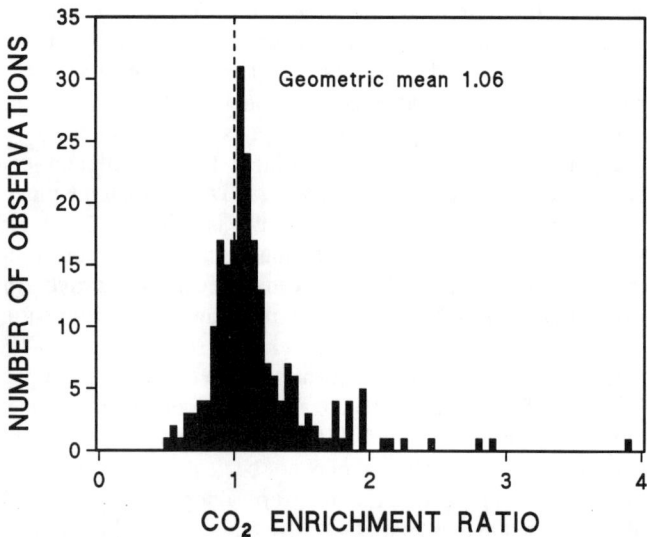

Fig. 18–3. Frequency distribution for the response of root-to-shoot ratios to CO_2 enrichment. The data are the root-to-shoot ratio of tree seedlings in elevated CO_2 divided by the ratio in ambient CO_2. The analysis was of a subset from a larger survey of 58 controlled-exposure studies covering 73 tree species (Wullschleger et al., 1995). The mean response of log-transformed data is a 6% increase in root-to-shoot ratio in elevated CO_2.

markedly different root system morphology than seedlings, there must be consideration of the differences between the responses of fine, short-lived roots and larger, woody, perennial roots. The responses to CO_2 enrichment of the woody root system have important implications for the overall C storage capacity of the tree, but the fine root responses should be more directly relevant to the issue of C-nutrient interactions.

An increase in fine root proliferation is particularly important in increasing the volume of soil exploited by a root system, as well as in increasing the size of the platform for rhizosphere processes. In an experiment with white oak (Norby et al., 1986a), fine roots were the plant component that exhibited the largest relative response to CO_2 enrichment. In yellow-poplar (Norby & O'Neill, 1991) fine roots increased with CO_2 enrichment to the same extent as coarse roots. Fine roots of 2-yr-old sweet chestnut seedlings were the most responsive plant component to CO_2 enrichment (El Kohen et al., 1992). In a model wet-tropical assemblage, the mass of fine roots (<1-mm diam.) increased 63% in elevated CO_2. The relative effect of CO_2 enrichment was less for increasingly larger roots (Körner & Arnone, 1992).

Symbiotic N_2 fixation is an important source of exogenous N in some forest systems and is a process that is highly dependent on the C cycle. Symbiotic N_2 fixation is stimulated in elevated CO_2 because of an increase in the number or size of nodules. As was previously demonstrated with annual legumes, nitrogenase activity (or N_2 fixation per unit nodule mass) usually is not specifically stimulated. Tree seedlings in which elevated CO_2 stimulated N_2 fixation include *Robinia pseudoacacia* L. (black locust), *Elaeagnus angustifolia* L. (Russian olive), *Alnus glutinosa* (L.) Gaertn. (European alder) (Norby, 1987), *A. rubra* Bong. (red alder) (Arnone & Gordon, 1990), and the tropical tree *Gliricidia sepium* (Thomas et al., 1991). Carbon dioxide enrichment resulted in increased plant N content in *A. rubra*, especially in nodulated seedlings supplied with fertilizer N. This result was interpreted as part of a positive feedback loop between N_2 fixation and photosynthesis in CO_2-enriched plants (Arnone & Gordon, 1990). Foliar N content in nodulated and fertilized *G. sepium* seedlings also increased with CO_2 enrichment, and a larger fraction of the leaf N pool was derived from N_2-fixation compared to plants grown in ambient CO_2 (Thomas et al., 1991). In all three of these studies, however, CO_2 enrichment stimulated growth more than N uptake despite the increased N_2-fixation; hence, foliar or whole-plant N concentration declined. Increased nodulation in elevated CO_2 may require increased P availability. For example, the nodule mass ratio in CO_2-enriched *R. pseudoacacia* increased with increasing whole-plant P/N ratio (R.J. Norby, unpublished data).

Nitrogen fixation by the lichen *Lobaria pulmonaria* (L.) Hoffm. also was stimulated by elevated CO_2 (Norby & Sigal, 1989). This lichen contributes a small but meaningful amount of fixed N to some forest systems, particularly the Douglas fir forests of the Pacific Northwest (Denison, 1979). While not a belowground process, this response is an example of a change in the C cycle (CO_2 enrichment) potentially altering a nutrient cycle. The ability to sequester nutrients from exogenous sources is low in most unmanaged ecosystems, so increased C sequestration will depend either on redistribution of nutrients within the ecosystem or changes in C-to-nutrient ratios (Rastetter et al., 1992). Mechanisms such as N_2 fixation that sequester exogenous nutrients should increase the capacity of most systems to store additional C.

While symbiotic N_2 fixation is relatively uncommon in forest trees, the mycorrhizal relationship between soil fungi and tree roots is essentially universal. Mycorrhizae form a fundamental link between the biotic portions of an ecosystem and the geochemical matrix, with a fast, positive feedback across this important linkage (Odom & Biever, 1984; O'Neill et al., 1991). As a major pathway for distribution of energy in terrestrial ecosystems, mycorrhizae can play an important role in mediating ecosystem homeostasis (Odom & Biever, 1984). If CO_2 affects mycorrhizal establishment or function, then we can expect effects on this linkage between C and nutrient cycles and higher-order ecosystem function (O'Neill et al., 1991).

Since mycorrhizal fungi are dependent on photosynthate from the plant for their C and energy supply, an increase in C supply to the plant could well stimulate the establishment of mycorrhizal roots, with beneficial consequences for the nutrient supply to the plant. Indeed, the rate of mycorrhizal colonization increased dramatically in *P. echinata* Mill. (shortleaf pine) and white oak when these ectomycorrhizal species were planted in nutrient-poor forest soil with indigenous mycorrhizae and grown in a CO_2-enriched atmosphere (O'Neill et al., 1987a). Carbon dioxide enrichment also stimulated mycorrhizal colonization of one population of *P. taeda* seedlings but not of a different population (Lewis et al., 1994). In contrast, CO_2 caused no difference in either the timing or extent of colonization of yellow-poplar roots with vesicular-arbuscular (VA) mycorrhizae. Apparently, the intensity of colonization by VA mycorrhizae was determined not by C supply, but by the availability of root tissue for infection (O'Neill et al., 1991). If this difference in response of ectomycorrhizae relative to VA mycorrhizae confers a competitive advantage to ectomycorrhizal hosts (by increasing nutrient acquisition, for example), then changes in forest composition and species distribution could occur (O'Neill et al., 1991).

The analysis of effects of CO_2 enrichment becomes more difficult when the analysis moves away from the plant in which the primary response (via photosynthesis) occurs. Atmospheric CO_2 concentration is unlikely to have any direct effects on soil-based processes, since the CO_2 concentration gradient from soil to air would change little even after a doubling of atmospheric CO_2 (Lamborg et al., 1983). Nevertheless, indirect effects of plant responses can still be expected in the soil. Carbon allocated to woody root systems is incorporated or stored in perennial tissue, incorporated into more ephemeral fine roots, transferred to fungal symbionts, respired, or exuded. The effect of CO_2 concentration on the allocation of C to all of these processes has rarely been studied directly. We pulse-labeled *P. echinata* seedlings with $^{14}CO_2$ and followed the allocation patterns over the next 3 d (Norby et al., 1987). There was no effect of the CO_2 in which the plants were growing on the percentage of $^{14}CO_2$ recovered in root systems. There was a change in the distribution of ^{14}C within the root system, however, with more C allocated to fine roots and less to coarse roots in the seedlings grown in elevated CO_2. There also was a suggestion that more C was released to the soil as organic exudates. Increased input of root-derived, ^{14}C-labeled organic material was suggested in a system with wheat (*Triticum aestivum* L.) plants in elevated CO_2 (Lekkerkerk et al., 1990). Because of the preference of microorganisms for the easily decomposed material, the decomposition of more resistant native soil organic matter was thought to be lower in elevated CO_2.

Organic exudates are a primary C source for the rhizosphere microbial community, and an effect of atmospheric CO_2 on exudation could drive changes in the composition and activity of this community, including microbes involved in nutrient cycling (Rogers et al., 1994). Such pathways may seem too obscure or trivial to be of concern when addressing large-scale global issues. Odom and Biever (1984) concluded, however, that these high quality—but often low quantity—energy flows will have the most important impact on primary productivity and homeostatis of ecosystems. Hence, it is important to recognize and investigate the little-known dissolved organic matter pathways and microbes that control energy and material allocations along these paths (Odom & Biever, 1984). Unfortunately, there have been very few studies in this area in relation to the potential effects of elevated CO_2. O'Neill et al. (1987b) measured the numbers of total bacteria, nitrifers, and phosphate-dissolving bacteria in the rhizosphere of yellow-poplar plants growing in ambient and elevated CO_2. The response was not as predicted; the numbers of phosphate-solubilizing and nitrite-oxidizing bacteria were lower in the rhizosphere of CO_2-enriched plants. In their experiment with model humid tropical assemblages, Körner and Arnone (1992) reported that the efflux of CO_2 from soil was almost double in elevated CO_2. While some of the increased CO_2 efflux was probably associated with the increased fine root mass, they concluded there also probably was a stimulation of microbial activity, supported by their observation of greater soil xylanase and protease activity in elevated CO_2. Increased root turnover and exudation in elevated CO_2 may have stimulated the breakdown of the thin layer of leaf and bark compost that was part of their artificial soil.

The result of these various belowground responses to CO_2 enrichment is that there is a potential for nutrient availability in soil to be enhanced. To assess whether increased nutrient availability leads to greater nutrient uptake by plants in elevated CO_2, it is necessary to analyze the nutrient content of the entire plant, because CO_2 enrichment can alter the distribution of nutrients within the plant. For example, the proportion of the total plant N pool in fine roots of white oak seedlings increased with CO_2 enrichment, while the proportion in the tap root decreased (Norby et al., 1986a). Williams et al. (1986) reported that with CO_2 enrichment, N and P pools decreased in size in the shoots of deciduous tree seedlings, while these pools increased in the roots. El Kohen et al. (1992), on the other hand, reported that CO_2 enrichment caused the pool size of N to increase in stems and leaf litter and decrease in coarse roots in fertilized sweet chestnut seedlings. In unfertilized seedlings the decrease in coarse root pool size was balanced by increases in the fine root and leaf litter N pools. In neither case did total pool size of N change with CO_2 concentration.

Since uptake of many nutrients across root membranes is an active, energy-demanding process of ion transport against a concentration gradient, an increased supply of carbohydrate to root systems of CO_2-enriched plants might support increased active ion uptake. There have been no studies investigating indirect effects of atmospheric CO_2 on the process of nutrient uptake at the root surface, although there have been detailed mechanistic studies of the role of soil CO_2 (Thibaud et al., 1990). Luxmoore et al. (1986) used a growth analysis approach to show that in *P. virginiana* Mill. (Virginia pine) seedlings specific nutrient uptake (i.e., nutrient uptake per unit root and time) was not altered by CO_2 enrichment. Nevertheless, the total uptake of some nutrients (including N, Ca, Fe, and Zn) was stimulated by CO_2 enrichment because of an increase in root weight integrated over time. Tschaplinski et al. (1993) reported that free amino acid concentrations were reduced in CO_2-enriched *P. taeda* seedlings and suggested that this might indicate a limitation of N assimilation in elevated CO_2.

Plant nutrients differ in their physical, chemical, and biochemical attributes, and it is not surprising that different nutrients show various responses to elevated CO_2. We categorized the response of different nutrients in white oak (Norby et al., 1986a) and yellow-poplar (O'Neill et al., 1987b) as showing either no increase in uptake and hence a decline in concentration (N, S, B), increased uptake commensurate with increased growth (P, K, Fe, Cu, Al), or an intermediate response in which there was an increase in uptake that was less than the increase in growth, leading to a decline in concentration (Ca, Mg, Mn, Zn, Sr). These different patterns of response of the potentially limiting macronutrients might be understood as representing a gradient from a supply-driven function (e.g., N, S) to a demand-driven function (K). In a low-N forest soil as was used in these experiments, N uptake was limited by N supply regardless of increased physiological demands imposed by CO_2 enrichment. Uptake of K, which was plentiful in the soil, was more likely to be controlled by plant demand through a tight regulation of membrane-bound transporters, and whole-plant concentration of K was maintained as growth increased. The intermediate response shown by the divalent cations (Ca, Mg) represented passive uptake of a nonlimiting nutrient that increased as the plants become larger. The case with P is most interesting. The pattern of response followed that of a demand-driven nutrient, yet the supply in the soil was limited. After analysis of the pools of available P in soil, plant, and soil leachate, we concluded that P availability had increased in elevated CO_2. We attributed this response to the increased fine root proliferation in elevated CO_2, which provided an expanded platform for root-influenced microbial and chemical reactions that control P availability (Norby et al., 1986a). Through this mechanism an incipient P deficiency in CO_2-enriched plants was avoided, a conclusion that could not have been reached in a system with an artificial rooting medium (e.g., Goudriaan & de Ruiter, 1983).

BELOWGROUND RESPONSES IN FIELD STUDIES

Experiments with individual plants in pots have been valuable for suggesting and describing many of the responses of root systems to elevated CO_2 and how those responses might interact with nutrient cycles. However, these experiments are limited in the extent to which they show how the responses actually will be manifested in a natural ecosystem, with competing root systems and other influences on both C and nutrient cycles. That there is a *potential* for CO_2 enrichment to increase root exudation and thereby increase nutrient availability and uptake cannot be considered as evidence that these processes occur in the field, or that if they do, that they are important in whole-system response. For example, increased exudation could stimulate root pathogens instead of bacteria involved in nutrient turnover, or if nutrient mineralization does increase because of CO_2-stimulated exudation, the newly available nutrients might leach from the system instead of being taken up by the plant (cf. Körner & Arnone, 1993). Even if the hypothesized process does result in increased nutrient supply, the additional amount might be a tiny and undetectable fraction of the variability associated with soil heterogeneity or soil moisture.

These considerations emphasize the importance of exploring CO_2 responses at a larger system level which integrates more components of ecosystem variability. Although there have been successful experiments investigating the responses to CO_2 enrichment of intact ecosystems, this has been attempted only in systems of small stature—the arctic tussock tundra (Oechel et al., 1991), the Chesapeake Bay salt marsh (Drake, 1992), and a tallgrass prairie (Owensby et al., 1993). No such study of an intact forest has been attempted because of technical and economic, as well as biological and ecological reasons. Nevertheless, several recent and ongoing studies have investigated the responses of individual trees or simple assemblages of trees.

A field study was initiated in Oak Ridge in 1989 to determine whether the previously described short-term CO_2 responses of yellow-poplar and white oak to CO_2 are sustained over several growing seasons under field conditions (Norby et al., 1992; Wullschleger & Norby, 1992, Wullschleger et al., 1992, Gunderson et al., 1993; Norby et al., 1993). Seedlings, which were grown from seed in growth chambers with regulated CO_2, were planted directly in the unamended soil (Captina Series, a Typic Fragiudult) within open-top field chambers and maintained for three or four growing seasons without supplemental irrigation or fertilization. The CO_2 in replicate chambers were ambient (+0, or about 355 µmol mol^{-1}), ambient + 150 µmol mol^{-1}, and ambient + 300 µmol mol^{-1}; the CO_2 treatments were maintained continuously throughout the growing seasons. Photosynthesis was stimulated by CO_2 enrichment and foliar respiration was lower. The effect of CO_2 on photosynthesis was sustained in both species throughout the experiment despite lower foliar N and chlorophyll concentrations, and there was no evidence of a loss in photosynthetic capacity (Gunderson et al., 1993). Nevertheless, there was no significant effect of CO_2 on yellow-poplar dry mass (Norby et al., 1992). White oak, on the other hand, had more than twice the dry mass in elevated CO_2 compared to ambient air after 4 yr. Most of the effect of CO_2 on growth in white oak occurred early in the 1st yr, and there was not a sustained effect of CO_2 on relative growth rate (Norby et al., 1993).

While the field experiment with yellow-poplar and white oak provided an unrestricted and unmanipulated rooting environment for the trees, extensive characterization of belowground responses was precluded. The data on belowground responses are essentially snapshots of the status of roots near the end of the experiment. Additional, although similarly limited, data on belowground responses in the field come from a horticultural experiment with sour orange (*Citrus aurantium* L.) trees (Idso et al., 1991) and a relatively short-term experiment with bigtooth aspen (*Populus grandidentata* Michx.). Other more recent field studies of tree responses to CO_2 (Strain & Thomas, 1992) will be focusing more on belowground responses and should greatly augment our understanding of the processes at this larger scale.

NUTRIENT LIMITATIONS TO GROWTH

The results of seedling studies have led us to the conclusion that nutrient deficiency does not preclude growth responses to elevated CO_2. Others have concluded that growth responses will occur only when nutrients are not limiting. Clearly, this remains an important issue to evaluate in the field. The question can be put: Is nutrient amendment necessary to sustain a growth response to CO_2 enrichment in a nutrient-limited habitat? Or, How does the capacity of an ecosystem to respond to CO_2 vary with its fertility? These questions are more difficult to answer than the simpler questions regarding the short-term response of seedlings, and they require field experiments of much larger scope than those that so far have been completed.

The current field CO_2 experiments do, however, give us fodder on which to speculate. For example, it is tempting to conclude that the large difference in response of sour orange and yellow-poplar trees to multiyear CO_2 enrichment was due to the important difference in nutrient regimes. Stem volume of the sour orange trees was 180% greater in elevated CO_2 (Idso et al., 1991). The yellow-poplar showed only a 35% increase in growth efficiency (annual growth per unit leaf area) and no significant increase in whole-tree mass (Norby et al., 1992). The sour orange trees were grown under horticultural conditions with fertilizer added "as needed," whereas the yellow-poplar trees were grown under conditions more typical of a forest, with no fertilizer additions and in a soil in which sorghum-sudangrass [*Sorghum bicolor* (L.) Moench] had been grown to lower nutrient content. Although it cannot be proven that the yellow-poplar were nutrient-limited, foliar and whole-plant N concentrations were lower in elevated CO_2. This difference in response between the two experiments could be taken as evidence that growth increases will not occur in elevated CO_2 unless nutrients are added.

Unfortunately, the analysis is not so simple. White oak trees at the same site exhibited a large increase in mass with CO_2 enrichment (more than double) after 4 yr under the same nutrient conditions as the yellow-poplar (Norby et al., 1993). White oak is less nutrient demanding than yellow-poplar, and there was no effect of CO_2 on foliar N concentration. A conclusion that might follow from this comparison is that the potential CO_2 response is controlled not by the nutrient characteristics of the site but the nutrient characteristic of a particular species (or individual) on that site. In support of this, bigtooth aspen increased in growth with CO_2 enrichment even in extremely nutrient depauperate sand (Zak et al., 1993).

An analysis of the growth dynamics of the white oak reveals further complexity about its CO_2 response, with additional implications for the tentative conclusion about nutrient control. The large growth effect was mostly due to an effect of elevated CO_2 on seedling establishment early in the experiment. The larger plants with greater leaf area continued to grow faster, as would be expected for open-grown trees, but there was not a sustained stimulation of relative growth rate, and growth efficiency increased about the same as in the yellow-poplar. A similar analysis could be applied to the sour orange trees. It may be, then, that the sustained responses of sour orange, yellow-poplar, and white oak to CO_2 enrichment were essentially similar regardless of nutrient regime, and the differences in tree mass after several years were due to as yet unexplained effects on seedling establishment or differences in experimental protocol.

ALTERATIONS IN CARBON BUDGET

These field studies have provided evidence for mechanisms by which CO_2 enrichment of trees could alter the C budget of soils. Fine root density (mass per unit soil surface area) of yellow-poplar and white oak (Norby et al., 1992, 1993) was significantly greater in elevated CO_2. Recognizing that fine root production cannot be assessed from a single measurement of root mass without detailed information on turnover rates (Vogt et al., 1989), it is reasonable to assume that fine root production increased with CO_2 enrichment. Fine root density of white oak roots at the end of the 4-yr experiment was 140% higher in elevated CO_2, approximately the same as the relative increase in whole-tree mass, so there was not necessarily a specific stimulation of fine root production. The relative increase in yellow-poplar fine root density with CO_2 enrichment (120%), however, was much greater than the effect of CO_2 concentration on aboveground production, implying a shift in C allocation from leaf production to fine root production, perhaps in response to the declining foliar N concentrations in the leaves (Norby et al., 1992).

After one growing season, the number of roots, root length, and root elongation of bigtooth aspen all were consistently greater in elevated CO_2 than in ambient CO_2, but only total root mass (which was 76–84% of total biomass) was significantly increased (Zak et al., 1993). There was no effect of CO_2 on aboveground growth. In contrast, Idso and Kimball (1992) reported that the increase in fine root mass of sour orange trees in elevated CO_2 was similar or less than the increase in aboveground biomass. One regression line described the relationship between seasonal fine root mass and tree basal area for both ambient and elevated CO_2. The aspen trees were grown under nutrient deficient conditions (even after fertilization) and the sour orange trees were well supplied with nutrients, perhaps explaining the difference in relative responsiveness of the fine roots.

What are the implications of increased fine root production in elevated CO_2? Compared to seedlings, fine roots of saplings and larger trees represent a much smaller proportion of the total mass of the tree, so increased fine root production does not imply increased C sequestration by the living tree. However, fine roots are the origin of much of the organic matter in soil (McClaugherty et al., 1984), and this could represent a route toward increased C sequestration in soil. Addressing this potentially important response of forests to CO_2 enrichment will require much more data on fine root production and decomposition, as well as on long-term changes in soil organic matter.

An increase in CO_2 efflux from soil, representing an integration of root and microbial respiration, also is evidence of an alteration in the belowground C cycle in elevated CO_2. A single measurement beneath the yellow-poplar trees suggested that CO_2 efflux was higher in the elevated CO_2 chambers (Norby et al., 1992), in concert with the increase in fine root density. An increased loss of C through fine root respiration could perhaps explain the absence of a significant increase in tree mass despite increased CO_2 uptake by the canopy. More extensive measurements beneath the white oak trees indicated that increased CO_2 efflux was a consistent and significant response to CO_2 enrichment. The proportionate increase in CO_2 efflux was similar to the calculated increase in total fine root respiration. Although specific fine root respiration rates actually were lower in elevated CO_2, the higher fine root density more than compensated (C.T. Nietch & R.J. Norby, unpublished data).

A greater fine root density in soil should provide an increased platform for microbial processes. Some preliminary indications of a response in the rhizosphere microbial community is provided by an analysis of the phospholipid ester-linked fatty acids of fine roots and associated rhizosphere (Ringelberg & White, 1994). There was no effect of CO_2 concentration on community structure in the white oak rhizosphere or on viable biomass per gram rhizosphere, although total microbial biomass presumably increased in concert with fine root biomass. Zak et al. (1993) measured significantly higher microbial C both in rhizosphere and bulk soil and higher labile C in rhizosphere soil of elevated CO_2 chambers with bigtooth aspen. Another observed microbial response is that mycorrhizal establishment apparently was enhanced by CO_2 enrichment in the ectomycorrhizal white oak, but not in the VA mycorrhizal yellow-poplar (O'Neill, 1994), much the same as was observed in the aforementioned growth chamber experiments (O'Neill et al., 1991).

NUTRIENT RELATIONS AND LITTER DECOMPOSITION

These alterations in the soil C cycle could have both direct and indirect implications for soil nutrient cycles. Increased fine root and mycorrhizal density could increase the capacity of the tree to take up limiting, nonmobile nutrients by increasing the volume of soil exploited. Increased C flux through soil could stimulate microbial processes controlling nutrient turnover. If increased fine root production is at the expense of leaf production (or leaf chemical composition), changes in litter decomposition and nutrient mineralization could result. There have been no field studies long enough or comprehensive enough to assess how a CO_2-induced change in the C cycle alters forest nutrient cycling, although there are some relevant observations that may help to form testable hypotheses.

The increase in microbial C in the CO_2-enriched bigtooth aspen system (Zak et al., 1993) was associated with a significant increase in N mineralization measured by short-term laboratory incubations. Zak et al. (1993) concluded that elevated CO_2 concentrations can have a positive feedback effect on soil C and N dynamics which lead to greater N availability. Since in their highly N-deficient system there was no response to CO_2 enrichment until N was added, they concluded that if C gain is severely N limited, then elevated atmospheric CO_2 will have limited impact on processes mediated by changes in soil C cycling.

If increased fine root proliferation and stimulated nutrient turnover lead to a higher capacity for nutrient uptake, the expected result should be a higher nutrient content in the plant, especially for a nutrient that might be limiting growth. In the yellow-poplar experiment, however, despite increased fine root density, there was no difference in whole-plant content of N, P (Table 18–1), or any other nutrient. Nitrogen content actually tended to decline with increasing CO_2 concentration. Whole-plant and foliar concentrations of N and P were significantly lower in elevated CO_2, but there was no significant effect on N concentration in live or dead fine roots.

One suggestion that frequently has been made about forest ecosystem responses to increased CO_2 is that declining N concentrations in plant tissue will slow decomposition rates, creating a negative influence on N availability and growth (Lamborg et al., 1983; Strain & Bazzaz, 1983; Mooney et al., 1987). There are virtually no data on leaf litter from trees grown in elevated CO_2 in the field to support or refute this idea (O'Neill, 1994), and only meager data from tree seedlings grown in pots (Norby et al., 1986b; Melillo et al., 1990; Coûteaux et al., 1991). In a growth chamber experiment with white oak seedlings, there was no strong evidence of an effect of CO_2 concentration on the chemical composition of senescent leaves (litter quality) that would be expected to provide a negative influence on decomposition rate and N availability (Norby et al., 1986b). However, in that study the leaves did not senesce in response to normal environmental cues (declining photoperiod and temperature), and the leaf litter probably was not representative of natural white oak litter. In another growth chamber experiment (R.J. Norby & J. Pastor, unpublished data) in which the white oak leaves senesced and abscised at the end of two growth cycles in response to declining photoperiod and temperature, the largest and most consistent effect of CO_2 on litter chemistry was on soluble phenolic content, which

Table 18–1. Amount and concentration of N and P in *Liriodendron tulipifera* saplings after three growing seasons in different concentrations of atmospheric CO_2. The trees were planted in the soil within open-top chambers and received no fertilizer during the experiment. Fine roots (<1-mm diam.) were not analyzed for P, so fine root P is not included in the total and whole-plant concentration.

Component	CO_2 enrichment, µmol mol^{-1}		
	+0	+150	+300
	N (g)		
Leaf	8.1	7.1	6.8
Stem	3.2	3.6	3.4
Coarse root	3.5	3.6	3.6
Fine root	0.7	0.7	0.7
Total	15.5	15.0	14.6
	P (g)		
Leaf	0.84	0.82	0.72
Stem	0.73	0.90	0.77
Coarse root	0.42	0.41	0.36
Total	1.99	2.12	1.85
	Whole-plant concentration (g kg^{-1})		
N	5.9	4.8	4.7
P	0.9	0.8	0.7

increased significantly with elevated CO_2 in both growth cycles. The concentration of N was not significantly affected, and the lignin-to-N ratio increased with CO_2 concentration in the first cycle only. The exponential decay coefficient (k), which reflects the change in lignin-to-N ratio (Melillo et al., 1982) was significantly smaller in elevated CO_2 for Cycle 1 litter only. These values of k correspond to a predicted 1st-yr decomposition rate (mass loss) of 63% for ambient-grown leaves and 59% for leaves grown in elevated CO_2.

White oak and yellow-poplar leaf litter from the field study also were analyzed. There were no significant effects of CO_2 on any measured aspect of leaf litter chemistry in white oak (Table 18–2). Nitrogen concentration, which was higher than in the leaves from the growth chamber experiment, tended to be slightly lower in the elevated CO_2 treatments, and lignin-to-N ratio tended to be higher, but the differences were small, not significant, and translated into small differences in predicted decomposition rates. The concentration of phenolics tended to be higher in elevated CO_2 both years, as was the case in the growth chamber study. In yellow-poplar litter, N concentration was significantly lower in CO_2-enriched leaves, but the differences were smaller than in the green leaves prior to autumn senescence. Lignin concentration also tended to be lower in CO_2-enriched litter, so there was no significant difference in lignin-to-N ratio.

These data do not support the premise that plant growth in elevated CO_2 lowers leaf litter quality, so there is no strong reason to predict that decomposition rates will decline. To determine decomposition rates directly, yellow-poplar leaf litter was enclosed in mesh bags and distributed in the litter layer of a natural yellow-poplar stand (O'Neill, 1994). The bags were withdrawn periodically to determine the mass loss over time, as well as changes in litter chemistry. Mass loss (decomposition) of the leaf litter from the different CO_2 regimes ranged from 48 to 54% after 18 mo and did not significantly differ by treatment. There also were no significant treatment differences in N dynamics of the decomposing litter. This data set does not support the assumption that leaves grown in high CO_2 will decompose and release N more slowly. Although the CO_2-enriched leaves had significantly lower N concentration when they were green, this difference did not transfer to a significant difference in leaf litter quality or decom-

Table 18–2. Chemical constituents (litter quality) of senesced leaves of trees grown in open-top chambers under different levels of CO_2 enrichment after 1 and 2 yr of treatment.

Species	CO_2 enrichment	1989			1990		
		N	Lignin	Lignin-N ratio	N	Lignin	Lignin-N ratio
	µmol mol^{-1}	—————— g kg^{-1} ——————					
Quercus alba	+0	10.0	136	13.7	9.3	132	14.2
	+150	8.7	139	16.2	8.1	134	16.5
	+300	9.2	140	16.8	8.4	132	15.7
Liriodendron tulipifera	+0	10.8	176	16.3	11.1	180	16.3
	+150	9.3	152	16.4	11.0	185	16.8
	+300	9.6	160	16.6	10.8	182	16.8

position (O'Neill, 1994). Similar measurements are being made on decomposing white oak litter.

The decomposition of fine roots and its impact on soil C and N cycles have barely been considered in CO_2 experiments. The N concentration of white oak and yellow-poplar fine roots has not differed between CO_2 regimes. Since N is thought not to retranslocate from fine roots as they senesce (Nambiar, 1987), this should indicate that N concentration would not differ in decomposing roots. This was the case in the yellow-poplar fine roots, but a full analysis of fine root litter quality or decomposition has not been attempted.

Changes in decomposition also could result from CO_2 effects on allocation patterns in trees. Since different plant tissues vary widely in their C-to-N ratios, a change in the proportion of the various tissues returned to the soil would likely have a much larger effect on C and N cycles than any direct effect of CO_2 on the chemistry of a specific tissue. For example, increased growth efficiency in high CO_2 leads to an increased proportion of woody tissue with a much higher C-to-N ratio than leaf litter. To determine the net result to C and N budgets in a forest will require longer term and larger-scale studies.

MODELING FOREST-SCALE CARBON-NUTRIENT INTERACTIONS

Ultimately, our interest in belowground responses arises from the need to predict the responses of forests to rising atmospheric CO_2 concentration over large spatial and temporal scales. As difficult as it has been to measure belowground responses in multi-year field experiments on individual trees, it has been even more difficult to initiate experiments on intact forest ecosystems. Hence, there must be a reliance on models of forest response to address these larger-scale questions. To be at all believable, the models should incorporate the best evidence about how important forest processes, resource interactions, and feedbacks will be influenced by a higher CO_2 concentration. A few models that address forest response to global change include consideration of the potential effects of elevated CO_2 on interactions between C and nutrient cycles. The lack of experimental data limits the scope of these efforts to the testing of different scenarios of response, but this approach can be valuable for identifying new measurements to be made in the field and developing more specific hypotheses to test.

Rastetter et al. (1992) described how the Marine Biological Laboratory's General Ecosystem Model (MBL-GEM) was used to examine the response of temperate hardwood forests to increased CO_2 concentration. Their underlying premise was that changes in C storage in terrestrial ecosystems can be attributed to changes in the total amount of N in the system, net shifts in N between vegetation and soils, and changes in the C-to-N ratio in vegetation and soils. The model provides a process-level description of ecosystems by simulating changes in the amounts of C and N in 11 organic matter pools plus inorganic soil N. In response to a doubling of atmospheric CO_2 concentration, the model temperate forest (as well as the arctic tundra) increased C storage without increasing total N. The critical assumption (although not stated as such) driving their simulated responses was that CO_2 stimulation of photosynthesis produced leaf litter with a high C-to-N ratio. This litter caused increased N immobilization into soil organic matter and amplified the N stress of the vegetation. The net result was a shift in N from vegetation to soil.

Pastor and Post (1988) have shown the importance of changes in litter quality in driving long-term changes in forest productivity following climate change. In their simulations of forest productivity with a forest stand competition model, LINKAGES, the direct responses of stand growth to increased soil water deficit (brought on by climate change) were amplified by alterations in N cycling arising from changes in litter quality. The litter quality feedback occurred through changes in forest composition resulting from differential sensitivity of species to moisture deficit. They made no assumptions about the influence of CO_2 concentration on litter quality. More recent simulations explored the consequences for forest C storage of several potential direct effects of CO_2 on different plant processes (Post et al., 1992). Increased fine root production in elevated CO_2 without changes in basal area increment or leaf production, a scenario based partially on experimental observations (Norby et al., 1992), resulted in small increases in soil organic matter, as well as increased aboveground production because of greater N availability. Although only certain aspects of ecosystem C and N dynamics were considered in these simulations, and many other processes and interactions were ignored, Post et al. (1992) showed that there is considerable potential for increasing C storage in forest ecosystems resulting from plant responses to elevated CO_2.

CONCLUSIONS

A few experimental studies on tree responses to elevated CO_2, conducted on potted seedlings, model ecosystems, or field-grown seedlings and saplings have presented enough data on belowground responses to be intriguing. The data are fragmentary at best, but generally support the concept that belowground responses to elevated CO_2 have the potential to modify C and nutrient cycles in forest soils. Since these alterations in C and nutrient cycles can provide feedbacks to tree growth and C storage, it becomes especially important that analyses of forest responses to a changing atmosphere at least recognize—if not incorporate—the importance of longer-term, higher-order ecological responses. Of particular importance are those belowground responses that occur at the interface of the C and nutrient cycles.

Returning to the three questions in this arena that were posed earlier, what can we now conclude? Elevated CO_2 does seem to increase C cycling rates (Question 1)—photosynthetic uptake of CO_2 is higher and remains higher even when foliar N concentrations are lower. The return of CO_2 from the soil to the atmosphere also is higher. The steps in between, however, remain vague—How much of the increased CO_2 efflux from soil is due to root respiration rate and how much is attributable to microbial processes? To what extent does the increased C cycling influence other processes? There are only hints that microbial populations or processes might be affected or that any such effects increase nutrient availability. Seedling studies have shown that tree growth responses to elevated CO_2 over a relatively short-time period (less than 1 yr) do not require increased nutrient availability, although some data sets might suggest otherwise (Question 2). There is not yet evidence that various hypothesized mechanisms which might increase nutrient availability have actually improved tree nutrition. Furthermore, observations about CO_2 responses of seedlings in different nutrient environments are difficult to extend to a forest because the dominant processes in tree nutrition change as a tree grows larger. Especially important in this regard is the issue of feedbacks on nutrient cycling caused by increased C sequestration (Question 3). Despite speculation to the contrary, there is no strong evidence that leaf litter decomposition will be slower in a high CO_2 world. Other C sequestration processes that might alter nutrient cycling have not been explored.

There are other tree and forest ecosystem processes besides the belowground ones considered here that will contribute to our understanding of forest responses to rising CO_2 concentration. Especially important are questions about biochemical and physiological adjustments in nutrient use, distribution, and retranslocation within trees. As the scope of the inquiry expands, we also must consider biotic interactions, such as the role of herbivores in effecting C and nutrient redistribution, and changes in forest composition due to differential sensitivity of trees species to elevated CO_2. Having stressed the importance of belowground responses in global change research, it is distressing that only fragmentary and ambiguous data can be presented. Nevertheless, the observations that have been made allow us to speculate on a somewhat more informed basis. More important, the incrementally increasing knowledge base allows us to ask more specific and relevant questions about the mechanisms and consequences of CO_2 effects on the C and nutrient cycles of forest soils. Now it is time to move from the speculation and sink the shovels into the soil!

ACKNOWLEDGMENTS

Research sponsored by the Global Change Research Program, Office of Health and Environmental Research, U.S. Dep. of Energy, under contract no. DE-ACO5-840R21400 with Martin Marietta Energy Systems, Inc., Publication no. 4163, Environmental Sciences Division, Oak Ridge National Laboratory. The submitted manuscript has been authored by a contractor of the U.S. government under contract no. DE-ACO5-840R21400. In accordance, the U.S. government retains a nonexclusive, royalty-free license to publish or reproduce the published form of this contribution, or allow others to do so, for U.S. government purposes.

REFERENCES

Arnone, J.A., III, and J.C. Gordon. 1990. Effect of nodulation, nitrogen fixation and CO_2 enrichment on the physiology, growth and dry mass allocation of seedlings of *Alnus rubra* Bong. New Phytol. 116:55–66.

Bazzaz, F.A. 1990. Response of natural ecosystems to the rising global CO_2 levels. Ann. Rev. Ecol. Syst. 21:167–196.

Conroy, J., E.W.R. Barlow, and D.I. Bevege. 1986. Response of *Pinus radiata* seedlings to carbon dioxide enrichment at different levels of water and phosphorus: Growth, morphology and anatomy. Ann. Bot. 57:165–177.

Conroy, J.P., P.J. Milham, and E.W.R. Barlow. 1992. Effect of nitrogen and phosphorus availability on the growth response of *Eucalyptus grandis* to high CO_2. Plant Cell Environ. 15:843–847.

Conroy, J.P., P.J. Milham, M. Mazur, and E.W.R. Barlow. 1990. Growth, dry weight partitioning and wood properties of *Pinus radiata* D. Don after 2 years of CO_2 enrichment. Plant Cell Environ. 13:329–337.

Coûteaux, M.-M., M. Mousseau, M.-L. Célérier, and P. Bottner. 1991. Increased atmospheric CO_2 and litter quality: Decomposition of sweet chestnut leaf litter with animal food webs of different complexities. Oikos 61:54–64.

Denison, W.C. 1979. *Lobaria oregana*, a nitrogen-fixing lichen in old-growth Douglas fir forests. p. 266–275. *In* J.C. Gordon et al. (ed.) Symbiotic nitrogen fixation in the management of temperate forests. Oregon State Univ. School Forestry, Corvallis, OR.

Drake, B.G. 1992. A field study of the effects of elevated CO_2 on ecosystem processes in a Chesapeake Bay wetland. Aust. J. Bot. 40:579–595.

Eamus, D., and P.G. Jarvis. 1989. The direct effects of increase in the global atmospheric CO_2 concentration on natural and commercial temperate trees and forests. Adv. Ecol. Res. 19:1–55.

El Kohen, A., H. Rouhier, and M. Mousseau. 1992. Changes in dry weight and nitrogen partitioning induced by elevated CO_2 depend on soil nutrient availability in sweet chestnut (*Castanea sativa* Mill). Ann. Sci. For. 49:83–90.

Goudriaan, J., and H.E. de Ruiter. 1983. Plant growth in response to CO_2 enrichment, at two levels of nitrogen and phosphorus supply. 1. Dry matter, leaf area and development. Neth. J. Agric. Sci. 31:157–169.

Gunderson, C.A., R.J. Norby, and S.D. Wullschleger. 1993. Foliar gas exchange responses of two deciduous hardwoods during three years of growth in elevated CO_2: No loss of photosynthetic enhancement. Plant Cell Environ. 16:797–807.

Idso, S.B., and B.A. Kimball. 1992. Seasonal fine-root biomass development of sour orange trees grown in atmospheres of ambient and elevated CO_2 concentration. Plant Cell Environ. 15:337–341.

Idso, S.B., B.A. Kimball, and S.G. Allen. 1991. CO_2 enrichment of sour orange trees: 2.5 years into a long-term experiment. Plant Cell Environ. 14:351–353.

Johnson, D.W., J.T. Ball, and R.F. Walker. 1994. Effects of CO_2 and nitrogen on nutrient uptake in ponderosa pine seedlings. Plant Soil (In press.)

Körner, C., and J.A. Arnone III. 1992. Responses to elevated carbon dioxide in artificial tropical ecosystems. Science (Washington, DC) 257:1672–1675.

Kramer, P.J. 1981. Carbon dioxide concentration, photosynthesis, and dry matter production. BioScience 31:29–33.

Lamborg, M.R., R.W.F. Hardy, and E.A. Paul. 1983. Microbial effects. p. 131–176. *In* E.R. Lemon (ed.) CO_2 and plants. Westview Press, Boulder, CO.

Lekkerkerk, L.J.A., S.C. Van de Geijn, and J.A. Van Veen. 1990. Effects of elevated atmospheric CO_2-levels on the carbon economy of a soil planted with wheat. p. 423–429. *In* A.F. Bouwman (ed.) Soils and the greenhouse effect. John Wiley & Sons, Chichester, England.

Lewis, J.D., R.B. Thomas, and B.R. Strain. 1994. Effect of elevated CO_2 on mycorrhizal colonization rates of loblolly pine (*Pinus taeda* L.) seedlings. Plant Soil 165:81–88.

Luxmoore, R.J. 1981. CO_2 and phytomass. BioScience 31:626.

Luxmoore, R.J., E.G. O'Neill, J.M. Ells, H.H. Rogers. 1986. Nutrient uptake and growth response of Virginia pine to elevated atmospheric carbon dioxide. J. Environ. Qual. 15:244–251.

McClaugherty, C.A., J.D. Aber, and J.M. Melillo. 1984. Decomposition dynamics of fine roots in forested ecosystems. Oikos 42:378–386.

McLaughlin, S.B., and R.J. Norby. 1991. Atmospheric pollution and terrestrial vegetation: Evidence of changes, linkages and significance to selection processes. p. 61–101. *In* G.E. Taylor et al. (ed.) Ecological genetics and air pollution. Springer-Verlag, New York.

Melillo, J.M., J.D. Aber, and J.F. Muratore. 1982. Nitrogen and lignin control of hardwood leaf litter decomposition. Ecology 63:621–626.

Melillo, J.M., T.V. Callaghan, F.I. Woodward, E. Salati, and S.K. Sinha. 1990. Effects on ecosystems. p. 282–310. *In* J.T. Houghton et al. (ed.) Climate change: The IPCC scientific assessment. Cambridge Univ. Press, Cambridge, England.

Mooney, H.A., P.M. Vitousek, and P.A. Matson. 1987. Exchange of materials between terrestrial ecosystems and the atmosphere. Science (Washington, DC) 238:926–932.

Mueller-Dombois, D. 1992. Potential effects of the increase in carbon dioxide and climate change on the dynamics of vegetation. Water Air Soil Pollut. 64:61–79.

Nambiar, E.K.S. 1987. Do nutrients retranslocate from fine roots? Can. J. For. Res. 17:913–918.

Norby, R.J. 1987. Nodulation and nitrogenase activity in nitrogen-fixing woody plants stimulated by CO_2 enrichment of the atmosphere. Physiol. Plant. 71:77–82.

Norby, R.J., C.A. Gunderson, S.D. Wullschleger, E.G. O'Neill, and M.K. McCracken. 1992. Productivity and compensatory responses of yellow-poplar trees in elevated CO_2. Nature (London) 357:322–324.

Norby, R.J., and E.G. O'Neill. 1989. Growth dynamics and water use of seedlings of *Quercus alba* L. in CO_2-enriched atmospheres. New Phytol. 111:491–500.

Norby, R.J., and E.G. O'Neill. 1991. Leaf area compensation and nutrient interactions in CO_2-enriched yellow-poplar (*Liriodendron tulipifera* L.) seedlings. New Phytol. 117:515–528.

Norby, R.J., E.G. O'Neill, W.G. Hood, and R.J. Luxmoore. 1987. Carbon allocation, root exudation, and mycorrhizal colonization of *Pinus echinata* seedlings grown under CO_2 enrichment. Tree Physiol. 3:203–210.

Norby, R.J., E.G. O'Neill, and R.J. Luxmoore. 1986a. Effects of atmospheric CO_2 enrichment on the growth and mineral nutrition of *Quercus alba* seedlings in nutrient-poor soil. Plant Physiol. 82:83–89.

Norby, R.J., E.G. O'Neill, S.D. Wullschleger, C.A. Gunderson, and C.T. Nietch. 1993. Growth enhancement of *Quercus alba* saplings by CO_2 enrichment under field conditions. Bull. Ecol. Soc. Am. 74(Suppl.):375.

Norby, R.J., J. Pastor, and J.M. Melillo. 1986b. Carbon-nitrogen interactions in CO_2-enriched white oak: Physiological and long-term perspectives. Tree Physiol. 2:233–241.

Norby, R.J., and L.L. Sigal. 1989. Nitrogen fixation in the lichen *Lobaria pulmonaria* in elevated atmospheric carbon dioxide. Oecologia 79:566–568.

Odom, E.P., and L.J. Biever. 1984. Resource quality, mutualism, and energy partitioning in food chains. Am. Nat. 124:360–376.

Oechel, W.C., G. Riechers, W.T. Lawrence, T.T. Prudhome, N. Grulke, and S.J. Hastings. 1991. Long-term *in situ* manipulation and measurement of CO_2 and temperature. Func. Ecol. 6:86–100.

O'Neill, E.G. 1994. Responses of soil biota to elevated atmospheric carbon dioxide. Plant Soil 165:55–65.

O'Neill, E.G., R.J. Luxmoore, and R.J. Norby. 1987a. Increases in mycorrhizal colonization and seedling growth in *Pinus echinata* and *Quercus alba* L. in an enriched CO_2 atmosphere. Can. J. For. Res. 17:878–883.

O'Neill, E.G., R.J. Luxmoore, and R.J. Norby. 1987b. Elevated atmospheric CO_2 effects on seedling growth, nutrient uptake, and rhizosphere bacterial populations of *Liriodendron tulipifera* L. Plant Soil 104:3–11.

O'Neill, E.G., R.V. O'Neill, and R.J. Norby. 1991. Hierarchy theory as a guide to mycorrhizal research on large-scale problems. Environ. Pollut. 73:271–284.

Owensby, C.E., P.I. Coyne, and L.M. Auen. 1993. Nitrogen and phosphorus dynamics of a tallgrass prairie ecosystem exposed to elevated carbon dioxide. Plant Cell Environ. 16:843–850.

Pastor J., and W.M. Post. 1988. Response of northern forests to CO_2-induced climatic change. Nature (London) 334:55–58.

Post, W.M., J. Pastor, A.W. King, and W.R. Emanuel. 1992. Aspects of the interaction between vegetation and soil under global change. Water Air Soil Pollut. 64:345–363.

Rastetter, E.B., R.B. McKane, G.R. Shaver, and J.M. Melillo. 1992. Changes in C storage by terrestrial ecosystems: How C-N interactions restrict responses to CO_2 and temperature. Water Air Soil Pollut. 64:327–344.

Ringelberg, D.B., and D.C. White. 1994. Analysis of the effect of increased CO_2 on the biomass, community structure and nutritional status of the rhizosphere microbiota of white oaks, long leaf pine and cottonwood. Rep. no. 33. Environ. Inst., Univ. Alabama, Tuscaloosa, AL.

Rogers, H.H., G.B. Runion, and S.V. Krupa. 1994. Plant responses to atmospheric CO_2 enrichment with emphasis on roots and the rhizosphere. Environ. Pollut. 83:155–189.

Stitt, M. 1991. Rising CO_2 levels and their potential significance for carbon flow in photosynthetic cells. Plant Cell Environ. 14:741–762.

Strain, B.R., and F.A. Bazzaz. 1983. Terrestrial plant communities. p. 177–222. *In* E.R. Lemon (ed.) CO_2 and plants. Westview Press, Boulder, CO.

Strain, B.R, and R.B. Thomas. 1992. Field measurements of CO_2 enhancement and climate change in natural vegetation. Water Air Soil Pollut. 64:45–60.

Thibaud, J.-B., V. Toulon, H. Sentenac, J.-C. Davidian, and C. Grignon. 1990. CO_2 high partial pressure in the rhizosphere affect mineral nutrition of plants: An approach to the mechanisms involved. Symbiosis 9:69–70.

Thomas, R.B., D.D. Richter, H. Ye, P.R. Heine, and B.R. Strain. 1991. Nitrogen dynamics and growth of seedlings of an N-fixing tree (*Gliricidia sepium* (Jacq.) Walp.) exposed to elevated atmospheric carbon dioxide. Oecologia 88:415–421.

Tschaplinski, T.J., R.J. Norby, and S.D. Wullschleger. 1993. Responses of loblolly pine seedlings to elevated CO_2 and fluctuating water supply. Tree Physiol. 13:283–296.

Vogt, K.A., D.J. Vogt, E.E. Moore, and D.G. Sprugel. 1989. Methodological considerations in measuring biomass, production, respiration and nutrient resorption for tree roots in natural ecosystems. p. 217–232. *In* J.G. Torrey and L.J. Winsnip (ed.) Applications of continuous and steady-state methods to root biology. Kluwer Acad. Publ., Dordrecht, the Netherlands.

Waring, R.H., and W.H. Schlesinger. 1985. Forest ecosystems. Concepts and management. Acad. Press, Orlando, FL.

Williams, W.E., K. Garbutt, F.A. Bazzaz, and P.M. Vitousek. 1986. The responses of plants to elevated CO_2 IV. Two deciduous-forest tree communities. Oecologia 60:454–459.

Wullschleger, S.D., and R.J. Norby. 1992. Respiratory cost of leaf growth and maintenance in white oak saplings exposed to atmospheric CO_2 enrichment. Can. J. For. Res. 22:1717–1721.

Wullschleger, S.D., R.J. Norby, and C.A. Gunderson. 1992. Growth and maintenance respiration in leaves of *Liriodendron tulipifera* L. saplings exposed to long-term carbon dioxide enrichment in the field. New Phytol. 121:515–523.

Wullschleger, S.D., W.M. Post, and A.W. King. 1995. On the potential for a CO_2 fertilization effect in forests: Estimates of the biotic growth factor based on 58 controlled-exposure studies. p. 85–107. *In* G.M. Woodwell and F.T. Mackenzie (ed.) Biotic feedbacks in the global climatic system: Will warming feed the warming? Oxford Univ. Press, New York.

Zak, D.R., K.S. Pregitzer, P.S. Curtis, J.A. Teeri, R. Fogel, and D.L. Randlett. 1993. Elevated atmospheric CO_2 and feedback between carbon and nitrogen cycles in forested ecosystems. Plant Soil 151:105–117.

19 Soil Organic Matter: A Link Between Forest Management and Productivity

Gray S. Henderson
University of Missouri
Columbia, Missouri

Soil organic matter is of scientific interest today because of two separate, but not unrelated issues. The first is its role as a source or sink for atmospheric carbon dioxide, and the second is its importance to long-term forest (or agricultural) productivity. The productivity issue is emphasized in this paper but much of the discussion is pertinent to the source/sink issue.

Long-term forest productivity concerns have emerged because of real or suspected declines in second rotation tree growth in various parts of the globe (Dyck & Skinner, 1990; Powers et al., 1990; Turner & Gessel, 1990). Two factors are commonly linked to long-term productivity, organic matter and its associated nutrients, and compaction, which is partially related to soil organic matter (Froehlich & McNabb, 1984).

Soil organic matter is thought to be in dynamic equilibrium between inputs and outputs. It accumulates to a "carbon carrying potential," or maximum, which is controlled by climate, topography and vegetation (Johnson & Kern, 1991). At equilibrium, the amount of atmospheric C transferred to the soil via vegetation is offset by decomposition and solubilization losses. Land management upsets this equilibrium by altering vegetation and influencing microclimate and soil properties affecting decomposition and solubilization. Simplified conceptual models of organic matter response were presented by Moore et al. (1981) (Fig. 19-1). The classic negative exponential response occurs upon agricultural disturbance when replacement of native vegetation by crops lowers inputs and increased decomposition rates accelerate outputs (Fig. 19-1A). Eventually the new vegetation (crop) inputs and decomposition losses reach equilibrium at a content smaller than the "native" level. If cultivated land is abandoned or revegetated the organic matter content rebuilds towards the native equilibrium content (Fig. 19-1A).

For disturbance associated with forest harvest and regrowth the organic matter response is somewhat different (Fig. 19-1B). Residue from slash and roots initially increases soil organic matter which subsequently declines as C inputs fall short of outputs. However, if revegetation is rapid this decline would

---

Copyright © 1995 Soil Science Society of America, 677 S. Segoe Rd., Madison, WI 53711, USA.
Carbon Forms and Functions in Forest Soils.

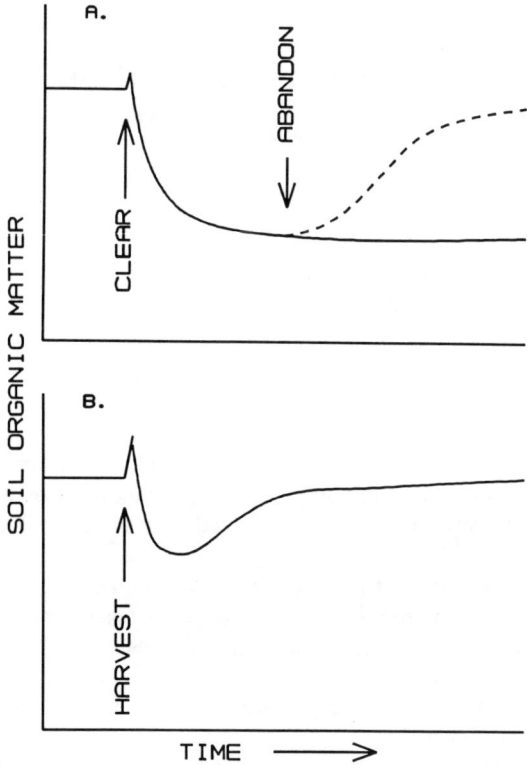

Fig. 19–1. Theoretical model of the time course of soil organic matter content as affected by cultivation (A) and forest harvest (B) (adapted from Moore et al., 1981).

be short-lived and as inputs approach and then exceed outputs, organic matter would again increase.

For both scenarios questions remain as to whether the equilibrium towards which the regrowing vegetation is building will be less than, the same, or more than the "native" content. The answer may depend on "damage" to the site during cultivation or harvest and/or the productivity of the regrowing vegetation, the annual input of organic matter to the soil. Can the new equilibrium level be increased by establishing non-native vegetation? Can the organic matter content, and therefore productivity, be permanently decreased by damage to soils? Regarding reforestation of abused lands, a key question is whether various soil properties dependent on or related to organic matter are so degraded that native equilibrium levels can not be re-established.

Considerable effort is being expended estimating global terrestrial C storage and quantifying its changes due to land management. Carbon storage in world soils is estimated at approximating 1500 to 1600 Pg (Post et al., 1990; Eswaran et al., 1993), about 75% of total terrestrial amounts. This estimate is based on areal estimates of soil categories composited at the suborder or order level and is limited to a 1-m depth. While such estimates delineate relative roles of soils and other ecosystem components in C storage, they are not particularly useful in assessing management or land use impacts on storage at specific sites and even less helpful in assessing resulting changes in long-term productivity.

For many soils the 1-m depth limitation is inappropriate both for estimating pools and assessing productivity (Hammer et al., 1995; see Chapter 10). Stone et al. (1993) and Eswaran et al. (1993) point out the significance of omitting deeper soil horizons for Spodosols and Histosols, respectively. Not all soils possess prominent organic-rich spodic horizons, but many accumulate organic matter at depths greater than 1 m. As Stone and Kalisz (1991) have laboriously documented, roots extend to great depths more often than many of us care to admit. These deep roots and their associated soil environs can not be ignored in site productivity assessments.

Even within 1 m, horizonation and organic matter distribution varies. While in some soils surface horizon processes predominate, equally important processes may be taking place deeper in others. It is important to consider these depth differences and how associated water storage, root distribution, nutrition, and other factors influence growth. Ultimately, this spatial variance should be considered relative to potential impacts of forest management on productivity.

This paper will not concentrate so much on specific sites but rather attempts to provide an overview of organic matter changes associated with various aspects of forest management and potential links to productivity. I will try to highlight what we know and do not know and hopefully will introduce a few topics to stimulate ideas for future examination.

FOREST MANAGEMENT, SOIL ORGANIC MATTER AND FOREST PRODUCTIVITY LINKAGES

Land management, soil organic matter and productivity are intimately linked (Fig. 19–2). Land management can directly influence the amount of soil organic matter. Most management activities reduce soil organic matter but additions also occur, and in forest management short-term or spatial enrichment is probable. Soil organic matter, through its effect on soil properties, can profoundly influence forest productivity and these relationships will be examined in this paper. In turn, the product of carbon dioxide fixation is partially returned to the soil. The amount

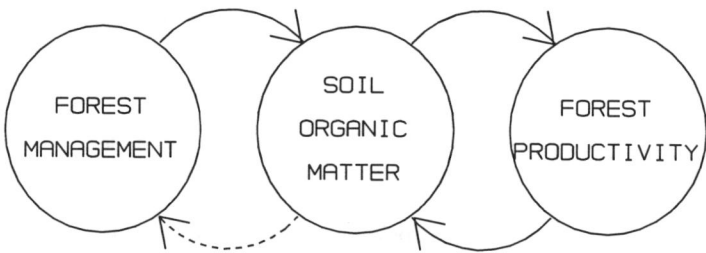

Fig. 19–2. Linkages among forest management, soil organic matter and forest productivity.

of input to the soil is obviously a direct function of productivity. Hence, a dynamic cycle exists between forest productivity and soil organic matter. Finally, it is possible for soil organic matter to influence management decisions. A manager may choose to consider soil organic matter content and its ramifications on future productivity in selecting the type of land management he/she follows. Certainly, in the realm of global climate change research one topic is management of forests for C sequestration. Discussions of methods to enhance reforestation, decrease forest degradation, and/or implement agroforestry are commonplace (Johnson & Kern, 1991). In forestry, management alternatives may consider concern for organic matter status. Alternatives include the type and timing of harvest, site preparation, type of species to replant, fertilization and maintenance schedules, or possibilities for reforestation of degraded lands.

To discuss linkages among forest management, soil organic matter and productivity, it is necessary to first summarize, without simplifying too much, the changes in vegetation, soil and site conditions associated with various management activities. The second step will relate these changes to soil properties influenced by organic matter and examine what is known and not known about their implications for long-term productivity. To facilitate this approach, forest management activities are considered in four categories: harvesting, site preparation, stand maintenance, and, reforestation and agroforestry.

FOREST HARVESTING INFLUENCES ON SOIL ORGANIC MATTER

There are two distinct aspects of a harvesting operation that influence future productivity. One involves the road and skid trail system necessary for extracting timber while the other considers harvesting patterns and biomass utilization levels. The two are not independent as harvest pattern influences the transportation network, but they can be conveniently separated for considering long-term productivity impacts.

Roads and Skid Trails

The density and placement of roads and skid trails or cable systems depends on timber volume (value), extraction system, and topography. There is an economic tradeoff between road miles and the cost of operating ground-based or aerial skidding systems (Pearce & Stenzel, 1972). Topography not only influences road location but also density due to the greater cost of road building in steep terrain. The acreage devoted to roads and skid trails reduces productivity. The objective in road construction is to produce a traffic-bearing surface and this is accomplished by removing organic matter and compacting mineral soil. Future productivity of these areas is zero, so the denser the road network the greater the reduction in productivity.

While not intended, compaction of skid trails is inevitable. Froehlich and McNabb (1984) concluded that compaction reduces growth for decades. When the skid trail network is unplanned or used without regard to soil trafficability, productivity on an area basis may be reduced by 15% or more. In addition, improper skid trail location can accelerate erosion resulting in further loss of productivity.

A rarely mentioned aspect of road location is the potential to alter soil water movement. On hillslopes, road construction often intercepts lateral soil water movement and diverts it to ditches and more direct routing to streams. The effect on productivity of lower slope positions that could utilize this water during periods of moisture stress is unknown. Overall, roads and skid trails together may reduce productivity enough to alter organic matter inputs and soil properties on a scale greater than the area they occupy.

Harvest Pattern and Intensity

The two variables of interest are the pattern of harvest and the degree of utilization of the felled trees. Harvests can be patterned as area clearcuts, selection cuts of various intensities, or strip cuts and small block cuts, which are alternating clearcuts. Utilization determines amount of biomass removal, whether only boles or the whole tree, including branches, foliage and even stumps and roots. In addition, distribution of the residue varies between operations. In one, limbing and slash disposal may result in relatively uniform residue distribution, whereas in another these activities may be concentrated at landings with corresponding depletion of organic matter over the larger area.

Expected changes in soil organic matter relate primarily to slash distribution and decomposition dynamics. Harvesting results in immediate inputs to the soil, both as aboveground woody and foliage material and belowground root biomass. The input quantity increases with cutting intensity and decreases with biomass utilization. Decomposition rates of the added and native organic matter increase in response to temperature and moisture increases at the harvest site. The temperature increase depends on the harvest pattern. Soil temperature in large clear cuts potentially increases more than in strip or block cuts that receive partial shading from adjacent uncut areas according to cut orientation. The shading of subsequent strip or block cuts is progressively reduced and is minimal for the last area as long as the time interval between cuts is relatively short. Selective cutting only partly removes the overstory and thus raises soil temperature the least.

While the forest floor and mineral soil surface experience accelerated drying following harvest, overall soil water contents increase following harvest because transpiration demands decrease (Brooks et al., 1991). The increase is directly related to percentage basal area removed (Douglas, 1983) and occurs primarily during the growing season (Douglas & Swank, 1975). Similar to temperature, soil water contents would increase most in large clearcuts and least in selective cuts with smaller cuts intermediate because of edge effects. The coincidental increase in soil water availability with elevated temperatures results in enhanced decomposition.

Harvesting also removes nutrients in proportion to the intensity of harvest and degree of biomass utilization (Leaf, 1979; Turner & Lambert, 1986; Compton & Cole, 1991). Additional nutrients may be lost after release by decomposition and during site preparation (Johnson et al., 1988; Mann et al., 1988).

SITE PREPARATION INFLUENCES ON SOIL ORGANIC MATTER

Site preparation refers to activities following harvest, usually clear cutting, that prepare the site for revegetation, commonly through planting or seeding a specific species. These activities produce various degrees of mineral soil displacement and/or reduction of surface litter and are practiced to facilitate revegetation and control competing vegetation. Practices include crushing or chopping residues, root raking, scalping, windrowing, subsoiling (ripping), bedding, and burning, and may be practiced in various combinations and intensities.

Potential changes, depending on the practices, include organic matter loss or incorporation into mineral soil, temperature increases resulting from loss of mulch effect, increase in bulk density, loss of desirable structure, decrease in infiltration rate resulting in increased erosion potential, volatilization of nutrients and C, mineral soil heating, and reduction in competing vegetation with accompanying lessened shade, water and nutrient uptake, and organic matter input.

PRODUCTIVITY LINKAGES TO CHANGES ASSOCIATED WITH HARVESTING AND SITE PREPARATION

The soil properties likely affected by changing organic matter status are nutrient content, bulk density and water content. Recent summaries of long-term forest productivity generally conclude that nutrient supply and compaction are the two factors most likely involved in second or third rotation decline (Powers et al., 1990; Dyck & Skinner, 1990; Johnson, 1992), while changes in water relations have received less emphasis. The following paragraphs explore what is known about how these properties influence forest productivity.

Nutrient Supply

Of course the most straightforward effect on nutrient supply is removal in products—the greater the removal the greater the nutrient loss. Beyond that, the potential for changing nutrient supply depends on organic matter decomposition rates, physical redistribution of nutrient capital, and losses through volatilization and erosion.

Decomposition rates are accelerated by harvesting and site preparation due to the combination of warmer temperatures, greater water availability and more organic substrate (Morris & Pritchett, 1983; Edwards & Ross-Todd, 1983). Whether accelerated decomposition results in nutrient losses depends on the combination of a site's resistance and resilience as shown in Fig. 19–3 (Henderson, 1985). Resistance is defined as the tendency to resist increased mineralization of nutrients. For comparable temperature and moisture conditions it is related to substrate C/N ratios, wider C/N ratios being less likely to mineralize nutrients and especially N. Burger and Pritchett (1984) point out, however, that mineralizable N incubations reflecting organic matter "quality" indicate N release potential better than C/N ratio. Resilience describes revegetation potential with those sites recovering faster being more resilient. These two factors describe the temporal pattern and relative quantities of nutrients lost in soil leachate following release from organic matter.

The greatest potential for loss occurs at sites with narrow C/N ratios, and presumably high N levels, and low revegetation rates. These conditions rarely occur unless revegetation is restricted such as with chemical weed control. The Hubbard Brook Watershed forest denudation study is the extreme example of these conditions (Likens et al., 1977). Most forest sites have greater resistance

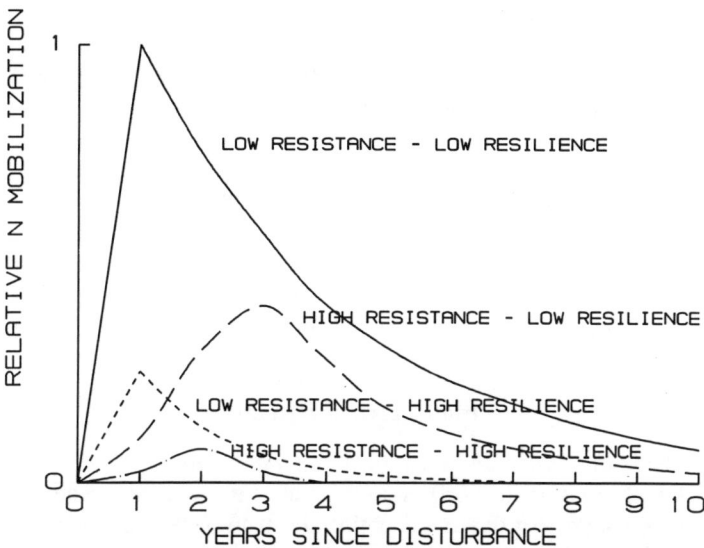

Fig. 19–3. Theoretical N mobilization from organic matter as influenced by four combinations of resistance to mineralization and resilience of vegetation regrowth (after Henderson, 1985).

(slower organic matter decomposition) and/or resilience (faster regrowth) than occurred in the Hubbard Brook experiment and sequester mineralized nutrients so that the leaching loss is minimal except for a short time following disturbance. The supporting evidence for describing patterns of resistance and resilience is found in studies of dissolved nutrients in stream water from disturbed sites (Henderson, 1985).

While the potential for dissolved loss is not great, a larger threat is loss of nutrients from surface horizons due to burning or erosion. Depending on intensity, burning not only volatilizes N and C from litter layers and surface mineral soils (Johnson, 1992), but also can lead to accelerated erosion. Ulery and Graham (1993) found redder color (Fe oxidation), 15 to 68% reductions in organic matter, and fusion of clay into larger particles in surface soils of severely burned areas that covered 1 to 2% of western conifer sites. Prescribed burns that are too intense not only leave the soil surface unprotected from raindrop impact, they also can destroy residual root binding and structure. The resulting loss of infiltration capacity due to surface sealing generates runoff which transports soil organic and mineral material and their associated nutrients (Brooks et al., 1991). While short-term elevated nutrient concentrations have been documented (Raison, 1979), the long-term effects of burning on nutrient supply and productivity are not generally known. However, Vance and Henderson (1984) have shown that 35 to 40 yr of annual or periodic burning in an oak (*Quercus* spp.)-hickory (*Carya* spp.) forest reduced mineralizable N without decreasing total N. This observation, like that of Burger and Pritchett (1984), indicates that organic matter quality is important to N supply and may change in response to disturbance.

Harvesting and site preparation methods that concentrate residue or redistribute surface soil, even though they are confounded with other factors, reduce productivity on some sites (Dyck et al., 1989; Powers et al., 1990). The amount of reduction depends on the proportion of the nutrients in the rearranged soil layers and intensity of the disturbance. If surface disturbance bares a significant area, accelerated erosion becomes a possibility.

To generalize, soils most at risk for productivity loss are those marginally supplied with nutrients and dependent on organic matter decomposition for resupply (Dyck & Skinner, 1990). In these cases, the removal, disturbance and leaching losses can be a significant amount of the nutrient supply required for the next rotation. Care is needed in managing these sites to maintain what nutrient capital does exist following harvest.

Subsoil Nutrient Dynamics

The above discussion and most of the literature stresses the importance of nutrient loss from surface horizons. Organic matter and associated nutrients deeper in the solum also deserve attention. In many forest soils and especially Spodosols, appreciable organic matter accumulations occur at depths of 1 to 2 m or more (Daniels et al., 1975; Stone et al., 1993). The dynamics of these accumulations and their role in productivity is rarely studied. How important are these horizons? How reactive are they?

McFee and Stone's (1965) study of Spodosols in the Adirondacks is often cited in textbooks to illustrate typical organic matter distribution. Their Fig. 3, showing change in B horizon organic matter as a function of stand age (time since disturbance by fire), is often overlooked. It indicates the dynamic connection of this subsurface horizon to the O horizon. The slope of their regression yields annual C and N sequestering of 110 kg/ha (98 lb/acre) and 2.8 kg/ha (2.5 lb/acre), respectively. If the regression is extrapolated to the year of disturbance, it suggests that 33600 kg/ha (30000 lb/acre) organic matter and 875 kg/ha (780 lb/acre) N had been mobilized from the B horizon. While this relationship is undoubtedly not linear, it does indicate the potential for alteration of subsoil organic matter. In contrast, Spodosols in North Carolina with organic matter accumulation to depths of 10 m are apparently not linked to surface horizon dynamics (Daniels et al., 1975; Holzhey et al., 1975). In fact these horizons appear to have ceased organic matter accumulation (Holzhey et al., 1975). This raises interesting questions about conditions influencing subsurface accumulation. For example, is immobilization related to a base cation gradient and as this gradient diminishes does organic matter accumulation cease? Or, are other processes such as silica-organo bonding important such as suggested by Holzhey et al. (1975)?

While Spodosols only account for about 2% of world soils (Eswaran et al., 1993), their subsoil C dynamics may indicate similar processes occur in more widely distributed soils. Do other soils, such as Alfisols and Ultisols, with less prominent subsoil organic matter accumulations behave similarly? If nutrients are lost from or physical properties altered in these horizons following disturbance, what are their prospects for recovery? Are the conditions that contributed to genesis of these horizons still operable today?

Bulk Density

Bulk density is closely related to other soil properties, porosity, aeration, and mechanical impedance, which can be altered by compaction. Collectively these properties influence root and aboveground growth (Sutton, 1991). Forest harvesting operations result in compacted soils. The degree of compaction varies with the harvesting system and the care with which the system is utilized. This compaction reduces productivity on an areal basis (Froehlich & McNabb, 1984) and thus potentially reduces future inputs to soil organic matter. Ripping is sometimes used to overcome subsoil compaction problems and increase seedling survival by creating desirable microsites (Sutton, 1991). Disking is used to reduce competition and surface compaction but may create conditions conducive to accelerated erosion.

In addition to this direct impact of harvest and site preparation on future productivity, accelerated decomposition reduces soil organic matter contents (Johnson, 1992) and potentially increases bulk density, thereby reducing porosity and aeration. Various studies have demonstrated reduced productivity at least partially related to bulk density, porosity, and aeration changes (Skinner et al., 1989; Dyck & Skinner, 1990; Sutton, 1991). In productivity index formulations bulk density is included as a function that progressively limits growth at values greater than 1.3 g/cm^3 (Henderson et al., 1990).

Bulk density can be altered directly by compaction and indirectly through organic matter loss. While it is known that factors such as texture, water content and organic matter affect susceptibility to compaction and the influence of bulk density on root growth (Greacen & Sands, 1980; Daddow & Warrington, 1983; Sutton, 1991), how well has this information been assembled for dissemination to harvest planners? Is it known how accelerated decomposition, especially in subsoils, influences bulk density and how this varies with soil type? What soils are most at risk? How long does it take to naturally re-establish lighter bulk densities and reverse associated productivity losses? Are there situations where the change is irreversible? These questions, and their answers, are important for long-term productivity maintenance and are beginning to be addressed systematically (Dyck & Mees, 1989; Powers et al., 1990). However, such studies require long-term commitment and dedication of both investigator and funding source (Burger & Powers, 1991), and we need to encourage gleaning as much information from them as possible, including data on rooting patterns and subsoil dynamics.

Water Relations

Harvesting and site preparation potentially influence productivity by changing infiltration capacity, water holding capacity, and/or water table level. In addition to accelerating nutrient depletion through erosion, reduced infiltration capacities also mean less water for recharging depleted storage. Where water is limiting, such as in marginal rainfall areas or on slopes without water tables within reach of roots, incomplete recharge could expand the period of water deficit. Reduced transpiration demands would offset this reduction in water availability, however, persistence of reduced infiltration after evapotranspiration demands are re-established could lead to productivity declines. Organic matter loss resulting in decreased porosity and water holding capacity could similarly induce greater drought stress and/or a shorter growing season.

Reports of water table rise and subsequent difficulties in reforestation are well known for level coastal plain soils (Pritchett & Fisher, 1987). Bedding is now commonly practiced to obtain plantation establishment and again lower water tables on these sites. What is the implication of drastically reducing the seasonal water table rise and fall on forest productivity and resulting soil organic matter replenishment? Is the soil volume exploited by roots ultimately reduced due to changes in soil organic matter distribution related to a higher water table?

In addition to water table changes, organic matter dynamics following harvest and site preparation could influence water relations in other ways that remain largely unexplored. If disturbance reduces subsurface organic matter to the point where densities or porosity restricts root growth, or if old root channels become occluded lessening the probability that new roots follow these channels, will productivity be reduced? Examples of root exploitation of water and nutrients below a restrictive layer are not uncommon (Armson, 1977). Root penetration of these layers is enhanced by old root channels which provide a conduit through the barrier. Does plantation forestry result in altered rooting patterns so these avenues are more limited and potential productivity is reduced? Stone and Kalisz (1991) compiled deep rooting observations for 211 species from throughout the world. What proportion of these observations were from plantations vs. naturally occurring stands? Gale and Grigal (1987) suggested genetic and successional controls exist for vertical root density. When compared to natural stands, do rooting habits of plantation species differ due to water table or density conditions? How important are changes in root distribution to productivity? If they are, could the root channels be maintained or the density changes minimized by another harvesting or site preparation method, such as selection cutting? These questions are unanswered and I know of no ongoing research on this subject.

STAND MAINTENANCE INFLUENCES ON SOIL ORGANIC MATTER

Stand maintenance activities include planting, competition control, thinning and fertilization.

Planting

Since planting follows soon after site preparation it might be more appropriate to consider it under that topic. However, the aspects of planting relevant to soil organic matter are more related to planning, selection of species and planting densities, than physical manipulation. Selection of specific species or varietal stock is presumably done to assure the best possible productivity and thus can influence organic matter dynamics. The greater the growth rate the sooner the site is reoccupied, the lesser the need for control of competing vegetation, the lower the decomposition loss due to temperature and moisture enhancement, and ultimately the possibility for greater biomass production. Hence the potential exists for maintaining or even enhancing soil organic matter contents after harvest and site preparation.

The rooting habit of the selected species may be genetically different from the species previously growing on the site (Gale & Grigal, 1987; Sutton, 1991). Does this alter the soil volume exploited by the plantation? Certainly annual agronomic crops have shallower root distributions than the native forests or prairies which they replaced. If the soil volume exploited is either increased or decreased, water and nutrient uptake are similarly affected. What is the implication of these changes for productivity? Should we encourage deep-rooted species or development of varieties with these traits?

Fine root mortality is another aspect to consider. In addition to establishing the importance of roots to C and N cycles, the recent literature also indicates species differences in the annual turnover rate (Joslin & Henderson, 1987) and even possible differences among mycorrhizal associations. What are the implications of fine root mortality differences among species for soil organic matter accumulation and hence long-term productivity?

Litter nutritional characteristics and hence decomposition rates of species differ. Classic (Lyford, 1963; Edwards et al., 1970) and contemporary (Fried et al., 1990) experiments have established relationships between organic matter distribution in organic and mineral horizons (mull vs. moder vs. mor) and the activities of organisms responding to litter nutrition. Tarrant et al. (1969), Youngberg and Wollum (1970), and Permar and Fisher (1983) have documented increased N contents in forest stands containing N-fixing trees or scrubs. In tropical agroforestry *Erythrina spp.* have been shown to increase K availability to annual bean (*Phaseolus* spp.) crops through uptake and cycling from deep soil horizons not exploited by the bean roots (D. Cass, 1986, personal communication). The extent to which long-term productivity might be altered by increased nutrient availability induced by various tree species remains a subject for future research.

Plantation spacing is another variable. Spacing differences influence temperature, moisture, and amount of litter as a function of the time required to achieve crown closure. These factors are strong determinants of decomposition rates and, other factors being equal, the sooner crown closure is obtained the sooner organic matter dynamics should regain preharvest conditions. Therefore, it is likely that closer spacings favor this re-establishment.

Competition Control

Herbicides are used to control competition for water, nutrients, and sunlight in order to increase seedling productivity. Depending on the zone of vegetation inhibition (general broadcast vs. selective spray application), potential changes include loss of organic input, decrease in shading leading to temperature increase, reduction in water use, and possibly an increase in erosion (Spittlehouse & Childs, 1990). These conditions are conducive to increased decomposition with accelerated nutrient mobilization as demonstrated by Neary et al. (1990) for slash (*Pinus elliottii* Engelm.) and loblolly pine (*P. taeda* L.). The mobilized nutrients can be taken up by vegetation or lost through leaching. How significant are these short-term effects on nutrient availability to long-term productivity? Do the short-term benefits offset any long-term detrimental impacts? Research is limited in this aspect of herbicide use.

Thinning

Precommercial and commercial thinning is commonly employed in plantation forestry. Changes associated with thinning are periodic pulses of organic matter input, short-lived alterations in soil water while evapotranspiration rates adjust, and, if products are removed, some nutrient loss and compaction (Pritchett & Fisher, 1987). Unless compaction occurs or whole tree utilization depletes nutrient capital (Malkonen, 1979), these changes are relatively minor and probably insignificant to long-term productivity.

Fertilization

Fertilization, normally with N or P, is obviously done to stimulate productivity. Johnson (1992) reviewed soil C response to fertilization and, as expected, found increases related to greater forest productivity. The distribution of organic matter between above- and belowground plant components also could be altered by nutrient availability. The implication to soil organic matter dynamics of such a redistribution, if it occurs, is largely unknown.

A largely unexplored aspect of forest fertilization is amelioration of subsoil acidity and/or density to promote deep rooting. Berry (1985) has shown an increase in deep rooting of loblolly pine in response to subsoiling in a compact soil. In my work with root ingrowth cores (Fig. 19–4) there has been a distinct root mass and length increase in response to Ca additions to acid soils. Could subsoiling with Ca injection be used to promote deep rooting and greater soil volume exploitation?

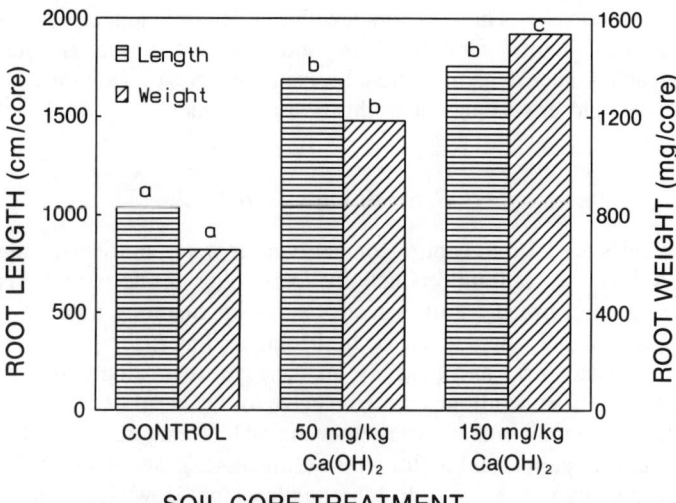

Fig. 19–4. Length and weight of Norway spruce roots growing into control and calcium amended soil cores after two growing seasons at Harvard Forest (G.S. Henderson, unpublished data).

PRODUCTIVITY LINKAGES TO ORGANIC MATTER CHANGES ASSOCIATED WITH STAND MAINTENANCE

Stand maintenance activities are designed to enhance productivity, or at least increase accumulation in usable wood products. For activities other than competition control, more residues should result, and thus input to soil organic matter will be greater, and the "healing" rate following disturbance may be shortened. The unknowns concern implications of species selection on exploitable soil volume and organic matter distribution, and ramifications of chemical weed control. Can we select species that penetrate dense or acid barriers in order to take advantage of underlying soil moisture and nutrient reserves that would enhance productivity? Are there soil treatments that would achieve the same end? In addition to the desirable result of promoting early seedling growth, does herbicide use produce undesirable effects that could be detrimental to productivity in the longer term?

REFORESTATION AND AGROFORESTRY EFFECTS ON SOIL ORGANIC MATTER

Since initiation of the 1985 Farm Bill, there has been activity in reforestation of marginal agriculture lands utilizing the incentives of the Conservation Reserve Program. For that matter, worldwide reforestation is a subject receiving increasing attention as issues of atmospheric C sequestration are raised. In addition, there is a small but growing interest in temperate zone agroforestry whereby agronomic crops and specialized tree crops are intercropped (Garrett & Kurtz, 1983). This is a subject area with which most foresters and soil scientists have little contact or, probably interest. However, there are interesting questions associated with these activities to which the answers are not known.

Reforestation and agroforestry can be expected to increase organic matter input to soils, alter temperature regimes, change moisture relationships, reduce erosion, retain solution nutrients, and enhance pesticide degradation. Henderson et al. (1991) postulated that these changes will increase soil organic matter storage, improve soil structure and bulk density conditions resulting in better water relations for both crops and trees, improve overall productivity, and improve surface and groundwater quality.

In fact, these improvements have not been demonstrated, especially for agroforestry. There is considerable opportunity for research on these subjects and many of us in the forest soils discipline, with our watershed and ecological emphasis, could make a substantial contribution. Ultimately, questions as to whether the benefits justify the costs in changing the way the agriculture community manages the land will be asked, and the answers will need to be forthcoming.

SUMMARY

What is known about linkages among forest management, soil organic matter and forest productivity can be grouped into established relationships, likely or possible relationships, and speculative relationships. These also outline challenges for future research.

Established Relationships

1. Roads, skid trails and landings result in compaction and productivity losses in direct proportion to area disturbed.
2. Harvesting removes nutrients as determined by intensity of harvest and biomass utilization. Productivity loss can result on soils with marginal nutrient contents.
3. Harvesting results in increased soil temperature and moisture and adds organic matter to soil. These factors lead to increased decomposition rates and potential for nutrient leaching.
4. Site preparation results in accelerated organic matter loss, soil displacement, and compaction and can lead to productivity loss.
5. Intensity of prescribed burning influences N content and erosion potential and, if uncontrolled, can lead to lost productivity.

Likely or Possible Relationships

1. Soil organic matter can be increased by enhancing productivity through fertilization, species selection, or agroforestry.
2. Organic matter contents and associated soil properties can be degraded to the point where re-establishment of productivity is unlikely, or at best a long-term process.
3. Organic matter loss raises bulk density and may alter rooting patterns and decrease productivity.
4. Denser plantation spacings result in decreased decomposition rates and, ultimately, retention of site productivity.
5. Species conversion may alter litter nutrient characteristics that influence patterns of organic matter accumulation and nutrient turnover and thereby influence productivity.
6. Agroforestry can reduce erosion, improve water relations and water quality, and result in overall productivity increases on marginal agricultural lands.

Speculative Relationships

The following are presented as questions because of the uncertainty of their influences on productivity.

1. Does diversion of subsurface soil water by roads reduce productivity of lower slope position sites?
2. Does subsoil C respond to surface disturbance in forest soils other than Spodosols? Can it be degraded resulting in loss of nutrients and desirable physical properties that enhance productivity?
3. Is subsurface organic matter accumulation controlled by cation gradients or other processes that are not as active after harvest (now) as when initial soil development took place?
4. Does short-rotation plantation forestry raise water table levels in aquic soils and does this reduce productivity by shrinking the soil volume exploitable by roots?

5. Does harvest intensity affect maintenance of root channels through restrictive soil horizons and thus exploitable soil volume and productivity?
6. Can species selection influence productivity through alteration of root depth or fine root turnover leading to improvement in soil properties?
7. Can forest fertilization be used to promote deep rooting and thus increase productivity through greater soil exploitation?
8. Does herbicide use to control competition extend conditions favoring decomposition and erosional loss and thus result in long-term productivity decline?
9. Does agroforestry promote conditions which enhance pesticide degradation?

REFERENCES

Armson, K.A. 1977. Forest soils: Properties and processes. Univ. Toronto Press, Toronto, Ontario, Canada.

Berry, C.R. 1985. Subsoiling and sewage sludge aid loblolly pine establishment on adverse sites. Reclam. Reveg. Res. 3:301–311.

Brooks, K.N., P.F. Ffolliott, H.M. Gregersen, and J.L. Thames. 1991. Hydrology and the management of watersheds. Iowa State Univ. Press, Ames, IA.

Burger, J.A., and R.F. Powers. 1991. Field designs for testing hypotheses in long-term productivity studies. p. 79–105. *In* W.J. Dyck and C.A. Mees (ed.) Long-term field trial to assess environmental impacts of harvesting. Proc. IEA/BE T6/A6 Workshop, Amelia Island, FL. February 1990. IEA/BE T6/A6 Rep. no. 5. FRI Bull. 161. For. Res. Inst., Rotorua, New Zealand.

Burger, J.A., and W.L. Pritchett. 1984. Effects of clearfelling and site preparation on nitrogen mineralization in a southern pine stand. Soil Sci. Soc. Am. J. 48:1432–1437.

Compton, J.E., and D.W. Cole. 1991. Impact of harvest intensity on growth and nutrition of successive rotations of Douglas fir. p. 151–161. *In* W.J. Dyck and C.A. Mees (ed.) Long-term field trial to assess environmental impacts of harvesting. Proc. IEA/BE T6/A6 Workshop, Amelia Island, FL. February 1990. IEA/BE T6/A6 Rep. no. 5. FRI Bull. 161. For. Res. Inst., Rotorua, New Zealand.

Daddow, R.L., and G.E. Warrington. 1983. Growth-limiting soil bulk densities as influenced by soil texture. Rep. no. WSDG-TN-00005. Watershed Systems Development Group, USDA-FS, Fort Collins, CO.

Daniels, R.B., E.E. Gamble, and C.S. Holzhey. 1975. Thick Bh horizons in the North Carolina coastal plain: I. Morphology and relation to texture and soil ground water. Soil Sci. Soc. Am. J. 39:1177–1181.

Douglas, J.E. 1983. The potential for water yield augmentation from forest management in the eastern United States. Water Resour. Bull. 19:351–358.

Douglas, J.E., and W.T. Swank. 1975. Effects of management practices on water quality and quantity: Coweeta Hydrologic Laboratory, North Carolina. p. 1–13. *In* Proc. Municipal watershed Manage. Symp. USDA-FS, Northeastern For. Exp. Stn., Broomhall, PA.

Dyck, W.J., and C.A. Mees. 1989. Research strategies for long-term site productivity. Proc. IEA/BE A3 Workshop, Seattle, WA. August 1988. IEA/BE A3 Rep. no. 8. FRI Bull. 152. For. Res. Inst., Rotorua, New Zealand.

Dyck, W.J., C.A. Mees, and N.B. Commerford. 1989. Medium-term effects of mechanical site preparation on radiata pine productivity in New Zealand—a retrospective approach. p. 79–92. *In* W.J. Dyck and C.A. Mees (ed.) Research strategies for long-term site productivity. Proc. IEA/BE A3 Workshop, Seattle, WA. August 1988. IEA/BE A3 Rep. no. 8. FRI Bull. 152. For. Res. Inst., Rotorua, New Zealand.

Dyck, W.J., and M.F. Skinner. 1990. Potential for productivity decline in New Zealand radiata pine forests. p. 318–332. *In* S.P. Gessel et al. (ed.) Sustained productivity of forest soils. Proc. North American For. Soils Conf., 7th, Vancouver, BC. 24–28 July 1988. Univ. British Columbia, Faculty of For. Publ., Vancouver, BC.

Edwards, C.A., D.E. Reichle, and D.A. Crossley. 1970. The role of soil invertebrates in turnover of organic matter and nutrients p. 147–172. *In* D.E. Reichle (ed.) Analysis of temperate forest ecosystems. Springer-Verlag, New York.

Edwards, N.T., and B.M. Ross-Todd. 1983. Soil carbon dynamics in a mixed deciduous forest following clear-cutting with and without residue removal. Soil Sci. Soc. Am. J. 47:1014–1021.

Eswaran, H., E. Van Den Berg, and P. Reich. 1993. Organic carbon in soils of the world. Soil Sci. Soc. Am. J. 57:192–194.

Fried, J.S., J.R. Boyle, J.C. Tappeiner II, and K. Cromack, Jr. 1990. Effects of bigleaf maple on soils in Douglas-fir forests. Can. J. For. Res. 20:259–266.

Froehlich, H.A., and D.H. McNabb. 1984. Minimizing soil compaction in Pacific Northwest forests. p. 159–192. *In* E.L. Stone (ed.) Forest soils and treatment impacts. Proc. North America For. Soils Conf., 6th, Knoxville, TN. 19–23 June 1983. Univ. Tennessee, Dep. Forestry, Wildlife and Fisheries Publ., Knoxville, TN.

Gale, M.R., and D.F. Grigal. 1987. Vertical root distributions of northern tree species in relation to successional status. Can. J. For. Res. 17:829–834.

Garrett, H.E., and W.B. Kurtz. 1983. Silvicultural and economic relations of integrated forestry-farming with black walnut. Agrofor. Syst. 1:245–256.

Greacen, E.L., and R. Sands. 1980. Compaction of forest soils: A review. Aust. J. Soil Res. 18:163–189.

Hammer, R.D., G.S. Henderson, R. Udawatta, and D.K. Brandt. 1995. Soil organic matter in the forest-prairie ecotone. p. 201–231. *In* W.W. McFee and J.M. Kelly (ed.) Carbon forms and functions in forest soils. SSSA, Madison, WI.

Henderson, G.S. 1985. Nutrient dynamics in disturbed forests and associated influences on stream chemistry. p. 55–65. *In* B.G. Blackmon (ed.) Proc. For. Water Quality: A Mid-South Symp., Hot Springs, AR. May 1986. Univ. Arkansas, Monticello, AR.

Henderson, G.S., R.D. Hammer, and D.F. Grigal. 1990. Can measurable soil properties be integrated into a framework for characterizing forest productivity? p. 137–154. *In* S.P. Gessel et al. (ed.) Sustained productivity of forest soils. Proc. North American For. Soils Conf., 7th, Vancouver, BC. 24–28 July 1988. Univ. British Columbia, Faculty For. Publ., Vancouver, BC.

Henderson, G.S., J. Krstansky, and D. Ramsey. 1991. The role of agroforestry in water quality: Our ongoing research program. p. 1–6. *In* Proc. Water Quality Conf., Columbia, Missouri. 10 May 1991. Missouri Agric. Exp. Stn., Univ. Missouri, Columbia, MO.

Holzhey, C.S., R.B. Daniels, and E.E. Gamble. 1975. Thick Bh horizons in the North Carolina coastal plain: II. Physical and chemical properties and rates of organic additions from surface sources. Soil Sci. Soc. Am. J. 39:1182–1187.

Johnson, D.W. 1992. Effects of forest management on soil carbon storage. Water Air Soil Pollut. 64:83–120.

Johnson, D.W., J.M. Kelly, W.T. Swank, D.W. Cole, H. Van Miegroet, J.W. Hornbeck, R.S. Pierce, and D.H. Van Lear. 1988. The effects of leaching and whole-tree harvesting on cation budgets of several forests. J. Environ. Qual. 17:418–426.

Johnson, M.G., and J.S. Kern. 1991. Sequestering carbon in soils: A workshop to explore the potential for mitigating global climate change. USEPA Rep. 600/3–91-031. USEPA Environ. Res. Lab., Corvallis, OR.

Joslin, J.D., and G.S. Henderson. 1987. Organic matter and nutrients associated with fine root turnover in a white oak stand. For. Sci. 33:330–346.

Leaf, A.L. (ed.). 1979. Impact of intensive harvesting on forest nutrient cycling. Proc. Workshop, Syracuse, NY. 13–16 Aug. 1979. State Univ. New York, Syracuse, NY.

Likens, G.E., F.H. Bormann, R.S. Pierce, J.S. Eaton, and N.M. Johnson. 1977. Biogeochemistry of a forest ecosystem. Springer-Verlag, New York.

Lyford, W.H. 1963. Importance of ants to brown podzolic soil genesis in New England. Harvard For. Pap. no. 7. Harvard Univ., Petersham, MA.

Malkonen, E. 1979. Biomass and nutrient removal in whole tree harvesting in some coniferous stands thinned for the first time. p. 405. *In* A.L. Leaf (ed.) Impact of intensive harvesting on forest nutrient cycling. Proc. Workshop, Syracuse, NY. 13–16 Aug. 1979. State Univ. New York, Syracuse, NY.

Mann, L.K., D.W. Johnson, D.C. West, D.C. Cole, J.W. Hornbeck, C.W. Martin, H. Riekirk, C. Smith, W.T. Swank, L.M. Tritton, and D.H. Van Lear. 1988. Effects of whole-tree and stem-only clearcutting on postharvest hydrologic losses, nutrient capital and regrowth. For. Sci. 34:412–428.

McFee, W.W., and E.L. Stone. 1965. Quantity, distribution, and variability of organic matter and nutrients in a forest podzol in New York. Soil Sci. Soc. Am. Proc. 29:432–436.

Moore, B., R.D. Boone, J.E. Hobbit, R.A. Houghton, J.M. Melillo, B.J. Peterson, G.R. Shaver, C.J. Vorosmarty, and G.M. Woodwell. 1981. A simple model for analysis of the role of terrestrial ecosystems in the global carbon budget. p. 365–385. *In* B. Bolin (ed.) Modelling the global carbon cycle: SCOPE 16. John Wiley & Sons, New York.

Morris, L.A., and W.L. Pritchett. 1983. Effects of site preparation on *Pinus elliottii-P. palustris* flatwoods forest soil properties. p. 243–251. *In* R. Ballard and S.P. Gessel (ed.) I.U.F.R.O. Symp. on Forest Site and Continuous Productivity. USDA-FS Gen. Tech. Rep. PNW-163. Pacific Northwest For. Range Exp. Stn. Portland, OR.

Neary, D.G., E.J. Jokela, N.B. Commerford, S.R. Colbert and T.E. Cooksey. 1990. p. 432–450. *In* S.P. Gessel et al. (ed.) Sustained productivity of forest soils. Proc. North American For. Soils Conf., 7th, Vancouver, BC. 24–28 July 1988. Univ. British Columbia, Faculty For. Publ., Vancouver, BC.

Pearce, J.K., and G. Stenzel. 1972. Logging and pulpwood production. John Wiley & Sons, New York.

Permar, T.A., and R.F. Fisher. 1983. Nitrogen fixation and accretion by wax myrtle (*Myrica cerifera*) in slash pine (*Pinus elliottii*) plantations. For. Ecol. Manage. 5:39–46.

Post, W.M., T.H. Peng, W.R. Emanuel, A.W. King, V.H. Dale and D.L. DeAngelis. 1990. The global carbon cycle. Am. Sci. 78:310–326.

Powers, R.F., D.H. Alban, R.E. Miller, A.E. Tiarks, C.G. Wells, P.E. Avers, R.G. Cline, R.O. Fitzgerald, and N.S. Loftus, Jr. 1990. Sustaining site productivity in North American forests: Problems and prospects. p. 49–70. *In* S.P. Gessel et al. (ed.) Sustained productivity of forest soils. Proc. North American For. Soils Conf., 7th, Vancouver, BC. 24–28 July 1988. Univ. British Columbia, Faculty For. Publ. Vancouver, BC.

Pritchett, W.L., and R.F. Fisher. 1987. Properties and management of forest soils. John Wiley & Sons, New York.

Raison, R.J. 1979. Modification of the soil environment by vegetation fires: A review. Plant Soil 51:73–108.

Skinner, M.F., G. Murphy, E.D. Robertson, and J.G. Firth. 1989. Deleterious effects of soil disturbance on soil properties and the subsequent early growth of second-rotation radiata pine. p. 201–211. *In* W.J. Dyck and C.A. Mees (ed.) Research strategies for long-term site productivity. Proc. IEA/BE A3 Workshop, Seattle, WA. August 1988. IEA/BE A3 Rep. no. 8. FRI Bull. 152. For. Res. Inst. Rotorua, New Zealand.

Spittlehouse, D.L., and S.W. Childs. 1990. Evaluating the seedling moisture environment after site preparation. p. 80–94. *In* S.P. Gessel et al. (ed.) Sustained productivity of forest soils. Proc. North American For. Soils Conf., 7th, Vancouver, BC. 24–28 July 1988. Univ. British Columbia, Faculty For. Publ., Vancouver, BC.

Stone, E.L., W.G. Harris, R.B. Brown, and R.J. Kuehl. 1993. Carbon storage in Florida Spodosols. Soil Sci. Soc. Am. J. 57:179–182.

Stone, E.L., and P.J. Kalisz. 1991. On the maximum extent of tree roots. For. Ecol. Manage. 46:59–102.

Sutton, R.F. 1991. Soil properties and root development in forest trees: A review. Ontario Inf. Rep. 0-X-413. For. Canada, Ontario Region, Sault Ste. Marie, Canada.

Tarrant, R.F., K.C. Lu, W.B. Bollen, and J.F. Franklin. 1969. Nitrogen enrichment of two forest ecosystems by red alder. Gen. Tech. Rep. PNW-6. USDA-FS, Portland, OR.

Turner, J., and S.P. Gessel. 1990. Forest productivity in the southern hemisphere with particular emphasis on managed forests. p. 23–39. *In* S.P. Gessel et al. (ed.) Sustained productivity of forest soils. Proc. North American For. Soils Conf., 7th, Vancouver, BC. 24–28 July 1988. Univ. British Columbia, Faculty For. Publ., Vancouver, BC.

Turner, J., and M.J. Lambert. 1986. Effects of forest harvesting nutrient removals on soil nutrient reserves. Oecologia 70:140–148.

Ulery, A.L., and R.C. Graham. 1993. Forest fire effects on soil color and texture. Soil Sci. Soc. Am. J. 57:135–140.

Vance, E.D., and G.S. Henderson. 1984. Soil nitrogen availability following long-term burning in an oak-hickory forest. Soil Sci. Soc. Am. J. 48:184–190.

Youngberg, C.T., and A.G. Wollum. 1970. Nonleguminous symbiotic nitrogen fixation. p. 385–395. *In* C.T. Youngberg and C.B. Davey (ed.) Tree growth and forest soils. Oregon State Univ. Press, Corvallis, OR.

20 Soil Carbon in Northern Forested Wetlands: Impacts of Silvicultural Practices

Carl C. Trettin

Oak Ridge National Laboratory
Oak Ridge, Tennessee

Martin F. Jurgensen, Margaret R. Gale, and James W. McLaughlin

Michigan Technological University
Houghton, Michigan

Wetlands are typified by a hydrologic regime that induces anoxia in the soil during the growing season, and by vegetation that is adapted to O_2-deficient soils. This combination of hydrology, soil, and vegetation gives rise to site conditions that greatly affect ecosystem processes and landscape functions. Wetlands are now widely recognized for the important functions and values they provide, and concern exists regarding the integrity of the wetland resource and the sustainability of those attributes (Conserv. Found., 1988). Soil C is inextricably linked to many wetland functions including element cycling and storage (Hemond, 1980; Clymo, 1984a; Verhoeven, 1986; Damman, 1990; Verry & Urban, 1992), water quality (Hemond, 1990; Marin et al., 1990; David & Vance, 1991), hydrology (Boelter & Verry, 1977; Ingram, 1983), and vegetation dynamics (Vasander, 1982; Laine et al., 1992). Carbon accumulation and storage in wetland soils also are particularly important ecosystem processes that affect the global C budget. Estimates of C storage in northern peatlands alone range from 257 Pg (Eswaran et al., 1993) to 455 Pg (Post et al., 1982; Gorham, 1991), which comprise 16 and 33% of the respective estimates of the global soil C pool. Eswaran et al. (1993) reported an additional 51 Pg C in mineral soils classified in the aquic suborders, which are characteristically wetland soils. These estimates of soil C in wetlands are constrained by inadequate data on wetland area, organic horizon depth, and soil bulk density (Schlesinger, 1986; Eswaran et al., 1993). However, they demonstrate the importance of wetland soils as a global C sink.

The proportion of northern wetlands which are forested is unknown; however, statistics from five countries demonstrate that forests comprise a significant

---

Copyright © 1995 Soil Science Society of America, 677 S. Segoe Rd., Madison, WI 53711, USA.
Carbon Forms and Functions in Forest Soils.

Table 20–1. Area of wetlands and forested wetlands occuring North America and Eurasia, and the proportion of commercial forest land occuring as wetlands.

Country	Total wetland area	Forested wetland area	Proportion of commercial forest land in wetlands	Source†
	*10^6 ha	*10^6 ha	%	
Sweden	8.4	3.4§	20	1
Finland	10.4‡	?	47	2
USA	42.3	21.1	9	3,4
USSR	?	245.0	22	5
Canada	1110.0‡	?	?	6

†1 = Hanell (1991), 2 = Paivanen (1991), 3 = Frayer (1991), 4 = Abernethy and Turner (1987), 5 = Vompersky (1991), 6 = Haavisto and Jeglum (1991).
‡Estimate is for peatland area only, it does not include mineral soil wetlands.
§Does not include noncommercial forest land.

proportion of the wetland resource, and that commercial wood product operations are an important use of those wetlands (Table 20–1). In the northern USA forests comprise 51 to 66% of the total wetland area (Cubbage & Flather, 1993). The wetland resource in individual states may comprise a larger proportion of the commercial forest land area than the national average; for example in Michigan, Minnesota, and Wisconsin the proportions are 22, 29, and 16%, respectively (Smith & Hahn, 1987, 1989; Hahn & Smith, 1987). Estimates of forested wetlands within the vast Canadian peatlands are not available, however Haavisto and Jeglum's (1991) assessment of commercial forests operations demonstrates that wetlands are important. Similarly, in Scandinavia and the former USSR, intensive management of forested wetlands has been an integral component of commercial forest operations (Hanell, 1991; Paivanen, 1991; Vompersky, 1991).

Management regimes on these northern, forested wetlands vary considerably, ranging from harvesting alone to intensive silvicultural systems designed to maximize wood fiber production. Because of concern about wood utilization efficiency and sustained site productivity, management of forested wetlands typically involves harvesting, site preparation, reforestation, and stand tending. Drainage systems also are a common part of forest management prescriptions, particularly in the southeastern USA, Finland, and the former USSR (Allen & Campbell, 1988; Paivanen, 1991; Vompersky, 1991). Management prescriptions for forested wetlands in the northern USA and Canada do not usually involve drainage, although there is ongoing research and applications using such systems (Jeglum & Overend, 1991).

Common silvicultural practices used in northern forested wetlands include clearcut harvesting (stem-only and whole-tree), site preparation (mounding), replanting, and fertilization. How do these practices affect soil C dynamics in wetlands? What are the associated effects on ecosystem processes? What is the resultant effect on soil C balance at different spatial scales? These questions and others related to the impacts of silvicultural operations on northern forested wetlands have not been adequately studied. Johnson (1992) recently completed a review of forest management impacts on soil C in forests, but only one of the approximately 100 papers reviewed addressed northern forested wetlands. Johnson (1992) reported that harvesting alone generally caused little net change in upland soil C, while more intensive disturbances associated with site preparation usually reduced soil C levels. Organic matter decomposition in wetland soils is sensitive to the same biotic and abiotic factors which affect upland soils. However, the effect of silvicultural disturbances in wetlands may not be the same because of differences in silvicultural practices and site conditions, especially the large C content and anoxic properties of wetland soils.

Forested wetlands are an important component of northern landscapes. Ecological processes which affect wetland functions (e.g., modification of hydrologic regimes, water quality, habitat, or C storage) and societal values (e.g., aesthetics, recreation, forest products) are sensitive to the biotic and abiotic properties of the wetland and its C cycle. Accordingly, an understanding about the effects of forest management practices on C dynamics is necessary if those practices are to be used in a way that does not degrade wetland functions and values. This paper has three objectives: (i) review the distribution and function of C in northern forested wetlands, (ii) review how different silvicultural practices affect soil C levels, and (iii) consider the potential for recovery of soil C following disturbance by silvicultural practices.

WETLAND PROPERTIES AND PROCESSES

There are many classification systems and terms which are used to describe forested wetlands (Mader, 1991). Unfortunately, none of these are ideal for characterizing soil C content or the affect of soil C on wetland functions, such as hydrology, water quality, or nutrient cycling. This is because classification systems either lack specificity with respect to soil and hydrologic properties, or consensus on definitions and nomenclature. Generally, wetlands are classified on the basis of landform, vegetation, depth of peat accumulation, and trophic status. For purposes of this paper we distinguish three categories of forested wetlands: (i) mineral soil, (ii) histic-mineral soil, and (iii) peatlands (Histosols). These categories reflect the degree to which the wetlands have accumulated soil C (Table 20–2).

Mineral soil wetlands commonly occur in landforms which have a fluctuating water table and extended aerobic periods during the growing season. Organic matter does not accumulate in these soils because decomposition is not sufficiently limited by saturation. Accordingly, C flux is approximately balanced between inputs and outputs. Mineral soil wetlands are morphologically similar to upland soils except for the presence of hydric soil indicators (e.g., low chroma mottles within the upper 25 cm, seasonal high water table or flooding). Peatlands (Histosols) are wetlands that have a thick (>40 cm) accumulation of surface organic matter. Peatlands have traditionally been characterized by the source of water which sustains the plant community. Fens are minerotrophic peatlands supplied from groundwater and precipitation. In contrast, bogs are ombrotrophic, and receive water primarily from precipitation (Table 20–2). These water quality and hydrologic characteristics of peatlands have a significant effect on the vegetation and peat type. However, the present vegetation does not necessarily reflect

Table 20–2. Classification and characteristics of northern forested wetlands reflecting accumulated soil C.

Wetland property or attribute	Mineral soil	Wetlands soil class — Histic-mineral soil	Peatlands (Histosols) — C-accumulating wetlands	Peatlands (Histosols)
Common wetland nomenclature†	Swamp, palustrine forest	Swamp, mire, palustrine forest	Fen, treed fen, palustrine forest	Bog, treed bog, palustrine forest
Common tree species	*Acer rubrum* L., *Ulmus americana*, *Fraxinus pennsylvanica* Marshall, *F. nigra* Marshall, *Populus tremula* L.	*Picea mariana* (Mill.) B.S.P., *Larix laricina* (Du Roi) K. Koch, *Pinus banksiana* Lamb., *Acer rubrum* L.	*Thuja occidentalis* L., *Acer rubra*, *Pinus sylvestris* (L.), *Picea abies* (L.) Karst	*Picea mariana* (Mill.) B.S.P., *Larix laricina* (Du Roi) K. Koch, *Pinus sylvestris* (L.)
		Landform/hydrology classification‡		
Limnogenous LF: riparian zones WS: flooding, groundwater, precipitation	X			
Topogenous LF: topographic depressions or lows WS: groundwater, precipitation	X	X	X	
Ombrogenous LF: raised organic soils WS: precipitation				X

Table 20–2 (continued).

Wetland property or attribute		Forested wetlands		
		Minerotrophic	Minerotrophic	Ombrotrophic
Water characteristics				
Trophic status				
Soil water pH§	5.0–6.5	4.0–5.5	5.7–7.0	3.8–4.1
Soil properties				
Organic horizon thickness	0–5 cm	5–40 cm	40+ cm	40+ cm
Primary component of the O horizon	Tree litter	Bryophytes, sedges, tree litter	Bryophytes, sedges, grasses, tree litter	Bryophytes, tree litter

†Nomenclature used to describe forested wetlands [Moore & Bellamy, 1974; Cowardin et al., 1979, Natl. Wetlands Working Group (NWWG), 1988].
X = Common property or characteristic.
‡Categorization of wetlands types by landform (LF) and source of water sustaining the wetland (WS) (adapted from Damman & French, 1987).
§From Boelter and Verry (1977), and Damman and French (1987).

the composition of accumulated organic matter because it is a function the current hydrologic conditions, which may not have been the same during successional development of the wetland (Swanson & Grigal, 1989).

Intermediate between mineral soil wetlands and peatlands are the histic-mineral soil wetlands, which have a histic epipedon 5 to 40 cm thick. These wetlands are nascent peatlands, functioning as a C sink and having hydrologic and biogeochemical processes that are strongly affected by organic matter accumulation. Swamp is a common term used in conjunction with this wetland soil type (Table 20–2). Histic-mineral soil wetlands are not generally considered a peatland (Histosol) until the accumulation of surface organic materials is greater than 40 cm (Soil Survey Staff, 1975; NWWG, 1988).

Carbon Distribution

Soil

Soil C represents the largest C pool in northern forested wetlands, ranging from 50 to 1300 t ha^{-1} (Fig. 20–1). The amount of C stored in a particular wetland depends on the C concentration, thickness, and bulk density of the soil horizons. The rate of C accumulation in peatlands is dependent on wetland productivity, stage of successional development, and organic matter decomposition (Clymo, 1984b). Average C accumulation rates in peatlands range from 0.1 to 0.8 t C ha^{-1} yr^{-1} (Fig. 20–1).

Soil C in forested peatlands and histic-mineral soil wetlands occurs primarily as partially decomposed organic matter. Accumulated organic matter (i.e., peat) in these wetlands has a direct effect on soil hydrologic properties (Boelter & Verry, 1977; Ingram, 1983; Glaser, 1987; Siegel & Glaser, 1987; Verry, 1988). The degree of peat decomposition is the critical parameter affecting water retention, bulk density, and hydraulic conductivity (Boelter, 1969) and C mineralization (Wieder & Yavitt, 1991; Hogg, 1993). Two distinct zones are recognized within a peatland to distinguish areas of biological activity, hydrology, and edaphic properties (Fig. 20–2). The acrotelm, or surface zone, is characterized by a variable soil water content, a relatively high hydraulic conductivity, and active nutrient cycling and decomposition. The thickness of the acrotelm is defined by the lower limit of natural groundwater fluctuations. In contrast, the catotelm is permanently saturated, the peat is generally highly decomposed, biological activity is very low, and water flow is minimal due to a very low hydraulic conductivity (Ingram, 1983). In histic-mineral soils the surface organic horizons would be analogous to the acrotelm. These terms are not used in reference to mineral soil wetlands.

Vegetation

Northern forested wetlands contain diverse plant communities, which have developed in response to climate, hydrology, and substrate conditions (Teskey & Hinkley, 1978; Hofstetter, 1983; Glaser, 1987; NWWG, 1988). The occurrence and productivity of these plant communities vary with respect to water chemistry, water flow, water table depth, and climate (Richardson, 1978; Damman & French, 1987; Damman, 1990). Tree species which are common to northern

SOIL CARBON IN NORTHERN FORESTED WETLANDS

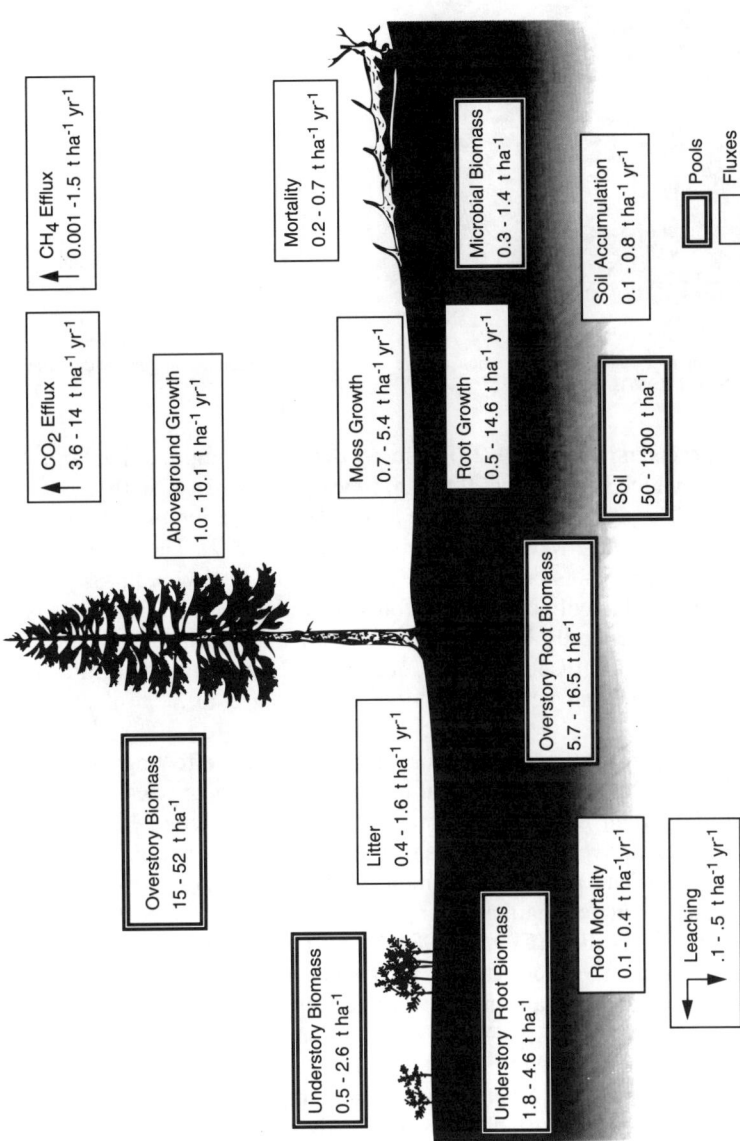

Fig. 20–1. Distribution and fluxes of C in northern forested wetlands. The values encompass the range reported by Reader and Stewart (1972), Reiners (1972), Parker and Schnieder (1975), Reader (1978), Richardson (1978), Brinson et al. (1981), Kenttämies (1981), Vasander (1982), Van Cleve et al. (1983), Grigal (1985), Grigal et al. (1985), Armentano and Menges (1986), Williams and Sparling (1988), Brock and Bregman (1989), Haland and Braekke (1989), Moore (1989), Rochefort et al. (1990), Gorham (1991), Kim and Verma (1992), Verry and Urban (1992), Bartlett and Harriss (1993).

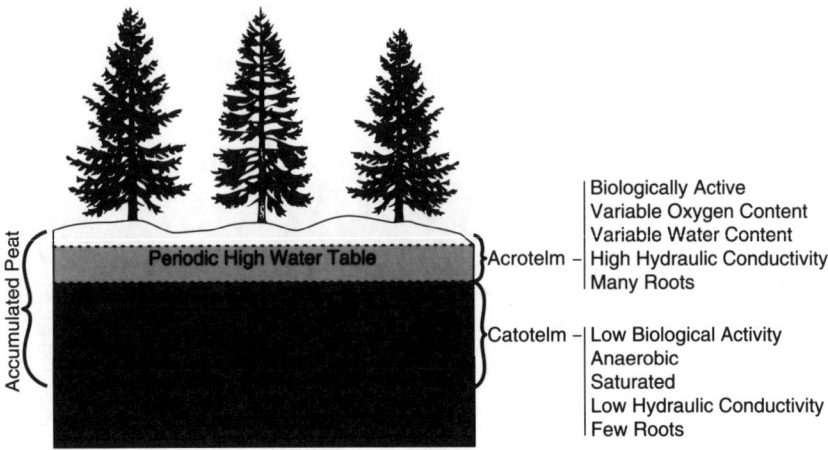

Fig. 20–2. Location and functional properties of the acrotelm and catotelm in peatlands (after Damman & French, 1987).

wetlands are presented in Table 20–2. The forest overstory comprises the largest biomass C pool, which is typically 20 to 100 times greater than C in the understory vegetation (Fig. 20–1). There is little available information on the belowground biomass in forested wetlands (Santantonio et al., 1977). The few studies conducted indicate that roots comprise 10 to 30% of the total C biomass pool. Net aboveground productivity of northern forested wetlands range from 1 to 10 t ha^{-1} yr^{-1}, which is generally lower than other temperate freshwater wetlands and upland forests (Richardson, 1978). Understory vegetation is a significant proportion of total wetland net productivity, comprising 5 to 50% of aboveground growth.

Sphagnum is perhaps the single most important plant affecting northern forested wetlands. While there are several hundred *Sphagnum* species, approximately 20 are important in the development of peatlands (Clymo, 1987). The occurrence of different *Sphagnum* species reflects microsites of varying pH conditions, available nutrients and water, the light (Andrus, 1986). *Sphagnum* affects wetland processes by increasing the water holding capacity of the peat (Boelter & Verry, 1977), acidification (Clymo, 1984a), and peat accumulation (Clymo, 1984b). The slow decomposition rate and high water retention characteristics of *Sphagnum* are important mechanisms of paludification and development of deep peatlands (Moore & Bellamy, 1974). The acidification and exchange capacity of *Sphagnum* affects plant distribution and nutrient cycling within the wetland (Hemond, 1980; Crum, 1988). *Sphagnum* also is an important component of net wetland productivity, often exceeding the productivity of trees (Grigal, 1985; Brock & Bregman, 1989).

Organic Matter Decomposition

The development of histic-mineral soil wetlands and peatlands requires that the rate of organic matter production exceed the rate of decay. Organic matter accumulation occurs until a steady state exists between production and decay within the organic horizons (Clymo, 1992). Soil water content, temperature, acidity, and substrate quality are the primary factors affecting the rate of organic matter decomposition in wetlands (Moore & Bellamy, 1974; Clymo, 1984b; Hogg, 1993). Estimates of long-term C accumulation rates, derived from ^{14}C dated peat cores, range from 8 to 41 g C m^{-2} yr^{-1}, with an average of 21 g C m^{-2} yr^{-1} (Tolonen et al., 1992). However, these accumulation rates do not account for sustained decay throughout the peat column. When this factor is considered, using Clymo's (1984b) decomposition model, the actual long-term C accumulation rate is 33% lower than the estimate derived from ^{14}C peat cores (Tolonen et al., 1992). Peatland surface accretion rates range from 0.1 to 4.3 mm yr^{-1}, with values commonly ranging between 0.5 to 2 mm yr^{-1} (Tolonen et al., 1992). Peat accumulation is not constant with time, the rate of accumulation declines as the thickness of peat increases (Clymo, 1984b).

Organic matter decomposition is most active in the surface soil or acrotelm, which has higher temperature and O_2 content, and a larger supply of labile C than deeper in the soil (Clymo, 1984b; Lieffers, 1988; Hogg, 1993). However, in peatlands having a hummock/hollow relief organic matter decomposition rate in the surface O horizons has been reported to be greater in the hummocks due to increased aeration (Farrish & Grigal, 1985), or less in hummocks due to higher acidity and water table fluctuations (Rochefort et al., 1990). Hogg et al. (1992) demonstrated that frequent wetting and drying cycles increase organic matter decomposition, as compared to extended dry periods alone.

Substrate quality and site conditions also are important factors affecting decomposition in forested wetlands. Hogg (1993) reported that CO_2 released from *Sphagnum* peat from near the surface was greater than older peat from 10- to 12-cm depth, when both peats were incubated under the same abiotic conditions. A primary factor contributing to the lower decay rate of deeper peat was its increased proportion of nonhydrolysable C (Jorgensen & Richter, 1992). Nutrient status of peatlands affects organic matter decomposition, with fens exhibiting higher decomposition rates than bogs (Farrish & Grigal, 1988). Interactions among wetland conditions and substrate quality also affect C mineralization from peat (Wieder & Yavitt, 1991).

Efflux of C from wetlands occurs both as gas and liquid, with gaseous outputs of C (CO_2 and methane [CH_4]) being the dominate pathway (Fig. 20–1). Soil temperature is a primary factor affecting organic matter decomposition. Correspondingly, CO_2 and CH_4 flux from wetlands is positively correlated to soil temperature (Kim & Verma, 1992; Bartlett & Harriss, 1993; Dunfield et al., 1993). Gas fluxes are highest during the summer, corresponding to the period of warmest soil temperatures. Although organic matter decomposition is low during the winter, it can be a significant proportion of the annual decay (Taylor & Jones, 1990; Braekke & Finer, 1990). Dise (1992) reported that CH_4 emissions during the winter can comprise 4 to 21% of the annual flux. Microbial population size and substrate availability also affect soil gas flux (Yavitt et al., 1987).

Soil aeration is another important factor affecting organic matter decomposition, and in peatlands it is strongly influenced by water table depth (Mannerkoski, 1985, p. 1–190). Soil aeration has a significant effect on the proportion of CO_2 and CH_4 in the total gas flux from a wetland (Moore & Knowles, 1989; Freeman et al., 1993). Methane oxidation is very rapid in aerated wetland soils, hence CH_4 emission is low when the water table recedes below the soil surface (Bartlett & Harriss, 1993). In contrast, CO_2 flux is stimulated when the water table is lowered (Kim & Verma, 1992; Freeman et al., 1993).

Carbon output in waters emanating from northern wetlands has long been considered to be high, but this C flux has not been adequately considered in estimating total C losses from wetlands, or in terms of watershed functions. Hemond (1990) hypothesized that riparian wetlands were the primary source of dissolved organic carbon (DOC) in streams in recently glaciated terrain. Other studies have recently established a relationship of DOC in streams with the occurrence of adjacent wetlands (Heikkinen, 1990; Koprivnjak & Moore, 1992). Organic acids comprise a significant proportion of the DOC, and are the principal anions providing the charge balance in waters from northern forested wetlands (Urban et al., 1987; Sallantaus, 1988). However, the transport of organic acids and associated cations are sensitive to the hydrologic flow path, especially when mineral soils are involved (Cronan, 1990; Heikkinen, 1990; David & Vance, 1991).

EFFECTS OF SILVICULTURAL PRACTICES

Wetness is a major constraint on silvicultural operations and vegetation productivity in northern wetlands. Accordingly, silvicultural practices are designed to ameliorate the wetness limitations. Drainage has been used extensively to manage water table levels, and it is most commonly used in conjunction with harvesting, site regeneration, and site improvement prescriptions (Paavilainen & Paivanen, 1988). The following discussion reviews the effects of those practices on soil C levels and the corresponding impact to soil and water processes.

Forest Drainage

Drainage systems are used in many forested wetlands in northern Europe and Asia (Heikurainen et al., 1978; Jeglum & Overend, 1991; Vompersky, 1991). Drainage has been a particularly important practice in Finland and the former USSR, where over 5.5 million hectares have been drained in each country (Paivanen, 1991; Vompersky, 1991). This practice has not been prevalent in northern forested wetlands of North America, although there is a history of drainage dating to the early 1900s (Zon & Averell, 1929) and considerable opportunities exist for expanded use in Canada (Haavisto & Jeglum, 1991). Silvicultural drainage systems typically use open ditches in a variety of configurations that lower the water table 30 to 73 cm below the soil surface (Paivanen & Wells, 1977; Terry & Hughes, 1978; Braekke, 1983).

Until recently the emphasis of studies on drained wetlands focused primarily on forest growth responses and nutrient cycling, with little regard to changes in soil C levels. Significant changes in soil C following drainage are expected due to induced changes in both biotic and abiotic conditions. However, only recently have studies of previously drained peatlands documented reductions in soil C (Table 20–3). These changes in soil C following drainage do not necessarily

Table 20-3. Change in soil C levels following silvicultural practices in northern forested wetlands.

Wetland type	Site characteristics	Silvicultural treatments	Assessment interval	Net change in soil C	Mean annual C loss	Source[†]
			yr	g C m^{-2}	g C m^{-2} yr^{-1}	
Peatlands						
Fen	Minerotrophic, tall sedge and tall sedge pine fens	Drainage	30	+1300–1800	+40–60	1
	Herb-rich pine fen, peat 1.7 m	Drainage	20	−650	−33	2
	Topogenous poor fen, peat 0.8–0.09 m.	Drainage, fertilization, planted	28	−3400[‡]	−120	3
Bog	Oligotrophic, low-sedge bog and cottongrass pine bog	Drainage	30	−1200 to 4800	−40 to −160	1
	Ombrotrophic bog	Drainage, fertilization	1	—	−700[§]	4
	Ombrotrophic bog, peat 2.7–3.7 m	Drainage, fertilization, seeded	27	−7850[‡]	−290	3
Histic-mineral soil	Conifer swamp, 15 cm *Sphagnum* epipedon	Whole-tree, harvest	1.5	−1800	−1200	5
	Conifer swamp, 15 cm *Sphagnum* epipedon	Whole-tree, harvest, mound, planted	1.5	−2200	−1500	5

[†] 1 = Laine et al. (1992), 2 = Sakovets and Germanova (1992), 3 = Braekke (1987), 4 = Silvola (1988), 5 = Trettin et al. (1992).
[‡] A factor of 0.5 was used to convert original data which was presented as organic matter to C.
[§] Based on CO_2 flux.

reflect total C loss or gain, but represent a net change that includes both inputs and outputs. Bogs generally exhibited greater soil C losses than fens, however, the cause of this difference has not been addressed. Drainage effects on soil C levels in forested, histic-mineral soil wetlands have not been documented, however, a reduction in soil C would be expected. Reductions in soil C from forest drainage is generally less than the 7.8 to 11.2 t ha^{-1} yr^{-1} which Armentano and Menges (1986) reported for agriculturally drained wetlands. Drainage does not always result in net reduction of soil C, although changes in soil C pools are difficult to detect when based on peat depth (Anderson et al., 1992). Increased soil C following drainage was documented in one study (Laine et al., 1992), which was attributed to improved site quality affects on belowground biomass production.

Pathways of C loss following drainage include both gas emissions and leaching. Carbon dioxide flux increases following drainage (Lahde, 1969; Silvola, 1986), while CH_4 flux is usually reduced (Bartlett & Harriss, 1993). Armentano and Menges (1986) reported C loss, as CO_2, from drained forested wetlands in the northern USA as approximately 2.8 t ha^{-1} yr^{-1}. That amount is within the range of C losses reported from forested peatlands which were measured by differences in soil C pools (Table 20–3). Increased DOC export from drainage of northern, forested peatlands also has been widely reported (Table 20–4; Heikurainen et al., 1978; Ahtiainen, 1988; Lundin, 1988; Heikkinen, 1990). However, reductions in DOC also have been found (Kenttamies, 1981; Hovi, 1988). Exports of DOC following drainage are a function of water balance alterations (Sallantaus, 1988). Drainage of fens reduces the normal influx of groundwater thereby reducing the total water flux through the wetland. In contrast, hydrologic inputs to bogs are not as affected by drainage because precipitation is the primary water source. The specific hydrologic response in drained wetlands is dependent on the drainage system design, regional hydrology, geomorphic setting, vegetation type, soil properties, and climate (Starr & Paivanen, 1986).

Reductions in soil C pools after draining peatlands, either as gases or DOC, is a direct result of increased organic matter decomposition due to changes in soil aeration and temperature regimes (Lieffers & Rothwell, 1987; Lieffers, 1988). Drainage systems increase the depth to water table thereby effectively increasing the volume of aerated soil (Munro, 1984; Mannerkoski, 1985, p. 1–190). The effects of drainage on soil temperature are varied, with reports of no change, to increases of 2 to 4°C in the upper 30 cm of peat (Table 20–5). Although either increased aeration or temperature alone would have a stimulatory effect on organic matter decomposition, the combined effect would have the most significant impact (Oades, 1988).

Harvesting

Wood fiber products in northern forested wetlands are usually harvested by clear-cutting, either using whole-tree or stem-only harvest systems. Clearcutting offers efficiencies in harvest operations, and it is an effective method for regenerating wetland forests. Both types of clearcutting systems reduce aboveground biomass, but whole-tree harvesting involves a greater removal of smaller

Table 20-4. Dissolved organic C export and change in water concentrations following silvicultural practices in northern forested wetlands.

Wetland type	Site description	Silvicultural treatments	Assessment interval	DOC export	Change in DOC†	Source‡
			yrs	g C m^{-2}	mg L^{-1}	
Peatlands						
Fen	Tall sedge fen, peat 3.2 m	Drainage	4	—	−0.5	1
	minerotrophic fen	Drainage	30	14	+18.0	2
		Undrained	30	8	—	2
Bog	Herb-rich pine fen	Drainage	20	—	+48.0§	3
	Ombrotrophic bog	Drained	30	16	+6	2
		Undrained	30	12		2
Histic-mineral soil	Conifer swamp, 15 cm *Sphagnum* epipedon	Whole-tree harvest	4	—	−2§	4
	Conifer swamp, 15 cm *Sphagnum* epipedon	Whole-tree harvest, mound, plant	4	—	+3§	4

†Change in concentration as compared to undisturbed reference site.
‡ 1 = Lundin (1988), 2 = Sallantaus (1992), 3 = Sakovets and Germanova (1992), 4 = J.W. McLaughlin (unpublished data).
§Soil water sampled at approximately 1-m depth.

Table 20–5. Change in midsummer soil temperature following silvicultural practices.

Wetland type	Site description	Silvicultural treatments	Assessment interval	Soil depth	Temperature change†	Source‡
			years	cm	°C	
Peatlands						
Fen	Small sedge mire, peat depth 2.4–3.2 m	Drainage	4	10, 20, 50, 75	0	1
	Treed fen, peat depth = 1.3 m	Drainage	2	10	+3	2
		Drainage	2	30	+2	2
	Conifer swamp, 6 cm histic epipedon	Drained,	11	5	+2.2	3
		Harvested	11	50	+2.5	3
Histic-mineral soil	Conifer swamp, 15 cm *Sphagnum* epipedon	Whole-tree Harvest	1	5	+2.6	4
			1	25	+2.7	4
	Conifer swamp, 15 cm *Sphagnum* epipedon	Whole-tree harvest, bed, plant	1	5	+5.8	4
			1	25	+4.6	4

† Mean temperature difference, over the assessment interval, between the disturbed and undisturbed reference site.
‡ 1 = Latja and Kurimo (1988), 2 = Lieffers (1988), 3 = Kubin and Kemppainen (1991), 4 = Trettin and Jurgensen (1992).

diameter material, sometimes including foliage (Mann et al., 1988). While many studies have been conducted comparing the effects of these two harvesting systems on biomass pools in upland soils we found no such studies on northern wetlands.

Musselman and Fox (1991) reported that clearcutting on all soils in the USA could reduce soil C pools 25 to 50%. Although a recent review of 13 studies showed that harvesting may either increase or decrease soil C in upland soils, the net effect was usually less than a 10% change in mineral soil C content (Johnson, 1992). Changes in the C content of the forest floor also increased or decreased depending on the quantity of logging residue. Effective maintenance of soil C would most likely occur for stem-only harvesting where a proportion of the overstory biomass remains on site. Although leaving woody residue may have a short-term effect on the size of the soil C pool, it is unlikely that the residue additions would change the soil C equilibrium levels, nor does it represent an effective mechanism for reducing nutrient losses (Fahey et al., 1991; Titus & Malcolm, 1992).

Only one study was found which considered the effect of harvesting alone on soil C pools in northern wetlands, since most silvicultural prescriptions in northern forested wetlands involve drainage, site preparation, or both (Table 20–3). Trettin et al. (1992) reported a 28% reduction in soil C 18 mo after whole-tree harvesting a histic-mineral soil swamp. The annualized C loss following harvesting alone was greater than that reported for peatlands which were drained (Table 20–3). However, these losses reported for peatlands represent a net change in soil C that includes C inputs and outputs over several decades; in contrast, the short-term response reported for the histic-mineral soil wetland involves C loss predominately.

The impact of harvesting on soil C in peatlands and histic-mineral soil wetlands is expected to be different from upland soils. Unlike upland forests, these forested wetlands are characterized by thick surface O horizons, which have developed because of slow organic matter decomposition rates. Changes in soil temperature, aeration, and nutrient availability after harvesting increase organic matter decomposition rates (Trettin & Jurgensen, 1992) and cause reductions in soil C levels. Higher soil temperature following harvesting is a primary factor contributing to increased organic matter decomposition (Table 20–5). This effect occurs rapidly following harvesting and can persist for more than 10 yr (Kubin & Kemppainen, 1991).

Water table depth greatly affects soil aeration, hence changes in the water table following harvesting also would impact organic matter decomposition. The affect of water table depth on aeration is largely controlled by the hydrologic regime, but it also varies temporally (Munro, 1984) and spatially within the wetland (Trettin, unpublished data). Depth to the water table in undrained wetlands, where the water table normally falls to at least 30 cm below the soil surface during the growing season, will typically decrease following harvesting (Verry, 1988). On these sites, anoxia induced by the higher water table would limit the rate of organic matter decomposition. In undrained wetlands where the water table normally remains within 30 cm of the soil surface, the amplitude of water table fluctuations after harvesting will increase, but the average depth will remain the same (Verry, 1986, 1988). In drained peatlands the water table typically rises following harvesting (Paivanen, 1982).

Mineralization of woody residue and soil organic matter following harvesting of forested wetlands increases DOC and mineral nutrient leaching (Ahtiainen, 1988; Hughes et al., 1990). The leached C and nutrients can have a direct effect on the productivity and acid-base chemistry of nearby streams. Huttunen et al. (1988) measured increases in both algal and bacteria primary production following clearcutting of forested wetlands in Finland, but no effect was found when an uncut buffer strip was left adjoining the stream. While DOC exports are an important pathway for C loss in all harvested peatlands, the flux may be dependent on the type of harvest system used. Whole-tree harvest removes more C as biomass, leaving less residue for subsequent decay. Stem-only harvest leaves more residue on the site, which is subject to decay and release as CO_2, CH_4 and DOC. Whether there is a difference in total C loss between these two types of harvesting systems in wetlands has not been studied.

Site Preparation

Site preparation is used on forested wetlands to establish sites for planting seedlings, and for the control of competing vegetation. Numerous site preparation systems are used on northern wetlands, which mostly involve some form of mounding (Sutton, 1993). Mounds are created either by disking or excavation. These mounds provide an elevated planting site and mitigate against increased water table levels following harvesting. Disked mounds (i.e., bedding) mix surface soil horizons into the center of the planting bed, while mounds created by excavation typically do not mix the surface soils (Attiwill et al., 1985; Sutton, 1993).

The degree of soil mixing as a result of site preparation greatly affects organic matter decomposition. Ross and Malcolm (1988) found in a laboratory study that mixing peat with mineral soil under both aerated and saturated conditions increased organic matter decomposition and nutrient mineralization in accordance to the degree of mixing and soil aeration levels. Results from a field study comparing harvesting and site preparation demonstrated that bedding caused greater reductions in soil C pools, as compared to harvesting alone (Table 20–3). These C losses occurred within the first 18 mo following harvest, and were a result of increased organic matter decomposition in the O horizons. Measurements 30 mo later showed no changes in soil C among site preparation treatments (J.W. McLaughlin, unpublished data). The loss of soil C as a result of bedding was greater than annual losses reported for drained peatlands or harvested histic-mineral soil wetlands (Table 20–3).

As was shown earlier for timber harvesting, soil temperature increases as a result of site preparation and is a primary factor affecting organic matter decomposition (Table 20–5). This is especially true for mounding, which increased the rate of soil warming in the spring and increased the maximum midsummer soil temperature in a histic-mineral soil wetland (Trettin & Jurgensen, 1992). Sutton (1993) also reported that mounding increased soil temperature.

The most pronounced effect of site preparation on organic matter decomposition occurs when both soil aeration and temperature are affected (Hogg et al.,

1992). Elevated soil temperature increased decomposition of peat under aerated conditions, but temperature did not have an obvious effect on organic matter decomposition when saturated peat was incubated in a laboratory study (Hogg et al., 1992). A similar temperature and water regime interaction on the cellulose decay was reported by Donnelly et al. (1990). Field measurements of organic matter decomposition following site preparation of a histic-mineral soil wetland found increased decomposition in the upper 25 cm of soil (Fig. 20–3) which was attributed to increased temperature and aeration (Trettin & Jurgensen, 1992). Harvesting alone did not increase soil temperature to the same extent as mounding, and the affect was evident in the lower decomposition rate. Increased organic matter decomposition as a result of mounding also have been reported for other northern (Sutton, 1993) and southern (Bridgham et al., 1991) forested wetlands.

RECOVERY OF SOIL CARBON

Increased organic matter decomposition rates following silvicultural practices reduces the ability of forested wetlands to serve as a C sink. In many cases, increased decomposition rates can transform a wetland from a C sink to a C source. Long-term accumulation rates of soil C in northern peatlands range from 8 to 41 g m^{-2} yr^{-1} (Tolonen et al., 1992). The time necessary to replace soil C loss

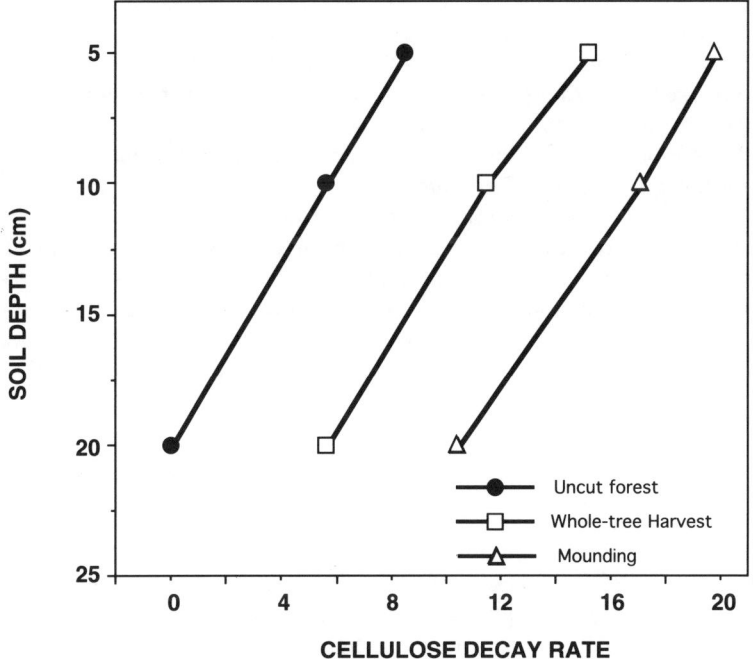

Fig. 20–3. Rate of cellulose decay in a histic-mineral soil swamp in northern Michigan 1 yr following whole-tree harvesting and whole-tree harvesting plus mounding. Cellulose decomposition rate was measured by Cotton Strip Assay (after Trettin & Jurgensen, 1992).

resulting from silvicultural practices could take decades to centuries. Reductions in soil C pools after harvesting range from approximately 1000 to 5000 g m^{-2} (Table 20–3). If postdisturbance accumulation rates were the same as predisturbance, the recovery time to return soil C to preharvest levels would range from 40 to 120 yr, using an accumulation rate of 23 g m^{-2} yr^{-1} (Gorham, 1991). However, it is unlikely that peat accumulation on disturbed wetlands would be the same as under predisturbance conditions because of changes in plant species composition and productivity, hydrology, and nutrient cycling. On drained peatlands the recovery of soil C may never occur because long-term CO_2 emissions (approx. 30 g m^{-2} yr^{-1}) exceeds average C accumulation rates (Gorham, 1991). Shifts in C allocation among the soil and vegetation will affect wetland functions, but there is little empirical data on which to assess this potential response.

The duration of impacts on soil C accumulation resulting from silvicultural practices is dependent on wetland properties and the type forest management practice. Results from several studies in northern peatlands demonstrated that C loss was still evident 20 to 30 yr after the wetland was drained (Table 20–3). Vompersky et al. (1992) suggested that improved plant growth following drainage may mitigate against loss of soil C. However, even if increased wetland productivity effectively reduces net C losses or perhaps even increases C storage on the site, any soil C loss should be expected to change soil hydrologic properties and nutrient cycling within the wetland. Only one study (Laine et al., 1992) has documented a net increase in soil C following drainage as a result of increased stand productivity. Braekke (1992) reported increased root growth following drainage and fertilization, but he did not determine the change in soil C levels. Sakovets and Germanova (1992) reported that total C content on a drained fen increased following drainage as a result of increased aboveground biomass production, although soil C levels actually decreased.

Mitigation of soil C loss by increased biomass production has only been demonstrated for fens. There are currently no studies which suggest that ombrotrophic wetlands have the potential for productivity gains to compensate for soil C loss. Changes in C allocation from soil C to above- and/or belowground biomass represents a functional change in the wetland that may temporarily balance the C flux and storage, but it is doubtful that increased C in biomass will replace the C accumulation function of peatlands and histic-mineral soils because cultivated biomass does not represent a long-term C sink (Vitousek, 1991).

CONCLUSIONS AND PERSPECTIVES

A significant, but yet uncertain, proportion of northern wetlands are forested, accumulate soil C, and are important for wood production in many countries. Carbon accumulation in these wetlands is a result of biomass production exceeding the rate of organic matter decomposition. Biotic and abiotic factors which affect organic matter accumulation include hydrology, geomorphic setting, microclimate, and vegetation composition and production. A unique aspect of C-accumulating wetlands is that the affect of those developmental factors on wetland functions change as organic matter accumulates. Accordingly, wetland functions, including soil water storage, modification of water quality, nutrient cycling, habitat conditions, and others, are sensitive to soil C levels. Carbon storage alone is an important wetland function that affects the global C budget.

The C balance in forested wetlands is sensitive to various silvicultural practices. However, information characterizing the impacts of silvicultural practices in terms of long-term C dynamics and wetland functions are very limited. Harvesting, site preparation, and drainage each cause a reduction in soil C. However, the magnitude of C loss is dependent on wetland type (bog, fen, swamp), management regime imposed, and time since disturbance. The actual loss of soil C caused by these silvicultural practices is difficult to ascertain with existing information. Most available data is from long-term studies that permit only an assessment of the net change in soil C storage. The loss of soil C following silvicultural practices occurs as CO_2, CH_4, and DOC. However, the relative importance of gaseous and aqueous pathways for C flux have not been adequately addressed. Increased soil temperature and aeration are the primary factors causing soil organic matter decomposition to increase following harvesting and site preparation. Carbon losses are greatest in drained and site-prepared wetlands, where aerated soil volume and water table fluctuations are the greatest.

In contrast, biomass production on wetlands sometimes increases following drainage and site preparation. That response has been postulated as a mechanism for mitigating soil C loss following drainage. Currently, only one study has measured an actual increase in soil C following drainage, all others have reported a net C loss. Increased C accumulation in biomass following other silvicultural practices (e.g., harvesting, site preparation) also may be considered to balance the loss of soil C. However, changes in C accumulation among different pools (e.g., biomass, soil) within the wetland will affect ecosystem functions.

Given the diversity of wetland types and the importance of these ecosystems for wood products, additional research is needed to determine the effects of silvicultural practices on soil C and associated ecosystem processes. Studies are needed to characterize the magnitude of soil C loss following silvicultural practices, and to determine the partitioning among the gas and aqueous pathways for C flux. These data are needed to determine the impacts on wetland functions, including, the global C balance, stream water quality, stream biota, hydrology, and element cycling and transport. Other studies are needed to further characterize the mechanisms controlling soil C loss, in particular the interaction between moisture and temperature regimes in field-scale experiments. Finally, long-term research is needed to assess the functional response of different wetland types to altered patterns of C allocation and storage following silvicultural practices. This information is particularly important for designing best management practices to sustain inherent wetland functions.

To address the growing public concern on the management of forested wetlands, regional, interdisciplinary studies are needed to enhance the knowledge-base regarding the effects of silvicultural practices. Ideally, such studies should have a common treatment design that is applied to regionally important wetland types. A good example of a comprehensive, integrated study is SUOSILMU, a Finnish program designed to determine the role of peatlands in the biogeochemical cycling of C (Kanninen & Anttila, 1992). In the USA, opportunities for integrated studies on managed wetlands are provided through research cooperatives (Shepard et al., 1993) and consortia (e.g., Consortium for Research on Southern Forested Wetlands). Expansion of these efforts, particularly in northern regions, is needed in order to provide an informed basis for decisions on the regulation and management of forested wetlands.

ACKNOWLEDGMENTS

We would like to thank the National Council of the Paper Industry on Air and Stream Improvement and the Lake States Forest Resources and Environmental Management Cooperative for supporting our northern, forested wetlands research. We also appreciate the comments and suggestions for improving this paper from two anonymous reviewers and the editors. Oak Ridge National Laboratory is managed by Martin Marietta Energy Systems, Inc., under contract DE-AC05-84OR21400 with the U.S. Dep. of Energy.

REFERENCES

Abernethy, Y., and R.E. Turner. 1987. U.S. forested wetlands: 1940–1980. BioScience 37:721–727.

Ahtiainen, M. 1988. Effects of clear-cutting and forestry drainage on water quality in the Nurmes-study. p. 206–219. In Proc. Int. Symp. Hydrology of Wetlands in Temperate and Cold Regions, Joensuu, Finland. 6–8 June. Vol. 1. No. 4/88. Publ. Acad., Helsinki, Finland.

Allen, H.L., and R.G. Campbell. 1988. Wet site pine management in the southeastern United States. p. 173–184. In D.L. Hook (ed.) The ecology and management of wetlands. Vol. 2. Croom Helm, London.

Anderson, A.R., D.G. Pyatt, J.M. Sayers, S.R. Blackhall, and H.P. Robinson. 1992. Volume and mass budgets of blanket peat in the north of Scotland. Suo 43:195–198.

Andrus, R.E. 1986. Some aspects of *Sphagnum* ecology. Can. J. Bot. 64:416–426.

Armentano, T.V., and E.S. Menges. 1986. Patterns of change in the carbon balance of organic soil-wetlands of the temperate zone. J. Ecol. 74:755–774.

Attiwill, P.M., N.D. Turvey, M.A. Adams. 1985. Effects of mound-cultivation (bedding) on concentration and conservation of nutrients in a sandy podzol. For. Ecol. Manage. 11:97–110.

Bartlett, K.B., and R.C. Harriss. 1993. Review and assessment of methane emissions from wetlands. Chromosphere 26:261–320.

Boelter, D.H. 1969. Physical properties of peat related to degree of decomposition. Soil Sci. Soc. Am. Proc. 28:433–435.

Boelter, D.H., and E.S. Verry. 1977. Peatland and water in the northern Lake States. USDA-FS, Gen. Tech. Rep. NC-31. North Central For. Exp. Stn., St. Paul, MN.

Braekke, F.H. 1983. Water table levels at different drainage intensities on deep peat in northern Norway. For. Ecol. Manage. 5:169–192.

Braekke, F.H. 1987. Nutrient relationships in forest stands: Effects of drainage and fertilization on surface peat layers. For. Ecol. Manage. 21:269–284.

Braekke, F.H. 1992. Root biomass changes after drainage and fertilization of a low-shrub pine bog. Plant Soil 143:33–43.

Braekke, F.H., and L. Finer. 1990. Decomposition of cellulose in litter layer and surface peat of low-shrub pine bogs. Scand. J. For. Res. 5:297–310.

Bridgham, S.D., C.J. Richardson, E. Maltby, and S.P. Faulkner. 1991. Cellulose decay in natural and disturbed peatlands in North Carolina. J. Environ. Qual. 20:695–701.

Brinson, M.M., A.E. Lugo, and S. Brown. 1981. Primary productivity, decomposition, and consumer activity in freshwater wetlands. Annu. Rev. Ecol. Syst. 12:123–161.

Brock, T.C.M., and R. Bregman. 1989. Periodicity in growth, productivity, nutrient content and decomposition of *Sphagnum recurvum var. mucronatum* in a fen woodland. Oecologia 80:44–52.

Clymo, R.S. 1984a. *Sphagnum*-dominated peat bog: A naturally acid ecosystem. Philos. Trans. R. Soc. London, B 305:487–499.

Clymo, R.S. 1984b. Limit of peat bog growth. Philos. Trans. R. Soc. London, B 303:605–654.

Clymo, R.S. 1987. Interaction of *Sphagnum* with water and air. p. 513–529. *In* T.C. Hutchinson and K.M. Meema (ed.) Effects of atmospheric pollutants on forests, wetlands, and agricultural ecosystems. Springer-Verlag, Berlin, Germany.

Clymo, R.S. 1992. Models of peat growth. Suo 43:127–136.

The Conservation Foundation. 1988. Protecting America's wetlands: An action agenda. Conserv. Found., Washington, DC.

Cowardin, L.M., V. Carter, F.C. Golet, and E.T. LaRoe. 1979. Classification of wetlands and deepwater habitats of the United States. FWS/OBS-79/31. U.S. Fish Wildlife Serv., Washington DC.

Cronan, C.S. 1990. Patterns of organic acid transport from forested watersheds to aquatic ecosystems. p. 245–260. *In* E.M. Perdue and E.T. Gjessing (ed.) Organic acids in aquatic ecosystems. John Wiley & Sons, Chichester, England.

Crum, H.A. 1988. A focus on peatlands and peat mosses. Univ. Michigan Press, Ann Arbor.

Cubbage, F.W., and C.H. Flather. 1993. Forested wetland area and distribution: A detailed look at the South. J. For. 91:35–40.

Damman, A.W.H. 1990. Nutrient status of ombrotrophic peat bogs. Aquilo Ser. Bot. 28:5–14.

Damman, A.W.H., and T.W. French. 1987. The ecology of peat bogs of the glaciated northeastern United States: A community profile. U.S. Fish Wildlife Serv. Biol. Rep. 85(7.16). U.S. Dep. Interior, Washington, DC.

David, M.B., and G.F. Vance. 1991. Chemical character and origin of organic acids in streams and seepage lakes of central Maine. Biogeochemistry 12:17–41.

Dise, N.B. 1992. Winter fluxes of methane from Minnesota peatlands. Biogeochemistry 17:71–83.

Donnelly, P.K., J.A. Entry, D.L. Crawford, and K. Cromack, Jr. 1990. Cellulose and lignin degradation in forest soils in response to moisture, temperature, and acidity. Microbiol. Ecol. 20:289–295.

Dunfield, P., R. Knowles, R. Dumont, and T.R. Moore. 1993. Methane production and consumption in temperate and subarctic peat soils: Response to temperature and pH. Soil Biol. Biochem. 25:321–326.

Eswaran, H., E. Van Den Berg, and P. Reich. 1993. Organic carbon in soils of the world. Soil Sci. Soc. Am. J. 57:192–194.

Fahey, T.J., P.A. Stevens, M. Hornung, and P. Rowland. 1991. Decomposition and nutrient release from logging residue following conventional harvest of Sitka spruce in north Wales. Forestry 64:289–301.

Farrish, K.W., and D.F. Grigal. 1985. Mass loss in a forested bog: Relation to hummock and hollow microrelief. Can. J. Soil Sci. 65:375–378.

Farrish, K.W., and D.F. Grigal. 1988. Decomposition in an ombrotrophic bog and a minerotrophic fen in Minnesota. Soil Sci. 145:353–358.

Frayer, W.E. 1991. Status and trends of wetlands and deepwater habitats in the conterminous United States, 1970's to 1980's. School Forestry Wood Products, Michigan Technol. Univ., Houghton, MI.

Freeman, C., M.A. Lock, and B. Reynolds. 1993. Fluxes of CO_2, CH_4 and N_2O from a Welsh peatland following simulation of water table draw-down: Potential feedback to climate change. Biogeochemistry 19:51–60.

Glaser, P.H. 1987. The ecology of patterned boreal peatlands of northern Minnesota: A community profile. U.S. Fish Wildlife Serv. Biol. Rep. 85(7.14). U.S. Dep. Interior, Washington, DC.

Gorham, E. 1991. Northern peatlands: Role in the carbon cycle and probable responses to climatic warming. Ecol. Monogr. 1:182–195.

Grigal, D.F. 1985. *Sphagnum* production in forested bogs in northern Minnesota. Can. J. Bot. 63:1204–1207.

Grigal, D.F., C.G. Buttleman, and L.K. Kernik. 1985. Biomass and productivity of the woody strata of forested bogs in northern Minnesota. Can. J. Bot. 63:2416–2424.

Haavisto, V.F., and J.K. Jeglum. 1991. Peatland potentially available for forestry in Canada. p. 30–37. *In* J.K. Jeglum and R.P. Overend (ed.) Peat and peatlands diversification and innovation, Quebec City, Quebec. 6–10 Aug. 1989. Vol. 6. Canadian Soc. Peat and Peatlands, Darmouth, Nova Scotia.

Hahn, J.T., and W.B. Smith. 1987. Minnesota's forest statistics, 1987: An inventory update. Gen. Tech. Rep. NC-118. USDA-FS, North Central For. Exp. Stn., St. Paul, NM.

Haland, B., and F.H. Braekke. 1989. Distribution of root biomass in a low-shrub pine bog. Scan. J. For. Res. 4:307–316.

Hanell, B. 1991. Peatland forestry in Sweden. p. 19–25. *In* J.K. Jeglum and R.P. Overend (ed.) Peat and peatlands diversification and innovation, Quebec City, Quebec. 6–10 Aug. 1989. Vol. 6. Canadian Soc. Peat and Peatlands, Darmouth, Nova Scotia.

Heikkinen, K. 1990. Nature of dissolved organic matter in the drainage basin of a boreal humic river in northern Finland. J. Environ. Qual. 19:649–657.

Heikurainen, L., K. Kenttamies, and J. Laine. 1978. The environmental effects of forest drainage. Suo 29:49–58.

Hemond, H.F. 1980. Biogeochemistry of Thoreau's Bog, Concord, Massachusetts. Ecol. Monogr. 50:507–526.

Hemond, H.F. 1990. Wetlands as the source of dissolved organic carbon to surface waters. p. 301–313. *In* E.M. Perdue and E.T. Gjessing (ed.) Organic acids in aquatic ecosystems. John Wiley & Sons, Chichester, England.

Hofstetter, R.H. 1983. Wetlands in the United States. p. 201–244. *In* A.J.P Gore (ed.) Mires: Swamp, bog, fen, and moor. Vol. 1. Elsevier Sci. Publ. Co., Amsterdam.

Hogg, E.H. 1993. Decay potential of hummock and hollow Sphagnum peats at different depths in a Swedish raised bog. Oikos 66:269–278.

Hogg, E.H., V.J. Lieffers, and R.W. Wein. 1992. Potential carbon losses from peat profiles; effects of temperature, drought cycles, and fire. Ecol. Applic. 2:298–306.

Hovi, A. 1988. Organic carbon dynamics in small brooks before and after forest drainage and clearcutting. p. 220–231. *In* Proc. Int. Symp. Hydrology Wetlands in Temperate and Cold Regions, Joensuu, Finland. 6–8 June. Vol. 1. No. 4/88. Publ. Acad. Finland, Helsinki, Finland.

Hughes, S., B. Reynolds, and J.D. Roberts. 1990. The influence of land management on concentrations of dissolved organic carbon and its effects on the mobilization of aluminum and iron in podzol soils in mid-Wales. Soil Use Manage. 6:137–144.

Huttunen, P., A.L. Holopainen, and A. Hovi. 1988. Effects of silvicultural measures on primary production in forest brooks. p. 239–248. *In* Proc. Int. Symp. Hydrology Wetlands in Temperate and Cold Regions, Joensuu, Finland. 6–8 June. Vol. 1. No. 4/88. Publ. Acad. Finland, Helsinki, Finland.

Ingram, H.A.P. 1983. Hydrology. p. 67–158. *In* A.J.P Gore (ed.) Mires: Swamp, bog, fen, and moor. Vol. I. Elsevier Sci. Publ. Co., Amsterdam.

Jeglum, J.K., and R.P. Overend (ed.) 1991. Peat and peatlands diversification and innovation, Quebec City, Quebec. 6–10 Aug. 1989. Canadian Soc. Peat Peatlands, Darmouth, Nova Scotia.

Johnson, D.W. 1992. Effects of forest management on soil carbon storage. Water Air Soil Pollut. 64:83–120.

Jorgensen, R.G., and G.M. Richter. 1992. Composition of carbon fractions and potential denitrification in drained peat soils. J. Soil Sci. 43:347–358.

Kanninen, M., and P. Anttila (ed.). 1992. The Finnish research programme on climate change. No. 3/92. Publ. Acad. Finland, Helsinki, Finland.

Kenttamies, K. 1981. The effects of water quality of forest drainage and fertilization in peatlands. Publ. Water Res. Inst., no. 43. Natl. Board Waters, Helsinki, Finland.

Kim, J., and S.B. Verma. 1992. Soil surface CO_2 flux in a Minnesota peatland. Biogeochemistry 18:37–51.

Koprivnjak, J.F., and T.R. Moore. 1992. Sources, sinks, and fluxes of dissolved organic carbon in subarctic fen catchments. Arctic Alpine Res. 24:204–210.

Kubin, E., and L. Kemppainen. 1991. Effect of clearcutting of boreal spruce forest on air and soil temperature conditions. Acta For. Fenn. 225:1–42.

Lahde, E. 1969. Biological activity in some natural and drained peat soils with special reference to oxidation-reduction conditions. Acta For. Fenn. 94:1–69.

Laine, J., H. Vasander, and A. Puhalainen. 1992. Effects of forest drainage on the carbon balance of mire ecosystems. p. 170–181. *In* Proc. 9th Int. Peat Congr., 9th, 22–26 June. Vol. 1. Uppsala, Sweden. Int. Peat Soc., Jyska, Finland.

Latja, A., and H. Kurimo. 1988. Temperature changes in the soil and close to the ground on wetlands drained for forestry. p. 46–51. *In* Proc. Int. Symp. Hydrology Wetlands Temperate and Cold Regions, Joensuu, Finland. 6–8 June. Vol. 1. No. 4/88. Publ. Acad. Finland, Helsinki, Finland.

Lieffers, V.J. 1988. *Sphagnum* and cellulose decomposition in drained and natural areas of an Alberta peatland. Can. J. Soil Sci. 68:755–761.

Lieffers, V.J., and R.L. Rothwell. 1987. Effects of drainage on substrate temperature and phenology of some trees and shrubs in an Alberta peatland. Can. J. For. Res. 17:97–104.

Lundin, L. 1988. Impacts of drainage for forestry on runoff and water chemistry. p. 197–205. *In* Proc. Int. Symp. Hydrology Wetlands Temperate Cold Regions, Joensuu. 6–8 June. Vol. 1. No. 4/88. Publ. Acad. Finland, Helsinki, Finland.

Mader, S.F. 1991. Forested wetland classification and mapping: A literature review. NCASI Tech. Bull. 606. Natl. Council Pap. Ind. Air Stream Improvement, New York.

Mann, L.K., D.W. Johnson, D.C. West, D.W. Cole, J.W. Hornbeck, C.W. Martin, H. Riekerk, C.T. Smith, W.T. Swank, L.M. Tritton, and D.H. Van Lear. 1988. Effects of whole-tree and stem-only clearcutting on postharvest hydrologic losses, nutrient capital, and regrowth. For. Sci. 34:412–428.

Mannerkoski, H. 1985. Effect of water table fluctuation on the ecology of peat soil. Publ. 7. Dep. Peatland For. Univ. Helsinki, Helsinki, Finland.

Marin, L.E., T.K. Kratz, and C.J. Bower. 1990. Spatial and temporal patterns in the hydrochemistry of a poor fen in northern Wisconsin. Biogeochemistry 11:63–76.

Moore, P.D., and D.J. Bellamy. 1974. Peatlands. Springer-Verlag, New York.

Moore, T.R. 1989. Growth and net production of *Sphagnum* at five fen sites, subarctic eastern Canada. Can. J. Bot. 67:1203–1207.

Moore, T.R., and R. Knowles. 1989. The influence of water table levels on methane and carbon dioxide emissions from peatland soils. Can. J. Soil Sci. 69:33–38.

Munro, D.S. 1984. Summer soil moisture content and the water table in a forested wetland peat. Can. J. For. Res. 14:331–335.

Musselman, R.C., and D.G. Fox. 1991. A review of the role of temperate forests in the global CO_2 Balance. J. Air Waste Manage. Assoc. 41:798–807.

National Wetlands Working Group. 1988. Wetlands of Canada. Ecol. Land Classification Ser., no. 24. Sustainable Devel. Branch. Environ. Canada, Ottawa, Ontario, and Polysci. Publ., Montreal, Quebec.

Oades, J.M. 1988. The retention of organic matter in soils. Biogeochemistry 5:35–70.

Paavilainen, E., and J. Paivanen. 1988. Use and management of forested hydric soils. p. 203–212. *In* D.L. Hook (ed.) The ecology and management of wetlands. Vol. 2. Croom Helm, London.

Paivanen, J. 1982. The effect of cutting and fertilization on the hydrology of an old forest drainage area. Folia For. 515:4–19.

Paivanen, J. 1991. Peatland forestry in Finland: Present status and prospects. p. 3–12. *In* J.K. Jeglum and R.P. Overend (ed.) Peat and peatlands diversification and innovation, Quebec City, Quebec. 6–10 Aug. 1989. Canadian Soc. Peat Peatlands, Darmouth, Nova Scotia.

Paivanen, J., and E.D. Wells. 1977. Guidelines for the development of peatland drainage systems for forestry in Newfoundland. Dep. Fish. Environ., Canadian For. Serv. Inf. Rep. N-X-156. St. Johns, Newfoundland.

Parker, G.R., and G. Schnieder. 1975. Biomass and productivity of an alder swamp in northern Michigan. Can. J. For. Res. 5:403–409.

Post, W.M., W.R. Emanuel, P.J. Zinke, and A.G. Stangenberger. 1982. Soil carbon pools and world life zones. Nature (London) 298:156–159.

Reader, R.J. 1978. Primary production in northern bog marshes. p. 53–62. *In* Freshwater wetlands. Acad. Press, New York.

Reader, R.J., and J.M. Stewart. 1972. The relationship between net primary production and accumulation for a peatland in southeastern Manitoba. Ecology 53:1024–1037.

Reiners, W.A. 1972. Structure and energetics of three Minnesota forests. Ecol. Monogr. 42:71–94.

Richardson, C.J. 1978. Primary productivity values in fresh water wetlands. p. 131–145. *In* P.E. Greeson et al. (ed.) Wetland function and values: The state of our understanding. AWRA, Minneapolis, MN.

Rochefort, L., D.H. Vitt, and S.E. Baley. 1990. Growth, production, and decomposition dynamics of *Sphagnum* under natural and experimentally acidified conditions. Ecology 71:1986–2000.

Ross, S.M., and D.C. Malcolm. 1988. Modeling nutrient mobilization in intensively mixed peaty heathland soil. Plant Soil 107:113–121.

Sakovets, V.V., and N.I. Germanova. 1992. Changes in carbon balance of forested mires in Karelia due to drainage. Suo 43:249–252.

Sallantaus, T. 1988. Water quality and man's influence on it. p. 80–98. *In* Proc. Int. Symp. Hydrology Wetlands Temperate Cold Regions, Joensuu, Finland. 6–8 June. Vol. 2. No. 4/88. Publ. Acad. Finland, Helsinki, Finland.

Sallantaus, T. 1992. Leaching in the material balance of peatlands—preliminary results. Suo 43:253–258.

Santantonio, D., R.K. Hermann, and W.S. Overton. 1977. Root biomass studies in forest ecosystems. Pedobiologia 17:1–31.

Schlesinger, W.H. 1986. Changes in soil carbon storage and associated properties with disturbance and recovery. p. 195–219. *In* J.R. Trabalka and D.E. Reichle (ed.) The changing carbon cycle—A global analysis. Springer-Verlag, New York.

Shepard, J.P., A.A. Lucier, and L.W. Haines. 1993. Industry and forest wetlands: Cooperative research initiatives. J. For. 91:29–33.

Siegel, D.I., and P.H. Glaser. 1987. Ground water flow in a bog-fen complex Lost River peatland, northern Minnesota. J. Ecol. 75:743–754.

Silvola, J. 1986. Carbon dioxide dynamics in mires reclaimed for forestry in eastern Finland. Ann. Bot. Fennici 23:59–67.

Silvola, J. 1988. Effect of drainage and fertilization on carbon and nutrient mineralization of peat. Suo 39:27–37.

Smith, W.B., and J.T. Hahn. 1987. Michigan's forest statistics, 1987: An inventory update. USDA-FS Gen. Tech. Rep. NC-112. North Central For. Exp. Stn., St. Paul, MN.

Smith, W.B., and J.T. Hahn. 1989. Wisconsin's forest statistics, 1987: An inventory update. USDA-FS Gen. Tech. Rep. NC-130. North Central For. Exp. Stn., St. Paul, MN.

Soil Survey Staff. 1975. Soil taxonomy: A basic system of soil classification for making and interpreting soil surveys. USDA-SCS Agric. Handb. 436. U.S. Gov. Print. Office, Washington, DC.

Starr, M.R., and J. Paivanen. 1986. Runoff response to peatland forest drainage in Finland: A synthesis. p. 43–50. *In* S.P. Gessel (ed.) Forest site productivity. Martinus Nijoff Publ., Dordrecht, the Netherlands.

Sutton, R.F. 1993. Mounding site preparation: A review of European and North American experience. New For. 7:151–192.

Swanson, D.K., and D.F. Grigal. 1989. Vegetation indicators of organic soil properties in Minnesota. Soil Sci. Soc. Am. J. 53:491–495.

Taylor, B.R., and H.G. Jones. 1990. Litter decomposition under snow cover in a balsam fir forest. Can. J. Bot. 68:112–120.

Terry, T.A., and J.H. Hughes. 1978. Drainage of excess water—why and how? p. 148–166. *In* W.E. Balmer (ed.) Proc. Soil Moisture—Site Productivity Symp., Myrtle Beach, SC. 1–3 Nov. 1977. Southeast For. Exp. Stn., Asheville, NC.

Teskey, R.O., and T.M. Hinkley. 1978. Impact of water level changes on woody riparian and wetland communities. U.S. Fish Wildlife Serv. Biol. Rep. FWS/OBS-78/88. U.S. Dep. Interior, Washington, DC.

Titus, B.D., and D.C. Malcolm. 1992. Nutrient leaching from the litter layer after clearfelling of sitka spruce stands on peaty gley soils. Forestry 65:389–416.

Tolonen, K., H. Vasander, A.W.H. Damman, and R.S. Clymo. 1992. Rate of apparent and true carbon accumulation in boreal peatlands. p. 319–333. *In* Proc. Int. Peat Congr., 9th, 22–26 June. Vol. 1. Uppsala, Sweden, Int. Peat Soc., Jyska, Finland.

Trettin, C.C., and M.F. Jurgensen. 1992. Organic matter decomposition response following disturbance in a forested wetland in northern Michigan, U.S.A. p. 392–399. *In* Proc. Int. Peat Congr., 22–26 June. Uppsala, Sweden. Int. Peat Soc., Jyska, Finland.

Trettin, C.C., M.R. Gale, M.F. Jurgensen, J.W. McLaughlin. 1992. Carbon storage response to harvesting and site preparation in a forested mire in northern Michigan, U.S.A. Suo 43:281–284.

Urban, N.R., S.J. Eisenreich, and E. Gorham. 1987. Proton cycling in bogs: Geographic variation in northeastern North America. p. 577–598. *In* T.C. Hutchinson and K.M. Meema (ed.) Effects of atmospheric pollutants on forests, wetlands, and agricultural ecosystems. NATO ASI Ser. Vol G16. Springer-Verlag, Berlin.

Van Cleve, K., L. Oliver, R. Schlentner, L.A. Viereck, and C.T. Dryness. 1983. Productivity and nutrient cycling in taiga forest ecosystems. Can. J. For. Res. 13:747–766.

Vasander, H. 1982. Plant biomass and production in virgin, drained, and fertilized sites in a raised bog in southern Finland. Ann. Bot. Fennici 19:103–125.

Verhoeven, J.T.A. 1986. Nutrient dynamics in minerotrophic peat mires. Aquat. Bot. 25:117–137.

Verry, E.S. 1986. Forest harvesting and water: the Lake States experience. Water Res. Bull. 22:1039–1047.

Verry, E.S. 1988. The hydrology of wetlands and man's influence on it. p. 41–61. *In* Proc. Int. Symp. Hydrology Wetlands Temperate Cold Regions, Joensuu, Finland. 6–8 June. Vol. 2. No. 4/88. Publ. Acad. Finland, Helsinki. Int. Peat Soc., Jyska, Finland.

Verry, E.S., and N.R. Urban. 1992. Nutrient cycling at Marcell Bog, Minnesota. Suo 43:147–154.

Vitousek, P.M. 1991. Can planted forests counteract increasing atmospheric carbon dioxide? J. Environ. Qual. 20:348–354.

Vompersky, S.E. 1991. Current status of forest drainage in the USSR and problems in research. p. 13–18. *In* J.K. Jeglum and R.P. Overend (ed.) Peat and peatlands diversification and innovation, Quebec City, Quebec. 6–10 Aug. 1989. Canadian Soc. Peat and Peatlands, Darmouth, Nova Scotia.

Vompersky, S.E., M.V. Smagina, A.I. Ivanov, and T.V. Glukhova. 1992. The effect of forest drainage on the balance of organic matter in forest mires. p. 17–22. *In* O.M. Bragg et al. (ed.) Peatland ecosystems and man: An impact assessment. Dep. Biol. Sci., Univ., Dundee, England.

Wieder, R.K., and J.B. Yavitt. 1991. Assessment of site differences in anaerobic carbon mineralization using reciprocal peat transplants. Soil Biol. Biochem. 23:1093–1095.

Williams, B.L., and G.P. Sparling. 1988. Microbial biomass carbon and readily mineralized nitrogen in peat and forest humus. Soil Biol. Biochem. 20:579–581.

Yavitt, J.B., G.E. Lang, and R.K. Wieder. 1987. Control of carbon mineralization to CH_4 and CO_2 in anaerobic, *Sphagnum*-derived peat from Big Run Bog, West Virginia. Biogeochemistry 4:141–157.

Zon, R., and J.L. Averell. 1929. Drainage of swamps and forest growth. Res. Bull. 89. Agric. Exp. Stn., Univ. Wisconsin, Madison.

21 Carbon Dynamics Following Clear-Cutting of a Northern Hardwood Forest

Chris E. Johnson and Charles T. Driscoll
Syracuse University
Syracuse, New York

Timothy J. Fahey
Cornell University
Ithaca, New York

Thomas G. Siccama
Yale University
New Haven, Connecticut

Jeffrey W. Hughes
University of Vermont
Burlington, Vermont

Concern regarding the increasing levels of greenhouse gases in the atmosphere has resulted in a renewed interest in the study of C cycling in terrestrial ecosystems. Recent research has focused on defining the global pools of C in soils and biomass and quantifying the potential effects of land management practices on terrestrial C sources and sinks (Detwiler, 1986; Lugo et al., 1986; Schlesinger, 1986; Goldin & Lavkulich, 1990). Forest clearing has been identified as a potential contributing factor to atmospheric CO_2 increases (Houghton et al., 1987, 1991), while the regrowth of forest lands has been suggested as a potential sink for atmospheric CO_2 (Marland, 1988; Vitousek, 1991; Lugo & Brown, 1992). Evaluating the role of forest management in the global C cycle requires both extensive analysis of patterns across biomes and comprehensive studies of effects at individual sites. To date, most of this research has focused on the possible changes in soil C pools caused by forestry practices. In two extensive studies, Allen (1985) and Johnson (1992) reviewed the literature on soil C loss after clear-cutting at sites in the Tropics and worldwide. Although it is not possible to draw

---

Copyright © 1995 Soil Science Society of America, 677 S. Segoe Rd., Madison, WI 53711, USA. Carbon Forms and Functions in Forest Soils.

clear conclusions from the data, it appears that C losses are more likely to occur in tropical soils than in temperate-zone soils.

From an ecosystem perspective, there are three major pathways for C loss during and following logging. Carbon may be exported from the ecosystem in biomass for use in wood products. The significance of this flux depends on the harvesting method used (i.e., selective cutting, stem-only clear-cutting, whole-tree harvesting) and the relative amount of C in forest biomass and soils. Accelerated loss of C via soil respiration also can be significant when postharvest conditions stimulate microbial activity. Finally, the loss of the forest canopy leads to increased water fluxes in soils, which may result in increased export of particulate and dissolved C in drainage waters. The loss of C through erosion is believed to be a major C flux associated with deforestation in the Tropics (Detwiler & Hall, 1988).

Following logging, C is sequestered in the ecosystem through photosynthetic assimilation. For a given site, forest management may thus result in net accumulation of C or loss of C depending on the balance between net photosynthesis and the various processes resulting in C export.

In Spodosols, the biogeochemistry of C and organic matter is central to soil development and soil properties, and is linked to the cycles of N, S, Al, Fe, Ca, and other major and trace elements. Thus, the disruption of the C cycle following clear-cutting or other forest management practices will have consequences beyond the direct impact on the C cycle. Indeed, even if logging has little effect on ecosystem C content there may be profound effects on the ecosystem due to redistribution of C within the soil. In Spodosols, which develop in environments of intense leaching, redistribution of C can be caused by translocation of organic matter in soil water from the O horizon into the mineral soil. Also, mechanical disturbance caused by logging machinery can result in dramatic changes in C distribution (Ryan et al., 1992). Through these processes, essential nutrients may be transported, along with C, down in the soil profile, away from the O horizon, an important source of nutrients for regrowing vegetation.

The effects of forest clearing on biogeochemical processes have been a focus of research in the northern hardwood forest ecosystem at the Hubbard Brook Experimental Forest in central New Hampshire for 25 yr (e.g., Bormann et al., 1968, 1974; Likens et al., 1970; Bormann & Likens, 1980; Martin et al., 1986). Previous manipulations of experimental watersheds at Hubbard Brook have included clear-cutting without biomass removal, stem-only clear-cutting, and strip-cutting. The present study began in 1983 with the goal of establishing a comprehensive long-term research effort aimed at determining the timing and mechanisms of nutrient reorganization following clear-cutting disturbance. In the winter of 1983–1984, Watershed 5 at Hubbard Brook was cut in a commercial whole-tree harvest. Since that time, we have been studying the changes in soils (Huntington & Ryan, 1990; Johnson et al., 1991a,b; Ryan et al., 1992), stream water chemistry (Lawrence et al., 1987; Fuller et al., 1988; Lawrence & Driscoll, 1988), soil solution chemistry (Fuller et al., 1987; Dahlgren & Driscoll, 1994), root dynamics (Fahey, et al., 1988; Fahey & Arthur, 1994), regrowing vegetation (Hughes & Fahey, 1988, 1992; Hughes et al., 1988; Mou et al., 1993), plant–animal interactions (Hughes & Fahey, 1991), and hydrology.

This contribution represents a synthesis and extension of many individual investigations of the dynamics of C in various ecosystem pools during the first 10 yr (1983–1992) of the Watershed 5 study. Our original research hypotheses regarding C dynamics were that whole-tree harvesting would result in: (i) decreased soil C pools, (ii) a large net loss of C from the forest floor (O horizon), (iii) decreased export of dissolved organic carbon (DOC) in stream waters due to coagulation with dissolved Al, (iv) increased respiratory losses of soil C, and (v) rapid recovery of C assimilation by plants.

SITE DESCRIPTION

The Hubbard Brook Experimental Forest is located in the southern White Mountain region (43°56' lat, 71°45'W long.), in the town of Woodstock, 22 km north of Plymouth, New Hampshire. Features of the experimental forest have been summarized in Likens et al. (1977) and Bormann and Likens (1980). In this paper, we present data from three experimental areas. Watershed 5, the clear-cut area, is a 22.5-ha gaged catchment which extends in elevation from about 510 to 750 m (Fig. 21–1). Watershed 6 (area = 13.2 ha), which serves as the biogeochemical reference watershed at Hubbard Brook, is adjacent to Watershed 5 and extends from about 540 to 790 m in elevation. Finally, the Bear Brook watershed is an ungaged area adjacent to Watershed 6. The experimental areas all have a generally southeastern exposure with slopes averaging about 13%.

The climate at Hubbard Brook is humid–continental, with short, cool summers, and cold winters. Average annual precipitation for Watersheds 5 and 6 is about 140 cm, of which 25 to 35% falls as snow (Federer et al., 1990). Each winter, a persistent snowpack develops to a depth of about 1.5 m. Mean annual runoff

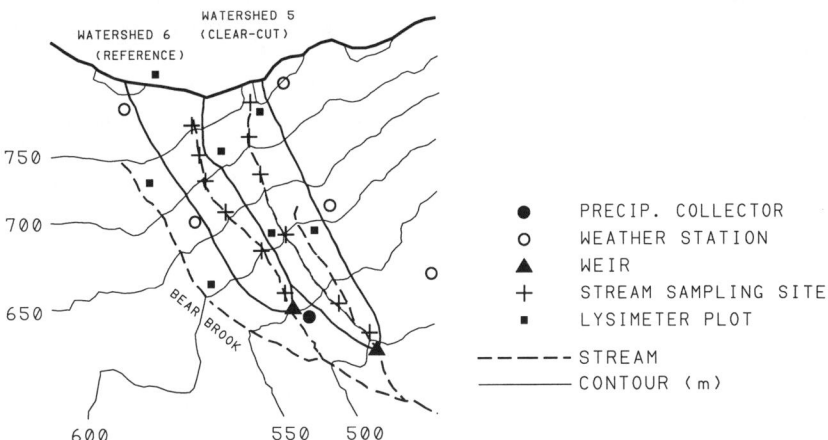

Fig. 21–1. Map of experimental Watersheds 5 and 6 at the Hubbard Brook Experimental Forest, New Hampshire.

from Watershed 6 is 90 cm, with about 50% occurring as snowmelt (Federer et al., 1990).

The Hubbard Brook Experimental Forest is underlain by gneissic rocks of the Rangely formation, which is extensively intruded by quartz monzonites of the Kinsman formation. Bedrock is covered by a mantle of felsic, base-poor glacial till, deposited during the Wisconsinan period ($\cong 14000$ yr ago), and ranging in depth from zero to several meters (Johnson et al., 1981). Soils developed in the till are highly acidic (pH<4.5; Johnson et al., 1991b) Haplorthods and Fragiorthods of the Tunbridge, Lyman, Berkshire, Skerry, and Becket series (Huntington et al., 1988; Johnson et al., 1991a), with an average solum depth of about 60 cm (Johnson et al., 1991a).

The dominant vegetation type at Hubbard Brook is the northern hardwood forest, which is dominated by American Beech (*Fagus grandifolia* Ehrh.), sugar maple (*Acer saccharum* Marsh.), and yellow birch (*Betula alleghaniensis* Britt.). At high elevation and in isolated pockets, the spruce-fir forest type is present, with red spruce (*Picea rubens* Sarg.), balsam fir (*Abies balsamea* [L.] Mill), and white birch (*B. papyrifera* var. *cordifolia* [Marsh.] Regel). The experimental area was extensively logged around 1910 to 1919. There is no evidence of fire disturbance.

EXPERIMENTAL METHODS

Whole-Tree Harvesting

In the autumn and winter of 1983–1984, all trees on Watershed 5 which were ≥5-cm diam.-at-breast-height (dbh) were cut with chain saws and feller-bunchers. Whole trees (boles and branches) from the lower two-thirds of Watershed 5 were removed by rubber-tired skidders during the winter of 1983–1984. Trees from most of the upper one-third were removed in the summer of 1984. Felled trees in one small area ($\cong 12\%$) were not cleared until 1985, and 3% of the watershed was not cleared at all. In addition, trees were never felled in a "buffer" strip along the watershed boundary.

The clear-cutting operation was intended to be a commercial whole-tree harvest, with very few restrictions imposed for scientific purposes. The cutting and clearing were performed by local operators, with the single restriction that disturbance was to be minimal in twelve 25- by 25-m "biomass plots." Thus, we view the results of the Watershed 5 experiment to be largely indicative of the consequences of commercial clear-cut logging of this landscape.

Biomass Estimation

In the summer of 1982, the dbh was measured for all trees on Watershed 5 with dbh ≥10 cm, and a subsample of trees with 2 cm ≤ dbh < 10 cm. Above- and belowground biomass was estimated using the dimensional analysis method of Whittaker et al. (1974), with revised allometric equations (T.G. Siccama, 1992, unpublished data). Revisions to the Whittaker et al. (1974) equations were based in part on allometric analyses performed on trees cut from Watershed 5.

The concentration of C in aboveground plant tissues was assumed to be 50% for budgetary calculations. The concentration of C in roots was assumed to be 52% (Fahey et al., 1988).

Biomass in trees removed from the 12 biomass plots was measured directly. After skidding, the trees were chipped and loaded into trucks. Total biomass was estimated by difference in weight of the empty and loaded trucks.

Aboveground biomass in regrowing vegetation was estimated in post-harvest Years 1 to 6 by collecting all herbs, shrubs, seedlings (woody plants <50-cm height), and saplings (>50 cm) in 100 1- by 1-m plots randomly located throughout Watershed 5. Because of the increasing density of saplings, sapling biomass in Year 7 was estimated by measuring the basal diameter of saplings with dbh >2 cm, and applying allometric equations developed from seven individuals of each major species. The dbh of saplings also was measured in Year 7 on 99 circular plots (5-m diam.). Belowground biomass was estimated from aboveground boimass by applying species-specific root-shoot ratios, estimated by excavation in Year 3 of regrowth (Mou et al., 1993).

Soils and Roots

Soils were intensively sampled on Watershed 5 prior to clear-cutting and in post-harvest Years 3 and 8 (1986 and 1991). Sixty soil pits were excavated in each sampling year, at sites randomly selected within six elevation strata. Details of the quantitative sampling method are available in Huntington et al. (1988) and Johnson et al. (1991a). Briefly, soils were collected in depth increments using a 0.5-m^2 (0.707- by 0.707-m) template. The Oi and Oe layers were collected as a single sample, followed by the Oa horizon. Mineral soil was sampled in three layers: 0 to 10, 10 to 20, and 20 cm to C or bedrock. Soils were passed through a 13-mm (0.5-in.) screen in the field. All rocks (>13 mm) and soil removed from the pits were weighed in the field. Mineral soils and thick Oa horizons were subsampled. Organic samples (Oi+Oe, Oa) were air dried and passed through a 5-mm stainless-steel sieve. Air-dried mineral soil samples were passed through a 2-mm stainless-steel sieve. Using this quantitative method, we developed precise estimates of soil masses on an areal basis (Johnson et al., 1991a), which could be used to estimate soil chemical pools. Standard errors for soil masses ranged from 5 to 10% of the mean for mineral soils, and 10 to 15% of the mean for the O horizon. Thus, we could detect changes in element pools of approximately 10 to 30% (Johnson et al., 1990). Woody roots were sorted from these samples to estimate lateral root biomass (see below). Many of the pits (29 in 1983, 48 in 1986, 59 in 1991) also were mapped and sampled by horizon using the criteria of the Soil Survey Staff (1975).

Soil C was measured using the method described in Huntington et al. (1988). About 10 to 20 g of soil was oven dried (105°C) and ground. A 2- to 50-mg subsample was then combusted in pure O_2, and the resultant gases analyzed chromatographically using a Carlo Erba Model 1500 (1983, 1986 samples) or EA 1108 (1991 samples) elemental analyzer (Carlo Erba Instruments/Fisons Instruments, Danvers, MA). About 10% of the samples were replicates or standards.

In addition to the quantitative pits, loss of soil organic matter from the forest floor (O horizons) was estimated in situ. Approximately 200 15- by 15-cm forest floor blocks were excavated in the early summer of 1984 using a wooden template (Johnson et al., 1982; Federer et al., 1993). The blocks were weighed moist and a subsample was collected for moisture determination. The samples were then enclosed in pouches made from fiberglass grain sacks and returned to Watershed 5 for incubation. The pouches effectively excluded entry of new roots. A nylon mesh screen was placed on the blocks to keep out fresh litter while allowing water infiltration. Blocks were resampled in postharvest Years 2, 3, 4, and 7.

Roots were collected from: (i) field screens, (ii) laboratory sieves, and (iii) dried soil samples. Root samples were sorted into five size classes (0.6–1.0, 1.0–2.5, 2.5–5.0, 5–10, and >10 mm), and weighed. Using the various soil mass estimates, total lateral root biomass was computed (Fahey et al., 1988; Fahey & Arthur, 1994).

Root decomposition rates were estimated by incubating fresh fine roots (<0.6 mm) in nylon bags within the forest floor at randomly selected locations within Watershed 5; small woody roots (0.6–10 mm) and large woody roots (10–100 mm) were tethered on nylon fishing line and placed in direct contact with soil approximately 10 cm below the O horizon. Samples were collected periodically for weight-loss determinations. Decay of large woody-roots also was inferred from differences in specific gravity between living roots and samples collected from a chronosequence of clearcuts at Hubbard Brook ranging in age from 10 to 19 yr. Also, samples of fresh large woody roots were trimmed at the edges, tethered with fishing line, and incubated in situ in Watershed 5 and measured for density and weight loss periodically after clearcutting. Details of the root decomposition research methods can be found in Fahey et al. (1988) and Fahey and Arthur (1994).

Solution Chemistry

Tension-free lysimeters were installed at four sites within Watershed 5 and three reference sites in the Bear Brook watershed in the fall of 1983. Lysimeters were placed near the bottom of the Oa, Bh, and within the Bs2 soil horizons at sites in the spruce-fir, high-elevation hardwood, and low-elevation hardwood zones (Fig. 21–1). Three replicate lysimeters were placed in each of the three horizons at the hardwood zone sites, while duplicates were placed in each horizon at spruce-fir zone sites. The depths of the lysimeters were 3 to 6 cm, 19 to 26 cm, and 40 to 49 cm below the ground surface for Oa, Bh, and Bs lysimeters, respectively. Two types of lysimeters were used: (i) 10-cm polyethylene funnels filled with acid-washed silica sand, and (ii) rectangular polyethylene containers (commercially available sandwich-holders, 20 by 3.5 by 14.5 cm) filled with acid-washed teflon beads. The lysimeters drained into 1- or 2-L polyethylene bottles. Through 1985, no statistically significant differences were detected between lysimeter types (Driscoll et al., 1988). Thus, samples were bulked in the field starting in 1986.

Stream samples were collected at 12 sites, 6 each along elevational gradients within Watersheds 6 and 5 (Fig. 21–1). Due to ephemeral conditions, the sample numbers vary between sampling sites. In general, fewer samples were obtained from higher-elevation sites during the summer period.

Stream sampling began in the summer of 1982, while lysimeter sampling began in the spring of 1984. All solution samples were collected in polyethylene bottles and kept refrigerated until analysis. Analyses for pH, dissolved inorganic carbon (DIC) and dissolved organic carbon (DOC) were usually performed within 3 d of collection. Samples were collected on approximately monthly intervals. The data reported here include samples collected through December 1992.

Stream and soil solution samples were filtered using glass-fiber filters. Dissolved inorganic carbon was measured by acidification, followed by infrared CO_2 detection. Dissolved organic carbon was analyzed (after removal of DIC) by ultraviolet-enhanced persulfate oxidation of organic matter, followed by infrared CO_2 detection (Dohrmann Instruments, Xertex Corp., Santa Clara, CA).

Carbon Dioxide Evolution

Evolution of CO_2 from the soil surface was measured using a soda-lime trap method (Edwards, 1982), with precautions to avoid bias as prescribed by Nadelhoffer and Raich (1992). Monthly measurements were made from May to November at 20 random positions along transects from about 500- to 650-m elevation in Watershed 5 and in the adjacent mature forest in 1991–1993. For each measurement, chambers were left in place for 48 h. Annual flux of CO_2-C was calculated by integrating under the curves for each site, assuming that mid-November values were maintained through the winter months.

Statistical Analysis

The Student's two-sample t-test was used to detect significant differences between preharvest and postharvest soil pools and concentrations. The Student's two-sample t-test also was used to test hypotheses concerning differences in mean C concentrations in soil solutions and stream waters between Watershed 5 and 6. Samples from replicate lysimeters were treated as independent observations. Tests involving solution concentrations were performed on nontransformed data. For computation of solution fluxes, solution concentrations were first volume-weighted, based on the total flow at the Watershed 5 or 6 weir for the sample month. There is a small, but significant, positive relationship between DOC and flow at Hubbard Brook (Lawrence & Driscoll, 1990).

We acknowledge that the experimental design used in this study suffers from pseudoreplication. This is an inherent problem with watershed manipulation studies, which can only be replicated at enormous expense (even then replication is less than perfect). As Hurlbert (1984) pointed out, the use of standard statistical methods to test for *treatment effects* is inappropriate in the presence of pseudoreplication. Therefore, in a strict sense, the statistical tests performed in this study only show the significance of *differences between means*. In interpreting these data, it is our contention that clear-cutting and associated recovery are the most important factors influencing the observed differences in the biogeochemistry of these adjacent watersheds.

RESULTS

In 1982, the total amount of forest biomass (live + dead) on Watershed 5 with dbh ≥ 2 cm was 269 Mg ha^{-1} (Table 21–1). According to Siccama et al. (1970), the biomass of herbs, shrubs, and trees <2 cm at Hubbard Brook is less than 2% of the total forest biomass. Aboveground, living biomass accounted for 77% of the total biomass pool, while roots contributed 15% and standing dead trees were 7% (Table 21–1). Bole wood and bark accounted for 71% of the aboveground biomass in living trees. Thus, whole-tree harvesting had a potential wood-fiber yield 41% higher than stem-only clear-cutting.

Sugar maple accounted for approximately 40% of the total aboveground living biomass, followed by American beech and yellow birch, with about 22% each (Table 21–1). The spruce-fir forest type accounted for 10% of the aboveground living biomass, with only about one-half of that fraction as conifers.

Of the 4610 Mg of estimated "harvestable" biomass on Watershed 5 in the summer of 1983, approximately 3690 Mg were actually removed from the ecosystem during the clear-cut (Table 21–2). The buffer strip and the steep area not skidded accounted for a total area of 2.6 ha. Thus, the allometrically estimated biomass export from areas which were felled and skidded was 185 Mg ha^{-1}. This value is in remarkably close agreement (<3% discrepancy) with biomass removal from the biomass plots, as measured by the weights of chip trucks (Table 21–2). It appears that the allometric equations relating dbh to biomass at Hubbard Brook can be used to estimate aboveground biomass pools extremely well.

Based on these values, the estimated C export in wood fiber was 93 Mg ha^{-1}. Residual biomass C pools include: 8 Mg ha^{-1} in trees in the buffer strip (Table 21–2), 20 Mg ha^{-1} in the roots of felled trees (Table 21–1), 4 Mg ha^{-1} in trees felled but not skidded (Table 21–2), and 8 Mg ha^{-1} in "slash," the twigs and limbs broken from trees during skidding (Table 21–2).

By 1990, the 7th yr after harvesting, the amount of living biomass on Watershed 5 had increased to 19.3 Mg ha^{-1}, 8% of the precut biomass pool (Table 21–3). Most of the accumulation was in trees, with shrubs and herbs accounting for only about 6% of biomass after 6 yr (Mou et al., 1993). Based on these data, the estimated C accumulation in forest biomass averaged 1.9 Mg ha^{-1} yr^{-1} in the first 3 yr, and 3.4 Mg ha^{-1} yr^{-1} after Year 4. Using the root-shoot ratios of Mou et al. (1993), the aboveground C accumulation averaged 1.3 Mg ha^{-1} yr^{-1} in Years 1 to 3 and 2.4 Mg ha^{-1} yr^{-1} in Years 4 to 7; belowground C accumulation averaged 0.6 Mg ha^{-1} yr^{-1} in Years 1 to 3, and 1.0 Mg ha^{-1} yr^{-1} in Years 4 to 7 (Table 21–3).

Three sets of data can be used to assess the fate of forest floor organic matter after clear-cutting. Seven years after logging, isolated forest floor blocks showed a total mass loss (dry basis) of 22.2% in the hardwood forest, and 27.2% in the spruce-fir zone (Fig. 21–2, weighted average = 23.2%). According to these data, the forest floor soil organic matter pool does not appear to have reached a minimum as of Year 7. Direct measurement of the forest floor C pool using the soil pit data showed an initial increase after 3 yr, followed by a decline by Year 8 (Table 21–3). The increase between 1983 and 1986 was not statistically significant. However, both the decline between 1986 and 1991 and the overall loss

Table 21–1. Biomass in trees (≥2-cm dbh) on Watershed 5 at the Hubbard Brook Experimental Forest in 1982, prior to whole-tree harvest clear-cutting.

Species	Bole wood	Bark	Branch	Total aboveground†	Root	Total live	Dead
				Mg ha^{-1}			
Sugar Maple	53.5	6.6	20.0	82.2	15.3	97.5	2.3
Beech	28.8	1.8	14.2	46.0	9.3	55.3	5.7
Yellow Birch	24.0	2.9	15.7	45.6	9.0	54.6	8.8
White Ash	4.7	0.29	2.7	7.9	1.2	9.1	0.1
Balsam Fir	4.2	0.57	1.6	6.8	2.3	9.1	1.1
Red Spruce	2.8	0.39	1.1	4.7	1.6	6.3	0.9
White Birch	5.6	0.70	3.4	10.0	2.1	12.1	0.4
Total‡	125.8	13.6	60.5	207.0	41.6	248.6	20.0

†Includes twigs and leaves.
‡Includes minor species.

Table 21–2. Exported and residual biomass on Watershed 5 at the Hubbard Brook Experimental Forest during whole-tree harvest clear-cutting.

Biomass component	Mass	Watershed area
	Mg	ha
1. 1982 Aboveground living biomass ($\geq$ 10-cm dbh)	4480	22.5
2. Estimated increment: 1982–1983	132	22.5
3. Harvestable Biomass (1 + 2)	4610	22.5
4. Trees left in "buffer strip"	368	1.8
5. Total biomass cut (3–4)	4250	20.7
6. Trees cut but not skidded	180	0.8
7. "Slash" remaining (1984)	380	19.9
8. Total estimated biomass removal [5 – (6 + 7)]	3690	19.9
Estimated removal (Mg ha^{-1})	185	
Measured removal (biomass plots, Mg ha^{-1})	180	

Table 21–3. Accumulation of biomass in vegetation on Watershed 5 following whole-tree harvest clear-cutting.

Year	Biomass		Increment	
	Aboveground	Belowground	Aboveground	Belowground
	Mg ha^{-1}		Mg ha^{-1} yr^{-1}	
1985	1.4	0.6	0.7	0.3
1986	4.0	1.7	2.6	1.1
1988	9.3	4.0	2.7	1.2
1990	13.5	5.8	2.1	0.9

(1983–1991) were statistically significant (Table 21–3). The total C loss from the forest floor, based on the soil pit data, was approximately 25%, a value similar to the mass loss observed in the in situ blocks. Finally, there were no significant differences in the concentration of DOC in Oa horizon soil water between lysimeters in Watershed 5 and the Bear Brook watershed during the first 3 yr after cutting (Fig. 21–4). However, DOC concentrations have been elevated in Watershed 5 since postharvest Year 4 (1987). High variability makes it difficult to detect even large changes. However, significantly higher concentrations were noted in Oa horizon soil water in postharvest Years 5 and 7 (1988, 1990).

There were no detectable changes in the pool of C in mineral soils following clear-cutting (Table 21–4). With this intensive sampling effort, we could have detected a change in total mineral soil C of 10 to 20% (Johnson et al., 1990). However, the C contents of the 10 to 20 cm and 20 cm to C mineral soil layers were significantly lower 8 yr after cutting than prior to cutting.

The mean C content of the whole solum (forest floor + mineral soil) did not show a significant change in the first 3 yr after logging. After 8 yr, an estimated decrease of 20 Mg ha^{-1} was significant at the $P \leq 0.10$ level, but not at the $P \leq 0.05$ level.

Carbon was redistributed within the mineral soil following clear-cutting. A C-rich (8.4%) A horizon, classified Ap (Federer, 1982), was observed in 25% of

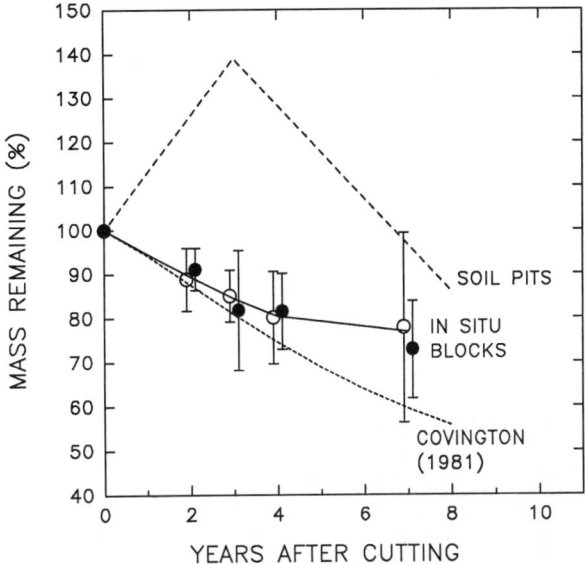

Fig. 21–2. Mass loss from forest floor blocks incubated on Watershed 5 after clear-cutting. Loss from blocks in the spruce-fir and hardwood zones are shown by filled and open circles, respectively. The weighted average for Watershed 5 is indicated by the solid line. Values based on quantitative soil pits and Covington (1981) also are shown.

the postharvest pits, while A horizons are rare in the undisturbed forest (Johnson et al., 1991a). Also, the concentration of C increased in the E horizon in the first 8 yr after logging (Table 21–5). The Bs2 horizon, which comprises the bulk of the solum, showed no significant change in C concentration.

Patterns in the concentrations of DOC in Bh and Bs soil solutions contrasted somewhat (Fig. 21–3). In the Bh horizon, DOC was elevated in Watershed 5 soil

Table 21–4. Soil C pools prior to (1983) and following (1986, 1991) whole-tree harvest clear-cutting on Watershed 5 at the Hubbard Brook Experimental Forest (1983, 1986 data from Huntington & Ryan, 1990). Sample size was 59 in 1983 and 1986, 60 in 1991.

| | Year | | |
Layer	1983	1986	1991
		Mg ha^{-1}	
Oi + Oe	11 (0.9)†	8.9 (1.6)	8.2 (1.2)
Oa	20 (2.5)	31 (8.5)	14 (2.2)
0–10 cm mineral	32 (1.2)	34 (1.5)	32 (1.5)
10–20 cm mineral	29 (1.4)	31 (1.9)	23 (1.4)*
20 cm–C horizon	82 (6.6)	75 (6.6)	63 (5.9)*
Total O horizon	30 (3.1)	39 (9.3)	22 (2.5)*
Total mineral soil	130 (8.1)	140 (8.6)	118 (6.2)
Whole solum	160 (8.7)	180 (12.7)	140 (6.7)

*Significantly different ($P \leq 0.05$) from the 1983 (preharvest) value.
†Values in parentheses are standard errors.

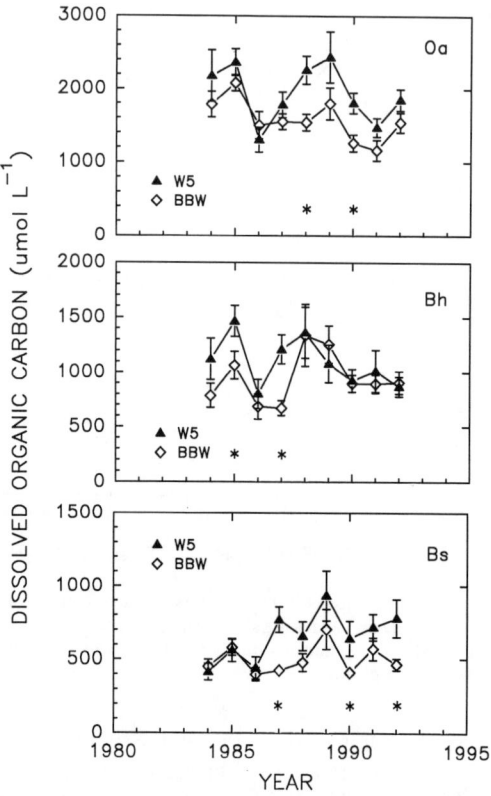

Fig. 21–3. Annual average dissolved organic carbon (DOC) concentrations in soil solutions collected from zero-tension lysimeters in Bear Brook Watershed (BBW, uncut) and Watershed 5 (W5, clear-cut). Error bars are used to denote one standard error. Significant differences ($P \leq 0.05$) between the two catchments are indicated by an asterisk (*).

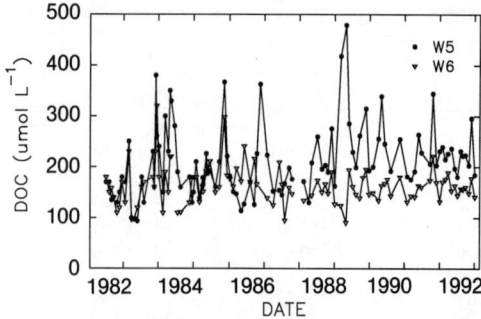

Fig. 21–4. Dissolved organic carbon (DOC) concentrations in stream samples collected at the outlets of Watershed 5 (W5, clear-cut) and Watershed 6 (W6, uncut) at the Hubbard Brook Experimental Forest.

Table 21–5. Soil C and organic matter concentrations (loss-on-ignition) by horizon on Watershed 5 prior to (1983) and following (1986, 1991) whole-tree harvest clear-cutting.

Horizon	C			Organic matter			C: OM		
	1983†	1986‡	1991	1983§	1986§	1991	1983¶	1986	1991
	%			%					
Oi + Oe	47.0	38.0*	45.1	87.1	73.4*	77.9*	0.56	0.52	0.58
Oa	30.0	22.0*	27.0*	51.6	41.6*	47.2	0.54	0.53	0.64
A / Ap	ND#	ND	8.5	14.8	24.6	16.9	ND	ND	0.48
E	2.5	ND	3.1	4.6	2.9*	6.4	0.59	ND	0.50*
Bh	6.8	ND	9.2*	13.8	14.5	19.7*	0.52	ND	0.46*
Bs1	6.3	ND	6.8	13.6	12.3	15.1	0.48	ND	0.45*
Bs2	3.6	ND	3.5	8.1	8.0	8.5	0.43	ND	0.41
C	1.4	ND	0.9*	3.8	3.6	2.7	0.37	ND	0.33

*Significantly different ($P \leq 0.05$) from the preharvest (1983) mean.
†Huntington et al. (1988).
‡Huntington & Ryan (1990).
§Johnson et al. (1991a).
¶Huntington et al. (1989).
ND = not determined.

solutions, relative to Bear Brook watershed, in the first 4 yr after clear-cutting, with significant differences detected in postharvest Years 2 and 4 (1984, 1987). After Year 4, the two watersheds showed nearly identical patterns. In the Bs horizon, elevated DOC concentrations were observed in Watershed 5 starting in postharvest Year 4 (1987), with significant differences noted in Years 4, 7 and 9 (1987, 1990, 1992). Since 1987, the annual mean DOC concentration in Bs soil solutions has, on average, been 50% greater in Watershed 5 than Bear Brook watershed. The average difference in mean annual DOC concentration over that period has been 240 μmol L^{-1}.

Using allometric relationships, the biomass in tree roots on Watershed 5 prior to clear-cutting was estimated to be 41.6 Mg ha^{-1} (Table 21–1). According to Whittaker et al. (1974), lateral root biomass accounts for 65% of total root biomass (the rest being root crowns). Thus, our allometric estimate of lateral root biomass is 27.0 Mg ha^{-1}. This value is remarkably close to the measured value for Watershed 5 (26.8 Mg ha^{-1}; Fahey et al., 1988). Based on a C concentration of 52% (Fahey et al., 1988), the C pool in belowground biomass prior to logging was 21.6 Mg ha^{-1}. In the precut forest, most of the biomass in woody roots was found in the larger-diameter size classes (<1 mm, 5%; 1–10 mm, 39%, >10 mm, 56%; Fahey et al., 1988). The rate of mass loss (and C loss) decreased with increasing root diameter, but the general patterns were similar for all size classes. Mass loss from small woody roots ranged from 20 to 28% in the first 2 yr after logging. For large woody roots, the loss after 5 yr ranged from 45 to 63% (Fahey & Arthur, 1994). The values for large woody roots are probably somewhat overestimated (Fahey & Arthur, 1994). For budget calculations, we assumed an average C loss from all decomposing roots of 30% after 3 yr, and 60% after 8 yr. The associated C flux to the soil or soil solution was 2.2 Mg ha^{-1} yr^{-1} during the first 3 yr, and 1.7 Mg ha^{-1} yr^{-1} during postharvest Years 4 to 8.

The response of DOC concentrations in stream water was delayed (Fig. 21–4). Prior to harvesting, the two streams had very similar DOC concentrations. There appears to have been an initial response immediately after harvesting, in the first half of 1984. Between 1985 and 1987, the two streams had similar DOC concentrations. Since the summer of 1987, Watershed 5 has had consistently higher DOC concentrations in stream water than Watershed 6 (Table 21–6). Significantly higher average stream DOC concentrations were observed in Watershed 5 in 1984 and 1988–1992.

The DOC concentrations can be coupled with water flow to compute DOC fluxes from the two catchments (Table 21–6). Increased hydrologic losses in the first several years after cutting, a commonly observed phenomenon (Hornbeck et al., 1970, 1986; Douglas & Swank, 1972), resulted in elevated DOC export from Watershed 5, relative to Watershed 6. In later years, water flows in the catchments were similar, but the higher DOC concentrations in Watershed 5 contributed to the continued elevated DOC fluxes. When summed for 1984–1992, the total export of DOC in Watershed 5 stream water since clear-cutting was about 57.4 kg ha^{-1} (or 33%) greater than for Watershed 6 stream water.

For the period 1991–1993, estimated efflux of CO_2 from the soil was 20% greater in the clear-cut watershed than in the adjacent, uncut forest. The average flux in Watershed 5 was 5.7 Mg C ha^{-1}, while in the uncut forest it was 4.7 Mg C ha^{-1}.

Table 21-6. Annual mean concentrations of DOC and estimated fluxes in stream water collected monthly at the outlets of Watersheds 5 (W5) and 6 (W6) between 1982 and 1992. Watershed 5 was clear-cut in the fall/winter of 1983–1984.

Year	DOC concentration			Flux†	
	W5	W6	Significance	W5	W6
	—— µmol L^{-1} ——			—— kg C ha^{-1} ——	
1982	155	145	—	12.3	12.9
1983	193	177	—	33.8	25.7
1984	223	150	*	16.8	12.7
1985	175	159	—	20.0	21.3
1986	178	189	—	16.6	15.8
1987	174	148	—	14.7	13.6
1988	199	154	*	28.1	12.5
1989	297	153	**	34.1	20.3
1990	248	159	**	30.0	21.8
1991	225	159	**	24.8	17.2
1992	221	161	**		
Total 1984–1992	216	159		231	174

†Fluxes computed for water years beginning June 1.

DISCUSSION

Soil Carbon Dynamics

Covington (1981) hypothesized that decreased litter inputs and accelerated decomposition following clear-cutting result in decreases up to 50% in the forest floor soil organic matter content in hardwood stands after about 15 yr. His observations, based on samples from a chronosequence of stands ranging in age from 3 to 70 yr have been critical in guiding research in forest floor dynamics in northern hardwood stands (Aber et al., 1978; Federer, 1984; Snyder & Harter, 1987). Data from the Watershed 5 study are in certain respects consistent with Covington's hypothesis, yet lead to the observation that physical site disturbance caused by logging machinery may be equally important to forest floor soil organic matter and C dynamics.

Forest floor blocks incubated in Watershed 5 after clear-cutting showed mass losses of up to 27% in the first 7 yr (Fig. 21–2). These values are probably overestimates because of disturbance associated with the original collection of the blocks. Also, new litter and roots were excluded from the blocks, resulting in further bias in the estimate of mass loss. Even with the overestimation of mass loss, the observed values are less than the rate reported by Covington (1981). Nevertheless, it is apparent from the in situ blocks that conditions in Watershed 5 have resulted in significant reductions in mass of the pre-existing forest floor. This observation is supported by the higher DOC concentrations observed in Oa horizon soil solutions on Watershed 5, relative to the Bear Brook watershed (Fig. 21–3). Also, respiration rates from the forest floor were greater in the clear-cut forest than in the reference forest.

Direct measurement using soil pits also revealed a significant 8-yr decline in the forest floor C pool (Table 21-4). This decrease was the result of decreases in both the concentration of C in O horizons (Table 21-5) and in forest floor mass (87 Mg ha^{-1} in 1983, 75 Mg ha^{-1} in 1991).

Based on the in situ block data, it is clear that decomposition is a major contributing factor to the decline in the forest floor C pool. However, it also is clear that physical disturbance plays an important role. Logging-related disturbance on Watershed 5 was extensive, perhaps amplified by the long, narrow shape of the catchment. In a transect study, Ryan et al. (1992) reported that only 35% of Watershed 5 was undisturbed by logging equipment, while 17% was depressed or scarified, and the remaining 48% of the catchment was scalped, rutted, or mounded. Evidence that this activity resulted in extensive mixing of mineral soil and forest floor includes: (i) significantly lower loss-on-ignition and C concentration in the forest floor following cutting (Table 21-5), (ii) significantly higher forest floor mass 3 yr after cutting (Johnson et al., 1991a), (iii) significantly higher exchangeable Al in the Oa horizon (Johnson et al., 1991b). Thus, C and organic matter from the forest floor was "exported" to the upper mineral soil as mineral-rich material was "imported" to the forest floor.

Mixing of mineral soil and forest floor also influenced C distribution in the mineral soil; in 25% of the post-harvest soil pits, an A horizon was observed. Further down in the mineral soil profile, the concentration of soil organic matter (as estimated by loss-on-ignition), decreased significantly in the E horizon in the first 3 yr after clear-cutting, while other horizons remained unchanged (Table 21-5: Johnson et al., 1991a). Between postharvest Years 3 and 8, loss-on-ignition increased significantly in the E, Bh, and Bs1 horizons, but remained unchanged in the Bs2 horizon, which comprises the bulk of the solum (Johnson et al., 1991a). Carbon concentration data were not available for soil horizon samples collected in 1986. However, a comparison of preharvest and Year 8 samples shows a significant increase in C concentration in the Bh horizon. Based on these data, it appears that soil organic matter and C were mobilized from the forest floor and upper mineral soil (E horizon) immediately after cutting (postharvest Years 1-3), while, in later years, soil organic matter and C were accumulated in the upper mineral soil (E, Bh, and Bs1).

Patterns in Stream and Soil Water Dissolved Organic Carbon

There are striking temporal trends in postharvest DOC concentrations in stream and soil water. The concentration of DOC in Bh horizon soil solutions in Watershed 5 were elevated relative to Bear Brook watershed in the first 4 yr after clear-cutting (Fig. 21-3). The opposite pattern was found in Bs horizon soil solutions, where DOC concentrations were similar immediately after cutting, but elevated in Watershed 5 starting in 1987. In Oa horizon soil solutions, DOC concentrations were generally higher in Watershed 5 throughout the first 9 yr, with the largest and most significant differences occurring in postharvest Years 5 to 7 (1988-1990). Thus, it appears that in the first 3 yr after clear-cutting, the dominant process in soil water was the increased downward transport of DOC from the upper mineral soil and immobilization in the lower mineral soil. This pattern is consistent with the observed decrease in the loss-on-ignition of the E horizon.

Starting approximately 4 yr after cutting, however, losses from the upper mineral soil (Bh) were no longer elevated in Watershed 5 relative to the Bear Brook watershed, while Oa DOC concentrations in Watershed 5 were increasing relative to Bear Brook. This change, coupled with the observed increases in the loss-on-ignition of E and Bh horizons between postharvest Years 3 and 8, indicate that immobilization of DOC in the upper mineral soil was an important DOC sink in later years of the study.

Soil water DOC concentrations in the Bs horizon were nearly identical in Watershed 5 and the Bear Brook watershed the first 3 yr after clear-cutting (Fig. 21–3). After postharvest Year 3 the DOC concentrations in Bear Brook watershed remained relatively constant, while DOC in Watershed 5 increased, with significant differences observed in 3 of the last 6 yr. In Watershed 5 during postharvest Years 5 to 9, the concentration of DOC in Bs horizon soil solutions averaged about 750 µmol L^{-1}, while in the Bh horizon the average DOC concentration was about 1050 µmol L^{-1}. In the Bear Brook Watershed in the same period, the average DOC concentrations in Bs and Bh horizon soil solutions were about 530 µmol L^{-1} and 1060 µmol L^{-1}, respectively. Thus, under undisturbed conditions, about 530 µmol L^{-1} of DOC was immobilized or oxidized in the lower mineral soil, neglecting evapotranspiration and other hydrologic effects, while in the clear-cut ecosystem, about 300 µmol L^{-1} of DOC was immobilized or oxidized.

We hypothesize that the difference in DOC behavior in Bs horizon soil solutions is due to abiotic processes in the soil. Specifically, the lower retention of DOC in Watershed 5 is consistent with the mechanism of DOC adsorption/desorption in Hubbard Brook soils. In a laboratory experiment, Romkens et al. (1993, unpublished data) observed that DOC concentrations in solutions equilibrated with Hubbard Brook Bs soils were highly pH dependent, with a minimum DOC concentration around pH 4.2, and increasing DOC at higher and lower pH values. Since clear-cutting, we have observed significant increases in the pH of Bs horizon soils (Johnson et al., 1991b) and soil waters (Dahlgren & Driscoll, 1994), which is consistent with the solution DOC concentrations.

Stream and Bs horizon soil solution DOC concentrations followed similar patterns. Stream DOC concentrations at the outlets of Watershed 5 and 6 were nearly equal prior to and immediately after clear-cutting, with the exception of 1984, the 1st yr postharvest (Fig. 21–4, Table 21–6). Significantly higher DOC concentrations were observed starting in the 5th yr after cutting. The elevated DOC concentrations in the Watershed 5 stream appear to have been relatively uniform temporally and spatially. An analysis of longitudinal trends showed that DOC concentrations were elevated after 1987 at all stream sampling sites within Watershed 5 except at the uppermost site, which drains the spruce-fir zone (Fig. 21–5). Also, DOC concentrations at the Watershed 5 outlet increased after 1987 in all seasons of the year (Fig. 21–6).

We suggest that the observed increases in stream DOC concentrations are largely the result of higher DOC concentrations in Bs horizon soil solutions. Several observations are consistent with this hypothesis. First, the temporal patterns in DOC concentration were similar between Bs soil solutions and stream waters. Second, the spruce-fir zone did not contribute to the increased stream DOC concentrations. Soils in the spruce-fir zone are often shallow, and it has

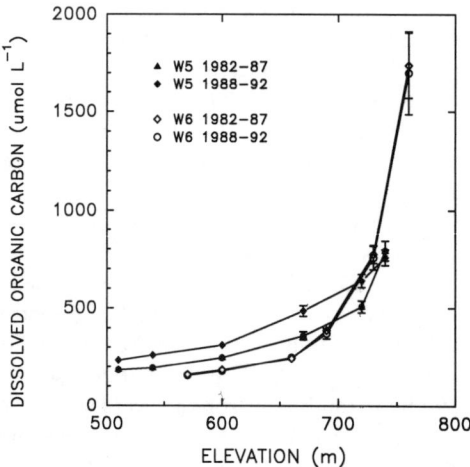

Fig. 21–5. Elevation trends in DOC concentrations in stream waters draining Watersheds 5 (W5, clear-cut) and 6 (W6, uncut). Error bars are used to indicate one standard error, and are often smaller than the symbol.

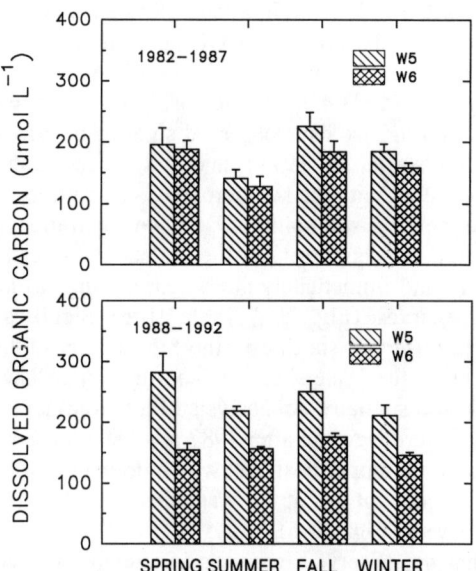

Fig. 21–6. Seasonal trends in stream water DOC concentrations in Watersheds 5 (W5, clear-cut) and 6 (W6, uncut): spring, April to May; summer, June to August; Fall, September to November; winter, December to March.

been suggested that the contribution of deeper soil waters to stream flow is less important in this area than in the hardwood zones (Lawrence et al., 1988; Lawrence & Driscoll, 1990). Thus, if processes deep in the soil are responsible for elevated DOC concentrations, the spruce-fir zone would be expected to show a smaller response. Third, the absence of a seasonal pattern in the DOC response is consistent with the hypothesis that the relevant process is abiotic.

Root Decomposition and Carbon Dynamics

Comprehensive studies of root turnover and decomposition are rare due to difficulties associated with sampling and analysis. However, it is clear that the release of C and nutrients from decaying roots is a quantitatively important flux in the early years after clear-cutting (Fahey et al., 1988; Fahey & Arthur, 1994). The estimated annual release of C by root decomposition (2.2 Mg ha^{-1} in postharvest Years 1–3, and 1.7 Mg ha^{-1} in Years 4–8) represents a large flux relative to solution fluxes (see next section). Therefore, the C released from roots must have been retained in the soil or lost from the ecosystem via respiration. It is not possible to separate the two processes quantitatively. Note that although the total estimated C release from roots after 8 yr (15.1 Mg ha^{-1}) represents a significant fraction (9%) of the total soil C pool (179 Mg ha^{-1}), it is not large enough to have been detected by our sampling method.

Based on the soil respiration data, CO_2 flux at the soil surface in the uncut forest at Hubbard Brook is about 4.7 Mg C ha^{-1} yr^{-1}, while the CO_2 flux in Watershed 5, 8 to 10 yr after clear-cutting was estimated to be 5.7 Mg C ha^{-1} yr^{-1}. This difference compares favorably to data presented by Hendricksen et al. (1989), who observed a difference of 0.6 Mg C ha^{-1} between clear-cut and uncut plots over a 100-d period in a northern mixed forest. It is clear that a modest increase in soil respiration may account for a significant portion of the C released from decomposing plant roots.

Carbon Budgets

Using the data from this study and other investigations at Hubbard Brook (Gosz et al., 1976; Likens et al., 1977; McDowell & Likens, 1988; Hughes & Fahey, 1994), we have constructed C budgets for uncut and clear-cut watershed ecosystems (Fig. 21–7, 21–8). A number of assumptions were made to facilitate the calculations: (i) the Bh and Bs horizon lysimeters drained the 0- to 10- and 10- to 20-cm mineral soil layers, respectively, (ii) biomass accumulated at the same rate in 1991 as in 1987 to 1991, (iii) water flows downward in the solum, evapotranspiration is distributed among the soil horizons according to the fine root distribution in the precut watershed (Fahey et al., 1988), (iv) stream degassing of CO_2 is the difference between DIC fluxes in Bs soil solutions and stream water, (v) The ratio of litter from living root systems to total living belowground biomass remained constant. Assumption (v) may result in underestimation of root litter in Watershed 5. Anecdotal information from in-growth cores and root screens indicates that by Years 5 to 8, root growth in Watershed 5 was about two-thirds of the growth in the uncut forest, while turnover rates were similar (T.J. Fahey, 1993, unpublished data).

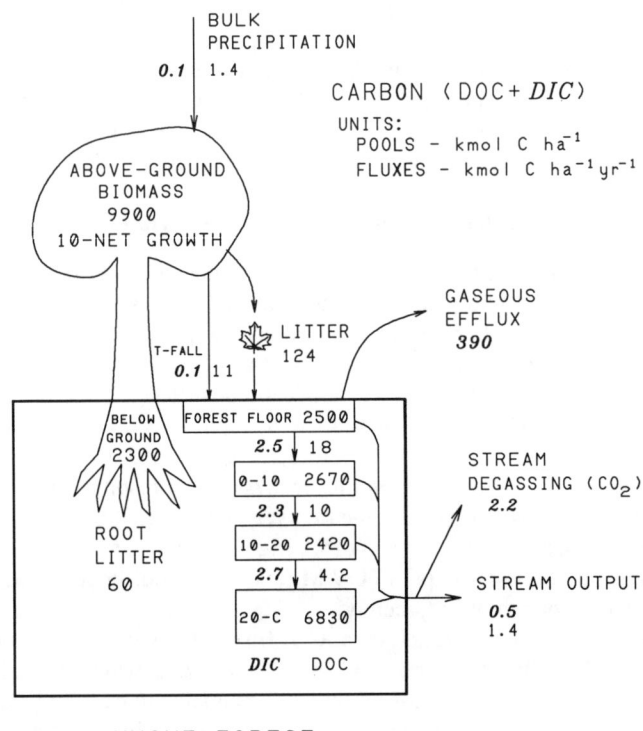

Fig. 21-7. Carbon budget for 1983 to 1992 in an uncut watershed ecosystem (Watershed 6) at the Hubbard Brook Experimental Forest. See text for details. Fluxes of DIC are shown using italics.

The C cycle can be divided into two subcycles (Fig. 21–7, 21–8). The fluxes associated with biologically mediated processes, which include plant growth, decomposition, and respiration, are large, on the order of hundreds of kmol C per hectare per year, while hydrologically related fluxes (precipitation, throughfall, soil solutions, streams) are much smaller, on the order of kmol C per hectare per year at most. Because the two subcycles are tightly linked, an integrative approach is necessary to evaluate responses to disturbance. However, detailed conclusions are compromised by the fact that hydrologic fluxes are quantitatively insignificant compared to biological fluxes, and uncertainty in biological terms overwhelms the hydrological fluxes.

For example, the budgets can be used to assess the status of C in the Hubbard Brook ecosystem. In the undisturbed forest, the total C input to the solum is about 215 kmol C ha^{-1} yr^{-1} (litter + throughfall + root litter), or 2.6 Mg ha^{-1} yr^{-1}. This value is less than the estimated total output from the solum, about 390 kmol C ha^{-1} yr^{-1} (gaseous loss + stream output + stream degassing = 394 kmol C ha^{-1} yr^{-1}), or 4.7 Mg ha^{-1} yr^{-1}. Thus, it appears that the net loss of C from the solum in the undisturbed forest is about 180 kmol C ha^{-1} yr^{-1} (2.2 Mg h^{-1} yr^{-1}).

CLEAR-CUTTING A NORTHERN HARDWOOD FOREST & CARBON DYNAMICS

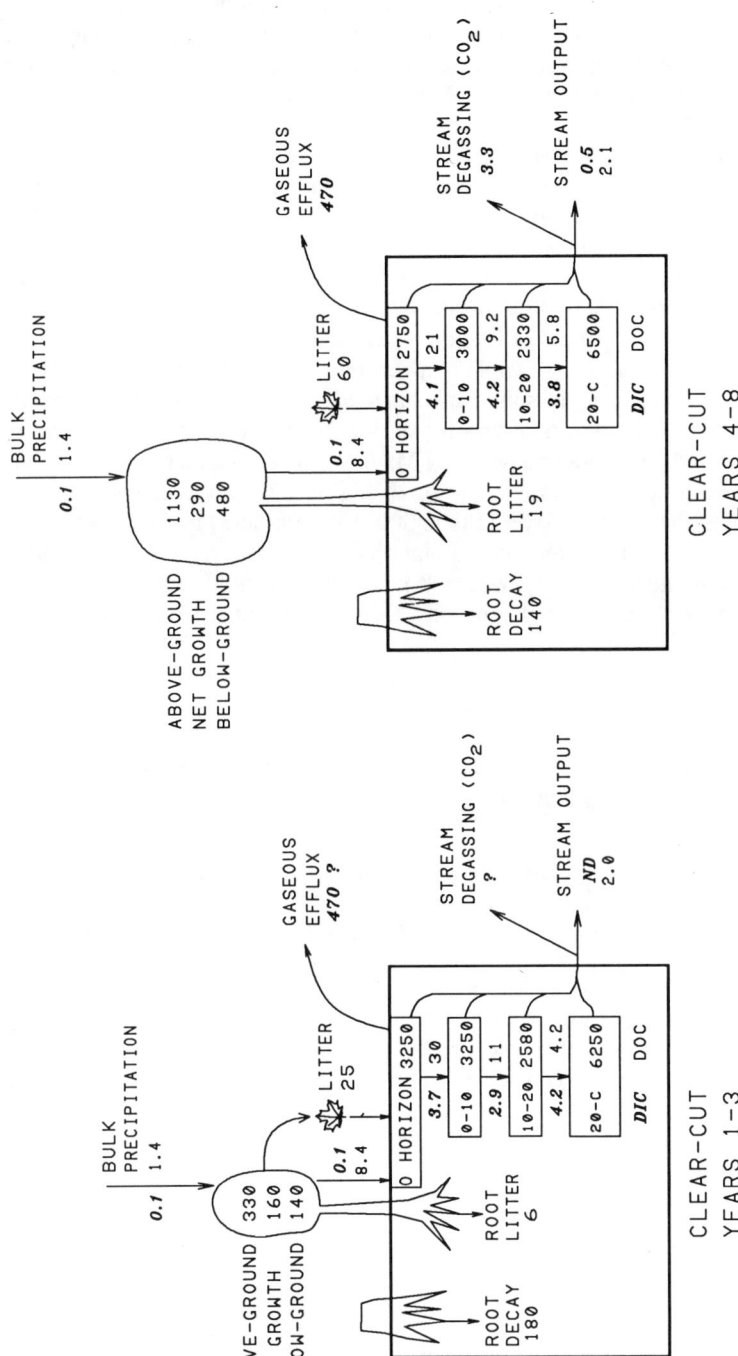

Fig. 21-8. Carbon budget for a clear-cut watershed ecosystem (Watershed 5) at the Hubbard Brook Experimental Forest. Terms are as in Fig. 21-7.

In the first 3 yr after clear-cutting, the litter inputs (above- and belowground) to the solum declined markedly, but root decomposition in the clear-cut watershed made up for the decrease. Thus, the total C input to the system in the first three postharvest years averaged 220 kmol C ha^{-1} yr^{-1}, or 2.6 Mg ha^{-1} yr^{-1} (litter + root litter + throughfall + root decomposition), while output increased to about 470 kmol C ha^{-1} yr^{-1} (5.6 Mg ha^{-1} yr^{-1}). From these estimates, we estimate that the net C loss from the solum increased to approximately 250 kmol C ha^{-1} yr^{-1} or 3.0 Mg ha^{-1} yr^{-1}. This is a large flux relative to solution fluxes, but impossible to detect through direct soil sampling.

In postharvest Years 4 to 8, input of leaf and root litter increased, while root decomposition decreased. As a result, the overall solum C inputs and outputs remained similar: total input of C to the solum averaged approximately 230 kmol C ha^{-1} yr^{-1} (2.8 Mg ha^{-1} yr^{-1}), while outputs were about 470 kmol C ha^{-1} yr^{-1} (5.6 Mg ha^{-1} yr^{-1}). Therefore, the solum appears to have been a net source of about 3.0 Mg ha^{-1} yr^{-1} of C in the first 8 yr after cutting. Note that the total C loss over 8 yr, as estimated from these budgets (24 Mg ha^{-1}), is very similar to the estimated loss of C from the solum (Table 21–4), 20 Mg ha^{-1}. However, this loss was not significant at the $P \leq 0.05$ level (although it was significant at the 0.10 level). If the trend of soil C loss continues, a significant decrease in solum C should be detected in our next soil sampling (1996). However, with the recovery of vegetation we anticipate that the difference in CO_2 effluxes between Watershed 5 and 6 will decrease.

The values used in the analysis of solum C balance must be viewed as highly speculative because of: (i) the lack of comprehensive data regarding gaseous CO_2 emission from the soil in the first 5 yr after logging, and (ii) the absence of information regarding root litter in the clear-cut watershed. However, the assertion that clear-cutting resulted in increased net loss of C from the soil is probably a safe conclusion.

Implications for Other Element Cycles

The observed patterns in C cycling following clear-cutting on Watershed 5 have important implications for the cycling of other elements, particularly metals. At Hubbard Brook and other recently glaciated sites in the northeastern USA, soil clay contents are low. As a result, most of the cation exchange capacity (CEC) of the soil is provided by deprotonated acidic functional groups associated with soil organic matter (Federer & Hornbeck, 1985). Also, pools of exchangeable nutrient cations such as Ca and Mg are quite small at Hubbard Brook. For example, the pool of exchangeable Ca in the solum is approximately 8 kmol ha^{-1} (Johnson et al., 1991b), while the amount of Ca in forest biomass is estimated to be approximately 17 kmol ha^{-1} (T.G. Siccama, 1992, unpublished data). Thus, the fate of soil organic matter after clear-cutting is critical to the availability of nutrients for regrowing vegetation.

In the case of Watershed 5, leaching of the forest floor has resulted in the downward migration of soil organic matter and associated nutrients. However, physical and natural processes have helped to retain soil organic matter in the top 20 cm of the solum (Table 21–4), thereby preserving nutrient pools. In addition, changes in the soil environment can alter the chemical properties of soil organic matter. We observed an increase in CEC per gram soil organic matter in Bh and Bs1 horizons in the first 3 yr after clear-cutting on Watershed 5 (Johnson et al., 1991b), resulting in increased exchangeable nutrient cations in those horizons, even though soil organic matter content did not change. Based on the observed changes in C distribution in the solum (Table 21–5), it is likely that the exchangeable nutrient concentrations in the upper mineral soil further increased between postharvest Years 3 and 8.

The cycles of several trace metals also are closely tied to the biogeochemical movement of C. In particular, Pb and Fe are known to form strong complexes with organic ligands in solution (Schnitzer & Skinner, 1967; Stevenson, 1982). In both cases, transport of the metal in the environment is essentially determined by the movement of organometallic complexes. At Hubbard Brook, significant correlations between DOC and Pb or Fe have been observed in stream and soil water (Driscoll et al., 1988; Fuller et al., 1988). In an investigation conducted between July 1984 and October 1985, Fuller et al. (1988) observed no significant response in soil solution (Oa and Bs) or stream water Pb and Fe concentrations on Watershed 5. Considering the strong correlation with DOC and the minimal change in DOC concentrations in those solutions (Fig. 21–3), the result was not surprising. However, given the delayed increase in DOC in soil solutions and stream water, it seems likely that there also have been delayed increases in trace metal concentrations and fluxes. Unfortunately, the monitoring of Fe and Pb was discontinued in 1985.

Many forest areas also serve as watersheds for municipal water supply. Our results suggest that extensive clear-cutting may result in prolonged elevated DOC concentrations. This may be problematic because: (i) elevated DOC concentrations may be accompanied by increased trace metal concentrations, (ii) DOC causes taste and odor problems in drinking water, and (iii) chlorination during water treatment results in the formation of carcinogenic trihalomethanes.

ACKNOWLEDGMENTS

This is a contribution of the Hubbard Brook Ecosystem Study. The Hubbard Brook Experimental Forest is administered by the USDA-FS, Northeast Forest Experiment Station, Radnor, Pennsylvania. Funding for this research was provided by the National Science Foundation, Division of Environmental Biology (in part through the Long-Term Ecological Research program) and the Andrew W. Mellon Foundation. We are grateful for the thoughtful reviews of two anonymous colleagues. We also thank J.J. Ruiz, E. Letvin, M. Aloi, R. Lynch, and R. Boes Romanowicz for their contributions.

REFERENCES

Aber, J.D., D.B. Botkin, and J.M. Melillo. 1978. Predicting the effects of different harvesting regimes on forest floor dynamics in northern hardwoods. Can. J. For. Res. 8:306–315.

Allen, J.C. 1985. Soil response to forest clearing in the United States and the tropics: Geological and biological factors. Biotropica 17:15–27.

Bormann, F.H., and G.E. Likens, 1980. Pattern and process in a forest ecosystem. Springer-Verlag, New York.

Bormann, F.H., G.E. Likens, D.W. Fisher, and R.S. Pierce. 1968. Nutrient loss accelerated by clearcutting of a forest ecosystem. Science (Washington, DC) 159:882–884.

Bormann, F.H., G.E. Likens, T.G. Siccama, R.S. Pierce, and J.S. Eaton. 1974. The export of nutrients and recovery of stable conditions following deforestation at Hubbard Brook. Ecol. Monogr. 44:255–277.

Covington, W.W. 1981. Changes in forest floor organic matter and nutrient content following clear cutting in northern hardwoods. Ecology 62:41–48.

Dahlgren, R.A., and C.T. Driscoll. 1994. The effects of whole-tree clear-cutting on soil processes at the Hubbard Brook Experimental Forest, New Hampshire, USA. Plant Soil 158:239–262.

Detwiler, R.P. 1986. Land use change and the global carbon cycle: The role of tropical soils. Biogeochemistry 2:67–93.

Detwiler, R.P., and C.A.S. Hall. 1988. Tropical forests and the global carbon cycle. Science (Washington, DC) 239:42–47.

Douglas, J.E., and W.T. Swank. 1972. Streamflow modification through management of eastern forests. USDA-FS Res. Pap. SE-94. USDA-FS, Southeast For. Exp. Stn., Asheville, NC.

Driscoll, C.T., R.D. Fuller, and D.M. Simone. 1988. Longitudinal variations in trace metal concentrations in a northern forested ecosystem. J. Environ. Qual. 17:101–107.

Edwards, N.T. 1982. The use of soda-lime for measuring respiration rates in terrestrial systems. Pedobiologia 23:321–330.

Fahey, T.J., and M.A. Arthur. 1994. Further studies of root decomposition following harvest of northern hardwoods forest. For. Sci. 40:618–629.

Fahey, T.J., J.W. Hughes, M. Pu, and M.A. Arthur. 1988. Root decomposition and nutrient flux following whole-tree harvest of northern hardwood forest. For. Sci. 34:744–768.

Federer, C.A. 1982. Subjectivity in the separation of organic horizons of the forest floor. Soil Sci. Soc. Am. J. 46:1090–1093.

Federer, C.A. 1984. Organic matter and nitrogen content of the forest floor in even-aged northern hardwoods. Can. J. For. Res. 14:763–767.

Federer, C.A., L.D. Flynn, C.W. Martin, J.W. Hornbeck, and R.S. Pierce. 1990. Thirty years of hydrometeorologic data at the Hubbard Brook Experimental Forest, New Hampshire. Gen. Tech. Rep. NE-141. USDA-FS, Northeast For. Exp. Stn., Radnor, PA.

Federer, C.A., D.E. Turcotte, and C.T. Smith. 1993. The organic fraction-bulk density relationship and the expression of nutrient content in forest soils. Can. J. For. Res. 23:1026–1032.

Federer, C.A., and J.W. Hornbeck. 1985. The buffer capacity of forest soils in New England. Water Air Soil Pollut. 26:163–173.

Fuller, R.D., C.T. Driscoll, G.B. Lawrence, and S.C. Nodvin. 1987. Processes regulating sulphate flux after whole-tree harvesting. Nature (London) 325:707–710.

Fuller, R.D., D.M. Simone, and C.T. Driscoll. 1988. Forest clearcutting effects on trace metal concentrations: Spatial patterns in soil solutions and streams. Water Air Soil Pollut. 40:185–195.

Goldin, A., and L.M. Lavkulich. 1990. Effects of historical land clearing on organic matter and nitrogen levels in soils of the Fraser lowland of British Columbia. Can. J. Soil Sci. 70:583–592.

Gosz, J.R., G.E. Likens, and F.H. Bormann. 1976. Organic matter and nutrient dynamics of the forest and forest floor in the Hubbard Brook Forest. Oecologia 22:305–320.

Hendricksen, O.Q., L. Chatarpaul, and D. Burgess. 1989. Nutrient cycling following whole-tree and conventional harvest in northern mixed forest. Can. J. For. Res. 19:725–735.

Hornbeck, J.W., R.S. Pierce, and C.A. Federer. 1970. Streamflow changes after forest clearing in New England. Water Resour. Res. 6:1124–1132.

Hornbeck, J.W., C.W. Martin, R.S. Pierce, F.H. Bormann, G.E. Likens, and J.S. Eaton. 1986. Clearcutting northern hardwoods: Effects on hydrologic and nutrient ion budgets. For. Sci. 32:667–686.

Houghton, R.A., R.D. Boone, J.R. Fruci, J.E. Hobbie, J.M. Melillo, C.A. Palm, B.J. Peterson, G.R. Shaver, G.M. Woodwell, B. Moore, D.L. Skole, and N. Myers. 1987. The flux of carbon from terrestrial ecosystems to the atmosphere in 1980 due to changes in land use: geographic distribution of the global flux. Tellus 39B:122–139.

Houghton, R.A., D.L. Skole, and D.S. Lefkowitz. 1991. Changes in the landscape of Latin America between 1850 and 1985. II. Net release of CO_2 to the atmosphere. For. Ecol. Manage. 38:173–199.

Hughes, J.W., and T.J. Fahey. 1988. Seed dispersal and colonization of a disturbed northern hardwood forest. Bull. Torrey Bot. Club 115:89–99.

Hughes, J.W., and T.J. Fahey. 1991. Availability, quality, and selection of browse by white-tailed deer after clearcutting. For. Sci. 37:261–270.

Hughes, J.W., and T.J. Fahey. 1992. Colonization dynamics of herbs and shrubs in a disturbed northern hardwood forest. J. Ecol. 79:605–616.

Hughes, J.W., and T.J. Fahey. 1994. Litterfall dynamics and ecosystem recovery during forest development. For. Ecol. Manage. 63:181–198.

Hughes, J.W., T.J. Fahey, and F.H. Bormann. 1988. Population persistence and reproductive ecology of a forest herb: Aster acuminatis. Am. J. Bot. 75:1057–1064.

Huntington, T.G., and D.F. Ryan. 1990. Whole tree harvesting effects on soil nitrogen and carbon. For. Ecol. Manage. 31:193–204.

Huntington, T.G., D.F. Ryan, and S.P. Hamburg. 1988. Estimating soil nitrogen and carbon pools in a northern hardwood forest ecosystem. Soil Sci. Soc. Am. J. 52:1162–1167.

Hurlbert, S.H. 1984. Pseudoreplication and the design of ecological field experiments. Ecol. Monogr. 54:187–211.

Johnson, A.H., T.G. Siccama, and A.J. Friedland. 1982. Spatial and temporal patterns of lead accumulation in the forest floor in the northeastern United States. J. Environ. Qual. 11:577–580.

Johnson, C.E., A.H. Johnson, and T.G. Huntington. 1990. Sample size requirements for the determination of changes in soil nutrient pools. Soil Sci. 150:637–644.

Johnson, C.E., A.H. Johnson, T.G. Huntington, and T.G. Siccama. 1991a. Whole-tree clear-cutting effects on soil horizons and organic-matter pools. Soil Sci. Soc. Am. J. 55:497–502.

Johnson, C.E., A.H. Johnson, and T.G. Siccama. 1991b. Whole-tree clear-cutting effects on exchangeable cations and soil acidity. Soil Sci. Soc. Am. J. 55:502–507.

Johnson, D.W. 1992. Effects of forest management on soil carbon storage. Water Air Soil Pollut. 64:83–120.

Johnson, N.M., C.T. Driscoll, J.S. Eaton, G.E. Likens, and W.H. McDowell. 1981. 'Acid rain', dissolved aluminum and chemical weathering at the Hubbard Brook Experimental Forest, New Hampshire. Geochim. Cosmochim. Acta 45:1421–1437.

Lawrence, G.B., and C.T. Driscoll. 1988. Aluminum chemistry downstream of a whole-tree harvested watershed. Environ. Sci. Technol. 22:1293–1299.

Lawrence, G.B., and C.T. Driscoll. 1990. Longitudinal patterns of concentration-discharge relationships in stream water draining the Hubbard Brook Experimental Forest, New Hampshire. J. Hydrol. 116:147–165.

Lawrence, G.B., R.D. Fuller, and C.T. Driscoll. 1987. Release of aluminum following whole-tree harvesting at the Hubbard Brook Experimental Forest, New Hampshire. J. Environ. Qual. 16:383–390.

Lawrence, G.B., C.T. Driscoll, and R.D. Fuller. 1988. Hydrologic control of aluminum chemistry in an acidic headwater stream. Water Resour. Res. 24:659–669.

Likens, G.E., F.H. Bormann, N.M. Johnson, D.W. Fisher, and R.S. Pierce. 1970. Effects of forest cutting and herbicide treatment on nutrient budgets in the Hubbard Brook watershed-ecosystem. Ecol. Monogr. 40:23–47.

Likens, G.E., F.H. Bormann, R.S. Pierce, J.S. Eaton, and N.M. Johnson. 1977. Biogeochemistry of a forested ecosystem. Springer-Verlag, New York.

Lugo, A.E., and S. Brown. 1992. Tropical forests as sinks of atmospheric carbon. For. Ecol. Manage. 54:239–255.

Lugo, A.E., A.J. Sanchez, and S. Brown. 1986. Land use and organic carbon content of some subtropical soils. Plant Soil 96:185–196.

Marland, G. 1988. The prospect for solving the CO_2 problem through global reforestation. Tech. Rep. 39. U.S. Dep. Energy, Office of Energy Res., Washington, DC.

Martin, C.W., R.S. Pierce, G.E. Likens, and F.H. Bormann. 1986. Clearcutting affects stream chemistry in the White Mountains of New Hampshire. Res. Pap. NE 579. USDA-FS, Northeastern For. Exp. Stn., Broomall, PA.

McDowell, W.H., and G.E. Likens. 1988. Origin, composition and flux of dissolved organic carbon in the Hubbard Brook valley. Ecol. Monogr. 58:177–195.

Mou, P., T.J. Fahey, and J.W. Hughes. 1993. Effects of soil disturbance on vegetation recovery and nutrient accumulation following whole-tree harvest of a northern hardwood ecosystem, Hubbard Brook Experimental Forest. J. Appl. Ecol. 30:661–675.

Nadelhoffer, K.J., and J.W. Raich. 1992. Fine root production estimates and belowground carbon allocation in forest ecosystems. Ecology 73:1139–1147.

Ryan, D.F., T.G. Huntington, and C.W. Martin. 1992. Redistribution of soil nitrogen, carbon and organic matter by mechanical disturbance during whole-tree harvesting in northern hardwoods. For. Ecol. Manage. 49:87–99.

Schlesinger, W.H. 1986. Changes in soil carbon storage and associated properties with disturbance and recovery. p. 194–220. *In* J.R. Trabalka and D.E. Reichle (ed.) The changing carbon cycle: A global analysis. Springer-Verlag, New York.

Schnitzer, M., and S.I.M. Skinner. 1967. Organometallic interactions in soils: 7. Stability constants of Pb^{2+}-, Ni^{2+}-, Mn^{2+}-, Co^{2+}-, Ca^{2+}-, and Mg^{2+}-fulvic acid complexes. Soil Sci. 103:247–252.

Siccama, T.G., F.H. Bormann, and G.E. Likens. 1970. The Hubbard Brook ecosystem study: Productivity, nutrients and phytosociology of the herbaceous layer. Ecol. Monogr. 40:389–402.

Snyder, K.E., and R.D. Harter. 1987. Forest floor dynamics in even-aged northern hardwood stands. Soil Sci. Soc. Am. J. 51:1381–1383.

Soil Survey Staff. 1975. Soil taxonomy: A basic system of soil classification for making and interpreting soil surveys. USDA-SCS Agric. Handb. 436. U.S. Gov. Print. Office, Washington, DC.

Stevenson, F.J. 1982. Humus chemistry: Genesis, composition, reactions. John Wiley & Sons, New York.

Vitousek, P.M. 1991. Can planted forests counteract increasing atmospheric carbon dioxide? J. Environ. Qual. 20:348–354.

Whittaker, R.H., F.H. Bormann, G.E. Likens, and T.G. Siccama. 1974. The Hubbard Brook Ecosystem Study: Forest biomass and production. Ecol. Monogr. 44:233–252.

22 Distribution of Carbon in a Piedmont Soil as Affected by Loblolly Pine Management

D.H. Van Lear and P.R. Kapeluck

Clemson University
Clemson, South Carolina

Melissa M. Parker

Texas Parks and Wildlife Department
Nacogdoches, Texas

Little information is available to accurately estimate C in soil and root system pools of Piedmont sites. Soils in the Piedmont physiographic province of the southeastern USA suffered severe erosion from agricultural activity prior to 1940 (Giddens & Garman, 1941). In the upper Piedmont of South Carolina where the present study was conducted, an average of 23 cm of topsoil, including associated organic matter, was lost between 1880 to 1930, and some sites lost as much as 60 cm (Trimble, 1974). As a result of these losses, these eroded sites were left with low quantities of soil C.

During the Great Depression of the 1930s, many farms in the Piedmont were abandoned as much of the farmland had become unproductive. About 12000 ha of this worn-out farmland was purchased by the Federal government in the mid-1930s and eventually became the Clemson Experimental Forest (Sorrells, 1984). Loblolly pine (*Pinus taeda* L.) plantations established on the Forest in the late 1930s now represent some of the oldest pine plantations in the South and afford an excellent opportunity to study their role in sequestering atmospheric C.

The objectives of this paper are to: (i) describe the distribution of C in soil and root-system biomass after 55 yr of pine plantation management on a poor-quality Piedmont site, (ii) model the increase in soil and root-system C following agricultural abandonment, and (iii) describe effects of harvest on soil and root-system C pools over a 13-yr period.

STUDY SITE DESCRIPTION

The study area has been described previously (Van Lear et al., 1984; Van Lear & Kapeluck, 1990). Briefly, the 55-yr-old plantation occupies an eroded

---

Copyright © 1995 Soil Science Society of America, 677 S. Segoe Rd., Madison, WI 53711, USA. Carbon Forms and Functions in Forest Soils.

phase of the Pacolet fine sandy loam (a clayey, kaolinitic, thermic Typic Kanhapludult). The major portion of the study site, based on plant and other site indicators, falls within the subxeric type in Jones' (1991) Landscape Ecosystem Classification System for the upper Piedmont. Slopes average about 13% and site index ranged from 21 to 24 m at 50 yr. The stand was planted at 2- by 2-m spacing in 1939 and was thinned lightly from below at ages 22 and 30 yr. Management records indicate that about 21 t ha^{-1} of stem biomass was removed in these thinnings.

Topography of the study site suggests a history of severe erosion, as evidenced by frequent deep gullies that function as semistable ephemeral stream channels. A 1939 photograph (Fig. 22–1) of a similar site on the Clemson Experimental Forest indicates the sparse quantity of early successional vegetation supported by the badly degraded old-fields at the time of plantation establishment.

Stemwood from a small (0.5-ha) watershed within the plantation was harvested in the winter of 1979–1980. The harvested watershed had been burned annually from 1977 to 1979 using low intensity prescribed fires to prepare a seedbed for natural regeneration and to control understory hardwoods. The stand was regenerated naturally to a dense stand of loblolly pine using clearcutting with seed in place (Cox & Van Lear, 1985). The regenerated natural stand was 13 yr old at this writing and served as the treated watershed, while an adjacent unharvested watershed in the 55-yr-old plantation served as the control.

Fig. 22–1. Eroded old field near study site in 1939.

METHODS

The distribution of C in the forest floor, mineral soil, and root system of the mature and regenerated stand was determined from several related studies conducted on the site over the past 13 yr. Carbon concentrations in all samples were analyzed on either a Perkin Elmer Model 240C (Perkin Elmer Elemental Analyzer, Model 240C, Norwalk, CT) at the Chemical Analysis Laboratory of the University of Georgia or a C analyzer (LECO Carbon Analyzer, Model CR-12, St. Joseph, MI) at the Coweeta Hydrologic Laboratory of the USDA Forest Service.

Forest Floor Carbon

The forest floor of the treated watershed was sampled just before and after harvest in 1979–1980 on 10 paired 0.09-m^2 quadrats. The forest floor was separated into O_i and O_e layers and oven dried at 70°C to a stable weight. Carbon content was determined from concentrations multiplied by ash-free weight of forest floor components.

Within 3 yr following harvest, the preharvest forest floor had decomposed to the point that it could not be sampled. Thirteen years later, the forest floor in the regenerated stand was similar in mass, thickness, and C content to that reported for a 15-yr-old loblolly pine stand on a similar site within 1 km of the study site (Van Lear & Goebel, 1976).

Soil Carbon

Twelve to twenty samples of the mineral soil matrix (exclusive of root channels, roots, and rocks) were collected at 3 to 5 permanent sampling locations from each watershed at three depths (0–10, 10–30, and 30–50 cm). At each sampling location, four corings per depth were composited in the field and taken to the laboratory for analysis. Samples were dried at 70°C to a constant weight, sieved through a 2-mm sieve, and pulverized in a hammermill prior to C analysis.

Soil matrix C was measured on the control watershed and the harvested watershed at time of harvest and 1, 2, 3 and 13 yr postharvest. Soil C concentration at the 50- to 100-cm depth was extrapolated from regressions of C concentration vs. soil depth using data from the 0- to 10-, 10- to 30-, and 30- to 50-cm depths. Data from both harvested and control watersheds were combined to develop the equation. Soil volumes (minus root volume, corrected for shrinkage, old root channels and rock volume) were calculated for each depth increment and multiplied by specific gravity to yield estimates of soil mass. Measured or estimated (50- to 100-cm depth) soil C concentrations were multiplied by the appropriate soil mass to yield C content.

Soil C was estimated to a depth of only 1 m because of the difficulty of sampling lower soil depths and because this depth corresponded to the depth of rooting on this site. Richter et al. (1994, see Chapter 11) have recently demonstrated that C is found, albeit in low concentrations, to depths of 6 m on some Piedmont soils. Therefore, the C pools reported here are conservative and representative of the rooting depth, rather than the C pool to bedrock.

The quantity of C in the soil matrix in 1939 at the time of agricultural abandonment was estimated from current C concentrations, excepting that of the 0- to 30-cm depth. We assumed that severe erosion during the 100 yr prior to plantation establishment removed most of the A and part of the thick B horizon and associated C. Present soil and topographic features, as well as photographs from 1939, verify this assumption. Thus, the soil near the surface at that time would have been impoverished of C and the relatively stable C concentrations of the subsoil (Emanuel et al., 1981) would have prevailed. Therefore, the current measured C concentration of the 30- to 50-cm depth was used to estimate the postagricultural content of the 0- to 30-cm depth. The C content of lower soil depths at that time was estimated from a regression equation developed from pooled concentration data from both treatments.

Increases in C content of the soil after plantation establishment were attributed to the virtual cessation of erosion under forest cover (Douglass & Van Lear, 1983) and incorporation of organic matter into the surface soil during stand development and management. Bulk density, used to calculate C content, varied little throughout the profile to a depth of 1 m and was assumed to be 1.5 g cm^{-3} in 1939 and at the present time. We assumed that root decomposition was the dominant source of increasing soil C during reforestation (Persson, 1980). Therefore, we used the pattern of increasing root density found by Coile (1936) for Piedmont loblolly pine over a 70-yr period to shape our model, using our estimated value for soil C in 1939 and our measured soil C value for the present 55-yr-old plantation to put limits on the curve.

The volume of soil to a depth of 1 m occupied by old root channels was estimated from analyses of 60 1-m^2 soil profiles of the site (McCollum, 1992). Outlines of root channels and rocks and/or rock cavities were traced onto acetate sheets. In the laboratory, sketches were measured and converted to area by use of the measuregraph of the Geographic Information System (Professional Map Analysis System, Version 2.0, Spatial Information Systems, Inc., Omaha, NE). These soil profiles were considered analogous to planar intersects used to measure volumes of forest fuels (Brown, 1974); therefore, root channel cross-sectional areas were converted to volumes.

Carbon Content of the Root System

Coarse root biomass, estimated when the plantation was 50 yr old, was determined by a combination of excavation and regression methods, while fine root biomass was estimated from soil cores (Kapeluck and Van Lear, 1995). The distribution of lateral roots in the stand was determined from counts made on 30 soil profiles extending to a 1-m depth, the maximum rooting depth for loblolly pine on this site (Kapeluck, 1988; McCollum, 1992).

Carbon sequestered in root systems was estimated from data on root distribution and biomass of loblolly pines on this subxeric site (Kapeluck, 1988; McCollum, 1992). Concentrations of C were determined for five size classes of pine roots (taproots, >50 mm, 25–50 mm, 6–25 mm, and <6 mm). The distribution of lateral root biomass was determined from the percentage of roots in each size class by depth from the soil profile data. These percentages were multiplied by the root biomass of each size class to allocate lateral root biomass to the appropriate soil depth.

Taproot biomass distribution was determined by dividing the volume of the average taproot [based on tree of average diameter breast height (DBH)] into sections corresponding with the depth of soil sampling (0–10, 10–30, 30–50, and 50–100 cm). The average taproot was considered cone-shaped and the sections viewed as frustrums of a cone. Percentage volume of each frustrum of the taproot was multiplied by the taproot biomass per hectare and its C content allocated to the appropriate soil depth.

Root Decomposition After Harvest

Root decomposition of harvested mature plantations was evaluated by sampling decomposing root systems of similar stands harvested previously on subxeric sites on the Clemson Forest. Specific gravity of three root samples for each of four diameter classes from stands harvested between 0 and 10 yr prior to sampling were determined by the volume displacement method, using melted paraffin to seal the sample (Barber & Van Lear, 1984). Changes in C content during this 10-yr decomposition period were determined from regressions of changes in biomass over time coupled with measured C concentrations.

RESULTS AND DISCUSSION

Distribution of Carbon in the Soil and Below-Stump Biomass of a 55-Year-Old Stand

Soil Carbon

About 76% of the belowground C (71.8 t ha^{-1}) was in the soil, including C in the forest floor and old root channels (Table 22–1). Because of higher C concentrations in the upper soil layers and the disproportionate distribution of root biomass, nearly three-fourths of the belowground C was within 30 cm of the surface.

Carbon concentrations in root channels (1.35%) under this loblolly pine stand were higher than those in the surrounding soil matrix, especially in the subsoil where C concentrations ranged from 0.60 to 0.08%. Root channels are the cavities of once-living roots that have decomposed and filled with soil and organic matter carried down from surface layers by soil fauna and gravity

Table 22–1. Distribution of belowground C in a 55-yr-old loblolly pine plantation on a poor site in the Piedmont of South Carolina.

Soil depth	Forest floor	Soil matrix	Root channels	Roots	Total
cm			t ha^{-1}		
Oi + Oe	11.9	—	—	—	11.9
0–10		10.0	2.1	6.7	18.8
10–30		13.5	1.9	5.8	21.2
30–50		8.6	0.9	2.8	12.3
50–100		5.1	0.6	1.9	7.6
Total	11.9	37.2	5.5	17.2	71.8

(Lutz & Chandler, 1955). Although the volume of root channels was only 2.7% of the total soil volume sampled, C in root channels was about 10% (5.5 t ha^{-1}) of soil C, excluding the root system.

Many root channels contained living roots. Root density was up to 100 times greater in root channels than in the soil matrix (McCollum, 1992). Root channels are preferred by living roots because they are often characterized by enhanced fertility, better aeration, and better moisture relations (Van Rees, 1984). Root channels provide increased spatial heterogeneity in the soil profile, contribute to increased infiltration and percolation of soil water, and enhance gas diffusion throughout the rooting zone (Lutz & Chandler, 1955). These channels facilitate root penetration of dense clay Piedmont subsoils.

The volume of soil occupied by old root channels is low. Although the vegetative complex of the Piedmont was historically dominated by *Quercus – Carya – Pinus* or *Quercus – Carya – Liriodendron* (Skeen et al., 1993), all except the deeper roots of these forests would have been obliterated by the 100 yr of row cropping and the accompanying loss of surface soil to erosion. Some of the surviving root channels are attributed to shortleaf (*P. echinata* Mill.) pine, which was a deep-rooted component of the preagricultural landscape, and to loblolly pines removed in earlier thinnings from the most recent stand. Both of these species have relatively deep roots systems. Pine charcoal found in root channels on the study area suggest that many root channels predate agriculture, since no burning has been conducted in the present unharvested loblolly pine stand.

Carbon Stored in Below-Stump Biomass

Carbon stored in the root system of the mature stand totaled 17.2 t ha^{-1}, 24% of the belowground C (Table 22–1). The general shape of the taproot (modeled as an inverted cone) and the unequal distribution of lateral roots (Fig. 22–2) account for the large quantity (73%) of root C in the upper 30 cm of the mineral soil profile.

About 56% of the root system carbon of this 55-yr-old pine stand was in the stump and taproot (Table 22–2). Fine pine roots, characterized by their rapid turnover (Ruark, 1993), contained 1.8 t ha^{-1} C. Because of the difficulty of extracting fine roots from the soil cores, this component of the root system is probably conservatively estimated.

It is important to note that site quality of the study area was poor. If ratios between root and total biomass are similar across site classes, much higher figures for root C would be expected on better sites. Wells and Jorgensen (1975) and Switzer and Nelson (1972) reported up to 2.5 times as much aboveground biomass for loblolly pine stands on good sites as measured here (Van Lear et al., 1984). These differences are due primarily to site quality, but also to the fact that the plantation area in the present study included a road and the stand had been previously thinned.

Effects of Fifty-Five Years of Plantation Management on Belowground Carbon

Continual cultivation dramatically reduces the organic matter content of noneroded Piedmont soils (Giddens, 1957). In steeper terrain, such as this study site, erosion following annual plowing also would have contributed to the

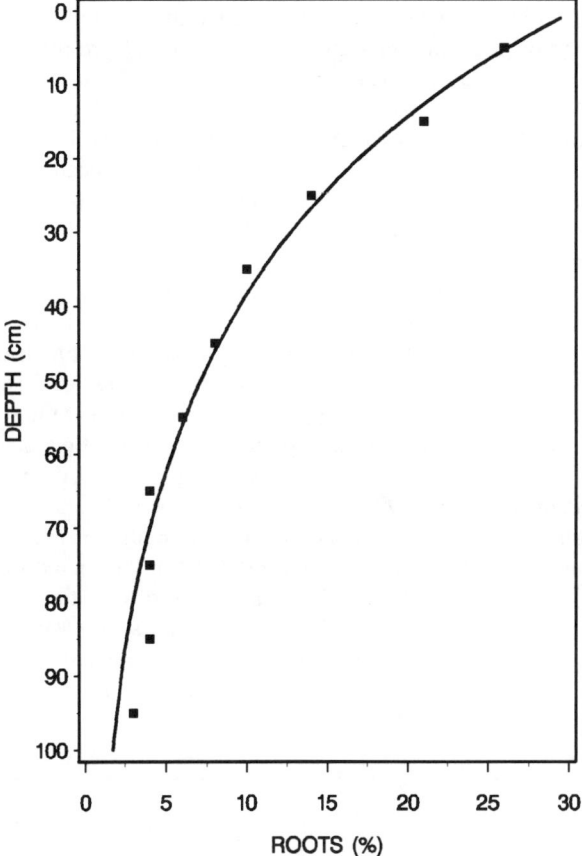

Fig. 22–2. Percentage of loblolly pine roots by depth on an eroded Piedmont site.

Table 22–2. Biomass and C content of the root system of a mature loblolly pine plantation on a poor Piedmont site.

		Root diameter size class (cm)			
Total	Taproot†	>5.0	2.5–5.0	0.6–2.5	<0.6
		t ha^{-1}			
Biomass	19.8	5.0	2.7	4.5	4.0	36.0
C	9.7	2.4	1.3	2.0	1.8	17.2

†Includes the stump to a 15-cm height.

removal of any accumulations of surface organic matter. On this site, only remnants of the A and E horizons remained following many decades of row cropping. At the time of agricultural abandonment in 1939, the C horizon was at the surface over most of the site. However, it would have been a C horizon without the benefits of 55 yr of forest regrowth. Therefore, current subsoil C concentrations from appropriate lower depths were used to estimate soil matrix C content in 1939 to be 11.6 t ha^{-1} at the termination of agricultural activity.

After 55 yr of plantation management consisting of fire protection and two light thinnings, soil matrix C had accumulated to 37.2 t ha^{-1}, a 222% increase since 1939. As explained in the methods, we modeled the C accumulation curve (Fig. 22–3) after the pattern of increasing root density found by Coile (1936) for Piedmont loblolly pine over a 70-yr period. Our model indicates a much greater accumulation rate for soil C during reforestation of this impoverished site than studies of reforestation of more fertile Piedmont sites (Schiffman & Johnson, 1988; Richter et al., 1994, see Chapter 11).

The forest floor of the 55-yr-old plantation contained 11.9 t ha^{-1} C, a quantity probably attained during the first two decades of the rotation. The forest floor of a 15-yr-old loblolly plantation growing on a similar site near the present study contained 10.2 t ha^{-1} C, based on a reported biomass of 20.8 t ha^{-1} (Van Lear & Goebel, 1976). These data support the contention that the forest floor in loblolly pine stands is primarily in an aggrading mode in the first 15 to 20 yr, after which it stabilizes as the rate of litter deposition approaches the rate of decomposition (Switzer & Nelson, 1972).

The accumulation of C in root biomass over a rotation is assumed to follow a pattern similar to that found by Coile (1936) for the increasing density of loblolly pine roots over time. Assuming near-zero root C at time of agricultural abandonment (reasonable since the fine roots of the sparse early successional vegetation would turn-over rapidly and erosion continued unchecked until reforestation), C has accumulated in the root systems of the plantation to a total of 17.2 t ha^{-1} at 55 yr (Fig. 22–4).

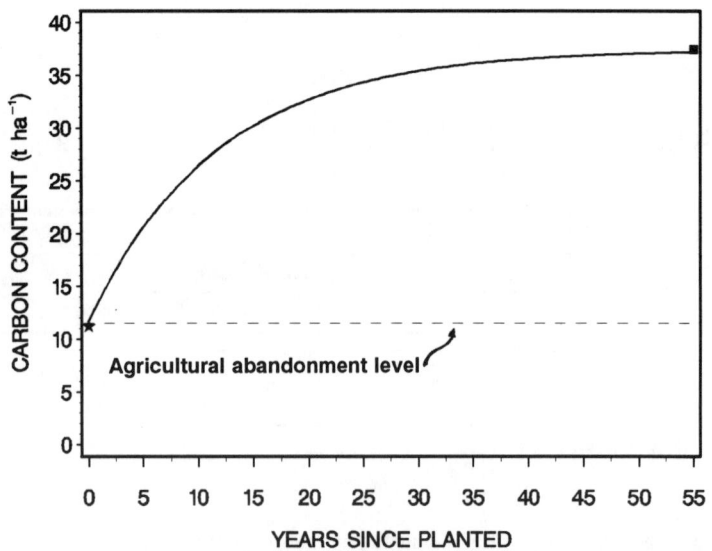

Fig. 22–3. Carbon accretion in the soil matrix during 55 yr of pine occupancy on a poor Piedmont site. The lower bound of the regression is an estimate while the upper bound was measured in the control watershed.

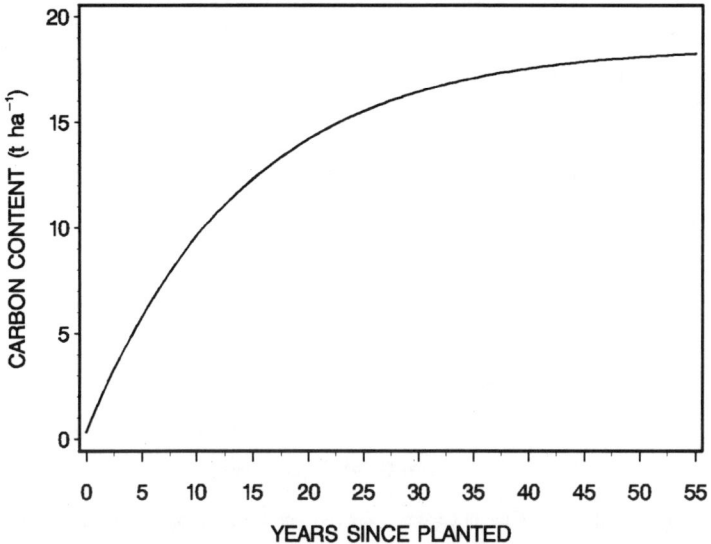

Fig. 22–4. Carbon accretion in the root systems of mature loblolly pine over a 55-yr period on a poor site in the Piedmont.

During the 55 yr of plantation management, C storage in the soil matrix, forest floor, and below-stump biomass increased 472% over soil matrix C levels at the time of agricultural abandonment. While this percentage increase is greater than the 235 and 187% increase in ecosystem storage of C under loblolly pine reforestation in a Virginia Piedmont study (Schiffman & Johnson, 1988) and another South Carolina Piedmont study (Richter et al., 1994, see Chapter 11), absolute quantities of C stored are lower. The percentage increase in the present study is high because estimates of soil matrix C at the time of plantation establishment are much lower than those in other Piedmont studies due to differences in prior erosion during the agricultural period. Both of these other studies were conducted on more level terrain and higher quality sites, which accounts for the greater initial C content of the soils and the greater C accumulation during reforestation in both phytomass (aboveground biomass was included in both of these cited studies, but not in the present study) and forest floor components. The higher initial C levels also largely explain the relatively small (<10%) increases in soil C storage in those studies vs. the 2.2-fold increase during reforestation in the present study.

Despite differences in absolute quantities of C stored and in the relative role of ecosystem components, all of these studies strongly indicate that Piedmont ecosystems have acted as net C sinks during reforestation following agricultural use.

Effects of Harvest on Belowground Carbon

Following harvest, the root system begins to decay almost immediately. Although different-sized roots decay at different rates depending on their size and degree of lignification, one overall root decomposition equation (Fig. 22–5) was

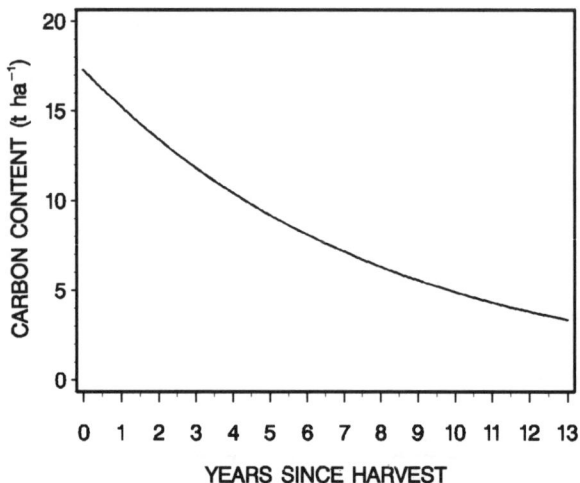

Fig. 22–5. The pattern of C decline in decomposing roots on poor Piedmont sites following harvest.

calculated for ease of interpretation. Assuming a decay rate based on the overall decay coefficient and using measured C concentrations, the decaying root system contained 53% of its original C after 5 yr and 19% after 13 yr. In addition to root decomposition following harvest, decomposing loblolly pine slash also contributes C to the soil matrix and accounts for an undetermined portion of the increase in soil matrix C observed after harvest. About 50% of woody logging slash is decomposed within 10 yr of harvest (Barber & Van Lear, 1984).

Low intensity burns had reduced the forest floor on the treated watershed to 17.8 t ha^{-1} (8.9 t ha^{-1} C) at time of harvest. Within 3 yr after harvest, the forest floor had decomposed to the point where it was minimal. This rapid decomposition is attributed to the increase in fine structural material after burning, fragmentation and mixing by soil animals, and to the higher solar radiation and temperatures reaching the harvested site. Following crown closure (4 yr) by the dense natural regeneration, the forest floor rapidly returned to near preharvest quantities.

The measured C content of the soil matrix increased for 2 yr after harvest (Fig. 22–6) as the root system, forest floor, and logging slash from the previous stand was fragmented and incorporated by soil fauna (Hole, 1981). However, by Year 3, soil C had fallen below preharvest levels, probably because labile forms of finely divided organic matter were mineralized or solubilized and leached from the profile. In addition, some of the native soil organic matter may have been mineralized due to the priming action of the residues incorporated after harvest (Giddens, 1957).

Soil matrix C soon reaccumulates through inputs from root decomposition (both from previous and current stand) and the forest floor. At age 7, the dense natural stand of 3 to 4 m tall saplings was precommercially thinned, which provided a large input of organic residues to the site. This pulse was not modeled into Fig. 22–6 because it was assumed that most of the trees removed in the precommercial thinning would have soon succumbed to natural mortality.

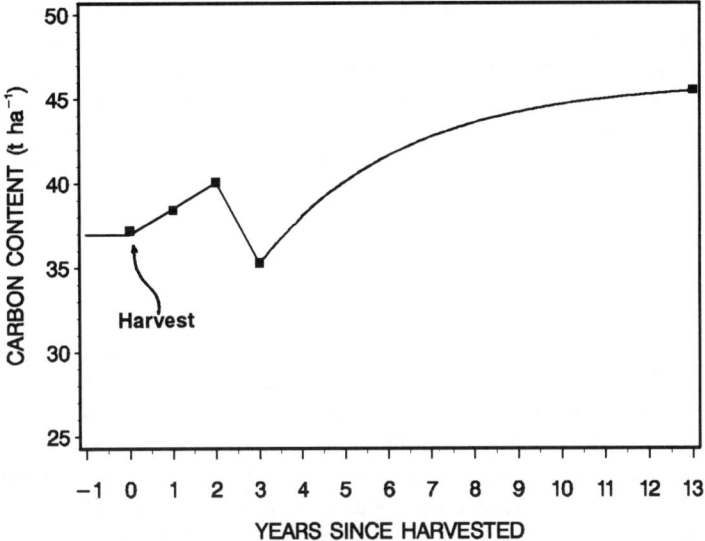

Fig. 22–6. The trend of soil matrix C following harvest on a poor Piedmont site.

Our model of the response of soil matrix C following harvest, based on empirical data, supports a similar theoretical model developed by Moore et al. (1981). This study indicates that within a relatively short time after harvest (13 yr in this case) soil matrix C had increased even above the assumed steady-state level of the mature plantation.

Carbon concentrations in both the 0- to 10- and 10- to 30-cm depths of the mineral soil matrix were significantly higher on the harvested site after 13 yr than in the 55-yr-old plantation (Table 22–3). The increase in soil C in the upper 30 cm may be a short-lived occurrence associated with the extremely high stem density and the corresponding high rates of root mortality in the naturally regenerated stand. In addition, detritus from the previous plantation is still contributing

Table 22–3. Carbon concentration in the soil matrix 13 yr after harvest of a mature loblolly pine plantation on a poor Piedmont site.

Soil depth	Control watershed	Harvested watershed
cm		%
0–10	0.97ax*	1.48ay
10–30	0.60bx	0.75by
30–50	0.37cx	0.35cx
50–100	0.08†	0.8

*Means within a column followed by the same letter (a, b, or c) are not significantly different at the $P = 0.05$ level. Means within rows followed by the same letter (x or y) are not significantly different at the 0.05 level.
†Values at the 50- to 100-cm depth were extrapolated and not analyzed statistically.

C as it decays. Debris inputs from the precommercial thinning 6 yr prior to final sampling also may have elevated soil matrix C levels in the short term.

Whether this higher level of soil matrix C is sustained throughout the second rotation can only be determined by subsequent sampling. However, since temperature and moisture, along with vegetation, are the primary determinants of soil C levels, it seems likely that soil C will once again decline to the levels found in the mature plantation.

SUMMARY

The distribution of C in belowground pools (soil and root system) was determined in a 55-yr-old loblolly pine plantation growing on a poor Piedmont site. Nearly 76% of the belowground C was in the soil (including the forest floor and old root channels), and about 24% was in the root system. Root channels occupied <3% of the soil volume but contained about 10% of the soil C. Soil matrix C increased 220% from the impoverished levels following agricultural abandonment of the severely eroded site. Carbon sequestered in the forest floor and root system totaled 29.1 t ha^{-1}. A portion of the mature plantation was harvested 13 yr ago, after which soil matrix C exhibited a short positive pulse, then a decline followed by a surge to a level exceeding, at least temporarily, that of the unharvested stand. Plantings of loblolly pine following agricultural abandonment not only slowed erosion, but also made significant contributions to storage of belowground C. Further, this study suggests that these plantations can be harvested and regenerated with only minor and temporary effects on storage of belowground C.

REFERENCES

Barber, B.L., and D.H. Van Lear. 1984. Weight loss and nutrient dynamics in decomposing woody loblolly pine logging slash. Soil Sci. Soc. Am. J. 48:906–910.

Brown, J.K. 1974. Handbook for inventorying downed woody material. Gen. Tech. Rep. INT-16. USDA-FS Int. For. and Range Exp. Stn., Ogden, UT.

Coile, T.S. 1936. Distribution of forest tree roots in North Carolina Piedmont soils. J. For. 35:4–39.

Cox, S.K., and D.H. Van Lear. 1985. Biomass and nutrient accretion on Piedmont sites following clearcutting and natural regeneration of loblolly pine. p. 501–506. *In* E. Shoulders (ed.) Proc. Biennial Southern Silvicultural Res. Conf., 3rd, Atlanta, GA, 7 8 Nov. 1984. Gen. Tech. Rep. S-54. USDA-FS South. For. Exp. Stn., New Orleans, LA.

Douglass, J.E., and D.H. Van Lear. 1983. Prescribed burning and water quality of ephemeral streams in the Piedmont of South Carolina. For. Sci. 29:181–189.

Emanuel, W.R., G.E.G. Killough, and J.S. Olson. 1981. Modeling the circulation of carbon in the world's terrestrial ecosystems. p. 335–353. *In* B. Bolin (ed.) Carbon cycle modeling. John Wiley & Sons, New York.

Giddens, J. 1957. Rate of loss of carbon from Georgia soils. Soil Sci. Soc. Am. Proc. 21:513–515.

Giddens, J., and W.H. Garman. 1941. Some effects of cultivation on the Piedmont soils of Georgia. Soil Sci. Soc. Am. Proc. 6:439–446.

Hole, F.D. 1981. The effects of animals on soil. Geoderma 25:75–112.

Jones, S.M. 1991. Landscape ecosystem classification for South Carolina. p. 59–68. *In* D.L. Mengel and D.T. Tew (ed.) Proc. Symp. Ecological Landscape Classification: Applications to Identify Productive Potential of Southern Forests, Charlotte, NC. Gen. Tech. Rep. SE-68. 7–9 January. USDA-FS Southeast For. Exp. Stn. Asheville, NC.

Kapeluck, P.R. 1988. Root biomass and nutrient content of a mature loblolly pine plantation on a Piedmont site. M.S. thesis. Clemson Univ., Clemson, SC.

Kapeluck, P.R., and D.H. Van Lear. 1995. A technique for estimating below-stump biomass of mature loblolly pine plantations. Can. J. For. Res. (In press.)

Lutz, H.J., and R.F. Chandler, Jr. 1955. Forest soils. John Wiley & Sons, New York.

McCollum, M.M. 1992. Density and distribution of loblolly pine (*Pinus taeda* L.) roots along an environmental gradient. M.S. thesis. Clemson Univ., Clemson, SC.

Moore, B., R.D. Boone, J.E. Hobbie, R.A. Houghton, J.M. Melillo, B.R. Petersen, G.R. Shaver, C.J. Vörösmarty, and G.M. Woodwell. 1981. A simple model for analysis of the role of terrestrial ecosystems in the global carbon budget. p. 365–386. *In* B. Bolin (ed.) Carbon cycle modelling. SCOPE Rep. 16, John Wiley & Sons, Chichester, England.

Persson, T. 1980. Depth and replacement of fine roots in a mature scots pine stand. Ecol. Bull. 32:251–260.

Richter, D.D., D. Markewitz, C.G. Wells, H.L. Allen, J. Duncombe, K. Harrison, P.R. Heine, A Sturnes, B. Urrego, and G. Bonai. 1994. Carbon cycling in a loblolly pine forest: Implications for the missing carbon sink and for the concept of soil. p. 233–252. *In* W.W. McFee and J.M. Kelly (ed.) Carbon forms and functions in forest soils. SSSA, Madison, WI.

Ruark, G.A. 1993. Modeling soil temperature effects on *in situ* decomposition rates for fine roots of loblolly pine. For. Sci. 39:118–129.

Schiffman, P.M., and W.C. Johnson. 1988. Phytomass and detrital carbon storage during forest regrowth in the southeastern United States Piedmont. Can. J. For. Res. 19:69–78.

Skeen, J.N., P. Doerr, and D.H. Van Lear. 1993. Oak-hickory-pine forests. p. 1–35. *In* W.H. Martin et al. (ed.) Biodiversity of the southeastern United States: Upland terrestrial communities. John Wiley & Sons, New York.

Sorrells, R.T. 1984. The Clemson Experimental Forest—Its first fifty years. Clemson Univ., College of For., Recreation Resour., Clemson, SC.

Switzer, G.L., and L.E. Nelson. 1972. Nutrient accumulation and cycling in loblolly pine (*Pinus taeda* L.) plantation ecosystems: The first 20 years. Soil Sci. Soc. Am. Proc. 36:143–147.

Trimble, S.W. 1974. Man-induced soil erosion on the Southern Piedmont—1700–1970. Soil Conserv. Soc. Am.

Van Lear, D.H., J.B. Waide, and M.J. Teuke. 1984. Biomass and nutrient content of a 41-year-old loblolly pine (*Pinus taeda* L.) plantation on a poor site in South Carolina. For. Sci. 30:395–404.

Van Lear, D.H., and N.B. Goebel. 1976. Leaf fall and forest floor characteristics in loblolly pine plantations in the South Carolina Piedmont. Soil Sci. Soc. Am. J. 40:116–119.

Van Lear, D.H., and P.R. Kapeluck. 1990. Nitrogen pools and processes during natural regeneration of loblolly pine. p. 234–252. *In* S.P. Gessel et al. (ed.) Sustained productivity of forest soils. Proc. 7th North American For. Soils Conf., Vancouver, Canada. 24–28 July 1988. Univ. British Columbia, Faculty of For. Publ., Vancouver, BC.

Van Rees, K.C.J. 1984. Root distribution of a slash pine plantation on a flatwoods spodosol. M.S. thesis. Univ. Florida, Gainesville, FL.

Wells, C.G., and J.R. Jorgensen. 1975. Nutrient cycling in loblolly pine plantations. p. 137–158. *In* B. Bernier and C.H. Winget (ed.) Forest soils and land management. Proc. 4th North American For. Soils Conf., Quebec, Canada. August 1973. Les Presses de L'Universite' Laval, Quebec, Canada.

23 The Role of Forest Soils in the Global Carbon Cycle

Alex F. Bouwman and Rik Leemans

*National Institute of Public Health and Environmental Protection
Bilthoven, the Netherlands*

The conversion of forests to agricultural, urban and other land use results in the release of carbon dioxide (CO_2) comparable in magnitude to the release of CO_2 by the combustion of fossil fuels (Bolin, 1977), which amounted to about 6 Pg C yr^{-1} in 1990 (WRI, 1992). Tropical forests are exploited by people for a variety of purposes, including timber and fuelwood extraction, shifting cultivation and clearing for permanent agriculture (Seiler & Crutzen, 1980). The main processes responsible for such CO_2 release after deforestation are burning and decomposition of aboveground biomass and soil organic matter, but each pathway has a different effect on the rate and timing of changes in C contents. The degree of C losses depends on the final land use and cover after clearing and the rate of reforestation. Globally, tropical forests are the major source of release of CO_2 (Lanly, 1982; Detwiler, 1986; Detwiler & Hall, 1988; Houghton et al., 1983, 1985, 1987). These forests currently cover large extents, but are mostly located in areas with ever-increasing population pressures and therefore high rates of deforestation.

Some sinks have been identified regionally within the terrestrial biosphere. Delcourt and Harris (1980) carefully analyzed past and recent forestry records and concluded that the C content for the temperate forests of the eastern USA has increased slightly since the 1950s and Kauppi et al. (1992) came to similar conclusions for Europe. However, Sedjo and Solomon (1989) indicated that this trend could never cope with the increasing anthropogenic CO_2 emissions. Recently, Tans et al. (1990) suggested through a simulation study that the terrestrial biosphere at temperate latitudes should be a major sink for atmospheric CO_2 sequestering 2 to 3.4 Pg C yr^{-1}. There are a number of potential processes that contribute to this sink, such as increased photosynthesis at higher atmospheric CO_2 concentrations (CO_2 fertilization), reforestation and accumulation of soil C.

Future global warming may increase the rates of soil organic matter decomposition. This may lead to increased soil C losses (Jenkinson et al., 1991). Shifts in vegetation zones caused by climate warming (Leemans, 1992) may lead to a

---

Copyright © 1995 Soil Science Society of America, 677 S. Segoe Rd., Madison, WI 53711, USA.
Carbon Forms and Functions in Forest Soils.

slight increase of 1% in the global soil C pool (Prentice & Fung, 1990), but Smith et al. (1992) show that this estimate is in fact the upper limit. Depending on the regional characteristics of the tropical forests it could be much less.

Global C fixation through photosynthesis by terrestrial plants and vegetation may be approximately 110 Pg C yr^{-1}, of which about 50% finally ends up in the soil compartment. Net primary production (NPP, the total amount of organic material synthesized in excess of that used in respiration) by terrestrial biota is equivalent to approximately 60 Pg C yr^{-1}, litterfall 40 to 50 Pg C yr^{-1}, and the estimated total stock of litter may be 60 Pg C (Ajtay et al., 1979). Estimates of the total size of the global biomass C pool range from approximately 560 Pg C (Ajtay et al., 1979; Goudriaan & Ketner, 1984) to 835 Pg C (Whittaker & Likens, 1975). The amounts of C stored in soils may range between 1000 and 3000 Pg C (Table 23–1). The other C pools are: atmosphere (720 Pg C), oceans (38000 Pg C), fossil C reserves (6000 Pg C) (Goudriaan & Ketner, 1984) and caliche (petrocalcic horizons in arid and semiarid regions, 800 Pg C) (Schlesinger, 1982). On a worldwide basis the amount of C in decaying plant litter and soil organic matter may well exceed the amount of C in live vegetation by a factor of two or more. Small changes in litter production or decomposition could therefore have large consequences for atmospheric CO_2 levels. Soils may form either significant sinks or additional sources for C.

In the first part of this paper we discuss estimates of global soil C pools. In the second part we will make an attempt to quantify so-called "equilibrium fluxes" of soil C, including soil organic matter accumulation, weathering of bedrock and riverine transport, and "transient" fluxes of soil C caused by land use changes. Finally, we will present an overview of the literature of the effects of climate change and increasing atmospheric CO_2 on soil C.

Table 23–1. Estimates of the pool of C in world soils and annual loss of soil C caused by deforestation.

Approach	Reference	Soil C pool	Annual loss of soil C
		Pg C	Pg C yr^{-1}
Vegetation types	Bolin (1977)	700	0.3
	Schlesinger (1977)	1456	0.9
	Bolin et al. (1979)	1672	
	Ajtay et al. (1979)	1635	
	Schlesinger (1984)	1515	0.8
	Detwiler (1986)		0.11–0.26
Soil types	Bohn (1976)	3000	
	Atjay et al. (1979)	2070	
	Post et al. (1982)	1395	
	Bohn (1982)	2200	
	Buringh (1984)	1477	1.54–5.4
	Bouwman (1990)	1700	
	Kimble et al. (1990)	1061	
	Adams et al. (1990)	1115	
	Eswaran et al. (1993)	1576	
Life zone groups	Post et al. (1982)	1395	
	Post et al. (1985)	1272	
	Zinke et al. (1984)	1300–1700	
Modeling	Meentemeyer et al. (1981)	1457	

GLOBAL SOIL CARBON POOLS

There have been several attempts to estimate the storage of soil organic matter for the different ecosystems of the world (Table 23–1). These include: (i) estimates based on vegetation types—this methodology assumes that soil C content is characteristic for each vegetation type. Total soil C content can then be computed when the extent of a vegetation type is known. The disadvantage of these estimates is that the variability of soil factors that are important for organic matter accumulation and decomposition is not considered. (ii) Estimates based on soil types—these estimates are based on generalized soil C contents based on Soil Survey Staff (1975) (Bohn, 1976; Eswaran et al., 1993) or a combination of generalized C contents of profiles and vegetation types (Buringh, 1984; based on Soil Survey Staff, 1975; Bouwman, 1990; based on FAO-Unesco, 1971–1982). (iii) Estimates based on climate classifications—the estimates by Post et al. (1982, 1985) and Zinke et al. (1984) are based on the Life Zone Classification of climates as developed by Holdridge (1967). Zinke et al. (1984) analyzed about 2700 soil profiles representing virtually every major ecosystem. They list C content (not including C stored in O horizons) and bulk density for each soil horizon or at standard depths for each location. (iv) Estimates based on modeling—Meentemeyer et al. (1981) applied a simple model to calculate soil C contents based on climate and litter input to soils.

It becomes clear from the studies presented in Table 23–1 that global soil organic matter could range from 1000 to 3000 Pg C. These estimates still exclude the amounts of inorganic C stored in soils, of which calcium carbonate in "calcic" horizons in deserts and semideserts, and charcoal are the major forms. Some 800 Pg C may be stored in caliche (calcic horizons) occurring in the deserts throughout the world.

Sanford et al. (1985) reported that the uppermost 1 m of the terra firme forest soils in Venezuela contain amounts of charcoal of up to 20 Mg ha^{-1}. Fossil charcoal may be related to recurrent wildfires from historical times (Jones & Chalone, 1991) and this illustrates the degree of recalcitrance of this type of C. Based on several assumptions Esser (1990) suggested that since 1860 about 8 Pg of charcoal C has been formed in grasslands and savannas and 1.5 Pg C in temperate forests. However, not all this charcoal finds its way into the soil since part is injected into the atmosphere and part is lost due to erosion.

FLUXES OF SOIL CARBON

We consider "equilibrium" fluxes and "transient" fluxes of soil C. Fluxes resulting from soil formation and weathering are considered equilibrium fluxes, because they may have been fairly constant in the last centuries to millennia. Riverine transport of C originating from forest soils also is considered to be an equilibrium flux, although it may change under the influence of man on a shorter time scale. Changes in soil C as a result of changes in land use (deforestation, reforestation) are transient, because they may change at a time scale of years to decades. Shifting cultivation is a process with changing intensity and extent at the global scale. Although shifting cultivation is a much more slowly changing process than permanent deforestation, we classify the associated soil C changes as transient fluxes.

Equilibrium Fluxes of Soil Carbon

Soil Carbon Accumulation

Long-term accumulation rates derived from analysis of 16 soil profiles between 3000 and 10000 yr old were used to estimate the maximum potential for the present accretion of soil C of 2.4 ± 0.7 g C m^{-2} yr^{-1} (Schlesinger, 1990). Applying this number to the global area of upland soils yields a global accumulation flux of 0.32 Pg C yr^{-1}. This is consistent with Goudriaan and Ketner (1984) who estimated an increase of humus plus charcoal C of 56 Pg C since 1780 based on both transient and equilibrium fluxes. The simulated equilibrium accumulation rate presented by Adams et al. (1990) amounts to 490 Pg since the last glacial maximum 18000 yr ago. This is much lower than the above 0.32 Pg yr^{-1}. Soil C accumulation rates depend strongly on soil forming process. The accumulation of C in the formation of humus rich B_h horizons in Spodosols may be up to 55 g C m^{-2} yr^{-1} (Harden et al., 1992), while C sequestration in Alifisols and Inceptisols may be much lower than the rate assumed by Schlesinger (1990), illustrating the variability.

Armentano and Menges (1986) estimated that the world's wetlands annually sequester 0.08 Pg C yr^{-1}, based on a wetland extent in temperate zones of 349 Mha. More recent global estimates of the wetland area add up to 530 to 570 Mha (Matthews & Fung, 1987; Aselmann & Crutzen, 1989) and this would lead to somewhat higher C accumulation rates.

It is assumed that the global soil C accumulation in upland and wetland soil of 0.4 Pg C yr^{-1} occurs mainly in forest soils (Table 23–2). As indicated above, soil carbonates form a significant global C pool. The estimated annual flux of CO_2 from the atmosphere to desert-soil carbonate of 0.023 Pg C yr^{-1} (Schlesinger, 1985, 1986) is negligible compared to the C accumulation rate estimated for global upland and wetland soils.

Weathering

Weathering of calcium carbonate is a soil forming process typical for all calcareous soils in areas with a surplus of rainfall over evapotranspiration. From data summarized in Van Breemen and Protz (1988) and Van Breemen and Feijtel (1990), we can conclude that removal of atmospheric CO_2 by dissolution of calcium carbonates is in the order of 4 to 30 g C m^{-2} yr^{-1}.

In noncalcareous areas the drainage flux of HCO_3^- is determined by the rate of silicate weathering. This depends strongly on the types and quantities of minerals present in the soil and the bedrock and on intensities of biological activities accelerating weathering processes. Generally the CO_2 sink associated with silicate weathering is much less than the one discussed for calcareous soils. Lowest rates are observed in acidic soils low in easily weatherable minerals (e.g., in Hubbard Brook watershed, 0.003 g C m^{-2} yr^{-1}), in areas with easily weatherable silicate rock values could range from 0.6 g C m^{-2} yr^{-1} (e.g., for pellitic schists in Maryland) to 12 g C m^{-2} yr^{-1} (very high precipitation areas in Cascade Mountains) (Van Breemen & Feijtel, 1990, and references therein).

Table 23–2. Summary of global estimates of soil C fluxes for forest soils.

	C loss	C accumulation	References
	— Pg C yr^{-1} —		
Equilibrium fluzes			
Soil C accumualtion		0.4	Schlesinger (1990)
Weathering		0.2–0.3	Van Breemen & Feijtel (1990)
Riverine transport	0.4		Meybeck (1982)
Transient fluxes			
Forest conversion to agriculture	0.3–0.7		This chapter
Shifting cultivation	0.1–0.3		This chapter
Forest conversion to "other land"	0.1		This chapter
Charcoal formation†		0.1–0.3	This chapter
Reforestation		0.3	This chapter
Global temperature rise	1.0		Jenkinson et al. (1991)

†Formation of charcoal caused by biomass burning, including the process of permanent deforestation, shifting cultivation and savanna burning.

Net removal of atmospheric CO_2 by drainage of water with dissolved CO_2 in excess of that dissolved in equilibrium with the atmospheric partial CO_2 pressure could be on the order of 0.1 to 10 g C m^{-2} yr^{-1} (Van Breemen & Feijtel, 1990). Part of this dissolved CO_2 could facilitate weathering processes in aquifers and the remainder is removed by degassing when groundwater enters open streams.

In summary, worldwide soils may remove 0.16 to 0.27 Pg C yr^{-1} of atmospheric CO_2 as HCO_3^- during weathering of the parent material (Holland, 1978). We assume that this removal occurs mainly in forest soils (Table 23–2).

The currently increasing atmospheric CO_2 concentration, associated global warming, shifts in vegetation patterns and changes in biological activity may increase the importance of weathering. There may be large areas of comparatively unweathered land at high latitudes that may act as sinks for C in the future (Bird et al., 1990).

Riverine Transport

Another potential sink for CO_2 may be export of dissolved organic C compounds from the solum. Concentrations below the rooting zone are generally 2 to 10 g C m^{-3}, with the highest values for Spodosols and Histosols, and the lowest values for C-rich soils and tropical soils high in Fe and Al; intermediate values would apply to most other temperate soils (Van Breemen & Feijtel, 1990). Based on annual percolation rates of between 100 and 1000 mm water, these concentrations imply a removal of organic dissolved C from soils on the order of 0.2 to 10 g C m^{-2} yr^{-1}. Part of this may be retained as solid humic material in deeper strata, part may be oxidized in oxygenated aquifers, and part may finally be transported by rivers.

Meybeck (1982) estimated a global riverine transport of 0.6 Pg C yr^{-1} including dissolved organic and inorganic organic matter, and 0.35 Pg C yr^{-1} of particulate organic and inorganic C. Organic C transport by rivers (dissolved and particulate) is about 0.4 Pg C yr^{-1} (Meybeck, 1982; Schlesinger & Melack, 1981). Part of this material is decomposed during transport, so that the amount of C reaching the oceans could be less.

Global estimates for soil C losses via rivers are extremely uncertain, and the role of the world's forests is even more uncertain. Based on estimates for broad ecosystems from Meybeck (1982) we estimate that riverine transport of organic C from forest soils could be 0.4 Pg C yr^{-1} (Table 23–2).

Transient Fluxes of Soil Carbon Caused by Land Use Changes

General

The annual loss of soil C caused by changes in land use can be considerable, although large variation exists in different parts of the world. The contribution of soils always must have been a significant part of the global C cycle. Buringh (1984) gives an estimate of 537 Pg C for total net soil C losses caused by deforestation since prehistoric times. Estimates of the current annual C release from soils due to land use changes for the 1980s are still uncertain (Table 23–1). Forest conversion and forest management may have a number of important consequences for soil organic matter: (i) secondary growth or revegetation of cleared areas, and in case of permanent conversion, the land use after clearing, determines the quantity and quality of organic material added to the soil each year (Lugo & Brown, 1986; Detwiler, 1986; Mann, 1985, 1986; Schlesinger, 1986). The lack of a vegetation cover after clearing could accelerate the decomposition of soil organic matter (Cunningham, 1963; Ellis & Graley, 1983), although the opposite also has been observed (Johnson, 1992). (ii) Soils under cultivation generally have higher bulk densities than soils under natural vegetation. This complicates comparisons between virgin and cultivated soils (Voroney et al., 1981; Mann, 1985; 1986; Johnson, 1992; Davidson & Ackerman, 1993). (iii) Burning of vegetation converts a portion of the biomass into inert charcoal (Seiler & Crutzen, 1980). (iv) Clearing of vegetation could cause significant losses of material from the O and A horizon caused by wind and rainfall erosion.

Tropical deforestation has increased from 11 to 15 Mha yr^{-1} between the early 1980s and 1990s (Table 23–3). The major part of this deforestation is occurring in Latin America (currently 7.4 Mha yr^{-1}), while current deforestation in Africa may be 4.1 Mha yr^{-1} and 3.9 Mha yr^{-1} in Asia (Table 23–4). At present close of 1% of the total tropical forest area is cleared annually. It is generally believed that deforestation rates peaked in 1987 for the Amazon region and deforestation rates have decreased steadily since then (WRI, 1992). This is in agreement with recent satellite measurements (Skole & Tucker, 1993). The global arable area has increased from 1330 to 1440 Mha between 1960 and 1990, an increase of about 4 Mha yr^{-1} (FAO, 1991). In the same period the global grassland area has increased from 3210 to 3400 Mha, while in all developing countries, the area of permanent pastures decreased from 2410 to 2130 Mha (FAO, 1991). Forest clearing for expansion of arable land may have been 7.9 Mha yr^{-1} and forest conversion to grasslands 1.5 Mha yr^{-1} in the early 1980s (Houghton et al., 1987) or maybe 12 Mha yr^{-1} and 2.5

Table 23-3. Estimates of global deforestation rates and shifting cultivation.

	Mha yr^{-1}	
Permanent clearing		
Lanly (1982)	11.3†	
FAO (1993)	15.4‡	
Shifting cultivation	Closed forests	Open forests
	Mha yr^{-1}	
Seiler & Crutzen (1980)	17.5–43.6	6.9–21.9
Detwiler et al. (1985)	18	18.6
Houghton et al. (1985)	13.5	

†For early 1980s.
‡For late 1980s.

Table 23-4. Comparison of regional reforestation and deforestation rates and areas of forest plantations [sources: FAO (1993) for deforestation; areas of forest plantations and net area planted for Africa, Latin America and Caribbean, Asia and Pacific from FAO (1993); net forest area planted for North America, Europe and former USSR from WRI (1992)].

Region	Deforestation	Net forest area planted†	Area of forest plantations
	Mha yr^{-1}		Mha
Africa	4.1	0.1	3.0
North America	0.0	1.7	ND‡
Latin America and Caribbean	7.4	0.3	8.6
Asia and Pacific	3.9	1.5	32.3
Europe	0.0	0.7	ND
Former USSR	0.0	3.2	ND
Global	15.4	7.4	43.9§

†The net forest area planted is taken as 70% of the reported areas planted to account for failed plantations (FAO, 1993)
‡ND = no data
§For the Tropics only.

Mha yr^{-1} for arable land and grassland expansion at present. Hence, part of the agricultural land is lost by land degradation or conversion to other uses, and loss of productivity is replaced by further clearing of remaining forests. Table 23-5 presents a matrix of global land transformations, including deforestation, expansion of agricultural land, land degradation, urbanization and other conversions, and reforestation. The estimated land degradation loss described in Table 23-5 could be close to the estimated loss due to mismanagement of 4 Mha yr^{-1} (Buringh & Dudal, 1987). The sum of the annual conversion of agricultural to nonagricultural land (3 Mha yr^{-1}) and forest conversion to "other land" (3 Mha yr^{-1}) is close to the estimate of 8 Mha yr^{-1} (Buringh & Dudal, 1987). Obviously, these global land use changes are still extremely uncertain.

Table 23–5. Assumed curent global annual land use changes [total deforestation rates are from FAO (1993) and conversion rates are updated for 1990 with data for the early 1980s reported by Houghton et al. (1987), reforestation rates from Table 23–4.

	Land degradation, urbanization and other conversions†	Deforestation	Reforestation‡	Total change
		Mha yr^{-1}		
Arable land	–6.0	12.0		4.0
Grassland	3.0	2.5		5.5
Forest		–15.5	7.5	–8.0
Other land§	3.0	3.0	–7.5	–1.5

†Rates of land degradation, urbanization and other conversions (e.g., arable land to grassland) are adapted to arrive close to the total changes reported by FAO (1991).
‡Reforestation is assumed to take place on "other land" and not on arable land or grassland.
§Including urban land.

Permanent Conversion of Forest to Arable Land and Grassland

The C content is usually lower in cultivated soils than in soils under forest or grass cover. There are a number of possible explanations for this: (i) much of the biomass produced on arable land is harvested, (ii) decomposition rates are higher in cultivated soils due to higher soil temperatures, (iii) the organic substances of crops are less resistant against decomposition (Schimel et al., 1985) and cultivation increases the accessibility of organic matter to microbial attack by disruption of soils structure (Voroney et al., 1981), (iv) reduced interception by the crop canopy causes increased infiltration. Soil C losses due to leaching are generally smaller in forests, (v) changes in microbial species composition may change such that the importance of fungi declines (Voroney et al., 1981).

A great number of authors have tried to quantify the effect of cultivation on soil C. Detwiler (1986) concluded for tropical regions that conversion of forests to arable land results in a decline of soil C of 40%, and that clearing followed by pasturing results in a loss of only 20%. Schlesinger (1986) found an average soil C loss of 21% for cultivated soils from the Americas, Africa, India, Thailand and New Guinea. Post and Mann (1990) analyzed over 1100 soil profiles and found that the amount of C or N lost after starting cultivation is a function of the initial amount stored in the soil. Average C loss for soils with high initial C was about 23%. Soils with very low initial C actually increased their C storage during cultivation. In soils with intermediate C contents, soils with higher C/N ratios lost more C than those with low C/N ratios. Jones (1972) calculated from data for a savanna in Guinea that soils under cultivation or under fallow following cultivation, have an average C content of slightly more than 50% of the original soil content. Lugo and Brown (1986) reported losses of 46% after clearing and 10 yr of cultivation of a Central American subtropical wet forest. This value increased to 70% after 100 yr of cultivation. In dry areas losses were usually lower. In all cases in relatively short periods of time (decades) most of the losses could be regained by the abandonment of cultivation and secondary forest succession or use as pasture. Johnson (1992) concluded that forest harvesting and subsequent cultivation result in losses of soil C on the order of 30 to 50% over a period of several decades. Davidson and Ackerman (1993) reviewed reports of paired comparisons of soil C in virgin and cultivated soils and concluded that between 20 and 40% of soil C is lost following cultivation, the best estimate being 30% for the entire soil column.

Using the matrix of land use changes in Table 23–5 and the method for estimating soil C losses proposed by Buringh (1984), the 1990 soil C loss would be 1.4 Pg C yr^{-1} for the 12 Mha yr^{-1} of newly formed arable land and 0.2 Pg yr^{-1} for the 2.5 Mha yr^{-1} of grassland expansion giving a total of 1.6 Pg C yr^{-1}. Buringh (1984) proposed to use soil C loadings in soils of 208 Mg C ha^{-1} for forests, 95 Mg C ha^{-1} for arable and 116 Mg C ha^{-1} for grasslands. The amounts of soil C assumed by Buringh (1984) for forests and the differences between forest soils and agricultural soils seem high compared to the numbers proposed by other authors (Post et al., 1982; Tinker & Imeson, 1990) and this may lead to overestimation of C loss.

Nearly all the soil C losses occur within the first 20 yr after clearing, and most occur within 5 yr (Davidson & Ackerman, 1993). Model simulations suggest that perhaps 100 yr are needed for a complete equilibrium (Van Veen & Paul, 1981). To estimate the transient flux we need to make an integration over time of soil C losses, including those caused by historic forest clearings. In the following model approach we use different model soil profiles to represent extreme cases of soil C loss. We included heavy and light textured Oxisols with about 150 and 105 T C ha^{-1}, respectively, and heavy and light textured Ultisols, with about 110 and 90 T C ha^{-1}, respectively. Ultisols and Oxisols are the most common soils in tropical forest regions. Based on the current conversion rate of 12 Mha yr^{-1} of forest to permanent arable land (Table 23–5), data for historic forest clearing rates (Houghton et al., 1983), and the soil model profiles presented in Bouwman (1989), we estimated soil C losses (Fig. 23–1 and Table 23–6). The C loss including historic clearings for the period 1900 to 1990 would be 0.3 to 0.7 Pg C yr^{-1}. The major C release comes from clearings of the last two to three decades. We omitted losses of soil C caused by forest conversion to grassland, because they are much lower than for forest conversion to arable land (Detwiler, 1986). Moreover, the forest conversion to grassland is much slower than the conversion to arable land (Table 23–6).

Charcoal may be formed during forest burning. Seiler and Crutzen (1980) estimated that the amount of charcoal C remaining after burning may be 20 to 30% of the C in the unburned aboveground biomass. Crutzen and Andreae (1990) improved this estimate by adding more observations and estimated that 5% of the C exposed or 10% of the C actually burned is converted to charcoal. The current burning of 200 to 700 Tg C yr^{-1} in permanent deforestation (Crutzen & Andreae, 1990) may result in 20 to 70 Tg C yr^{-1} of charcoal formation. Direct injection of charcoal into the atmosphere may be only 1% of the charcoal formed (Suman, 1988) or 18% of the total particulate matter in smoke (Crutzen & Andreae, 1990). Jordan (1985) reported that after burning initial losses of ash and topsoil are heavy. Burning mainly occurs towards the end of the dry season, when it is difficult to establish a suitable crop cover necessary to protect the soil surface. Due to its low density charcoal easily becomes airborne after burning (Seiler & Crutzen,

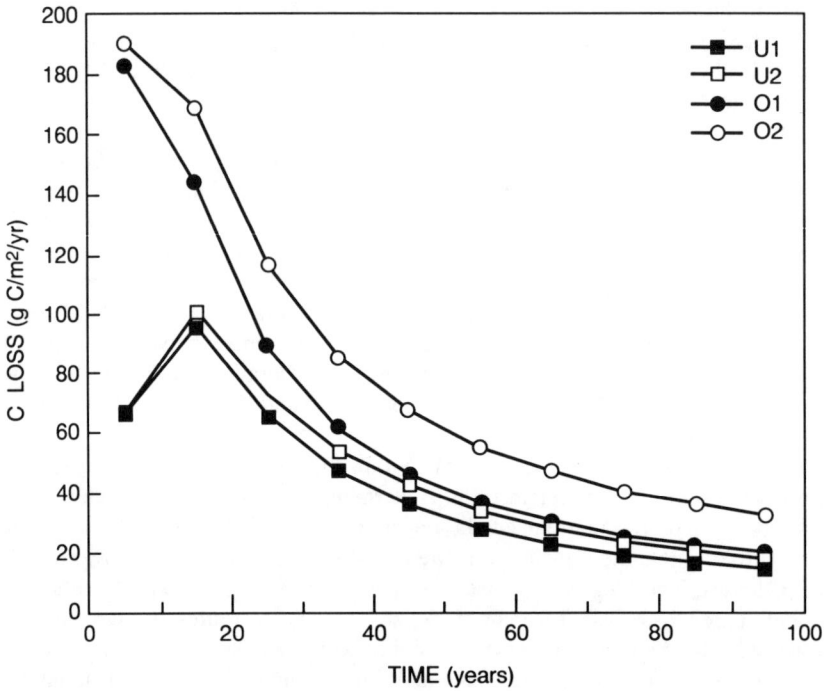

Fig. 23–1. Modeled 10-yr average C flux caused by loss of soil organic matter from soils under permanent cultivation. Model profiles U1 and U2 are respectively light textured and heavy textured Ultisols for tropical seasonal forest. Model profiles O1 and O2 are respectively light-textured and heavy-textured Oxisols for tropical rain forest. The 0- to 0.8-m model profiles have three layers. The 100-yr simulation resulted in a loss of initial soil organic C of about 45% for the U model profiles and 60% for the O model profiles (adapted from Bouwman, 1989).

1980). Up to 50% of the charcoal formed may be lost by soil run-off (Bouwman, 1989).

Shifting Cultivation and Savanna Burning

Shifting cultivation seldom results in a substantial soil organic matter depletion; usually soil C contents are maintained at 75% of natural equilibrium levels (Nye & Greenland, 1960), but during decreasing rehabilitation periods soil C contents could drop to 50% of the original levels (Sanchez, 1976). It is probable that with increasing population pressure shifting cultivation intensity increases with subsequent decreasing fallow periods and a larger proportion of shifting cultivation that is converted to permanent agriculture (Houghton et al., 1985). Both trends result in soil C loss (Nye & Greenland, 1960; Sanchez, 1976).

Seiler and Crutzen (1980) estimated that the total area under shifting cultivation is 300 to 500 Mha. They assumed a crop/fallow ratio of 1:10 and an average area cleared per person of 0.1 to 0.2 ha yr^{-1}. The estimated population of shifting cultivators for the early 1980s of 200 million then results in a total area

Table 23–6. Conversion of tropical and temperate forests to permanent agriculture and associated C loss caused by accelerated soil organic matter decomposition. Deforestation estimates are adapted from Houghton et al. (1983). C fluxes from soils are from Fig 23–1.

	1901–1940	1941–1960	Period 1961–1970	1971–1980	1981–1990	Total 1990
Tropical permanent forest conversion to arable land						
			Mha yr^{-1}			
	3	5	9	10	12	
Soil C loss in 1990 from areas cleared in period indicated						
			g C m^{-2} yr^{-1}			
Low estimate†	20	40	70	100	70	
High estimate†	50	80	120	170	190	
Soil C loss in 1990; × area cleared; × number of years per period						
			Tg C yr^{-1}			
Low estimate†	20	40	60	100	80	300
High estimate†	60	80	110	170	230	650

†High and low estimates refer to C losses from heavy textured Oxisols (O2) and light textured Ultisols (U1), respectively (Fig 23–1).

cleared annually of 20 to 60 Mha yr^{-1} (Table 23–3). The different shifting cultivation systems have been extensively described (Nye & Greenland, 1960; Sanchez, 1976; Jordan, 1985), but no improved quantitative and geographic data on areas cleared other than those presented in Table 23–3 are currently available (Meyers, 1991). Therefore it still proves very difficult to give reliable estimates of soil C fluxes. The model described in Bouwman (1989) predicts a decline of soil C to 70 to 80% of initial levels after 100 yr of shifting cultivation (Fig. 23–2). Assuming that soil erosion plays no role in shifting cultivation and based on shifting cultivation of about 40 Mha yr^{-1} for the early 1980s, and clearings in other periods calculated as a function of population in developing countries, the soil C loss in 1990 is 0.1 to 0.3 Pg C yr^{-1} (Table 23–7).

Significant amounts of charcoal may build up during shifting cultivation. The assumption that 10% of the material burned remains as charcoal (Crutzen & Andreae, 1990) in shifting cultivation systems leads to a global amount of charcoal formation of 50 to 100 Tg C yr^{-1} or 100 to 200 g C m^{-2} yr^{-1}. This estimate is in line with the estimated 70 g C m^{-2} yr^{-1} for burning in Central America (Suman, 1988) and the 500 g C ha^{-1} of "unhumified" C—including charcoal—added to the soil compartment during shifting cultivation in each rotation period (Nye & Greenland, 1964). Accounting for water and wind erosion losses, the in situ accumulation of soil charcoal is probably much less.

Savannas and bushland are ecosystems with an abundant grass cover intermingled with trees or shrubs covering about 1900 Mha worldwide. Savannas may be burned every 1 to 4 yr during the dry season with highest frequency in humid savannas (Crutzen & Andreae, 1990). This process may not be a net source of atmospheric CO_2, but considerable amounts of charcoal may be formed during burning. The amount of C burned in this process is 300 to 1600 Tg C yr^{-1}, of which 10% or 30 to 160 Tg C yr^{-1} may be converted into charcoal.

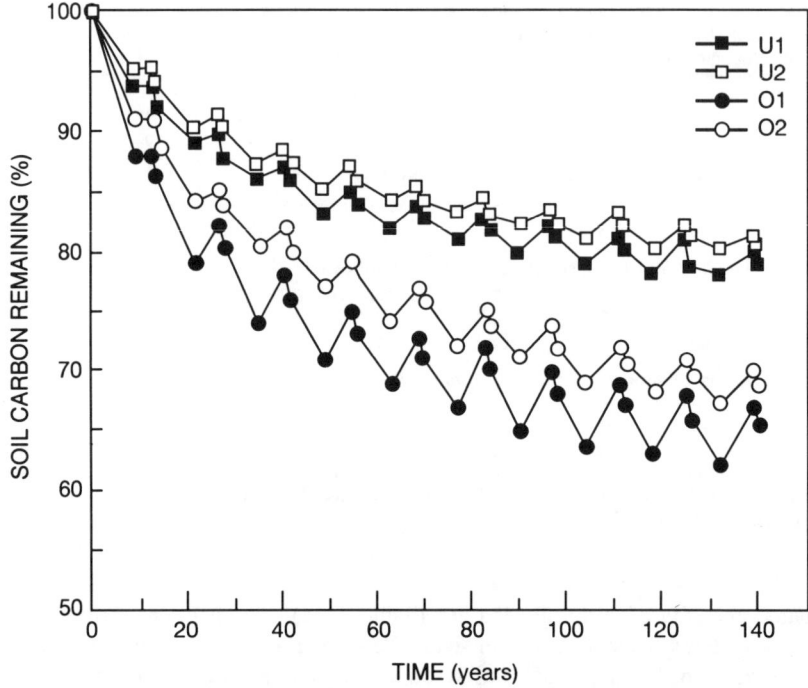

Fig. 23–2. Simulated soil C loss from soils under shifting cultivation for soil model profiles U1, U2, O1 and O2 (see Fig. 23–1). Two years of cropping is followed by 12 yr of regeneration of the forest. Annual additions of C during the two cropping years are 120 and 69 g C m^{-2} yr^{-1}, for litter and root material, respectively. For the rain forest sites (O) forest litter C input increases from 150 g C m^{-2} in the 3rd yr to 550 g C m^{-2} after 14 yr of regeneration. Root C inputs increase from 33 to 120 g C m^{-2} between these years. For seasonal forest sites (U) forest litter inputs increases from 140 to 280 and root C inputs from 85 to 130 g C m^{-2} yr^{-1}. At the beginning of each rotation assumed initial inputs from clearing residues were 500 g C m^{-2}.

Other Forest Clearing Processes

It is difficult to say what use the 3 Mha of cleared forest land which is converted to "other land" may have (Table 23–5). Certainly it means degradation of forests, and this may lead to a decline in soil C. If we follow Buringh's (1984) approach, the 3 Mha yr^{-1} of forest conversion to "other land" would lead to losses of 0.5 Pg C yr^{-1}. Buringh (1984) used soil C loadings of 208 Mg ha^{-1} for forests and 30 Mg C ha^{-1} for "other lands." The soil C loading assumed for other lands seems rather low and that of forests high compared to Post et al. (1982) and Tinker and Imeson (1990), and this may lead to overestimation of the C loss. The global C loss resulting from forest conversion to "other land" could amount to 0.1 Pg C yr^{-1} based on the low estimate for C losses (Table 23–6), assuming that part of the "other lands" are degraded forests or fallow lands with lower soil C contents than forest soils.

The estimated global soil C loss caused by permanent conversion to agriculture, shifting cultivation and forest conversion to other uses is 0.5 to 1.1 Pg C yr^{-1}

Table 23–7. Increase of the area of open and closed forests cleared in shifting cultivation and associated C loss caused by accelerated soil organic matter decomposition. Areas cleared in shifting cultivation are calculated from an assumed 40 Mha yr^{-1} clearing for the early 1980s (Table 23–3). Cleared areas in other periods were calculated as a function of population in developing countries from WRI (1988). Carbon dioxide fluxes from soils are calculated from Fig. 23–2 and Bouwman (1989).

	1900	1901–1942	Period 1943–1970	1971–1984	1985–1990	Total 1990
Tropical forest clearing in shifting cultivation						
	Total		Increase in annual cleared area			Total
	Mha		——— Mha yr^{-1} ———			Mha
	12	5	12	11	5	45
Soil C loss in 1990 from deforestation in period indicated						
			g C m^{-2} yr^{-1}			
Low estimate†		10	10	30	40	
High estimate†		20	30	60	100	
Soil C loss in 1990; × increase in area cleared; × number of years per period						
			Tg C yr^{-1}			
Low estimate†		20	30	50	10	110
High estimate†		40	100	90	30	270

†High and low estimates refer to C losses from heavy–textured Oxisols (O2) and light textured Ultisols (U1), respectively (Fig 23–2).

(Table 23–2), which is in good agreement with Schlesinger (1984), who reported a loss of 0.8 Pg yr^{-1} of soil C.

Reforestation

The worldwide net reforestation rate is 7.4 Mha yr^{-1} (Table 23–4), including the establishment of plantations for industrial and nonindustrial purposes. This estimate does not include regeneration of old tree crops through either natural regeneration or forest management. However, some countries may have reported such regeneration as reforestation (WRI, 1992). Double counting also may occur when failed plantations are replanted. Many trees also are planted for nonindustrial purposes, such as village wood lots. Reforestation data often exclude this component (WRI, 1992). It is obvious that deforestation and reforestation are occurring in different areas, with reforestation mostly in temperate regions (Table 23–4).

Where former agricultural land is reverted to forest, soil C and C stored in the O horizon usually increase substantially. In a review Johnson (1992) concluded that in most studies a significant accumulation of C in soil and O horizon was found, with increases of 35% to even a doubling. In some cases soil C decreased, but this decrease was more than offset by increases in C storage in the O horizon.

As most of the plantations are in temperate climates, a C loading in soils before plantation of 50 Mg C ha^{-1} (50% of the forest soil C) and equilibrium soil C in the mature forest of 100 Mg C ha^{-1} (for temperate forests, see Post et al., 1982; Tinker & Imeson, 1990) were used. We assume that the accumulation of 50

Mg C ha^{-1} occurs in three decades. About 60% of the 44 Mha of forest plantation in the Tropics has been established during 1981 to 1990 (FAO, 1993). Therefore, we assume that all reported tropical plantations (Table 23–4) are less than 30 yr old, so that the accumulation is perhaps 0.07 Pg C yr^{-1}. There are no data available for temperate regions (WRI, 1992). The current rate of net forest plantation in temperate regions exceeds that in the Tropics by a factor of three (Table 23–4). Assuming that the total temperate plantation area with ages of <30 yr is also three times greater than in the Tropics, a very tentative estimate of the global soil C accumulation rate in recent forest plantations is 0.3 Pg yr^{-1}.

Forest Management

Detwiler et al. (1985) and Johnson (1992) concluded that soil C losses result from soil use, not from clearing per se. Johnson (1992) reviewed several studies reporting soil C changes after harvesting of forests. Harvesting alone had a considerable effect on C in the forest floor mass, causing either increases or decreases depending on how much slash was left behind. The effects of harvesting on soil C varied from site to site, with observed increases, decreases or no effects. The majority of the studies reported no effects or only slight decreases in soil C. The overall conclusion was that little (10%) or no change of soil C can be expected to occur after forest harvesting alone. Therefore, managed forests, currently covering 970 Mha worldwide (WRI, 1992) may have no net global effect on the C cycle (Detwiler, 1986).

Natural Disturbances

Lugo and Brown (1986) indicated that forested landscapes are frequently disturbed by, e.g., fires and storms and this may affect the soil C fluxes. There may be loss of C (including soil C) shortly after disturbances and accumulation of plant and soil C over longer periods (<300 yr). However, data on frequency and magnitude of "natural" forest disturbances are insufficient to produce a global estimate of the effects on soil C.

Nitrogen Deposition

Peterson and Melillo (1985) estimated that the potential C storage in forests caused by N loading from fossil fuel combustion could be 0.1 Pg C yr^{-1}. If this storage is occurring, most of it could be in temperate forests, and about half in forest soils. These calculations were based on a release of 20 Tg N yr^{-1} from fossil fuel combustion, of which 30% was assumed to reach forests via atmospheric transport and precipitation. Peterson and Melillo (1985) assumed that only 60% of the N input is retained. A more recent estimate of the N release from fossil fuel combustion is 37 Tg N yr^{-1} (Jaffe, 1992), so that the above potential C storage could be higher. Part of the deposition of 89 Tg N yr^{-1} as NH_3 and NH_4^+ also could reach forests, and add to the potential C storage. The C accumulation due to eutrophication may not have been realized, because some of the N inputs may be lost through denitrification, and some forests may not be N limited and would not store more C with higher N loading. In addition, as ecosystems become richer in N, the N use efficiency may decrease. The response to N also may be limited because increases in N deposition are accompanied by increases in H$^+$, sulphate, heavy metals and other toxins (Peterson & Melillo, 1985).

Deposition of N is not spatially uniform. In regions receiving high N inputs from deposition, dramatic C sequestration should be observed. Increased standing wood volume has been reported for many European countries (Kauppi et al., 1992; Huettl & Zoettl, 1992) and North America (Delcourt & Harris, 1980; Sampson, 1992). The causes of this increase are still uncertain. They may include N deposition, but also CO_2 fertilization and management. As the effect of N deposition on soil C in forest soils is still very uncertain, no estimate is presented in Table 23–2.

Transient Fluxes of Soil Carbon Caused by Climate and Vegetation Change

General

The decomposition of soil organic matter often shows an initial period of rapid loss of labile constituents such as soluble carbohydrates, followed by an extended period of ever-decreasing decomposition rates during which the more recalcitrant components such as lignin are oxidized. Minderman (1968) was among the first authors who showed that it is necessary to know the decomposition rates of the separate chemical constituents for estimating the decomposition rate of a mass of organic matter. Modern soil organic matter turnover models use different pools of soil organic matter, whereby inputs of organic matter are converted into microbial biomass and pools of soil organic matter with different "turnover" times. Transformations follow first-order rate kinetics, and the change in soil C as a result of all transformations can be calculated from the sum of all the C pools. In mathematical form these models can be written as

$$d\ C(p)/D\ t = -k \times M \times T \times C(p) + A(p) \qquad [1]$$

where $C(p)$ is the C in soil organic matter Pool p, t is time, k is the first-order transformation rate constant (t^{-1}), M and T are rate reduction coefficients used to simulate suboptimal conditions for soil moisture status and temperature, respectively (no dimension), and $A(p)$ is the periodic addition of organic matter to Pool p. Several models of this type exist (Van Veen & Paul, 1981; Jenkinson et al., 1987, p. 1–8; Parton et al., 1987). Other turnover models have coupled C and N cycles (McGuire et al., 1992).

Equation [1] describes the effect of the major factors that determine soil organic matter dynamics: litter input to soils, temperature and moisture. Differences in soil texture can be included in the transformation rate constant k (Parton et al., 1987). For complete analysis of the effect of climate and vegetation change, the specific changes in the above factors should be known. As we will discuss below, with the current capabilities and knowledge, only partial studies of these effects have been performed.

Global Warming

The five compartment Rothamsted soil model described by Jenkinson et al. (1987, p. 1–8) was extended to the global scale to study the effect of temperature change. Arbitrarily an annual input of "inert" C of 300 g C m^{-2} yr^{-1} was assumed for all freely drained soils of the world and clay content was assumed constant for all soils. Jenkinson et al. (1991) recognized that the value of 0.25 for the ratio of easily decomposable and recalcitrant organic matter in the annual inputs used could be too high for coniferous forests. For a global temperature rise of 0.03°C yr^{-1} the estimated annual release of soil C as CO_2 was 1 Pg C y^{-1} (or 61 Pg over a period of 60 yr) (Table 23–2). These model results also suggest that decomposition in the wet tropical forest zone is not affected by a 25% increase in precipitation since the soil will always be wet enough to proceed at its maximum permitted rate. In contrast, in the cool temperate steppe zone a 25% decrease in precipitation spread uniformly over the year should increase the stock of C considerably, by retardation of decomposition by lack of moisture.

Global warming enhances decomposition and additional N may be mineralized and made available to plant growth. The C/N ratio of forest soils is typically much lower than that of trees. Hence, enhanced decomposition and associated N release may cause storage of C in tree tissue that exceeds soil C loss. In general, ecosystems with large woody components have a higher capacity to increase C storage by redistributing nutrients from soils to vegetation than nonwoody ecosystems (Petersen & Melillo, 1985; Raststetter et al., 1992). Recently developed models with coupled C and N cycling predict an increase in NPP for a doubling of CO_2 and climate change (McGuire et al., 1992; Melillo et al., 1993). For tropical and dry temperate ecosystems, predicted increases in NPP are caused mainly by the effects of elevated CO_2, while in northern and moist temperate ecosystems NPP increases are dominated by the effects of increased N availability which more than offset the higher plant respiration caused by higher temperatures (Melillo et al., 1993). In many northern and temperate ecosystems NPP is limited by the availability of inorganic N and here the effects of increased N availability may be stronger than in some tropical ecosystems where P may be limiting growth (Melillo et al., 1993; Raststetter et al., 1992; McGuire et al., 1992).

Global warming may cause a significant loss of soil C (Table 23–2). However, redistribution of C and N from soils to vegetation may account for a significant C accumulation in vegetation.

Climate Induced Vegetation Change

The models described above do not consider the redistribution and change of vegetation that may be induced by climate change. Terrestrial vegetation is a large, dynamic pool of C in direct exchange with the atmosphere. This pool is sensitive to climate, and projected CO_2-induced climate changes are large enough to cause significant responses from vegetation at seasonal, interannual, decadal, and longer time scales. Coarse scale vegetation patterns (biomes) can easily be determined using climatic data. Early analyses have used the Holdridge Life Zone Classification system (Emanuel et al., 1985a,b; Leemans, 1992; Cramer & Leemans, 1993) or the Köppen-climate classification (Guetter & Kutzbach, 1990). These models determine the boundaries of biomes by using two or more climatic indices based on monthly temperature and precipitation. A more recent development has moved away from these simple climate classifications and defined biomes with a more stringent ecophysiological justification (e.g., Box, 1981; Woodward, 1987). This comprehensive approach resulted in the BIOME model (Prentice et al., 1992) that is mainly based on physiological considerations of the distribution of functional groups of plant-species and designed especially for climate change and C-cycle applications. The model takes into account species response to extreme temperatures, growing season characteristics and local moisture conditions. The latter is based on soil properties obtained from the FAO soil databases (Zobler, 1986).

Some analyses have used the above models to show shifts in global C storage due to climatic change. The common methodology is to run a model for both current climatic conditions and doubled CO_2 climate conditions. The current climate is based on extensive collections of climate records (e.g., Leemans & Cramer, 1991), while the doubled CO_2 conditions are most often derived from simulations with General Circulation Models (GCM's) for the atmosphere (e.g., Mitchell et al., 1990).

Emanuel et al. (1985a,b) first showed the potential shifts in vegetation as a result of climatic change. In their analyses, the global C content of the terrestrial biosphere decreased, but this was mainly an artifact of the climatic scenario used. They considered temperature increase only and, through an increase of evapotranspiration this led to a large shift from forested areas to grasslands and deserts. More comprehensive analyses with both changing temperature and precipitation showed that these vegetation shifts should increase the global C storage in vegetation and soils (Prentice & Fung, 1990; Smith et al., 1992). Smith et al. (1992) analyzed four different GCM scenarios. All scenarios result in an increase in forested areas, although they are different both regionally and in magnitude. There is a poleward shift of the forested zones, with an increase in the areal extent of tropical forests and a shift of the boreal forest zone into the region currently occupied by tundra. Terrestrial C storage increased from 8.5 to 180 Pg above the estimates for present conditions. A similar increase has actually occurred since the last glacial maximum (Adams et al., 1990), although the actual magnitude is uncertain (Prentice & Fung, 1990).

Recent analyses with the BIOME model (Prentice et al., 1994) supported an increasing aboveground C content of 300 to 700 Pg since the last glacial maximum (Prentice et al., 1993). Klein-Goldewijk et al. (1994) listed an increase of 60 Pg C and 50 Pg C between now and the Year 2035 for the IPCC-A and IPCC-D scenarios (Houghton et al., 1992), respectively. Both studies do not account for changes in different feedbacks processes such as CO_2 fertilization on plant growth and the response of soil respiration to climatic change. Changes in soil C are more difficult to determine and could, depending on the climate scenario, decrease by 100 Pg or increase by 41 Pg (Turner & Leemans, 1993) in the coming four to five decades. The analyses by Klein-Goldewijk et al. (1994) and Turner and Leemans (1993) encompass larger uncertainties than long-term simulations (Prentice et al., 1992). This is mainly due to the disadvantages of equilibrium models to assess the impact of rapid climate change. On time scales of decennia (IPCC-scenarios) the turnover time of soil organic matter used in these models is relatively long and the equilibrium soil C contents do not change drastically. For the difference between the last glacial maximum and current conditions this turnover time is relatively short and equilibrium models are appropriate tools to assess changes in C content. The experiments for doubled CO_2 climate are, however, appropriate to get an impression for the direction and the potential feedbacks between soils, vegetation and the atmosphere.

CONCLUSIONS

We have reviewed the state-of-the-art understanding of C content and fluxes between the terrestrial biosphere and the atmosphere, with emphasis on forest soils. We distinguish "equilibrium fluxes," that are relatively constant on a long time scale (centuries–millennia), and "transient fluxes," that are determined by magnitude and type of land use change and that are variable over a short time scale (years–decades). Finally we discussed the effects of climate change and potential shifts in vegetation patterns on soil C.

Equilibrium fluxes of soil C are not influenced significantly by changes in land use. We conclude that the equilibrium soil C fluxes between the atmosphere and soils and between soils and aquatic systems cannot be disregarded in global C cycle studies. The transient soil C fluxes that we have attempted to quantify include those associated with land use changes and climate change. During the last centuries deforestation (forest conversion to arable land, pastures and other land uses, and shifting cultivation) has and is still affecting major changes in soil C contents over much of the tropical and temperate regions (Table 23–2). Due to the slow pedogenic processes, these land use changes are likely to continue to dominate the C fluxes between the different compartments over the coming decade.

Changes in soil C also may be caused by shifts in vegetation patterns induced by climate change. Only on a time scale of decennia to centuries could climate change and the resulting shifting vegetation zones become the dominant control of the global C cycle. When this happens the boreal zone will play a major role, because more of the climate change is predicted to occur here. Over the next century boreal soils could constitute a major source of CO_2, comparable in magnitude to current deforestation in the Tropics (Billings, 1987).

The estimates of the different fluxes summarized in Table 23–2 cannot be simply added because of the differences in time scales and because of the impossibility to estimate their relative uncertainty. Although the sum of estimates of the sink candidates in Table 23–2 is considerable, it is uncertain whether or not the processes listed account for a significant part of "the missing sink" of atmospheric C at the global scale.

Increased temperatures may lead to redistribution of C and nutrients from soil to vegetation, resulting in net accumulation in the soil–vegetation system due to much higher C:nutrient ratios in vegetation than in soils. At the same time, the changing atmospheric composition (e.g., CO_2 fertilization, N deposition) may affect C fluxes and pools. The combined effect of changes in temperature and precipitation, CO_2 concentration, and N deposition is poorly understood, but these mechanisms could form a significant global C sink.

Several models have been used to predict vegetation changes, but they are based on limited understanding of the functioning of soils and the biosphere under changing conditions. Their validation and calibration are hampered by insufficient primary data sets, such as actual land cover (Leemans & van den

Born, 1994). Our understanding will only improve considerably with the creation of comprehensive, high-resolution databases on soil properties, climate, land use and vegetation. Remote sensing, for example, can facilitate the generation of some of these databases, but the majority of the work to be done is dependent on time and labor intensive field inventories. As discussed above some of this research is underway, but the planned publication dates exceed the current research needs with many years. Until then, only creative analyses of the scanty available data will have to suffice to increase our understanding.

ACKNOWLEDGMENTS

We are grateful to two anonymous reviewers. Their comments have lead to major improvements of this paper. Preparation of this paper was under MAP Project no. 481 507 (Global Biosphere) of the National Institute of Public Health and Environmental Protection (RIVM).

REFERENCES

Adams, J.M., H. Faure, L. Faure-Denard, J.M. McGlade, and F.I. Woodward. 1990. Increases in terrestrial carbon storage from the last glacial maximum to the present. Nature (London) 348:711–714.

Ajtay, G.L., P. Ketner, and P. Duvigneaud. 1979. Terrestrial primary production and Phytomass. p. 129–181. *In* B. Bolin et al. (ed.) The global carbon cycle. John Wiley & Sons, New York.

Armentano, T.V., and E.S. Menges. 1986. Patterns of change in the carbon balance of organic soil-wetlands of the temperate zone. J. Ecol. 74:755–774.

Anselmann, I., and P.J. Crutzen. 1989. Global distribution of natural freshwater wetlands and rice paddies: Their net primary productivity, seasonality and possible methane emissions. J. Atmos. Chem. 8:307–358.

Billings, W.D. 1987. Carbon balance of Alaskan tundra and taiga ecosystems: Past, present and future. Quat. Sci. Rev. 6:165–177.

Bird, M., B. Fyfe, A. Chivas, and F. Longstaffe. 1990. deep weathering at extra-tropical latitudes: A response to increased atmospheric CO_2. p. 383–389. *In* A.F. Bouwman (ed.) Soils and the greenhouse effect. John Wiley & Sons, Chichester, England.

Bohn, H.L. 1976. Estimate of organic carbon in world soils. Soil Sci. Soc. Am. J. 40:468–470.

Bohn, H.L. 1982. Estimates of organic carbon in world soils. II. Soil Sci. Soc. Am. J. 46:1118–1119.

Bolin, B. 1977. Changes of land biota and their importance to the carbon cycle. Science (Washington, DC) 196:613–615.

Bolin, B., E.T. Degens, P. Duvigneaud, and S. Kempe. 1979. The global carbon cycle. p. 1–56. *In* B. Bolin et al. (ed.) The global carbon cycle. John Wiley & Sons, New York.

Bouwman, A.F. 1989. Modelling soil organic matter decomposition and rainfall erosion in two tropical soils after forest clearing for permanent agriculture. Land Degrad. Rehabil. 1:125–140.

Bouwman, A.F. 1990. Exchange of greenhouse gasses between terrestrial ecosystems and the atmosphere. p. 61–127. *In* A.F. Bouwman (ed.) Soils and the greenhouse effect. John Wiley & Sons, Chichester, England.

Box, E.O. 1981. Macroclimate and plant forms: An introduction to predictive modeling in phytogeography. Dr. W. Junk Publ., the Hague, the Netherlands.

Buringh, P. 1984. Organic carbon in soils of the world. p. 91–109. *In* G.M. Woodwell (ed.) The role of terrestrial vegetation in the global carbon cycle. Measurement by remote sensing. John Wiley & Sons, New York.

Buringh, P., and R. Dudal. 1987. Agricultural land use in space and time. p. 9–43. *In* M.G. Wolman and F.G.A. Fournier (ed.) Land transformations in agriculture. SCOPE Vol. 32, John Wiley & Sons, Chichester, England.

Cramer, W., and R. Leemans. 1993. Assessing impacts of climate change on vegetation using climate classification systems. p. 190–217. *In* A.M. Solomon and H.H. Shugart (ed.) Vegetation dynamics and global change. Chapman and Hall, New York.

Crutzen, P.J., and M.O. Andreae. 1990. Biomass burning in the tropics: impact on atmospheric chemistry and biogeochemical cycles. Science (Washington, DC) 250:1669–1678.

Cunningham, W. 1963. The effect of clearing on a tropical forest soil. J. Soil Sci. 14:334–345.

Davidson, E.A., and I.L. Ackerman. 1993. Changes in soil carbon inventories following cultivation of previously untilled soils. Biogeochemistry 20:161–193.

Delcourt, H.R., and W.F. Harris. 1980. Carbon budget of the Southeastern U.S. biota: Analysis of historical change in trend from source to sink. Science (Washington, DC) 210:321–323.

Detwiler, R.P. 1986. Land use change and the global carbon cycle: The role of tropical soils. Biogeochemistry 2:67–93.

Detwiler, R.P., and C.A.S. Hall. 1988. Tropical forests and the global carbon cycle. Science (Washington, DC) 239:42–47.

Detwiler, R.P., C.A.S. Hall, and P. Bogdonoff. 1985. Land use change and carbon exchange in the tropics: II. Estimates for the entire region. Environ. Manage. 9:335–344.

Ellis, R.C., and A.M. Graley. 1983. Gains and losses in soil nutrients associated with harvesting and burning Eucalypt rain forest. Plant Soil 74:437–450.

Emanuel, W.R., H.H. Shugart, and M.P. Stevenson. 1985a. Climatic change and the broad-scale distribution of terrestrial ecosystems complexes. Clim. Change 7:29–43.

Emanuel, W.R., H.H. Shugart, and M.P. Stevenson. 1985b. Response to comment: Climatic change and the broad-scale distribution of terrestrial ecosystems complexes. Clim. Change 7:457–460.

Esser, G. 1990. Modeling global terrestrial sources and sinks of CO_2 with special reference to soil organic matter. p. 247–262. In A.F. Bouwman (ed.) Soils and the greenhouse effect. John Wiley & Sons, Chichester, England.

Eswaran, H., E. Van Den Berg, and P. Reich. 1993. Organic carbon in soils of the world. Soil Sci. Soc. Am. J. 57:192–194.

Food and Agriculture Organization. 1991. AGROSTAT PC. Computerized Information Series 1/3. Land use. FAO Publ. Div., U.N. FAO, Rome.

Food and Agriculture Organization. 1993. Forest resources assessment 1990. Tropical countries. FAO For. Pap. No. 112. FAO, Rome.

Food and Agriculture Organization-Unesco. 1971–1982. Soil map of the world 1:5,000,000. U.N. FAO, Rome.

Goudriaan, J., and P. Ketner. 1984. A simulation study for the global carbon cycle, including man's impact on the biosphere. Clim. Change 6:167–192.

Guetter, P.J., and J.E. Kutzbach. 1990. A modified Köppen classification applied to model simulations of glacial and interglacial climates. Clim. Change 16:193–215.

Harden, J.W., E.T. Sundquist, R.F. Stallard, and R.K. Mark. 1992. Dynamics of soil carbon during deglaciation of the Laurentide ice sheet. Science (Washington, DC) 258:1921–1924.

Holdridge, L.R. 1967. Life zone ecology. Tropical Science Center, San José, Costa Rica.

Holland, H.D. 1978. The chemistry of the atmosphere and oceans. John Wiley & Sons, New York.

Houghton, J.T., B.A. Callander, and S.K. Varney (ed.). 1992. Climate change 1992: The supplementary report to the IPCC scientific assessment. Intergovernmental Panel on Climate Change, Supplementary Rep. Cambridge Univ. Press, Cambridge, England.

Houghton, R.A., R.D. Boone, J.R. Fruci, J.E. Hobbie, J.M. Melillo, C.A. Palm, B.J. Peterson, G.R. Shaver, G.M. Woodwell, B. Moore, and D. Skole. 1987. The flux of carbon from terrestrial ecosystems to the atmosphere in 1980 due to changes in land use: Geographic distribution of the global flux. Tellus 39:122–139.

Houghton, R.A., R.D. Boone, J.M. Melillo, C.A. Palm, G.M. Woodwell, N. Myers, B. Moore III, and D.L. Skole, 1985. Net flux of carbon dioxide from tropical forests in 1980. Nature (London) 316:617–620.

Houghton, R.A., J.E. Hobbie, J.M. Melillo, B. Moore, B.J. Peterson, G.R. Shaver, and G.M. Woodwell. 1983. Changes in the carbon content of terrestrial biota and soils between 1860 and 1980: A net release of CO_2 to the atmosphere. Ecol. Monogr. 53:235–262.

Huettl, R.F., and H.W. Zoettl. 1992. Forest fertilization: Its potential to increase the CO_2 storage capacity and to alleviate the decline of the global forests. Water Air Soil Pollut. 64:229–249.

Jaffe, D.A. 1992. The nitrogen cycle, p. 263–284. In S.S. Butcher et al. (ed.) Global biogeochemical cycles. Acad. Press, London.

Jenkinson, D.S., D.E. Adams, and A. Wild. 1991. Model estimates of CO_2 emissions from soil in response to global warming. Nature (London) 351:304–307.

Jenkinson, D.S., P.B.S. Hart, J.H. Rayner, and L.C. Parry. 1987. Modelling the turnover of organic matter in long-term experiments at Rothamsted. INTECOL Bull. Vol. 15. Int. Assoc. Ecol. Athens, GA.

Johnson, D.W. 1992. Effects of forest management on soil carbon storage. Water Air Soil Pollut. 64:83–120.

Jones, M.J. 1972. The organic matter content of the savanna soils of West Africa. J. Soil Sci. 24:42–53.

Jones, T.P., and W.G. Chalone. 1991. Fossil charcoal, its recognition and paleoatmospheric significance. Global Planet. Change 97:39–50.

Jordan, C.F. 1985. Nutrient cycling in tropical forest ecosystems: Principles and their application in management and conservation. John Wiley & Sons, Chichester, England.

Kauppi, P.E., K. Mielikäuainen, and K. Kuusela. 1992. Biomass and carbon budget of European Forests, 1971 to 1990. Science (Washington, DC) 256:70–74.

Kimble, J.M., H. Eswaran, and T. Cook. 1990. Organic carbon on a volume basis in tropical and temperate soils. p. 248–253. *In* M. Koshimo et al. (ed.) Trans. 14th Int. Cong. Soil Sci., Kyoto, Japan. 12–18 Aug. 1990., Vol. 5. Int. Soil Sci. Soc., Kyoto, Japan.

Klein-Goldewijk, K., J.G. van Minne, G.J.J. Kreileman, M. Vloedbeld, and R. Leemans. 1994. Simulating the carbon flux between the terrestrial environment and the atmosphere. Water Air Soil Pollut. 76:199–230.

Lanly, J.P. 1982. Tropical forest resources. For. Pap. no. 30. U.N.FAO, Rome.

Leemans, R. 1992. Modelling ecological and agricultural impacts of global change on a global scale. J. Sci. Ind. Res. 51:709–724.

Leemans, R., and W.P. Cramer. 1991. The IIASA database for mean monthly values of temperature, precipitation and cloudiness on a global terrestrial grid. RR-91-18. Int. Inst. Appl. Syst. Analyses, Laxenburg, Austria.

Leemans, R., and G.J. van den Born. 1994. Determining the potential distribution of natural vegetation, crops and agricultural productivity. Water Air Soil Pollut. 76:133–161.

Lugo, A.E., and S. Brown. 1986. Steady state terrestrial ecosystems and the global carbon cycle. Vegetatio 68:83–90.

Mann, L.K. 1985. A regional comparison of carbon in cultivated and uncultivated alfisols and mollisols in the central United States. Geoderma 36:241–253.

Mann, L.K. 1986. Changes in soil carbon storage after cultivation. Soil Sci. 142:279–288.

Matthews, E., and I. Fung. 1987. Methane emission from natural wetlands: Global distribution, area, and environmental characteristics of sources. Global Biogeochem. Cycles 1:61–86.

McGuire, A.D., J.M. Melillo, L.A. Joyce, D.W. Kicklighter, A.L. Grace, B. Moore III, and C.J. Vorosmarty. 1992. Interactions between carbon and nitrogen dynamics in estimating net primary productivity for potential vegetation in North America. Global Biogeochem. Cycles 6:101–124.

Meentemeyer, V., E.O. Box, M. Folkoff, and J. Gardner. 1981. Climatic estimation of soil properties; soil pH, litter accumulation and soil organic content. Bull. Ecol. Soc. Am. 62:104.

Melillo, J.M., A.D. McGuire, D.W. Kicklighter, B. Moore III, C.J. Vorosmarty, and A.L. Schloss. 1993. Global climate change and terrestrial net primary production. Nature (London) 363:234–240.

Meybeck, M. 1982. Carbon, nitrogen and phosphorous transport by world rivers. Am. J. Sci. 282:401–450.

Meyers, N. 1991. Tropical forest: Present status and future outlook. Clim. Change 19:3–32.

Minderman, G. 1968. Addition, decomposition and accumulation of organic matter in forests. J. Ecol. 56:355–362.

Mitchell, J.F.B., S. Manabe, V. Meleshko, and T. Tokioka. 1990. Equilibrium climate change—and its implications for the future. p. 131–172. *In* J.T. Houghton et al. (ed.) Climate change: The IPCC scientific assessment. Cambridge Univ. Press, Cambridge, England.

Nye, P.H., and D.J. Greenland. 1960. The soil under shifting cultivation. Tech. Commun. 51. Commonwealth Bureau Soils, Harpenden, England.

Nye, P.H., and D.J. Greenland. 1964. Changes in the soil after clearing tropical forest. Plant Soil 21:101–112.

Parton, Q.J., D.S. Schimel, C.F. Cole, and D.S. Ojima. 1987. Analysis of factors controlling soil organic matter levels in Great Plains grasslands. Soil Sci. Soc. Am. J. 51:1173–1179.

Peterson, B.J., and J.M. Melillo. 1985. The potential storage of carbon caused by eutrophication of the biosphere. Tellus 37B:117–127.

Post, W.M., W.R. Emanuel, P.J. Zinke, and A.G. Staugenberger. 1982. Soil carbon pools and world life zones. Nature (London) 298:156–159.

Post, W.M., and L.K. Mann. 1990. Changes in soil organic carbon and nitrogen as a result of cultivation. p. 401–406. *In* A.F. Bouwman (ed.) Soils and the greenhouse effect. John Wiley & Sons, Chichester, England.

Post, W.M., J. Pastor, P.J. Zinke, and A.G. Stangenberger. 1985. Global patterns of soil nitrogen storage. Nature (London) 317:613–616.

Prentice, I.C., W. Cramer, S.P. Harrison, R. Leemans, R.A. Monserud, and A.M. Solomon. 1992. A global biome model based on plant physiology and dominance, soil properties and climate. J. Biogeogr. 19:117–134.

Prentice, I.C., M.T. Sykes, M. Lautenschlager, S.P. Harrison, O. Denissenko, and P.J. Bartlein. 1994. Modeling global vegetation patterns and terrestrial carbon storage at the last glacial maximum. Glob. Ecol. Biogeogr. Lett. 3:67–76.

Prentice, K.C., and I.Y Fung. 1990. The sensitivity of terrestrial carbon storage to climate change. Nature (London) 346:48–51.

Raststetter, E.B., R.B. McKane, G.R. Shaver, and J.M. Melillo. 1992. Changes in C storage by terrestrial ecosystems: How C-N interactions restrict responses to CO_2 and temperature. Water Air Soil Pollut. 64:327–344.

Sampson, R.N. 1992. Forestry opportunities in the United States to mitigate the effects of global warming. Water Air Soil Pollut. 64:157–180.

Sanchez, P.A. 1976. Properties and management of soils in the tropics. John Wiley & Sons, New York.

Sanford, R.L., J. Saldarriaga, K.E. Clark, C. Uhe, and R. Herrerra. 1985. Amazon rainforest fires. Science (Washington, DC) 227:53–55.

Schimel, D.S., D.C. Coleman, and K.A. Horton. 1985. Soil organic matter dynamics in paired rangeland and cropland toposequences in North Dakota. Geoderma 36:201–214.

Schlesinger, W.H. 1977. Carbon balance in terrestrial detritus. Ann. Rev. Ecol. System. 8:51–81.

Schlesinger, W.H. 1982. Carbon storage in the caliche of arid soils. A case study from Arizona. Soil Sci. 133:247–255.

Schlesinger, W.H. 1984. Soil organic matter: a source of atmospheric CO_2. p. 111–127. *In* G.W. Woodwell (ed.) The role of terrestrial vegetation in the global carbon cycle. John Wiley & Sons, New York.

Schlesinger, W.H. 1985. The formation of caliche in soils of the Mojave Desert, California. Geochim. Cosmochim. 49:57–66.

Schlesinger, W.H. 1986. Changes in soil carbon storage and associated properties with disturbance and recovery. p. 194–220. *In* J.R. Trabalka and D.E. Reichle (ed.) The changing carbon cycle. A global analysis. Springer Verlag, New York.

Schlesinger, W.H. 1990. Evidence from chronosequence studies for a low carbon-storage potential of soils. Nature (London) 348:232–234.

Schlesinger, W.H., and J.M. Melack. 1981. Transport of organic carbon in the world's rivers. Tellus 33:172–187.

Sedjo, R.A., and A.M. Solomon. 1989. Climate and forests. p. 105–119. *In* N.J. Rosenberg et al. (ed.) Greenhouse warming: Abatement and adaptation. Resour. Future, Washington, DC.

Seiler, W., and P.J. Crutzen. 1980. Estimates of gross and net fluxes of carbon between the biosphere and the atmosphere from biomass burning. Clim. Change 2:207–247.

Skole, D., and C. Tucker. 1993. Tropical deforestation and habitat fragmentation in the Amazon: Satellite data from 1978 to 1988. Science (Washington, DC) 260:1905–1910.

Smith, T.M., R. Leemans, and H.H. Shugart. 1992. Sensitivity of terrestrial carbon storage to CO_2 induced climate change: Comparison of four scenarios based on general circulation models. Clim. Change 21:367–384.

Soil Survey Staff. 1975. Soil taxonomy: A basic system of soil classification for making and interpreting soil surveys. USDA-SCS Agric. Handb. 436. U.S. Gov. Print. Office, Washington DC.

Suman, D.O. 1988. The flux of charcoal to the troposphere during the period of agricultural burning in Panama. J. Atmos. Chem. 6:21–34.

Tans, P.P., I.Y. Fung, and T. Takahashi. 1990. Observational constraints on the global atmospheric CO_2 budget. Science (Washington, DC) 247:1431–1438.

Tinker, P.B., and P. Imeson. 1990. Soil organic matter and biology in relation to climate change. p. 71–87. *In* H.W. Scharpenseel et al. (ed.) Soils on a warmer earth. Elsevier, Amsterdam.

Turner, D.P., and R. Leemans. 1993. Equilibrium analysis of projected climate change effects on the global soil organic matter pool. p. 59–63. *In* T. Vinson and T.P. Kolchugina (ed.) Carbon cycling in boreal forest and subarctic ecosystems: Biospheric responses and feedbacks to global climate change. Environ. Res. Lab., USEPA, Corvallis, OR.

Van Breemen, N., and T.C.J. Feijtel. 1990. Soil processes and properties involved in the production of greenhouse gases, with special reference to soil taxonomic systems. p. 195–223. *In* A.F. Bouwman (ed.) Soils and the greenhouse effect. John Wiley & Sons, Chichester, England.

Van Breemen, N., and R. Protz. 1988. Rates of calcium carbonate removal from soils. Can. J. Soil Sci. 68:449–454.

Van Veen, J.A., and E.A. Paul. 1981. Organic carbon dynamics in grassland soils. 1. Background information and computer simulation. Can. J. Soil Sci. 61:185–201.

Voroney, R.P., J.A. van Veen, and E.A. Paul. 1981. Organic C dynamics in grassland soils. 2. Model validation and simulation of the long-term effects of cultivation and rainfall erosion. Can. J. Soil Sci. 61:211–224.

Whittaker, R.H., and G.E. Likens. 1975. The biosphere and man. p. 305–328. *In* H. Lieth and R.H. Whittaker (ed.) Primary production of the biosphere. Springer Verlag, Heidelberg, Germany.

Woodward, F.I. 1987. Climate and plant distribution. Cambridge Univ. Press, Cambridge, England.

World Resources Institute. 1988. World Resources 1988–89. An assessment of the resource base that supports the global economy. WRI-Int. Inst. Environ. Develop., Basic Books Inc., New York.

World Resources Institute. 1992. World resources 1992–1993. A report of the World Resources Institute. Oxford Univ. Press, New York.

Zinke, P.J., A.G. Stangenberger, W.M. Post, W.R. Emanuel, and J.S. Olson. 1984. Worldwide organic soil carbon and nitrogen data. ORNL/TM-8857, Oak Ridge Natl. Lab., Oak Ridge, TN.

Zobler, R. 1986. A world soil file for global climate modeling. NASA Tech. Memor. No. 87802. NASA, New York.

24 Comparison of Carbon Accumulation in Douglas Fir and Red Alder Forests

Dale W. Cole, Jana E. Compton, and R.L. Edmonds
University of Washington
Seattle, Washington

Peter S. Homann
Oregon State University
Corvallis, Oregon

H. Van Miegroet
Utah State University
Logan, Utah

Soil organic C or organic matter has been studied under various plant communities. In comparisons of similar soils having either grassland or forest vegetation, surface horizons of the forest soil may have lower levels of soil C (Ugolini & Schlichte, 1973; Almendinger, 1990) or higher levels (Jenny, 1980). Even between forest types, soil C may vary. Alban (1982) found the forest floor in a *Populus tremuloids* stand had 10 to 40% less organic matter than in *Picea glauca*, *Pinus resinosa* or *Pinus banksiana* stands of the same age. Surface soil (0–12 cm) in a *Picea abies* forest had a significantly higher concentration of organic matter than did the soil from *Quercus rubra* or *Pinus strobus* (Challinor, 1968). In comparisons of *Pinus radiata* and eucalyptus forests in Australia, Turner and Kelly (1985) found higher concentrations of C in the surface mineral soil of one pine plantation at only one of two sites. In another *P. radiata*–eucalyptus comparison, Turner and Lambert (1988) found greater soil C in a pine plantation in a low-fertility site, but less soil C in a high-fertility site.

The most dramatic change in soil C occurs with N_2-fixing vegetation, which generally cause a substantial (20–100%) increase in soil C during their lifespan (Johnson, 1992). Actinorhizal N_2-fixing alders (*Alnus* spp.) cause significant and rapid changes to other soil properties including an increase in soil N (Van Cleve et al., 1971; Bormann & DeBell, 1981; Brozek, 1990), a decrease in soil pH (Crocker & Major, 1955; Ugolini, 1968; Bormann & DeBell, 1981; Van Miegroet & Cole, 1985), an increase in exchangeable Al and changes in P availability (Cole et al., 1990). A comparison of red alder (*A. rubra* Bong.) and Douglas fir (*Pseudotsuga menziesii* [Mirb.] Franco) established on the same soil series at the Thompson Research Center in western Washington, has shown that red alder can significantly increase soil C and N relative to Douglas fir, while it decreases soil pH and P availability (Cole et al., 1990). Both red alder and Douglas fir are common species in lowland areas throughout western Washington, Oregon and British Columbia, especially after disturbance. The changes in soil properties associated with red alder may alter site productivity and decomposition rates and hence affect the C accumulation potential at the site.

Although a rotation of red alder can alter numerous soil properties (Cole et al., 1990), the following questions still remain concerning the factors responsible for such changes, the stability of these changes over time, and the effect they have on productivity and C accumulation remain:

1. What are the mechanisms responsible for the greater soil C accumulation under the N-rich red alder forest stands compared to Douglas fir—a greater rate of detritus production, a slower rate of decomposition, or both?
2. What is the effect of previous vegetation (and associated ecosystem properties) on future C dynamics and storage in forest plantations?
3. Does soil organic C under alder change rapidly in response to harvesting and plantation establishment?

To provide answers we compared organic matter production and storage in adjacent stands of mature alder and Douglas fir. Changes to the soil organic pool and litter production were then followed through harvest and the first 7 yr after plantations of red alder and Douglas fir had been re-established at these sites. In this paper we examine the shifts in organic accumulation and production during these stages of stand development and discuss the processes causing these differences.

SITE DESCRIPTION AND METHODS

The research was conducted in adjacent stands of red alder and Douglas fir at the Thompson Research Center, Cedar River Watershed, Washington. This site is located 56 km southeast of Seattle at 220 m elevation in the foothills of the Cascade Mountains. The climate is maritime with cool, dry summers and moderate, wet winters. The recorded mean annual temperature for the area is 9.8°C, with average monthly temperatures of 2.8 and 16.8°C in January and July. Mean annual precipitation is 130 cm, 75% falling as rain between October and March (Cole & Gessel, 1968).

Fifty-Year-Old Alder and Douglas Fir Stands

The Douglas fir plantation was established in 1931 after a series of slash fires followed the logging of the original old-growth Douglas fir–western hemlock (*Tsuga heterophylla* [Raf.] Sarg.) forest between 1910 and 1920 (Turner et al., 1976). The understory vegetation consists mainly of salal (*Gaultheria shallon*

Pursh.), Oregon grape (*Berberis nervosa* [Pursh.] Nutt.), and bracken fern (*Pteridium aquilinum* Kuhn var. *pubescens* Underw.). There also are several species of mosses, predominantly *Hylocomium* spp. and *Eurynchium oreganum* (Sull.) Jaeg. Adjacent areas not planted with Douglas fir or destroyed by fires became established with red alder over the following 10 yr. These areas developed into mature red alder stands (average age 46 in 1985). The understory under this forest type is more prominent and consists mainly of a dense growth of sword fern (*Polystichum munitum* [Kaulf.] Presl.) and bracken fern intermixed with some Oregon grape. For simplicity, these two stands are referred to as 50-yr-old stands throughout the rest of this paper.

The soil supporting both the red alder and the Douglas fir stands belongs to the Alderwood series (loamy-skeletal, mixed, mesic Vitrandic Durochrept), developed from ablation till overlying compacted basal till at approximately the 1-m depth. Due to the presence of the compacted basal till, drainage is restricted and a perched water table is common during winter. On a mass basis, the sand-sized particles (0.063–2 mm) contain approximately 35% quartz, 10% K-feldspar, and 30% plagioclase (Na/Ca molar ratio of approximately 3:2). Vermiculite dominates the clay-sized particles (<2 μm), with kaolinite, chlorite and mica also present. A comparison of soil from the two plots indicates similar mineralogy at both sites (April & Newton, 1992). Soil from both plots have similar percentages of sand, silt and clay, and bulk soil chemistry (total Fe, Mn, Ti, Ca, K, P, Si, Al, Mg, and Na). Differences in clay minerals and light and heavy minerals are within the error ranges. These similarities indicate that the glacial material underlying the stands is fairly homogeneous and that it is reasonable to assume that the differences in C accumulation between these stands are largely due to species effects.

In this study, a 20- by 20-m control plot was established in 1979 in both the Douglas fir and red alder stands. The two plots are about 200 m apart. Slope is nearly level in both plots. To compare stand C accumulation, C content of various ecosystem components and detrital production were measured in these plots. The sampling procedures for the 50-yr-old stands are summarized from Homann et al. (1992). All living trees were numbered and their diameters and heights were measured every year or every other year through 1988. Woody tissue biomass for live trees was estimated from the 1985 diameter measurements and regression equations (Dice, 1970; Gholz et al., 1979). Woody tissue was categorized as: branches, bole bark, bole wood, and woody roots. Fine roots were not quantified. Woody tissue growth was determined for each live tree as the change in woody tissue biomass between measurements. Woody tissue mortality was defined as the transfer at tree death of live woody tissue to a detrital pool of coarse woody debris; this latter pool was not quantified but was considered to include standing dead trees, fallen logs, and large woody pieces (>6 cm in diam.) lying on the surface of the O horizon. Average rates of woody tissue growth and mortality were determined for the 9-yr period from 1979 to 1988.

Total foliar mass in red alder was determined as 1.12 times foliar litterfall mass, based on a study at this site by Turner et al. (1976). Because red alder is deciduous, current foliage is equivalent to total foliage. Total foliar mass in

Douglas fir was determined as 5.88 times foliar litter, based on the studies of Marshall and Waring (1986) and Heilman and Gessel (1963a,b). Douglas fir retains its needles for approximately 6 yr on sites of similar productivity (Cole, 1981). Current foliage mass was assumed to be equivalent to annual foliar litter. Nonwoody (fine) root mass and production were not determined.

Understory vegetation was collected in July 1986 from four 1- by 1-m areas in each plot. Samples were taken at periodic intervals across the plots. Aboveground parts of all living understory vegetation, including moss, were removed from this area. Organic (O) horizon was collected from a 0.5- by 0.5-m area within each understory subplot and separated into (i) wood and (ii) litter plus humus. Wood included all woody material within the O horizon plus woody material <6 cm in diameter on the surface. Wood >6 cm in diameter on the surface was not measured. Samples were dried (75°C) and weighed.

Mineral soil was sampled by depth intervals (0–7, 7–15, 15–30 and 30–45 cm) in June 1985 in five soil pits per plot. Soil was air dried and sieved (<2 mm). Soil bulk density and gravel content (>2 mm) were determined by depth in areas adjacent to the plots (2 measurements per stand in 1980 plus 8 measurements per stand in 1984).

Litterfall was collected continuously in 0.5- by 0.5-m frames with 1-mm nylon mesh bottoms (4 per plot) from September 1985 through August 1988. Litter was removed monthly, oven dried (75°C), sorted into foliage and nonfoliar components, and weighed.

To assess the rate of leaf litter decomposition, fresh litter was collected from each mature stand in August through October 1986 on large plastic sheets placed beneath the forest canopies. Air-dried Douglas fir needle litter or red alder leaf litter equivalent to 5 g oven-dried (75°C) mass was placed in 18- by 18-cm bags made of 1-mm mesh nylon (Nitex; Tetko Inc., Elmsford, NY). This fabric allows movement of water through the litter during decomposition. Larger mesh size did not significantly increase decomposition (R.L. Edmonds, unpublished data). In February 1987, the bags were placed on the surface of the O horizons in the stands from which the litter had been collected. In an area adjacent to each study plot, litterbags were laid out 0.5 m apart along four (nearly) parallel transects that were 0.5 to 1 m apart. Litterbags were numbered consecutively along transects and randomly selected for pick up. For each stand, 10 bags were immediately collected to serve as initial replicates, and 9, 10, or 15 bags were collected after 0.3, 1, 2, 4, and 6 yr. Roots that grew into the bags were removed to the extent possible, and the remaining material was dried at 65 to 75°C and weighed. After 6 yr, samples contained roots that could not be separated from the remaining litter, hence values for 6 yr overestimate remaining mass.

Conversion Plot Establishment and Sampling

To test the influence of harvesting and forest species conversion on soil C levels, we clearcut a 1.0-ha area in the previously described Douglas fir and alder stands in September 1984. Stems were removed by overhead cable yarding to minimize soil disturbance. Foliage, branches, and tops were left on the site to minimize nutrient loss associated with the harvesting operation. In

February 1985, one-half of each harvested plot was then replanted with Douglas fir and half with red alder seedlings, yielding four 0.5-ha forest conversion plots:

1. Previously Douglas fir, planted with Douglas fir, referred to as "fir-to-fir."
2. Previously Douglas fir, planted with red alder, referred to as "fir-to-alder."
3. Previously red alder, planted with Douglas fir, referred to as "alder-to-fir."
4. Previously red alder, planted with red alder, referred to as "alder-to-alder."

Figure 24–1 illustrates the layout of these plots. Eight 15- by 15-m subplots were delineated in each conversion area. Douglas fir seedlings (2 yr old) came from a nursery and red alder (height approximately 0.3 m) from a nearby gravel roadbed. Prior to harvesting, in July 1984, understory and mineral soils were sampled. Understory biomass was sampled from two 1- by 1-m collection areas per conversion subplot (n = 16), dried at 75°C, and weighed. Three 0.3- by 0.3-m soil pits were excavated per subplot and mineral soil was sampled by depth (0–15, 15–30, and 30–45 cm). Samples were composited by subplot yielding eight samples at each depth per plot. Soils were air dried and sieved (<2 mm), and C was determined from this <2-mm fraction.

Understory and O horizon mass within the plots was determined on 8 Aug. 1988. The understory was collected from a 1- by 1-m area in each of the eight subplots per plot. After understory removal, the forest floor was sampled from a 0.5- by 0.5-m metal frame placed in the center of the understory collection area. Samples of O horizon and understory were dried at 75°C for 21 and 5 d, respectively. All O horizon samples were ground separately to <2 mm for C determination.

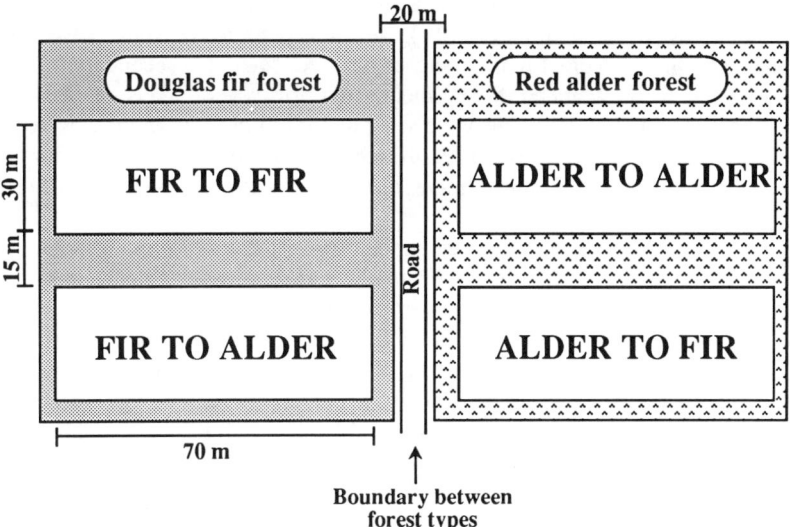

Fig. 24–1. Douglas fir and red alder conversion plots at the Thompson Research Center. Each of the four individual plots is 0.5 ha.

In November 1991, the O horizon was sampled from two randomly located 0.3- by 0.3-m areas within each of the eight subplots. Soil pits were excavated and mineral soil was sampled by depth intervals 0–15, 15–30, 30–45 cm). Samples were composited by subplot ($n = 8$). Soils were air dried, sieved to <2 mm and stored at room temperature until analysis.

The height of seedlings in these plots was measured annually from the time of planting in 1985 through the 1992 growing season. Diameters were measured annually in the young alder stands beginning in 1988 and in the Douglas fir stand in 1990, as trees were large enough to measure diameters. Biomass regressions for the aboveground components of the conversion plot stands were developed by destructive harvest for each plot in 1988 and 1992. Trees were sampled across the range of tree heights and diameters within the plots. On 2 Oct. 1988, the aboveground parts of eight seedlings were collected from the buffer strips in each of the fir-to-alder and alder-to-alder plots. The buffer strips have the same planting density as the subplots and should have biomass distribution similar to subplot trees. The seedlings were divided into stems and branches plus leaves, and weighed. Subsamples were taken for dry mass determination (75°C). Selected branches were stripped of their leaves, and the branch and stem weighed separately. The dry mass was then determined (75°C). For the young fir plots (fir-to-fir and alder-to-fir), 1988 biomass regressions are taken from Brozek (1990). Thirteen 6-yr-old Douglas fir were randomly chosen from each of the young fir plots. Trees were separated into current and older needles, current and older twigs, and stems. These components were dried at 75°C and weighed.

On 11 to 14 Aug. 1992, six trees were selected from the buffer strips of each conversion plot. Each tree was divided into stem and branches plus leaves and weighed. A subsample from the midpoint of each third of the stem was dried and weighed. For the young alder stands (fir-to-alder and alder-to-alder), leaves were removed from a subsample of branches from each third of the tree. For the young fir stands (fir-to-fir and alder-to-fir), secondary branches plus leaves were removed from six to eight main branches per tree. Main and secondary branches plus leaves were weighed fresh, then dried for several days at 75°C and reweighed. Secondary branches were separated into branches and leaves, redried at 75°C and reweighed. Subsamples of each component were taken for dry mass determination (75°C). Biomass regressions constructed in 1988 and 1992 are shown in Appendix 24–1.

To assess the rate of litter decomposition in these conversion plots, fresh leaf litter was collected from the alder-to-alder and fir-to-alder stands in October and November 1989, using the same process we used to collect fresh litter from the mature forests. Litter was not collected from the young fir stands. Air-dried litter equivalent to 5 g of oven-dried (75°C) mass was placed in 15- by 15-cm bags constructed of 1-mm mesh nylon. On 11 Feb. 1990, the bags were placed on the surface of the O horizon in the stand from which the litter had been collected as well as in the reciprocal stand. Five replicate bags from each treatment were collected after 1 and 2 yr. The contents were removed and dried at 75°C. Mass remaining was determined after 1 and 2 yr as well as decomposition constants (k values). The k values (yr^{-1}) were calculated with the following equation: $X/X_0 = e^{-kt}$, where X = mass at time t, and X_0 = initial mass (Olson, 1963).

Analytical Procedures and Statistics

Total C concentration was determined on all mineral soil and O horizon samples by LECO total C analyzer (LECO model 176-100C determinator, St. Joseph, MI). Total C concentration is presented on an oven-dried (105°C) basis for mineral soil samples and on a 75°C basis for O horizon samples. These soils are noncalcareous and hence total mineral soil C is considered to be equivalent to soil organic C (Nelson & Sommers, 1982). Carbon contents (kg C ha^{-1}) were calculated by multiplying C concentrations by mineral soil masses (kg ha^{-1}). Mineral soil masses (<2 mm) were determined from bulk density and coarse fragment (>2-mm) measurements. Organic horizon C content (kg C ha^{-1}) was estimated from Eq. [1], which corrects O horizon samples for mineral soil contamination.

$$\text{O horizon C content (kg C ha}^{-1}) = \frac{C_{sample} - C_{mineral}}{C_{organic} - C_{mineral}} \times C_{organic}\left(\text{mg C g}^{-1}\right) \times \text{O horizon mass }\left(\text{kg ha}^{-1}\right) \quad [1]$$

where C_{sample} is the C concentration measured in the O horizon sample, $C_{mineral}$ is the highest concentration in the surface mineral soil horizon for that plot and $C_{organic}$ is 500 g kg^{-1} (the highest value for any O horizon sample).

Carbon in the overstory and understory vegetation was estimated by multiplying 500 g C kg^{-1} organic matter times the mass of the pool on an areal basis. Carbon flux in litterfall was calculated assuming that the C concentration was 500 g kg^{-1} organic matter. Annual tree woody C accumulation in the 50-yr-old stands was calculated as the stem, branch, and woody root biomass in 1988 minus the biomass in 1979 divided by 9 yr. Annual woody tissue growth in the conversion plots was calculated as the difference between stem plus branch biomass after 1988 growing season and stem plus branch biomass after 1992 divided by 4 yr.

Statistical comparisons of the mean C contents and concentrations are valid only in determining whether there are significant differences between plots. Since there is no replication of the entire experimental design at a second site, it is not possible to apply the results of this experiment across the landscape, since inherent soil differences may be confounded with treatment effects. However, this is the problem with many ecosystem-level or watershed-level studies because they are difficult and expensive to replicate. Consequently, the results of this study are limited to this site, but may be representative of what could happen at other sites. For measurements where there were replicate samples and samples were analyzed separately, mean C concentrations and contents were compared between plots by either a t-test (Freund, 1988) or one-way ANOVA (Wilkinson, 1989). These comparisons were possible for the understory, O horizon and soil C concentrations and for the litterfall rates. Differences among mean k values for the conversion plot litter decomposition study were determined with two-way ANOVA (Wilkinson, 1989).

RESULTS AND DISCUSSION

Carbon Accumulation in Fifty-Year-Old Red Alder and Douglas Fir Stands

After approximately 50 yr of stand development, there were greater soil C concentrations and more total C storage in the alder ecosystem than in the Douglas fir ecosystem (Tables 24–1 and 24–2). The difference in total C storage is a result of higher accumulation of C in the O horizon and mineral soil profile of the alder stand, which more than offsets the higher C content of the Douglas fir vegetation. The aboveground C accumulation by Douglas fir was approxi-

Table 24–1. Mean soil C concentrations (with standard deviation in parentheses) in the 50-yr-old Douglas fir and red alder stands. Means were compared between plots at a given horizon by a two-tailed two-sample t-test (Freund, 1988).

Horizon	Douglas fir	Red alder
	mg C kg^{-1} soil	
O horizon	403 (38.5) NS‡§	381 (51.6) NS
Wood†	456 (61.5) NS	492 (52.7) NS
0–7 cm	80.0 (41.5)****§	118 (41.5)****§
7–15 cm	45.8 (13.4)****§	62.4 (14.6)****§
15–30 cm	32.9 (12.1)*	56.7 (12.2)*
30–45 cm	28.2 (12.3) NS	40.5 (8.7) NS

*,****Significantly different at the 0.05 and 0.10 levels, respectively.
†Includes wood in O horizon and wood <6 cm on surface.
‡NS = not significantly different.
§n = 4 for O horizon, n = 6 for wood, n = 5 for mineral soil. (This applies to all values in table.)

Table 24–2. Stand C distribution in the 50-yr-old Douglas fir and red alder stands. Tree and understory C concentrations were assumed to be 500 mg C kg^{-1}. Where possible, variability between plots was assessed and means compared via two-tailed, two-sample t-test (Freund, 1988).

Component	Douglas fir	Red alder
	kg C ha^{-1}	
Total tree	158 000	136 000
Aboveground	136 000	113 000
Leaf	3 100	1 750
Understory	1 550 (500)*†	3 000 (1 050)*
O horizon	9 470 (235)***	30 300 (640)***
Wood	6 020 (1 130)**	9 300 (1 250)**
Mineral soil		
0–15 cm	37 700	50 200
15–30 cm	21 900	32 700
30–45 cm	20 100	21 600
Total soil	79 700	105 000
Total stand	255 000	284 000

*,**,***Significant at the 0.05, 0.01 and 0.001 probability levels, respectively.
†Numbers in parentheses are standard deviations (n = 4).

mately 12% higher than alder, while the mineral soil C accumulation was 30% higher, and the O horizon C is 200% higher under alder.

The higher amount of C in O horizon and mineral soil under red alder is a function of higher production of detritus, lower decomposition rate, or both. Understanding how the differences arose requires estimates of both above- and belowground detrital production and decomposition over the history of the stand. Belowground information is unavailable, although current aboveground litterfall and leaf–litter decomposition measurements provide important insights. The red alder stand currently produces about three times as much overstory leaf litterfall (1560 vs. 525 kg C ha^{-1} yr^{-1}) and two times as much nonleaf litterfall as the Douglas fir stand (680 vs. 365 kg C ha^{-1} yr^{-1}; Table 24–3). This is consistent with the general trend that deciduous species produce more leaf litter than coniferous species (Cole & Rapp, 1980). Further, the understory contributes considerable litter. For the same stands in 1975, Douglas fir understory litterfall was 15% of understory biomass (Turner, 1975), and red alder understory litterfall was 21% of understory biomass (Turner et al., 1976). On the basis of understory C content (Table 24–2) and these estimated turnover rates, understory litterfall was 630 and 233 kg C ha^{-1} yr^{-1} in the red alder and Douglas fir stands, respectively. Thus, total aboveground litterfall under red alder is about 2.6 times that under Douglas fir (2870 vs. 1120 kg C ha^{-1} yr^{-1}).

In addition to greater litterfall deposition, differential decomposition rates may also explain the greater amount of C within the O horizon and mineral soil of the red alder stand. Using a mass-balance approach, we determined turnover time for C in the O horizon. Dividing O horizon C content by the litterfall rate, we obtain a turnover time of 8.5 yr for the Douglas fir O horizon and 10.6 yr for the alder O horizon. Assuming that the O horizon is in steady-state with respect to litterfall, the red alder O horizon is turning over more slowly than the Douglas fir O horizon.

This finding is supported by the long-term (6-yr) decomposition study. Decomposition of leaf litter proceeded more rapidly for red alder than Douglas fir during the 1st yr (Fig. 24–2). During the 2nd yr, Douglas fir litter lost mass more quickly, confirming the trend observed by Edmonds (1980). Beyond 2 yr,

Table 24–3. Rate of tree woody tissue C accumulation and litterfall in the 50-yr-old Douglas fir and red alder stands. Woody tissue and litter C concentrations were assumed to be 500 mg C kg^{-1}. Where possible, variability between plots was assessed and means compared via two-sample t-test (Freund, 1988). Numbers in parentheses are standard deviations about the mean.

Stand	Douglas fir	Red alder
	kg C ha^{-1} yr^{-1}	
Tree woody tissue production†	2955	2960
Tree woody tissue morality‡	1880	1570
Tree leaf litterfall§	525 (41)***	1560 (95)***
Tree nonleaf litter	365 (125)**	680 (95)**
Understory litterfall	233	630

,*Significant at the 0.01 and 0.005 levels, respectively.
†Mean change in above- and belowground woody biomass 1979 to 1988.
‡Mean annual biomass of trees that died between 1979 and 1988.
§Litterfall was collected September 1985 to August 1988 ($n = 4$ for each plot).

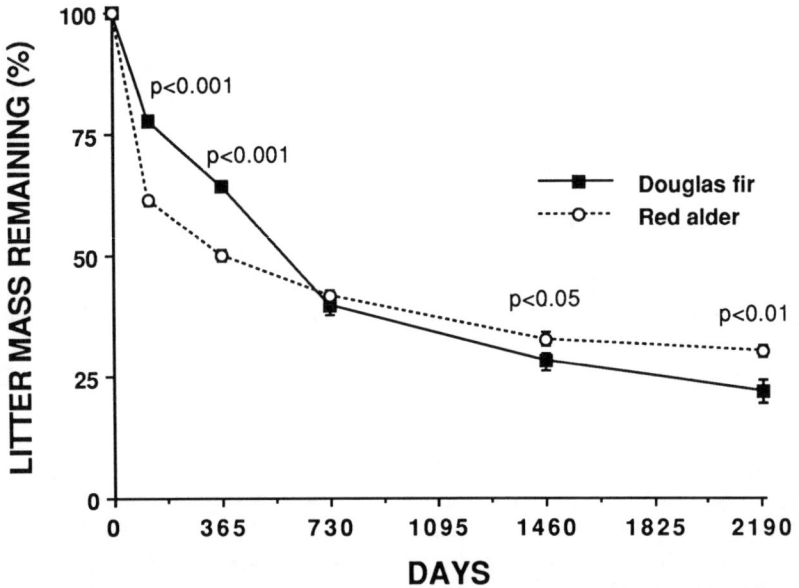

Fig. 24–2. Decomposition rates for Douglas fir and red alder litter, expressed as a percentage of remaining litter mass. Error bars are standard error of mean (n = 9–15). Some errors bars are smaller than symbols. Those months where means were significantly different between alder and Douglas fir are shown by probability (P) values at the given level.

Douglas fir continued to lose mass more quickly, resulting in only 21% of Douglas fir litter mass remaining at 6 yr compared with 30% for red alder. This loss demonstrates that short-term litterbag studies (1–2 yr) are not adequate for predicting long-term dynamics of detritus.

Comparison of red alder and Douglas fir leaf-litter dynamics suggests the greater detrital production under alder is initially more important than the slower decomposition rate in enhanced accumulation of C in the O horizon. As evident in the ratio between the litter decomposition rates of these two species, there is a significant change in the decomposition pattern over this period. The initial rapid decomposition of alder (Fig. 24–3) results in a decrease in the decomposition ratio of red alder compared to Douglas fir from 3.0 to 2.4 after 365 d. With the decomposition rate of alder litter rapidly diminishing after this initial 730- to 1460-d period (2–4 yr), the ratio increases to 4.1 by the end of 2192 d (6 yr). The data suggest that for red alder, litterfall production plays the dominant role in initial C accumulation in the forest floor, producing approximately 2.6 times more than Douglas fir (Table 24–3). However, with time, the relatively slow decomposition rate of alder litter becomes increasingly important in explaining C accumulation. For example, by 6 yr the three times greater litter production under alder as compared to Douglas fir has resulted in only 1.5 times more C accumulation in the forest floor. This study did not determine how much O horizon and soil C can be attributed to production and how much to decomposition at stand maturity. However, the trend discussed above and illustrated in Fig. 24–3 indi-

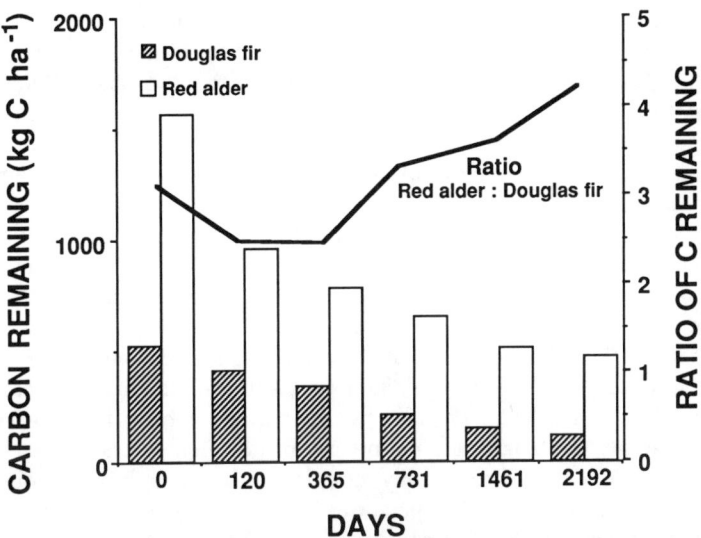

Fig. 24–3. Decrease of Douglas fir and red alder litter mass and ratio of the decomposition rate of red alder to Douglas fir over a 2200-d (6-yr) period. Masses were determined by multiplying the percentage remaining at each time interval (Fig. 24–2) by the annual leaf litter deposition rate (Table 24–3). Ratio shown is the mass of red alder litter remaining divided by the mass of Douglas fir litter remaining.

cates that decomposition plays a major role in explaining the high C accumulation under alder. However, nonfoliar overstory litter and understory litter also could be a factor in this relationship. Red alder twigs and wood decomposed more quickly than Douglas fir wood after 5 yr (Edmonds et al., 1986). Unfortunately, we have no information about understory litter decomposition, which may play an important role in site C accumulation dynamics, based on the magnitude of understory litterfall.

There are several explanations for the litter decomposition trends noted under red alder (initial rapid decomposition followed by slower decomposition after 2 yr). The results of Parsons et al. (1990) with aspen leaf litter indicate most of the mass lost in the first 6 mo of decomposition can be attributed to loss of water-soluble material. In our study, water-soluble material initially constituted 21% of the Douglas fir litter and 20% of the red alder litter (P.S. Homann, unpublished data); therefore the much more rapid loss of mass from the red alder litter during the first 4 mo (Fig. 24–2) cannot be attributed to greater initial amounts of soluble material. Lignin has been implicated in decreasing mass-loss rates and N implicated in increasing rates (Cromack, 1973; Fogel & Cromack, 1977; Fyles & McGill, 1987; Melillo et al., 1982; White et al., 1988). Similar lignin contents of red alder and Douglas fir litter (21 and 18%, respectively; Edmonds, 1980) indicate it also is not a key factor in our comparison. However, red alder litter had four times the N concentration of the Douglas fir (2.25 and 0.54%, respectively; Homann & Cole, 1990), which may explain the greater mass loss during the 1st yr of decomposition.

Over longer periods of decomposition, higher N contents may retard rather than enhance mass loss. Berg (1986) proposed that N enhances the rate of mass loss of readily decomposable portions of detritus (extractives and cellulose), but decreases the rate of more recalcitrant fractions (lignin). Our results are consistent with this proposal, given the rapid mass loss of N-rich alder followed by a subsequent much slower decomposition rate. Red alder leaves decomposed slower than other leaf types with a similar initial ratio of litter lignin to N, over a 3-yr period (Harmon et al., 1990). Camiré et al. (1991) speculated that slower-than-expected decomposition of black alder (*A. glutinosa*) roots was due to high levels of N in the roots allowing formation of N-lignin derivatives. Tannin content also may be important in controlling mass loss (summarized by Anderson, 1991) and may be a factor in slower long-term decomposition of alder litter. Fyles and Fyles (1993) found that Douglas fir leaf litter had 6% tannins and red alder leaf litter had 21% tannins. Tannins have been found to reduce biodegradability and humification of organic matter (Benoit & Starkey, 1968a,b; Zucker, 1983). Other factors may contribute to the trends we observed. Microenvironmental factors (temperature, moisture) may have differed between the two stands and affected decomposition rates. The litter layer under alder tends to be wetter and colder in the winter months and drier in the summer months. Both factors would restrict the rate of decomposition, independent of litter chemistry, although the importance of these factors has not been tested.

Changes in Soil and O-Horizon C Following Harvesting and Plantation Establishment

Soil C concentration and content before harvesting (1984) and 7 yr after harvesting and replanting are reported in Table 24–4. In the former Douglas fir site, with initially lower C content, harvesting and re-establishment of Douglas fir had minimal impact on the C supply consistent with the literature reviewed by Johnson (1992). Replanting the former Douglas fir site with alder resulted in an apparent 27% increase in the soil C content after only 7 yr. In the former alder site, where accumulation of soil C was highest prior to harvesting, there was a 25% decrease in soil C content 7 yr after harvest whether the site was replanted with alder or Douglas fir. This decrease took place across all soil depths. The area retaining the highest soil C level was the former alder-to-alder plot. This finding is consistent with the observations discussed previously—inputs of C are highest and decomposition is lowest under alder.

Unfortunately the O horizon mass was not determined before harvest of the conversion plots in 1984, thus making it difficult to follow the direct effect of harvest on decomposition rates in the surface organic horizon. To make this comparison between preharvest to postharvest C contents in the O horizon, we determined the mass of the O horizon from adjacent 50-yr-old Douglas fir and red alder stands in 1986 (Fig. 24–4). The fir-to-alder plot showed the largest increase in O horizon C, while the alder-to-fir plot showed the largest decrease. This result is consistent with those reported above for the C content in the soil profile. The C content of the O horizon at the former alder sites decreased 7 yr following harvest, although the change was not large for the alder-to-alder plot.

Results of the decomposition experiment conducted with red alder litter in the conversion plots are shown in Table 24–5. The results are consistent with the

Table 24-4. Soil C concentration and content in conversion plots before (1984) and after (1991) harvest and planting. Mean C concentration was compared between 1984 and 1991 for given plot depth by independent samples t-test using SYSTAT (Wilkinson, 1989).

Site	Depth	C concentration		C content	
		1984	1991	1984	1991
	cm	g C kg^{-1}		kg C ha^{-1}	
Fir-to-fir	0–15	47.0 (9.5)†	51.1 (20.5)	21 900	23 800
	15–30	32.5 (9.1)	31.7 (14.4)	9 590	9 350
	30–45	30.2 (6.5)*	22.9 (5.8)*	8 970	6 800
	Total			40 400	39 900
Fir-to-alder	0–15	38.3 (10.0)	49.1 (26.7)	22 700	29 100
	15–30	26.4 (10.5)	33.7 (19.1)	10 500	13 400
	30–45	20.9 (5.74)	25.8 (11.4)	9 280	11 500
	Total			42 500	54 000
Alder-to-fir	0–15	80.6 (16.1)	74.2 (12.9)	39 400	36 300
	15–30	49.6 (12.5)	43.6 (14.7)	13 800	12 100
	30–45	48.7 (15.5)*	33.6 (11.3)*	15 300	10 500
	Total			68 400	58 900
Alder-to-alder	0–15	109 (26.4)‡	81.7 (29.4)‡	66 000	49 300
	15–30	68.1 (15.2)	61.8 (23.4)	25 400	23 100
	30–45	67.4 (27.2)*	36.8 (19.5)*	17 600	9 600
	Total			109 000	82 000

*, ‡ Significant at the 0.05 and 0.10 levels, respectively.
† Standard deviations are shown in parentheses ($n = 8$).

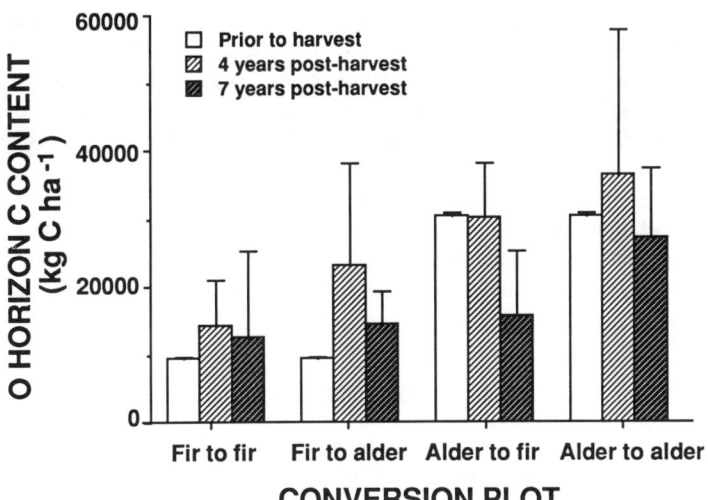

Fig. 24-4. Comparison of preharvest to postharvest C content (kg ha^{-1}) in the O horizon of the conversion plots. Preharvest C content is from adjacent 50-yr-old stands in 1986. Error bars are one standard deviation of the mean (preharvest, $n = 4$; 1988 and 1991, $n = 8$).

Table 24–5. Decomposition rate constants (k) and percentage of alder leaf mass remaining after 1 and 2 yr in the young alder stands. Leaf origin refers to the stand from which the leaves were collected (leaf composition varies). Location refers to the stand within which the litterbags were placed (decomposition environment varies). Probability values (P) are the result of a two-way ANOVA on k using SYSTAT (Wilkinson, 1989).

Leaf origin	Location	Year 1 % remaining	k†	Year 2 % remaining	k†
Fir-to-alder	Fir-to-alder	44	−0.82	44	−0.42
Fir-to-alder	Alder-to-alder	48	−0.74	41	−0.45
Alder-to-alder	Fir-to-alder	54	−0.62	52	−0.33
Alder-to-alder	Alder-to-alder	54	−0.62	46	−0.39
Treatment effects on k†			P		
Leaf origin			0.002		0.010
Location			0.339		0.097
Origin× location			0.316		0.616

†Rate constant (k) determined for fraction remaining = e^{-kt} (t in yr) February 1990 to February 199 and February 1991 to February 1992.

findings for the 50-yr-old stand in that alder litter lost 50% of its mass in the 1st yr but shows little mass change in the 2nd yr. This rapid 1st yr weight loss may be explained by the faster decomposition of intervein leaf material compared with veins, or because relatively soluble materials are rapidly released under alder, leaving more resistant material.

This experiment also showed that litter from the two alder stands decomposed at different rates. By placing the alder leaves from one stand in the same stand (e.g., fir-to-alder leaves on fir-to-alder forest floor) and in the other stand (e.g., fir-to-alder leaves on alder-to-alder forest floor), we could separate the effects of litter quality (which stand the leaves came from) from decomposition environment (light, temperature, and moisture levels on the forest floor of the stand where they came from). The fir-to-alder litter decomposed at a significantly faster rate, irrespective of where the leaves were placed. Litter from the fir-to-alder stand had significantly higher concentrations of N, P, K, and Mg (Compton, 1990) and thus could be considered to have better quality. Litter from both stands decomposed slightly faster in the alder-to-alder stand after 2 yr, perhaps because the lower aboveground biomass allowed more light to reach the forest floor, creating slightly more favorable decomposition conditions. Even within a species, litter quality, decomposition rate, and hence O horizon C storage can be affected by site conditions. Tree nutrition also may be playing an important role in these differences.

Growth Rate of Young Alder and Douglas Fir

A striking difference in growth rates was observed when alder and Douglas fir were replanted on the conversion plots (Fig. 24–5a,b). In both cases the lowest height and diameter growth occurred when a species was replanted on a site formally occupied by the same species. This effect was particularly evident for red alder, where the average height of alder-to-alder trees was nearly 4 m less than found for fir-to-alder trees. This reduction in growth may be the result of the decrease in the availability of soil P that developed during the previous alder rotation (Compton, 1990).

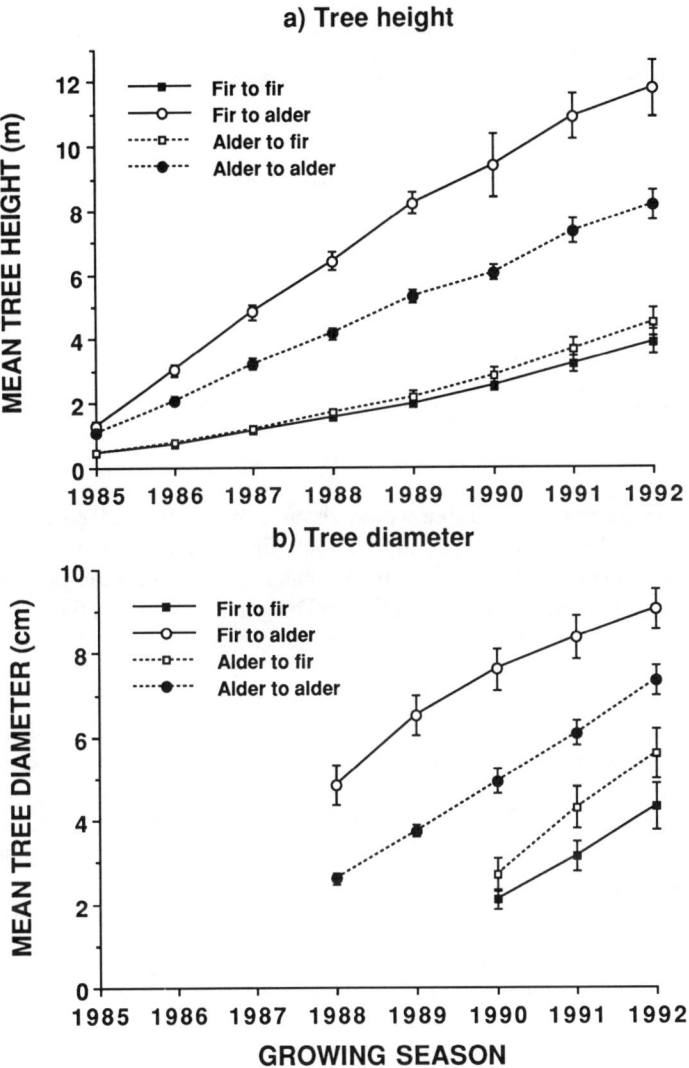

Fig. 24–5. Mean height (*a*) and diameter at 1.4 m (*b*) of red alder and Douglas fir in the conversion study plots from the time of establishment in 1985 to age of eight in 1992. Error bars indicate standard deviations of mean and may be smaller than symbols.

In the case of Douglas fir, we believe that the lower growth rate of the fir-to-fir plot is due to the N nutrition problem prevalent throughout the region (Stegemoeller & Chappell, 1990). Douglas fir is more productive on the former alder site because the N deficiency was somewhat corrected from symbiotic fixation and N accumulation that took place during the previous rotation. There may have been an even greater difference between the alder-to-fir vs. the fir-to-fir sites had it not been for the extensive presence of competing vegetation on the nutrient-rich, alder-to-fir site.

Table 24–6. Tree C content, woody tissue C accumulation and litterfall rates in the 8-yr-old Douglas fir and red alder conversion plots. Woody tissue C accumulation is the mean annual change in stem and branch biomass from 1988 to 1992. Litterfall in young Douglas fir stands was estimated from leaf mass in 1991 assuming that 20% of all foliage is lost each year. Carbon concentration in tissue was assumed to be 500 mg C kg^{-1} organic matter.

Plot	C content		Woody C growth†	Litterfall	
	Tree	Understory		Leaf	Nonleaf
	kg C ha^{-1}			kg C ha^{-1} yr^{-1}	
Fir-to-fir	3 850	2 350	648	112	ND†
Fir-to-alder	26 100	1 230	4 040	1 624 (146)‡	467 (287)
Alder-to-fir	4 080	2 360	696	105	ND
Alder-to-alder	11 900	1 013	2 350	1 691 (237)	29 (19)

†ND = not determined.
‡Litterfall in young alder stands was collected May 1992 to January 1993; standard deviations in parentheses (n = 4).

Carbon accumulation in tree woody tissue closely follows the height results (Table 24–6). The greatest accumulation took place in the fir-to-alder plot. After 8 yr, C accumulation was over five times greater on the fir-to-alder plot (4040 kg C ha^{-1}) than when the site was replanted to Douglas fir (fir-to-fir; 648 kg C ha^{-1}). Carbon accumulation was nearly two times greater on the fir-to-alder plot (4040 kg C ha^{-1}) than on the alder-to-alder plot (2350 kg C ha^{-1}). Leaf mass is much greater under alder, but leaf mass as a percentage of tree biomass is much smaller under alder (10%) relative to Douglas fir (26 and 22%, fir-to-fir and alder-to-fir, respectively). As expected, because of greater production and annual turnover of alder leaves, the litterfall return was substantially higher (approximately 16-fold) at the sites replanted to alder than at those replanted to Douglas fir. At this relatively young age, there is little in the way of litterfall return under Douglas fir, even though leaf mass makes up >20% of its biomass. In contrast, the deciduous characteristic of red alder results in the annual return of its entire leaf biomass.

Relative Importance of the Understory Species in Conversion Plot Carbon Accumulation

In the four conversion plots, the understory species played roles of varying importance in supplying C to the site (Fig. 24–6). At the time of harvest there was approximately twice as much understory vegetation in the red alder stand as compared to Douglas fir, and may have presented in more competition to the young trees growing on the former alder plots. Four years after harvesting, the presence of understory species had increased greatly for all conversion plots evidently due to the open canopy and the presence of adequate light. After 7 yr, the understory biomass greatly decreased except where Douglas fir had been replanted to Douglas fir and the canopy had not yet closed.

CONCLUSIONS

Carbon accumulation is markedly different in alder and Douglas fir ecosystems. After 50 yr, alder accumulated approximately 30% more C in the O hori-

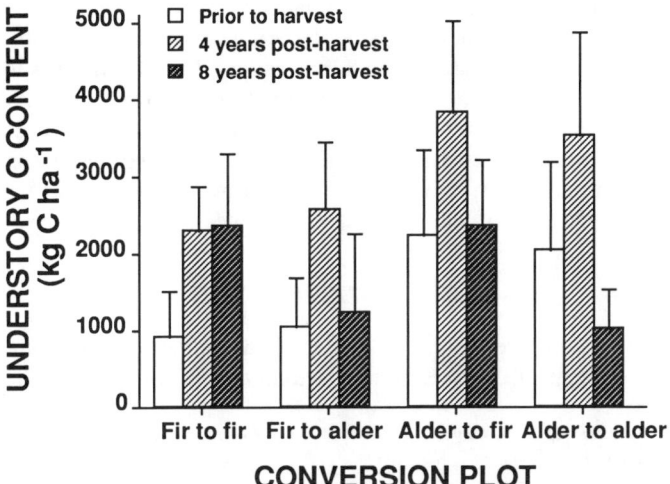

Fig. 24-6. Mean understory C content in conversion plots prior to harvest (1984) and 4 and 8 yr after plantation establishment. Error bars are one standard deviation above the mean.

zon and soil profile than did Douglas fir. In addressing Question 1 (What are the mechanisms responsible for the greater soil accumulation under the red alder forest stand compared to Douglas fir?) we found that greater soil C accumulation with alder is due both to higher litterfall rate and to slower rate of litter decomposition. We estimate from a 6-yr study of leaf litter decomposition that approximately one-third of this extra accumulation in the O horizon under alder is due to a slower rate of decomposition of the litter material and two-thirds to the higher rate of litter production. Although the higher rate of litter production is clearly a result of the deciduous nature of alder, it is not as evident why alder litter material should decompose more slowly than that of Douglas fir after the first 2 yr. Factors could include initial presence of secondary organic compounds resistant to decomposition, nutritional deficiencies or toxicities that directly affect the decomposing organisms, stabilizing N-humus interactions or enhanced microbial synthesis of organic compounds resistant to decomposition.

Previous vegetation on a site clearly influences the future site productivity (Question 3). The capacity of alder to accumulate C is greatly diminished in a second rotation alder stand. In the first 7 yr, the productivity of second rotation alder is approximately one-half that of the first rotation. In addition, a significant amount of the additional C that accumulated in the O horizon and mineral soil to the 45-cm depth during the first rotation of alder decreased 7 yr after harvesting and replanting. In contrast, where Douglas fir was the first rotation species, the soil C level did not decrease after harvesting and has increased slightly in the 7 yr since it was replanted with red alder. Whether O horizon C content decreased after clearcutting depended on the species planted. In addressing Question 2 (Does soil organic C under red alder change rapidly in response to harvesting and plantation establishment?) soil C significantly decreased in clearcut red alder sites, but increased rapidly when a fir site was planted with red alder.

It is clear that the alder rotation dramatically increased ecosystem C levels above that under Douglas fir, but it is not clear whether this accumulation will be sustained in the second rotation. Our early results (7 yr of growth) suggest that species rotation provided the best mechanism to maximize C accumulation. It certainly represents the best mechanism to maximize production by either alder or Douglas fir. Future effort should focus on understanding the reasons for the differences in productivity, and whether nutrient availability (N for Douglas fir and P for red alder) can offset the negative influence of the previous rotation.

ACKNOWLEDGMENTS

The Thompson Site Research Center is maintained by the University of Washington College of Forest Resources within the City of Seattle's Cedar River

App. 24–1. Biomass regression equations developed for the conversion plots in 1988 and 1992 by destructive sampling.

Plot	Component	X	Slope	Intecept	r^2
		1988 biomass regression equations			
Fir-to-fir	Stem	Ht†	0.151	−0.535	0.89
	New needle	Ht	0.0344	−0.099	0.74
	Old needle	Ht	0.021	−0.057	0.74
Alder-to-fir	Stem	Ht	0.115	−0.275	0.53
	Branch	Ht	0.022	−0.029	0.35
	Leaf	Ht	0.010	0	0.15 NS
Fir-to-alder	Stem+branch	Ht × dbh^2	0.0714	0.135	0.98
	Leaf	Ht × dbh^2	0.0138	0.0250	0.97
Alder-to-alder	Stem+branch	Ht × dbh^2	0.0594	0.356	0.98
	Leaf	Ht × dbh^2	0.0096	0.0577	0.97
		1992 biomass regression equations			
Fir-to-fir	Stem	Ht × dbh^2	0.0362	0.195	0.99
	Branch	Ht × dbh^2	0.0201	0.025	0.99
	Leaf	Ht × dbh^2	0.0172	0.1852	0.94
Alder-to-fir	Stem	Ht × dbh^2	0.0228	0.412	0.99
	Branch	Ht × dbh^2	0.0173	−0.076	0.99
	Leaf	Ht × dbh^2	0.0131	−0.0448	0.99
Fir-to-alder	Stem	dbh^2	1.140	−1.972	1.00
	Branch	dbh^2	0.6998	−1.494	1.00
	Leaf	dbh^2	0.1837	−0.3597	0.97
Alder-to-alder	Stem	dbh^2	0.8398	−0.4992	0.99
	Branch	dbh^2	0.7169	−1.7122	0.96
	Leaf	dbh^2	0.1736	−0.2532	0.98

†The equation is in the format:

$$\text{Mass (kg tree}^{-1}) = (\text{slope} \times X) + \text{intercept}$$

where X, slope and intercept are shown in the table (Ht = height in ft, dbh = diameter at breast height in in.).

Watershed. This project was initiated as part of the Integrated Forest Study on the effects of acid deposition, funded by the Electric Power Research Institute and coordinated by Oak Ridge National Laboratory. Completion of the project was funded by the National Council of the Paper Industry for Air and Stream Improvement. We thank Robert Gonyea, Bert Hasselberg and Mike Johnson for field assistance, Jay Kuhn and Jacquie Fenning for laboratory analyses and two anonymous reviewers for valuable suggestions.

REFERENCES

Alban, D.H. 1982. Effects of nutrient accumulation by aspen, spruce and pine on soil properties. Soil Sci. Soc. Am. J. 46:853–861.

Almendinger, J.C. 1990. The decline of soil organic matter, total-N and available water capacity following the late-holocene establishment of jack pine on sandy Mollisols, north-central Minnesota. Soil Sci. 150:680–694.

Anderson, J.M. 1991. The effects of climate change on decomposition processes in grassland and coniferous forests. Ecol. Appl. 1:326–347.

April, R., and R. Newton. 1992. Mineralogy and mineral weathering. p. 378–425. In D.W. Johnson and S.E. Lindberg (ed.) Atmospheric deposition and forest nutrient cycling: A synthesis of the integrated forest study. Springer-Verlag, New York.

Benoit, R.E., and R.L. Starkey. 1968a. Enzyme inactivation as a factor in the inhibition of decomposition of organic matter by tannins. Soil Sci. 105:203–208.

Benoit, R.E., and R.L. Starkey. 1968b. Inhibition of decomposition of cellulose and some other carbohydrates by tannins. Soil Sci. 105:291–296.

Berg, B. 1986. Nutrient release from litter and humus in coniferous forest soils—A mini review. Scand. For. Res. 1:359–369.

Bormann, B.T., and D.S. DeBell. 1981. Nitrogen content and other soil properties related to age of red alder stands. Soil Sci. Soc. Am. J. 45:428–432.

Brozek, S. 1990. Effect of soil changes caused by red alder (*Alnus rubra*) on biomass and nutrient status of Douglas-fir (*Pseudotsuga menziesii*) seedlings. Can. J. For. Res. 20:1320–1325.

Camiré, C., B. Côté and S. Brulotte. 1991. Decomposition of roots of black alder and hybrid poplar in short-rotation plantings: Nitrogen and lignin control. Plant Soil 138:123–132.

Challinor, D. 1968. Alteration of surface soil characteristics by four tree species. Ecology 49:286–290.

Cole, D.W. 1981. Nitrogen uptake and translocation by forest ecosystems. In F.E. Clark and T. Rosswall (ed.) Terrestrial nitrogen cycles. Ecol. Bull. (Stockholm): 33:219–232.

Cole, D.W., J. Compton, H. Van Miegroet, and P.S. Homann. 1990. Changes in soil properties and site productivity caused by red alder. Water Air Soil Pollut. 54:231–246.

Cole, D.W., and S.P. Gessel. 1968. Cedar River Research: A means for studying the nutrient cycling in forest ecosystems. For. Resour. Monogr. Univ. Washington College For. Resour., Seattle, WA.

Cole, D.W., and M.R. Rapp. 1980. Elemental cycling in forest ecosystems. p. 341–409. In D. Reichle (ed.) Dynamic properties of forest ecosystems—IBP Synthesis. Vol. 23. Cambridge Univ. Press, Cambridge, England.

Compton, J.E. 1990. Phosphorus distribution and cycling under red alder: Implications for growth and nutrition of second rotation alder. M.S. thesis. Univ. Washington, Seattle, WA.

Crocker, R.L., and J. Major. 1955. Soil development in relation to vegetation and surface age at Glacier Bay, Alaska. J. Ecol. 43:427–448.

Cromack, K., Jr. 1973. Litter production and litter decomposition in a mixed hardwood watershed and in a white pine watershed at Coweeta Hydrologic Station, North Carolina. Ph.D. diss. Univ. Georgia, Athens (Diss. Abstr. 73-31874).

Dice, S.F. 1970. The biomass and nutrient flux in a second-growth Douglas-fir ecosystem. Ph.D. diss. Univ. Washington, Seattle (Diss. Abstr. 70-19621).

Edmonds, R.L. 1980. Litter decomposition and nutrient release in Douglas-fir, red alder, western hemlock, and Pacific silver fir ecosystems in western Washington. Can. J. For. Res. 10:327–337.

Edmonds, R.L., D.J. Vogt, D.H. Sandberg, and C.H. Driver. 1986. Decomposition of Douglas-fir and red alder wood in clear-cuttings. Can. J. For. Res. 16:822–831.

Fogel, R., and K. Cromack, Jr. 1977. Effect of habitat and substrate quality on Douglas fir litter decomposition in western Oregon. Can. J. Bot. 55:1632–1640.

Freund, J.E. 1988. Modern elementary statistics. Prentice-Hall, Englewood Cliffs, NJ.

Fyles, J.W., and I.H. Fyles. 1993. Interaction of Douglas-fir with red alder and salal foliage litter during decomposition. Can. J. For. Res. 23:358–361.

Fyles, J.W., and W.B. McGill. 1987. Decomposition of boreal forest litters from central Alberta under laboratory conditions. Can. J. For. Res. 17:109–114.

Gholz, H.L., C.C. Grier, A.G. Campbell, and A.T. Brown. 1979. Equations for estimating biomass and leaf area of plants in the Pacific Northwest. For. Res. Lab. Res. Pap. 41. Oregon State Univ., Corvallis, OR.

Harmon, M.E., G.A. Baker, G. Spycher, and S.A. Greene. 1990. Leaf-litter decomposition in the *Picea/Tsuga* forests of Olympic National Park, Washington, U.S.A. For. Ecol. Manage. 31:55–66.

Heilman, P.E., and S.P. Gessel. 1963a. Nitrogen requirements and the biological cycling of nitrogen in Douglas-fir stands in relationship to the effects of nitrogen fertilization. Plant Soil 18:386–402.

Heilman, P.E., and S.P. Gessel. 1963b. The effects of nitrogen fertilization on the concentration and weight of nitrogen, phosphorus, and potassium in Douglas-fir trees. Soil Sci. Soc. Am. Proc. 27:102–105.

Homann, P.S., H. Van Miegroet, D.W. Cole, and G.V. Wolfe. 1992. Cation distribution, cycling, and removal from mineral soil in Douglas-fir and red alder forests. Biogeochemistry 16:121–150.

Homann, P.S., and D.W. Cole. 1990. Sulfur dynamics in decomposing forest litter: Relationship to initial concentration, ambient sulfate, and nitrogen. Soil Biol. Biochem. 22:621–628.

Jenny, H. 1980. The soil resource. Springer-Verlag, New York.

Johnson, D.W. 1992. Effects of forest management on soil carbon storage. Water Air Soil Pollut. 64:83–120.

Marshall, J.D., and R.H. Waring. 1986. Comparison of methods of estimating leaf area index in old-growth Douglas-fir. Ecology 67:975–979.

Melillo, J.M., J.D. Aber, and J.F. Muratore. 1982. Nitrogen and lignin control of hardwood leaf litter decomposition dynamics. Ecology 63:621–626.

Nelson, D.W., and L.E. Sommers. 1982. Total carbon, organic carbon and organic matter. p. 539–579. *In* A.L. Page et al. (ed.) Methods of soil analysis. Part 2. Agron. Monogr. 9. ASA and SSSA, Madison, WI.

Olson, J.S. 1963. Energy storage and the balance of producers and consumers in ecological systems. Ecology 44:322–331.

Parsons, W.F.J., B.R. Taylor, and D. Parkinson. 1990. Decomposition of aspen (*Populus tremuloides*) leaf litter modified by leaching. Can. J. For. Res. 20:943–951.

Stegemoeller, K.A., and H.N. Chappell. 1990. Growth response of unthinned and thinned Douglas-fir stands to single and multiple applications of nitrogen. Can. J. For. Res. 20:343–349.

Turner, J. 1975. Nutrient cycling in a Douglas-fir ecosystem with respect to age and nutrient status. Ph.D. diss. Univ. Washington, Seattle (Diss. Abstr. 75-28453).

Turner, J., D.W. Cole, and S.P. Gessel. 1976. Mineral nutrient accumulation and cycling in a stand of red alder (*Alnus rubra*). J. Ecol. 64:965–974.

Turner, J., and J. Kelly. 1985. Effect of radiata pine on soil chemical characteristics. For. Ecol. Manage. 11:257–270.

Turner, J., and M.J. Lambert. 1988. Soil properties as affected by *Pinus radiata* plantations. N.Z. J. For. Res. 18:77–91.

Ugolini, F.C. 1968. Soil development and alder invasion in a recently deglaciated area of Glacier Bay, Alaska. p. 115–140. *In* J.M. Trappe et al. (ed.) Biology of alder. U.S. For. Serv. Pacific Northwest For. Range Exp. Stn. Portland, OR.

Ugolini, F.C., and A.K. Schlichte. 1973. The effect of Holocene environmental changes on selected western Washington soils. Soil Sci. 116:218–227.

Van Cleve, K., L.A. Viereck, and R.L. Schlentner. 1971. Accumulation of nitrogen in alder (*Alnus*) ecosystems near Fairbanks, Alaska. Art. Alp. Res. 3:101–114.

Van Miegroet, H., and D.W. Cole. 1985. Acidification sources in red alder and Douglas-fir soils—Importance of nitrification. Soil Sci. Soc. Am. J. 49:1274–1279.

White, D.L., B.L. Haines, and L.R. Boring. 1988. Litter decomposition in southern Appalachian black locust and pine-hardwood stands: Litter quality and nitrogen dynamics. Can. J. For. Res. 18:54–63.

Wilkinson, L. 1989. SYSTAT: The system for statistics. SYSTAT, Inc. Evanston, IL.

Zucker, W.V. 1983. Tannins: Does structure determine function? An ecological perspective. Am. Nat. 121:335–365.

25 Modeling Carbon and Nitrogen Dynamics in Western Red Cedar and Western Hemlock Forests

Rodney J. Keenan and J.P. (Hamish) Kimmins

University of British Columbia
Vancouver, Canada

John Pastor

University of Minnesota
Duluth, Minnesota

The storage and flux of C are important components in the functioning of forest ecosystems (Waring & Schlesinger, 1985), and these are often intimately coupled with the cycling of nutrients (Attiwill, 1986). For example, in temperate forests N generally limits productivity when water is not limiting, and its availability in the soil depends largely on the rate at which N is mineralized during decomposition of organic matter returned to the forest floor in above- and belowground litterfall. When climatic conditions are constant, this is mainly determined by the chemistry of the C associated with the N in litterfall (Meentmeyer, 1978). Thus, there are strong feedbacks between the way in which different species store and utilize the C fixed in photosynthesis, the rate at which N is released in the forest floor during decomposition, N uptake, and, ultimately, to level of production that the forest sustains (Hobbie, 1992). These feedbacks can lead to alternative stable states in the composition of vegetation and N supply in the soil, even when abiotic conditions such as climate, soil parent material and topography are quite similar, in both forest (Flanagan & Van Cleve, 1983; Nadelhoffer et al., 1983; Pastor et al., 1984; Zak et al., 1986), and grassland ecosystems (Wedin & Tilman, 1990).

The landscape of the Pacific coast of North America is dominated by coniferous species. Individual trees grow to large sizes and large amounts of biomass accumulate in forest stands (Waring & Franklin, 1979). However, there is considerable variation between these conifer species in maximum size and age, and in their susceptibility to attack by insects and diseases. This variation is indicative of differences in growth patterns and C allocation to secondary chemicals in the production process. There also are differences in the growth response of each

---

Copyright © 1995 Soil Science Society of America, 677 S. Segoe Rd., Madison, WI 53711, USA. Carbon Forms and Functions in Forest Soils.

species to varying levels of N availability, although this has been less intensively investigated. Despite the economic importance of—and scientific interest in—these forests, there has been little consideration of the way in which these differences in chemistry and response to N availability might express themselves in ecosystem functioning and affect community structure through the kinds of feedback mechanisms described above.

The forests of northern Vancouver Island are in the northern latitudinal range of these Pacific coastal forests. They are composed largely of western redcedar (*Thuja plicata* Donn), western hemlock [*Tsuga heterophylla* (Raf.) Sarg.] and amabilis fir [*Abies amabilis* (Dougl.) Forbes], with occasional Sitka spruce [*Picea sitchensis* (Bong.) Carr.] on the outer coast, and lodgepole pine (*Pinus contorta* var. *contorta*) on more poorly drained sites. In the gently undulating landscape on the eastern side of the north island there has been little recent wildfire, and the predominant disturbances to the forest canopy are windstorms. On well-drained to somewhat-imperfectly drained sites the forests are either dominated by old-growth western redcedar with a lesser component of western hemlock (the cedar–hemlock type); or more uniform, second-growth western hemlock and amabilis fir, that largely originated after a catastrophic windstorm in 1906 (the hemlock–amabilis type). The more open cedar–hemlock stands have a major understorey component dominated by the ericaceous shrub salal (*Gaultheria shallon* Pursh). The locality is described in more detail by Prescott et al. (1995, see Chapter 17). Mean annual precipitation is about 1700 mm, and mean annual temperature is 7.9°C. The two forest types occupy similar topographic positions. Soils are medium-textured, humo-ferric podzols (Agriculture Canada, 1987)—haplorthods in the U.S. system, underlain by periglacial and fluvio-glacial tills derived from acid igneous substrates. These forests contain around 250 Mg ha^{-1} of C and 2000 kg ha^{-1} of N in detrital biomass (Keenan et al., 1993). However, the availability of N in the forest floor of the cedar-dominated, old-growth cedar–hemlock stands is considerably lower than in hemlock-dominated, hemlock–amabilis stands (Prescott et al., 1993), and this contributes to the slower growth of planted and naturally regenerated seedlings following clear-cutting on the cedar–hemlock sites (Weetman et al., 1989a,b).

Western red cedar and western hemlock differ in their maximum life span and maximum size, and in their resistance to insects and pathogens (Minore, 1983). Western red cedar can live up to 1000 yr, and can attain a maximum diameter of 250 to 300 cm, while hemlock rarely survives more than 400 yr, and reaches a maximum diameter of about 120 cm (Franklin & Dyrness, 1973; Keenan, 1993). These differences in size and life history are associated with different growth rates and regeneration dynamics in this environment, and they indicate differences in wood and foliar chemistry between the two species.

The objective of this paper was to investigate the extent to which lower nutrient availability in the forest floors of cedar–hemlock stands could be explained by differences in species chemistry. The study tested the LINKAGES model in this environment, using it to simulate biomass accumulation and N availability over time, given constant climate and mineral soil conditions, for forests dominated by two species: western red cedar, and western hemlock.

Because of the time scales involved in ecosystem development, it is difficult to investigate such a hypothesis using traditional empirical approaches, such as a controlled experiment, as has been done in more rapidly developing vegetation types like grasses (Wedin & Tilman, 1990). The stand history of these long-lived species also is difficult to reconstruct, and no study sites were available that could have been used to investigate forest floor properties with varying proportions of each species in different stages of development. Computer modeling provides a way of overcoming some of these investigative difficulties. In developing a model, current knowledge and ideas about the way ecosystems develop and function are organized in a logical framework, and the implications of this empirical and theoretical understanding can be investigated. Hypotheses can be investigated that are difficult or impossible to test through empirical means (Yarie, 1990). The model, in effect, is a kind of experimental microcosm in which different scenarios can be explored using assumptions based on current (albeit imperfect) knowledge of ecosystem functioning. However, models are not reality, and should not be considered a replacement for empirical approaches to investigation and hypothesis testing.

Models can depict the functioning of ecosystems at various scales of space and time (Kimmins, 1988). In this study, the ecosystem properties of interest were those that develop over decades or millennia, in a space of the order of a single forest stand. Because forests in this locality have been the subject of relatively little scientific investigation, it was important to use a simple model that described the functioning of the ecosystem using the dynamics of a few key components for which calibration data were available. In longer-time step models, the dynamics of larger, slower-moving components can integrate the dynamics of smaller components that turn over more rapidly (Bunnell, 1989). For example, tree diameter growth integrates the kinetics of water and nutrient uptake, photosynthesis, and C allocation; and the nutrient dynamics in decomposing litter integrate the activities of various fungal and microbial decomposers. By carefully choosing the components included in the simulation, a relatively simple model structure can be developed that has a considerable degree of descriptive realism.

MODEL STRUCTURE AND CALIBRATION

The model used in this study was LINKAGES (Pastor & Post, 1985, 1986), a simulation model of forest growth and nutrient cycling based on the JABOWA model (Botkin et al., 1972; Botkin, 1993). Both LINKAGES and JABOWA were developed to simulate forest growth and community dynamics in the northeastern USA. The model LINKAGES includes more detailed representations of climate, and soil N and water availability, and the way they affect the growth of different tree species, than JABOWA. It explicitly represents the feedback between species chemistry, N availability, and forest production that may control species composition in these forests. It requires a relatively simple set of calibration data, and can simulate the development of both even-aged, single species, and mixed-age, mixed-species stands found in the study area.

The model LINKAGES consists of three basic components: production, climate and decomposition. The production component is the single-tree, nonspatial model construct of JABOWA (Shugart, 1984) which has been used extensively to simulate tree growth and community dynamics in many different forest types around the world (for a full description see Botkin, 1993). It has previously been applied, with some success, to the Douglas fir ecosystems of the Pacific Northwest (Dale et al., 1986). In this component, tree establishment, growth and mortality are simulated on a small area plot. Individual trees establish [as saplings with a diameter at breast height (dbh) chosen stochastically from between 1 and 3 cm] at a user-specified rate, if light and moisture conditions are suitable for the species. These established individuals increment in diameter on an annual time step at a rate determined by the potential maximum diameter increment under optimal conditions (a function of the maximum age and maximum diameter of each species), and modified according to the simulated availability of light, water and nutrients, whichever is most limiting, for each individual. These modifiers are calculated from species-specific coefficients of response functions that are entered in the input file. Tree height is a parabolic function of diameter. Mortality is simulated in two ways: (i) exogenous mortality is simulated by killing a small proportion of trees each year, so that 1% of trees reach the potential maximum age for their species; and (ii) within stand competition is simulated using a flag for slow growth. In this study a diameter increment less than 10% of the potential maximum was considered slow growth, and the probability of mortality in a given year increased to 0.2 for trees with more than 10 consecutive years of slow growth.

Light at any level in the canopy is a function of the foliage biomass above that level, and is determined from allometric relationships between diameter and foliar biomass, and assuming that all foliage for an individual is situated at the top of the tree, and spread across the entire plot. Available moisture is calculated in the climate component. The mean and the standard deviation of monthly temperature and precipitation for the study area are read into the model as input, and normally distributed, random values are selected to simulate an annual climate. Thornthwaite and Mather's monthly actual evapotranspiration (AET) is calculated according to an approximation function (Pastor & Post, 1984), and combined with soil moisture-holding capacity (from soil texture) to determine the proportion of the growing season that soil moisture falls below field capacity. This value is used to reduce diameter growth.

The availability of N, the nutrient assumed to be limiting tree growth, is calculated in the decomposition component. Foliar, root and twig litterfall are calculated for each year in the production component from foliar biomass and foliage retention time. Woody litterfall is determined by tree mortality. The decomposition and N dynamics of these annual inputs of each type, and species, of litter are monitored each year as separate cohorts. Foliar and fine root litter lose mass as a function of the simulated climate (annual AET) and their lignin/N ratio, according to a relationship developed by Pastor and Post (1985). Woody litter cohorts lose mass at user-specified annual rates. Once a cohort reaches a certain percentage mass remaining (a function of the initial lignin concentration, DeHaan, 1977, p. 21–30), it is transferred to a combined pool of humus, which loses mass, and N, at a constant rate (1% yr^{-1} in the original model, changed to 2% in this study) or well-decayed wood which loses mass at 1.5% per year.

The immobilization or mineralization of N is simulated for each cohort using a linear relationship between the mass remaining and the N concentration in the remaining material (Aber & Melillo, 1980), of the form

$$\text{percent N} = a + b(\text{percentage mass remaining}) \qquad [1]$$

The coefficients of this relationship are specified as input for each litter type. A similar relationship is used to model the dynamics of lignin. Lignin and N concentrations of each cohort change during the course of decomposition, and therefore these are dynamic components that are recalculated annually in the model. These changes affect the mass loss in the following year. The net mobilization of N is calculated from the sum of immobilization or mobilization in each cohort, and the mobilization from humus and well-decayed wood. A proportion of immobilization is satisfied by throughfall (16% of litterfall N) and external inputs (1 kg ha^{-1} yr^{-1} in this region).

Parameters used in the growth component of the model are shown in Table 25–1. Maximum age, dbh, and height were either obtained from the literature (Franklin & Dyrness, 1973), or from measurements of stand structure in the study area (Keenan, 1993). The coefficients B2 and B3 determine the shape of the diameter growth curve. These were calculated following the selection of the value for G, the growth scaling parameter in JABOWA (Botkin, 1993). This parameter was determined for each species in an iterative process, by applying various values and using the one that gave the best fit to stand growth from permanent plots with a site index of 25 m (at breast height age 50, J. Barker, 1992,

Table 25–1. Parameters used for each species in the growth component of the model.

Parameter	Western red cedar	Western hemlock
Maximum age (yrs)	750	400
Maximum dbh (cm)	250	122
Maximum height (m)	50	50
B2†	0.08	0.27
B3†	38.90	81.21
G‡	100	190
Stemwood biomass slope	2.081	2.257
Stemwood biomass intercept	−2.0927	−2.172
Foliage biomass trees<30 cm dbh (slope)	1.922	2.218
Foliage biomass, trees <30 cm dbh (intercept)	−3.007	−4.130
m^2 leaf area/kg foliage	0.0044	0.0044
Shade tolerance, a	10.59	10.20
Shade tolerance, b	0.062	0.038
Shade tolerance, c	9.31	10.15
N tolerance, a	0.998	1.258
N tolerance, b	0.174	−8.808
N tolerance, c	−0.028	−0.01
Foliage retention time (yr)	4	4

† Parameters for shaping diameter growth curve.
‡ Parameter for determining diameter growth.

personal communication). Foliage and stemwood biomass relationships were taken from Gholz et al. (1979), Dale and Hemstrom (1984), and Feller (1992). In simulating the forests of northern Vancouver Island there were a number of simplifying factors. First, the object was to simulate nutrient availability under constant climate and mineral soil conditions for forests dominated by two species, western red cedar, and western hemlock. Amabilis fir was a minor component in the stands where detailed investigations were undertaken, and fir generally has similar life history and foliar characteristics to hemlock. Initial model runs using climatic data from the locality with a relatively fine-textured soil indicated that the soil moisture content remained above field capacity throughout the growing season. This was consistent with field observations, and therefore it was assumed that there was no moisture limitation on tree growth. The two species are close to the center of their geographic range in this locality, so no temperature limitations were imposed.

The limiting factors on tree growth were therefore assumed to be light and N availability. In the model, tree species are assigned shade tolerance categories based on the extent to which growth is reduced in relation to the proportion of above canopy light reaching the tree. In applying the JABOWA model to the Pacific Northwest, Dale and Hemstrom (1984) developed relationships between foliage biomass and leaf area, and leaf area and percentage of above-canopy light. These relationships conformed fairly well to recent data from Vancouver Island (Smith, 1993), and were used to determine the proportion of above canopy light (PACL) at various heights in the canopy. A recent study of shade tolerance of cedar and hemlock in coastal British Columbia by Carter and Klinka (1992) provided the following relationship between PACL and relative growth rate, and the parameters for each species

$$RG = [a \, (1 - \exp^{-b(\text{PACL})})]c \qquad [2]$$

where RG = relative growth rate; and a, b, and c are parameters that determine the shape of the growth response. Parameters for the two species on fresh sites were used in the model.

The response to N availability function of the two species was more problematical. In LINKAGES, the Mitterlich relationships developed by Aber et al. (1979) for hardwood and conifer species in northeastern USA were used to describe three different classes of response to varying levels of annual net N mineralization (not including tree uptake). However, there have been few studies of annual N mineralization in the Pacific Northwest, and those have indicated considerably lower annual rates than in the forests where Aber et al. (1979) developed their relationships. In addition, there has been little broadscale investigation of the response of these two species to fertilization, or to varying levels of available N. For western hemlock there was a general pattern of site index increasing across an N status (measured as total and anaerobically mineralized N) gradient from poor to medium, with no further increase on sites classified as N rich (Kayahara et al., 1993, unpublished data). The response of cedar is more variable, and may depend to some degree on the extent of nitrification (Krajina et al., 1973; Minore, 1983).

There have been a number of studies in the study area of response to fertilization, and of the growth of seedlings planted on the two site types. These indicated that hemlock responds strongly to fertilization on the cedar–hemlock sites but not on the hemlock–amabilis sites, and that hemlock grows much more rapidly on the hemlock–amabilis than the cedar–hemlock sites. Growth response to fertilization on the hemlock–amabilis sites was generally small. The response of cedar planted on the hemlock–amabilis sites, and to fertilization, is much lower (Weetman et al., 1990; Thompson et al., 1993, unpublished data). Thus it appears that N availability in the two forest types may lie across the poor to medium threshold for hemlock described by Kayahara et al. (1993). A similar pattern may well apply for cedar, but the limit below which N availability has to fall before growth is significantly reduced is below the levels found in these two forest types.

Thus, although it appeared that the Mitterlich function was an appropriate descriptor of the growth response to N availability for the two species, the problem was to determine appropriate parameters. This was done using height growth data from an experiment where the two species were planted on clearcuts of the two forest types, and assuming annual N availability was 20 kg ha^{-1} yr^{-1} on the cedar–hemlock sites (Weetman et al., 1990) and 50 kg ha^{-1} yr^{-1} on the hemlock–amabilis sites. With no measured estimate or best guess available, the latter figure was somewhat arbitrary. The difference, compared to cedar–hemlock sites, is greater than that reported for KCl-extractable N, but less than that reported for N mineralized in 40-d aerobic incubations (Prescott et al., 1993). Parameters were developed based on the assumption that cedar growth ceased at zero net mineralization, while hemlock growth did so at 10 kg ha^{-1} yr^{-1}. Parameters for the relationship are shown in Table 25–1, and the response functions in Fig. 25–1.

The parameters used in the decomposition component of the model are shown in Table 25–2. Initial N contents were measured during an associated study (Keenan, 1993), and initial lignin values were from Harmon et al. (1990). The parameters for the relationships between N concentration vs. percentage organic matter remaining are from unpublished studies (R. Edmonds, 1992, personal communication). The slope of the line was kept the same between the two species, but the intercept was adjusted to reflect the differences in initial N concentration. Parameters for the lignin vs. organic matter remaining relationship were derived in a similar way.

Woody debris is a major component of these systems and, to more realistically reflect their N dynamics, the model was modified to include a larger number of woody litter types which have different rates of decay according to the species. Rates of decay have been reported for logs of these species in similar ecosystems (Graham & Cromack, 1982; Sollins et al., 1987). These values were doubled to obtain the numbers reported in Table 25–2 to take into account Sollins' (1981) argument that those estimates, based on changes in density, do not take into account fragmentation losses. The parameters for N and lignin dynamics in decaying wood were derived from measurements of wood in various stages of decay in the study area (Keenan et al., 1993).

The model simulated forest growth and nutrient dynamics on a 0.2-ha plot. The model has stochastic elements for calculating climate and tree mortality, and allows for the simulation of up to 100 plots with different random effects. Results

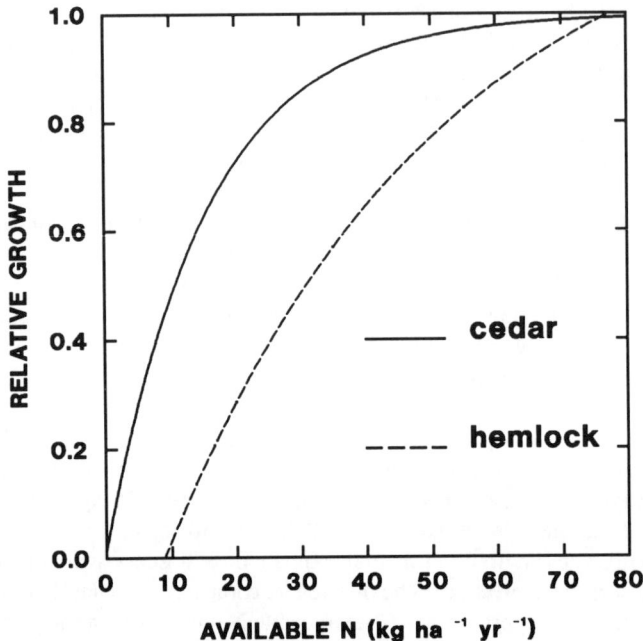

Fig. 25–1. Growth response functions for western redcedar and western hemlock to available N. Calculated according to the equation: Relative growth = $a[1 - 10^{-c(N+b)}]$, where N is available N, and a, b and c are the N response parameters specified in Table 25–1.

are then reported as a mean value with 95% confidence limits. Climate is relatively consistent in this locality, and simulation of 10 plots produced output with confidence limits within 10% of the mean for most variables. Model runs were done on a IBM 386/25 Mhz PC. The model output was varied from that described by Pastor and Post (1985) to get a more comprehensive picture of the simulated forest floor, and to obtain diameter-class distributions of simulated stands. A component was included to simulate a windstorm disturbance, in which the probability of death was high for big trees and low for small trees.

The model was initialized with a mass of humus and N specified in the input file. This represents a starting humus pool that loses mass and mineralizes N at a constant, relatively slow, rate of 2% per year. Thus, initial N can only be varied by varying the size and N concentration of this humus pool. Once a forest becomes established and a forest floor of decomposing cohorts is built up, most of the mineralized N comes from these decomposing cohorts. Initializing the model with a low amount of soil N required a considerable period for cedar (the more low-N tolerant species) to become established and develop a forest floor. Consequently, the first 2000 yr or so of all simulations should be considered an initialization stage and largely ignored for the evaluation of model performance. All simulations were run with a fine-textured mineral soil (field capacity of 38.3 cm, wilting point of 20 cm).

Table 25–2. Parameters used in decomposition component.

Litter type	Initial N	N parameters intercept, slope	Initial lignin	Lignin parameters intercept, slope	Mass loss when litter becomes humus or decayed wood	Annual mass loss
	g kg^{-1}		g kg^{-1}		%	
Cedar foliage	4.0	0.0191, 0.0186	231	0.664, 0.673	50	NA†
Hemlock foliage	7.5	0.0261, 0.186	240	0.673, 0.433	49	NA
Fine roots	4.6	0.0163, 0.0117	283	NA†	58	10
Cedar wood >60 cm dbh	0.6	0.0033, 0.0027	NA†	NA	50	0.6
Cedar wood <60 cm dbh	0.6	0.0033, 0.0027	NA	NA	50	1.2
Hemlock wood >30 cm dbh	0.6	0.0033, 0.0027	NA	NA	50	1.6
Hemlock wood <30 cm dbh	0.6	0.0033, 0.0027	NA	NA	50	2.3
Twigs	3.0	0.0195, 0.0157	NA	NA	61	10
Well-decayed wood	3.0	0.005, 0.002	NA	NA	25	1.5

†NA = not applicable

Simulations

Four scenarios were investigated with the model:

1. Cedar–hemlock type. Cedar and hemlock were grown together on soil with an initial soil N of 2 Mg/ha. Cedar was established at one seedling per plot every 4 yr and hemlock at one seedling every year.
2. Hemlock–amabilis type. Hemlock established at five seedlings per plot per year, with an initial soil N of 4 Mg/ha. Simulated windthrow occurred every 300 yr (stems and foliage of dead trees were added to the forest floor).
3. Hemlock only, no disturbance. Five seedlings were established per plot per year, with an initial soil N of 4 Mg/ha.
4. Cedar only, no disturbance. Five seedlings were established per plot per year, with an initial soil N of 4 Mg/ha.

The first two scenarios were designed to simulate the two forest types found in the region. A higher rate of seedling establishment in the hemlock–amabilis than the cedar–hemlock type largely reflects the presence of salal in the cedar–hemlock type, which was thought to limit the rate of recruitment in the cedar–hemlock type, while seedling establishment in hemlock–amabilis stands without salal is quite prolific (Keenan, 1993). It has been postulated that the hemlock–amabilis forest type has been subject to repeated windthrow (Lewis, 1982), so this kind of event was included in the hemlock–amabilis simulation. The model only simulates the dynamics of tree species, understorey species such as salal were not included. The latter two scenarios were more hypothetical and were designed to demonstrate the effect of the different species with other things (establishment, mortality and initial N) being held equal.

In order to determine the management consequences of the differences in nutrient availability which result from stands being dominated by the different species, four further scenarios were investigated. These involved running the cedar–hemlock and the hemlock–amabilis scenarios for 3000 yr, clearcutting the simulated stands (i.e., all trees were killed, stemwood was not added to the forest floor, but foliage was) and planting cedar or hemlock plantations at 3000 stems per hectare. These plantations were then grown for two 100-yr rotations.

RESULTS

Diameter distributions for the two species, once the model had equilibrated, for simulated cedar–hemlock stands are shown in Fig. 25–2. These are compared to the average values measured in three stands. The shape of the simulated and actual distributions are quite similar. There were some gaps in the measured cedar distribution compared with simulated one, that may indicate that disturbance not accounted for in the simulations has disrupted the regular patterns of establishment and mortality. The hemlock distributions suggest that annual recruitment may be higher than the one stem per plot in the simulation, and that the potential maximum diameter used may be a little high for the species in this type.

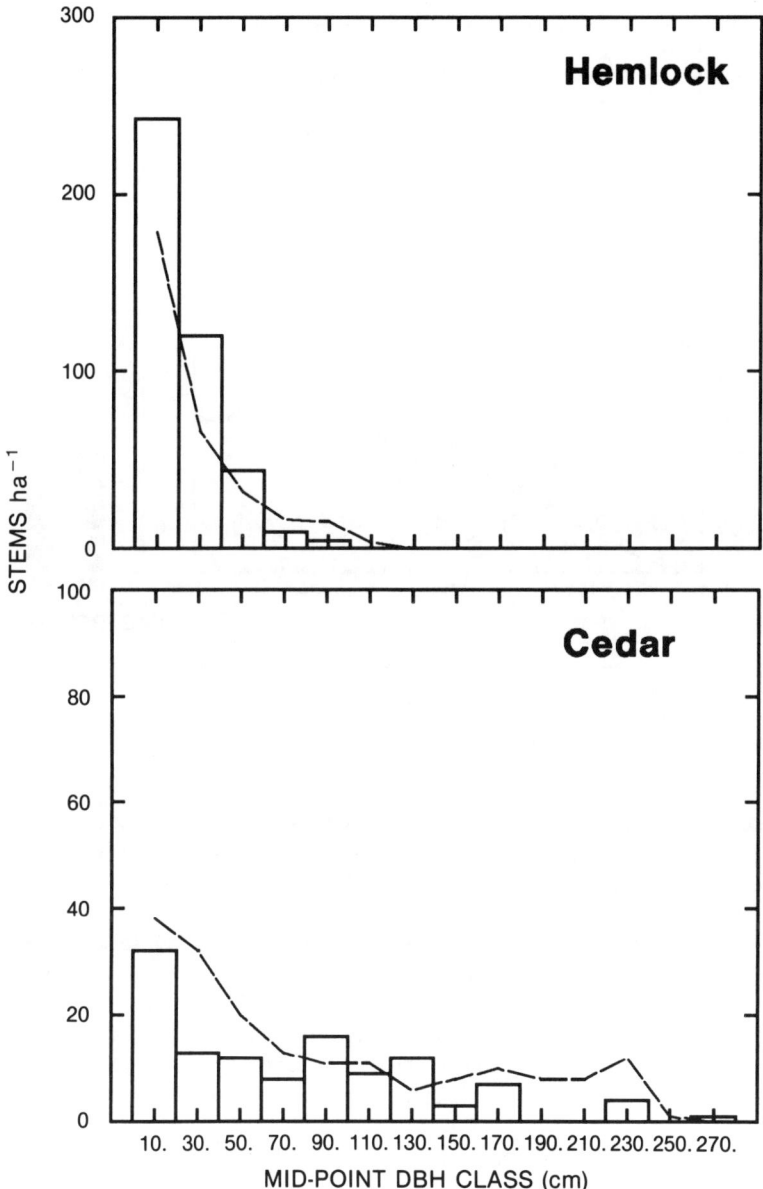

Fig. 25–2. Diameter distributions for western hemlock and western redcedar in cedar–hemlock stands. Bars are values obtained from measured stands, dotted lines are model output.

Aboveground live biomass simulated in the four scenarios, and "measured" biomass amounts for the two forest types (calculated using measured diameters and the same allometric relationships used in the model) are shown in Fig. 25–3. Available N is shown in Fig. 25–4, and forest floor biomass in Fig. 25–5. Values for all variables stabilized fairly early in the simulation for the single species

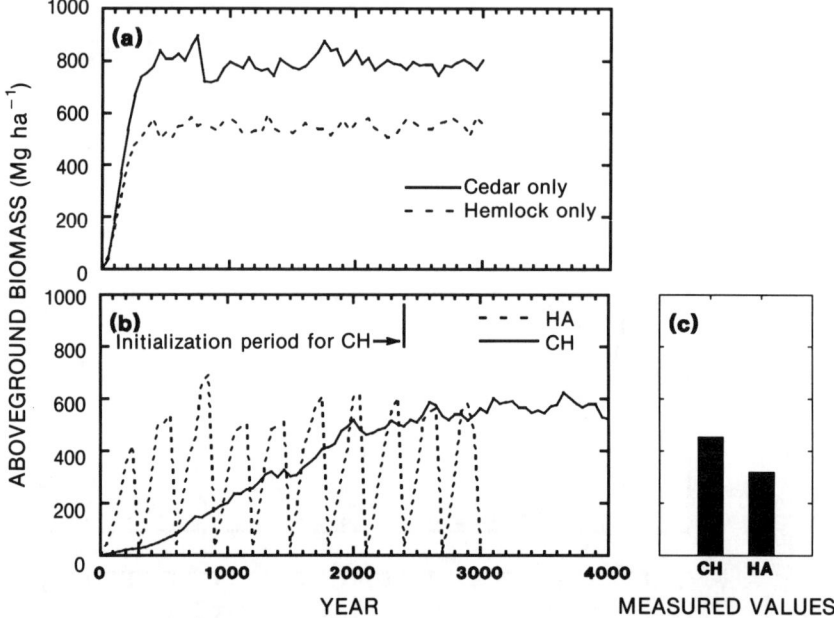

Fig. 25–3. Simulated aboveground biomass (Mg ha^{-1}) for four different modeling scenarios. (*a*) Cedar and hemlock only, both with high initial N. (*b*) Cedar–hemlock (CH, with low initial N), and hemlock–amabilis (HA, with high initial N). (*c*) Mean aboveground biomass calculated from three stands in the cedar–hemlock and hemlock–amabilis forest types.

scenarios, but it took 2000 to 3000 yr to reach a stable equilibrium in the simulation of the cedar–hemlock type. This is because of the low initial N and the low recruitment in this scenario, which meant that it took some time before a closed canopy nutrient cycle was established.

Aboveground, live biomass in the simulated cedar–hemlock stand (Fig. 25–3b) varied between 550 and 630 Mg ha^{-1}, slightly higher than the mean "measured" value from three stands of about 560 Mg ha^{-1}. The division of the biomass between the two species is not shown in the figure, but it also was relatively stable over the last 1000 yr of the 4000-yr simulation with hemlock comprising about 25% of the biomass. In the stands measured about 22% of the basal area was hemlock and fir.

Aboveground live biomass in the simulated hemlock–amabilis stands varied widely because of the periodic disturbance, peak values were between 500 and 600 Mg ha^{-1} at a stand age of 300 yr, and the biomass of simulated 100-yr-old stands varied between 300 and 400 Mg ha^{-1}. This compared quite well with the 320 Mg ha^{-1} measured in the stands disturbed by the major windstorm about 85 yr ago. Biomass in the hemlock-only scenario (Fig. 25–3a), with no disturbance, stabilized between 500 and 550 Mg ha^{-1}, and in the cedar-only scenario at around 800 Mg ha^{-1}. Cedar biomass was greater because it can attain larger sizes, and therefore it can accumulate a considerably greater mass in each stem than hemlock.

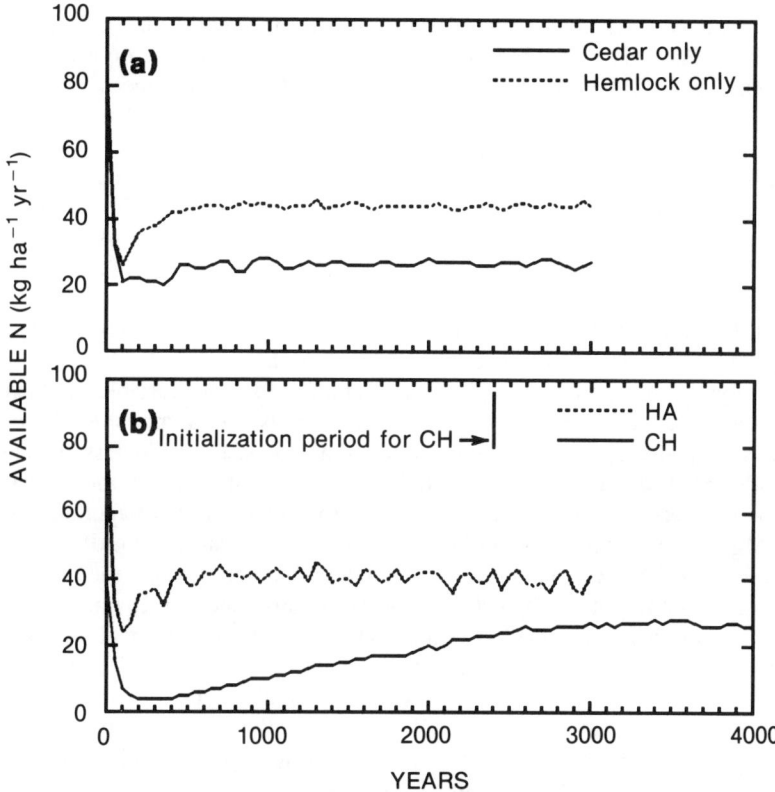

Fig. 25–4. Simulated N availability (kg ha^{-1} yr^{-1}) for four different modeling scenarios (*a*) Cedar and hemlock only (both with high initial N). (*b*) Cedar–hemlock (CH, with low initial N), and hemlock–amabilis (Ha, with high initial N).

Simulated available N (Fig. 25–4) stabilized at about 27 kg ha^{-1} yr^{-1} in the cedar–hemlock scenario, and varied between 37 and 43 kg ha^{-1} yr^{-1} in the hemlock–amabilis scenario. The variation in the hemlock–amabilis scenario was largely brought about by the flush of N in the litter deposited on the forest floor in the simulated windthrow. Although they both began with similar high levels of N, the single species simulations showed similar differences in nutrient availability after about 500 yr of simulation. Available N in the cedar-only scenario was almost identical of the equilibrium value in the cedar–hemlock scenario—27 kg ha^{-1} yr^{-1}. In the hemlock-only simulation available N stabilized at about 44 kg ha^{-1} yr^{-1}. As stated previously, no field estimates of annual N mineralization were available with which to compare these simulated estimates. However, the simulated values for the cedar–hemlock stands were within the range of 20 to 30 kg ha^{-1} yr^{-1} estimated by Weetman et al. (1990), and those from all simulations are within the range reported for coastal forests in the Pacific Northwest (Edmonds et al., 1990). The differential between the two forest types is similar to the differences in N availability measured in seedling bioassays, and KCl extractions by Prescott et al. (1993).

Simulated forest floor accumulation under the different scenarios is shown in Fig. 25–5. The total forest floor mass in the cedar–hemlock scenario is about 500 Mg ha^{-1}, about 140 Mg ha^{-1} less than the 640 Mg ha^{-1} measured in the cedar–hemlock stands (Keenan et al., 1993). Interestingly, this underestimate was about the same amount that the model overestimates live biomass. This could have been due to the effects of the 1906 windstorm, which resulted in higher mortality among larger trees than the constant exogenous mortality simulated in the model, with less live, and more detrital, biomass. It is therefore possible that the model relationships and parameters are simulating organic matter dynamics more accurately than might be suggested by these discrepancies. The proportion of the forest floor contained in the woody and nonwoody L, and F and H layers (H layer in the U.S. soil system) corresponds fairly well to the measured values, a further indication that the input rates and decay parameters for these variables in this forest type are set at appropriate levels. Similar conclusions can be drawn from the figures from the hemlock–amabilis simulation. Forest floor biomass fluctuates considerably because of the simulated disturbance, but the forest floor biomass at 100 yr after each windstorm (marked with arrows) is a similar magnitude to the 440 Mg ha^{-1} measured in the 85-yr-old stands. In the cedar–hemlock type, simulated F and H layer was larger in the simulation than measured in the field.

Impact of varying N availability, caused by differences in characteristics of species comprising the previous stand, on simulated growth of plantations of cedar and hemlock is shown in Fig. 25–6. Cedar growth was only slightly affected by lower N availability on the cedar–hemlock type, predicted biomass accumulation over 100 yr is about 310 Mg ha^{-1} on the cedar–hemlock type, and 330 Mg ha^{-1} on the hemlock–amabilis type. Current growth projections developed by forest managers in this area (of stemwood volume, which we multiplied by the density of cedar to calculate mass, and by 1.2 to get gross bole biomass and foliage), indicate that cedar stands contain 275 Mg/ha on poorer sites and 350 Mg/ha on richer ones (J. Barker, 1992, personal communication). Simulated growth of hemlock, on the other hand, was markedly lower on the cedar–hemlock type. Biomass after 100 yr was 230 Mg ha^{-1}, compared with 380 Mg ha^{-1} on the hemlock–amabilis type. These values are lower than those projected by managers for site indices of 20 and 30 (440 and 540 Mg ha^{-1}, respectively). It is interesting to note that the simulated growth of hemlock improves in the second rotation, as it begins to influence N availability.

DISCUSSION

These results suggest that the modified LINKAGES model, using a relatively simple calibration data set, provided reasonable simulations of population dynamics, organic matter turnover and nutrient dynamics in this environment. Simulated values of live and detrital biomass accumulation were generally within +/− 20% of measured mean values, and the predicted values for annual N mineralization (available N) were within the range of field estimates of average N availability.

Growing the two species in single species stands, with high initial N, resulted in lower N availability in the forest floor of cedar-dominated stands than those

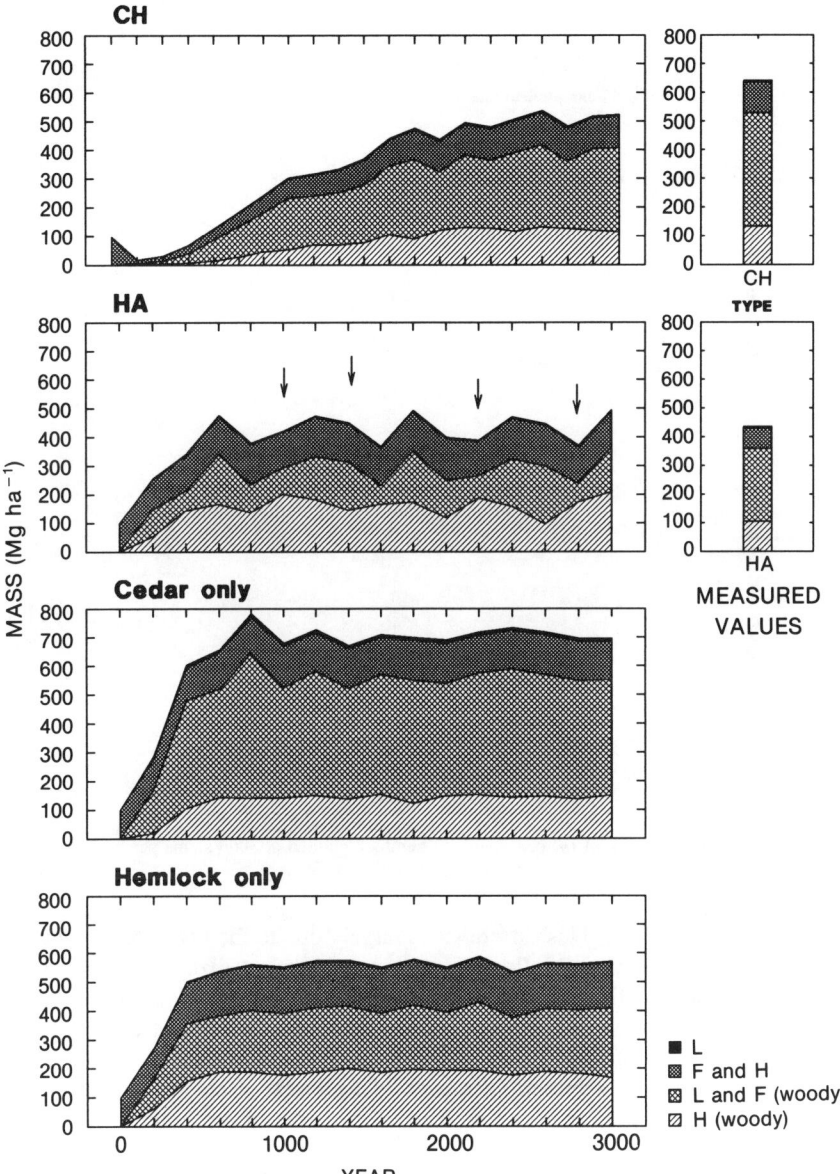

Fig. 25–5. Simulated forest floor accumulation by forest floor layers for the cedar–hemlock (CH) and hemlock–amabilis (HA) scenarios, and mean measured forest floor accumulation in three stands of the two forest types. Simulated forest floor accumulation for the cedar-only and hemlock-only scenarios. L and F (woody) includes woody debris and partly decomposed wood in the forest floor, H (woody) is well-decomposed wood.

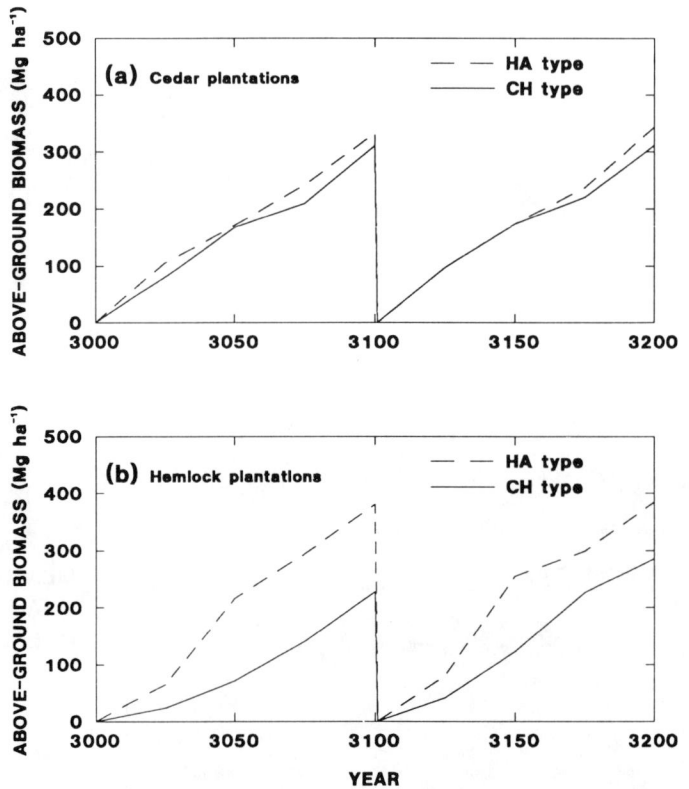

Fig. 25-6. Simulated growth of (a) cedar and (b) hemlock planted at 3000 stems per hectare on the cedar–hemlock, and the hemlock–amabilis forest types.

dominated by hemlock. This difference is largely due to the lower N concentration of cedar foliage, which influenced the rate of N mineralization in two ways. There was a lower initial pool of N for the same quantity of foliage, and the simulated decay rate of litter was slower, because the lignin/N ratio of cedar foliage was higher. N concentrations used in the model did not differ greatly (0.40% for cedar, and 0.75% for hemlock), however, N feedback mechanisms caused the relatively small variation to have a marked effect on decomposition rate and, consequently, on N mineralization. Actual evapotranspiration in this cool, moist environment was comparatively nonlimiting, and the lignin/N ratio had a large effect, especially when compared with drier and warmer, or colder environments (Meentemeyer, 1978; Pastor & Post, 1985). Also, both species had comparatively high lignin concentrations (23% for cedar, 24% for hemlock), and small variations in N concentration had a considerable impact on lignin/N ratio. These simulated differences in decay rate were supported by litter bag studies using foliage of the two species (Keenan, 1993).

Lower N availability in the forest floors under cedar compared to hemlock is contrary to previously reported studies of the two species. Cedar has previously been considered a "calcium pump," capable of raising the forest floor pH (Krajina, 1969), and allowing the development of microbially mediated decomposition, compared with fungal-dominated turnover under hemlock (Turner & Franz, 1985). However, these latter results were from forests with a drier, more continental climate than the one in this study, and Stone (1975) argues that effects of cedar on pH and C levels largely disappear at higher levels of precipitation. It has been argued that cedar prefers N available in the form of nitrate rather than ammonium, which may explain the general lack of response to fertilization and to planting on the hemlock–amabilis sites. However, there is no evidence in this locality to indicate that the presence of cedar can lead to greater rates of nitrification, as found in other areas (Turner & Franz, 1985).

Because cedar has a different response function to the availability of N than hemlock, these differences in N availability affect the growth of simulated plantations of each species differently, with cedar growing more slowly than hemlock on the hemlock type, but more rapidly on the cedar–hemlock type. There is a small decline in production once cedar becomes established on a type with a higher available N; however this does not lead to an ongoing feedback between declining foliar N concentration and productivity, and ultimately to zero production, because of the way litter N is included in the model. Initial litter N and lignin concentrations are fixed attributes of the species, and do not fluctuate with varying levels of N availability. Many ecologists and foresters would suggest that the N concentration in foliar litter can vary considerably, for example when trees are supplied with additional N in fertilizer. However, under N-limited conditions the concentration of N in litterfall will be a relatively constant attribute of the species, determined by the extent to which trees can retranslocate nutrients prior to senescence. This "base-level" of N is a function of the chemical composition of leaf structural components (maybe "lignin-bound" N; Berg & Theander, 1984), and this will vary between species, and in turn determine the effect of species on N availability. Because the N concentration can vary, it is therefore important to determine the species' N concentrations used in the model using foliar litter from N-limited sites.

If foliar litter contains more N than this base structural amount, then this extra N consists of more labile proteins and amino acids, and it is logical to assume that if the species is not actively retranslocating these compounds prior to senescence then its growth is not limited by N. If this is the case it also is likely that the decomposers are not N limited, and the foliage also may not decompose any faster; in fact, decomposition can be inhibited by high N concentrations (Berg & Tamm, 1991). If the N-response function in the model is calibrated correctly, increases in litterfall N for the same species on more N rich sites should not result in any growth increase because they will occur at, or above, the N "saturation" level for the species.

The idea that species characteristics can influence nutrient availability, and consequently, forest composition, is not new. Much of the early work investigating differences between mull and mor humus forms arrived at this conclusion (Rommell, 1935; Handley, 1954). However, the extension of these findings to the kind of logical conclusion described by Gosz (1981), where feedbacks between litter quality, nutrient availability, and production can lead to alternative positive or negative spirals in ecosystem productivity may not be entirely appropriate. The assumptions in LINKAGES suggest that these feedbacks are bounded at the lower limit by the degree to which a species can retranslocate N, which varies with the species and the chemical structure of its supporting tissues, and at the upper limit by the level at which growth no longer responds to increasing N availability. There is considerable variation among conifers in this upper limit, but it is relatively low compared to deciduous species. Thus, instead of Gosz's (1981) spirals, what may occur are shifts in rates of N cycling and forest productivity as forest composition changes from one set of species to another. These shifts occur in the LINKAGES model, given the interaction of life history characteristics, environment, and disturbance events, and lead to alternative stable equilibria in the absence of disturbance or climatic shifts. These are probably more common ecological states than the continuously declining or aggrading conditions that Gosz's (1981) conceptual model implies. Climatic conditions also are important in determining the extent to which species influence N availability, because of the varying influence of the lignin/N ratio in different climates.

A further factor, not varied in the model simulations, that could increase the simulated differences in N availability between the two types was the rate of humus turnover. This is specified as input to the model and was set at 2% per year for both types. However, the results in Fig. 25-5 suggest that while this figure is appropriate for the cedar–hemlock type, it may be low for the hemlock-dominated stands. This is further supported by the results of standard chemical and nuclear magnetic resonance (NMR) imaging analyses of the forest floors in the two types (de Montigny et al., 1993), which indicate that there are some important differences in the chemical composition of the humus between the two types, perhaps due to tannins originating in salal, which is leading to N being immobilized in stable tannin–protein complexes.

Two other factors have the potential to further reduce the nutrients available for the growth of planted seedlings, and were not included in the management scenarios used to produce Fig. 25-6. The first is the shrub salal. It is the dominant component of the understorey in the cedar–hemlock stands, but a relatively minor component of the hemlock–amabilis stands, and it rapidly forms a continuous cover on the cedar–hemlock sites following clear-cutting. Measurements in clearcuts indicate that salal can attain preharvest levels of aboveground biomass (about 4 Mg ha^{-1}) within 5 yr after cutting (Messier & Kimmins, 1991; Keenan, 1993). Allocation aboveground to leaves rather than stem is about 75% in the clearcuts, but only 25% in the uncut stands. Salal also allocates a considerably greater proportion belowground in the clear-cut areas. Taking just the aboveground leaf component of 3 Mg ha^{-1} at 5 yr, and multiplying by the leaf N concentration of 1% (Keenan, 1993), indicates that salal can take up at least 6 kg ha^{-1} yr^{-1} of N during the early years after cutting. This would decrease the N available for conifer growth to about 20 kg ha^{-1} yr^{-1}, further reducing seedling growth. In contrast, the hemlock–amabilis type is rapidly dominated by coniferous advanced growth and fireweed (*Epilobium angustifolium*) after clear-cutting. Fireweed dies off much more rapidly than salal, and its litter decomposes relatively rapidly, so nutrients immobilized in early succession in the hemlock–amabilis type become available more rapidly to the coniferous regeneration.

Second, the management practice of slash burning could have a significant impact on nutrient availability in the short-term. During the modeling, the importance of maintaining the actively decomposing layer of forest floor became apparent. The majority of available N was released during the first few years of foliar litter decomposition. Following this, the rate of N release was determined by humus turnover, which was quite slow. If tree establishment was not rapid enough for them to take advantage of the nutrient flush created by a disturbance and rapidly recreate a forest nutrient cycle, then growth was very slow and the establishment of a fully stocked stand took a long time. This was part of the reason the cedar–hemlock simulation took so long to reach a steady state. Burning of logging slash and the actively decomposing L layer following clearcutting can produce a small, short-term increase in N-mineralization, but it also can remove a considerable portion of the labile pool of N in these ecosystems, and may substantially reduce long-term N availability. Thus, while the large amounts of humus in these ecosystems gives the impression that they are well buffered against disturbance impacts such as fire, these considerable stores of N are not readily mineralized. The contrary conclusion, that ecosystems containing large accumulations of detrital biomass have low amounts of available nutrients, may well be a more general one, and managers should take every effort to conserve the *available* nutrient pool in these forests. High detrital accumulations, particularly in temperate environments such as the one in this study, probably indicate that much of the organic C and N in the system either begins in, or can quickly become part of, chemical complexes that are resistant to decomposition. This phenomenon probably also is a major contribution to the differences in N availability between mull and mor humus forms (Handley, 1954).

The forest floor conditions created by cedar may lead to the development of conditions in which cedar is able to grow more rapidly than hemlock despite lower N availability. This suggests that, in the absence of a catastrophic disturbance, the cedar–hemlock type is a self-sustaining system in which hemlock will remain a smaller component. However, this also will depend on the relative abilities of the two species to establish under a canopy. If hemlock is more successful at this stage of its life history, then it may eventually achieve a greater dominance. It has been speculated that the cedar–hemlock type is the climatic "climax" for this locality, eventually replacing hemlock-dominated stands (Lewis, 1982). However, the mechanism for the transition has not previously been identified. Both species similarly high shade tolerance (Carter & Klinka, 1992), and both are capable of establishing on organic substrates, such as decaying logs (Keenan, 1993). However, the mechanism could be the difference in growth response to N availability. Salal may mediate this development by invading later stages of the hemlock stands once gaps begin to form in their canopy, and sufficient light is available at the forest floor. Salal would lessen the dominance of hemlock by competing with the hemlock advanced growth for light and nutrients, while cedar, once established, is more tolerant of lower nutritional conditions. However, once the cedar–salal association has developed it is likely to take a particularly severe disturbance to allow hemlock to re-establish dominance, because of the lasting effect that the association has on forest floor nutrient mineralization.

There is considerable potential to improve various aspects of this investigation. The response functions for each species to varying rates of annual N mineralization need to be developed in a more thorough way, using growth information for the two species planted on a wider range of sites, perhaps using data from fertilization trials. The rates of turnover of the various forest floor components should be investigated further, and there was no field information available on the decay rates and nutrient dynamics in decomposing fine roots. Differences in root nutrient content and decay rates could make a further substantial contribution to the observed differences in N availability. To more accurately depict these ecosystems it would be valuable to include the dynamics of understorey species (particularly salal), because the biomass of the understorey and its effect on light available for seedling establishment is probably an important determinant of the dynamics of these ecosystems. To further understand the mechanisms involved in the efficiency of nutrient uptake and nutrient use by different species, and to simulate how these might affect the competitive dynamics among the different species, a more mechanistic model of nutrient cycling would be required.

CONCLUSIONS

The results of this study tend to support the hypothesis that the characteristics of the dominant coniferous species, particularly lower foliar N concentration, can explain much of the difference in N availability between the two forest types. The interaction of coniferous litter and salal exudates may be contributing further to lower rates of N availability in the cedar–hemlock type, by forming stable protein-tannin complexes deeper in the humus profile. The management practice of slash burning may further lower nutrient availability by burning off much of the foliar litter fraction containing most of the labile N. The small amount of N mineralized during the burn is rapidly taken up by salal. The lower N availability caused by these four factors has a greater impact on the productivity of more nutrient demanding species such as hemlock and spruce than it does on the growth of western red cedar.

ACKNOWLEDGMENTS

This study was funded by the Canadian Natural Sciences and Engineering Research Council, and Western Forest Products Ltd., Macmillan Bloedel Ltd., and Fletcher-Challenge Canada Ltd. through a Collaborative Research and Development Grant, and through the Donald S. McPhee fellowship awarded to Rod Keenan. Earlier application of LINKAGES to Pacific Northwest forests, upon which this work was partly based, was made possible by funds from the U.S. Forest Service, through the Center for Streamside Studies and the College of Forestry, University of Washington.

REFERENCES

Aber, J.D., and J.M. Melillo. 1980. Litter decomposition: measuring relative contributions of organic matter and nitrogen to forest soils. Can. J. Bot. 58:416–421.

Aber, J.D., D.B. Botkin, and J.M. Melillo. 1979. Predicting the effects of different harvesting regimes on productivity and yield in northern hardwoods. Can. J. For. Res. 9:10–14.

Agriculture Canada. 1987. The Canadian system of soil classification. 2nd ed. Publ. 1646. Agric. Canada, Ottawa, Canada.

Attiwill, P.M. 1986. Interactions between carbon and nutrients in forest ecosystems. Tree Physiol. 2:389–399.

Berg, B., and C.O. Tamm. 1991. Decomposition and nutrient dynamics of litter in long-term optimum nutrition experiments. I. Organic matter decomposition of Norway spruce needle litter. Scand. J. For. Res. 6:305–321.

Berg, B., and O. Theander. 1984. The dynamics of some nitrogen fractions in decomposing Scots pine needles. Pedobiologia 27:161–167.

Botkin, D.B. 1993. Forest dynamics. Oxford Univ. Press, Oxford, England.

Botkin, D.B., J.F. Janak, and J.R. Wallis. 1972. Some ecological consequences of a computer model of forest growth. J. Ecol. 60:849–872.

Bunnell, F.L. 1989. Alchemy and uncertainty: What good are models? USDA-FS Gen. Tech. Rep. PNW-232. USDA-FS, Portland, OR.

Carter, R.E., and K. Klinka. 1992. Variation in shade tolerance of Douglas fir, western hemlock and western red cedar in coastal British Columbia. For. Ecol. Manage. 55:87–105.

Dale, V.H., and M. Hemstrom. 1984. Climacs: A computer model of forest stand development for western Oregon and Washington. USDA-FS Res. Pap. PNW-327. USDA-FS, Portland, OR.

Dale, V.H., M. Hemstrom, and J. Franklin. 1986. Modeling the long-term effects of disturbances on forest succession, Olympic Peninsula, Washington. Can. J. For. Res. 16:56–67.

DeHaan, S., 1977. Soil organic matter studies. Vol. 1. Int. Atomic Energy Agency, Vienna, Austria.

de Montigny, L.E., C.M. Preston, P.G. Hatcher, and I. Kögel-Knabner. 1993. Comparison of humus horizons from two ecosystem phases on northern Vancouver Island using ^{13}C CPMAS NMR spectroscopy and CuO oxidation. Can. J. Soil Sci. 73:9–25.

Edmonds, R.L., D. Binkley, M.C. Feller, P. Sollins, A. Abee, and D.D. Myrold. 1990. Nutrient cycling: Effects on productivity of northwest forests. p. 17–35. In D.A. Perry et al. (ed.) Maintaining the long-term productivity of Pacific Northwest forest ecosystems. Timber Press, Portland, OR.

Feller, M.C. 1992. Generalized versus site-specific biomass regression equations for *Pseudotsuga menziesii* var. menziesii and *Thuja plicata* in coastal British Columbia. Bioresour. Technol. 39:9–16.

Flanagan, P.W., and K. Van Cleve. 1983. Nutrient cycling in relation to decomposition and organic matter quality in taiga ecosystems. Can. J. For. Res. 13:795–817.

Franklin, J.F., and C.T. Dyrness. 1973. Natural vegetation of Washington and Oregon. USDA-FS Gen. Tech. Rep. PNW-8. Pacific Northwest For. and Range Exp. Stn., Portland, Oregon.

Gholz, H.L., C.C. Grier, A.G. Campbell, and A.T. Brown. 1979. Equations for estimating biomass and leaf area of plants in the Pacific Northwest. For. Res. Lab. Pap. 41. Oregon State Univ., Corvallis, OR.

Gosz, J.R. 1981. Nitrogen cycling in coniferous forest ecosystems. Ecol. Bull. (Stockholm) 33:405–426.

Graham, R.L., and K. Cromack Jr. 1982. Mass, nutrient content and decay rate of dead boles in rainforest of Olympic National park. Can. J. For. Res. 12:511–521.

Handley, W.R.C. 1954. Mull and mor formation in relation to forest soils. Br. For. Commun. Bull. 23. HMSO, London.

Harmon, M.E., G.A. Baker, G. Spycher, and S.E. Greene. 1990. Leaf-litter decomposition in the *Picea/Tsuga* forest of Olympic National Park, Washington, U.S.A. For. Ecol. Manage. 31:55–66.

Hobbie, S.E. 1992. Effects of plant species on nutrient cycling. Trends Ecol. Evol. 7:336–339.

Keenan, R.J. 1993. Structure and function of western red cedar and western hemlock forests on Northern Vancouver Island, Ph.D. diss. Univ. British Columbia, Vancouver, BC.

Keenan, R.J., C.E. Prescott, and J.P. Kimmins. 1993. Mass and nutrient content of the woody debris and forest floor in western redcedar and western hemlock forests on northern Vancouver Island. Can. J. For. Res. 23:1052–1059.

Kimmins, J.P. 1988. Community organization: Methods of study and prediction of the productivity and yield of forest ecosystems. Can. J. Bot. 66:2654–2672.

Krajina, V.J. 1969. Ecology of forest trees in British Columbia. Ecol. Western N. Am. 2:1–146.

Krajina, V.J., S. Madoc-Jones, and G. Mellor. 1973. Ammonium and nitrate in the nitrogen economy of some conifers growing in Douglas-fir communities of the Pacific Northwest of America. Soil Biol. Biochem. 5:143–147.

Lewis, T. 1982. Ecosystems of block 4, TFL 25. Int. Rep. Western For. Prod. Ltd, Vancouver, BC.

Meentemeyer, V. 1978. Macroclimate and lignin control of litter decomposition rates. Ecology 59:465–472.

Messier, C., and J.P. Kimmins. 1991. Above and below-ground vegetation recovery in recently clearcut and burned sits dominated by *Gaultheria shallon* in coastal British Columbia. For. Ecol. Manage. 46:275–294.

Minore, D. 1983. Western redcedar: A literature review. USDA-FS Gen. Tech. Rep. PNW-150. USDA-FS, Portland, OR.

Nadelhoffer, K.J., J.D. Aber, and J.M. Melillo. 1983. Leaf-litter production and soil organic matter dynamics along a nitrogen availability gradient in Southern Wisconsin. Can. J. For. Res. 13:12–21.

Pastor, J., J.D. Aber, C.A. McClaugherty, and J.M. Melillo. 1984. Above-ground production and N and P cycling along a nitrogen-availability gradient on Blackhawk Island, Wisconsin. Ecology 65:256–268.

Pastor, J., and W.M. Post. 1984. Calculating Thornthwaite and Mather's actual evapotranspiration using an approximating function. Can. J. For. Res. 14:466–467.

Pastor, J., and W.M. Post. 1985. Development of a linked forest productivity-soil-process model. Environ. Sci. Div. Publ. no. 2455. Oak Ridge Natl. Lab., Oak Ridge, TN.

Pastor, J., and W.M. Post. 1986. Influence of climate, soil moisture, and succession on forest carbon and nitrogen cycles. Biogeochemistry 2:3–27.

Prescott, C.E., M.A. McDonald, and G.F. Weetman. 1993. Differences in availability of N and P in the forest floors of adjacent stands of western redcedar-western hemlock and western hemlock-amabilis fir on northern Vancouver Island. Can. J. For. Res. 23:605–610.

Prescott, C.E., L.E. deMontigny, C.M. Preston, R.J. Keenan, and G.F. Weetman. 1995. Carbon chemistry and nutrient supply in cedar-hemlock and hemlock-amabilis fir forest floors. p. 377–396. *In* W.W. McFee and J.M. Kelly (ed.) Carbon forms and functions in forest soils. SSSA, Madison, WI.

Rommell, L.G. 1935. Ecological problems of the humus layer in the forest. Mem. 170. Cornell Univ. Agric. Exp. Stn., New York.

Shugart, H.H. 1984. A theory of forest dynamics. Springer-Verlag, New York.

Smith, N.J. 1993. Estimating leaf area index and light extinction coefficient in stands of Douglas-fir (*Pseudotsuga menziessi*). Can. J. For. Res. 23:317–321.

Sollins, P. 1981. Input and decay of coarse woody debris in coniferous stands in western Oregon and Washington. Can. J. For. Res. 12:18–28.

Sollins, P., S.P. Cline, T. Verhoeven, D. Sachs, and G. Spycher. 1987. Patterns of log decay in Douglas-fir forests. Can. J. For. Res. 17:1585–1595.

Stone, E.L. 1975. Effects of species on nutrient cycles and soil change. Phil. Trans. R. Soc. London, B271:149–162.

Turner, D.P., and E.H. Franz. 1985. The influence of western hemlock and western redcedar on microbial numbers, nitrogen mineralization and nitrification. Plant Soil 88:259–267.

Waring, R.H., and J.F. Franklin. 1979. Evergreen coniferous forests of the Pacific Northwest. Science (Washington, DC) 204:1380–1386.

Waring, R.H., and W.H. Schlesinger. 1985. Forest ecosystems: Concepts and management. Acad. Press, San Diego, CA.

Wedin, D.A., and D. Tilman. 1990. Species effects on nitrogen cycling: A test with perennial grasses. Oecologia 84:433–441.

Weetman, G.F., R. Fournier, J. Barker, E. Schnorbus-Panozzo, and A. Germain. 1989a. Foliar analysis and response of fertilized chlorotic Sitka spruce plantations on salal dominated cedar-hemlock cutovers on Vancouver Island. Can. J. For. Res. 19:1501–1511.

Weetman, G.F., R. Fournier, J. Barker, E. Schnorbus-Panozzo, and A. Germain. 1989b. Foliar analysis and response of fertilized chlorotic western hemlock and western redcedar reproduction on salal dominated cedar-hemlock cutovers on Vancouver Island. Can. J. For. Res., 19:1512–1520.

Weetman, G.F., R. Fournier, E. Schnorbus-Panozzo, and J. Barker. 1990. Post-burn nitrogen and phosphorus availability of deep humus soils in coastal British Columbia cedar/hemlock forests and the use of fertilization and salal eradication to restore productivity. p. 451–459. *In* S.P. Gessel et al. (ed.) Sustained productivity of forest soils. Proc. 7th N. Am. For. Soils Conf., Faculty For., Univ. British Columbia, Vancouver, BC.

Yarie, J. 1990. Role of computer models in predicting the consequences of management on forest productivity. p. 3–18. *In* W.J. Dyck and C.A. Mees (ed.) Impact of intensive harvesting on forest site productivity. For. Res. Inst. Bull. 159. Rotorua, NZ.

Zak, D.R., K.S. Pretzinger, and G.E. Host. 1986. Landscape variation in nitrogen mineralization and nitrification. Can. J. For. Res. 16:1258–1263.

26 Carbon and Nitrogen Dynamics in Oak Stands along an Urban-Rural Gradient

Richard V. Pouyat
USDA-FS, NEFES
c/o Institute of Ecosystem Studies
Millbrook, New York

M.M. Carreiro
Fordham University
Bronx, New York

M.J. McDonnell, S.T.A. Pickett, and P.M. Groffman
Institute of Ecosystem Studies
Millbrook, New York

Robert W. Parmelee
Ohio State University
Columbus, Ohio

Kimberly E. Medley
Miami University
Miami, Ohio

W.C. Zipperer
USDA-FS
Northeast Forest Experiment Station
Syracuse, New York

In North America, landscapes have been modified and fragmented extensively by urbanization (Godron & Forman, 1983; Zipperer et al., 1990). In the USA, urban areas (those with human populations of at least 620 individuals/km^2) have been expanding at a rate of 1.3 million acres (526000 ha) per year (USDA, 1982). Urban areas differ from less densely populated, or rural areas, in several

---

Copyright © 1995 Soil Science Society of America, 677 S. Segoe Rd., Madison, WI 53711, USA. Carbon Forms and Functions in Forest Soils.

environmental factors that have the potential to influence biogeochemical processes, such as C and N dynamics. Environmental factors that can influence biogeochemical processes in urban areas include a modified mesoclimate, increased concentrations of atmospheric pollutants, modified disturbance regimes, and compositional changes of plant and animal species due to introductions of non-native species and habitat fragmentation (Fig. 26–1). In addition to these environmental factors, plant litter quality, or the decomposability of litter, potentially can be altered by urban pollution (e.g., Findlay & Jones, 1990; Jordan et al., 1991).

The study of forest ecosystems in urban areas is important, not only because of the spatial extent of urbanization, but because such areas provide an excellent opportunity to study how changes in temperature, atmospheric chemistry, and species composition interact to affect biogeochemical processes, such as nutrient cycling (McDonnell & Pickett, 1990). For example, older cities generally have more heavy industrial activity and are often characterized by having higher deposition of heavy metals and warmer temperatures than surrounding areas. Newer, less industrialized cities, while having warmer temperatures than surrounding areas, may not experience the deposition levels of an older industrial city. This provides an opportunity to compare how such contrasting combinations of factors affect the C and N dynamics of ecosystems.

In this paper we consider the unique abiotic and biotic aspects of forest ecosystems embedded within urban and suburban areas and describe how urban environments can be used as study subjects of multiple anthropogenic impacts on forest ecosystems. We then present a case study to describe our approach to the investigation of C and N dynamics in forest stands along a transect extending from New York City to northwestern Connecticut. Finally, we discuss the significance of the results in relation to the structure and function of forest ecosystems exposed to urban environments and multiple stresses at local and regional scales.

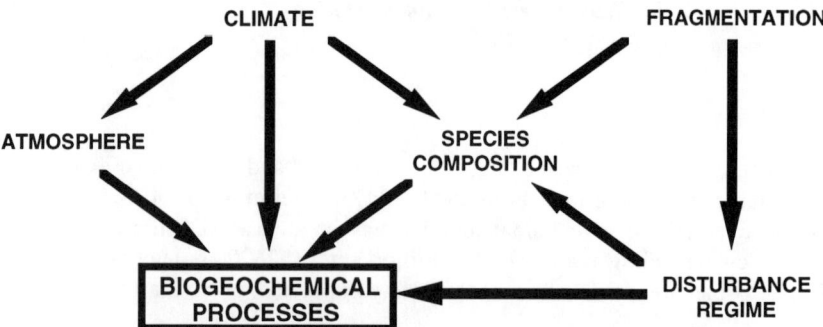

Fig. 26–1. Conceptual model of environmental factors that are expected to influence biogeochemical processes in urban and suburban areas. Individual environmental factors are described in text.

URBAN ENVIRONMENTS

Defined in geographers' terms, urbanization is a process in which landscapes become extensively modified due to increased human habitation and increased per capita energy consumption (McDonnell & Pickett, 1990). Despite urbanization, patches of forest and other unmanaged plant communities often are found in an urban matrix (Dickinson, 1966; Stearns & Montag, 1974; McDonnell & Pickett, 1990). Such forest stands have many of the same attributes of nonurban forests, including closed canopies and a forest floor (Pouyat & Zipperer, 1992). However, in spite of similar structural appearance, these stands may differ functionally (Stearns & Montag, 1974; Bornkamm et al., 1982; Baker, 1986; Seinfeld, 1989; Graedel & Crutzen, 1989), due to a complex of environmental changes caused by urbanization (McDonnell, 1988). These novel combinations of factors can alter soil chemical and physical properties that may in turn affect soil community structure, and over time ecosystem processes and forest vigor (Bormann, 1985; Smith, 1990; Pouyat, 1991).

Perhaps the most significant effect of urban environments on forest soil properties is the modification of soil temperature and moisture regimes. Urban environments are characterized by localized increase in temperature known as the "heat island" effect (Landsberg, 1981; Oke, 1990) and localized increases in precipitation (Changnon, 1980). Urban heat islands are caused both by a reduction in evapotranspiration and an increase in heat storage by urban structures (Oke, 1990). These changes in the flux of latent heat increase both the maximum and minimum temperatures of urban environments. Temperature differences between the urban core and outlying rural areas generally range from 3 to 8°C and usually are greatest in early evening. A rise in average minimum temperature increases the number of frost-free days in urban areas. This extends the growing season for plants in urban areas relative to the surrounding countryside and can affect decomposition rates in forest soils. Increases in ambient temperature also can enhance O_3 concentrations (Candelino & Chameides, 1990).

Increased combustion and construction in urban areas produce large quantities of condensation nuclei that can increase precipitation downwind of urban sources (Changnon, 1980; Oke, 1990). The resultant increase in heavy rainfall can enhance water availability for plants and soil organisms in well-drained forest soils. However, the potential for wetter soils in urban forests can be counteracted by the tendency of soils to form hydrophobic surfaces (Craul, 1985; White & McDonnell, 1988). Hydrophobic soil surfaces reduce water infiltration rates that can lower soil water potential (DeVano, 1971).

Urban environments usually have higher depositional fluxes of atmospheric chemicals than rural environments. Forest stands in urban areas receive relatively high amounts of heavy metals, N, S, and base cations in wet and dry atmospheric deposition (Ruppert, 1975; McColl & Bush, 1978; Gatz, 1991) due to fossil fuel combustion, emissions from heavy industry, and temperature inversions (Seinfeld, 1989). As a result, urban forest soils typically have high levels of metal cations in the O and A horizons (Parker et al., 1978; Pouyat & McDonnell, 1991).

The effects of N and S inputs on the chemistry of forest soils depends greatly on the soil's buffering capacity (Frink & Voigt, 1976). In Europe, the long-term input of S and N acidic compounds has increased soil acidity and depleted the cation content of poorly buffered soils (e.g., Ulrich et al., 1980). However, inputs of base cations associated with acidic deposition can compensate for a portion of cation losses due to leaching (Driscoll et al., 1989). In addition to acidifying soils, long-term chronic deposition of N compounds can increase N availability in forest soils to a point where the biological demand will be exceeded, or "saturated" (Agren & Bossatta, 1988; Skeffington & Wilson, 1988; Aber et al., 1989).

Urban forest ecosystems differ from rural forest ecosystems in species composition because of the naturalization of non-native plant and animal species (Bagnall, 1979; Airola & Bucholz, 1984; Hobbs, 1988; Rudnicky & McDonnell, 1989; McDonnell & Pickett, 1990). Most non-native species that have become naturalized in urban areas were either introduced intentionally as ornamentals or were introduced accidently as a consequence of human activity (Gilbert, 1989). However, forest species composition also is influenced by the fragmented nature of urban landscapes (Davis & Glick, 1978). The fragmentation of forest or other vegetation cover decreases patch size and thus interior habitat. This increases the isolation of patches and increases the susceptibility of the forest community to the immigration of plants and animals from adjacent areas (Janzen, 1983). As a result of species compositional changes, nutrient fluxes can be modified in urban forest ecosystems (Groffman et al., 1994).

URBAN ENVIRONMENTS AS ECOSYSTEM MANIPULATIONS

Multiple stresses, both anthropogenic (e.g., pollution stress) and natural (e.g., drought stress), have been hypothesized as causing forest decline in some parts of the USA and Europe (Hinrichsen, 1986; Cowling, 1989; Likens, 1989). Measuring forest ecosystem response to stress is problematic, since many environmental factors affect tree health (Hinrichsen, 1986; Cowling, 1989; Tamm, 1989) and manipulative experiments on the ecosystem level are difficult to implement (Tamm, 1989).

One approach to studying the effects of multiple stress on forest ecosystems is to compare forest stands chronically exposed to anthropogenic impacts with similar forests that are more remote from human influences. Comparisons of this kind are feasible in urban landscapes and adjacent undeveloped areas. By comparing forest stands in urban areas to similar forest stands in rural areas, the suite of environmental factors associated with urban environments can serve as analogues for multiple stresses imposed by regional air pollution and changes in climate. Since urbanization can be considered a treatment that exposes entire ecosystems to altered environments, urban environments can be viewed as "ecosystem level manipulations." This permits replication of sites on landscape or regional scales with rural stands acting as "controls" for urban and suburban stands. By selecting forest stands exposed to different levels of abiotic factors and by manipulating these factors in field and lab experiments, we can potentially match the ecological responses to their respective causes.

CASE STUDY: THE NEW YORK URBAN-RURAL GRADIENT EXPERIMENT

Approach

To study the influence of urban environments on forest ecosystem structure and function, we have utilized the environmental gradient paradigm as a research tool (McDonnell & Pickett, 1990). Whittaker (1967) introduced the concept of gradient analysis as a methodology to examine and explain species composition of communities along gradients of environmental factors. Since that time, Keddy (1989) and Vitousek and Matson (1991) effectively have studied natural gradients, such as gradients of soil moisture and salinity, to understand the relationships between environmental variation and community structure. The gradient paradigm assumes that environmental variation is ordered in space. The spatial pattern of the environment constrains the structural and functional components of ecosystems, such that a coexisting pattern emerges along with various ecosystem attributes (McDonnell & Pickett, 1990; McDonnell et al., 1993).

Environmental gradients also may occur when human population densities, human-built structures, and human-generated processes vary spatially on a landscape. Environmental gradients, determined largely by the degree to which the landscape is urbanized, are called urban-rural gradients (McDonnell & Pickett, 1990). Urban-rural environmental gradients are those that express the contrast between environments determined largely by urban land use (e.g., residential, commercial, and industrial), compared to other environments determined by agriculture, wilderness, contiguous forest or some combination of these.

To study the contrast in environments that occurs among various land uses, we established a transect along an urban-rural land use gradient that intersects similar ecosystem types. We use a hypothetical landscape to illustrate our approach (Fig. 26–2). In this landscape an array of soil types creates a spatial pattern of soil chemical and physical characteristics (Fig. 26–2a). Soils have formed in the landscape under the constraints of soil forming factors, such as climate, parent material, relief and biota (Jenny, 1941). To hold the chemical and physical environmental variation in the soil to a minimum, a transect is placed on the landscape in which stands are located on sites having similar soil type, such that in the absence of urbanization the soil-forming factors remain essentially constant (Fig. 26–2b). Based on the assumption that the factors are constant, variation in the physical and chemical properties of the soils between sites should be negligible. However, if the hypothetical landscape is modified by urbanization, such that the transect now cuts through an array of land use types (e.g., the New York City metropolitan area), a new level of variation appears, which can be attributed to an urban effect (Fig. 26–2c). In such a case, the transect represents an urban-rural environmental gradient.

Study Area

We have been investigating ecosystem structure and function in forest ecosystems along an urban-rural land use gradient in the New York City metropolitan area for the last 7 yr (White & McDonnell, 1988; McDonnell & Pickett, 1990; Pouyat & McDonnell, 1991; Pouyat, 1992; McDonnell et al., 1993;

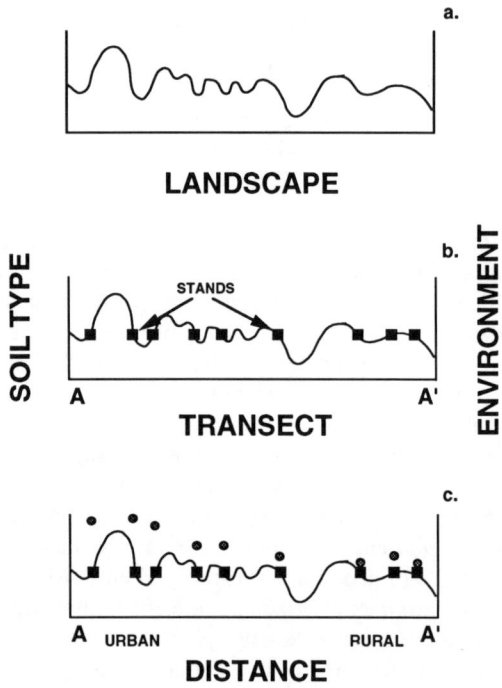

Fig. 26–2. A hypothetical landscape is used to show soil environmental changes that occur on a transect (A to A') that traverses an array of soil types (*a*). Sampling locations (solid squares) are stratified along this transect, so that study locations are situated on similar soil types (*b*). Along the transect soil chemical and physical properties are measured at each study site (*c*). If the measured properties (shaded circles) deviate from a line of slope, 0, along an urban-rural land use gradient, then a soil environmental gradient exists.

Groffman et al., 1994). The body of past and current research includes detailed characterization of the gradient including quantification of land use characteristics and soil chemical and physical properties as well as ecological analysis of plant community structure, soil C and N dynamics, soil organism abundances and trace gas fluxes.

The study region includes contrasting land uses extending from Bronx County, New York (New York City) to sites in Westchester County, New York, and Litchfield County, Connecticut (McDonnell & Pickett, 1990). Within the region, human population density ranges from 10 individuals per square kilometer in Litchfield to 10000 individuals per square kilometer in New York City. The climate of the region is characterized by warm humid summers and cold winters, average annual air temperatures range from 12.5°C in New York City to 8.5°C in northwestern Connecticut (NOAA, 1985). Much of this difference in air temperature is attributed to the heat island effect associated with New York City (Bornstein, 1968). Precipitation is distributed evenly throughout the year for the entire study area and ranges from an annual average of 108 cm in New York City to 103 cm in northwestern Connecticut (NOAA, 1985). The region receives acid deposition transported from other areas of the USA (Wolff et al., 1979) that adds to local sources of heavy metals (Homolya & Lambert, 1981) and acid deposition (Volchok & Bogen, 1971). Ozone levels in New York City rise intermittently above 60 to 170 ppb (60–170 pg/L) (New York State Dep. of Environ. Conserv., 1989), which exceeds 4-hr plant damage thresholds set by USEPA (Smith, 1990).

The Bronx, New York, and the study area to the north constitute the southern portion of the Northeastern Upland Physiographic Province (Broughton et al., 1966). The bedrock consists of highly metamorphosed and dissected crystalline rocks that are composed of schist, granite and gneiss (Schuberth, 1968). Soils in the study area are classified as Typic or Lithic Dystrochrepts, coarse-loamy, mixed, mesic subgroups. The soil types included in the study are well-drained, moderate to shallow, sandy loam soils situated on gently sloping terrain (Hill et al., 1980).

METHODS

We established a transect in 1988 as a 20-km wide by 130-km long belt from highly urbanized Bronx County, New York to rural Litchfield County, Connecticut. To establish an anthroposequence, we selected nine study sites along the transect with each site consisting of three to four individual stands. A total of 31 stands at one time or another were used in our research, depending on the focus of a specific study. For most statistical procedures, a mean was computed for each site using stand data (n = 3 or 4). The stands were selected using the following criteria. (i) location on upland sites on either of two soil series (Hollis and Charlton) both of which are classified to the same family (Gonick et al., 1970), (ii) oak-dominated forest with *Quercus rubra* L. or *Q. velutina* Lam. as major components of the overstory; (iii) minimum stand age of 70 yr, and (iv) no visual evidence of disturbance and no documented evidence of human impact for at least 70 yr. By design, our transect did not include any stands with a significant proportion of non-native tree species.

We collected data to quantify various land use features related to urbanization along the transect. We measured the total length of major roads and highways and percentage urban land use from aerial photographs and U.S. Geological Survey topographic maps for 16 km^2 land units that were situated along the transect. We determined population density and traffic volume for the same land area using U.S. Bureau of Census county subdivision data (U.S. Bureau of Census, 1990) and New York and Connecticut traffic volume reports (New York State Dep. of Transportation, 1990; Connecticut Dep. of Transportation, 1990). We used these measures of urbanization as independent variables in correlations with various soil properties and C and N cycling processes measured in the stands. Soil sampling procedures and laboratory methods for chemical and physical analyses, and the measurement of C and N cycling rates are described in Pouyat and McDonnell (1991) and Pouyat (1992).

Before the correlation with land use data, the soil data were submitted to a Principal Component Analysis using the SAS package (SAS Inst., 1987) to generalize the soil variables and to ordinate sites. We also used the geographical data to cluster the study sites into urban, suburban, and rural land use types that were considered treatment levels in a one-way analysis of variance (ANOVA). Data were transformed appropriately to reduce heteroscedasticity using the Box-Cox transformation tests (Box & Cox, 1982). Tukey Studentized Range Test and Hochberg's Method for unequal sample sizes (SAS Inst., 1987) were used to determine significant differences between land use type means at $P < 0.05$.

RESULTS AND DISCUSSION

Soil Properties

We observed marked changes in chemical properties of mineral soil (10-cm depth) and forest floor along the transect. We ordinated 20 forest soil variables (Table 26–1) using Principal Components Analysis (Fig. 26–3). The first four components of the Principal Component Analysis accounted for 90% of the variation in soil chemical properties. The first principal component, which accounted for 42% of the variation, had positive loadings corresponding to high concentrations of heavy metals, especially Pb, Cu, and Ni, high total soluble salt concentrations, high organic matter concentration, high concentrations of base cations, and a slight increase in soil acidity. Sites located closer to the urban core had positive loadings on the first principal component, while sites located beyond 30 km of the urban core had negative loadings (Fig. 26–3). A curvilinear relationship best explained the first principal component as a function of distance from the urban core and was highly significant (Fig. 26–4a). An ANOVA supports the results obtained by the Principal Component Analysis (Table 26–2).

These results indicate that a soil environmental gradient exists along the transect. In contrast, little of the variation in soil texture and bulk density was

Table 26–1. Soil properties measured and methods of analysis for soil samples collected (10-cm depth) along an urban-rural transect in the New York City metropolitan area (spring 1989).

Soil property	Method	Reference
Organic matter	Loss on ignition	Wilde et al., 1972
pH, in water	2:1	Wilde et al., 1972
pH, $CaCl_2$	0.01 M $CaCl_2$	McLean, 1982
Electrical conductivity	1:1	Rhoades, 1982
Exchange acidity	$BaCl_2$	Thomas, 1982
Total C and N	Flash combustion	Carlo Erba, 1986
Cu, acid-soluble	HNO_3 digest	Friedland, 1985
Ni, acid-soluble	HNO_3 digest	Friedland, 1985
Mn, acid-soluble	HNO_3 digest	Friedland, 1985
Co, acid-soluble	HNO_3 digest	Friedland, 1985
Zn, acid-soluble	HNO_3 digest	Friedland, 1985
Pb, acid-soluble	HNO_3 digest	Friedland, 1985
Cr, acid-soluble	HNO_3 digest	Friedland, 1985
P, available	NaOAc extract	Chapman, 1965
K, available	NaOAc extract	Chapman, 1965
Mg, available	NaOAc extract	Chapman, 1965
Ca, available	NaOAc extract	Chapman, 1965
Fe, available	NaOAc extract	Chapman, 1965
Al, available	NaOAc extract	Chapman, 1965

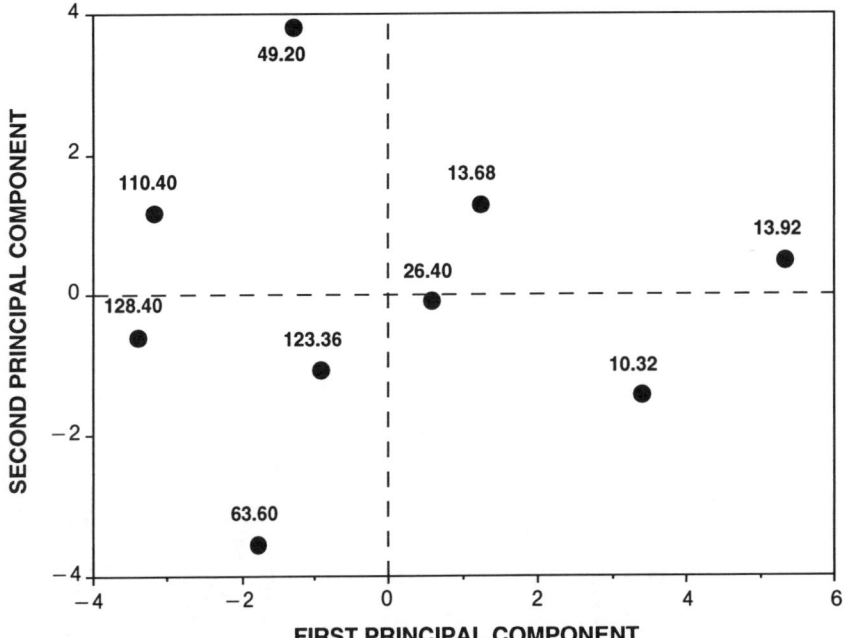

Fig. 26–3. Scatter plot of the first and second principal component scores of soil chemical properties measured along an urban-rural transect in the New York City metropolitan area. Distance to the urban core is given in kilometers for each site ($n = 9$). Sites that are greater than 30 km from the urban core have negative scores, and sites less than 30 km from the urban core have positive scores for the first principal component axis (adapted from Pouyat, 1995).

Table 26–2. Means of soil chemical properties and organic matter (10-cm) by land use type along an urban-rural transect in the New York City metropolitan area (spring 1989). Values are the mean of five rural, two suburban, and two urban sites (adapted from Pouyat, 1995).

Soil properties	Means by type			P value
	Rural	Suburban	Urban	ANOVA
pH, equivalents	249.0a[†]	272.0ab	430.0b	0.036
conductivity, S m^{-1}	0.081a	0.115b	0.131c	0.005
Cu, mg kg^{-1}	14.5a	22.9ab	34.3b	0.012
Ni, mg kg^{-1}	15.4	18.0	24.2	0.200
Pb, mg kg^{-1}	27.8a	75.8a	115.0c	0.001
Ca, mg kg^{-1}	45.8	122.3	160.9	0.100
Mg, mg kg^{-1}	11.9	21.1	30.9	0.057
K, mg kg^{-1}	49.7	58.0	69.8	0.286
N, g kg^{-1}	0.21	0.21	0.29	0.174
Organic matter, %	7.5a	8.4ab	10.8b	0.041

[†] Means with different letters are significantly different at $P < 0.05$.

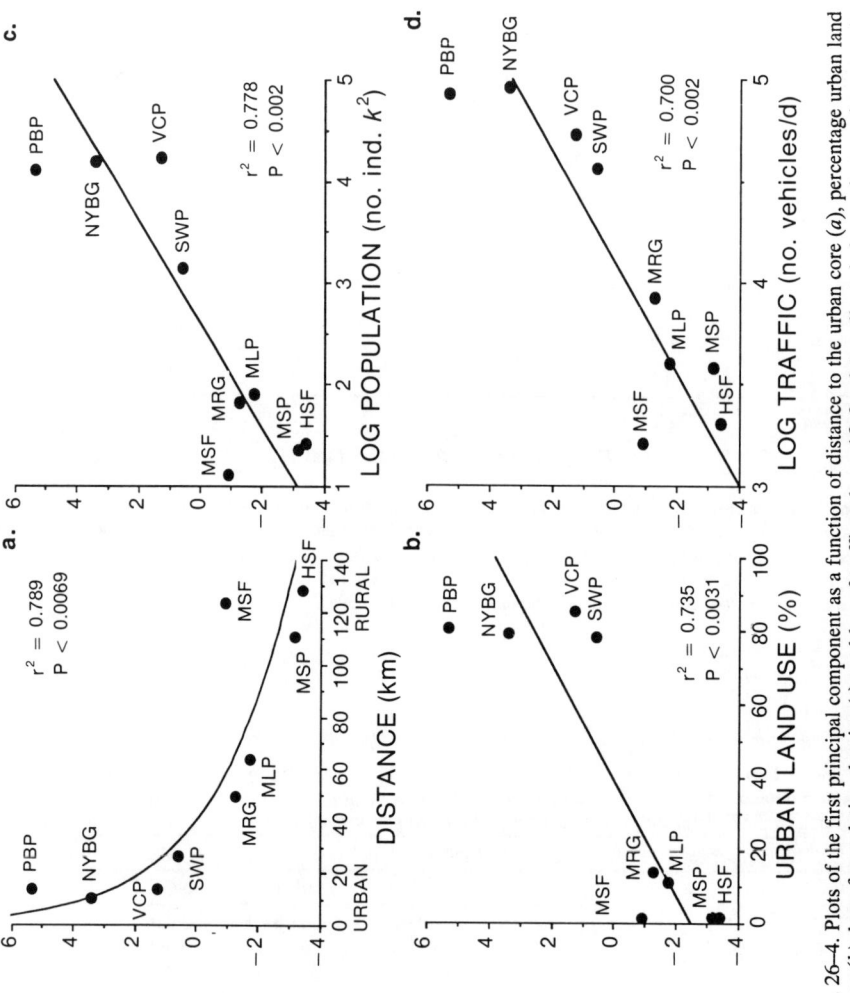

Fig. 26-4. Plots of the first principal component as a function of distance to the urban core (a), percentage urban land use (b), log of population density (c), and log of traffic volume (d) for data collected along urban-rural transect in New York City metropolitan area. Sites are indicated by three-letter abbreviations adjacent to the corresponding symbol in each plot. Values are the means of three sites (36 composited soil samples per site, adapted from Pouyat, 1995).

Table 26–3. Correlation matrix between land use features and physical properties of soils sampled along an urban-rural transect in the New York City metropolitan area; $n = 9$ (adapted from Pouyat, 1995).

Property	Population	Urban	Distance	Principal component
Bulk Density, kg m^{-3}	–0.123[†]	–0.196	0.088	–0.349
	0.752[‡]	0.613	0.821	0.358
Sand, %	–0.398	–0.519	0.268	–0.497
	0.289	0.152	0.485	0.174
Silt, %	0.453	0.525	–0.253	0.570
	0.220	0.147	0.511	0.109
Clay, %	0.284	0.452	–0.270	0.356
	0.458	0.222	0.483	0.347

[†] Pearson correlation coefficient.
[‡] P value.

correlated with the measures of urbanization or distance to the urban core (Table 26–3). These results support our supposition that we were able to select similar soil types along the transect using soil survey maps, though more detailed soil profile descriptions are needed to verify the soil type of each of the 31 stands. Soil pedon descriptions of the Hollis and Charlton soil series and additional soil information can be found in Pouyat (1992).

To investigate potential causes of this gradient in soil chemical properties, we investigated the relationship between the first principal component and the various quantitative characteristics of land use type (Fig. 26–4). Of the land use measures we quantified (population density, road density, percentage urban land use) traffic volume was the best predictor of the variation in soil chemical properties measured along the transect (Fig. 26–4d). The close correlation between traffic volume and soil chemistry suggests that automobiles are an important source of pollutants to forest stands in the New York City metropolitan area. Since all of the sampling locations in our study were at least 30 m from the nearest road, the depositional pattern of these pollutants are not confined to areas directly adjacent to roads.

Soil-Carbon and Nitrogen Dynamics

Our observation of a soil environmental gradient led us to postulate that biotic responses also would vary along the transect. We hypothesized that soil fungi and invertebrate abundances would be lower at the urban end relative to the rural end of the gradient, due to their sensitivity to pollution (Tyler et al., 1989). Total hyphal length of litter fungi was greater toward the rural end of the gradient (Fig. 26–5). We also found declining microarthropod abundance at the urban end of the gradient and observed fewer nematodes in the middle range of the gradient with the highest numbers in rural forests (Pouyat, 1992). Abundances of both microinvertebrates and especially litter fungi, were negatively correlated with heavy metal concentrations (Fig. 26–6). A somewhat surprising result was the observation that non-native species of earthworms were abundant in urban stands but were not present in rural stands (Pouyat, 1992). The high earthworm activity observed in the urban stands are noteworthy, since the entire study area

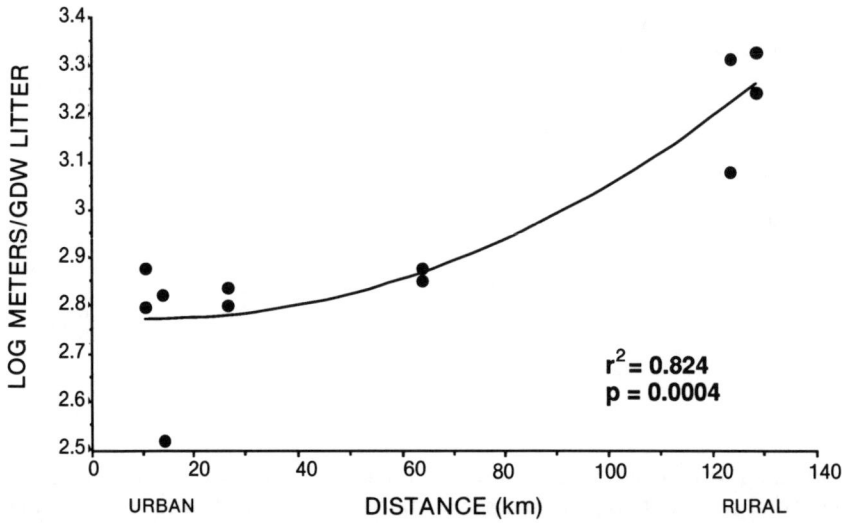

Fig. 26–5. Plot of mean fungal densities (log meters of hyphae; $n = 4$ per stand) in oak litter of 12 stands in January 1991 as a function of distance to the urban core along an urban-rural transect in the New York City metropolitan area (Carreiro & Pouyat, 1993).

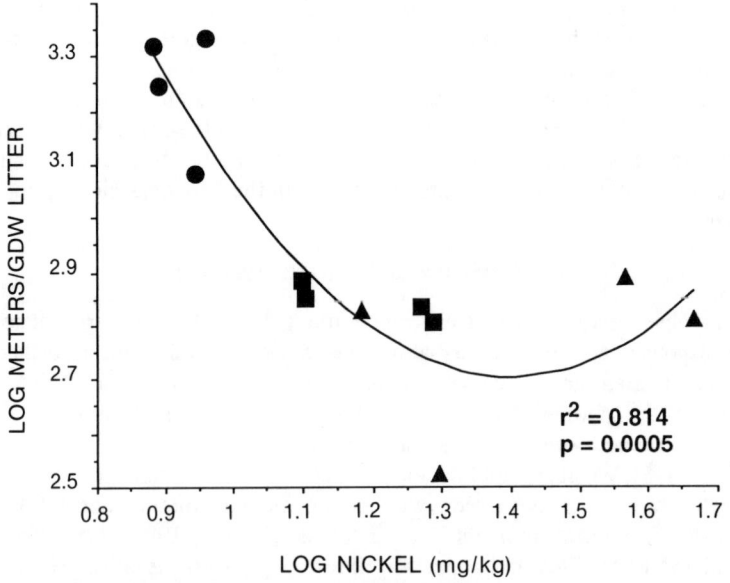

Fig. 26–6. Mean fungal densities (log meters of hyphae; $n = 4$ per stand) in oak litter of 12 stands in January 1991 regressed on forest floor Ni concentrations (log mg kg^{-1}; $n = 4$) along an urban-rural transect in the New York City metropolitan area. Rural stands are circles, suburban stands are squares, and urban stands are triangles (Carreiro & Pouyat, 1993).

was previously glaciated and it is believed that native earthworm populations do not occur in these areas (Gates, 1976). The lack of earthworm activity in the rural stands would support this view. The existence of European lumbricids in the urban rather than in rural stands, however, is an interesting result and perhaps explained by the close association these exotic species have with human activity (Schwert, 1990).

The dynamics of litter in forest ecosystems integrates the effects of resource quality, site environment and activities of decomposer organisms (Swift et al., 1979). We therefore hypothesized that the measured changes in soil properties and soil populations along the transect would result in differences in leaf litter decomposition rates. Mass loss rates of a reference litter (*Acer rubrum* L.) along the transect were significantly higher ($r^2 = 0.71$; $P < 0.005$) at the urban end of the transect than at the rural end of the transect (Pouyat, 1992). These results were counterintuitive since microinvertebrate and litter fungi abundances were lower in the urban than in the rural stands. We therefore believe the difference in mass loss rate of the reference litter was due to fragmentation by earthworms and to higher temperatures at the urban end of the gradient (Pouyat, 1992).

However, litter quality can be potentially altered by urban atmospheric pollution. Therefore, to test for potential differences in litter quality, we reciprocally transplanted red oak litter between urban and rural stands. As with the decomposition of the reference litter, the urban sites exhibited higher decomposition rates than the rural sites regardless of litter type (Fig. 26–7). However, the rural

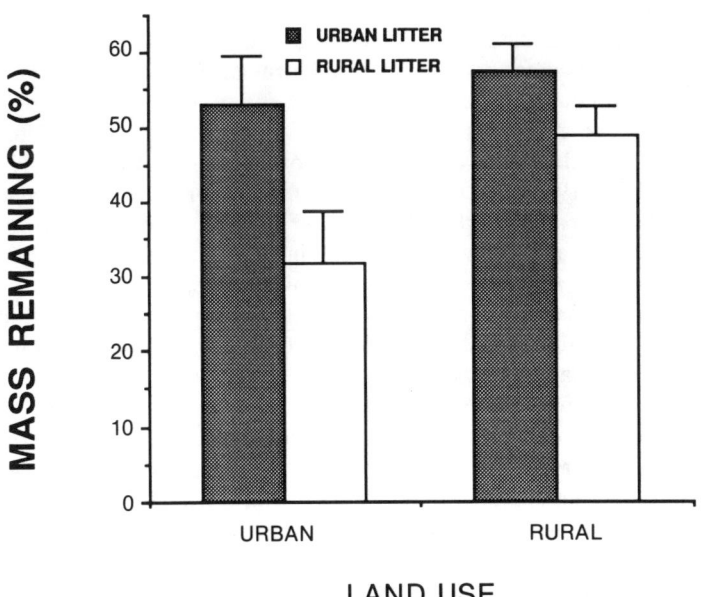

Fig. 26–7. Mean (standard error) percentage mass remaining of reciprocal transplants of oak litter after a 20-mo field incubation period along an urban-rural transect in the New York City metropolitan area. Shaded bars represent urban litter and open bars represent rural litter. Values are the means of four stands (16 litter bags for each sampling period; adapted from Pouyat, 1992).

oak litter decomposed more rapidly than the urban litter regardless of site, which suggests that the urban litter was of poorer quality than the rural litter (Fig. 26–7). These results were reproduced using the same litter sources in controlled laboratory incubations (Carreiro & Pouyat, 1993). The difference in leaf litter quality within the same plant species may be due to leaf damage by high O_3 concentrations found in the New York City metropolitan area (New York State Dep. of Environ. Conserv., 1989). Findlay and Jones (1990) reported that leaf exposure to high concentrations of O_3 reduced subsequent decomposition rates of poplar (*Populus deltoides* Bact.) leaves in laboratory incubations, suggesting that poplar litter quality is reduced when exposed to high ozone concentrations.

Input of more recalcitrant leaf litter should over time increase total soil C pools and reduce N cycling rates in the urban stands relative to rural stands. However, higher decomposition rates, and potentially higher rates of N deposition in these urban stands, should decrease total soil C pools and increase N cycling rates relative to rural stands (Groffman et al., 1994). To quantify the net response of these factors, we measured net potential N-mineralization rates of A horizon soil in laboratory incubations. We found that potential N-mineralization rates were higher in the urban stands than in the rural stands (Pouyat, 1992). These results were substantiated using reciprocally transplanted soil cores incubated in buried bags (Pouyat, 1992). Soil cores taken from urban areas accumulated more inorganic N (NH_4^+ and NO_3^-) than soil cores taken from rural areas regardless of site (Fig. 26–8). In addition, the soil cores incubated at the urban sites, regardless of where the cores originated, had higher accumulations of inorganic N than cores incubated at the rural sites (Fig. 26–8). In urban and rural sites, nitrification accounted for close to 50% of the total N mineralization measured in urban soil cores compared to 20% for rural soil cores regardless of incubation location (Pouyat, 1992).

Both the laboratory incubation and buried bag methods test for labile fractions of soil organic matter and are indicators of the availability of N in soil (Binkley & Hart, 1989). Therefore, the different responses in net N mineralisation between urban and rural forest stands can be due to changes in the quality of soil organic matter and the availability of N in the A horizon. In addition, buried bag methods are sensitive to variations in soil temperature while soil moisture content is held essentially constant throughout the incubation period (Binkley & Hart, 1989). Therefore, we attribute differences in field rates of net N mineralization between urban and rural sites to differences in temperature caused by the urban heat island effect (Pouyat, 1992).

However, the net N mineralization results contradict the litter transplant results, which indicated urban litter is more recalcitrant than rural litter. Typically, poor quality litter input either decreases the rate at which labile C mineralizes N or increases the amount of organic matter transferred to recalcitrant pools, or both (Lamb, 1975). Indeed, preliminary measurements of soil C pools along the urban-rural transect suggests that recalcitrant C pools are higher in urban forest soils relative to rural forest soils (Groffman et al., 1994). High net N mineralization rates in the urban forest soils may be caused by increases in N availability by earthworm activity and by greater inputs of atmospheric N. Earthworms ingest a large proportion of the total organic matter in surface soils and through their frag-

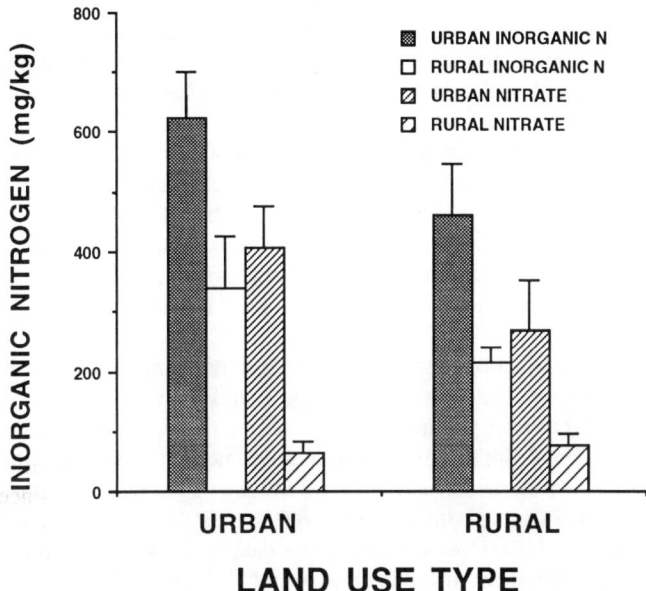

Fig. 26–8. Mean (standard error) accumulation of net inorganic N ($NH_4^+ + NO_3^-$) and nitrate expressed on an ash-free weight basis for reciprocal transplants of soil cores after an 8-mo field incubation period. Shaded bars represent the net amount of inorganic N that was mineralized in urban soil cores and clear bars represent the net amount of inorganic N that was produced in rural soil cores. Dense hash bars represent the net amount of nitrate accumulated in urban soil cores and less dense hash bars represent the net amount of nitrate accumulated in rural soil cores. Values are the means of four stands (16 soil cores for each sampling period; adapted from Pouyat, 1992).

menting, and cast-forming activities stimulate microbial activity, thereby increasing N availability (Shaw & Pawlak, 1986, Scheu, 1987). An increase in available N can result in more rapid and efficient utilization of litter by decomposers (Chapman et al., 1988) even if litter quality is more recalcitrant (Seastedt, 1984).

We are continuing our investigation along the current transect to test these hypotheses by conducting more detailed studies of soil N mineralization and earthworm dynamics, measurements of variation in N deposition and analysis of O_3 effects on plant growth and litter quality. To investigate the importance of individual environmental factors, we are manipulating these factors in lab and field experiments. We are particularly interested in establishing additional urban-rural transects in New York City and other metropolitan areas in North America to perform comparative studies.

SUMMARY AND CONCLUSION

Our analysis of environmental monitoring data and our investigations of abiotic environmental factors identified patterns in soil chemistry, air temperature and air quality along an urban-rural transect extending from New York City to western Connecticut. We present evidence suggesting that these abiotic factors are affecting soil and plant biota and biogeochemical processes in oak forest ecosystems located along the transect. Changes in plant litter quality, possibly from O_3 damage, heavy metal effects on litter fungi and the introduction of non-native species of earthworms are biotic responses that we hypothesize are influencing C and N dynamics in the urban forest stands. Our results suggest that the urban environment in the New York City metropolitan area exhibits a suite of abiotic and biotic factors that increase recalcitrant C pools in forest soils, and simultaneously accelerate N fluxes relative to rural environments. Changes in N availability, in turn, should have an influence on net primary productivity, litter quality, and, at longer time scales, plant species composition in forest ecosystems.

We conclude that the deposition of pollutants at the urban end of the gradient appears to be occurring at landscape scales, since all of the stands were greater than 30 m from the nearest road, a zone that spatially defines the "roadside environment" (Smith, 1976). The measurement of urban soil effects outside the roadside environment is an important finding of our research, since the urban environment may significantly affect forest stands that are not directly adjacent to a major road. Therefore, although more data is needed from other metropolitan areas, the differences in soil chemistry and C and N pools measured along this transect may be regarded as an "urban" regional phenomenon and not just a local phenomenon found only near roads. The "urban phenomenon" can serve as an unplanned manipulation that ecologists can utilize for studying the responses of forest ecosystems to multiple stress factors. Heavy metal contamination of forest soils, O_3 damage effects on plants, elevated air and soil temperatures, increased N deposition, and species compositional changes are all potential environmental factors associated with urban environments that can be regarded as "ecosystem level" treatments inviting further investigation.

ACKNOWLEDGMENTS

We would like to thank Michelle Conant, Gina Garcia, Jenifer Haight, Judy Koch, Leslie Likens, Scott Robison, Ethan Santos, and Heather Toyryla for their assistance with field sampling and laboratory analysis. Funding support came from the USDA-FS, Northern Global Change Program and Research Work Unit (NE-4952), Syracuse, New York, and by the Dewitt-Wallace Readers' Digest Fund. A contribution to the program of the Institute of Ecosystem Studies. The use of trade, firm, or corporation names in this publication is for the information and convenience of the reader. Such use does not constitute an official endorsement or approval by the USDA or the Forest Service of any product or service to the exclusion of others that may be suitable.

REFERENCES

Aber, J.D., K.J. Nadelhoffer, P. Steudler, and J.M. Melillo. 1989. Nitrogen saturation in northern forest ecosystems. BioSci. 39:378–386.

Agren, G.I., and E. Bosatta. 1988. Nitrogen saturation of terrestrial ecosystems. Environ. Pollut. 54:185–197.

Airola, J.M., and K. Bucholz. 1984. Species structure and soil characteristics of five urban forest sites along the New Jersey Palisades. Urban Ecol. 8:149–164.

Bagnall, R.G. 1979. A study of human impact on an urban forest remnant: Redwood Bush, Tawa, near Wellington, New Zealand. N.Z. J. Botany 17:117–126.

Baker, H.G. 1986. Patterns of plant invasion in North America. p. 44–57. *In* H.A. Mooney and J.A. Drake (ed.) Ecology of biological invasions of North America and Hawaii. Springer-Verlag, New York.

Binkley, D., and S.C. Hart. 1989. The components of nitrogen availability assessment in forest soils. Adv. Soil Sci. 10:57–112.

Bormann, F.H. 1985. Air pollution and forests: An ecosystem perspective. BioSci. 35:434–441.

Bornkamm, R.D., J.A. Lee, and M.R.D. Seaward. 1982. Urban ecology. Blackwell Sci. Publ., Oxford, England.

Bornstein, R.D. 1968. Observations of the urban heat island effect in New York City. J. Appl. Meterol. 7:575–582.

Box, G.E.P., and D.R. Cox. 1982. An analysis of transformations revisited, rebutted. J. Am. Stat. Assoc. 77:209–210.

Broughton, J.G., D.W. Fisher, Y.W. Isachsen, and L.V. Richard. 1966. Geology of New York: A short account. Leaflet no. 20. Univ. New York, State Educ. Dep., New York State Museum and Sci. Ser. Educ., Albany, NY.

Candelino, C.A., and W.L. Chameides. 1990. Natural hydrocarbons, urbanization, and urban ozone. J. Geophys. Res. 95:13971–13979.

Carlo Erba. 1986. Nitrogen analyzer 1500 instruction manual. Carlo Erba, Milan, Italy.

Carreiro, M.M., and R.V. Pouyat. 1993. Variation in oak leaf litter quality and its role in decomposition along an urban-rural land use gradient. Bull. Ecol. Soc. Am. 74:187.

Chagnon, S.A., Jr. 1980. Evidence for urban and lake influences on precipitation in the Chicago area. J. Appl. Meteorol. 19:1137–1159.

Chapman, H.D. 1965. Cation exchange capacity. p. 891–901. *In* C.A. Black (ed.) Methods of soil analysis. Agron. Monogr. 9. ASA and CSSA, Madison, WI.

Chapman, K., J.B. Whittaker, and O.W. Heal. 1988. Metabolic and faunal activity in litters of tree mixtures compared with pure stands. Agric. Ecosys. Environ. 24:33–40.

Connecticut Department of Transportation. 1990. Traffic log of state numbered routes and roads, 1989. Div. Planning Inventory and Data and USDOT, Federal Highway Comm. Connecticut Dep. Transportation, Storrs, CT.

Cowling, E.B. 1989. Recent changes in chemical climate and related effects on forests in North America and Europe. Ambio 18:167–171.

Craul, P.J. 1985. A description of urban soils and their desired characteristics. J. Arboricult. 11:330–339.

Davis, A.M., and T.F. Glick. 1978. Urban ecosystems and island biogeography. Environ. Conserv. 5:299–304.

DeBano, L.f. 1971. The effects of hydrophobic substances on water movement in soil during infiltration. Soil Sci. Soc. Am. Proc. 35:340–343.

Dickinson, R.E. 1966. The process of urbanization. p. 463–478. *In* F.F. Darling and J.P. Milton (ed.) Future environments of North America. Natural History Press, Garden City, New York.

Driscoll, C.T., G.E. Likens, L.O. Hedin, J.S. Eaton, and F.H. Bormann. 1989. Changes in the chemistry of surface waters: 25-year results at the Hubbard Brook Experimental Forest, New Hampshire. Environ. Sci. Technol. 23:137–143.

Findlay, S.E.G., and C.J. Jones. 1990. Exposure of cottonwood plants to ozone alters subsequent leaf decomposition. Oecologia 82:248–250.

Friedland, A.J. 1985. Trace metal accumulation, distribution and fluxes in forests of the northeastern United States. Ph.D. diss. Univ. Pennsylvania, Philadelphia (Diss. Abstr. 85-15377).

Frink, C.R., and G.K. Voigt. 1976. Potential effects of acid precipitation on soils in the humid temperate region. USDA-FS Gen. Tech. Rep. no. NE23. USDA-FS, Broomall, PA.

Gates, G.E. 1976. More on oligochaete distribution in North America. Megadrilogica 2:1–8.

Gatz, D.F. 1991. Urban precipitation chemistry: A review and synthesis. Atmos. Environ. 25:1–15.

Gilbert, O.L. 1989. The ecology of urban habitats. Chapman and Hall, New York.

Godron, M., and R.T.T. Forman. 1983. Landscape modification and changing ecological characteristics. p. 12–28. *In* H.A. Mooney and M. Godron (ed.) Disturbance and ecosystems. Springer-Verlag, New York.

Gonick, W.N., A.E. Shearin, and D.E. Hill. 1970. Soil survey of Litchfield County, Connecticut. USDA-SCS. U.S. Gov. Print. Office, Washington DC.

Graedel, T.E., and P.J. Crutzen. 1989. The changing atmosphere. Sci. Am. 261:58–68.

Groffman, P.M., R.V. Pouyat, M.J. McDonnell, S.T.A. Pickett, and W.C. Zipperer. 1994. Carbon pools and trace gas fluxes in urban forest soils. Adv. Soil Sci. (In press.)

Hill, D.E., E.H. Sautter, and W.N. Gonick. 1980. Soils of Connecticut. Connecticut Agric. Exp. Stn. Bull. 787.

Hinrichsen, D. 1986. Multiple pollutants and forest decline. Ambio 15:258–265.

Hobbs, E. 1988. Using ordination to analyze the composition and structure of urban forest islands. For. Ecol. Manage. 23:139–158.

Homolya, J.B., and S. Lambert. 1981. Characterization of sulfate emissions from non-utility boilers firing low sulphur residual oils in New York City, U.S.A. J. Air Pollut. Control Assoc. 31:139–143.

Janzen, D.H. 1983. No park is an island: Increase in interference from outside as park size decreases. Oikos 41; 402–410.

Jenny, H. 1941. Factors of soil formation. McGraw-Hill Book Co., New York.

Jordon, D.N., T.H. Green, A.H. Chappelka, B.G. Lockaby, R.S. Meldahl, and D.H. Gjerstad. 1991. Response of total tannins and phenolics in loblolly pine foliage exposed to ozone and acid rain. J. Chem. Ecol. 17:505–513.

Keddy, P.A. 1989. Competition. Chapman and Hall, New York.

Lamb, D. 1975. Patterns of nitrogen mineralization in the forest floor of stands of *Pinus radiata* on different soils. J. Ecology 63:615–625.

Landsberg, H.E. 1981. The urban climate. Acad. Press, New York.

Likens, G.E. 1989. Some aspects of air pollutant effects on terrestrial ecosystems and prospects for the future. Ambio 18:172–178.

McColl, J.G., and D.S. Bush. 1978. Precipitation and throughfall chemistry in the San Francisco Bay area. J. Environ. Quality 7:352–357.

McDonnell, M.J. 1988. A forest for New York. The Public Garden, New York 3:28–31.

McDonnell, M.J., and S.T.A. Pickett. 1990. The study of ecosystem structure and function along urban-rural gradients: An unexploited opportunity for ecology. Ecology 71:1231–1237.

McDonnell, M.J., S.T.A. Pickett, and R.V. Pouyat. 1993. The application of the ecological gradient paradigm to the study of urban effects. p. 175–189. *In* M.J. McDonnell and S.T.A. Pickett (ed.) Humans as components of ecosystems: Subtle human effects and the ecology of populated areas. Springer-Verlag, New York.

McLean, E.O. 1982. Soil pH and lime requirement. p. 199–223. *In* A.L. Page et al. (ed.) Methods of soil analysis. Part 1. ASA and SSSA, Madison, WI.

National Oceanic and Atmospheric Administration. 1985. Climates of the states. Vol. 2. 3rd ed. Gale Res. Co., Detroit, MI.

New York State Department of Environmental Conservation. 1989. Air quality report, ambient air monitoring system, annual 1988. Div. Air Resour.

New York State Department of Transportation. 1990. 1989 Traffic volume report. Planning Div., New York State Dept. Transportation, Dept. Environ. Conserv., Albany, NY.

Oke, T.R. 1990. The micrometeorology of the urban forest. Quat. J. R. Meteorol. Soc. 324:335–349.

Parker, G.R., W.W. McFee, and J.M. Kelly. 1978. Metal distribution in forested ecosystems in urban and rural northwestern Indiana. J. Environ. Qual. 7:337–342.

Pouyat, R.V. 1991. The urban-rural gradient: An opportunity to better understand human impacts on forest soils. p. 212–218. *In* Proc. Soc. Am. For. 1990 Annu. Conv. Washington, DC. 27 July–1 Aug. 1990.

Pouyat, R.V. 1992. Soil characteristics and litter dynamics in mixed deciduous forests along an urban-rural gradient. Ph.D. diss. Rutgers Univ., New Brunswick, NJ (Diss. Abstr. 92-32946).

Pouyat, R.V., M.J. McDonnell, and S.T.A. Pickett. 1995. Soil characteristics of oak stands along an urban-rural land use gradient. J. Environ. Qual. (In press.)

Pouyat, R.V., and M.J. McDonnell. 1991. Heavy metal accumulation in forest soils along an urban to rural gradient in southern NY, USA. Water Air Soil Pollut. 57–58:797–807.

Pouyat, R.V., and W.C. Zipperer. 1992. The uses and management of woodlands in urban and suburban environments. p. 26–29. *In* P. Rodbell (ed.) Proc. Natl. Urban For. Conf., 5th, Los Angeles, CA. 12–17 Nov. 1991. Am. For. Assoc., Washington, DC.

Rhoades, J.D. 1982. Soluble salts. p. 167–179. *In* A.L. Page et al. (ed.) Methods of soil analysis. Part 1. Agron. Monogr. 9. ASA and SSSA, Madison, WI.

Rudnicky, J.L., and M.J. McDonnell. 1989. Forty-eight years of canopy change in a hardwood-hemlock forest in New York City. Bull. Torrey Bot. Club 116:52–64.

Ruppert, H. 1975. Geochemical investigations on atmospheric precipitation in a medium-sized city (Göttingen, Germany). Water Air and Soil Pollut. 4:447–460.

SAS Institute. 1987. SAS Statistics users' guide. SAS Institute, Inc., Cary, NC.

Scheu, S. 1987. Microbial activity and nutrient dynamics in earthworms casts (Lumbricidae). Biol. Fert. Soils 5:230–234.

Schuberth, C.J. 1968. The geology of New York City and environs. Natural History Press, New York.

Schwert, D.P. 1990. Oligochaeta: Lumbricidae. p. 341–356. In D.L. Dindal (ed.) Soil biology guide. John Wiley & Sons, New York.

Seastedt, T.R. 1984. The role of microarthropods in decomposition and mineralization processes. Ann. Rev. Entomol. 29:25–46.

Seinfeld, J.H. 1989. Urban air pollution: State of the science. Science (Washington, DC) 243:745–752.

Shaw, C., and J. Pawluk. 1986. Faecal microbiology of *Octolasion tyrtaeum*, *Aporrectodea turgida* and *Lumbricus terrestris* and its relation to the carbon budgets of three artificial soils. Pedobiologia 29:377–389.

Skeffington, R.A., and Wilson, E.J. 1988. Excess nitrogen deposition: Issues for consideration. Environ. Pollut. 54:159–184.

Smith, W.H. 1976. Lead contamination of the roadside ecosystem. J. Air Pollut. Control Assoc. 26:753–766.

Smith, W.H. 1990. Air pollution and forests: Interaction between air contaminants and forest ecosystems. 2nd ed. Springer-Verlag, New York.

Stearns, F., and T. Montag. 1974. The urban ecosystem: A holistic approach. Dowden, Hutchinson and Ross, Inc., Stroudsburg, PA.

Swift, M.J., O.W. Heal, and J.M. Anderson. 1979. Decomposition in terrestrial ecosystems. Univ. California Press, Berkeley, CA.

Tamm, C.O. 1989. Comparative and experimental approaches to the study of acid deposition effects on soils as substrate for forest growth. Ambio 18:184–191.

Thomas, G.W. 1982. Exchangeable cations. p. 159–165. In A.L. Page et al. (ed.) Methods of soil analysis. Part 1. ASA, Madison, WI.

Tyler, G., A.M. Balsberg Pahlsson, G. Bengtsson, D. Bääth, and L. Tranvik. 1989. Heavy-metal ecology of terrestrial plants, microorganisms and invertebrates. Water Air Soil Pollut. 47:189–215.

Ulrich, B., R. Mayer, and P.K. Khann. 1980. Chemical changes due to acid deposition in a loess derived soil in Central Europe. Soil Sci. 120:193–199.

U.S. Bureau of Census. 1990. Census user's guide. U.S. Dep. Commerce, U.S. Gov. Print. Office, Washington, DC.

U.S. Department of Agriculture. 1982. Natural resource inventory. Bull. 756. SCS and Iowa State Statistical Lab., Washington, DC.

Vitousek, P.M., and P.A. Matson. 1991. Gradient analysis of ecosystems. p. 287–298. In J.J. Cole et al. (ed.) Comparative analysis of ecosystems: Patterns, mechanisms and theories. Springer-Verlag, New York.

Volchok, H.L., and D. Bogen. 1971. Trace metals-fallout in New York City. Science (Washington, DC) 156:1487–1489.

White, C.S., and M.J. McDonnell. 1988. Nitrogen cycling processes and soil characteristics in an urban versus rural forest. Biogeochemistry 5:243–262.

Whittaker, R.H. 1967. Gradient analysis of vegetation. Biol. Rev. 49:207–264.

Wilde, S.A., G.K. Voigt, and J.C. Iyer. 1972. Soil and plant analysis for tree culture. Oxford and IBH Publ. Co., New Delhi, India.

Wolff, G.T., P.J. Lioy, H. Golub, and J.S. Hawkins. 1979. Acid precipitation in the New York metropolitan area: Its relationship to meteorological factors. Environ. Sci. Technol. 13:209–212.

Zipperer, W.C., R.L. Burgess, and R.D. Nyland. 1990. Patterns of deforestation and reforestation in different landscape types in central New York. For. Ecol. Manage. 36:103–117.

27 A Perspective on the Evolution of Forest Soil Science[1]

Robert F. Chandler, Jr.
International Rice Research Institute
Clermont, Florida

I am grateful for the opportunity to join a group of forest soil scientists after 45 yr of absence from this field of activity. Because of my considerable age and a lack of familiarity with recent advances, I—no doubt—should have declined the invitation to make a summary statement at the close of this gathering. The urge to accept was great, however, for I was extremely curious to find out what was going on in forest soil research today, compared with what we were doing 50 yr ago.

Obviously, it would be impossible for me to present a comprehensive summary of 28 oral presentations and 34 poster displays. Rather, I'll touch on what we forest soil scientists were doing in the 1930s and 1940s and then make comparisons between those activities and some of the more important conclusions that can be drawn from the papers and exhibits of this conference.

In the earlier days of forest soil research we were very much interested in the role of soil organic matter in maintaining soil productivity. We recognized it, particularly in light-textured soils, as the principal source of cation exchange capacity and soil moisture retention. It improved soil structure and served as a source of energy for soil microorganisms. We conducted many studies that supported such relationships (Chandler, 1939; Heiberg & Chandler, 1941).

We examined root distribution in the horizons of Spodosols (then called podzols), finding high concentrations of fine roots in the O and Bhs (Then A_o and B_h) horizons where much of the organic matter accumulated (Hopkins, 1939).

We conducted mineralogical studies in mature Spodosols in the boreal coniferous forests of the northeastern USA, showing the breakdown of certain primary minerals in the acid E (then A2) horizon (Richard & Chandler, 1944). In general, the ferromagnesian minerals, especially hornblende, were altered by chemical weathering while such minerals as quartz, zircon and garnet were resistant. This weathering was induced, of course, by the organic acids originating in the deep, highly acid mor humus layer at the soil surface. Below the E horizon the

---

[1]Dr. Chandler was Pack Professor of Forest Soils at Cornell University 1935 to 1947, and is Director, Emeritus, of the International Rice Research Institute.

weathering of minerals decreased progressively with depth and there was little alteration in the glacial till, C horizon.

We conducted much research on the mineral nutrient content of forest tree foliage, realizing as we do today that nutrient cycling is essential for forest tree productivity (Lutz & Chandler, 1946). I believe we paid more attention to Ca than is the case today. We found that the foliar Ca content was largely a function of tree species rather than site, except on highly acid soils (less than pH 4.5). The Ca content of all species that were sampled, when expressed in chemical equivalents, exceeded the oxalate content. The Ca content of foliage varied among species from as high as 4.0% in red cedar (*Juniperus virginiana* L.) and basswood (*Tilia americana* L.) to less than 0.60% in eastern hemlock [*Tsuga canadensis* (L.) Carr].

The effect of foliar Ca content on surface soil pH could be readily demonstrated. For example, in central New York state, we sampled the humus layer under an isolated 100-yr-old hemlock tree growing in a hardwood forest composed largely of sugar maple (*Acer saccharum* Marsh.) and beech (*Fagus grandifolia* Ehrh.), with an admixture of black birch (*Betula lenta* L.), black cherry (*Prunus serotinia* L.), and hop hornbeam (*Ostrya virginiana* Mill.) on a soil derived from acid glacial till. Although the trees had been harvested at least twice, the land had never been cultivated or pastured. We took surface soil samples successively from outside the spread of the hemlock branches toward the trunk and found that the pH ranged from about 5.5 under the hardwood trees to as low as 3.2 near the bole of the hemlock. In contrast, we sampled the soil under a 40-yr-old red cedar tree in an abandoned field in Connecticut and found that the pH of the surface soil in the grass-covered area was 5.5 but reached a level of 7.0 under the red cedar tree. Thus a high-Ca tree increases soil pH on acid soils while a low-Ca tree decreases it. I use the two extremes here simply to illustrate the principle. Because, in general, highly acid forest soils are less productive than moderately acid soils, we argued that trees with a foliar Ca content of over 2.0% were "site improvers," while those with less than 1.0% had a detrimental effect on site quality. Of course, trees influence soils and soils influence trees. There is a tendency for the "site improvers" to occur more frequently on the better soils and for the low-Ca species (such as hemlock, spruce and pine) to predominate on the more acid soils where they can compete successfully with the high-Ca species.

With respect to the other nutrients, we found that there was a tendency for the Mg content of the leaves to be positively correlated with the Ca content, although the actual amounts were much lower. The N and K contents seemed to be largely associated with the available N and K levels in the soil, rather than being a function of tree species.

We valued humus layer type because it told us much about the soil as a whole and its productivity. I understand that the system of classification of humus layer types has been changed several times since 1941 when Professor Heiberg of Syracuse University and I published a paper on a revised nomenclature of forest humus layers for the northeastern USA (Heiberg & Chandler, 1941). I could not help but note how little mention was made of humus layer type in the papers presented at this conference.

We counted the mesofauna responsible for much of the early disintegration of the leaf and other litter on the soil surface. In highly productive soils on farm woodlots in central New York state, we found that there were as many as 2.5 million earthworms per hectare. They were so active feeding on the leaves and incorporating them into the soil that at the end of a summer season very little leaf litter from the previous year's leaf fall remained on the soil surface, and the topsoil had an excellent crumb structure to a depth of at least 15 cm (Eaton & Chandler, 1942). We found that earthworms had a distinct preference for leaves of some species and a dislike for those of others. For example, they would eat leaves of basswood (*T. americana* L.), white ash (*Fraxinus americana* L.) and sugar maple but wouldn't touch red oak (*Quercus rubra* L.) leaves. In the highly acid Spodosols of the coniferous boreal forests there were no earthworms, but mites, millipedes, springtails and other fauna took care of the early stages of leaf and needle litter disintegration, followed by the activity of bacteria and fungi (Lutz & Chandler, 1946).

We were interested in the role of ectomycorrhizae in forest tree nutrition, especially with regard to P availability to tree roots. Earl Stone, who is present today, chose to work with mycorrhizae for his doctoral dissertation (Stone, 1950).

Enough of the past. It is time to turn our attention to what is taking place these days, based on what has been presented at this conference.

The principal difference between studies of organic matter in forest soils 50 yr ago and today is that the earlier work was confined largely to the effect of total soil organic matter. Today, however, with advanced analytical methods, one can measure many more fractions of soil organic matter than was possible 50 yr ago and attempt to relate certain of these fractions to site productivity. I am not qualified to pass judgment on this subject. I did note that several scientists stressed the importance of using nuclear magnetic resonance for identifying the various fractions of soil organic matter.

In several papers it was reported that there was a substantial contribution to soil organic matter from the death of fine roots and from root exudates, subjects that received little attention in the early days.

A number of studies showed that in the sandy, light-textured Spodosols of the southeastern USA developed on flat topography, the organic matter penetrates to a greater depth than was formerly appreciated. In such soils, samples should be taken to a depth of 2 m or more, if the total organic matter content is to be measured. Formerly, most sampling went to a depth of only 1 m or less.

The theme of several papers and posters stressed the sequential stages of soil organic matter decomposition. First there is a loss of carbohydrates, next, lignin, and last, a loss of the waxes. One paper used the terminology for the fractions as labile, semilabile and recalcitrant. Another scientist termed the three fractions as active, slow and passive. The first fraction shows considerable seasonal variability and is believed to be associated with short-term productivity. The second stage reflects longer-term productivity, while the last stage (recalcitrant or passive) is a stable measure of productivity over time. Another author asserted that the compositional changes during breakdown must arise from two processes: (i) the selective preservation of recalcitrant organic compounds in the original undecomposed materials, and (ii) the in situ synthesis of organic compounds by microorganisms.

Several participants emphasized the significance of organic matter on nutrient supply and cycling, pointing out the importance of organic forms of N, P, and S.

The effect of applying sewage sludge to reforested areas previously cultivated was reported in several papers. Sludge applications accelerated the decomposition rate of soil organic matter, but the net effect was to increase total soil organic matter content. In some cases the growth rate of the treated stands was more than 100% greater than that of the controls.

In one experiment, artificially elevated atmospheric CO_2 increased the rate of photosynthesis, decreased foliar respiration, enhanced root growth, and raised the rate of CO_2 efflux from the soil. There was little increase in total biomass, which may have been due to N deficiency in the soil.

Work in New Hampshire studying the changes in amount of composition of soil organic matter during an 8-yr period following clear-cutting showed that the total amount of soil organic matter remained essentially unchanged, principally because the death of roots compensated for the reduced contribution from leaf litter. The principal activity was a downward movement of humic substances and acid functional groups in the soil, causing an increase in the cation exchange capacity in the Bh and Bs horizons. Another participant agreed that the increases in soil organic matter following clear-cutting are caused primarily by root death and decomposition.

Several scientists reported on the magnitude of the C sink in forest ecosystems. A study in South Carolina, for example, showed that 34 yr after reforestation with pine (*Pinus* spp.) on formerly cultivated land, the amount of C accumulated was about 134000 kg/ha, averaging approximately 3900 kg/ha/yr. About 75% of this was in the aboveground biomass, 12% in roots, 11% in the forest floor and only 1% in the surface 60 cm of soil (perhaps deeper sampling would have revealed that there was more C in the mineral soil).

Another study in South Carolina determined the C distribution in mineral soil, root systems and forest floor in a mature, intensively managed loblolly pine (*P. taeda* L.) plantation. The investigators found that C in the mineral soils totaled 51000 kg/ha, that in the root systems amounted to 36000 kg/ha and that in the forest floor, 2000 kg/ha.

In general, studies of the amount of leaf litter deposited annually ranged from 2500 to 3500 kg/ha, agreeing with the work done 50 or more years earlier. The quantities of nutrients deposited by leaf litter agreed with our early results, but more attention these days is given to the amounts contributed by throughfall as well as litterfall.

Studies of the contribution of soil respiration to atmospheric CO_2 have given variable results, depending on the environment, nature of the forest stand, etc. Data obtained in the northeastern USA revealed that about 350 g of C per square meter per year were emitted from the soil, including that from the forest floor. Approximately 40% came from surface layers (mostly from decomposing leaf litter), and 60% from the soil (including root respiration). The statement was made that, on a global basis, the efflux of CO_2 from the soil and litter contributes over 12 times as much CO_2 to the atmosphere as does that from fossil fuel combustion.

Various studies showed that, in general, the downward movement of organic matter is greater in southern soils than in northern ones, and that in wetland soils high amounts of organic matter pass into the groundwater and eventually into streams.

As was true 50 yr ago, many of the current investigations point out that N is the principal nutrient limiting tree growth. The high C/N ratio in the acid humus layer of northern soils causes much of the N to be unavailable for plant use. In our early work with forest tree fertilization we found that in order to get a growth response it was necessary to apply much larger doses of nitrogenous fertilizer than would be necessary for agricultural crops. This, of course, was because so much of the N was tied up in the organic matter and in the microorganisms whose growth and activity were limited by N.

In the field trip during this conference we saw vivid demonstrations of the growth response to N and P applications, as well as that of herbicides, on pine plantations in Florida. A growth response was obtained with smaller fertilizer applications than would be necessary in northern forests because in the warmer climate the organic matter decomposed more rapidly and the C/N ratio was much lower.

Many of the participants mentioned that there is a poor correlation between total soil organic matter and site quality. This has always been true, of course, and should be expected. Many factors influence soil productivity. For example, compare the conditions around tree roots in a northern coniferous forest growing on a light-textured soil derived from glacial till overlain with a thick highly acid mor humus layer, to those under a hardwood forest farther south on a less acid silt loam soil, where earthworms and other fauna, as well as microorganisms, are highly active in decomposing the leaf litter. It is possible that *total* amount of organic matter in the two soil profiles is similar, but in the latter condition there is a rapid turnover of organic matter and a quick release of N and other nutrients, and tree growth is much faster.

To end this inadequate summary, there were a number of rather obvious thoughts that came to mind as I listened to the papers. They are presented briefly below:

Although many of the soils of the northern USA and Canada are classified as Spodosols, so are vast areas in the sandy flatlands of the southeastern USA. We must continue to recognize the differences between the soils in these two geographic areas. The Spodosols of the north are mostly developed from glacial deposits less than 15000 yr old, while the southern Spodosols are much older. The humus layer types, the vegetation, the weathering of soil minerals, the depth of tree rooting and the penetration of soil organic matter are completely different in the two types.

Modern methods of forest harvesting employ heavy machinery that compacts the soil. Furthermore, site preparation for reforestation disturbs the soil greatly. Forest soil scientists should continue to explore the effects of these practices on soil properties and determine what sets the proper balance between soil protection and economic return.

From the discussions of fire and its impact on tree growth, it would appear that controlled burning has an entirely different effect in different forest environments. The results obtained in coniferous forests in Oregon, for instance, are completely different from those in a loblolly pine plantation in Florida. The work in Oregon showed that up to 36000 kg/ha of organic matter from the humus layers and surface mineral soil can be lost by burning, with a decrease in tree growth rate. Certainly more studies are needed on the impact of controlled burning, especially in western USA.

Although the effect of increasing CO_2 on global warming is yet unknown, it seems to be an indisputable fact that the CO_2 content of the atmosphere is increasing and will continue to do so for some time to come. Forest scientists have an opportunity to study the impact of higher concentrations of CO_2 on tree growth, especially differences among species. It would seem that the relative response to solar radiation and atmospheric CO_2 concentration of a shade-tolerant tree, such as eastern hemlock, might be quite different from that of a shade-intolerant species, such as eastern white pine (*P. strobus* L.).

Forests continue to be cleared to allow for the growing of food to feed the world's increasing population. Thus, globally speaking, we are losing C from the soil. In many areas in Third World countries the time-honored practice of shifting cultivation is no longer a viable one. Because of population pressures the forest fallow period becomes too short to regenerate soil fertility. It is imperative not only that we make every effort to stem the tide of the burgeoning human population, but find suitable ways to manage tropical soils on a sustainable basis as a substitute for shifting cultivation.

In the final analysis, forest trees, like agricultural crops, require water, carbon dioxide, solar radiation, and mineral nutrients for growth. The objective of forest soil research will continue to be to find ways of managing forest ecosystems to keep these factors at optimum levels in a sustainable system.

REFERENCES

Chandler, R.F., Jr. 1939. Cation exchange properties of certain forest soils in the Adirondack section. J. Agric. Res. 59:491–505.

Eaton, T.H., Jr., and R.F. Chandler. 1942. The fauna of the forest humus layers in New York, Cornell Agric. Exp. Stn. Mem. 247.

Heiberg, S.O., and R.F. Chandler, Jr. 1941. A revised nomenclature of forest humus layers for the northeastern United States. Soil Sci. 52:87–99.

Hopkins, H.T., Jr. 1939. Root distribution of forest trees in the central Adirondack Region. M.S. thesis, Cornell Univ., Ithaca, NY.

Lutz, H.L., and R.F. Chandler, Jr. 1946. Forest soils. John Wiley & Sons, New York.

Richard, J.A., and R.F. Chandler, Jr. 1944. Some chemical and physical properties of mature podzol profiles. Soil Sci. Soc. Am. Proc. 8:379–383.

Stone, E.L. 1950. Some effects of mycorrhizae on the phosphorus nutrition of Monterey pine seedlings. Soil Sci. Soc. Am. Proc. 14:340–345.